stereochemistry of organometallic and inorganic compounds 5

Chains, Clusters, Inclusion Compounds, Paramagnetic Labels, and Organic Rings

stereochemistry of organometallic and inorganic compounds 5

Chains, Clusters, Inclusion Compounds, Paramagnetic Labels, and Organic Rings

Edited by

PIERO ZANELLO

Università di Siena, Dipartimento di Chimica, Pian dei Mantellini 44, 53100 Siena, Italy

ELSEVIER

Amsterdam — London — New York — Tokyo 1994

ELSEVIER SCIENCE B.V.
Sara Burgerhartstraat 25
P.O. Box 211, 1000 AE Amsterdam, The Netherlands

Library of Congress Cataloging-in-Publication Data

Chains, clusters, inclusion compounds, paramagnetic labels, and organic rings / edited by Piero Zanello.
p. cm. -- (Stereochemistry of organometallic and inorganic compounds ; 5)
Includes bibliographical references and index.
ISBN 0-444-81581-3 (acid-free)
1. Organometallic compounds. 2. Complex compounds. 3. Cyclic compounds. I. Zanello, Piero. II. Series.
QD481.S763 no. 5
[QD411]
541.2'23 s--dc20
[541.2'242]
93-42633
CIP

ISBN: 0-444-81581-3

This book is printed on acid-free paper.

Printed in The Netherlands

PREFACE

Volume 5 of *Stereochemistry of Organometallic and Inorganic Compounds* appears with a delay of about three years. Due to personal circumstances, Professor Ivan Bernal, who was the founder of this Series, has been forced to renounce to his task. I was asked to conclude the editing of the present Volume, which prematurely closes a Series planned to cover an important piece of *stereochemistry*. I beg preliminarily the readers' forgiviness for my inexperienced editorship, since the refinement and authoritativeness of Professor Bernal will be profoundly missed.

In Chapter 1, Drs. Averbuch-Pouquot and Durif accurately depict the *state of art* of the cyclic condensation of phosphates. Most of us have a rather precise idea of the extraordinary and multiple structural arrangements that the SiO_4 units assume in *silicates*, and of their practical appliances; we now learn that the PO_4 units are equally able to give interesting aggregates in *polyphosphates*, especially in *cyclophosphates*. We hope that the profound knowledge of the structural and chemical aspects of polyphosphates brought on by Averbuch-Pouquot and Durif may stimulate new studies in the field (for instance, the dramatic problem of *eutrophication* immediately comes to mind).

The belief that metal cluster compounds are able to act as electron sponges is deep-rooted. In Chapter 2, I have tried to show that such a generalization may lead to misconceptions, in that multiple electron transfers which do not cause framework destruction in metal-carbonyl complexes are not so common. In addition, in those cases where chemically reversible redox changes occur, we have examined in detail the structural consequences accompanying such electron addition/removal processes. Coupling electrochemical measurements to structural parameters may afford a more precise bonding description of this class of compounds, which have significant effects on catalysis and material sciences.

In Chapter 3, Professor Harada describes the encapsulation of organometal complexes in structurally suitable organic receptors (namely, *cyclodextrins*). This has led to pioneering work devoted to correlate the geometries of inorganic molecules with those of organic substrates. Supramolecular Chemistry, the most recent branch of Coordination Chemistry, will find uses in the study of *inclusion compounds*, particularly in analysing the bonding forces which keep together stable organic and stable inorganic molecules.

EPR techniques are ideally suited to characterize paramagnetic inorganic molecules (*e.g.*, magnetic properties and bonding parameters). Nevertheless, obtaining stereochemical informations from EPR spectra is by no means an easy matter. Professors S.S.Eaton and G.R.Eaton have found that introduction of a paramagnetic nitroxyl group into a metal complex having unpaired electrons leads to an internal "electron ping pong" between the two paramagnetic centres. The analysis of such an electron flow, depicted in Chapter 4, gives invaluable informations about the structure, conformation, or isomer constitution of the molecules. Since the distance as well as the molecular shape of the framework separating the two paramagnetic centres can be finely tuned by properly varying either the ligand in the metal complex or the complexity of the nitroxyl radical itself, it is evident that a more complete body of structural information will be gained by this unpaired-electron labelling method.

The remaining two Chapters intend to show how the stereochemical organization of organic assemblies may be properly addressed by appending to them selected metal complexes. In Chapter 5, Dr. A.Heumann describes the use of *palladium* complexes to promote homo- and hetero-cyclization of organic chains according to well determined stereochemical pathways. In Chapter 6, Dr. M.Uemura reviews the use of arenetricarbonyl*chromium* complexes to obtain a wide range of stereodefined classes of benzene derivatives. The clarity and completeness of these two Chapters will bring organic chemists' attention to this kind of *inorganic* approach to stereoselective *organic* syntheses.

To conclude, my gratitude goes to Professor I.Bernal for his friendly invitation to contribute repeatedly to this Series, as well as to the Administrative Editor of ELSEVIER for entrusting me with this editorial task. Likewise, I am indebted to all the co-authors of the present Volume for their spirit of co-operation, particularly for their patience to update their manuscripts.

List of Contributors

Marie Thérèse Averbuch-Pouchot — Centre National de la Recherche Scientifique, Laboratoire de Cristallographie, associé à l'Université J.Fourier, 166X - 38042 Grenoble Cédex, France

André Durif — Centre National de la Recherche Scientifique, Laboratoire de Cristallographie, associé à l'Université J.Fourier, 166X - 38042 Grenoble Cédex, France

Gareth R.Eaton — Department of Chemistry, University of Denver, Denver, Colorado 80208, U.S.A.

Sandra S.Eaton — Department of Chemistry, University of Denver, Denver, Colorado 80208, U.S.A.

Akira Harada — Department of Macromolecular Science, Faculty of Science, Osaka University, Japan

Andreas Heumann — Ecole Nationale Supérieure de Synthèses, de Procédés et d'Ingénierie Chimiques d'Aix-Marseille Faculté de St-Jérome, 13397 Marseille Cedex 13, France

Motokazu Uemura — Faculty of Science, Osaka City University, Sugimoto 3-3-138, Sumiyoshi-ku, Osaka 558, Japan

Piero Zanello — Dipartimento di Chimica, Università di Siena, Piano dei Mantellini, 44, 53100 Siena, Italy

Table of Contents

Chapter 1

CRYSTAL CHEMISTRY OF CYCLOPHOSPHATES

M.T.Averbuch-Pouchot and A.Durif

Chapter 2

STEREOCHEMICAL ASPECTS OF THE REDOX PROPENSITY OF HOMOMETAL CARBONYL CLUSTERS

P.Zanello

Chapter 3

PREPARATION, PROPERTIES, AND STEREOCHEMICAL ASPECTS OF INCLUSION COMPOUNDS OF ORGANOMETALLIC COMPLEXES WITH CYCLODEXTRINS

A. Harada

Chapter 4

METAL-NITROXYL INTERACTIONS AS PROBES OF STEREOCHEMISTRY

S.S.Eaton and G.R.Eaton

Chapter 5

RECENT DEVELOPMENTS ON STEREOCONTROL *VIA* (η^6-ARENE)CHROMIUM COMPLEXES

M.Uemura

Chapter 6

THE STEREOCHEMISTRY OF PALLADIUM CATALYZED CYCLIZATION REACTIONS

A.Heumann

CRYSTAL CHEMISTRY OF CYCLOPHOSPHATES

M.T. Averbuch-Pouchot and A. Durif

CRYSTAL CHEMISTRY OF CYCLOPHOSPHATES

M.T. AVERBUCH-POUCHOT and A. DURIF

Centre National de la Recherche Scientifique
Laboratoire de Cristallographie, associé à l'Université J. Fourier
166X - 38042 Grenoble Cédex, France.

1. INTRODUCTION

Cyclophosphates constitute an important part of the condensed phosphate chemistry. This chemistry has been very long to develop and, even today, is still relatively poor if compared with that of silicates, for instance.

Let us start explaining what is commonly called a condensed phosphate. We can very simply say, before proceeding into the subject with more details, that any phosphoric anion in which exists a P-O-P bond is a condensed phosphoric anion. Another simple way to define these salts is to say that any phosphoric anion corresponding to a formula characterized by a ratio P/O larger than 1/4 is a condensed one. It can be noticed that this last definition is a direct consequence of the first one, since the basic unit of all phosphoric anions is the PO_4 tetrahedron.

These P-O-P bonds can be obtained by various processes ; one of the simplest is the reorganization of two molecules of a monohydrogenomonophosphate after the elimination of a water molecule :

```
    O               O                O       O
    |               |                |       |
O - P - O-H + H-O - P - O ----> O -  P - O - P - O + H2O
    |               |                |       |
    O               O                O       O
```

This scheme of condensation illustrates the most common way to prepare, for instance, tetrasodium diphosphate :

$$2Na_2HPO_4 \text{ -------> } Na_4P_2O_7 + H_2O$$

This condensation phenomenon can generate a great number of phosphoric anions with various geometries and it is today recognized that no classification of condensed phosphates is possible unless supported by the geometry of the anions.

1.1. Classification of Condensed Phosphates

In the present state of the crystal chemistry of condensed phosphates, one observes, for the anions, three very different types of condensation geometries.

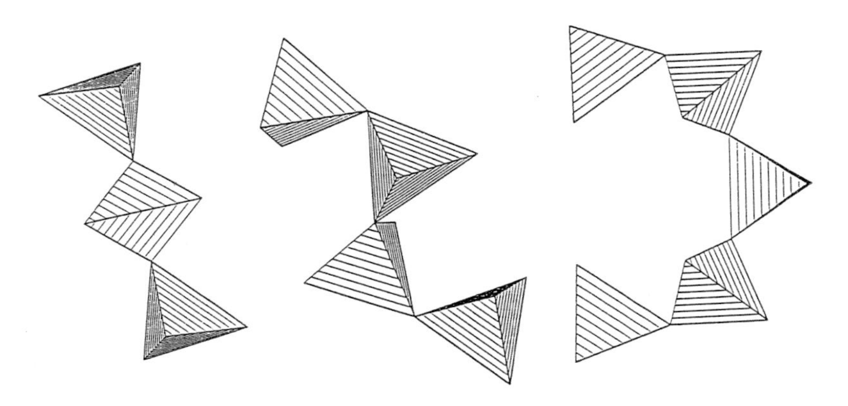

Figure 1.1 Some examples of polyphosphate anions.

The first one corresponds to a progressive linear linkage of PO_4 tetrahedra sharing one or two of their oxygen atoms. Fig. 1.1 reports some examples of such condensed anions. The corresponding phosphates are usually called *polyphosphates*. The general formula for this type of anions is given by : $(P_nO_{3n+1})^{(n+2)-}$.
Phosphorus atoms, belonging to PO_4 tetrahedra sharing two of their oxygen atoms with neighboring PO_4, are usually called "internal" phosphorus, the other ones "terminal" phosphorus.

The second type of condensation is a cyclic one, leading to the formation of $(P_nO_{3n})^{n-}$ rings. Today, phosphoric rings are known for n = 3, 4, 5, 6, 8, 10 and 12. The corresponding phosphates are named *cyclophosphates*. Fig. 1.2 illustrates some ring-anions.

If in the first two types of condensation, one PO_4 tetrahedron shares one or two of its oxygen atoms with the neighboring PO_4 groups, things are different for the third type of condensation observed in a class of P_2O_5-rich phosphates, called *ultraphosphates*. Here, in the anion, some PO_4 tetrahedra share three of their oxygen atoms with the neighboring PO_4 groups. This type of condensation leads to very various anion geometries : finite groups, infinite ribbons, infinite layers or three-dimensional networks. The phosphorus atom of a PO_4 tetrahedron sharing three of its oxygen atoms with adjacent tetrahedra is named "branching" phosphorus. Fig. 1.3 gives the representation of an ultraphosphate anion.

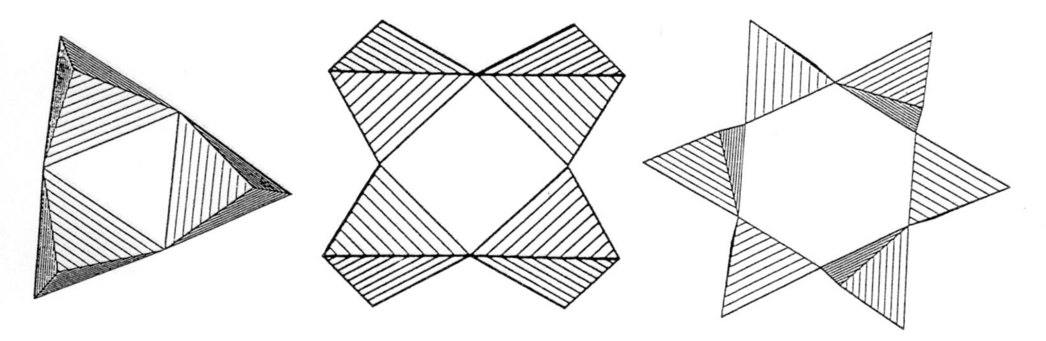

Figure 1.2 Some examples of ring-anions.

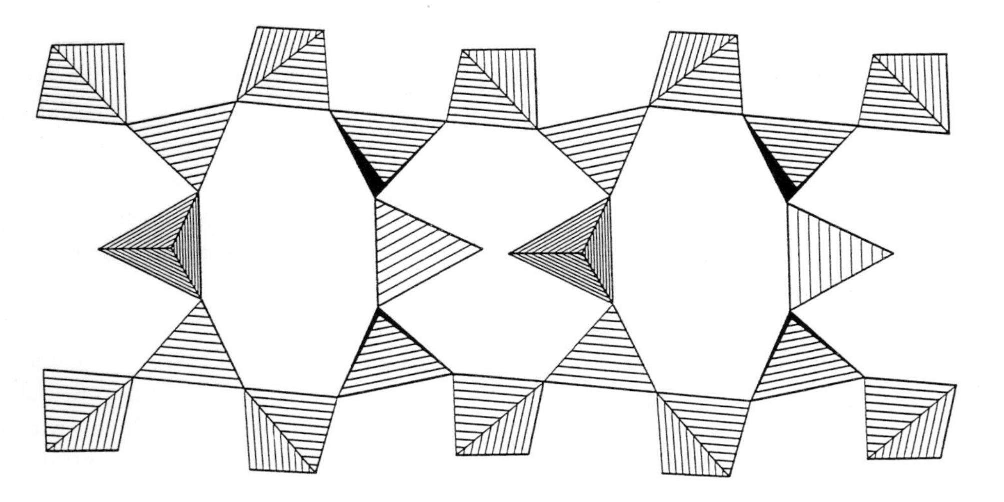

Figure 1.3 An ultraphosphate anion as observed in SmP_5O_{14}.

1.2. Nomenclature of Condensed Phosphates

Along years, the nomenclature used for condensed phosphates has been very often confusing. In the few next lines, we simply report what is today commonly accepted for the appellation of the condensed phos-phates.

In the case of phosphates containing anions corresponding to the first type of condensation geometry, the general appellation is *polyphos-phates*, but with some variations according to the degree of condensation. As we said above, in these salts the general formula of the phosphoric anion is : $(P_nO_{3n+1})^{(n+2)-}$. For small degrees of condensation (2<n<5), the usual name is *oligophosphates* and the today accepted nomenclature is the following one :

- for n = 2, (the P_2O_7 anion), the corresponding salts have for a long time been called "pyrophosphates", it is now the rule to say *diphosphates.*

- for n = 3, (the P_3O_{10} anion), the presently accepted appelation is *triphosphates* , but "tripolyphosphates" is still erroneously used.

- for n = 4 and 5, (the P_4O_{13} and P_5O_{16} anions), the denominations *tetraphosphates* and *pentaphosphates* are now preferred to the previously used "tetrapolyphosphates" and "pentapolyphosphates".

- when n is very large, the formula of the anion becomes close to PO_3 and its geometry is that of an infinite chain. Phosphates containing this type of anion are commonly called *long-chain polyphosphates.*

Phosphates, containing cyclic anions of general formula P_nO_{3n}, have for a long time been called "metaphosphates" (trimetaphosphates, tetrametaphosphates...). Actually it is convenient to employ more descriptive words such as : *cyclotriphosphates* for phosphates containing P_3O_9 anions, *cyclotetraphosphates, cyclopentaphosphates* for phosphates including larger ring-anions : P_4O_{12}, P_5O_{15}...

For *ultraphosphates*, corresponding to the third type of condensation, no systematic or logical nomenclature exists. This lack is a source of frequent confusions, since some of them have been erroneously designated by physicists or crystallographers according to the number of phosphorus atoms in the formula unit. So, some rare-earth ultraphosphates, with LnP_5O_{14} as general formula, are commonly known as "pentaphosphates", a denomination now used for the oligophosphates containing the P_5O_{16} anion. The restricted number of well characterized ultraphosphates can probably explain the absence of any coherent nomenclature and classification for this kind of compounds, still not extensively investigated.

1.3. Short Survey of Condensed Phosphate History

The present knowledge about the geometry of condensed phosphoric anions is the fruit of a long period of confusion spreading over more than one century and half. Since the fundamental discovery of Graham [1] and his early attempt to classify phosphates into three main groups, ortho-, pyro- and metaphosphates, a great amount of studies have been published dealing mainly with the various possibilities of condensation for "metaphosphates". For the longest part of this period, chemical analysis being the only way to characterize a compound, the degrees of condensation of the various forms of $NaPO_3$, for example, were guessed rather than really established through the determination of the formula of precipitated double salts, $BaNaP_3O_9.4H_2O$, for instance. By this way and through many experiments, Fleitmann and Henneberg [2] came to the conviction of the existence of a trimeric form of $NaPO_3$ called trimetaphosphate. More than twenty five years later, Lindbom [3] suggested a cyclic formulation for the corresponding anion.

Later on, the development of cryoscopic methods, the introduction of the techniques of chromatography by Ebel [4] and mainly the great possibilities offered by X-ray diffraction analysis clarified progressively this chapter of Inorganic Chemistry.

Several well-documented surveys of the development of the condensed phosphate chemistry have been published during the past forty years by Topley [5], Van Wazer [6], Thilo [7] and Kalliney [8]. Unfortunately, these reviews were all published before the huge expansion of X-ray structural analysis and so, cannot report many fundamental results obtained through this technique.

1.4. General Organization of the Review

Each chapter will contain, first, a short introduction dealing with the development along years of the corresponding matter, followed by a survey of the present state of chemistry in this field. This part does not pretend to be exhaustive, but tries to report as completely as possible what is today known about crystalline materials characterized with a good degree of certainity. This chemical part will be always complemented by updated tables reporting, when they are available, the main crystallographic features of the described materials.

When general methods of preparation exist, they will be described in a separate section intitled "Chemical Preparation of...". But, if a reported chemical preparation appears as being non-conventional, it will be described in the survey of "Present State of... Chemistry".

An important part of each chapter intitled "Some Atomic Arrangements of..." is devoted to the description of some structure types considered as the most representative within the family described.

A detailed study of the various geometries and internal symmetries of the ring anions will terminate each chapter.

Some abreviations will be commonly used along this review : S. G. for space-group, Z for the number of formula units in the unit cell, m.p. for melting-point. The units used are : ångströms for lengths, decimal degrees for angles and Kelvin degrees for temperatures.

A good number of data indispensable for the discussion of the ring-anion geometries are not always given in publications (P-P-P angles for instance), so these values have been recalculated by the reviewers from the atomic parameters found in literature.

Most of the drawings have been performed using a polyhedral representation of the ring anions. This process seems to us the most pedagogical one in a review mainly devoted to the geometry of tetrahedral anions. All the polyhedral drawings have been made using the STRUPLO system [9].

REFERENCES

1 T. Graham, *Phil. Trans. Roy. Soc.*, A123 (1833) 253.

2 T. Fleitmann and W. Henneberg, *Ann.*, 65 (1848) 304.

3 C.G. Lindbom, *Ber.*, 8 (1875) 122.

4 J.P. Ebel, *Bull. Soc. Chim. Fr.*, (1968) 1663.

5 B. Topley, *Quart. Rev.*, 3 (1949) 345.

6 J.R. Van Wazer, *Phosphorus and its Compounds,* Vol. 1, Interscience, New-York, 1966.

7 E. Thilo, *Adv. Inorg. Chem. Radiochem.*, 4 (1962) 1.

8 S.Y. Kalliney, *Topics in Phosphorus Chemistry*. Vol. 7 : Cyclophosphates, Interscience, New-York, 1972, pp. 255-309.

9 R.X. Fischer, *J. Appl. Cryst.*, 18 (1985) 258.

2. CYCLOTRIPHOSPHATES

During the past century, several chemists suspected the existence of cyclic phosphoric anions and produced several crystalline compounds possessing this type of anions. For instance, sodium cyclotriphosphate seems to have been prepared as early as 1833 by Graham [1]. In the middle of the past century, Fleitmann and Henneberg [2], using the Graham's principles, prepared several salts and double salts, today recognized as cyclotriphosphates. They also suggested the trimeric nature of the anion. During the second part of the past century some other chemists as Lindbom [3], Tammann [4, 5], Von Knorre [6] described a good number of additional cyclotriphosphates. It must be recognized that most of them have been, more or less recently, reproduced and clearly characterized as cyclotriphosphates. During the first half of our century, this part of chemistry did not develop significantly because no proper starting material was available in convenient amounts and purity for syntheses. The chemical preparations of sodium cyclotriphosphate, the parent compound of almost all the compounds prepared during the past century, were uneasy, tedious and in many cases not reproducible. Nevertheless, as early as 1938, Boullé [7] succeeded in preparing silver cyclotriphosphate monohydrate, $Ag_3P_3O_9.H_2O$, and showed this compound can be used through a metathesis reaction to synthesize water-soluble cyclotriphosphates. This reaction has, since this date, been extensively used and is today known as the Boullé's process. This process will be described in details in the part devoted to the general methods of preparation of cyclotriphosphates. It was not before the early fifties that a careful study of the thermal reorganization of NaH_2PO_4 by Thilo and Grunze [8] opened the way to a reproducible production of pure $Na_3P_3O_9$ according to the following scheme :

$$3NaH_2PO_4 \text{ -------> } Na_3P_3O_9 + 3H_2O$$

The first structural evidence for the cyclic nature of the anion, already clearly suggested by Lindbom [3], was reported by Eanes and Ondik [9] in 1965, when performing the crystal structure of $LiK_2P_3O_9.H_2O$.

During the past 25 years more than one hundred cyclotriphosphates have been characterized.

2.1. Present State of the Cyclotriphosphate Chemistry

2.1.1. Cyclotriphosphates of monovalent cations

- *$Li_3P_3O_9.3H_2O$*

Up to now, the only reported lithium cyclotriphosphate is the trihydrate $Li_3P_3O_9.3H_2O$. Its first characterization was reported by Eanes [10] and its atomic arrangement was later on determined by Masse *et al* [11, 12]. Grenier and Durif [13] observed that the trihydrate is stable up to 403K and that the release of water molecules occurs between 403 and 423K. At temperatures higher than 423K, it slowly transforms into the long-chain polyphosphate. This compound is normally prepared by the Boullé's metathesis reaction [7]. The hydrolytic decyclization kinetics of this salt in LiOH aqueous solution, in a temperature range 313-353K, has been studied by Sotnikova-Yuzhik *et al* [14]. The decyclization scheme is :

$$Li_3P_3O_9 + 2LiOH \text{ ------> } Li_5P_3O_{10} + H_2O$$

A crystalline triphosphate, $Li_5P_3O_{10}.5H_2O$, is so obtained with a good yield.

No anhydrous lithium cyclotriphosphate has been observed during the process of thermal dehydration-condensation of LiH_2PO_4 (see "Cyclo-hexaphosphates" chapter).

- $Na_2HP_3O_9$

Two crystalline forms of $Na_2HP_3O_9$ have been reported by Griffith [15]. The products obtained by this author were always polycrystalline. According to the author, form (I) is not water soluble, while form (II) is. Crystals of $Na_2HP_3O_9$ (I) have been obtained by Averbuch-Pouchot *et al* [16] during a study of the P_2O_5-SrO-Na_2O system. The authors report the chemical preparation and give a detailed description of the atomic arrangement of this salt, which, up to now, seems to be the only example of an acidic cyclotriphosphate.

- $Na_3P_3O_9.nH_2O$

A detailed crystal-chemical study of $Na_3P_3O_9$ and of its various hydrates has been performed by Ondik and Gryder [17]. $Na_3P_3O_9$, $Na_3P_3O_9.H_2O$, $Na_3P_3O_9.3/2H_2O$, $Na_3P_3O_9.3H_2O$ and $Na_3P_3O_9.6H_2O$ are clearly characterized and for the first time, unit-cell dimensions for these compounds are reported. Chemical preparations and conditions of stability of the various hydrates are described. The authors observed neither the existence of the hemihydrate nor that of polymorphs for $Na_3P_3O_9$ as well as for $Na_3P_3O_9.H_2O$, as previously reported in chemical literature. They also noticed a stricking analogy between unit-cell parameters of $Na_3P_3O_9$ and its monohydrate.

Crystal structure determinations of $Na_3P_3O_9$ and $Na_3P_3O_9.H_2O$ were performed some years later by Ondik [18]. The two atomic arrangements are closely related, as previously suggested by the analogy of their unit-cell dimensions [17]. In spite of this similarity, the water molecule of the monohydrate is not of a zeolitic nature but belongs to the coordination polyhedra of the sodium atoms.

Crystal structure of $Na_3P_3O_9.6H_2O$ has been performed by Tordjman and Guitel [19]. A detailed investigation of the dehydration of this salt has been reported by Prodan and Pytlev [20].

- $Ag_3P_3O_9.nH_2O$

As early as 1848, the existence of $Ag_3P_3O_9.H_2O$ was reported by Fleitmann and Henneberg [2]. These authors also reported the existence of the anhydrous salt, up to now not confirmed. Later, Lindbom [3] and Von Knorre [6)]characterized also the monohydrate, but Boullé [7] was the first to describe a reproducible process for the preparation of this salt and to demonstrate its possible use as a good starting material for the preparation of water soluble cyclotriphosphates. Since this publication, this salt has been extensively used and what is today known as the Boullé's process is described in "Chemical Preparation" section. The first crystallographic investigations of $Ag_3P_3O_9.H_2O$ have been performed simultaneously by Eanes [21] and Grenier [22] and its atomic arrangement determined by Bagieu-Beucher *et al* [23].

- $K_3P_3O_9$ and $(NH_4)_3P_3O_9$

Both $K_3P_3O_9$ and $(NH_4)_3P_3O_9$ have been prepared by the Boullé's process and investigated by Grenier [22] and Grenier and Durif [24]. The two salts are isotypic. This type of atomic arrangement has been determined by Bagieu-Beucher *et al* with the potassium salt [25].

$K_3P_3O_9$ transforms irreversibly into a long-chain polyphosphate on heating :

$$K_3P_3O_9 \dashrightarrow 3KPO_3$$

Excepted for the ammonium salt, this fact is common to all monovalent cation cyclotriphosphates. In the case of the potassium salt, the effect of various factors on the kinetics of this thermal conversion have been investigated by Dombrovskii and Koval [26].

- $Rb_3P_3O_9.H_2O$ and $Cs_3P_3O_9.H_2O$

$Rb_3P_3O_9.H_2O$ and $Cs_3P_3O_9.H_2O$, are isotypic. Grenier [22] was the first to describe chemical preparation of these two phos-phates by the Boullé's process and to report their main crystallographic features. Crystal structure for this type of compounds was later determined by Tordjman *et al* [27] using the caesium salt.

- $Tl_3P_3O_9$

$Tl_3P_3O_9$ was first described by Grenier [22] and its crystal structure solved by Boudjada [28].

- $LiK_2P_3O_9.H_2O$

Eanes and Ondik [9] prepared $LiK_2P_3O_9.H_2O$ by using the Boullé's process ; they report a complete determination of the atomic arrangement.

- $Na_2KP_3O_9$

The $NaPO_3$-KPO_3 phase-equilibrium diagram was first investigated by Tamman and Ruppelt [29]. They found this diagram to be of the eutectic type, nevertheless, they reported the existence of $Na_2KP_3O_9$. Then, Morey [30] redetermined this diagram and found $Na_3K(PO_3)_3$ to be the only compound in this system. Griffith and Van Wazer [31] characterized this last compound as a cyclotriphosphate. Bukhalova and Mardirosova [32] did not confirmed the Morey's result and found $Na_2KP_3O_9$ as the only compound in the system. Finally, this salt appears as a congruent melting compound (m.p. = 832K) in the $NaPO_3$-KPO_3 phase-equilibrium diagram

- $Na_2RbP_3O_9$

The salt $Na_2RbP_3O_9$, which was characterized by Cavero-Ghersi and

Durif [35] during the elaboration of the $NaPO_3$-$RbPO_3$ phase-equilibrium diagram (Fig. 2.2), is isotypic with $Na_2KP_3O_9$ and appears as an incongruent melting compound decomposing at 785K.

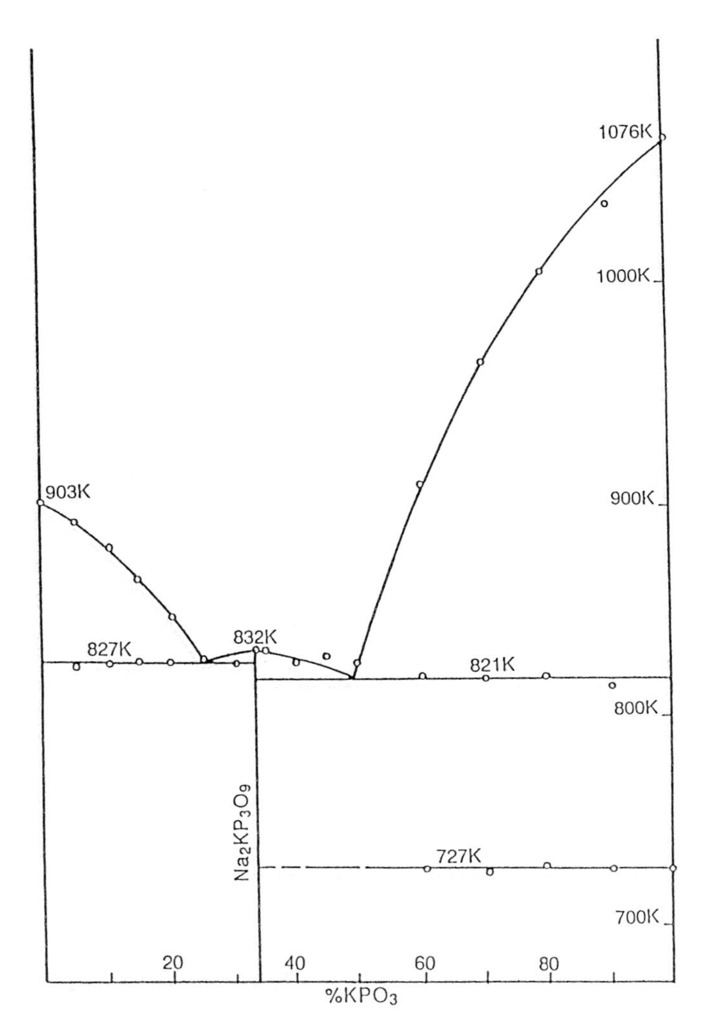

Figure 2.1 The $NaPO_3$-KPO_3 phase-equilibrium diagram as revised by Cavero-Ghersi and Durif [33].

- ***$Na_2TlP_3O_9$ and $Na_2NH_4P_3O_9$***

$Na_2TlP_3O_9$ and $Na_2NH_4P_3O_9$ have been prepared by Cavero-Ghersi [36] and identified as isotypes of $Na_2KP_3O_9$.

In spite of numerous analogies in the geometry of condensed silicate and phosphate anions, very few cases of isomorphism between silicates

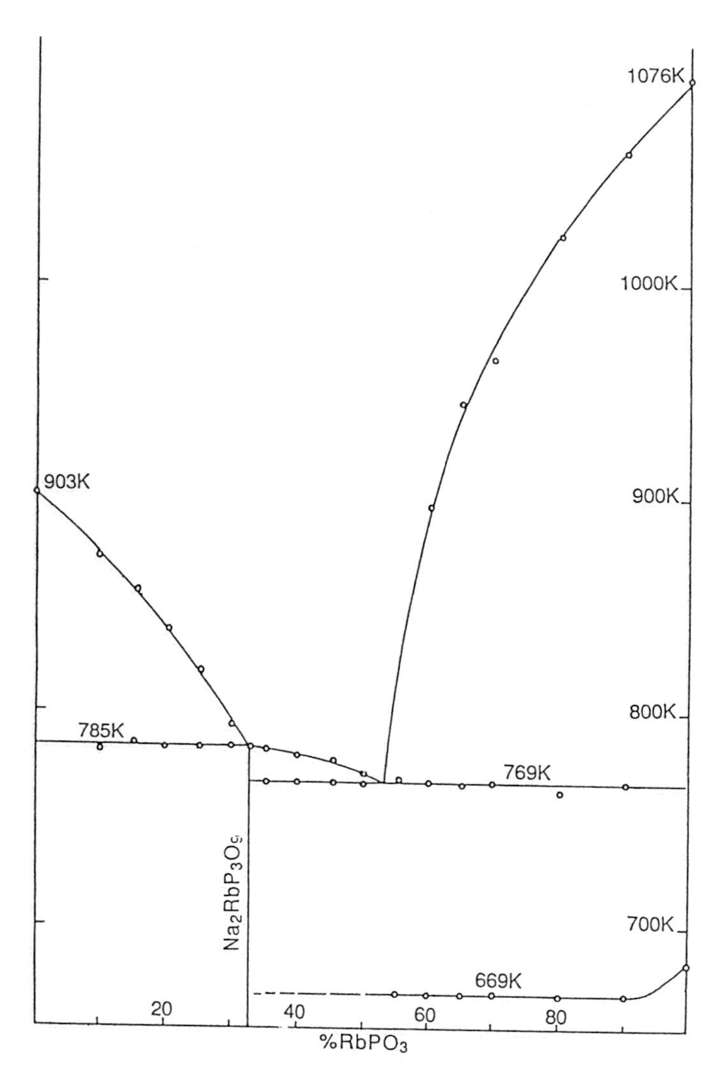

Figure 2.2 The $NaPO_3$-$RbPO_3$ phase-equilibrium diagram [35].

and phosphates have been reported. So we think it is worth noticing that four of the compounds described above, $Na_2KP_3O_9$, $Na_2RbP_3O_9$, $Na_2TlP_3O_9$ and $Na_2NH_4P_3O_9$ are isotypic of some margarosanite-type silicates, like $Ca_2BaSi_3O_9$, $Ca_2PbSi_3O_9$...

- $Na_2CsP_3O_9.2H_2O$

$Na_2CsP_3O_9.2H_2O$ was first observed during attempts to prepare caesium cyclotriphosphate by the Boullé's process with a silver cyclotriphosphate contaminated by sodium. Its crystal structure has been performed by Boudjada [28].

- ***$Na_2LiP_3O_9.4H_2O$***

The mixed compound, $Na_2LiP_3O_9.4H_2O$, was accidentally obtained for the first time during a preparation of lithium cyclotriphosphate by ion-exchange resins. A reproducible preparation was later reported by Averbuch-Pouchot and Durif [37] who performed an accurate determination of its atomic arrangement.

Table 2.1
Crystal data for monovalent cation cyclotriphosphates.

Formula	a α	b β	c (Å) γ (°)	S. G.	Z	Ref.
$Li_3P_3O_9.3H_2O$	12.511(4)	12.511(4)	5.594(2)	R3	3	10-12
$Na_2HP_3O_9$	7.788(5) 116.69(5)	7.809(5) 103.41(5)	7.129(5) 81.94(5)	$P\bar{1}$	2	16
$Na_3P_3O_9$	7.928(2)	13.214(3)	7.708(2)	Pmcn	4	18
$Na_3P_3O_9.H_2O$	8.500(1)	13.189(1)	7.558(1)	Pmcn	4	18
$Na_3P_3O_9.6H_2O$	7.826(4) 112.79(3)	9.530(5) 106.10(3)	10.828(5) 88.55(3)	$P\bar{1}$	2	19
$Ag_3P_3O_9.H_2O$	7.78(1) 114.70(5)	9.24(1) 91.00(5)	7.78(1) 114.70(5)	$P\bar{1}$	2	21-23
$K_3P_3O_9$	11.074(8)	11.965(9) 102.18(3)	7.350(6)	$P2_1/n$	4	22, 24, 25
$(NH_4)_3P_3O_9$	11.515(6)	12.206(8 101.63(5)	7.699(4)	$P2_1/n$	4	22, 24
$Rb_3P_3O_9.H_2O$ *	10.240(12) 96.72(5)	7.778(8) 69.32(5)	7.813(8) 96.23(5)	$P\bar{1}$	2	22
$Cs_3P_3O_9.H_2O$	10.610(6) 96.64(8)	7.966(4) 68.84(8)	8.172(5) 95.42(8)	$P\bar{1}$	2	22, 27
$Tl_3P_3O_9$	12.035(6)	18.74(1) 90.63(4)	8.940(3)	$P2_1/a$	8	22, 28

* Triclinic setting was recalculated for comparison with $Cs_3P_3O_9.H_2O$.

The main crystallographic features for monovalent and mixed monovalent cation cyclotriphosphates are given in Tables 2.1 and 2.2, respectively.

Table 2.2
Crystal data for mixed monovalent cation cyclotriphosphates.

Formula	a α	b β	c (Å) γ (°)	S. G.	Z	Ref.
$LiK_2P_3O_9.H_2O$	8.669(3)	14.497(4) 99.72(5)	7.634(6)	$P2_1/c$	4	9
$Na_2KP_3O_9$	6.886(2) 110.07(2)	9.494(3) 104.69(2)	6.797(2) 86.68(2)	$P\bar{1}$	2	33, 34
$Na_2RbP_3O_9$	7.010(2) 108.98(5)	9.542(3) 104.26(5)	6.783(3) 87.37(5)	$P\bar{1}$	2	35
$Na_2TlP_3O_9$	6.977 108.70	9.511 104.40	6.787 86.82	$P\bar{1}$	2	36
$Na_2NH_4P_3O_9$	6.918 106.87	9.412 106.87	7.006 88.09	$P\bar{1}$	2	36
$Na_2CsP_3O_9.2H_2O$	11.393(6)	13.188(8) 102.31(4)	7.622(3)	$P2_1/a$	4	28
$Na_2LiP_3O_9.4H_2O$	6.905(5) 95.00(5)	9.346(5) 104.36(5)	9.876(5) 107.75(5)	$P\bar{1}$	2	37

2.1.2. Cyclotriphosphates of divalent cations

Some anhydrous cyclotriphosphates of divalent cations have been reported in the chemical literature, but their existence has not yet been confirmed. Up to now, only cadmium, calcium, manganese, strontium, barium and lead cyclotriphosphate hydrates have been clearly characterized.

- $Cd_3(P_3O_9)_2.14H_2O$

The first evidence for the existence of a cadmium cyclotriphosphate was reported by Averbuch-Pouchot and Durif [38] when describing the chemical preparation by the Boullé's process and the main crystallographic data for the cadmium cyclotriphosphate tetradecahydrate : $Cd_3(P_3O_9)_2.14H_2O$. Later on, its atomic arrangement was determined by Averbuch-Pouchot *et al* [39] and re-examined at low temperature by Averbuch-Pouchot *et al* [40]. The physical properties of this salt have been carefully investigated by Michot [41], Simonot-Grange [42] and Simonot-Grange and Michot [43-44]. These authors showed that, under dynamical vacuum, eight of the fourteen water molecules can be removed without any alteration of the atomic framework and that this phenomenon is reversible. They also investigated the decondensation scheme of the tetradecahydrate in wet atmospheres at various temperatures.

- $Cd_3(P_3O_9)_2.10H_2O$

For some reasons not yet clearly understood, the Boullé's process, leading normally to the tetradecahydrate at room-temperature, sometimes produces the decahydrate $Cd_3(P_3O_9)_2.10H_2O$. Crystals, so obtained, are invariably twinned. Good quality crystals, suitable for a structural study, have been prepared accidentally during experiments for the preparation of an yttrium-lithium-cadmium cyclotriphosphate. The crystal structure determined by Averbuch-Pouchot *et al* [46] shows that two of the ten water molecules are of a zeolitic nature.

No relationship exists between the structures of the two hydrates, explaining why the decahydrate was never observed during dehydration experiments on the tetradecahydrate.

- $Ca_3(P_3O_9)_2.10H_2O$ and $Mn_3(P_3O_9)_2.10H_2O$

$Ca_3(P_3O_9)_2.10H_2O$ was first apparently prepared by Boullé [7] as an example for the use of his metathesis reaction. The author attributed to this salt the formula of an enneahydrate. The existence of a manganese cyclotriphosphate with 9 or 11 water molecules has been reported at the end of the past century by Tammann [5] and by Von Knorre [6]. El-Horr and Durif [47] prepared these two salts by reaction of $H_3P_3O_9$ with the corresponding carbonates and showed they are isotypes of the cadmium decahydrate salt. An heptahydrate of calcium, isotypic with the strontium salt described below, has been prepared by Durif [45], but its preparation is not reproducible.

- $Sr_3(P_3O_9)_2.7H_2O$

Ropp *et al* [48] claimed for the preparation of an anhydrous strontium cyclotriphosphate, but up to now, there is no structural evidence for the existence of such a salt. The only reported cyclotriphosphate of strontium is the hydrate $Sr_3(P_3O_9)_2.7H_2O$, described by Durif *et al* [49]. The authors report its chemical preparation by the Boullé's process and its main crystallographic features. The atomic arrangement of this salt was later determined by Tordjman *et al* [50].

- $Ba_3(P_3O_9)_2.6H_2O$ and $Ba_3(P_3O_9)_2.4H_2O$

Up to now, there is no evidence for the existence of the anhydrous barium cyclotriphosphate reported by Fleitmann and Henneberg [2], but two hydrates are well characterized.

$Ba_3(P_3O_9)_2.6H_2O$ was originally prepared through the Boullé's process [7] by Grenier and Martin [51]. A detailed structural characterization was later reported by Masse *et al* [52]. Two of the six water molecules are of a zeolitic nature. Some time after, Averbuch-Pouchot and Durif [53] optimized a chemical preparation of this hydrate not involving the Boullé's process. The thermal behavior of this salt has been carefully investigated by Tacquenet [54] and by Thrierr-Sorel *et al* [55].

$Ba_3(P_3O_9)_2.4H_2O$ was prepared by Averbuch-Pouchot and Durif [53] when investigating in details the reactions between $Na_3P_3O_9$ and $BaCl_2$ solutions at various concentrations. The authors reported a reproducible process for producing the tetrahydrate as polycrystalline samples or as single crystals. By adding slowly a very concentrated aqueous solution of $BaCl_2$ (15 g of $BaCl_2.2H_2O$ in 35 cm^3 of water) to a concentrated aqueous solution of $Na_3P_3O_9$ (3 g of $Na_3P_3O_9$ in 25 cm^3 of water), without any mechanical stirring, one observes almost immediately the formation of a large amount of small tufts of lath-like needles. After filtration to eliminate the mother liquor, an equivalent volume of water is added to the solid phase, which within a few minutes is transformed into small crystals of $Ba_3(P_3O_9)_2.4H_2O$. By performing the same experiment with less concentrated solutions, the same quantities of starting materials being respectively dissolved in 50 cm^3 of water, one observes after some hours the formation of large crystals of the tetrahydrate. In this case, the preparation may be contaminated by 10 to 20% of large elongated prisms of the hexahydrate. All the experiments have been run at room temperature. Crystals of $Ba_3(P_3O_9)_2.4H_2O$ appear as stout monoclinic prisms stable at room temperature. The authors [53] report a description of the atomic arrangement.

For these two hydrates, firing at 873K for some hours leads to the polyphosphate :

$$Ba_3(P_3O_9)_2.4(6)H_2O \text{ -------> } 3\ Ba(PO_3)_2 + 4(6)H_2O$$

- $Pb_3(P_3O_9)_2.3H_2O$

$Pb_3(P_3O_9)_2.3H_2O$ has been reported for the first time in 1848 by Fleitman and Henneberg [1]. Later, Durif and Brunel-Laügt [56] improved its chemical preparation and reported its main crystallographic features. The crystal structure has been determined by Brunel-Laügt *et al* [57]. The thermal dehydration of this salt has been investigated by Kuz'menkov *et al* [58].

Various other divalent cation cyclotriphosphates have been reported at the end of the past century : $Zn_3(P_3O_9)_2.9H_2O$, $Cu_3(P_3O_9)_2.9H_2O$ and $Co_3(P_3O_9)_2.9H_2O$ by Tammann [5], $Fe_3(P_3O_9)_2.12H_2O$ by Lindbom [3], but their existence has not yet been confirmed.

- $Ba_2Zn(P_3O_9)_2.10H_2O$

$Ba_2Zn(P_3O_9)_2.10H_2O$ has been prepared by Durif *et al* [59] using the Boullé's process. The authors report a complete description of its atomic arrangement. This salt is up to now the only mixed divalent cation cyclotriphosphate reported in chemical literature.

The main crystallographic features of the divalent cation cyclotriphosphates are reported in Table 2.3.

Table 2.3
Crystal data for divalent cation cyclotriphosphates.

Formula	a α	b β	c (Å) γ (°)	S. G.	Z	Ref.
$Cd_3(P_3O_9)_2.14H_2O$	12.228(3)	12.228(3)	5.451(3)	$P\bar{3}$	1	38-40
$Cd_3(P_3O_9)_2.10H_2O$	9.424(8)	17.87(1) 107.72(1)	7.762(7)	$P2_1/n$	2	46
$Ca_3(P_3O_9)_2.10H_2O$	9.332(7)	18.13(1) 106.69(5)	7.841(5)	$P2_1/n$	2	47
$Mn_3(P_3O_9)_2.10H_2O$	9.219(4)	17.733(8) 107.37(2)	7.644(3)	$P2_1/n$	2	47
$Sr_3(P_3O_9)_2.7H_2O$	16.05(1)	12.33(1)	10.87(1)	Pnma	4	49, 50
$Ba_3(P_3O_9)_2.6H_2O$	7.547(4) 108.58(8)	11.975(6) 100.35(8)	13.068(8) 95.54(8)	$P\bar{1}$	2	51, 52
$Ba_3(P_3O_9)_2.4H_2O$	16.09(1)	8.368(5) 95.38(5)	7.717(3)	C2/m	2	53
$Pb_3(P_3O_9)_2.3H_2O$	11.957(5)	11.957(5)	12.270(5)	$P4_12_12$	4	56, 57
$Ba_2Zn(P_3O_9)_2.10H_2O$	26.52(3)	7.625(5) 100.93(5)	12.92(1)	C2/c	4	59

2.1.3. Divalent-monovalent cation cyclotriphosphates

M^{II}- M^{I} cyclotriphosphates presently investigated can be roughly classified into two groups according to their chemical formulas :

$$M^{II}M^{I}_4P_3O_9.nH_2O \text{ and } M^{II}M^{I}P_3O_9.nH_2O$$

It can be noticed that lithium has never been observed in this series of compounds.

Most of the anhydrous salts have been characterized during the elaborations of M^IPO_3-$M^{II}(PO_3)_2$ phase-equilibrium diagrams and prepared as single crystals by various flux methods, while the great majority of the hydrates has been synthesized by the Boullé's process [7] or by classical methods of aqueous chemistry.

Anhydrous divalent-monovalent cation cyclotriphosphates

- $CaNa_4(P_3O_9)_2$ and $CdNa_4(P_3O_9)_2$

Griffith [60] reported a chemical preparation for the calcium salt and by various experiments brought the proof for the cyclic nature of its anion. Grenier *et al* [61] revised the $NaPO_3$-$Ca(PO_3)_2$ phase-equilibrium diagram, previously established by Morey [62] and reported crystal data for $CaNa_4(P_3O_9)_2$ which is a congruent melting compound (m.p. = 1006K).

The cadmium salt has been characterized by Averbuch-Pouchot and Durif [63] during the elaboration of the $NaPO_3$-$Cd(PO_3)_2$ phase-equilibrium diagram ; it melts congruently at 968K. From the crystal data reported by the authors, this salt seems isotypic with the corresponding calcium salt.

- $PbNaP_3O_9$ and $PbNa_4(P_3O_9)_2$

The phase-equilibrium diagram $NaPO_3$-$Pb(PO_3)_2$ established by Mahama *et al* [64] shows the existence of only one compound, $PbNa_4(P_3O_9)_2$, melting congruently at 913K ; its crystal structure was afterwards determined by Averbuch-Pouchot and Durif [65]. During attempts to prepare crystals of $PbNa_4(P_3O_9)_2$, the authors discovered another cyclotriphosphate, $PbNaP_3O_9$, not observed in the phase-equilibrium diagram and identified it as an isotype of the corresponding barium salt described just below.

- $BaNaP_3O_9$ and $BaNa_4(P_3O_9)_2$

Two compounds, $BaNaP_3O_9$ and $BaNa_4(P_3O_9)_2$, appear in the $NaPO_3$-$Ba(PO_3)_2$ phase-equilibrium diagram (Fig. 2.3) elaborated by Martin and Durif [66], the first one as a congruent melting compound (m.p. = 939K), the second one as a non-congruent melting compound decomposing at 956K. The existence of the first one was quoted in 1848 by Fleitmann and Henneberg [2] and its atomic arrangement determined by Martin and Mitschler [67]. The main crystallographic features for $BaNa_4(P_3O_9)_2$ and its chemical preparation have been reported by Averbuch-Pouchot and Durif [65], who identified this salt as an isotype of $PbNa_4(P_3O_9)_2$.

- $CdAgP_3O_9$

Chemical preparation and crystal data have been reported for $CdAgP_3O_9$ by Pouchot *et al* [68], who recognized it as an isotype of the mineral benitoite, $BaTiSi_3O_9$. $CdAgP_3O_9$ appears as a non-congruent melting compound decomposing at 923K in the $AgPO_3$-$Cd(PO_3)_2$ phase-equilibrium diagram elaborated by Averbuch-Pouchot [69].

- $BaAgP_3O_9$

Crystal data for $BaAgP_3O_9$, prepared by heating at 623K the tetrahydrate described below, have been reported by Durif and Averbuch-

Pouchot [70]. It is isotypic of $BaNaP_3O_9$.

- $SrAgP_3O_9$

The $Sr(PO_3)_2$-$AgPO_3$ phase diagram has been established by Savenkova *et al* [71]. The non-congruent melting compound, $SrAgP_3O_9$, observed in this system is said to be a cyclotriphosphate, but no structural investigation has confirmed this assumption.

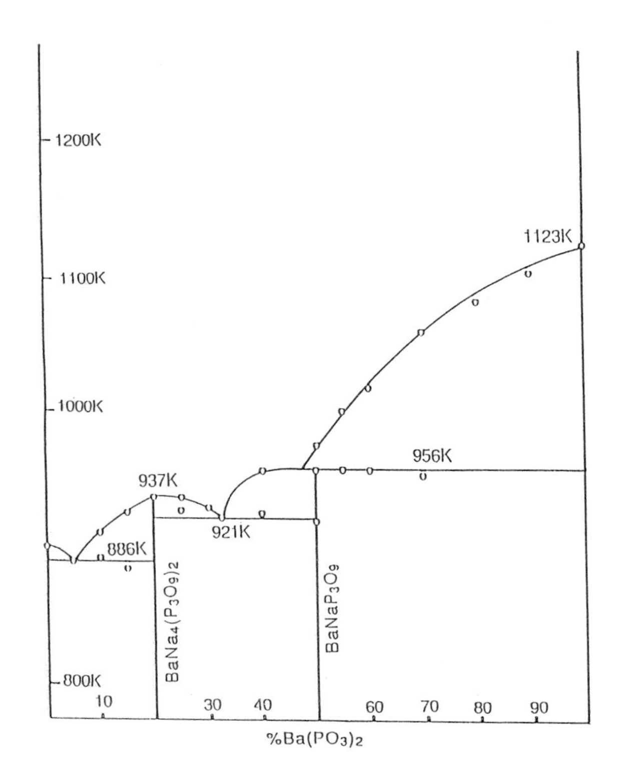

Figure 2.3 The $NaPO_3$-$Ba(PO_3)_2$ phase-equilibrium diagram [66].

- $CaKP_3O_9$ and $MgKP_3O_9$

The two salts, $CaKP_3O_9$ and $MgKP_3O_9$, were first characterized by Andrieu and Diament [72] during the elaborations of the KPO_3-$Mg_2P_4O_{12}$ and KPO_3-$Ca(PO_3)_2$ phase-equilibrium diagrams. They were soon recognized as isotypic with benitoite by Andrieu *et al* [73]. The KPO_3-$Mg_2P_4O_{12}$ phase-equilibrium diagram was subsequently revised by Averbuch-Pouchot *et al* [74]. This updated diagram, reported in Fig. 2.4, shows that $MgKP_3O_9$ is a congruent melting compound (m.p. = 1176K), as $CaKP_3O_9$ (m.p. = 1123K). Masse *et al* [75] refined the crystal structure of the magnesium salt. In addition, an orthorhombic form of $CaKP_3O_9$ has been prepared and described by Masse *et al* [76]. This second form is a distorsion of the benitoite arrangement.

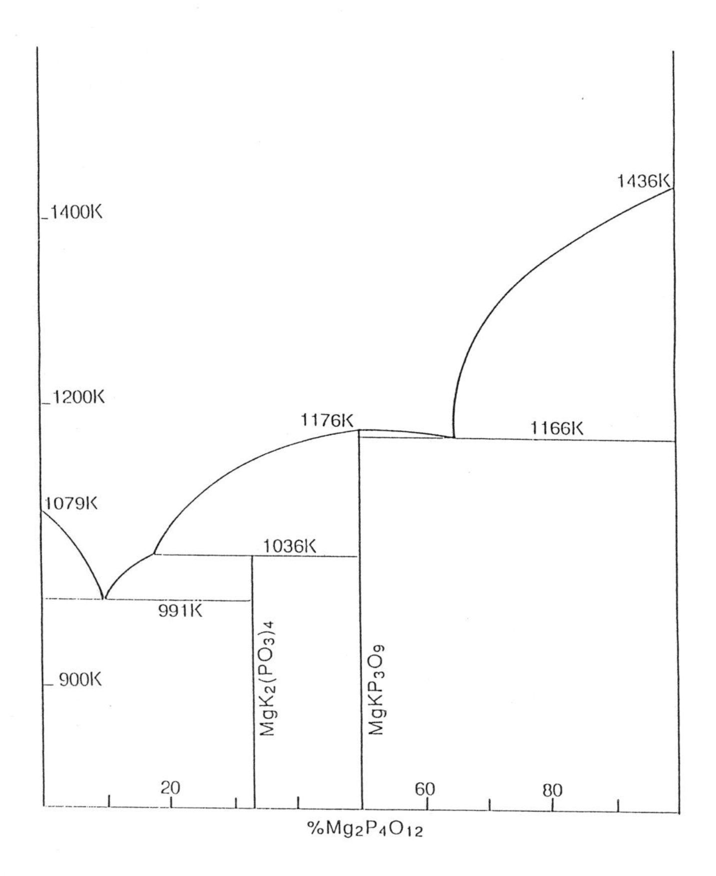

Figure 2.4 The KPO_3-$Mg_2P_4O_{12}$ phase-equilibrium diagram revised by Averbuch *et al* [74].

- $ZnKP_3O_9$, $CoKP_3O_9$, $MnKP_3O_9$ and $CdKP_3O_9$

The four cyclotriphosphates, $ZnKP_3O_9$, $CoKP_3O_9$, $MnKP_3O_9$ and $CdKP_3O_9$, have been prepared by Andrieu *et al* [73] and identified as isotypes of benitoite.

The zinc salt was said to be the only intermediate compound in the KPO_3-$Zn(PO_3)_2$ phase-equilibrium diagram elaborated by Krivovyazov *et al* [77]. This assumption was not confirmed during a reinvestigation of this system by Averbuch-Pouchot *et al* [74], who identified in the diagram a second compound, $ZnK_2(PO_3)_4$, a long-chain polyphosphate. $ZnKP_3O_9$ is a congruent melting compound (m.p. = 1111K).

The cobalt salt appears also as a congruent melting compound (m.p.

= 1069K) in the KPO_3-$Co_2P_4O_{12}$ phase-equilibrium diagram elaborated by Rakotomahanina-Rolaisoa [78].

The cadmium salt is a congruent melting compound (m.p. = 1021K) in the KPO_3-$Cd(PO_3)_2$ phase-equilibrium diagram (Fig. 2.5) elaborated by Mermet *et al* [79].

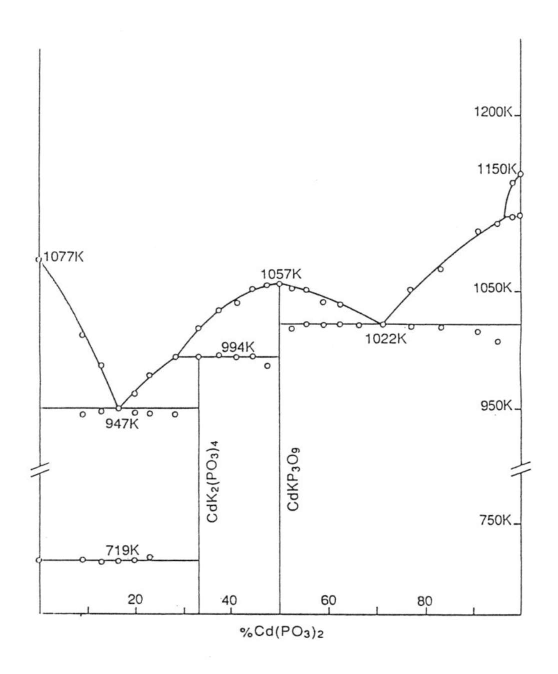

Figure 2.5 The KPO_3-$Cd(PO_3)_2$ phase-equilibrium [79].

- $HgKP_3O_9$

$HgKP_3O_9$ has been obtained during an investigation of the HgO-K_2O-P_2O_5 system by Averbuch-Pouchot and Durif [80]. The authors report a complete determination of the atomic arrangement showing the isotypy of this salt with $BaNaP_3O_9$.

- $M^{II}NH_4P_3O_9$ *(M^{II} = Zn, Co, Ca, Cd, Mg and Mn)*

Six cyclotriphosphates with the general formula $M^{II}NH_4P_3O_9$ have been synthesized by thermal methods and recognized as being isotypic with benitoite by Masse *et al* [75]. An accurate refinement of the crystal structure of the calcium-ammonium salt has been performed by Prisset [81]. Two of them, $MgNH_4P_3O_9$ and $CaNH_4P_3O_9$, are polymorphic. The magnesium salt can also crystallize as an orthorhombic compound. Its atomic arrangement determined by Grenier and Masse [82], a distorsion of the benitoite structure, is isotypic with the orthorhombic form of $CaKP_3O_9$.

The second form of the calcium salt is monoclinic. Its structure, first determined by Grenier and Masse [82] and later reinvestigated by Masse *et al* [76], is also a distorsion of the benitoite structure.

- $MgRbP_3O_9$

The cyclotriphosphate $MgRbP_3O_9$ is the high temperature form of the long-chain polyphosphate, $MgRb(PO_3)_3$, characterized during the elaboration of the $RbPO_3$-$Mg_2P_4O_{12}$ phase-equilibrium diagram by Rakotomahanina-Ralaisoa [83]. It is isotypic with the orthorhombic form of $MgNH_4P_3O_9$.

- $CdRbP_3O_9$

Chemical preparation of $CdRbP_3O_9$ has been described by Pouchot *et al* [68]. The crystal data reported by the authors show its isotypy with benitoite. This salt is a congruent melting compound (m.p. = 1024K) in the $RbPO_3$-$Cd(PO_3)_2$ phase-equilibrium diagram established by Mermet *et al* [79].

- $CaRbP_3O_9$

$CaRbP_3O_9$ appears as a congruent melting salt (m.p. = 1180K) in the $RbPO_3$-$Ca(PO_3)_2$ phase-equilibrium diagram elaborated by Henry and

Table 2.4

The common hexagonal space-group is $P\bar{6}2c$ for the 15 cyclotriphosphates isotypic with benitoite. Their unit cells reported in the present table contain two formula units.

Formula	a	c (Å)	Ref.
$CdAgP_3O_9$	6.622(2)	9.921(5)	68
$CaKP_3O_9$	6.795(1)	10.336(1)	73
$MgKP_3O_9$	6.605(1)	9.772(1)	73
$ZnKP_3O_9$	6.606(1)	9.743(1)	73
$CoKP_3O_9$	6.637(1)	9.795(1)	73
$MnKP_3O_9$	6.686(1)	9.958(1)	73
$CdKP_3O_9$	6.780(1)	10.148(1)	73
$ZnNH_4P_3O_9$	6.718(3)	9.819(5)	75
$CoNH_4P_3O_9$	6.695(3)	9.819(5)	75
$CaNH_4P_3O_9$	6.887(3)	10.448(5)	75, 81
$MgNH_4P_3O_9$	6.698(3)	9.831(5)	75
$MnNH_4P_3O_9$	6.771(3)	10.026(5)	75
$CdNH_4P_3O_9$	6.870(3)	10.233(5)	75
$CdRbP_3O_9$	6.858(2)	10.211(5)	68
$CdTlP_3O_9$	6.845(2)	10.154(5)	68

Durif [84]. From the crystal data reported by the authors, this compound is isotypic with the monoclinic form of $CaNH_4P_3O_9$.

- $CdCsP_3O_9$ and $CaCsP_3O_9$

The cadmium salt, $CdCsP_3O_9$, has been observed by Averbuch-Pouchot [85] during the elaboration of the $CsPO_3$-$Cd(PO_3)_2$ phase-equilibrium diagram. It melts congruently at 996K. Its atomic arrangement has been determined by Averbuch-Pouchot and Durif [86], confirming its isotypy with the orthorhombic form of $MgNH_4P_3O_9$. The calcium salt, $CaCsP_3O_9$, prepared and described by Masse *et al* [76], belongs to the same structure type.

- $SrCsP_3O_9$

Tokman and Bukhalova [87] established the $CsPO_3$-$Sr(PO_3)_2$ phase-equilibrium diagram. The non-congruent melting compound, $SrCsP_3O_9$, observed in this system is said to be a cyclotriphosphate, but no structural work confirmed this assumption.

- $MgTlP_3O_9$ and $MgTl_4(P_3O_9)_2$

$MgTlP_3O_9$ and $MgTl_4(P_3O_9)_2$ were characterized during the elaboration of the $TlPO_3$-$Mg_2P_4O_{12}$ phase-equilibrium diagram by Rakotomahanina-Rolaisoa *et al* [88]. Both salts are non-congruent melting compounds decomposing at 763K for $MgTl_4(P_3O_9)_2$ and 1031K for $MgTlP_3O_9$. This last compound is isotypic with the orthorhombic form of $MgNH_4P_3O_9$. A model for the atomic arrangement of $MgTl_4(P_3O_9)_2$ has been proposed by Rakotomahanina-Rolaisoa [78].

- $CoTl_4(P_3O_9)_2$ and $ZnTl_4(P_3O_9)_2$

The salts, $CoTl_4(P_3O_9)_2$ and $ZnTl_4(P_3O_9)_2$, were characterized during the elaborations of the $TlPO_3$-$Co_2P_4O_{12}$ and $TlPO_3$-$Zn(PO_3)_2$ phase-equilibrium diagrams by Rakotomahanina-Rolaisoa *et al* [88] for the first one, and Averbuch-Pouchot [89] for the second one. Both are non-congruent melting compounds, decomposing respectively at 754 and 713K. They are isotypic of the corresponding magnesium salt.

- $CdTlP_3O_9$ and $CdTl_4(P_3O_9)_2$

Chemical preparation of $CdTlP_3O_9$ have been reported by Pouchot *et al* [68]. The crystal data reported by the authors shows the isotypy of this compound with benitoite. The second compound, $CdTl_4(P_3O_9)$, not observed in the first version of the $TlPO_3$-$Cd(PO_3)_2$ phase-equilibrium diagram elaborated by Averbuch-Pouchot [69] was clearly characterized in the revised version of this diagram made by the same author [90] and identified as an isotype of the corresponding magnesium salt. $CdTlP_3O_9$ and $CdTl_4(P_3O_9)_2$ appear as non-congruent melting compounds, decomposing respectively at 962K and 749K.

Table 2.5
Main crystal data for $M^{II}M^{I}P_3O_9$ cyclotriphosphates.

Formula	a α	b β	c (Å) γ (°)	S. G.	Z	Ref.
$PbNaP_3O_9$	10.873(5)	12.11(5)	5.664(2)	$P2_12_12_1$	4	64
$BaNaP_3O_9$	11.134(8)	12.320(8)	5.802(3)	$P2_12_12_1$	4	66, 67
$BaAgP_3O_9$	11.05(1)	12.28(1)	5.903(5)	$P2_12_12_1$	4	70
$CaKP_3O_9$	7.316(4)	12.44(2)	9.955(6)	Pmcn	4	76
$HgKP_3O_9$	11.164(6)	12.46(1)	5.622(2)	$P2_12_12_1$	4	80
$MgNH_4P_3O_9$	7.22(1)	12.04(2)	9.33(1)	Pmcn	4	82
$CaNH_4P_3O_9$	7.446(5)	12.461(10)	10.050(10) 90.11(10)	$P2_1/n$	4	76, 82
$MgRbP_3O_9$	7.177(3)	11.981(4)	9.339(3)	Pmcn	4	83
$CaRbP_3O_9$	7.545(5)	12.51(2)	9.745(8) 90.00(5)	$P2_1/n$	4	84
$CdCsP_3O_9$	7.507(3)	12.68(2)	9.533(8)	Pmcn	4	85, 86
$CaCsP_3O_9$	7.523(4)	12.77(2)	10.056(6)	Pmcn	4	76
$MgTlP_3O_9$	7.185(5)	12.05(1)	9.331(5)	Pmcn	4	88
$CaTlP_3O_9$	7.471(5)	12.52(1)	9.913(5) 90.00(5)	$P2_1/n$	4	88

- ***$CaTl_4(P_3O_9)_2$ and $CaTlP_3O_9$***

$CaTl_4(P_3O_9)_2$ and $CaTlP_3O_9$ were first observed by Rakotomahanina-Rolaisoa *et al* [88] when elaborating the $TlPO_3$-$Ca(PO_3)_2$ phase-equilibrium diagram. $CaTl_4(P_3O_9)_2$ is a non-congruent melting compound isotypic of $MgTl_4(P_3O_9)_2$, decomposing at 793K, while $CaTlP_3O_9$, isotypic of the monoclinic form of $CaNH_4P_3O_9$, melts congruently at 1089K.

- ***$HgNa_2(NH_4)_2(P_3O_9)_2$***

The salt, $HgNa_2(NH_4)_2(P_3O_9)_2$, has been characterized during a study of the P_2O_5-HgO-Na_2O-$(NH_4)_2O$ system by Averbuch-Pouchot and Durif [91]. The authors report a complete determination of the atomic arrangement. The unit-cell dimensions of this compound are close by those measured for $CaNa_4(P_3O_9)_2$ and $CdNa_4(P_3O_9)_2$.

Tables 2.4, 2.5 and 2.6 report the main crystal data for all the anhydrous divalent-monovalent cation cyclotriphosphates presently known.

Table 2.6
Main crystal data for $M^{II}M^{I}_{4}P_3O_9$ cyclotriphosphates.

Formula	a α	b β	c (Å) γ (°)	S. G.	Z	Ref.
$CaNa_4(P_3O_9)_2$	13.248(5)	8.120(3) 94.65(5)	14.384(6)	Cc or C2/c	4	61
$CdNa_4(P_3O_9)_2$	13.05(1)	7.952(3) 94.23(1)	14.19(1)	Cc or C2/c	4	63
$PbNa_4(P_3O_9)_2$	7.268(4) 121.52(5)	8.151(5) 102.06(5)	7.851(5) 73.00(5)	$P\bar{1}$	1	64, 65
$BaNa_4(P_3O_9)_2$	7.313(4) 121.38(5)	8.237(5) 102.66(5)	7.865(5) 72.23(5)	$P\bar{1}$	1	65, 66
$MgTl_4(P_3O_9)_2$	7.191(3)	7.191(3)	19.55(1)	$P3_1c$	2	88
$CoTl_4(P_3O_9)_2$	7.220(3)	7.220(3)	19.62(1)	$P3_1c$	2	88
$ZnTl_4(P_3O_9)_2$	7.213(2)	7.213(2)	19.625(15)	$P3_1c$	2	89
$CdTl_4(P_3O_9)$	7.320(5)	7.320(5)	19.81(1)	$P3_1c$	2	90
$CaTl_4(P_3O_9)_2$	7.389(3)	7.389(3)	19.99(1)	$P3_1c$	2	88
$HgNa_2(NH_4)_2(P_3O_9)_2$	13.524(8)	8.362(5) 92.58(5)	14.390(8)	C2/c	4	91

Hydrated divalent-monovalent cation cyclotriphosphates

- $CaNaP_3O_9.3H_2O$ and $SrNaP_3O_9.3H_2O$

The characterization of the two salts, $CaNaP_3O_9.3H_2O$ and $SrNaP_3O_9.3H_2O$, has been early reported on 1875 by Lindbom [3]. Reproducible preparations and crystal data have been published by Durif [92] for the calcium salt and Martin and Durif [66] for the strontium one. The two salts are not isotypic. The atomic arrangement of $SrNaP_3O_9.3H_2O$ was determined by Zilber *et al* [93] ; that of $CaNaP_3O_9.3H_2O$ is up to now unknown. Influence of water vapor on the opening of the P_3O_9 ring during the thermal dehydration of $CaNaP_3O_9.3H_2O$ has been investigated by Simonot-Grange and Jamet [94].

- $BaNaP_3O_9.4H_2O$ and $BaNaP_3O_9.3H_2O$

The tetrahydrate, $BaNaP_3O_9.4H_2O$, was first prepared by Fleitmann and Henneberg [2]. Chemical preparation and crystallographic data for this salt have been reported by Martin and Durif [66] ; it is isotypic of the corresponding silver-barium salt described below. The trihydrate $BaNaP_3O_9.3H_2O$ has been obtained by crystallization at 333K of an aqueous solution of the tetrahydrate. Averbuch-Pouchot and Durif [95] reported a complete description of its atomic arrangement.

- $BaAgP_3O_9.4H_2O$

The hydrate, $BaAgP_3O_9.4H_2O$, prepared by the Boullé's process, has been described by Durif and Averbuch-Pouchot [70] and its crystal structure determined by Seethanen *et al* [96].

- $BaKP_3O_9.H_2O$

$BaKP_3O_9.H_2O$ was first characterized by Martin [97] and subsequently, its atomic arrangement was determined by Seethanen and Durif [98].

- $CaNH_4P_3O_9.3H_2O$

$CaNH_4P_3O_9.3H_2O$ has been prepared by Masse *et al* [76] and recognized as an isotype of $CaNaP_3O_9.3H_2O$. An hydrated calcium-ammonium cyclotriphosphate has been observed by Feldman and Grunze [99]. Its state of hydration is not well defined and according to the authors the ratio Ca/NH_4 is slightly different of unity.

- $BaNH_4P_3O_9.H_2O$, $BaCsP_3O_9.H_2O$ and $BaTlP_3O_9.H_2O$

The three salts, $BaNH_4P_3O_9.H_2O$, $BaCsP_3O_9.H_2O$ and $BaTlP_3O_9.H_2O$, have been prepared by using the Boullé's process. The preparation of $BaNH_4P_3O_9.H_2O$ was described by Durif *et al* [100] who report a complete description of the atomic arrangement. The caesium salt has been prepared by Masse and Averbuch-Pouchot [101]. From the crystal data, this salt appears as isotypic with $BaNH_4P_3O_9.H_2O$. Durif *et al* [100] reported chemical preparation and crystal data for $BaTlP_3O_9.H_2O$, showing its isotypy with the two former salts. A second form of this salt, this time isotypic with the barium-potassium salt, has been described by Martin [97].

- $NiNa_4(P_3O_9)_2.6H_2O$ and $NiAg_4(P_3O_9)_2.6H_2O$

Chemical preparations and crystal structure determinations of the two isotypic salts, $NiNa_4(P_3O_9)_2.6H_2O$ and $NiAg_4(P_3O_9)_2.6H_2O$, have been reported by Jouini and Dabbabi [102-104].

- $CuNa_4(P_3O_9)_2.4H_2O$

Chemical preparation of $CuNa_4(P_3O_9)_2.4H_2O$ is easily performed by mixing aqueous solutions of $Na_3P_3O_9$ and $CuCl_2$ in a stoichiometric ratio and by evaporating slowly at room temperature. The crystal structure have been performed by Durif and Averbuch-Pouchot [105].

- $NiK_4(P_3O_9)_2.7H_2O$ and $CoK_4(P_3O_9)_2.7H_2O$

The two isotypic salts, $NiK_4(P_3O_9)_2.7H_2O$ and $CoK_4(P_3O_9)_2.7H_2O$, prepared by the Boullé's process have been described by Seethanen *et al* [106]. The authors report a complete description of the atomic arrangement for the nickel compound.

- ***$CuK_4(P_3O_9)_2.4H_2O$ and $Cu(NH_4)_4(P_3O_9)_2.4H_2O$***

Both $CuK_4(P_3O_9)_2.4H_2O$ and $Cu(NH_4)_4(P_3O_9)_2.4H_2O$ have been prepared by using the Boullé's metathesis reaction. The atomic arrangement of the potassium salt has been determined by Durif and Averbuch-Pouchot [107]. From the crystal data, the two salts are isotypic.

- ***$Ni(NH_4)_4(P_3O_9)_2.4H_2O$ and $Co(NH_4)_4(P_3O_9)_2.4H_2O$***

The atomic arrangement of the nickel salt, $Ni(NH_4)_4(P_3O_9)_2.4H_2O$,

Table 2.7
Main crystal data for $M^{II}M^{I}_4P_3O_9.nH_2O$ cyclotriphosphates.

Formula	a α	b β	c (Å) γ (°)	S. G.	Z	Ref.
$NiNa_4(P_3O_9)_2.6H_2O$	9.186(2) 89.17(1)	8.020(2) 102.89(1)	6.838(1) 98.03(1)	$P\bar{1}$	1	102, 104
$NiAg_4(P_3O_9)_2.6H_2O$	9.209(3) 89.15(2)	8.053(3) 102.94(1)	6.841(2) 97.24(1)	$P\bar{1}$	1	102, 103
$CuNa_4(P_3O_9)_2.4H_2O$	7.907(5) 102.46(5)	8.364(5) 97.89(5)	7.122(5) 84.04(5)	$P\bar{1}$	1	105
$NiK_4P_3O_9)_2.7H_2O$	23.03(1)	11.882(4)	8.732(4)	Fm2m	4	106
$CoK_4(P_3O_9)_2.7H_2O$	22.98(1)	11.869(2)	8.751(2)	Fm2m	4	106
$CuK_4(P_3O_9)_2.4H_2O$	8.510(5)	14.303(8) 96.51(2)	8.487(5)	$P2_1/a$	2	107
$Cu(NH_4)_4(P_3O_9)_2.4H_2O$	8.60(1)	14.65(2) 96.66(8)	8.79(1)	$P2_1/c$	2	107
$Ni(NH_4)_4(P_3O_9)_2.4H_2O$	8.645(2)	14.698(3) 95.89(2)	8.774(2)	$P2_1/c$	2	102, 108
$Co(NH_4)_4(P_3O_9)_2.4H_2O$	8.612(2)	14.698(3) 95.67(1)	8.809(2)	$P2_1/c$	2	109
$CdK_4(P_3O_9)_2.2H_2O$	9.235(5) 96.38(1)	7.599(4) 103.90(1)	7.148(4) 102.06(1)	$P\bar{1}$	1	110
$ZnK_4(P_3O_9)_2.4H_2O$	12.444(4)	10.978(2) 124.41(2)	9.624(3)	C2/m	2	111
$CoRb_4(P_3O_9)_2.6H_2O$	13.216(3)	11.059(2) 126.71(1)	10.026(2)	C2/m	2	112
$NiCs_4(P_3O_9)_2.6H_2O$	19.992(4)	6.500(2)	18.445(4)	$Pca2_1$ Pcam	4	102

has been determined by Jouini and Dabbabi [108]. In addition, magnetic properties and IR spectrum are reported by the authors. Chemical preparation and crystal structure of $Co(NH_4)_4(P_3O_9)_2.4H_2O$ have been described by Belkhiria *et al* [109]. These two salts are isotypic with the corresponding copper-ammonium and copper-potassium compounds [107].

- $CdK_4(P_3O_9)_2.2H_2O$

Chemical preparation of $CdK_4(P_3O_9)_2.2H_2O$ by the Boullé's process was described by Averbuch-Pouchot [110]. The author performed a complete determination of the atomic arrangement.

- $ZnK_4(P_3O_9)_2.6H_2O$ and $CoRb_4(P_3O_9)_2.6H_2O$

From the crystal data and the atomic arrangements, the two salts $ZnK_4(P_3O_9)_2.6H_2O$ and $CoRb_4(P_3O_9)_2.6H_2O$ appear as isotypic. The zinc salt was first prepared and investigated by Seethanen *et al* [111]. The

Table 2.8
Main crystal data for $M^{II}M^{I}P_3O_9.nH_2O$ cyclotriphosphates.

Formula	a α	b β	c (Å) γ (°)	S. G.	Z	Ref.
$CaNaP_3O_9.3H_2O$	14.923(8)	14.923(8)	10.148(10)	$P6_3$...	2	92
$SrNaP_3O_9.3H_2O$	16.167(8)	12.013(5)	10.645(5)	Pnma	8	66, 93
$BaNaP_3O_9.4H_2O$	21.33(8)	7.01(1) 122.18(10)	18.26(8)	C2/c	8	66
$BaNaP_3O_9.3H_2O$	7.067(3) 116.46(5)	9.071(3) 95.97(5	9.906(4) 74.03(5)	$P\bar{1}$	2	95
$BaAgP_3O_9.4H_2O$	21.35(3)	7.163(3) 121.72(5)	18.35(2)	C2/c	8	70, 96
$BaKP_3O_9.H_2O$	7.34(1)	17.77(2) 95.24(5)	7.18(1)	$P2_1/n$	4	97, 98
$CaNH_4P_3O_9.3H_2O$	14.76(5)	14.76(5)	9.932(6)	$P6_3$...	8	76
$BaNH_4P_3O_9.H_2O$	11.70(1)	12.12(1) 101.05(5	7.559(5)	$P2_1/n$	4	100
$BaCsP_3O_9.H_2O$	11.764(8)	12.292(8) 101.16(5)	7.681(5)	$P2_1/n$	4	101
$BaTlP_3O_9.H_2O$(I)	11.76(1)	12.33(1) 100.92(5)	7.537(8)	$P2_1/n$	4	100
$BaTlP_3O_9.H_2O$(II)	7.39(1)	17.72(2) 95.45(5)	7.21(1)	$P2_1/n$	4	97

authors described this compound as a tetrahydrate. Subsequently, Belkhiria *et al* [112] reported a detailed determination of the atomic arrangement of the cobalt salt which appears as an hexahydrate. In this last salt, two of the six water molecules are weakly bonded and have relatively high thermal factors. This fact might explain the wrong assumption of the first authors [111] for the zinc salt.

- $NiCs_4(P_3O_9)_2.6H_2O$

Jouini and Dabbabi [102] reported chemical preparation and crystal data for $NiCs_4(P_3O_9)_2.6H_2O$ whose atomic arrangement is still unknown.

Tables 2.7 and 2.8 give the main crystallographic features for hydrated divalent-monovalent cation cyclotriphosphates.

2.1.4. Trivalent and trivalent-monovalent cation cyclotriphosphates

This class of compounds has not yet been extensively investigated and only a few number of them are today well characterized.

- $LnP_3O_9.nH_2O$

Serra and Giesbrecht [113] were the first to describe three rare-earth salts of general formula $TP_3O_9.3H_2O$ with T = La, Ce and Nd. Later on, Birke and Kempe [114-116] described chemical preparations and thermal behavior for $PrP_3O_9.4H_2O$, $LaP_3O_9.4H_2O$ and $ErP_3O_9.4.4H_2O$. Bagieu-Beucher and Durif [117] prepared a series of three $TP_3O_9.3H_2O$ compounds, with T = La, Ce and Pr and reported their crystal data. Crystal structure was determined by Bagieu-Beucher *et al* [118] with the cerium salt. The main crystallographic features of these three salts are reported in Table 2.9. It is not possible to assert that these salts are similar to those reported in [114-116] because of the lack of crystal data in these studies.

Table 2.9
The common hexagonal space-group is $P\bar{6}$ for the three $TP_3O_9.3H_2O$ cyclotriphosphates whose unit-cells reported in the present table contain one formula-unit.

Formula	a	c (Å)
$LaP_3O_9.3H_2O$	6.785(5)	6.112(5)
$CeP_3O_9.3H_2O$	6.770(5)	6.079(5)
$PrP_3O_9.3H_2O$	6.743(5)	6.048(5)

The thermal behavior of these three hexagonal $TP_3O_9.3H_2O$ compounds has been carefully investigated by Gobled [119]. They are not stable in air and decompose irreversibly according to the following scheme :

$$TP_3O_9.3H_2O \text{ -----> } TPO_4 + P_2O_5 + 3H_2O$$

This investigation has been run mainly with the lanthanum salt, the only one stable for more than a year.

Two other rare-earth compounds, $NdP_3O_9.5H_2O$ and $SmP_3O_9.5H_2O$, have been described by Bagieu-Beucher [120]. All attempts to produce single crystals of a proper size for a structural study failed. According to the author, $P2_1/a$ is the most probable space-group. Their monoclinic unit-cell dimensions are reported below :

$NdP_3O_9.5H_2O$	a = 13.954	b = 11.500	c = 7.745 Å	β = 105.83°
$SmP_3O_9.5H_2O$	a = 13.889	b = 11.496	c = 7.719	β = 105.79

- ***$M^{III}Na_3(P_3O_9)_2.9H_2O$*** *($M^{III}$ = Sm, Eu, Gd, Dy, Ho, Er, Y and Bi)*

The bismuth salt, $BiNa_3(P_3O_9)_2.9H_2O$, was first prepared by Bagieu-Beucher and Durif [121] by adding solid bismuth nitrate or chloride to a saturated solution of $Na_3P_3O_9$. In such a process, the crystal growth is very rapid. In some experiments, crytals up to 1 mm long were obtained within half an hour. These authors report a complete description of the atomic arrangement. This bismuth salt is not stable for long times. This unstability can probably be explained by the fact that some of the water molecules are of a zeolitic nature. A series of seven compounds, isotypic with $BiNa_3(P_3O_9)_2.9H_2O$, have been prepared by Bagieu-Beucher [122]. Their chemical preparations are identical to that described for the bismuth salt. They also are not stable for long times. The unit-cell dimensions for these eight compounds are reported in Table 2.10.

Table 2.10

Unit-cell dimensions for the $M^{III}Na_3(P_3O_9)_2.9H_2O$ compounds. $R\bar{3}c$ is the common rhomboedral space-group and Z = 18.

Formula	a	c (Å)	Ref.
$BiNa_3(P_3O_9)_2.9H_2O$	30.845(15)	13.085(3)	121
$SmNa_3(P_3O_9)_2.9H_2O$	31.07(1)	13.119(5)	122
$EuNa_3(P_3O_9)_2.9H_2O$	31.05(6)	13.075(6)	122
$GdNa_3(P_3O_9)_2.9H_2O$	30.99(1)	13.007(5)	122
$DyNa_3(P_3O_9)_2.9H_2O$	30.99(1)	12.951(4)	122
$HoNa_3(P_3O_9)_2.9H_2O$	30.95(1)	12.901(4)	122
$ErNa_3(P_3O_9)_2.9H_2O$	30.99(1)	12.887(4)	122
$YNa_3(P_3O_9)_2.9H_2O$	30.93(1)	12.904(5)	122

2.1.5. Organic and alkali-organic cation cyclotriphosphates

Recently, the reviewers extended the Boullé's metathesis reaction to the syntheses of organic cation or mixed metal-organic cation cyclotriphosphates. So, glycinium, guanidinium, methylammonium, ethylenediammonium, potassium-ethylenediammonium and isopropylammonium cyclotriphosphates have been clearly characterized. All these salts are stable in normal conditions. In all cases, the atomic arrangements have been accurately determined. Their main crystal features are reported in Table 2.11.

2.1.6. Cyclotriphosphate adducts with telluric acid

P_3O_9 ring anions appear also in adducts between monovalent cation cyclotriphosphates and telluric acid. These compounds are easily prepared by slow evaporation, at room-temperature, of aqueous solutions of the alkali cyclotriphosphates and telluric acid, prepared in a stoichiometric ratio. All their atomic arrangements have been described and, in all cases, the phosphoric ring anions and the $Te(OH)_6$ groups coexist as indepen-

Table 2.11
Main crystal data for organic and alkali-organic cation cyclotriphosphates.

Formula	a α	b β	c (Å) γ (°)	S. G.	Z	Ref.
$(EDA)_3(P_3O_9)_2$	15.558(8) 104.14(5)	10.450(6) 102.73(5)	7.639(4) 86.71(5)	$P\bar{1}$	2	123
$(IPA)_3P_3O_9$	25.22(2)	12.22(2) 123.90(2)	15.45(2)	C2	8	124
$K(EDA)P_3O_9$	20.850(8)	9.044(4)	11.653(5)	Ccca	8	125
$(MA)_3P_3O_9$	12.144(7)	5.361(5) 97.32(8)	7.203(7)	$P2_1/n$	4	126
$(Gly)_3P_3O_9$	12.223(8)	14.52(1) 100.47(5)	10.229(7)	$P2_1/c$	4	127
$(Gua)_3P_3O_9.4H_2O$	12.140(8)	15.183(8) 97.49(5)	10.706(5)	$P2_1/n$	4	128

EDA = $[NH_3\text{-}(CH_2)_2\text{-}NH_3]^{2+}$ (ethylenediammonium)
IPA = $[(CH_3)_2\text{-}CH\text{-}NH_3]^+$ (isopropylammonium)
MA = $(CH_3\text{-}NH_3)^+$ (methylammonium)
Gly = $(NH_3\text{-}CH_2\text{-}COOH)^+$ (glycinium)
Gua = $[C(NH_2)_3]^+$ (guanidinium)

dent units. As for all the other classes of phosphates forming adducts with telluric acid, no lithium salt has been observed during these investigations.

Crystal data for $Te(OH)_6.K_3P_3O_9.2H_2O$ and $Te(OH)_6.2Na_3P_3O_9.6H_2O$ have been reported by Boudjada [129]. Their atomic arrangements were subsequently determined by Boudjada *et al* [130, 131].

The following ammonium, rubidium, caesium and thallium adducts with telluric acid, corresponding to various formula, have been prepared and investigated from a structural point of view :

- $Te(OH)_6.2(NH_4)_3P_3O_9$ by Boudjada *et al* [132]
- $Te(OH)_6.Rb_3P_3O_9.H_2O$ by Boudjada and Durif [133]
- $Te(OH)_6.Cs_3P_3O_9.H_2O$ by Averbuch-Pouchot [134]
- $Te(OH)_6.2Tl_3P_3O_9$ by Boudjada [28].

In addition, two mixed-monovalent cation cyclotriphosphate-telluric acid adducts, $Te(OH)_6.Cs_2NaP_3O_9$ and $Te(OH)_6.Na_3P_3O_9.K_3P_3O_9$, have been studied by Boudjada [28] and Averbuch-Pouchot and Durif [135], respectively.

The main crystallographic features for cyclotriphosphate-telluric acid adducts are given in Table 2.12.

Table 2.12
Main crystal data for cyclotriphosphate adducts with telluric acid.

Formula	a α	b β	c (Å) γ (°)	S. G.	Z	Ref.
$Te(OH)_6.K_3P_3O_9.2H_2O$	15.57(2)	7.438(6) 107.16(9)	14.85(1)	$P2_1/c$	4	130
$Te(OH)_6.2Na_3P_3O_9.6H_2O$	11.67(1)	11.67(1)	12.12(1)	$P6_3/m$	2	131
$Te(OH)_6.2(NH_4)_3P_3O_9$	11.16(1)	11.16(1)	17.86(1)	$R\bar{3}$	3	132
$Te(OH)_6.Rb_3P_3O_9.H_2O$	15.564(6)	8.376(4) 113.33(2)	13.705(4)	$P2_1/a$	4	133
$Te(OH)_6.Cs_3P_3O_9.H_2O$	7.279(2)	13.984(8) 90.42(2)	17.071(4)	$P2_1/c$	4	134
$Te(OH)_6.2Tl_3P_3O_9$	11.168(1)	11.168(1)	11.733(3)	$P6_3/m$	2	28
$Te(OH)_6.Cs_2NaP_3O_9$	12.946(1)	9.174(1) 107.60(1)	13.406(1)	C2/c	4	28
$Te(OH)_6.Na_3P_3O_9.K_3P_3O_9$	18.42(1)	10.644(5) 119.76(5)	12.348(8)	C2/c	4	135

2.2. Chemical Preparations of Cyclotriphosphates

We report below some processes commonly used for the preparation of cyclotriphosphates. Most of these preparations involve the use of sodium cyclotriphosphate as starting material. This material can now be prepared in large amounts, with a good state of purity and high yields.

2.2.1. Chemical preparation of sodium cyclotriphosphate

By heating NaH_2PO_4, in a temperature range of 803-823K, for at least five hours, one obtains $Na_3P_3O_9$ according to the following scheme :

$$3NaH_2PO_4 \text{ ------> } Na_3P_3O_9 + 3H_2O$$

The so obtained cyclotriphosphate is sometimes contaminated by small amounts (mostly less than 1%) of some insoluble sodium phosphates, mainly long-chain polyphosphates. A dissolution in water followed by filtration eliminates these polyphosphates. After some days, crystals of the hexahydrate $Na_3P_3O_9.6H_2O$ appear in the resulting solution, kept at room temperature. Firing the hexahydrate at 623K leads to pure $Na_3P_3O_9$. A similar scheme of cyclization has been investigated for other alkali metals and for ammonium ion but the obtained yields are up to now deceiving.

2.2.2. The Boullé's process

The Boullé's process is a metathesis reaction widely used for the preparation of water-soluble cyclotriphosphates. The starting material is silver cyclotriphos-phate monohydrate, $Ag_3P_3O_9.H_2O$. This salt, sparingly water-soluble, is easily prepared by adding an aqueous solution of silver nitrate (~ M/10) to an aqueous solution of sodium cyclotriphosphate of approximately the same concentration. The reaction is :

$$Na_3P_3O_9 + 3AgNO_3 \text{ -----> } Ag_3P_3O_9.H_2O + 3NaNO_3$$

It is recommended to use an excess of silver salt to avoid the formation of a sodium contaminated compound. The precipitation is achieved within one day. The precipitate appears as nice monoclinic crystals, non-light sensitive and stable for years at room temperature.

A typical run of the Boullé's process is now exemplified for the chemical preparation of potassium cyclotriphosphate, $K_3P_3O_9$. A slurry of $Ag_3P_3O_9.H_2O$ in water is slowly added with an aqueous solution of potassium chloride in the stoichiometric ratio. The reaction is :

$$Ag_3P_3O_9.H_2O + 3KCl \text{ ------> } K_3P_3O_9 + 3AgCl + H_2O$$

After about one hour of mechanical stirring, the silver chloride is removed by filtration and the resulting solution is evaporated to obtain crystals of

$K_3P_3O_9$ or added with ethanol until the precipitation of polycrystalline potassium cyclotriphosphate. These last operations must be run at room or lower temperature in order to avoid the hydrolysis of the ring.

Many water-soluble cyclotriphosphates have been prepared by this process.

2.2.3. More classical methods

Classical methods of aqueous chemistry are in many cases available for the production of cyclotriphosphates. $CaNaP_3O_9.3H_2O$ is, for instance, very easily prepared by adding, in the proper ratio, a M/10 aqueous solution of $CaCl_2$ to an aqueous solution of $Na_3P_3O_9$ of same concentration. $CaNaP_3O_9.3H_2O$ is then precipitated by adding ethyl alcohol to the resulting solution.

2.2.4. Thermal methods

Many cyclotriphosphates have been prepared through what is commonly called "thermal methods". A typical example is the preparation of most of the $M^IM^{II}P_3O_9$ benitoite-like compounds. They are very easily obtained with a good purity by heating, at relatively low temperatures (many times less than 773K), a stoichiometric mixture of $(NH_4)_2HPO_4$ with the corresponding carbonates. The reaction scheme is for instance :

$$3(NH_4)_2HPO_4 + CoCO_3 + 1/2K_2CO_3 \text{ ------> } CoKP_3O_9 + 3/2CO_2 + 6NH_3 + 9/2H_2O$$

The starting mixture must be slowly heated up to the optimum temperature with frequent homogeneization grindings.

2.2.5. Flux methods

This type of process, very common in solid-state chemistry, has been frequently used, mainly for growing single-crystals. In the case of phosphates, the basic principle of this process is to use a starting mixture containing a large excess of H_3PO_4 or $(NH_4)_2HPO_4$. The composition of the starting mixture and the optimum temperature to be used are parameters difficult to determine. The optimization of a reproducible flux process for synthetizing a given material is in most cases a boring and long enterprise. After some hours or some days of heating at the appropriate temperature, a crystalline formation occurs into the melt. Then, the excess of the phosphoric flux is removed with hot water and the crystalline part separated by filtration. Because of this last necessity, this process, evidently, does not apply to the preparation of water-soluble phosphates.

2.2.6. Ion-exchange resins

The cyclotriphosphoric acid, $H_3P_3O_9$, can be obtained by using ion-exchange resins ; Amberlite IR220 or IRN77 have been extensively used. An aqueous solution of $Na_3P_3O_9$ is slowly passed through a column of resins and the so produced $H_3P_3O_9$ is immediately neutralized by the required carbonates or hydroxides. $Na_2LiP_3O_9$, $Mn_3(P_3O_9)_2.10H_2O$, $Ca_3(P_3O_9)_2.10H_2O$, $K_3P_3O_9$, for instance, have been prepared by this process.

2.2.7. Non-conventional methods

Very unconventional methods of preparation are sometimes reported with more or less details in chemical literature. Most of them have been elaborated by crystallographers in search for single-crystal production. These methods are reported in the part of the section "Present State of..." devoted to the given compound.

2.3. Some Atomic Arrangements of Cyclotriphosphates

In this section, we describe some structures of cyclotriphosphates selected among the most representative in this field.

2.3.1. Lithium cyclotriphosphate trihydrate

$Li_3P_3O_9.3H_2O$ [10-13] is the only cyclotriphosphate crystallizing with this trigonal type of atomic arrangement. Its unit-cell dimensions are :

a = 12.511(4), c = 5.594(2) Å in the hexagonal description,
a = 7.460 Å, α = 113.97° for the rhomboedral setting.

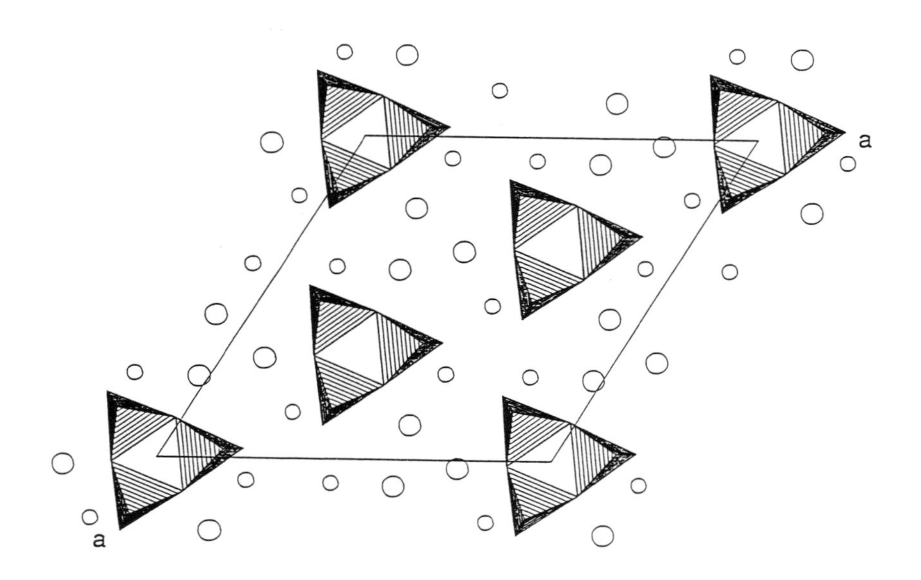

Figure 2.6 Projection along the c axis of the crystal structure of $Li_3(P_3O_9)_2.3H_2O$. Small empty circles are lithium atoms, the larger ones water molecules.

The space-group is R3, with Z = 3 for the hexagonal setting. The atomic arrangement, in projection along the c axis, is represented in Fig. 2.6.

The P_3O_9 ring anion is located around the threefold axis and so has a three-fold internal symmetry, relatively rare in the crystal chemistry of cyclotriphosphates.

Lithium atoms have a tetrahedral coordination made by three oxygen atoms and one water molecule. These $LiO_3(H_2O)$ tetrahedra form by sharing corners infinite chains spiraling around one set of 3_1 helical axes. Inside these $LiO_3(H_2O)$ tetrahedra the Li-O distances spread between 1.916 and 2.036 Å ($Li-H_2O$) and the O-Li-O angles between 99.7 and 120.0°.

2.3.2. Disodium hydrogen cyclotriphosphate

$Na_2HP_3O_9$ [16] is triclinic, $P\overline{1}$, with Z = 2 and the following unit-cell dimensions :

$$a = 7.788(5), \quad b = 7.809(5), \quad c = 7.129(5)\ \text{Å}$$
$$\alpha = 116.69(5), \quad \beta = 103.41(5), \quad \gamma = 81.94(5)^\circ$$

As shown in Fig. 2.7, which illustrates a projection, along the a axis, of the atomic arrangement, P_3O_9 ring anions connected by hydrogen

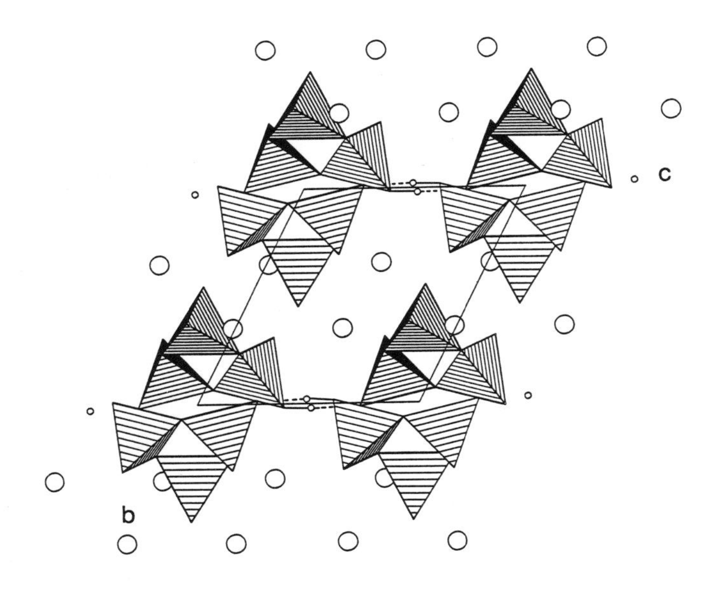

Figure 2.7 Projection, along the a axis, of the crystal structure of $Na_2HP_3O_9$. The small empty circles are hydrogen atoms, larger ones sodium atoms.

bonds form rows parallel to the c direction. The numerical values measured for these H-bonds : O-H = 0.86, H·····O = 1.59 Å and O-H·····O = 173° show they are very strong and probably explain the relatively low solubility of this salt. Inside this anionic array the P_3O_9 group has no internal symmetry and is so built by three crystallographically independent PO_4 tetrahedra. Cohesion between these rows of P_3O_9 is made by distorted NaO_6 octahedra, themselves interconnected by edge sharing as to form monodimensional arrays also parallel to the c direction. Inside the NaO_6 coordination polyhedra Na-O distances spread from 2.309 to 2.508 Å. This salt is today the only example of an acidic cyclotriphosphate.

2.3.3. Anhydrous sodium cyclotriphosphate

$Na_3P_3O_9$ [18] is orthorhombic with the following unit-cell dimensions

$$a = 7.928(2),\ b = 13.214(3),\ c = 7.708(2)\ \text{Å}$$

The space group is Pmcn and Z = 4. Fig. 2.8 reports the projection of

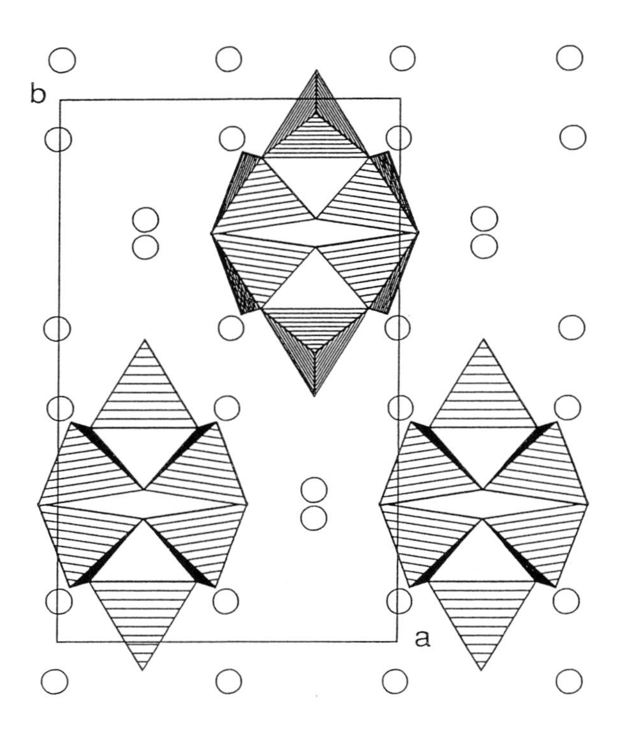

Figure 2.8 Projection, along the c axis, of the atomic arrangement of $Na_3P_3O_9$. Empty circles are sodium atoms. Mirror planes are in x = 1/4 and 3/4.

this atomic arrangement along the c axis. The P_3O_9 anion has a mirror symmetry and according to the authors a strong pseudo-3m symmetry, probably induced by the pseudo-hexagonality of the metrics of the unit cell. Two crystallographic independent sodium atoms are present in this arrangement ; one of them is located in the mirror plane. Both sodium atoms are fivefold coordinated and through edge and corner sharing these NaO_5 polyhedra build a three dimensional network.

The corresponding monohydrate, $Na_3P_3O_9.H_2O$, has an almost similar atomic arrangement.

2.3.4. Barium cyclotriphosphate tetrahydrate

$Ba_3(P_3O_9)_2.4H_2O$ [53] is monoclinic, C2/m, with Z = 2 and the following unit-cell dimensions :

$$a = 16.09(1),\ b = 8.368(5),\ c = 7.717(3)\ \text{Å},\ \beta = 95.38(5)°$$

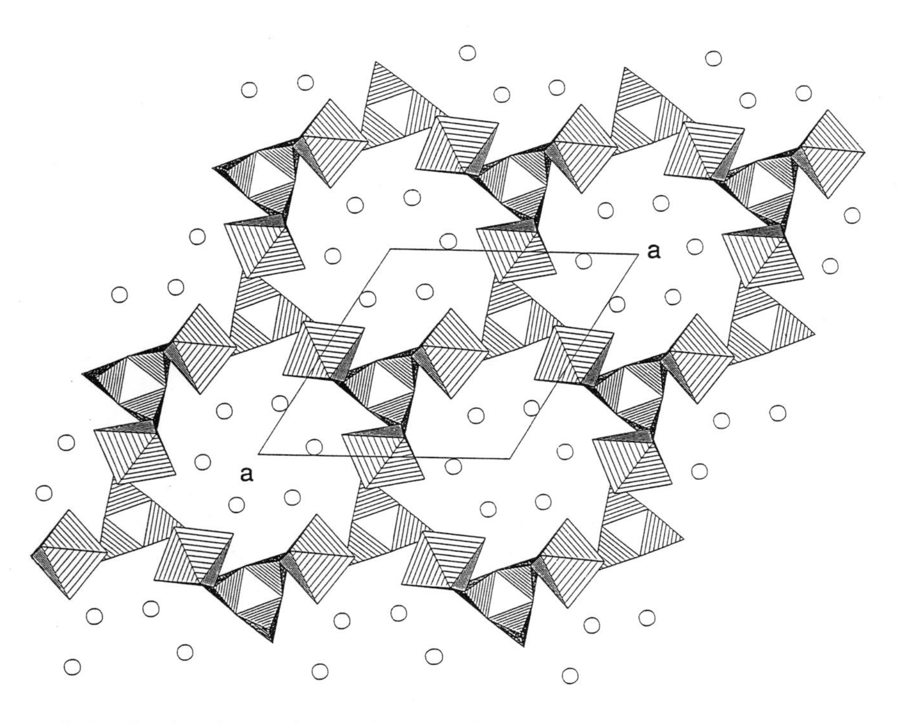

Figure 2.9 Projection, along the c axis, of the atomic arrangement of $Ba_3(P_3O_9)_2.4H_2O$. Small empty circles are barium atoms, large ones water molecules. The pairs of barium atoms located close by the inversion centres correspond to the statistical ones.

As shown in Fig. 2.9, which illustrates a projection of the structure along the c axis, the ring anion has a mirror symmetry. All the barium atoms are located in the mirror planes in y = 0 and 1/2, but their crystallographic situations are different. Four of them occupy a fourfold position and have a tenfold coordination made by eight oxygen atoms and two water molecules. The two remaining ones are distributed statistically on a fourfold position close to an inversion centre. They are located in a large oblongue centro-symmetrical cavity and have a sevenfold coordination built by five oxygen atoms and two water molecules. All the water molecules are involved in associated cation coordinations.

2.3.5. Cadmium cyclotriphosphate tetradecahydrate

$Cd_3(P_3O_9)_2.14H_2O$ [38-40] is trigonal, $P\overline{3}$, with Z = 1 and the following unit-cell dimensions :

$$a = 12.228, \quad c = 5.451 \text{ Å}$$

As shown in Fig. 2.10, the P_3O_9 ring anions are located around the

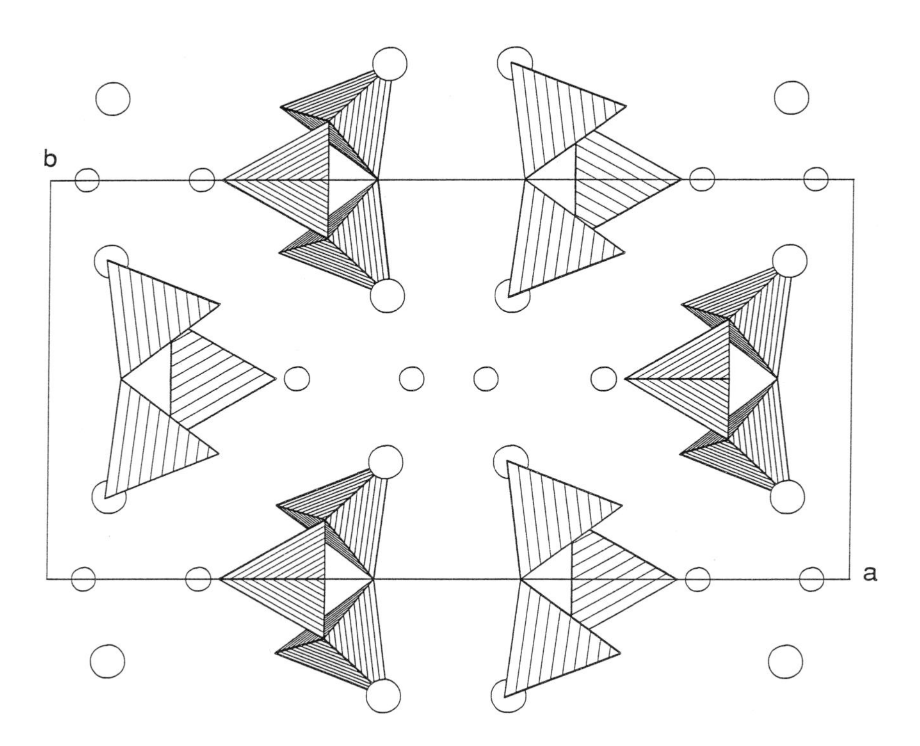

Figure 2.10 Projection of the crystal structure of $Cd_3(P_3O_9)_2.14H_2O$ along the c axis. Empty circles represent the zeolitic water molecules. The octahedra connecting the P_3O_9 groups are the $CdO_4(H_2O)_2$ polyhedra.

internal threefold axes at about z = 1/2 and so have a threefold internal symmetry. The cadmium atom is situated on the inversion center in 1/2,1/2,0 and its coordination is made by four oxygen atoms and two water molecules, forming an almost regular octahedron. Each cadmium octahedron shares two oxygen atoms with its two adjacent P_3O_9 groups. Such a network generates wide hexagonal channels around the $\overline{3}$ axes. The eight water molecules not involved in the cadmium coordination were not localized during the first crystal structure investigation, run at room temperature [39]. A reinvestigation of the structure at 93K [40] shows that six of the eight zeolitic water molecules are lining the channels, while the last two ones which could not be localized even at this temperature are probably located on the $\overline{3}$ axis with very high thermal factors. The six well localized zeolitic water molecules are arranged as to build an almost regular hexagon located around the threefold axis in approximately z = 0.5. Inside this hexagon the O(W)-O(W) distance is 2.812 Å at room temperature and decreases to 2.799 Å at 93K.

Inside the $CdO_4(H_2O)_2$ octahedron the Cd-O distances spread between 2.233 to 2.327 Å at room temperature and from 2.221 to 2.325 Å at 93K.

2.3.6. Calcium-ammonium cyclotriphosphate

$Ca(NH_4)P_3O_9$ [76, 81] is hexagonal, $P\overline{6}c2$, with Z = 2 and the following unit-cell dimensions :

$$a = 6.887(3), \quad c = 10.448(5) \text{ Å}$$

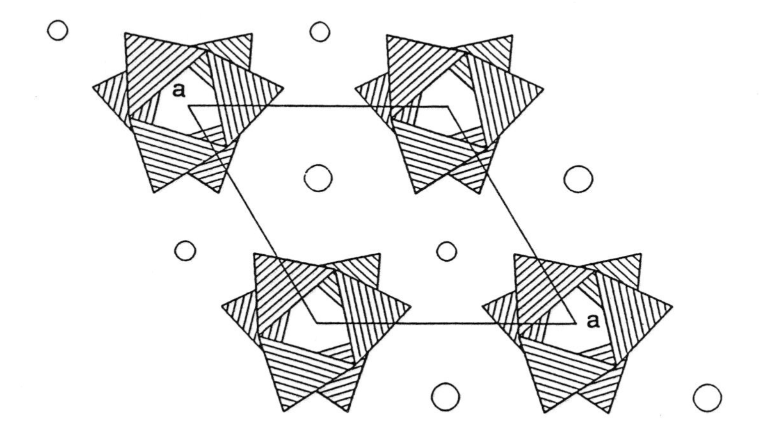

Figure 2.11 Projection, along the c axis, of the atomic arrangement of $CaNH_4P_3O_9$. The small empty circles represent calcium atoms ; the larger ones the ammonium groups.

It has the well-known structure of the mineral benitoite, $BaTiSi_3O_9$, like fourteen other cyclotriphosphates already reported in Table 2.4. As depicted in Fig. 2.11, the ring anions located around the $\overline{6}$ axis in z = 1/4 and 3/4 have the 3/m internal symmetry.

The divalent atoms, located on one of the threefold internal axes in z = 0 and 1/2, have a regular octahedral coordination, while the monovalent cations sited on the other internal threefold axis at the same heigths have a more distorted neighboring. So, the structure can be described as built by layers of phosphoric groups alternating along the c axis with layers of associated cations.

2.3.7. Nickel-potassium cyclotriphosphate heptahydrate

$NiK_4(P_3O_9)_2.7H_2O$ [106] is face-centred orthorhombic, Fm2m, with Z = 4 and the following unit-cell dimensions :

$$a = 23.03(1),\ b = 11.882(4),\ c = 8.732(4)\ Å$$

The ring anion has a mirror symmetry, with one phosphorus, two external oxygen and one bonding oxygen atoms in the mirror plane. The nickel atoms located in 0,0,0 have an octahedral coordination made by six water molecules. As shown by Fig. 2.12, these $Ni(H_2O)_6$ octahedra do not share any edge or corner. Within a range of 3 Å, the potassium atoms have a sixfold coordination. Through edge and corner-sharing, these KO_6 polyhe-

Figure 2.12 Projection, along the b axis, of the atomic arrangement of $NiK_4(P_3O_9)_2.7H_2O$. The empty circles are potassium atoms. The hatched octahedra represent the $Ni(H_2O)_6$ groups.

dra build layers parallel to the (a, c) planes. Such a layer is reported in Fig. 2.13. This organization was erroneously described by the authors in the original publication [106].

The corresponding cobalt salt is isotypic.

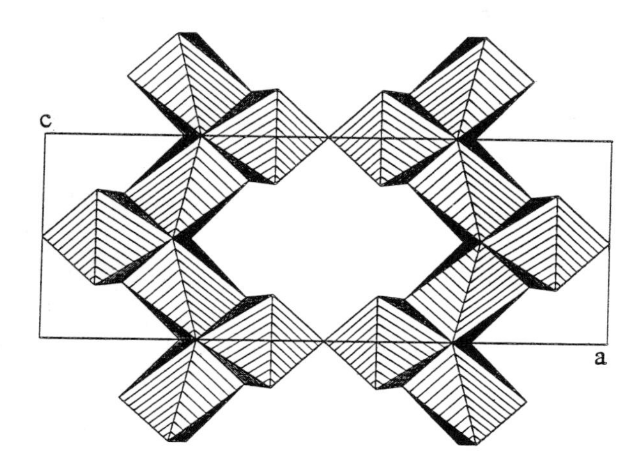

Figure 2.13 Projection, along the b axis, of a layer of edge and corner-sharing KO_6 octahedra in $NiK_4(P_3O_9)_2.7H_2O$.

2.3.8. Cerium cyclotriphosphate trihydrate

$CeP_3O_9.3H_2O$ [118] is hexagonal, $P\overline{6}$, with Z = 1 and the following unit-cell dimensions :

$$a = 6.770(5), \quad c = 6.079(5) \text{ Å}$$

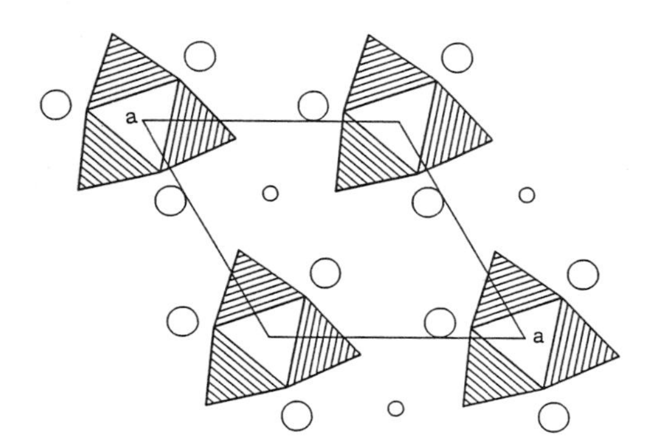

Figure 2.14 Projection of the atomic arrangement of $CeP_3O_9.3H_2O$ along the c axis. Small empty circles are cerium atoms, larger ones are water molecules.

Fig. 2.14 reports the projection, along the c axis, of the atomic arrangement of $CeP_3O_9.3H_2O$. The P_3O_9 ring anion located around the $\bar{6}$ axis has a planar configuration with a 3/m internal symmetry, quite comparable with that already observed in the cyclotriphosphates isotypic of benitoite. The cerium atom, located on one of the $\bar{6}$ internal axes, has a very regular ninefold coordination, made by six oxygen atoms and three water molecules.

2.3.9. Sodium cyclotriphosphate-telluric acid adduct hexahydrate

$Te(OH)_6.2Na_3P_3O_9.6H_2O$ [131] is hexagonal, $P6_3/m$, with Z = 2 and the following unit-cell dimensions :

$$a = 11.67(1), \quad c = 12.12(1) \text{ Å}$$

Fig. 2.15 reports a projection of the atomic arrangement along the c axis. The structure can be described as a succession of two kinds of alternate planes perpendicular to the c axis. Planes of the first type, located in z = 1/4 and 3/4, contain the sodium atoms and the water molecules, while z = 0 and 1/2 planes contain the $Te(OH)_6$ groups and the

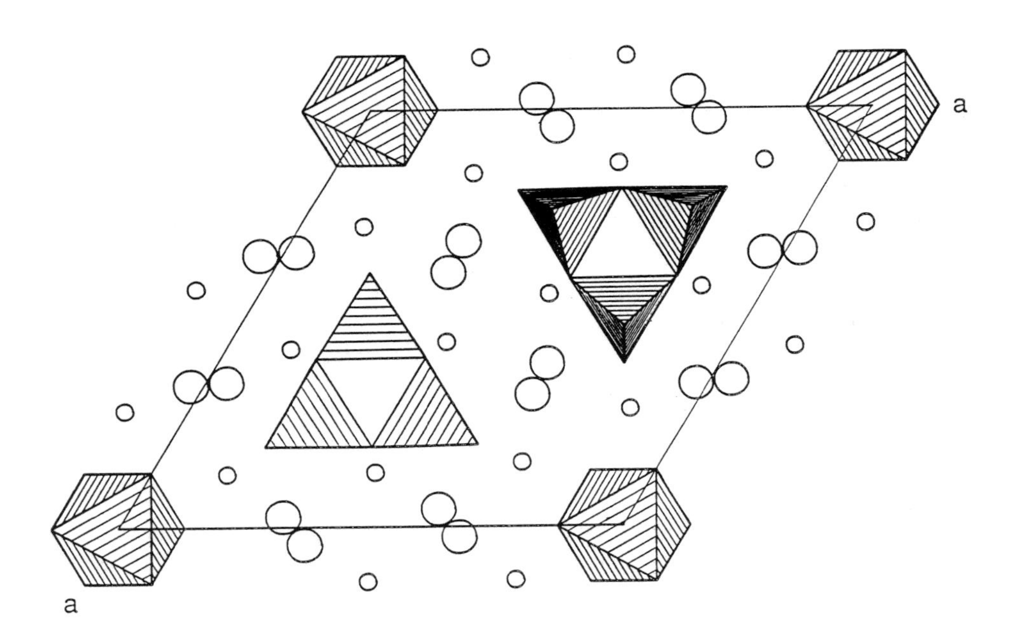

Figure 2.15 Projection, along the c axis, of the atomic arrangement of $Te(OH)_6.2Na_3P_3O_9.6H_2O$. Due to the presence of mirror planes in z = 1/4 and 3/4 two $Te(OH)_6$ groups and two P_3O_9 groups are superimposed in projection. Small empty circles are sodium atoms, larger ones are water molecules.

phosphoric ring-anions. These last groups, located around the $\overline{6}$ internal axes, have a threefold internal symmetry. The two independent sodium atoms have very distorted octahedral coordinations made by four oxygen atoms and two water molecules.

The $Te(OH)_6$ group is as usual built by six OH disposed around the tellurium atom as to build an almost regular octahedron.

2.4. The P_3O_9 Ring Anions

P_3O_9 anions can adopt various internal symmetries, 2, m, 3, 3/m, but an inspection of the fifty-six accurately investigated examples here reported, shows that the great majority of them, forty-one, have no internal symmetry. Among the remaining ones, two have a twofold symmetry, six a mirror symmetry, five a threefold and two a 3/m symmetry.

No correlation can be established between the nature of the associated cations and the internal symmetry of the ring inside the atomic arrangement.

This survey of the P_3O_9 ring geometry is given in the form of 4 tables (Tables 2.13, 2.14, 2.15 and 2.16) listing numerical values of what we consider as the main geometrical features of a phosphoric ring : the P-P distances and the P-P-P and P-O-P angles.

In the following lines, we report some averaged values for bond-distances or bond-angles measured in P_3O_9 ring anions. Some data published with a very poor degree of accuracy have not been taken into account in these calculations.

In addition, in many cases, the authors do not report the P-P distances and the P-P-P angles. These values have been recalculated by the reviewers from the original data.

The *P-P-P angles*, strictly equal to 60° in the rings having a 3 or 3/m internal symmetry, never depart significantly from this value. We report below the ranges of values observed for the various categories of rings :

$58.1 < P\text{-}P\text{-}P < 62.6°$	in rings with a mirror symmetry
$59.7 < P\text{-}P\text{-}P < 60.7$	in rings with a twofold symmetry
$58.6 < P\text{-}P\text{-}P < 61.4$	in rings with no internal symmetry.

The estimated standard deviation can be evaluated to 0.1°.

The *P-P distances* spread within the range, $2.818 < P\text{-}P < 2.974$ Å with an overall average of 2.890 Å.

We report below the ranges of values measured in each category of internal symmetry and the corresponding average inside this category :

3/m	2.856 < P-P < 2.963 Å	average 2.909 Å
3	2.877 < P-P < 2.915	2.897
2	2.864 < P-P < 2.901	2.887
m	2.826 < P-P < 2.974	2.898
no	2.818 < P-P < 2.946	2.887

For the numerical values given for P-P distances, the estimated standard deviation can be evaluated at 0.005, but for most of the recent data this value is close by 0.001 and sometimes lower.

Table 2.13

Main geometrical features for P_3O_9 ring anions with 3/m, 3, 2 or m internal symmetries.

Formula [Ref.]	P-P (Å)	P-O-P (°)	P-P-P (°)	Formula [Ref.]	P-P (Å)	P-O-P (°)	P-P-P (°)
Rings with a 3/m symmetry							
$CaNH_4P_3O_9$ [81]	2.963	137.1	60.0	$CeP_3O_9.3H_2O$ [118]	2.856	130.7	60.0
Rings with a 3 symmetry							
$Li_3P_3O_9.3H_2O$ [11]	2.915	130.9	60.0	$Cd_3(P_3O_9)_2.14H_2O$ [40]	2.911	132.2	60.0
$Te(OH)_6.2Na_3P_3O_9.6H_2O$ [131]	2.877	125.6	60.0	$Te(OH)_6.2(NH_4)_3P_3O_9$ [132]	2.896	128.3	60.0
$Te(OH)_6.2Tl_3P_3O_9$ [28]	2.886	127.4	60.0				
Rings with a m symmetry							
$Na_3P_3O_9$ [18]	2x2.887	2x126.3	59.9	$Na_3P_3O_9.H_2O$ [18]	2x2.873	2x125.4	59.2
	2.892	128.0	2x60.1		2.837	121.5	2x60.4
$Ba_3(P_3O_9)_2.4H_2O$ [53]	2x2.912	2x129.9	58.06	$CdCsP_3O_9$ [86]	2x2.864	2x126.1	62.6
	2.826	123.4	2x60.97		2.974	135.3	2x58.7
$NiK_4(P_3O_9)_2.7H_2O$ [106]	2x2.893	2x126.6	61.2	$CoRb_4(P_3O_9)_2.6H_2O$ [112]	2x2.924	2x131.7	61.4
	2.944	131.6	2x59.4		2.988	136.0	2x59.3
Rings with a 2 symmetry							
$K(EDA)P_3O_9$ [125]	2x2.901	2x127.6	60.0	$Te(OH)_6.Cs_2NaP_3O_9$ [28]	2x2.864	2x124.4	60.7
	2.900	129.1	2x60.0		2.893	126.5	2x59.7

EDA = $[NH_3-(CH_2)-NH_3]^{2+}$ (ethylenediammonium)

Table 2.14

Main geometrical data for P_3O_9 rings with no internal symmetry (first part).

Formula [Ref.]			Formula [Ref.]		
P-P (Å)	P-O-P (°)	P-P-P (°)	P-P (Å)	P-O-P (°)	P-P-P (°)
$LiK_2P_3O_9.H_2O$ [9]			$Na_2HP_3O_9$ [16]		
2.902	131.4	60.5	2.854	124.8	60.8
2.901	128.3	59.7	2.890	128.1	59.6
2.921	127.1	59.8	2.853	124.5	59.6
$Na_3P_3O_9.6H_2O$ [19]			$K_3P_3O_9$ [25]		
2.880	128.3	60.1	2.890	125.6	61.1
2.898	126.9	59.7	2.877	124.9	59.2
2.892	126.4	60.3	2.930	129.6	59.7
$Tl_3P_3O_9$, two independent rings in the unit cell [28]					
2.85	122.0	61.0	2.90	134.0	59.2
2.96	128.0	57.5	2.88	126.0	60.7
2.97	132.0	61.4	2.86	132.0	60.1
$Cs_3P_3O_9.H_2O$ [27]			$Na_2KP_3O_9$ [34]		
2.902	122.6	59.6	2.880	124.7	59.4
2.935	130.6	60.8	2.911	127.4	60.8
2.900	129.8	59.6	2.871	124.5	59.7
$CsNa_2P_3O_9.2H_2O$ [28]			$Na_2LiP_3O_9.4H_2O$ [37]		
2.880	126.8	59.7	2.884	127.5	60.5
2.866	126.1	60.2	2.896	128.0	60.1
2.880	126.9	60.2	2.864	126.1	59.4
$Cd_3(P_3O_9)_2.10H_2O$ [46]			$Sr_3(P_3O_9)_2.7H_2O$ [50]		
2.909	131.3	58.7	2.940	130.4	59.7
2.869	127.2	61.3	2.868	124.7	58.9
2.946	133.4	60.0	2.893	126.7	61.4
$Ba_3(P_3O_9)_2.6H_2O$, two independent rings in the unit cell [52]					
2.892	126.7	59.4	2.873	126.1	60.1
2.870	125.6	60.4	2.888	127.7	60.2
2.896	128.8	60.2	2.884	127.1	59.7
$Pb_3(P_3O_9)_2.3H_2O$ [57]			$Ba_2Zn(P_3O_9)_2.10H_2O$ [59]		
2.83	122.0	59.8	2.871	129.8	60.8
2.83	122.0	60.4	2.899	128.8	60.1
2.85	128.0	59.8	2.918	123.5	59.1
$PbNa_4(P_3O_9)_2$ [65]			$BaNaP_3O_9$ [67]		
2.878	126.4	60.3	2.856	128.5	60.4
2.894	127.5	60.1	2.886	123.0	59.3
2.898	126.7	59.6	2.890	126.9	60.3
$HgKP_3O_9$ [80]			$Hg(NH_4)_2Na_2(P_3O_9)_2$ [91]		
2.847	128.3	61.0	2.840	123.5	62.0
2.903	124.2	60.5	2.863	125.1	58.6
2.918	128.6	58.6	2.939	131.2	59.4

Table 2.15

Main geometrical features for P_3O_9 ring anions with no internal symmetry (second part).

Formula [Ref.]	P-P (Å)	P-O-P (°)	P-P-P (°)
$SrNaP_3O_9.3H_2O$ [93]			
	2.879	126.8	59.9
	2.888	127.5	60.2
	2.877	127.5	59.9
$BaNH_4P_3O_9.H_2O$ [100]			
	2.899	130.0	60.3
	2.872	126.6	59.4
	2.900	129.7	60.3
$NiNa_4(P_3O_9)_2.6H_2O$ [104]			
	2.899	129.0	59.5
	2.866	124.8	59.7
	2.860	125.8	60.8
$CuK_4(P_3O_9)_2.4H_2O$ [107]			
	2.878	126.3	60.7
	2.926	129.1	59.1
	2.909	131.6	60.2
$Co(NH_4)_4(P_3O_9)_2.4H_2O$ [109]			
	2.894	128.0	60.5
	2.907	129.2	60.0
	2.921	130.7	59.5
$BiNa_3(P_3O_9)_2.9H_2O$ [121]			
	2.848	124.5	59.5
	2.826	123.0	59.8
	2.818	123.2	60.6
$BaNaP_3O_9.3H_2O$ [95]			
	2.888	127.1	60.8
	2.882	128.8	59.5
	2.921	124.9	59.7
$NiAg_4(P_3O_9)_2.6H_2O$ [103]			
	2.877	128.8	59.7
	2.863	126.0	59.9
	2.858	126.1	60.4
$CuNa_4(P_3O_9)_2.4H_2O$ [105]			
	2.892	127.5	60.3
	2.895	129.2	59.9
	2.908	126.6	59.8
$Ni(NH_4)_4(P_3O_9)_2.4H_2O$ [108]			
	2.910	128.3	60.4
	2.910	130.6	59.8
	2.926	128.9	59.8
$CdK_4(P_3O_9)_2.2H_2O$ [110]			
	2.973	137.2	59.1
	2.930	129.2	59.7
	2.913	132.4	61.2

For the P-O-P angles, the overall average value is 127.9° with an estimated standard deviation evaluated at 0.1°. We report below the averages for the various categories of internal symmetries of the rings :

3/m	average 133.9
3	128.9
2	126.6
m	128.2
no	127.5

So, for this kind of very small rings we do not observe large deviations from the average values as we shall do when examining the larger ones, mainly P_6O_{18} and P_8O_{24}, since evidently, the geometrical strains decrease with the ring size.

Table 2.16

Main geometrical features for P_3O_9 ring anions with no internal symmetry (third part).

Formula [Ref.]			Formula [Ref.]		
P-P (Å)	P-O-P (°)	P-P-P (°)	P-P (Å)	P-O-P (°)	P-P-P (°)
$(EDA)_3(P_3O_9)_2$, two independent rings in the unit cell [123]					
2.907	129.7	59.9	2.918	131.0	60.2
2.895	127.6	60.3	2.902	128.6	60.2
2.890	127.9	59.7	2.919	130.5	59.6
$(IPA)_3(P_3O_9)_2$, two independent rings in the unit cell [124]					
2.874	128.3	59.6	2.875	128.4	59.8
2.866	128.8	59.8	2.866	126.8	60.1
2.895	129.5	60.6	2.872	128.5	60.0
$(MA)_3P_3O_9$ [126]			$(Gly)_3P_3O_9$ [127]		
2.851	124.9	60.2	2.865	125.8	60.0
2.874	127.0	59.5	2.883	128.3	60.3
2.876	126.9	60.3	2.872	128.7	59.7
$(Gua)_3P_3O_9.2H_2O$ [128]			$Te(OH)_6.K_3P_3O_9.2H_2O$ [130]		
2.884	127.2	59.0	2.894	129.1	60.2
2.902	124.5	60.8	2.905	128.6	59.9
2.851	128.6	60.2	2.895	127.5	59.9
$Te(OH)_6.Rb_3P_3O_9.H_2O$ [133]			$Te(OH)_6.Cs_3P_3O_9.H_2O$ [134]		
2.889	127.0	60.6	2.837	127.2	60.6
2.902	126.7	59.9	2.859	122.6	59.3
2.920	129.8	59.5	2.873	124.8	60.1

EDA = $[NH_3-(CH_2)_2-NH_3]^{2+}$ (ethylenediammonium)
MA = $(CH_3-NH_3)^+$ (methylammonium)
IPA = $[(CH_3)_2-CH-NH_3]^+$ (isopropylammonium)
Gly = $(NH_3-CH_2-COOH)^+$ (glycinium)
Gua = $[C(NH_2)_3]^+$ (guanidinium)

REFERENCES

1 T. Graham, *Phil. Trans. Roy. Soc., A123* (1833) 253.

2 T. Fleitmann and W. Henneberg, *Ann., 65* (1848) 304.

3 C.G. Lindbom, *Ber., 8* (1875) 122.

4 G. Tammann, *Z. phys. Chem., 6* (1890) 122.

5 G. Tammann, *J. prakt. Chem., 45* (1892) 417.

6 G. Von Knorre, *Z. anorg. Chem., 24* (1900) 369.

7 A. Boullé,
C. R. Acad. Sci., *206* (1938) 517.

8 E. Thilo and H. Grunze,
Z. anorg. allg. Chem., *281* (1955) 263.

9 E.D. Eanes and H.M. Ondik,
Acta Cryst., *15* (1962) 1280.

10 E.D. Eanes,
N.B.S., Monograph 25, Section 2, p 20.

11 R. Masse, J.C. Grenier, G. Bassi and I. Tordjman,
Z. Kristallogr.,*137* (1973) 17.

12 R. Masse, J.C. Grenier, G. Bassi and I. Tordjman,
Cryst. Struct. Comm., *1* (1972) 239.

13 J.C. Grenier and A. Durif,
Z. Kristallogr., *137* (1973) 10.

14 V.A. Sotnikova-Yuzhik, G.V. Peslyak and E.A. Prodan,
Russ. J. inorg. Chem., *32* (1987) 1505.

15 E.J. Griffith,
J. Am. Chem. Soc., *78* (1956) 3867.

16 M.T. Averbuch-Pouchot, J.C. Guitel and A. Durif,
Acta Cryst., *C39* (1983) 809.

17 H.M. Ondik and J.W. Gryder,
J. Inorg. Nucl. Chem., (1960) 240.

18 H.M. Ondik,
Acta Cryst., *18* (1965) 226.

19 I. Tordjman and J.C. Guitel,
Acta Cryst., *B32* (1976) 1871.

20 E.A. Prodan and S.I. Pytlev,
Izv. Akad. Nauk SSSR, Neorg. Mat., *19* (1983) 639.

21 E.D. Eanes,
private communication in Crystal Data, J.C.D.P.D.S., Vol. 2, 1973.

22 J.C. Grenier,
Bull. Soc. fr. Minér. Crist., *96* (1973) 171.

23 M. Bagieu-Beucher, A. Durif and J.C. Guitel,
Acta Cryst., *B31* (1975) 2264.

24 J.C. Grenier and A. Durif,
Rev. Chim. Minér., *9* (1972) 351.

25 M. Bagieu-Beucher, I. Tordjman, A. Durif and J.C. Guitel,
ActaCryst., *B32* (1976) 1427.

26 N.M. Dombrovskii and V.A. Koval,
Izv. Akad. Nauk SSSR, Neorg. Mat., *12* (1976) 738.

27 I. Tordjman, R. Masse and J.C. Guitel,
Acta Cryst., *B33* (1977) 585.

28 N. Boudjada,
Thesis, Univ. of Grenoble, France, 1985.

29 G. Tammann and A. Ruppelt,
Z. anorg. allg. Chem., *197* (1931) 65.

30 G. Morey,
J. Am. Chem. Soc., *76* (1954) 4724.

31 E.J. Griffith and J.R. Van Wazer, *J. Am. Chem. Soc.*, *77* (1955) 4222.

32 G.A. Bukhalova and I.V. Mardirosova, *Russ. J. Inorg. Chem.*, *11* (1966) 495.

33 C. Cavero-Ghersi and A. Durif, *C. R. Acad. Sci.*, *278C* (1974) 459-461.

34 I. Tordjman, A. Durif and C. Cavero-Ghersi, *Acta Cryst.*, *B30* (1974) 2701.

35 C. Cavero-Ghersi and A. Durif, *C. R. Acad. Sci.*, *280C* (1975) 579.

36 C. Cavero-Ghersi, *Thesis*, Univ. of Grenoble, France, 1975.

37 M.T. Averbuch-Pouchot and A. Durif, in preparation.

38 M.T. Averbuch-Pouchot and A. Durif, *Z. Kristallogr.*, *135* (1972) 318.

39 M.T. Averbuch-Pouchot, A. Durif and I. Tordjman, *Cryst. Struct.Comm.*, *2* (1973) 89.

40 M.T. Averbuch-Pouchot, A. Durif and J.C. Guitel, *Acta Cryst.*, *B32* (1976) 1533.

41 D. Michot, *Thesis*, Univ. of Dijon, France, 1975.

42 M.H. Simonot-Grange, *J. Solid State Chem.*, *46* (1983) 76.

43 M.H. Simonot-Grange and D. Michot, *Phosphorus*, *6* (1976) 103.

44 M.H. Simonot-Grange and D. Michot, *Phosphorus and Sulfur*, *4* (1978) 35.

45 A. Durif, in preparation.

46 M.T. Averbuch-Pouchot, A. Durif and J.C. Guitel, *Acta Cryst.*, *B32* (1976) 1894.

47 N. El-Horr and A. Durif, *C. R. Acad. Sci.*, *296-II* (1983) 1185.

48 R.C. Ropp, M.A. Aia, C.W.W. Hoffman, T.J. Veleker and R.W. Mooney, *Anal. Chem.*, (1959) 1164.

49 A. Durif, M. Bagieu-Beucher, C. Martin and J.C. Grenier, *Bull. Soc. fr. Minér. Crist.*, *95* (1972) 146.

50 I. Tordjman, A. Durif and J.C. Guitel, *Acta Cryst.*, *B32* (1976) 205.

51 J.C. Grenier and C. Martin, *Bull. Soc. fr. Minér. Crist.*, *98* (1975) 107-110.

52 R. Masse, J.C. Guitel and A. Durif, *Acta Cryst.*, *B32* (1976) 1892.

53 M.T. Averbuch-Pouchot and A. Durif, *Z. Kristallogr.*, *174* (1986) 219.

54 D. Tacquenet,
Thesis, Univ. of Dijon, France, 1978.

55 A. Thrierr-Sorel, D. Tacquenet and M.H. Simonot-Grange,
Phosphorus and Sulfur, *8* (1980) 73-78.

56 A. Durif and M. Brunel-Laügt,
J. Appl. Cryst., *9* (1976) 154.

57 M. Brunel-Laügt, I. Tordjman and A. Durif,
Acta Cryst., *B32* (1976) 3246.

58 M.I. Kuz'menkov, V.N. Makatun and N.V. Semenova,
Russ. J. Inorg.Chem., *24* (1979) 1151.

59 A. Durif, M.T. Averbuch-Pouchot and J.C. Guitel,
Acta Cryst., *B31* (1975) 2680.

60 E.J. Griffith,
Inorg. Chem., *2* (1962) 962.

61 J.C. Grenier, C. Martin and A. Durif,
Bull. Soc. fr. Minér. Crist., *93* (1970) 52.

62 G.W. Morey,
J. Am. Chem. Soc., *74* (1952) 5783.

63 M.T. Averbuch-Pouchot and A. Durif,
Mat. Res. Bull., *4* (1969) 859.

64 I. Mahama, M.T. Averbuch-Pouchot and J.C. Grenier,
C. R. Acad. Sci., *280C* (1975) 1105.

65 M.T. Averbuch-Pouchot and A. Durif,
Z. Kristallogr., *164* (1983) 307.

66 C. Martin and A. Durif,
Bull. Soc. fr. Minér. Crist., *95* (1972) 149.

67 C. Martin and A. Mitschler,
Acta Cryst., *B28* (1972) 2348.

68 M.T. Pouchot, I. Tordjman and A. Durif,
Bull. Soc. fr. Minér. Crist., *89* (1966) 405.

69 M.T. Averbuch-Pouchot,
C. R. Acad. Sci., *268C* (1969) 1253.

70 A. Durif and M.T. Averbuch-Pouchot,
J. Appl. Cryst., *9* (1976) 247.

71 M.A. Savenkova, I.V. Mardirosova and V.A. Matrosova,
Izv. Akad. Nauk SSSR, Neorg. Mat., *12* (1976) 1324.

72 R. Andrieu and R. Diament,
C. R. Acad. Sci., *259* (1964) 4708.

73 R. Andrieu, R. Diament, A. Durif, M.T. Pouchot and D. Tranqui,
C. R. Acad. Sci., *262B* (1966) 718.

74 M.T. Averbuch-Pouchot, C. Martin, E. Rakotomahanina-Rolaisoa and A. Durif,
Bull. Soc. fr. Minér. Crist., *93* (1970) 282.

75 R. Masse, J.C. Grenier, M.T. Averbuch-Pouchot, D. Tranqui and A. Durif,
Bull. Soc. fr. Minér. Crist., *90* (1967) 158.

76 R. Masse, A. Durif and J.C. Guitel,
Z. Kristallogr., *141* (1975) 113.

77 E.L. Krivovyazov, K.K. Palkina and N.K. Voskresenskaya, *Dokl. Akad. Nauk SSSR, 174* (1967) 610.

78 E. Rakotomahanina-Rolaisoa, *Thesis,* Univ. of Grenoble, France, 1972.

79 A. Mermet, M.T. Averbuch-Pouchot and A. Durif, *Bull. Soc. fr. Minéral. Crist., 92* (1969) 87.

80 M.T. Averbuch-Pouchot and A. Durif, *Acta Cryst., C42* (1986) 930.

81 J.L. Prisset, *Thesis*, Univ. of Grenoble, France, 1982.

82 J.C. Grenier and R. Masse, *Bull. Soc. fr. Minér. Crist., 91* (1968) 428.

83 E. Rakotomahanina-Ralaisoa, *Bull. Soc. fr. Minér. Crist., 95* (1972) 143.

84 Y. Henry and A. Durif, *Bull. Soc. fr. Minér. Crist., 92* (1969) 484.

85 M.T. Averbuch-Pouchot, *C. R. Acad. Sci., 269C* (1969) 26.

86 M.T. Averbuch-Pouchot and A. Durif, *Acta Cryst., B33* (1977) 3114.

87 I.A. Tokman and G.A. Bukhalova, *Izv. Akad. Nauk SSSR, Neorg. Mat., 11* (1975) 1654.

88 E. Rakotomahanina-Rolaisoa, Y. Henry, A. Durif and C. Raholison, *Bull. Soc. fr. Minér. Crist., 93* (1970) 43.

89 M.T. Averbuch-Pouchot, *Bull. Soc. fr. Minér. Crist., 95* (1972) 558.

90 M.T. Averbuch-Pouchot, *Thesis,* Univ. of Grenoble, France, 1974.

91 M.T. Averbuch-Pouchot and A. Durif, *Acta Cryst., C42* (1986) 932.

92 A. Durif, *C. R. Acad. Sci., 275C* (1972) 1379.

93 R. Zilber, I. Tordjman, A. Durif and J.C. Guitel, *Z. Kristallogr., 140* (1974) 350.

94 M.H. Simonot-Grange and P. Jamet, *Phosphorus and Sulfur, 3* (1977) 197.

95 M.T. Averbuch-Pouchot and A. Durif, *Acta Cryst., C43* (1987) 390.

96 D. Seethanen, A. Durif and J.C. Guitel, *Acta Cryst., B32* (1977)2716.

97 C. Martin, *Thesis*, Univ. of Grenoble, France, 1972.

98 D. Seethanen and A. Durif, *Acta Cryst., B34* (1978) 1091.

99 W. Feldman and I. Grunze, *Z. anorg. allg. Chem., 360* (1968) 225.

100 A. Durif, C. Martin and G. Bassi, *Bull. Soc. fr. Minér. Crist., 98* (1975) 19.

101 R. Masse and M.T. Averbuch-Pouchot, *Mat. Res. Bull.*, *12* (1977) 13.

102 A. Jouini and M. Dabbabi, *C. R. Acad. Sci.*, *301-II* (1985) 1347.

103 A. Jouini and M. Dabbabi, *Rev. Chim. Minér.*, *23* (1986) 776.

104 A. Jouini and M. Dabbabi, *Acta Cryst.*, *C42* (1986) 268.

105 A. Durif and M.T. Averbuch-Pouchot, *Z. anorg. allg. Chem.*, *514* (1984) 85.

106 D. Seethanen, I. Tordjman and M.T. Averbuch-Pouchot, *Acta Cryst.*,*B34* (1978) 2387.

107 A. Durif and M.T. Averbuch-Pouchot, *Acta Cryst.*, *C43* (1987) 819.

108 A. Jouini and M. Dabbabi, *Bull. Soc. Chim. Tunisie*, *2* (1987) 29.

109 M.S. Belkhiria, M. Ben-Amara and M. Dabbabi, *Acta Cryst.*, *C43* (1987) 609.

110 M.T. Averbuch-Pouchot, *Acta Cryst.*, *B34* (1978) 20.

111 D. Seethanen, A. Durif and M.T. Averbuch-Pouchot, *Acta Cryst.*, *B34* (1978) 14.

112 M.S. Belkhiria, M. Dabbabi and M. Ben-Amara, *Acta Cryst.*, *C43* (1987) 2270.

113 O.A. Serra and E. Giesbrecht, *J. Inorg. Nucl. Chem.*, *30* (1968) 793.

114 P. Birke and G. Kempe, *Z. Chem.*, *13* (1973) 151.

115 P. Birke and G. Kempe, *Z. Chem.*, *13* (1973) 65.

116 P. Birke and G. Kempe, *Z. Chem.*, *13* (1973) 110.

117 M. Bagieu-Beucher and A. Durif, *Bull. Soc. fr. Minér. Crist.*, *94* (1971) 440.

118 M. Bagieu-Beucher, I. Tordjman and A. Durif, *Rev. Chim. Minér.*, *8* (1971) 753.

119 D. Gobled, *Thesis*, Univ. of Dijon, France, 1973.

120 M. Bagieu-Beucher, *Thesis*, Univ. of Grenoble, France, 1980.

121 M. Bagieu-Beucher and A. Durif, *Z. Kristallogr.*, *178* (1987) 239.

122 M. Bagieu-Beucher, *C. R. Acad. Sci.*, *308-II* (1989) 377.

123 M.T. Averbuch-Pouchot, A. Durif and J.C. Guitel, *Acta Cryst.*, *C45* (1989) 1320.

124 M.T. Averbuch-Pouchot, A. Durif and J.C. Guitel, *Acta Cryst.*, *C44* (1988) 1907.

125 M.T. Averbuch-Pouchot and A. Durif, *Acta Cryst., C44* (1988) 1909.

126 M.T. Averbuch-Pouchot, A. Durif and J.C. Guitel, *Acta Cryst., C44* (1988) 97.

127 M.T. Averbuch-Pouchot, A. Durif and J.C. Guitel, *Acta Cryst., C44* (1988) 99.

128 M.T. Averbuch-Pouchot, A. Durif and J.C. Guitel, *Acta Cryst.*, in press.

129 N. Boudjada, *Z. anorg. allg. Chem., 477* (1981) 225.

130 N. Boudjada, M.T. Averbuch-Pouchot and A. Durif, *Acta Cryst., B37* (1981) 647.

131 N. Boudjada, M.T. Averbuch-Pouchot and A. Durif, *Acta Cryst., B37* (1981) 645.

132 N. Boudjada, A. Boudjada and J.C. Guitel, *Acta Cryst., C39* (1983) 656.

133 N. Boudjada and A. Durif, *Acta Cryst., B38* (1982) 595.

134 M.T. Averbuch-Pouchot, *Acta Cryst., C44* (1988) 1166.

135 M.T. Averbuch-Pouchot and A. Durif, *Acta Cryst., C43* (1987) 1653.

3. CYCLOTETRAPHOSPHATES

Because of the lack of proper structural characterization, the development of cyclotetraphosphate chemistry has been, at its beginning, marked by many controversies mainly based on the degree of condensation of the phosphoric anion.

As far as we know, Maddrell [1] has been the first to prepare cyclotetraphosphates. He synthesized a series of compounds of general formula $MO.P_2O_5$ (M = Mn, Cu, Co, Ni, Mg...), called by him "metaphosphates", which, later on, were recognized as cyclotetraphosphates. Some years later, several chemists, Fleitmann [2, 3], Glätzel [4], Tammann [5] and Warschauer [6] resumed and developed this investigation. These last authors using the copper cyclotetraphosphate, $Cu_2P_4O_{12}$, performed the synthesis of the corresponding sodium salt through the following reaction :

$$Cu_2P_4O_{12} + 2Na_2S \longrightarrow Na_4P_4O_{12} + 2CuS$$

Then, using the so produced sodium cyclotetraphosphate tetrahydrate, they synthesized numerous other compounds, mainly alkali and double salts. Structural investigations made during the past thirty years confirmed that most of these compounds are really cyclotetraphosphates. Fleitmann, in the absence of techniques for a proper determination of the condensation state, used the appellation "dimetaphosphates" to designate these salts. It seems it is not before Glätzel [4] that the tetrameric and cyclic nature of the anion was suspected. Then the denomination "tetrametaphosphates" created by Glätzel [4] was used by chemists, Warschauer [6] for instance, but unfortunately not always correctly applied.

In spite of being tedious and very time-consuming, the process described above for the preparation of the sodium salt has been for a long time the only possible one for the production of this starting material. One must wait the middle of the present century for the elaboration of a more convenient process based upon the low-temperature (273K) hydrolytic degradation of P_4O_{10}, which leads, with very high yields, to cyclotetraphosphoric acid, according to the following scheme :

$$P_4O_{10} + 2H_2O \text{ -----> } H_4P_4O_{12}$$

This reaction was mentioned as early as 1934 by Travers and Chu [7] and then by Topley [8], Raistrick [9], Bell *et al* [10] and finally carefully analysed and optimized by Thilo and Wicker [11].

The first structural proof of the cyclic nature of the P_4O_{12} anion was reported in 1937 by Pauling and Sherman [12] when they performed the crystal structure determination of the aluminium salt : $Al_4(P_4O_{12})_3$. Today, more than ninety cyclotetraphosphates are clearly characterized.

3.1. Present State of the Cyclotetraphosphate Chemistry

3.1.1. Cyclotetraphosphates of monovalent cations

- $Li_4P_4O_{12}.nH_2O$

The first experimental evidence for the existence of a lithium cyclotetraphosphate was reported by Grunze and Thilo [13] and Grunze [14] during their studies of the thermal condensation of LiH_2PO_4. This salt recrystallizes as a tetrahydrate. In a recent study of such a thermal condensation, Schülke and Kayser [15] characterized two forms for the anhydrous salt as well as, among the recrystallized salts, an octohydrate, an hexahydrate, two forms of tetrahydrate and two forms of dihydrate.

A crystal-chemistry study, performed by Grenier and Durif [16], concluded that lithium cyclotetraphosphate freshly prepared by the Boullé's process is an hexahydrate. From thermogravimetric experiments run with various aged specimens, these authors suspected that some of the water molecules must be of a zeolitic nature, and they depart with temperature, according to the following scheme :

$$Li_4P_4O_{12}.6H_2O \xrightarrow{343K} Li_4P_4O_{12}.4H_2O \xrightarrow{443K} Li_4P_4O_{12} \xrightarrow{623K} LiPO_3$$

In the first structural study run by Averbuch-Pouchot and Durif [17] on crystals prepared at room temperature according to the process described by Grenier and Durif [16], lithium cyclotetraphosphate is given as a pentahydrate ; two of the five water molecules being of a zeolitic nature. The very high thermal factors observed for the two non-bonded water molecules correspond probably to partly occupied positions leading to a chemical composition close by that of the tetrahydrate.

- $Na_4P_4O_{12}.nH_2O$

The anhydrous form of sodium cyclotetraphosphate, the monohydrate and two forms of the tetrahydrate are now well characterized. It has been claimed that a decahydrate crystallizes below 293K [18] but no structural data have been reported for this salt.

Single crystals of $Na_4P_4O_{12}$ and $Na_4P_4O_{12}.H_2O$ have been obtained by Wiench and Jansen [19] upon crystallization from aqueous solutions added with higher alcohols. The monohydrate exists in the temperature range 373-393K. The transition from the monohydrate to the anhydrous salt occurs topo-chemically and is reversible. A detailed description of the atomic arrangement of the monohydrate is reported by the authors. The anhydrous salt is isotypic with the monohydrate. This phenomenon has to be compared to the similar analogy observed between anhydrous sodium cyclotriphosphate and its monohydrate.

As we already said, $Na_4P_4O_{12}.4H_2O$ has been prepared by several chemists during the past century and at the beginning of the present one [2-6]. As early as 1937, Bonneman [20], using the Warschauer's process [6], prepared sodium cyclotetraphosphate tetrahydrate and determined its molecular weight by a cryoscopic procedure. He also produced the first X-ray diffraction diagrams for both the anhydrous salt and one form of the tetrahydrate.

Two forms of the tetrahydrate have been reported [10] : one is monoclinic, the second one triclinic. The form normally obtained at room temperature is the monoclinic one. The triclinic one is obtained when the crystallization is run at 353-363K.

Andress *et al* [21] described chemical preparation of the monoclinic form by the Warschauer's process and were the first to report unit-cell dimensions and space group for this salt. A detailed description for the preparation and the purification of $Na_4P_4O_{12}.4H_2O$ by the same procedure was published by Barney and Gryder [22].

Bell *et al* [10] investigated the low-temperature hydration of P_4O_{10} and were probably the first to describe a chemical preparation for

$Na_4P_4O_{12}.4H_2O$ based on the hydrolysis of P_4O_{10} into $H_4P_4O_{12}$ (see the begining of this Section).

Crystal structure of the monoclinic form has been performed by Ondik *et al* [23], while that of the triclinic form has been solved by Ondik [24]. These two structures have been later on re-examined by Averbuch-Pouchot and Durif [25] in order to localize the H-atoms.

- $Na_2H_2P_4O_{12}$ or $Na_2P_4O_{11}$

Chemical preparation of $Na_2H_2P_4O_{12}$ has been described by Griffith [26] by reaction of H_3PO_4 on NaH_2PO_4 at 673K :

$$2NaH_2PO_4 + 2H_3PO_4 \dashrightarrow Na_2H_2P_4O_{12} + 4H_2O$$

According to the author, the melting or, more probably, the decomposition point of $Na_2H_2P_4O_{12}$ is close by 673K. A more detailed preparation of this salt was afterwards reported by the same author [27]. This salt is almost insoluble in water and has a fibrous crystal habit. These two characteristics are rather surprising for an alkali cyclotetraphosphate. To explain these properties, Gryder *et al* [28] performed a crystallographic investigation of this material. They found the so-called $Na_2H_2P_4O_{12}$ to be an intimate intergrowth of two distinct crystalline forms, both monoclinic, and from various considerations they concluded that the anionic framework must be built up by chains of P_4O_{12} rings sharing an oxygen atom, thus leading to the formula of an ultraphosphate : $Na_2P_4O_{11}$. On the other hand, Jarchow [29] proposed a structural model based on the Griffith's $Na_2H_2P_4O_{12}$ formula. Due to the poor quality of the experimental data, this work does not bring conclusive solution and needs revision.

- $K_4P_4O_{12}.2H_2O$ and $K_4P_4O_{12}.4H_2O$

The existence of two varieties of $K_4P_4O_{12}.4H_2O$ and two varieties of the anhydrous salt reported by Van Wazer [30] has not been confirmed. Up to now, only one form of the dihydrate and one form of the tetrahydrate are well characterized. Averbuch-Pouchot and Durif reported chemical preparations and accurate determinations of the atomic arrangements of both dihydrate [25] and tetrahydrate [31].

- $(NH_4)_3HP_4O_{12}.H_2O$ and $(NH_4)_4P_4O_{12}$

According to Waerstad *et al* [32], $(NH_4)_3HP_4O_{12}.H_2O$ can be obtained by adding P_4O_{10} to a concentrated NH_4OH solution at a temperature lower than 278K. Unit-cell dimensions and an indexed powder diagram are reported by the authors.

$(NH_4)_4P_4O_{12}$ was first investigated by Romers *et al* [33] and discussed by Andress and Fischer [34]. The first valuable description of its atomic arrangement was reported by Romers *et al* [35]. This work was later on improved by Cruickshank [36] and a re-examination of the crystal structure, including the localization of the H-atoms, was reported by Koster and Wagner [37].

- $Rb_4P_4O_{12}.4H_2O$ and $Cs_4P_4O_{12}.4H_2O$

The two isotypic compounds, $M_4P_4O_{12}.4H_2O$ (M = Rb, Cs) have been prepared and characterized by Averbuch-Pouchot and Durif [38]. There is no evidence for the existence of other hydrates or of the anhydrous salts. It must be noticed that by a careful hydrolysis of the rubidium salt, at room

Table 3.1

Main crystallographic features for cyclotetraphosphates of monovalent cations.

Formula	a α	b β	c (Å) γ (°)	S. G.	Z	Ref.
$Li_4P_4O_{12}.5H_2O$	17.073(8)	17.029(8) 127.32(1)	13.554(6)	C2/c	8	17
$Na_4P_4O_{12}$	13.808(2)	13.633(2)	6.027(2)	$P2_12_12_1$	4	19
$Na_4P_4O_{12}.H_2O$	13.654(2)	13.475(3)	6.291(3)	$P2_12_12_1$	4	19
$Na_4P_4O_{12}.4H_2O$ (I)	6.652 103.40	9.579 106.98	6.320 93.28	$P\overline{1}$	1	24, 25
$Na_4P_4O_{12}.4H_2O$ (II)	9.667	12.358 92.58	6.170	$P2_1/a$	2	23, 25
$Na_2H_2P_4O_{12}$	18.74	14.79	7.03	Pbnm	8	29
$K_4P_4O_{12}.2H_2O$	8.153(4) 97.33(5)	8.222(4) 95.46(5)	11.154(8) 88.92(5)	$P\overline{1}$	2	25
$K_4P_4O_{12}.4H_2O$	9.061(3)	9.061(3)	10.284(5)	$I\overline{4}$	2	31
$(NH_4)_3HP_4O_{12}.H_2O$	11.144(4)	11.322(3) 90.67(3)	10.514(5)	$P2_1/n$	4	32
$(NH_4)_4P_4O_{12}$	10.433(2)	10.871(3)	12.588(3)	Cmca	4	33-37
$Rb_4P_4O_{12}.4H_2O$	9.163(3)	9.163(3)	21.356(8)	$P4_1$	4	38
$Cs_4P_4O_{12}.4H_2O$	9.466(5)	9.466(5)	21.933(9)	$P4_1$	4	38
$Tl_4P_4O_{12}$	7.635	7.635	11.087	$P\overline{4}2_1c$	2	44

temperature, Averbuch-Pouchot and Durif [39] prepared a new form of rubidium dihydrogen monophosphate, closely related to the monoclinic form of CsH_2PO_4 and identical to the intermediate form of RbD_2PO_4. The hydrolysis scheme is :

$$Rb_4P_4O_{12} + 4H_2O \dashrightarrow 4RbH_2PO_4$$

- $Tl_4P_4O_{12}$

The thermal dehydation of TlH_2PO_4 and of the corresponding arsenate has been studied by Dostal *et al* [40] and by Dostal and Kocman [41, 42). During these investigations, the authors characterized $Tl_4P_4O_{12}$ as a water soluble cyclotetraphosphate identical to $Tl_4As_4O_{12}$ and studied the solid solutions of the two salts. $Tl_4P_4O_{12}$ converts into a long-chain polyphosphate at about 690K. Its atomic arrangement has been determined by Fawcett *et al* [43, 44].

Table 3.1 reports the main crystallographic features of the thirteen presently investigated cyclotetraphosphates of monovalent cations.

- $Na_2M_2P_4O_{12}.2H_2O$ (M = K, NH_4 and Rb)

Two crystalline forms of $Na_2K_2P_4O_{12}.2H_2O$ have been prepared by Averbuch-Pouchot and Durif [25] : one is triclinic, the other one tetragonal. The authors report complete structural studies for the two forms.

Table 3.2

Main crystallographic features for cyclotetraphosphates of mixed monovalent cations.

Formula	a α	b β	c (Å) γ (°)	S. G.	Z	Ref.
$Na_2K_2P_4O_{12}.2H_2O$ (I)	11.366(8) 90.07(5)	7.908(5) 106.85(5)	7.929(5) 95.66(5)	$P\bar{1}$	2	25
$Na_2K_2P_4O_{12}.2H_2O$ (II)	7.928(5)	7.928(5)	21.66(2)	$P4_1$	4	25
$Na_2(NH_4)_2P_4O_{12}.2H_2O$	11.547(8) 89.76(4)	8.012(6) 106.22(3)	8.044(5) 94.78(3)	$P\bar{1}$	2	25
$Na_2Rb_2P_4O_{12}.2H_2O$	11.577(9) 89.79(5)	8.006(6) 106.58(4)	8.032(7) 95.19(4)	$P\bar{1}$	2	25
$Na_3CsP_4O_{12}.3H_2O$	11.39(1)	10.92(1) 95.24(5)	11.81(1)	$P2_1/c$	4	45
$Na_3CsP_4O_{12}.4H_2O$	14.50(2)	7.804(3)	7.006(3)	Imm2	2	45

$Na_2(NH_4)_2P_4O_{12}.2H_2O$ and $Na_2Rb_2P_4O_{12}.2H_2O$ have been synthesized during the same investigation and are isotypic of the triclinic form of the sodium-potassium salt.

- $Na_3CsP_4O_{12}.4H_2O$ and $Na_3CsP_4O_{12}.3H_2O$

When an aqueous solution of $Na_4P_4O_{12}$, acidified by a small amount of $H_4P_4O_{12}$, is added with caesium carbonate and then, kept at room temperature, several successive precipitations occur. The first one corresponds to the monoclinic form of $Na_4P_4O_{12}.4H_2O$, the second one to $Na_3CsP_4O_{12}.4H_2O$ and the last one to $Na_3CsP_4O_{12}.3H_2O$. Averbuch-Pouchot and Durif [45] reported chemical preparations and crystal structures of these last two hydrates.

Table 3.2 summarizes the main crystallographic features of the presently investigated cyclotetraphosphates of mixed monovalent cations.

3.1.2. Cyclotetraphosphates of divalent cations

Anhydrous divalent-cation cyclotetraphosphates

- $M_2P_4O_{12}$ (M = Zn, Ni, Co, Fe, Mg, Cu, Mn, Cd)

The eight cyclotetraphosphates $M_2P_4O_{12}$ (M = Zn, Ni, Co, Fe, Mg, Cu, Mn, Cd) possess the same atomic arrangement. Some of them have been characterized during the past century. For instance, $Cu_2P_4O_{12}$ has been described as early as 1847 by Maddrell [1] and successively by Fleitmann [2, 3] and Warschauer [6]. Some of these authors somewhat suspected the tetrameric nature of the anion. More recently, pioneering work in this field has been done by Thilo and Grunze [46]. They studied the thermal condensation of $M(H_2PO_4)_2$ monophosphates :

$$2M(H_2PO_4)_2 \longrightarrow M_2P_4O_{12} + 4H_2O$$

and they showed that the resulting compounds were isomorphous from X-ray experiments and very probably cyclotetraphosphates from paper chromatography. By spectrometric investigations, Steger [47] and Steger and Simon [48] confirmed these results.

Later on, unit-cell parameters of this series of compounds were reported by Beucher [49], Beucher and Grenier [50] and Laügt *et al* [51] and their isotypy was confirmed. The atomic arrangement was finally determined by Laügt *et al* [52] using the copper salt. With the exception of the cadmium salt, these compounds are prepared by using thermal or flux methods. In most cases, the crystals produced by flux methods are twinned.

This series of compounds have been the object of a great number of chemical, crystallographic and physical studies which will be reported below.

The thermal evolution of $Mn(H_2PO_4)_2.2H_2O$ and $Mg(H_2PO_4)_2.2H_2O$ leading to the formation of $Mg_2P_4O_{12}$ and $Mn_2P_4O_{12}$ has been studied in details by Shchegrov *et al* [53].

The mechanism of the thermal dehydration-cyclization of $Mg(H_2PO_4)_2.2H_2O$ leading to magnesium cyclotetraphosphate has been investigated by Serazetdinov *et al* [54] both in vacuum and at atmospheric pressure. Under vacuum, they observed the formation of an amorphous phase, at about 773K, before the crystallization of the cyclotetraphosphate. This phase does not appear when operating at atmospheric pressure.

A similar study has been performed by Bekturov *et al* [55]. Two crystalline forms of $Mg_2P_4O_{12}$ have been characterized as final condensation products.

Chemical preparation of $Mn_2P_4O_{12}$ using manganese carbonate and ammonium dihydrogeno-monophosphate as starting materials has been analysed by Bukhalova *et al* [56].

$Fe_2P_4O_{12}$ was prepared for the first time by Bagieu-Beucher *et al* [57]. The authors report crystal data showing the isotypy of this salt with the copper compound. This compound has also been characterized by Genkina *et al* [58] during a study of the $LiF-Fe_2O_3(FeO)-P_2O_5-H_2O$ system under hydrothermal conditions at 703K and 1000 atmospheres.

Surface tension of molten $Zn(PO_3)_2$ has been investigated by Krivovyazov *et al* [59]. According to these authors, the melting point is observed at 1135(2)K and the transformation from $Zn_2P_4O_{12}$ to $Zn(PO_3)_2$ occurs at 913K.

$Cd_2P_4O_{12}$ is a special case in this series. It cannot be prepared by a flux method similar to that used for the other $M_2P_4O_{12}$ salts. Its existence was reported for the first time by Thilo and Grunze [46] during a study of the thermal condensation of $Cd(H_2PO_4)_2.2H_2O$. These authors give the following scheme for that condensation:

$$Cd(H_2PO_4)_2.2H_2O \xrightarrow{373K} Cd(H_2PO_4)_2 \xrightarrow{453K} CdH_2P_2O_7 \xrightarrow{473K} Cd_2P_4O_{12}$$

According to these authors, the cyclotetraphosphate is transformed into a long-chain polyphosphate at about 873K.

Some years later, the same study was re-examined by Ropp and Aia [60] by differential thermal analysis. For the first two steps, they found transformation temperatures relatively close by those reported by Thilo and Grunze [46] : *i. e.* 403K to 433K instead of 373K and 483K instead of 453K. For the next step, corresponding to the condensation of the diphosphate into the cyclotetraphosphate, they give a temperature of 683K. According to the same authors, the so obtained cyclotetraphosphate transforms at 823K into a tetraphosphate : $Cd_3P_4O_{13}$. The existence of this last compound has not yet been confirmed.

In spite of these data, the preparation of pure specimens of $Cd_2P_4O_{12}$ was not possible until the process described by Laügt *et al* [51]. These authors used as starting material a not yet characterized hydrate of cadmium cyclotetraphosphate prepared at room temperature by action of $H_4P_4O_{12}$ on cadmium carbonate. By heating slowly this hydrate up to 573K and keeping it at this temperature for 30 min, they obtained pure specimens of $Cd_2P_4O_{12}$, not contaminated by the cadmium long-chain polyphosphate. In contradiction with the experiments of the previous authors, they observed that a prolongated heating at 573K transforms irreversibly $Cd_2P_4O_{12}$ into the long-chain cadmium polyphosphate. It is worth reporting that by firing, at temperatures higher than 673K, a phospho-arsenate $Cd(H_2P_{1-x}As_xO_4)_2.2H_2O$ (x~0.5), the same authors obtained a well crystallized mixture of $Cd_2P_4O_{12}$ and $Cd(AsO_3)_2$ without any contamination by the long-chain polyphosphate.

A good number of crystallographic investigations have been performed on this series of compounds :

(i) the crystal structure of $Mg_2P_4O_{12}$ has been refined from single crystal X-ray data by Nord and Lindberg [61];

(ii) the crystal structure of $Co_2P_4O_{12}$ has been refined by Nord [62] using neutron diffraction and a powdered specimen ;

(iii) the $Ni_2P_4O_{12}$ has been prepared as powder sample and its crystal structure refined by neutron diffraction by Nord [63]. The same author [64] investigated $NiCoP_4O_{12}$ and $NiZnP_4O_{12}$ samples by neutron diffraction showing that in both the solid solutions, the nickel atom preferentially enters the centrosymmetrical crystallographic site ;

(iv) the crystal structure of $Fe_2P_4O_{12}$ has been performed by Genkina *et al* [65]. The authors confirm the isotypy with the copper salt.

Except for the cadmium salt, a high-pressure form was observed for the $M_2P_4O_{12}$ compounds by Bagieu-Beucher *et al* [57]. At first, this high-pressure form was thought to be also a cyclotetraphosphate, but it was later recognized as a long-chain polyphosphate by Averbuch-Pouchot *et al* [66].

The magnetic properties of $M_2P_4O_{12}$ (M = Mn, Co, Ni, Cu) have been investigated by Gunsser *et al* [67].

- $Ca_2P_4O_{12}$

The salt $Ca_2P_4O_{12}$ has been obtained by Schneider and Jost [68] during a study of the thermal behavior of the tetrahydrate. The authors report only its unit-cell dimensions.

Table 3.3 summarizes the main crystallographic features for anhydrous divalent cation cyclotetraphosphates.

Table 3.3
Main crystal data for anhydrous cyclotetraphosphates of divalent cations.

Formula	a α	b β	c (Å) γ (°)	S. G.	Z	Ref.
$Cu_2P_4O_{12}$	12.56(1)	8.088(6) 118.58(2)	9.574(8)	C2/c	4	50, 52
$Mg_2P_4O_{12}$	11.77(1)	8.287(4) 118.87(2)	9.949(10)	C2/c	4	50
$Ni_2P_4O_{12}$	11.65(1)	8.241(4) 118.46(2)	9.857(6)	C2/c	4	50
$Co_2P_4O_{12}$	11.815(5)	8.310(8) 118.68(2)	9.339(10)	C2/c	4	50
$Zn_2P_4O_{12}$	11.78(1)	8.302(6) 118.81(2)	9.927(8)	C2/c	4	50
$Mn_2P_4O_{12}$	12.08(1)	8.471(6) 119.29(2)	10.171(8)	C2/c	4	50
$Fe_2P_4O_{12}$	11.952(2)	8.359(2) 118.76(5)	9.932(2)	C2/c	4	57, 65
$Cd_2P_4O_{12}$	12.319(5)	8.631(3) 119.33(5)	10.382(6)	C2/c	4	51
$Ca_2P_4O_{12}$	8.02(1) 97.4(1)	10.42(1) 109.8(1)	7.20(1) 90.4(1)	P1 or $P\bar{1}$	2	68

Hydrated divalent-cation cyclotetraphosphates

- $M_2P_4O_{12}.8H_2O$ (M = Zn, Mg, Fe, Co, Cu)

$Zn_2P_4O_{12}.8H_2O$ was at first characterized by Averbuch-Pouchot [69], who reported its chemical preparation and a complete description of its atomic arrangement. Later on, the corresponding magnesium, iron, cobalt and copper salts were prepared and recognized as isotypes of $Zn_2P_4O_{12}.8H_2O$ by Foumakoye [70].

- $Mn_2P_4O_{12}.10H_2O$, $Ni_2P_4O_{12}.10H_2O$ and $Cd_2P_4O_{12}.8H_2O$

The three hydrates $Cd_2P_4O_{12}.8H_2O$ and $M_2P_4O_{12}.10H_2O$ (M = Mn, Ni) have been characterized by Lavrov *et al* [71]. They are prepared by action of $H_4P_4O_{12}$ on an aqueous solution of the corresponding divalent cation perchlorates.

- $Ca_2P_4O_{12}.4H_2O$ and $Ca_2P_4O_{12}.H_2O$

The crystal structure of $Ca_2P_4O_{12}.4H_2O$ was performed by Schneider *et al* [72]. In spite of a close similarity with the corresponding lead salt, the authors could not conclude to their isotypy after a comparison of the two atomic arrangements. The existence of a second form of $Ca_2P_4O_{12}.4H_2O$ has been reported by Schülke [73], but, up to now, no crystal data are reported.

The thermal evolution of the monoclinic, $P2_1/n$, $Ca_2P_4O_{12}.4H_2O$ has been investigated by Schneider and Jost [68, 74]. The following scheme of evolution is reported by the authors :

$$Ca_2P_4O_{12}.4H_2O \xrightarrow{393K} Ca_2P_4O_{12}.H_2O \xrightarrow{493K} Ca_2P_4O_{12} \xrightarrow{733K} \text{amorphous phase}$$

$$\text{amorphous phase} \xrightarrow{793K} x\,Ca(PO_3)_2 \xrightarrow{813\text{-}893K} \beta\,Ca(PO_3)_2 \xrightarrow{1123K} \beta\,Ca_2P_2O_7$$

Each intermediate phase is crystallographically connected to the β-$Ca(PO_3)_2$ long-chain polyphosphate. It is of interest to notice that the intermediate amorphous phase does not interrupt the orientated course of the reaction. The crystallographic orientation relations between educts and products are reported for all steps of the reaction as well as the unit cells for each cyclotetraphosphate. The last step corresponds to a loss of P_2O_5.

- $Sr_2P_4O_{12}.6H_2O$

The hydrate $Sr_2P_4O_{12}.6H_2O$ has been obtained by Durif and Averbuch-Pouchot [75] by action of $H_4P_4O_{12}$ on a solution of strontium nitrate. The authors performed a complete determination of its atomic arrangement.

- $Pb_2P_4O_{12}.4H_2O$, $Pb_2P_4O_{12}.3H_2O$ and $Pb_2P_4O_{12}.2H_2O$

The tetrahydrate $Pb_2P_4O_{12}.4H_2O$ has been carefully investigated by Worzala [76], who reports a detailed determination of its atomic arrangement. This salt is of fundamental interest for the preparation, by a process of dehydration and recondensation, of lead cyclooctophosphate, a fundamental starting material for the production of other cyclo-octaphosphates (see the corresponding chapter). The dihydrate $Pb_2P_4O_{12}.2H_2O$ which constitutes the first step in this process, has been obtained as single crystals by Worzala [78]. They are formed topotactically by heating the tetrahydrate at 373K. The author reports a complete description of the atomic arrangement. The trihydrate $Pb_2P_4O_{12}.3H_2O$ has been prepared by Klinkert and Jansen [77] by using a gel technique. The authors report a detailed determination of the atomic arrangement. This hydrate does not appear in the process of recondensation of lead tetra-hydrate.

Table 3.4
Main crystal data for hydrated divalent cation cyclotetraphosphates.

Formula	a α	b β	c (Å) γ (°)	S. G.	Z	Ref.
$Zn_2P_4O_{12}.8H_2O$	8.610(5) 96.09(5)	7.137(5) 105.99(5)	7.108(5) 100.49(5)	$P\bar{1}$	1	69
$Mg_2P_4O_{12}.8H_2O$	8.740(5) 96.37(5)	7.139(5) 106.38(5)	7.103(5) 100.60(5)	$P\bar{1}$	1	70
$Fe_2P_4O_{12}.8H_2O$	8.643(5) 96.31(5)	7.175(5) 105.44(5)	7.114(5) 100.72(5)	$P\bar{1}$	1	70
$Co_2P_4O_{12}.8H_2O$	8.768(5) 92.43(5)	6.855(5) 105.83(5)	7.274(5) 104.44(5)	$P\bar{1}$	1	70
$Cu_2P_4O_{12}.8H_2O$	8.526(5) 93.20(5)	7.058(5) 105.58(5)	7.155(5) 100.59(5)	$P\bar{1}$	1	70
$Ca_2P_4O_{12}.4H_2O$	7.668(1)	12.895(1) 107.00(1)	7.144(1)	$P2_1/n$	2	68, 72
$Ca_2P_4O_{12}.H_2O$	7.72(1) 95.9(1)	10.52(1) 105.1(1)	7.15(1) 83.9(1)	P1 or $P\bar{1}$	2	68
$Sr_2P_4O_{12}.6H_2O$	6.644(3) 101.62(5)	7.365(4) 109.98(5)	8.618(4) 95.65(5)	$P\bar{1}$	2	75
$Pb_2P_4O_{12}.4H_2O$	8.07(2)	11.76(3) 108.2(3)	7.50(2)	$P2_1/n$	2	76
$Pb_2P_4O_{12}.3H_2O$	7.864(3) 97.42(2)	9.144(3) 100.63(2)	10.216(3) 114.92(2)	$P\bar{1}$	2	77
$Pb_2P_4O_{12}.2H_2O$	8.02(2) 98.8(2)	10.58(2) 108.7(2)	7.53(2) 82.6(3)	$P\bar{1}$	2	78

Table 3.4 reports the main crystallographic features for hydrated divalent cation cyclotetraphosphates.

3.1.3. Cyclotetraphosphates of divalent-monovalent cations

Anhydrous divalent-monovalent cation cyclotetraphosphates

- $Zn_4Na_4(P_4O_{12})_3$, $Co_4Na_4(P_4O_{12})_3$ and $Ni_4Na_4(P_4O_{12})_3$

Chemical preparations for the three isotypic compounds $M_4Na_4(P_4O_{12})_3$ (M = Zn, Co, Ni) have been reported by Averbuch-Pouchot and Durif [79]. The zinc salt has been used by the authors to determine the atomic arrangement. They discuss the close relationship between this structure and that of $Al_4(P_4O_{12})_3$.

- $SrNa_2P_4O_{12}$

$SrNa_2P_4O_{12}$ was not observed in the $NaPO_3$-$Sr(PO_3)_2$ phase-equilibrium diagram elaborated by Martin [80]. It has been characterized for the first time by Averbuch-Pouchot and Durif [81] during attempts to prepare single crystals of $SrNa_4(P_3O_9)_2$ by a flux method. The authors report its chemical preparation and a complete description of its atomic arrangement.

- $SrK_2P_4O_{12}$

In the KPO_3-$Sr(PO_3)_2$ phase-equilibrium diagram elaborated by Bukhalova and Tokman [82], $SrK_2P_4O_{12}$, the only compound observed in this system, is given as a non-congruent melting salt in contradiction with the revised diagram (Fig. 3.1) reported by Martin [80] in which $SrK_2P_4O_{12}$ appears as a congruent melting compound (m.p. = 977K). Chemical preparation and crystal data of this salt have been reported by Durif *et al* [83]

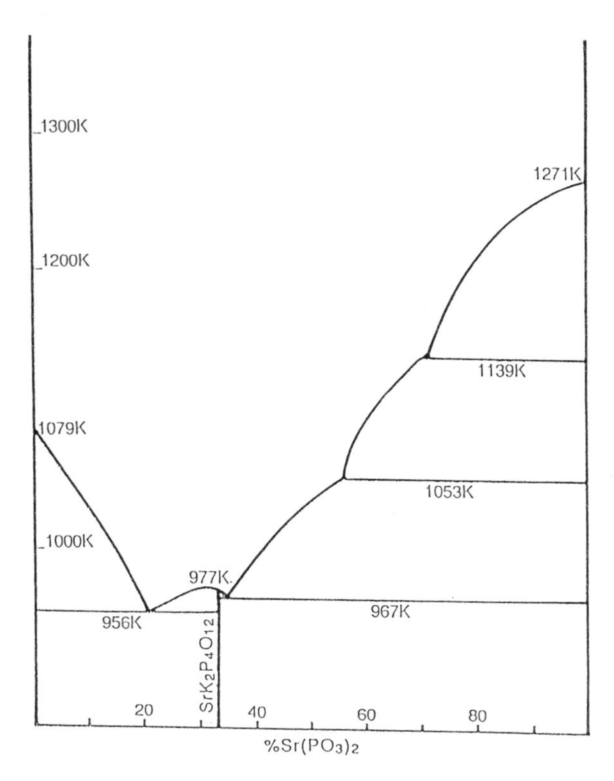

Figure 3.1 The $Sr(PO_3)_2$-KPO_3 phase-equilibrium diagram as revised by Martin [80].

and its crystal structure performed by Tordjman *et al* [84]. Eight tetragonal $M^{II}M^{I}_{2}P_4O_{12}$ cyclotetraphosphates have been clearly characte-rized as isotypes of this salt.

- $CaK_2P_4O_{12}$

Cavero-Ghersi and Durif [85] reported chemical preparation and crystal data for $CaK_2P_4O_{12}$, recognized as isotypic with $SrK_2P_4O_{12}$. At 973K, it decomposes as follows :

$$CaK_2P_4O_{12} \text{ -------> } CaKP_3O_9 + KPO_3$$

This explains why this salt has not been observed during the study of the KPO_3-$Ca(PO_3)_2$ phase-equilibrium diagram established almost simultaneously by Gill and Taylor [86] and Andrieu and Diament [87].

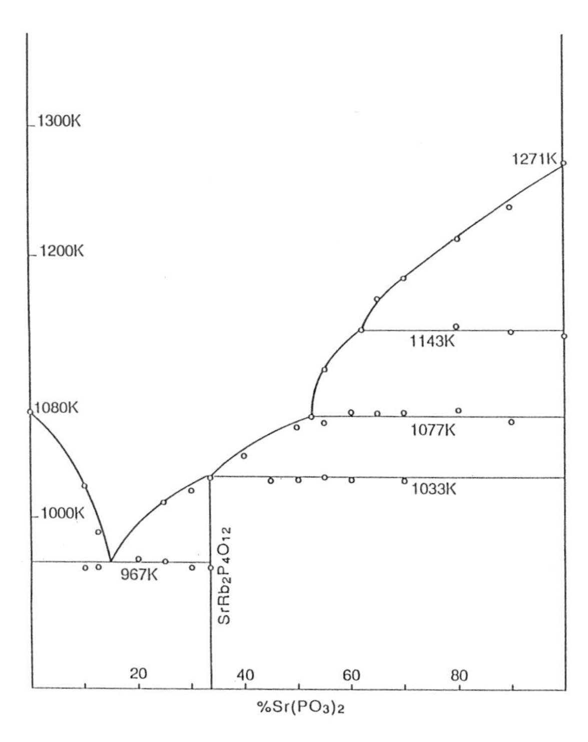

Figure 3.2 The $Sr(PO_3)_2$-$RbPO_3$ phase-equilibrium diagram. Two phase transformations of $Sr(PO_3)_2$ appear on the right side of the diagram.

- $PbK_2P_4O_{12}$

Existence and crystallographic investigations of $PbK_2P_4O_{12}$ have been reported by Cavero-Ghersi and Durif [85]. This compound, isotypic with $SrK_2P_4O_{12}$, appears as the low-temperature form of the $PbK_2(PO_3)_4$

long-chain polyphosphate observed by Mahama *et al* [88] during the elaboration of the KPO_3-$Pb(PO_3)_2$ phase-equilibrium diagram. At 810K, it transforms irreversibly into the long-chain polyphosphate :

$$PbK_2P_4O_{12} \text{ ------> } PbK_2(PO_3)_4$$

This transformation has been reinvestigated by Schneider [89] in dry and moist atmospheres. In dry atmosphere, he confirmed the previous result but in moist atmosphere he observed the following transformation :

$$PbK_2P_4O_{12} \text{ -----> } Pb_2P_2O_7 + \text{amorphous phase}$$

- $Sr(NH_4)_2P_4O_{12}$, $SrTl_2P_4O_{12}$ and $SrRb_2P_4O_{12}$

The three salts $SrM_2P_4O_{12}$ (M = NH_4, Tl, Rb) have been described by Durif *et al* [83] as isotypes of $SrK_2P_4O_{12}$. The last two species appear as the only compounds existing in the $RbPO_3$-$Sr(PO_3)_2$ and $TlPO_3$-$Sr(PO_3)_2$

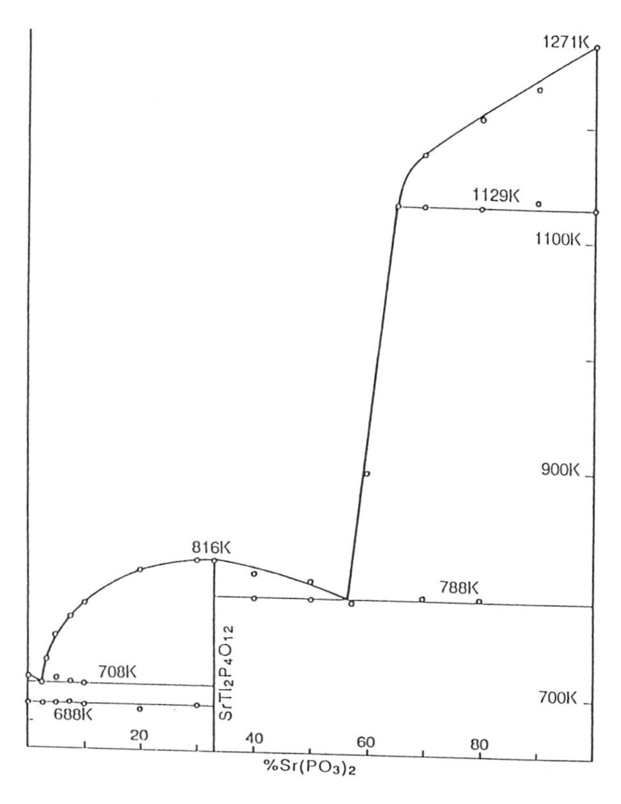

Figure 3.3 The $Sr(PO_3)_3$-$TlPO_3$ phase-equilibrium diagram [80].

phase-equilibrium diagrams elaborated by Martin [80] which is represented in Figs. 3.2 and 3.3. $SrTl_2P_4O_{12}$ is a congruent melting compound (m.p. = 816K), while $SrRb_2P_4O_{12}$ decomposes at 1033K.

- $Pb(NH_4)_2P_4O_{12}$, $PbRb_2P_4O_{12}$ and $PbTl_2P_4O_{12}$

Cavero-Ghersi and Durif [85] reported chemical preparations and crystal data for the three $PbM_2P_4O_{12}$ .compounds (M = NH_4, Tl, Rb) isotypic with $SrK_2P_4O_{12}$.

Table 3.5

Main crystallographic features for anhydrous divalent-monovalent cation cyclotetraphosphates.

Formula	a α	b β	c (Å) γ (°)	S. G.	Z	Ref.
$Zn_4Na_4(P_4O_{12})_3$	14.580	14.580	14.580	$I\bar{4}3d$	4	79
$Co_4Na_4(P_4O_{12})_3$	14.593	14.593	14.593	$I\bar{4}3d$	4	79
$Ni_4Na_4(P_4O_{12})_3$	14.446	14.446	14.446	$I\bar{4}3d$	4	79
$SrNa_2P_4O_{12}$	9.838(5)	9.838(5)	5.003(3)	P4/nbm	2	81
$SrK_2P_4O_{12}$	7.445(5)	7.445(5)	10.17(2)	$I\bar{4}$	2	83, 84
$CaK_2P_4O_{12}$	7.364(2)	7.364(2)	9.899(5)	$I\bar{4}$	2	85
$PbK_2P_4O_{12}$	7.434(2)	7.434(2)	10.208(4)	$I\bar{4}$	2	85
$Sr(NH_4)_2P_4O_{12}$	7.575(5)	7.575(5)	10.26(2)	$I\bar{4}$	2	83
$SrTl_2P_4O_{12}$	7.608(5)	7.608(5)	10.25(2)	$I\bar{4}$	2	83
$SrRb_2P_4O_{12}$	7.585(5)	7.585(5)	10.28(2)	$I\bar{4}$	2	83
$Pb(NH_4)_2P_4O_{12}$	7.550(2)	7.550(2)	10.350(8)	$I\bar{4}$	2	85
$PbRb_2P_4O_{12}$	7.567(2)	7.567(2)	10.357(8)	$I\bar{4}$	2	85
$PbTl_2P_4O_{12}$	7.591(3)	7.591(3)	10.356(8)	$I\bar{4}$	2	85
$Sr_3Cs_4H_2(P_4O_{12})_3$	15.455(5)	15.455(5)	15.455(5)	$I\bar{4}3d$	4	91

- $SrCs_2P_4O_{12}$ and $Sr_3Cs_4H_2(P_4O_{12})_3$

$SrCs_2P_4O_{12}$ appears as a non-congruent melting salt in the phase diagram $CsPO_3$-$Sr(PO_3)_2$ elaborated by Tokman and Bukhalova [90]. Up to now, there is no structural evidence to confirm the ring nature of the anion assumed by the authors. $Sr_3Cs_4H_2(P_4O_{12})_3$ was obtained by Averbuch-Pouchot and Durif [91] during an attempt to produce $SrCs_2P_4O_{12}$. It appears as a sparingly water soluble salt. A detailed determination of its atomic arrangement is reported by the authors, who discuss the close similarity of this structure with that of $Al_4(P_4O_{12})_3$.

Table 3.5 reports the main crystallographic features for anhydrous divalent-monovalent cation cyclotetraphosphates.

Hydrated divalent-monovalent cation cyclotetraphosphates

- $CaNa_2P_4O_{12}.11/2H_2O$ and $SrNa_2P_4O_{12}.6H_2O$

Chemical preparations and crystal structures of $CaNa_2P_4O_{12}.11/2H_2O$ and $SrNa_2P_4O_{12}.6H_2O$ have been reported by Averbuch-Pouchot and Durif [95] and Durif *et al* [93], respectively. No relationship exists between these two atomic arrangements. The crystal structure determination of the calcium salt provided the first example of a cyclotetraphosphate containing two crystallographic independent P_4O_{12} groups.

- $NiK_2P_4O_{12}.7H_2O$ and $Ni(NH_4)_2P_4O_{12}.7H_2O$

The two isotypic salts $NiM_2P_4O_{12}.7H_2O$ (M = K, NH_4) have been prepared by Jouini *et al* [94]. The atomic arrangement, determined by the authors with the ammonium salt, shows that three of the water molecules are of a zeolitic nature.

- $NiK_2P_4O_{12}.5H_2O$ and $CoK_2P_4O_{12}.5H_2O$

The two isotypic salts $MK_2P_4O_{12}.5H_2O$ (M = Ni, Co) have been

Table 3.6

Main crystallographic features for hydrated divalent-monovalent cation cyclotetraphosphates.

Formula	a α	b β	c (Å) γ (°)	S. G.	Z	Ref.
$CaNa_2P_4O_{12}.11/2H_2O$	27.88(10)	7.536(5)	7.378(5)	Pma2	4	92
$SrNa_2P_4O_{12}.6H_2O$	7.332(5)	7.663(5)	14.408(8)	I2mm	2	93
$NiK_2P_4O_{12}.7H_2O$	13.820(3) 98.13(4)	9.640(5) 97.31(4)	7.450(2) 102.85(4)	$P\bar{1}$	2	94
$Ni(NH_4)_2P_4O_{12}.7H_2O$	13.841(3) 98.05(4)	9.621(5) 97.25(4)	7.482(2) 103.01(4)	$P\bar{1}$	2	94
$CoK_2P_4O_{12}.5H_2O$	12.955(2)	16.294(2) 92.17(1)	7.432(2)	$P2_1/c$	4	95
$Ca_4K_4(P_4O_{12})_3.8H_2O$	20.38(1)	12.683(5) 89.31(5)	7.830(2)	$P2_1/a$	2	92
$Ca(NH_4)_2P_4O_{12}.2H_2O$	16.783(10)	10.888(6) 90.82(8)	7.913(2)	$P2_1/n$	4	96, 97

obtained by dehydration of the corresponding heptahydrates. The crystal structure determination has been performed by Jouini *et al* [95] using the cobalt salt. In addition, a detailed IR study of both compounds is reported by the authors and a tentative explanation for the dehydration and reorganization from the heptahydrate is proposed.

- $Ca_4K_4(P_4O_{12})_3.8H_2O$

Chemical preparation and crystal structure of $Ca_4K_4(P_4O_{12})_3.8H_2O$ have been reported by Averbuch-Pouchot and Durif [92].

- $Ca(NH_4)_2P_4O_{12}.2H_2O$

$Ca(NH_4)_2P_4O_{12}.2H_2O$ was first prepared by Cavero-Ghersi [96], who reported its main crystallographic features. Later on, its crystal structure has been determined by Tordjman *et al* [97].

Table 3.6 reports the main crystallographic features for hydrated divalent-monovalent cation cyclotetraphosphates.

3.1.4. Cyclotetraphosphates of trivalent cations

Up to day, all the well characterized trivalent cation cyclotetraphosphates belong to a type of structure, first investigated by Hendricks and Wyckoff [98] and described in 1937 by Pauling and Sherman [12] for the aluminium salt, $Al_4(P_4O_{12})_3$. This study was the first structural proof for the existence of cyclic phosphoric anions.

- $Fe_4(P_4O_{12})_3$ and $Cr_4(P_4O_{12})_3$

During an investigation of the Al_2O_3-P_2O_5 and Fe_2O_3-P_2O_5 systems, D'Yvoire [99] identified $Fe_4(P_4O_{12})_3$ as an isotype of the aluminium salt. $Cr_4(P_4O_{12})_3$ was characterized by Rémy and Boullé [100] when investigating the Cr_2O_3-P_2O_5 system.

Table 3.7

The common cubic space-group is $I\bar{4}3d$ for the six compounds whose unit cells reported below contain four formula units.

Formula	a (Å)	Ref.
$Al_4(P_4O_{12})_3$	13.730	12, 98, 107
$Cr_4(P_4O_{12})_3$	13.912	100
$Ti_4(P_4O_{12})_3$	13.82	101
$Fe_4(P_4O_{12})_3$	14.013(7)	99
$Sc_4(P_4O_{12})_3$	14.363(5)	102-105
$Yb_4(P_4O_{12})_3$	14.66(1)	106

- ***$Ti_4(P_4O_{12})_3$***

Chemical preparation and unit-cell dimensions of $Ti_4(P_4O_{12})_3$ have been reported by Liebau and Williams [101].

- ***$Sc_4(P_4O_{12})_3$***

Bagieu-Beucher [102] reported chemical preparation and crystal data for $Sc_4(P_4O_{12})_3$. The crystal structure was refined by Bagieu-Beucher and Guitel [103], after a first unsuccessful attempt performed by Wappler [104]. $Sc_4(P_4O_{12})_3$ was reinvestigated by Mezentseva *et al* [105], who reported density, crystallographic, optical and infra-red data.

- ***$Yb_4(P_4O_{12})_3$***

Chudinova [106] described the chemical preparation of $Yb_4(P_4O_{12})_3$ and gave an indexed powder X-ray diagram.

Crystal data for these six compounds are given in Table 3.7.

3.1.5. Cyclotetraphosphates of trivalent-monovalent cations

Since the characterization of stable rare-earth ultraphosphates, LnP_5O_{14}, by Beucher [108] and the discovery of their application as efficient laser materials, a great number of investigations have been performed in the field of Ln_2O_3-P_2O_5 or Ln_2O_3-M_2O-P_2O_5 systems. Most of the new materials characterized during these studies were long-chain polyphosphates with the exception of some $LnMP_4O_{12}$ compounds we report in this Section.

- ***$PrNH_4P_4O_{12}$ and $NdNH_4P_4O_{12}$***

Chemical preparation of monoclinic $PrNH_4P_4O_{12}$ and $NdNH_4P_4O_{12}$ have been reported by Masse *et al* [109]. The authors performed a detailed determination of the atomic arrangement of the praseodymium salt and showed the isotypy of the two compounds.

- ***$CeNH_4P_4O_{12}$***

The cyclotetraphosphate $CeNH_4P_4O_{12}$ is dimorphous. One of its forms is cubic, while the other one is monoclinic, isotypic of $PrNH_4P_4O_{12}$.

Vaivada and Konstant [110] reported the chemical preparation of $CeNH_4P_4O_{12}$ and studied its thermal evolution by DTA and X-ray diffraction. At 873K, the compound decomposes according to the following scheme :

$$2CeNH_4P_4O_{12} \text{ ------> } Ce(PO_3)_3 + CeP_5O_{14} + 2NH_3 + H_2O$$

The same authors [111] investigated the reaction between CeO_2 and $(NH_4)H_2PO_4$ between 293 and 1173K. For starting ratios P/Ce > 5, mainly $CeNH_4P_4O_{12}$, $Ce(NH_4)_2(PO_3)_5$ and CeP_5O_{14} are obtained. For higher CeO_2 concentrations the most stable phases are CeP_2O_7 and $Ce(PO_3)_3$. Rzaigui and Ariguib [112] described chemical preparations for the two crystalline forms of $CeNH_4P_4O_{12}$ and reported their crystal data. Rzaigui *et al* [113] reported an accurate structure determination for the atomic arrangement of the cubic form showing its structural analogy with $Al_4(P_4O_{12})_3$.

- $LaNH_4P_4O_{12}$

$LaNH_4P_4O_{12}$ has been characterized by Chudinova *et al* [114] as an intermediate compound during a study of the preparation of lanthanum polyphosphate using $NH_4H_2PO_4$ and lanthanum oxide or lanthanum oxalate as starting materials. The authors did not report any crystal data for this salt.

- $NdKP_4O_{12}$, $HoKP_4O_{12}$ and $EuKP_4O_{12}$

Tarasenkova *et al* [115] characterized $NdKP_4O_{12}$ during a study of the Nd_2O_3-K_2O-P_2O_5-H_2O system between 573 and 773K. They reported unit-cell dimensions and morphology showing the isotypy of this salt with $PrNH_4P_4O_{12}$. Litvin *et al* [116] described synthesis, crystal structure and properties of single crystals of $NdKP_4O_{12}$. The crystal structure of $HoK(PO_3)_4$ has been established by Palkina *et al* [117], while the synthesis of $EuKP_4O_{12}$ is described by Chudinova *et al* [118]. Both compounds are isotypic of $PrNH_4P_4O_{12}$.

- $NdRbP_4O_{12}$

Like $CeNH_4P_4O_{12}$, $NdRbP_4O_{12}$ is dimorphous. The crystal structure of the monoclinic form has been refined by Koizumi and Nakano [119]. The authors report the process used for crystal growth. The cubic form has been prepared by Dorokhova *et al* [120] who report a detailed determination of the atomic arrangement.

Table 3.8

The common cubic space-group is $I\bar{4}3d$ for the three compounds whose unit cells, reported below contain 12 formula units.

Formula	a (Å)	Ref.
$CeNH_4P_4O_{12}$	15.23(1)	112, 113
$NdCsP_4O_{12}$	15.223(3)	123
$NdRbP_4O_{12}$	15.241(7)	120

- $SmRbP_4O_{12}$

Palkina *et al* [121] described $SmRbP_4O_{12}$ and analyzed the structural differences in some isotypic compounds.

Byrappa and Litvin [122] have investigated the Rb_2O-Ln_2O_3-P_2O_5-

H_2O systems within the temperature range 573-1073K. A composition diagram showing the fields of crystallization for the different phases is given. The crystal chemistry of both $LnRbP_4O_{12}$ and $LnRb(PO_3)_4$ compounds is discussed.

- *$NdCsP_4O_{12}$*

The cubic atomic arrangement of $NdCsP_4O_{12}$ has been determined by Palkina *et al* [123].

Crystal growth, crystal structures and infra-red spectra of both $LnCs(PO_3)_4$ and $LnCsP_4O_{12}$ compounds have been discussed by Byrappa *et al* [124].

Table 3.9

Unit-cell dimensions of monoclinic trivalent-monovalent cyclotetraphosphates. Space-group is C2/c and Z = 4.

Formula	a	b β (°)	c (Å)	Ref.
$PrNH_4P_4O_{12}$	7.916(5)	12.647(10) 110.34(8)	10.672(9)	109
$NdNH_4P_4O_{12}$	7.881(8)	12.55(1) 110.80(10)	10.65(1)	109
$CeNH_4P_4O_{12}$	7.930(3)	12.634(5) 110.05(3)	10.699(5)	112
$NdKP_4O_{12}$	7.888	12.447 112.70	10.770	115, 116
$HoKP_4O_{12}$	7.798(1)	12.310(1) 112.63(1)	10.511(1)	117
$EuKP_4O_{12}$	7.76	12.03 112.5	10.96	113
$NdRbP_4O_{12}$	7.845(2)	12.691(3) 112.34(1)	10.688(3)	119
$SmRbP_4O_{12}$	7.868(2)	12.735(3) 111.25(2)	10.589	121

- *$InNaP_4O_{12}$*

During an investigation of the P_2O_5-Na_2O-In_2O_3-H_2O system, in a temperature range 423-773K, Avaliani *et al* [125] characterized the new diphosphate $InNa(H_2P_2O_7)_2$ and identified a new cyclotetraphosphate $InNaP_4O_{12}$, during its thermal evolution :

$$InNa(H_2P_2O_7)_2 \xrightarrow{588\text{-}678K} InNaP_4O_{12} \xrightarrow{973\text{-}1023K} In(PO_3)_3 + \text{melt}$$

Crystal data for cubic and monoclinic compounds, described in this section, are reported in Tables 3.8 and 3.9.

- $ErKP_4O_{12}.6H_2O$

Palkina *et al* [126] prepared several $LnMP_4O_{12}.nH_2O$ compounds by reaction of the monovalent cation cyclotetraphosphate with the proper rare-earth nitrate. $ErKP_4O_{12}.6H_2O$ has been investigated in detail by the authors, who report a complete description of the atomic arrangement of this monoclinic, C2/c, compound whose dimensions of the tetramolecular unit cell are :

$$a = 8.643(2),\ b = 12.015(8),\ c = 14.909(5)\ \text{Å},\ \beta = 90.65(5)^\circ$$

3.1.6. Cyclotetraphosphates of tetravalent cations

Only two tetravalent cation cyclotetraphosphates are known.

- $SnP_4O_{12}.4H_2O$

The chemical preparation of $SnP_4O_{12}.4H_2O$ has been described by Nariai *et al* [127]. The authors investigated the effect of water vapor on the thermal decomposition of this salt.

- UP_4O_{12}

Linde *et al* [128] proposed a crystal structure for this compound whose unit-cell dimensions are reported below :

$$a = 8.145(3),\quad b = 8.272(3),\quad c = 8.653(3)\ \text{Å}$$
$$\alpha = 117.83(3),\quad \beta = 117.24(3),\quad \gamma = 91.02(3)^\circ$$

The space group is $P\bar{1}$ with Z = 2.

3.1.7. Cyclotetraphosphate adducts with telluric acid

Results in the field of cyclotetraphosphate adducts with telluric acid are relatively poor if compared with those obtained with the cyclotriphosphate-telluric acid adducts we described before. Only three adducts have been clearly characterized. Their atomic arrangements have been determined accurately and show that, as in all the other class of adducts between telluric acid and phosphates, the $Te(OH)_6$ octahedra and the phosphoric groups coexist as independent units.

$2Te(OH)_6.(NH_4)_4P_4O_{12}.2H_2O$ was described by Durif *et al* [129].

$Te(OH)_6.K_4P_4O_{12}.2H_2O$ and $Te(OH)_6.Rb_4P_4O_{12}.2H_2O$ are isotypic and have been reported by Averbuch-Pouchot and Durif [130, 131]. The crystal structure of the potassium salt has been determined.

Crystallographic features for these adducts are given in Table 3.9.

Table 3.9

Main crystallographic features for adducts between telluric acid and alkali cyclotetraphosphates.

Formula	a α	b β	c (Å) γ (°)	S.G.	Z	Ref.
$2Te(OH)_6.(NH_4)_4P_4O_{12}.2H_2O$	11.845(6) 66.28(5)	8.554(5) 95.91(5)	7.433(5) 76.00(5)	$P\bar{1}$	1	129
$Te(OH)_6.K_4P_4O_{12}.2H_2O$	9.731(5)	11.43(1) 99.45(5)	17.16(1)	C2/c	4	130, 131
$Te(OH)_6.Rb_4P_4O_{12}.2H_2O$	10.049(3)	11.73(1) 99.16(5)	17.34(1)	C2/c	4	131

3.1.8. Organic and metal-organic cyclotetraphosphates

Recently, a good number of organic and metal-organic cation cyclotetraphosphates have been characterized. Two rather similar methods have been used for their preparations :

(i) action of $H_4P_4O_{12}$ acid, obtained from $Na_4P_4O_{12}.4H_2O$ by ion-exchange resins, on the appropriate amino-compound ;

(ii) addition of the proper quantity of P_4O_{10} to an iced aqueous solution of the amine.

In both cases, the reaction is of the type :

$$4\ R\text{-}NH_2 + H_4P_4O_{12} \text{------>} (R\text{-}NH_3)_4P_4O_{12}$$

Most of the compounds we report below are stable for years in normal conditions. In all cases, the atomic arrangements have been determined very accurately.

Glycinium cyclotetraphosphate, $(NH_3CH_2COOH)_4P_4O_{12}$, was described by Averbuch-Pouchot *et al* [132] and ethanolammonium cyclotetraphosphate, $[NH_3(CH_2)_2OH]_4P_4O_{12}$ by Averbuch-Pouchot *et al* [133].

An investigation of the $H_4P_4O_{12}$-$NH_2(CH_2)_2NH_2$-MO-H_2O systems has been recently performed. A survey of these studies has been reported by Averbuch-Pouchot *et al* [134], showing that the twelve compounds characterized during these investigations belong to four new structure-types :

(i) $(C_2N_2H_{10})_2P_4O_{12}.H_2O$ described by Averbuch-Pouchot *et al* [135] ;

(ii) a series of seven isomorphous compounds of general formula

$M(C_2N_2H_{10})_3(P_4O_{12})_2.14H_2O$ (M = Cu, Zn, Mg, Cd, Co, Ni, Mn), described by Averbuch-Pouchot and Durif [136) ;

(iii) $Ca(C_2N_2H_{10})P_4O_{12}.15/2H_2O$ investigated by Averbuch-Pouchot *et al* [137] ;

(iv) $Sr(C_2N_2H_{10})P_4O_{12}.5H_2O$ prepared and investigated by Bagieu-Beucher *et al* [138].

Some ethylenediammonium-alkali cyclotetraphosphates have also been described :

(i) two isomorphous compounds, $Na_2(C_2N_2H_{10})P_4O_{12}.2H_2O$ and $K_2(C_2N_2H_{10})P_4O_{12}.2H_2O$ prepared by Bdiri and Jouini [139]. This structure type has been determined by Jouini [140] using the sodium salt ;

(ii) $NaLi(C_2N_2H_{10})P_4O_{12}.3H_2O$ synthesized and investigated by Bdiri and Jouini [141].

Table 3.10
Main crystallographic features for organic-cation cyclotetraphosphates.

Formula	a α	b β	c (Å) γ (°)	S. G.	Z	Ref.
$(Gly)_4P_4O_{12}$	7.988(5) 111.64(5)	8.449(5) 105.27(5)	9.739(5) 99.40(5)	$P\bar{1}$	1	132
$(EA)_4P_4O_{12}$	10.746(8)	10.746(8)	10.071(8)	$I\bar{4}$	2	133
$(EDA)_2P_4O_{12}.H_2O$	13.168(9)	8.599(6)	15.152(10)	Pcca	4	135
$(DAP)_2P_4O_{12}.2H_2O$	14.627(1)	13.391(1)	10.068(1)	Pbca	4	142
$(DETA)_2P_4O_{12}.4H_2O$	7.966(1)	22.830(2) 116.1(1)	7.708(1)	$P2_1/n$	2	142, 143

Gly : $(NH_3CH_2COOH)^+$ (glycinium)
EA : $[NH_3(CH_2)_2OH]^+$ (ethanolammonium)
EDA : $[NH_3(CH_2)_2NH_3]^{+2}$ (ethylenediammonium)
DAP : $[NH_3(CH_2)_3NH_3]^{+2}$ (1-3 diammoniumpropane)
DETA : $[NH_3(CH_2)_2NH(CH_2)_2NH_3]^{+2}$ (diethylenetriammonium)

Lastly, 1-3 diammoniumpropane, $(C_3N_2H_{12})_2P_4O_{12}.2H_2O$, and di-ethylene-triammonium cyclotetraphosphate, $(C_4N_3H_{15})_2P_4O_{12}.4H_2O$, have been respectively described by Bdiri and Jouini [142, 143].

Tables 3.10 and 3.11 report the main crystal data for the organic cation and metal-organic cation cyclotetraphosphates, respectively .

Table 3.11

Main crystallographic features for metal-organic cation cyclotetraphosphates.

Formula	a α	b β	c (Å) γ (°)	S. G.	Z	Ref.
$Cu(EDA)_3(P_4O_{12})_2.14H_2O$	13.162(8)	13.301(8) 106.69(1)	12.308(8)	$P2_1/n$	2	136
$Zn(EDA)_3(P_4O_{12})_2.14H_2O$	12.902(8)	13.187(8) 106.04(1)	12.303(8)	$P2_1/n$	2	136
$Mg(EDA)_3(P_4O_{12})_2.14H_2O$	13.101(8)	13.292(8) 108.20(1)	12.465(8)	$P2_1/n$	2	136
$Cd(EDA)_3(P_4O_{12})_2.14H_2O$	13.193(8)	13.309(8) 107.13(1)	12.466(8)	$P2_1/n$	2	136
$Co(EDA)_3(P_4O_{12})_2.14H_2O$	13.146(8)	13.296(8) 106.77(1)	12.258(8)	$P2_1/n$	2	136
$Ni(EDA)_3(P_4O_{12})_2.14H_2O$	12.916(8)	13.189(8) 106.94(1)	12.363(8)	$P2_1/n$	2	136
$Mn(EDA)_3(P_4O_{12})_2.14H_2O$	13.098(8)	13.278(8) 107.15(1)	12.424(8)	$P2_1/n$	2	136
$Ca(EDA)P_4O_{12}.15/2H_2O$	14.611(3)	18.709(3)	7.81(2)	Pnma	4	137
$Sr(EDA)P_4O_{12}.5H_2O$	17.863(15)	15.317(13)	13.109(10)	Pbca	8	138
$Na_2(EDA)P_4O_{12}.2H_2O$	7.797(1)	14.657(1) 91.39(1)	12.916(1)	C2/c	4	139, 140
$K_2(EDA)P_4O_{12}.2H_2O$	8.172(1)	14.690(1) 91.19(1)	13.421(1)	C2/c	4	139
$NaLi(EDA)P_4O_{12}$ $3H_2O$	13.135(3)	7.737(2)	15.478(4)	Pcca	4	141

EDA : $[NH_3(CH_2)_2NH_3]^{+2}$ (ethylenediammonium)

3.2. Chemical Preparations of Cyclotetraphosphates

We report, here, the most common processes used for the preparation of cyclotetraphosphates. When compared to what we described for the chemical preparations of cyclotriphosphates two things must be noticed :

(i) as for cyclotriphosphates and other cyclophosphates, the most important starting material for syntheses is the sodium salt. But in the case of cyclotetraphosphates, this compound cannot be prepared by thermal condensation of NaH_2PO_4 through the hypothetical reaction scheme :

$$4NaH_2PO_4 \text{ -------->} Na_4P_4O_{12} + 4H_2O$$

(ii) the Boullé's metathesis reaction using the silver salt as starting material has almost never been used for the production of cyclotetra-phosphates.

As for the cyclotriphosphates, the most used starting material for these preparations is the sodium salt, here $Na_4P_4O_{12}.4H_2O$. For a long time, the only cyclotetraphosphates possible to prepare in a reproducible way were the members of the $M_2P_4O_{12}$ series (M = Cu, Mg, Co...). They are simply prepared by a thermal method leading to well crystallized samples. A mixture of phosphoric acid and of the proper bivalent cation salt is prepared with an excess of phosphoric acid and then heated for approximately one day at a temperature of 627-733K. After the apparition of a crystalline mass inside this bath, the excess of phosphoric flux is removed by hot water and the obtained crystalline $M_2P_4O_{12}$ is separated by filtration. One of these compounds, $Cu_2P_4O_{12}$, has been extensively used up to the end of the fifties to prepare the sodium salt through the process we describe below.

3.2.1. Preparation of $Na_4P_4O_{12}.4H_2O$ by exchange reaction

Finely divided $Cu_2P_4O_{12}$ is added to an aqueous solution of Na_2S. The mixture is mechanically stirred up to the end of the following reaction :

$$Cu_2P_4O_{12} + 2Na_2S \text{ ------>} Na_4P_4O_{12} + 2CuS$$

The temperature of the mixture must be kept close by 273K during the reaction in order to avoid the hydrolysis of the P_4O_{12} groups. Then, the insoluble copper sulfide is removed by filtration and the resulting solution is evaporated at room-temperature or added with alcohol or sodium chloride to precipitate $Na_4P_4O_{12}.4H_2O$.

3.2.2. Preparation of $H_4P_4O_{12}$ by hydrolysis of P_4O_{10}

When added to iced water, P_4O_{10} decomposes according to the following equation :

$$P_4O_{10} + 2H_2O \text{ ----->} H_4P_4O_{12}$$

so producing cyclotetraphosphoric acid. Very good yields (> 75%) can be obtained.

If a stoichiometric quantity of P_4O_{10} is added to an iced aqueous solution of Na_2CO_3 or $NaHCO_3$, one directly obtains a solution of sodium cyclotetraphosphate. Yields are also very good.

Most of the recently characterized alkali or mixed-alkali cyclotetraphosphates and a good number of organic cation derivatives have been prepared by using this process.

3.2.3. Thermal processes

Many anhydrous cyclotetraphosphates have been prepared by heating, at relatively low temperatures (623-723K), a stoichiometric mixture of the appropriate starting materials. For instance, $SrK_2P_4O_{12}$ can be easily obtained according to the following reaction :

$$SrCO_3 + 4(NH_4)_2HPO_4 + K_2CO_3 \dashrightarrow SrK_2P_4O_{12} + 2CO_2 + 8NH_3 + 6H_2O$$

Most of the $M^{II}K_2P_4O_{12}$ compounds were simply prepared as polycrystalline samples by using this process.

3.2.4. Flux methods

Most of the cyclotetraphosphates characterized during the elaboration of the M^IPO_3-$M^{II}(PO_3)_2$ phase-equilibrium diagrams and, in general, all insoluble cyclotetraphosphates have been synthesized as single-crystals by using flux methods similar to those described for the preparations of cyclotriphosphates. As said before, the main feature of this process is the use of a large excess of a phosphoric flux. A typical run of such a process used for the synthesis of $SrNa_2P_4O_{12}$ is reported below. In 8.6 cm^3 of H_3PO_4 (85%), 0.5 g of $SrCO_3$ and 3.5 g of Na_2CO_3 are added. The resulting mixture is then heated at 623K for one day. After removing the excess of phosphoric flux by hot water, a crystalline formation containing more than 80% of $SrNa_2P_4O_{12}$ is gathered.

3.2.5. Classical methods

As for cyclotriphosphates, classical methods of aqueous chemistry have been frequently used. For instance, $Sr(NH_4)_2P_4O_{12}$ is easily prepared by mixing N/10 aqueous solutions of $(NH_4)_4P_4O_{12}$ and $SrCl_2$. The same process has been used for the preparation of most of the cyclotetraphosphates reported in Table 3.6.

3.2.6. Ion-exchange resins

Cyclotetraphosphoric acid produced by passing an aqueous solution of $Na_4P_4O_{12}$ through a column of ion-exchange resins is commonly used. For instance, calcium and strontium cyclotetraphosphates have been synthesized by adding so prepared cyclotetraphosphoric acid to aqueous solutions of the corresponding nitrates. Usually, the so produced acid is added immediately to an appropriate mixture of carbonates or hydroxides in order to avoid a possible hydrolysis.

3.2.7. Non-conventional methods

When given by the authors, they are reported in the "Present State" Section.

3.3. Some Atomic Arrangements of Cyclotetraphosphates

3.3.1. Sodium cyclotetraphosphate monohydrate

$Na_4P_4O_{12}.H_2O$ is orthorhombic, $P2_12_12_1$, [19] with Z = 4 and the following unit-cell dimensions :

a = 13.654(2), b = 13.475(3), c = 6.291(3) Å

Fig. 3.4 is a projection, along the c axis, of the atomic arrangement. Within a range of 3 Å, the four independent sodium atoms have six neighbors. The NaO_6 polyhedra establish the cohesion between the P_4O_{12} groups. According to the authors, these last groups, in spite of having no internal symmetry, are very regular with, a very strong pseudo-D_{2d} conformation. Their mean planes are almost perpendicular to the c axis. The crystal structure of the anhydrous salt is similar.

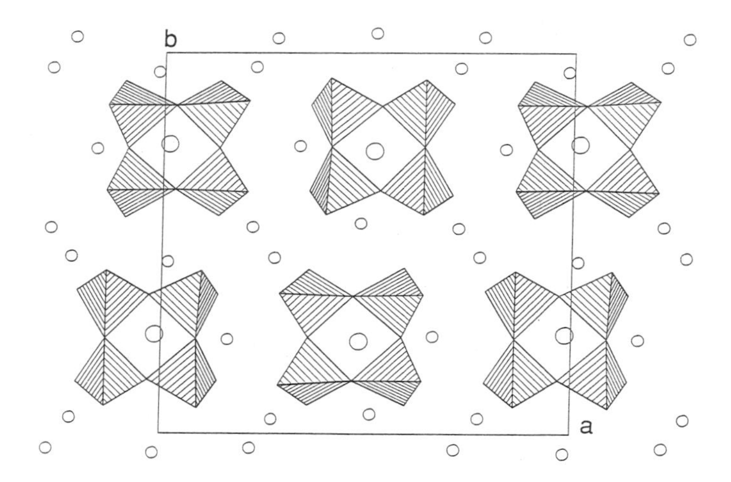

Figure 3.4 Projection of the atomic arrangement of $Na_4P_4O_{12}.H_2O$ along the c axis. Small empty circles represent sodium atoms, while large ones are water molecules.

3.3.2. Ammonium cyclotetraphosphate

$(NH_4)_4P_4O_{12}$ [33-37] is orthorhombic, Cmca, with Z = 4 and the following unit-cell dimensions :

a = 10.433(2), b = 10.871(3), c = 12.588(3) Å

Fig. 3.5 reports the projection of the atomic arrangement along the b axis. The ring anion has a 2/m symmetry ; two of its phosphorus atoms are located in the mirror plane, while the two last ones lie on a twofold axis. These ring anions have their centres in the planes y = 0 and y = 1/2 and form, in these planes, a pseudo square-centred arrangement. The ammonium groups are approximately located in planes y = 1/4 and 3/4, eight of them are located in the mirror planes and the last eight ones on the twofold axes. So, the atomic arrangement may be considered as formed by NH_4 and P_4O_{12} layers alternating in the b direction.

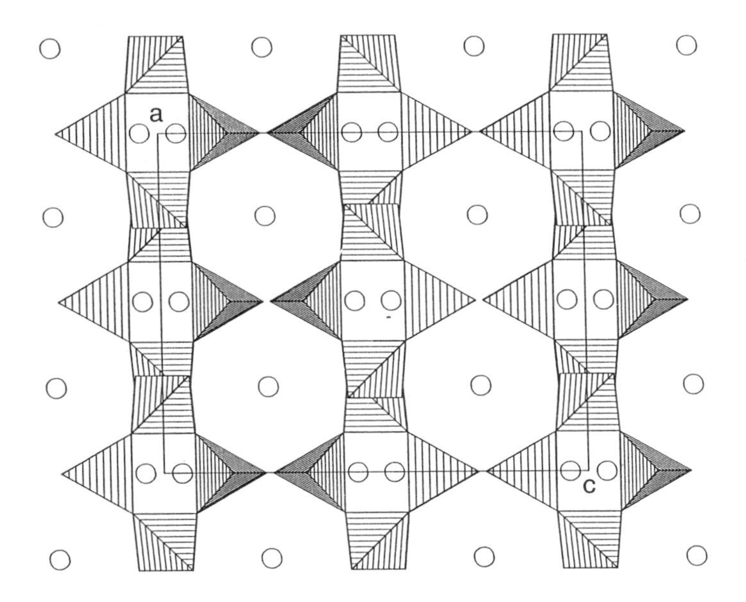

Figure 3.5 Projection of the atomic arrangement of $(NH_4)_4P_4O_{12}$ along the b axis. Empty circles represent the ammonium groups.

3.3.3. Thallium cyclotetraphosphate

$Tl_4P_4O_{12}$ [43,44] is tetragonal, $P\bar{4}2_1c$, with Z = 2 and the following unit-cell dimensions :

$$a = 7.635(5), \quad c = 11.087(7)\ \text{Å}$$

The ring-anions being centred around the $\bar{4}$ axes, in 1/2,1/2,0 and 0,0,1/2, have the $\bar{4}$ internal symmetry and are built by only one crystallographically independent PO_4 tetrahedron. The P_4O_{12} anions form layers spreading through the crystal structure, in planes z = 0, and z = 1/2. The two independent thallium atoms are located on binary axes and, within a range of 3.5 Å, have eight and tenfold coordinations. Fig. 3.6 reports a projection of this atomic arrangement along the c axis.

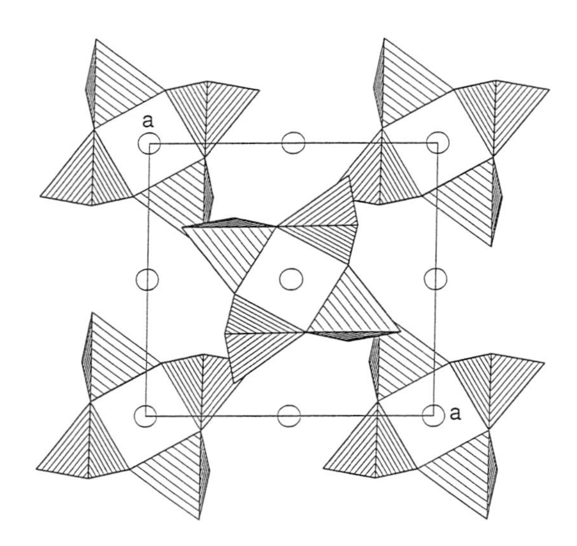

Figure 3.6 Projection of the atomic arrangement of $Tl_4P_4O_{12}$ along the c axis. Empty circles are thallium atoms.

3.3.4. Zinc cyclotetraphosphate octohydrate

$Zn_2P_4O_{12}.8H_2O$ [69] is triclinic, $P\overline{1}$, with Z = 1 and the following unit-cell dimensions :

$$a = 8.610(5),\ b = 7.137(5),\ c = 7.108(5)\ Å$$
$$\alpha = 96.09(5),\ \beta = 105.99(5),\ \gamma = 100.49(5)^\circ$$

The atomic arrangement of this salt is given in Fig. 3.7 in projection along the a axis. As shown, the structure can be described as layers of independent ZnO_6 octahedra spreading in (a, b) planes, forming a two-dimensional face-centered network. These layers are linked each other by the P_4O_{12} groups. The P_4O_{12} ring-anions and the two kinds of ZnO_6 octahedra are centrosymmetrical. One of the zinc polyhedron is built up by four oxygen atoms and two water molecules, while the second one is made by four water molecules and two oxygen atoms. A $ZnO_4(H_2O)_2$ octahedron shares four oxygen atoms with two adjacent P_4O_{12} groups, while a $ZnO_2(H_2O)_4$ shares two. In addition, a water molecule, not involved in the zinc coordination, is located between the ZnO_6 layers. A complete description of the hydrogen bond network is given by the author.

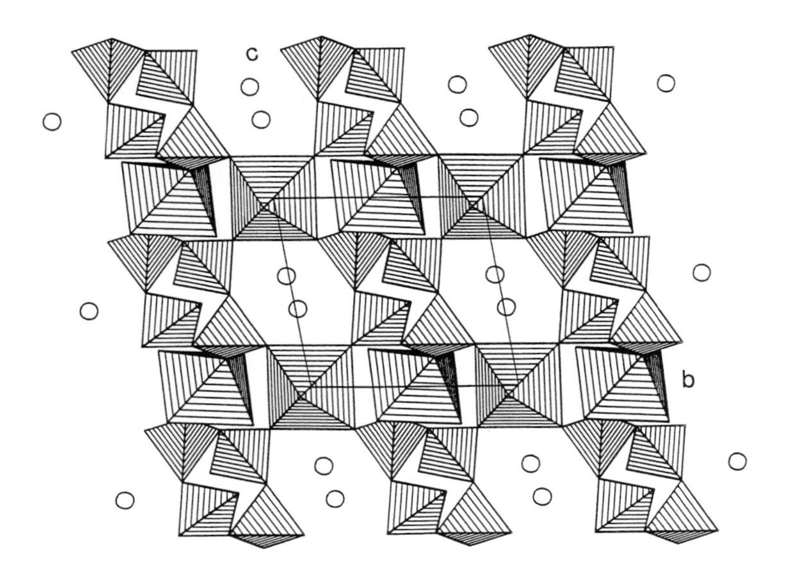

Figure 3.7 Projection, along the a direction, of the atomic arrangement of $Zn_2P_4O_{12}.8H_2O$. Hatched octahedra represent the $ZnO_4(H_2O)_2$ groups and the empty circles the non-bonded water molecules. The hydrogen atoms have been omitted.

3.3.5. Strontium-sodium cyclotetraphosphate

$SrNa_2P_4O_{12}$ [81] is tetragonal, P4/nbm, with Z = 2 and the following unit-cell dimensions :

$$a = 9.838(5), \quad c = 5.003(3) \text{ Å}$$

Up to now, this salt is the only example of a $MNa_2P_4O_{12}$ compound. Fig. 3.8 reports the projection of its atomic arrangement along the c axis. All the associated cations are located in plane z = 1/2. Sodium atoms, located on a site of 2/m symmetry, are surrounded by distorted oxygen octahedra having a pseudo-threefold axis parallel to the c direction, while strontium atoms located on a site of 42 symmetry are on the centres of square antiprisms, sharing four of their corners with the four neighboring NaO_6 octahedra. These polyhedra build infinite layers perpendicular to the c axis. The P_4O_{12} ring-anions, located around the $\bar{4}$ axes in z = 0, are so half-way between the associated cation polyhedra layers ; they have a $\bar{4}$ 2m internal symmetry.

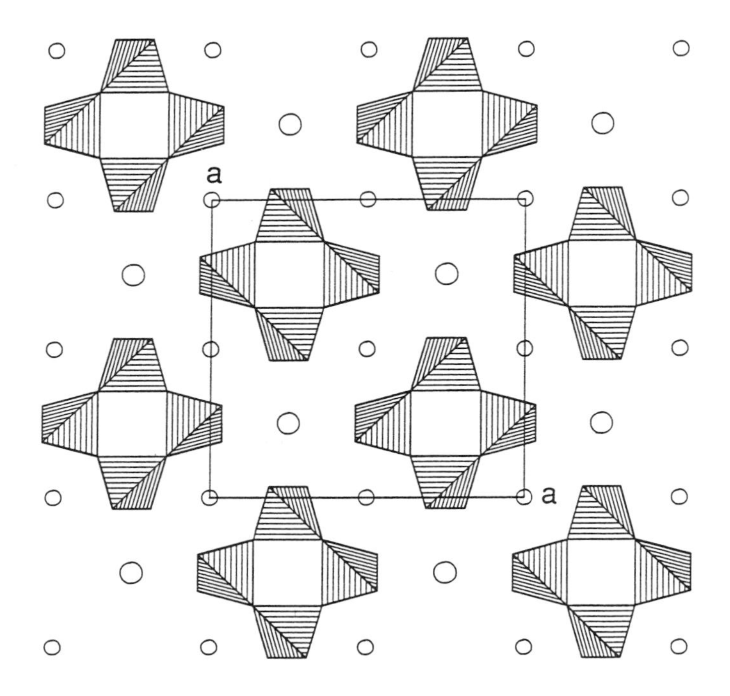

Figure 3.8 Projection of the atomic arrangement of $SrNa_2P_4O_{12}$ along the c axis. Larger empty circles are strontium atoms, the smaller ones sodium atoms

3.3.6. Strontium-potassium cyclotetraphosphate

$SrK_2P_4O_{12}$ [84] is tetragonal, $I\bar{4}$, with Z = 2 and the following unit-cell dimensions :

$$a = 7.445(5), \quad c = 10.17(2) \text{ Å}$$

Fig. 3.9 reports the projection of the atomic arrangement along the c axis. Each unit cell contains two P_4O_{12} groups built around the $\bar{4}$ axes and centred in 0,0,0 and 1/2,1/2,1/2. These rings, having the $\bar{4}$ internal symmetry, are built by only one independent PO_4 tetrahedron ; they alternate with potassium atoms along the $\bar{4}$ axes. The strontium atoms, in 0,1/2,1/4 and 0,1/2,3/4, are located on another series of $\bar{4}$ axes. The strontium atom has eight and the potassium atom ten oxygen neighbors. A strong structural analogy exists between the present arrangement and that previously described for $Tl_4P_4O_{12}$. Up to now, nine cyclotetraphosphates crystallize with this structure type (Table 3.4).

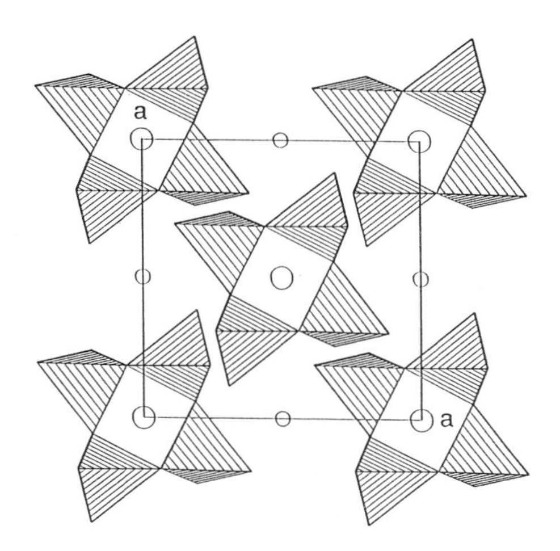

Figure 3.9 Projection, along the c axis, of the atomic arrangement of $SrK_2P_4O_{12}$. Small empty circles represent strontium atoms, large ones potassium atoms.

3.3.7. Calcium-sodium cyclotetraphosphate hydrate

$CaNa_2P_4O_{12}.11/2H_2O$ [92] is orthorhombic, Pma2, with Z = 4 and the following unit-cell dimensions :

$$a = 27.88(10),\ b = 7.536(5),\ c = 7.378(5)\ \text{Å}$$

Fig. 3.10 reports a projection of the atomic arrangement along the c axis. The main feature of this structure is the coexistence of two crystallographically independent P_4O_{12} anions with different internal symmetries. One of them has a twofold internal symmetry, the other one a mirror symmetry. In addition this salt provided the first example of a P_4O_{12} group with the m symmetry. As it can be seen, the mirror plane contains two opposite bonding oxygen atoms. The two independent calcium atoms have an eightfold coordination made by four oxygen atoms and four water molecules, while the two independent sodium atoms are located inside distorted octahedra built by four oxygen atoms and two water molecules. Inside these various polyhedra the Na-O distances spread between 2.284 and 2.705 Å and the Ca-O distances between 2.345 and 2.525 Å. All the water molecules are involved in the coordination polyhedra of the two associated cations.

This compound is one of the rare example of cyclotetraphosphate containing two P_4O_{12} ring anions with different internal symmetries.

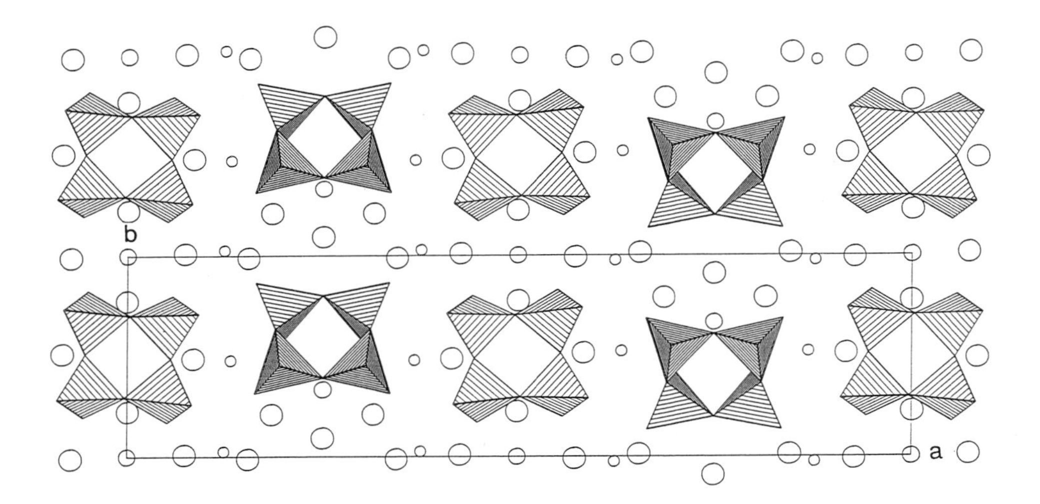

Figure 3.10 Projection of the atomic arrangement of $CaNa_2P_4O_{12}.11/2H_2O$ along the c axis. Smallest empty circles are sodium atoms, intermediate ones , calcium atoms and the largest ones represent water molecules.

3.3.8. Calcium-potassium cyclotetraphosphate octahydrate

Crystals of the monoclinic, $P2_1/a$, double-salt, $Ca_4K_4(P_4O_{12})_3.8H_2O$, [92] have the following dimolecular cell :

$$a = 20.38(1), \ b = 12.683(5), \ c = 7.830(2) \text{ Å}, \ \beta = 89.31(5)°$$

As in the previous example, the main feature of this atomic arrangement is the coexistence of two crystallographically independent P_4O_{12} rings. The first one has no internal symmetry, while the second one is centro-symmetrical. A projection of a part of this arrangement is illustrated in Fig. 3.11. The two independent calcium atoms have a sevenfold coordination, involving six oxygen atoms and one water molecule, with Ca-O distances ranging from 2.339 to 2.471 Å. Within a range of 3.5 Å, one of the potassium atom has a sevenfold coordination including two water molecules and the second a sixfold coordination including also two water molecules. Inside these KO_n polyhedra the K-O distances spread between 2.687 and 3.027 Å.

The four crystallographically independent water molecules are involved in the associated cation coordination polyhedra.

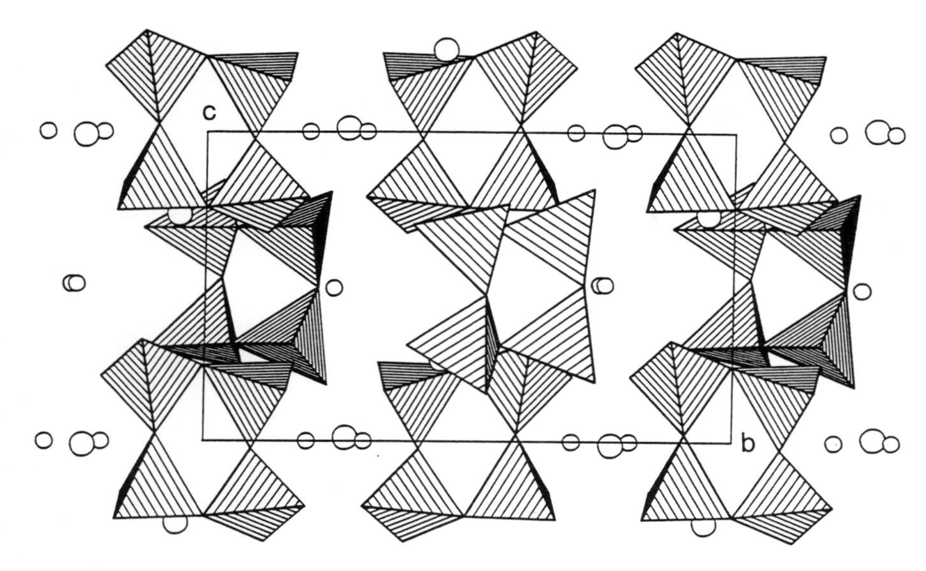

Figure 3.11 Projection of a part of the atomic arrangement of $Ca_4K_4(P_4O_{12})_3.8H_2O$ along the a axis. Small empty circles represent calcium atoms, larger ones are potassium atoms. Water molecules have been omitted.

3.3.9. Erbium-potassium cyclotetraphosphate hexahydrate

$ErKP_4O_{12}.6H_2O$ [126] is monoclinic, C2/c, and has the following unit cell :

a = 8.643(2), b = 12.015(8), c = 14.909(5) Å, β = 90.65(5)°

containing four formula units. Fig. 3.13, a projection of this structure along the b axis, shows the layer organization of this arrangement. Planes z = 0 and 1/2, containing the centrosymmetrical P_4O_{12} ring-anions, alternate with planes containing erbium and potassium atoms in z = 1/4 and 3/4. Erbium and potassium atoms, superimposed in the drawing, alternate along rows parallel to the b axis. The ErO_8 polyhedron is a bicapped distorted trigonal prism, made by four oxygen atoms and four water molecules. The irregular KO_8 polyhedron is also built by four oxygen atoms and four water molecules. These two types of polyhedra share edges as to form chains parallel to the b axis.

The water molecules are arranged as to build channels parallel to the b direction. Inside these channels are located the cation rows. All these water molecules take part to the associated cation coordination polyhedra.

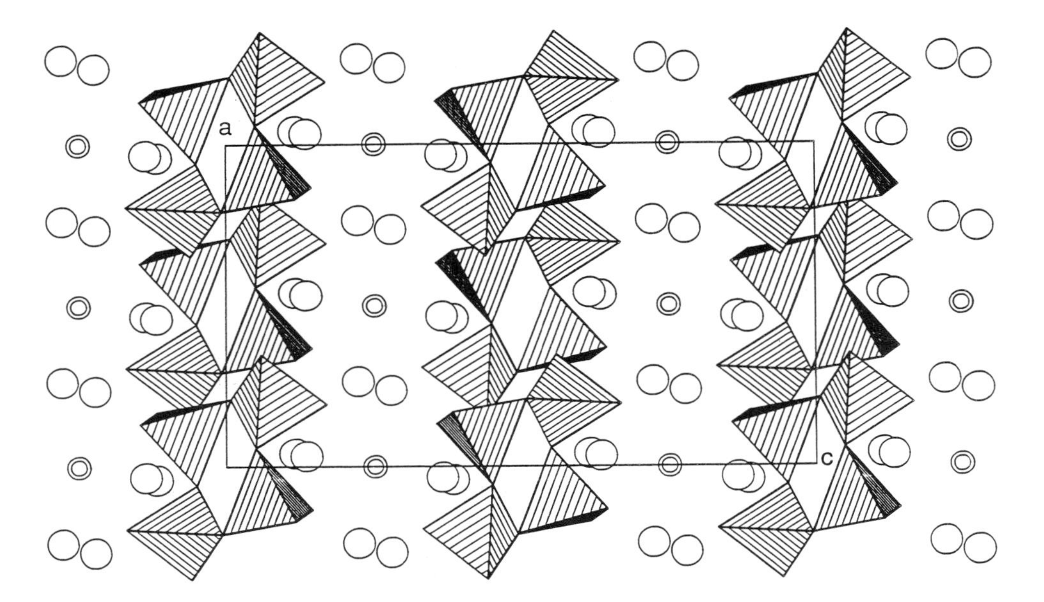

Figure 3.12 Projection of the atomic arrangement of $ErKP_4O_{12}.6H_2O$ along the b axis. The largest empty circles represent water molecules, the smallest ones are erbium atoms and the intermediate ones are potassium atoms.

3.4. The P_4O_{12} Ring Anions

One can imagine many geometrical conformations for the P_4O_{12} ring anion but we shall limit the discussion to the presently observed geometries. Today, forty-nine P_4O_{12} groups have been investigated with high or acceptable accuracies. As for the cyclotriphosphates, we report their main geometrical features, P-P distances, P-O-P and P-P-P angles in the form of some tables (3.12, 3.13, 3.14 and 3.15).

Up to now, if classified according to their geometries, eight types of P_4O_{12} rings are observed with $\bar{1}$, 2, 2/m, m, mm, $\bar{4}$, $\bar{4}$2m or no internal symmetry. The number of representatives in each group is very variable ; only one ring is observed with a 2/m symmetry while seventeen have a $\bar{1}$ symmetry.

If one excludes some rare exceptions, the P-P-P angles, strictly equal to 90° in rings with mm or $\bar{4}$2m symmetries, never depart significantly from this value. Their overall average value is 88.3°. For the calculation of this average, rings with $\bar{1}$, m and 2/m internal symmetries

Table 3.12
Main geometrical features of centrosymmetrical P_4O_{12} rings.

Formula [Ref.] P-P (Å)	P-O-P (°)	P-P-P (°)	Formula [Ref.] P-P (Å)	P-O-P (°)	P-P-P (°)
$Na_4P_4O_{12}.4H_2O$, monoclinic [25]			$Na_4P_4O_{12}.4H_2O$, triclinic [25]		
2.915	134.5	97.8	2.981	137.3	96.1
2.964	130.6	82.2	2.905	129.4	83.9
$CsNa_3P_4O_{12}.3H_2O$, two independent rings in the unit cell [45]					
2.972	136.0	93.8	2.978	137.7	86.1
2.988	136.5	86.2	2.991	138.4	93.9
$Cu_2P_4O_{12}$ [52]			$Zn_2P_4O_{12}.8H_2O$ [69]		
2.958	136.3	83.9	2.910	131.1	82.9
2.960	137.7	96.1	2.981	137.4	97.1
$Sr_2P_4O_{12}.6H_2O$ [75]			$Ca_2P_4O_{12}.4H_2O$ [72]		
2.951	131.6	81.5	2.876	135.3	98.3
2.916	129.3	98.5	2.955	126.8	81.7
$Ca_4K_4(P_4O_{12})_3.8H_2O$ [92]			$PrNH_4P_4O_{12}$ [109]		
2.956	136.9	95.0	2.953	138.2	86.6
2.966	133.4	85.0	2.985	134.4	93.4
$ErKP_4O_{12}.6H_2O$ [126]			$(NH_4)_4P_4O_{12}.2Te(OH)_6.2H_2O$ [129]		
2.912	130.5	96.7	2.961	136.5	86.6
2.930	133.0	83.3	2.767	124.1	93.4
$(Gly)_4P_4O_{12}$ [132]			$(DETA)_2P_4O_{12}.4H_2O$ [143]		
2.986	137.5	84.2	2.922	131.6	81.6
2.941	132.7	95.8	2.941	133.9	98.4
$Pb_2Cs_3(P_4O_{12})(PO_3)_3$, two independent rings in the unit cell [144]					
2.955	134.2	83.8	2.937	132.4	99.2
2.952	135.0	96.2	2.905	130.4	80.8
$KAl_2H_2(P_3O_{10})(P_4O_{12})$ [145]					
2.953	136.2	84.6			
2.966	137.8	95.4			

DETA : $[NH_3(CH_2)_2NH(CH_2)_2NH_3]^{+2}$ (diethylenetriammonium)
Gly : $(NH_3CH_2COOH)^+$ (glycinium)

have not been taken into account since in these cases the two different P-P-P angles are complementary. Angles of rings with mm and $\bar{4}2m$ have also been eliminated of this calculation, since as we already said strictly equal to 90°, for symmetry reasons. So, the average P-P-P value reported above is strongly weighted by values observed in rings of low internal symmetry. The largest P-P-P angle (97.8°) has been observed in the monoclinic form of $Na_4P_4O_{12}.4H_2O$ [25]. As for cyclotriphosphates, the estimated standard deviations for the reported P-P-P values can be evaluated to less than 0.1°, but for recent structural determination this same value is very often less than 0.03°.

Table 3.13
Main geometrical features for P_4O_{12} rings with no internal symmetry.

Formula [Ref.]	P-P (Å)	P-O-P (°)	P-P-P (°)
$Na_4P_4O_{12}.H_2O$ [19]			
	2.964	135.9	90.5
	3.015	137.4	90.2
	2.993	136.2	90.2
	2.999	132.5	89.1
$Cs_4P_4O_{12}.4H_2O$ [38]			
	2.953	133.0	89.2
	2.968	133.4	88.8
	2.943	131.6	89.6
	2.951	131.2	89.0
$Na_2K_2P_4O_{12}.2H_2O$, triclinic [25]			
	2.910	128.3	85.3
	2.916	128.5	84.7
	2.933	129.9	85.2
	2.925	129.2	85.0
$Cu(EDA)P_4O_{12}.14H_2O$ [136]			
	2.945	131.8	90.4
	2.929	130.4	88.1
	2.975	135.4	89.5
	2.945	132.3	89.0
$Pb_2P_4O_{12}.3H_2O$ [65]			
	2.912	131.5	90.0
	2.898	130.5	90.0
	2.907	131.3	89.6
	2.921	130.0	90.0
$Ca_4K_4(P_4O_{12})_3.8H_2O$ [92]			
	2.906	128.4	89.8
	2.892	131.2	89.2
	2.915	128.0	89.4
	2.901	128.2	89.8
$K_4P_4O_{12}.2H_2O$ [25]			
	2.973	133.4	84.5
	2.925	130.0	84.8
	2.923	130.0	85.0
	2.948	131.0	83.5
$Na_2K_2P_4O_{12}.2H_2O$, tetragonal [25]			
	2.928	128.9	85.1
	2.912	129.9	84.7
	2.937	128.7	84.9
	2.916	128.2	85.4
$Ni(NH_4)_2P_4O_{12}.7H_2O$ [94]			
	2.943	134.1	86.7
	2.944	132.7	91.4
	2.937	133.7	86.8
	2.946	133.5	91.2
$Sr(EDA)P_4O_{12}.5H_2O$ [138]			
	2.968	134.2	89.3
	2.917	129.4	87.9
	2.902	127.7	90.8
	2.925	130.6	89.0
$CoK_2(P_4O_{12}).5H_2O$ [95]			
	2.922	129.3	90.2
	2.953	131.9	89.4
	2.925	131.0	90.5
	2.935	131.1	89.8
$Ca(NH_4)_2(P_4O_{12}).2H_2O$ [97]			
	2.904	134.4	88.0
	2.947	129.5	90.2
	2.891	127.7	89.3
	2.893	127.3	91.0

EDA : $[NH_3(CH_2)_2NH_3]^{+2}$ (ethylenediammonium)

The P-P distances in cyclotetraphosphates are quite comparable to those observed in all the other kinds of cyclophosphates with an overall average of 2.938 Å. They spread for cyclotetraphosphates within the following range :

$$2.867 < \text{P-P} < 3.015 \text{ Å}$$

The estimated standard deviations for the reported P-P distance values can be evaluated in general at 0.005, but often close by 0.001 for the most recent data.

The P-O-P angles also do not depart siginificantly from their general average : 131.8° and spread from 124.1 to 139.8°.

Table 3.14

Main geometrical features for P_4O_{12} ring anions with higher symmetries (first part).

Formula [Ref.]				Formula [Ref.]			
	P-P (Å)	P-O-P (°)	P-P-P (°)		P-P (Å)	P-O-P (°)	P-P-P (°)
Rings with a 2 symmetry							
$Li_4P_4O_{12}.5H_2O$, two independent rings in the unit cell [17]							
	2.941	133.6	90.1		2.895	128.7	89.4
	2.914	130.3	89.7		2.916	130.2	90.6
$K_4P_4O_{12}.Te(OH)_6.2H_2O$ [130]				$(EDA)_2P_4O_{12}.H_2O$ [135]			
	2.908	129.7	84.6		2.963	133.9	87.7
	2.909	129.8	84.4		2.958	133.0	90.0
$Na_2(EDA)P_4O_{12}.2H_2O$ [140]				$NaLi(EDA)P_4O_{12}.3H_2O$ [141]			
	2.950	131.4	88.7		2.918	132.1	89.9
	2.949	132.8	88.5		2.919	132.1	89.9
$CaNa_2P_4O_{12}.11/2H_2O$ [92]							
	2.989	138.3	90.3				
	2.939	132.3	89.6				
Rings with a $\bar{4}$ symmetry							
$K_4P_4O_{12}.4H_2O$ [31]				$Zn_4Na_4(P_4O_{12})_3$ [79]			
	2.944	132.6	90.0		2.887	128.8	88.4
$Cs_4Sr_3H_2(P_4O_{12})_3$ [91]				$Ce(NH_4)P_4O_{12}$ [113]			
	2.905	129.3	89.8		2.867	125.1	89.3
$Sc_4(P_4O_{12})_3$ [103]				$Tl_4P_4O_{12}$ [44]			
	2.961	137.0	80.4		2.979	133.2	86.5
$(EA)_4P_4O_{12}$ [133]							
	2.935	131.9	89.8				

EDA : $[NH_3(CH_2)_2NH_3]^{+2}$ (ethylenediammonium)
EA : $[NH_3(CH_2)_2OH]^+$ (ethanolammonium)

As for cyclotriphosphates, no correlation could be established between the ring symmetry and the nature of the associated cations.

P_4O_{12} anions are also found in some mixed-anion phosphates as $Pb_2Cs_3(P_4O_{12})(PO_3)_3$ [144] and $KAl_2H_2(P_3O_{10})(P_4O_{12})$ [145]. This type of phosphates will be examined in a special chapter but, the numerical data for the corresponding rings have been included in Table 3.12.

Table 3.15

Main geometrical features for P_4O_{12} ring anions with higher symmetries (second part).

Formula [Ref.]	P-P (Å)	P-O-P (°)	P-P-P (°)	Formula [Ref.]	P-P (Å)	P-O-P (°)	P-P-P (°)
Rings with a 2/m symmetry							
$(NH_4)_4P_4O_{12}$ [37]							
	2.926	131.2	97.7				
			82.3				
Ring with a m symmetry							
$Ca(EDA)P_4O_{12}.15/2H_2O$ [137]				$CaNa_2P_4O_{12}.11/2H_2O$ [92]			
	2.935	130.9	90.2		2.904	135.3	90.5
	2.917	132.6	89.8		2.930	131.4	89.5
	2.951	134.3			2.953	131.3	
Ring with a mm symmetry							
$CsNa_3P_4O_{12}.4H_2O$ [45]				$SrNa_2P_4O_{12}.6H_2O$ [93]			
	3.012	129.8	90.0		2.971	135.1	90.0
	2.929	139.8			2.994	138.3	
Ring with a $\bar{4}2m$ symmetry							
$SrNa_2P_4O_{12}$ [81]							
	2.956	132.3	90.0				

EDA : $[NH_3(CH_2)_2NH_3]^{+2}$ (ethylenediammonium)

REFERENCES

1 R. Maddrell, *Liebigs Ann.*, 61 (1847) 53.

2 T. Fleitmann, *Ann. Chem.*, 72 (1849) 228.

3 T. Fleitmann, *Pogg. Ann.*, 78 (1849) 233 and 338.

4 A. Glatzel, *"Inaugural Dissertation"*, Würzburg, 1880.

5 G. Tammann, *J. prakt. Chem.*, 45 (1892) 417.

6 F. Warschauer, *Z. anorg. Chem.*, 36 (1903) 137.

7 A. Travers and Y.K. Chu, *C. R. Acad. Sci.*,198 (1937) 2169.

8 B. Topley, *Quart. Rev.*, 3 (1949) 345.

9 B. Raistrick,
Discussions Faraday Soc., 5 (1949) 234.

10 R.N. Bell, L.F. Audrieth and O.F. Hill,
Ind. Eng. Chem., 44 (1952) 568.

11 E. Thilo and W. Wicker,
Z. anorg. allg. Chem., 277 (1952) 27.

12 L. Pauling and J. Sherman,
Z. Krist., 96 (1937) 481.

13 H. Grunze and E. Thilo,
Z. anorg. allg. Chem., 281 (1955) 284.

14 H. Grunze,
Angew. Chem., 67 (1955) 408.

15 U. Schülke and R. Kayser,
Z. anorg. allg. Chem., 531 (1985) 167.

16 J.C. Grenier and A. Durif,
Z. Kistallogr., 137 (1973) 10.

17 M.T Averbuch-Pouchot and A. Durif,
Acta Cryst., C42 (1986) 129.

18 J.R. Van Wazer,
Phosphorus and its Compounds, Vol. 1, Interscience, New York, 1966, p701.

19 D.M. Wiench and M. Jansen,
Monat. Chem., 114 (1983), 699.

20 P. Bonneman,
C. R. Acad. Sci., 204 (1937) 865.

21 K.A. Andress, W. Gehring and K. Fischer,
Z. anorg. Chem., 260 (1949), 331.

22 D.L. Barney and J.W. Gryder,
J. Am. Chem. Soc., 77 (1955) 3195.

23 H.M. Ondik, S. Block and C.H. Mac Gillavry,
Acta Cryst., 14 (1961) 555.

24 H.M. Ondik,
Acta Cryst., 17 (1964) 1139.

25 M.T. Averbuch-Pouchot and A. Durif,
J. Solid State Chem., 58 (1985) 119.

26 E.G. Griffith,
J. Am. Chem. Soc., 76 (1954) 5892.

27 E.G. Griffith,
J. Am. Chem. Soc., 78 (1956) 3867.

28 G.W. Gryder, G. Donnay and H.M. Ondik,
Acta Cryst., 11 (1958) 38.

29 O.H. Jarchow,
Acta Cryst., 17 (1964) 1253.

30 J.R. Van Wazer,
Phosphorus and its Compounds, Vol. 1, Interscience, New York, 1966, p 699.

31 M.T. Averbuch-Pouchot and A. Durif, *Acta Cryst.*, C41 (1985) 1564.

32 K.R. Waerstad, G.H. McClellan, A.W. Frazier and R.C. Sheridan, *J. Appl. Cryst.*, 2 (1969) 306.

33 C. Romers, J.A.A. Ketelaar and C.H. Mac Gillavry, *Nature*, 164 (1949) 960.

34 K.A. Andress and K. Fischer, *Acta Cryst.*, 3 (1950) 399.

35 C. Romers, J.A.A. Ketelaar and C. H. Mac Gillavry, *Acta Cryst.*, 4 (1951) 114.

36 D.W.J. Cruickshank, *Acta Cryst.*, 17 (1964) 675.

37 D.A. Koster and A.J. Wagner, *J. Chem. Soc.*, A (1970) 435.

38 M.T. Averbuch-Pouchot and A. Durif, *Acta Cryst.*, C42 (1986) 131.

39 M.T. Averbuch-Pouchot and A. Durif, *Acta Cryst.*, C41 (1985) 665.

40 K. Dostal, V. Kocman and V. Ehrenbergrova, *Z. anorg. allg. Chemie*,367 (1969) 80.

41 K. Dostal and V. Kocman, *Z. anorg. allg. Chemie*, 367 (1969) 92.

42 K. Dostal and V. Kocman, *Z. Chem.*, 5 (1965) 344.

43 J.K. Fawcett, V. Kocman, S.C. Nyburg and R.J. O'Brien, *Chem.Comm.*, 18 (1970) 1213.

44 J.K. Fawcett, V. Kocman and S.C. Nyburg, *Acta Cryst.*, B30 (1974) 1979.

45 M.T. Averbuch-Pouchot and A. Durif, *J. Solid State Chem.*, 60 (1985) 13.

46 E. Thilo.and I. Grunze, *Z. anorg. allg. Chem.*, 290 (1957) 209.

47 E. Steger, *Z. anorg. allg. Chem.*, 294 (1958) 146.

48 E. Steger and A. Simon, *Z. anorg. allg. Chem.*, 294 (1958) 1.

49 M. Beucher, Dissertation, Univ. of Grenoble, France, 1968.

50 M. Beucher and J. C. Grenier, *Mat. Res. Bull.*, 3 (1968) 643.

51 M. Laügt, A. Durif and M.T. Averbuch-Pouchot, *Bull. Soc. fr. Minér. Crist.*, 96 (1973) 383.

52 M. Laügt, J.C. Guitel, I. Tordjman and G. Bassi, *Acta Cryst.*, B28 (1972) 201.

53 L.N. Shchegrov, N.M. Antraptseva and I.G. Ponomareva, *Izv. Akad. Nauk SSSR, Neorg. Mat.*, 25 (1989) 308.

54 D.Z. Serazetdinov, E.V. Poletsev and Yu.A. Kushnikov, *Russ. J.Inorg. Chem.*,12 (1967) 1599.

55 A.B. Bekturov, D.Z. Serazetdinov, Yu.A. Kushnikov, E.V. Poleaev and S.M. Divnenko,
Russ. J. Inorg. Chem., 12 (1967) 1242.

56 G.A. Bukhalova, I.G. Rabkina and I.V. Mardirosova,
Russ. J. Inorg.Chem., 20 (1975) 332.

57 M. Bagieu-Beucher, M. Gondrand and M. Perroux,
J. Solid State Chem., 19 (1976) 353.

58 E.A. Genkina, N.S. Triodina, O.K. Mel'nikov and B.A. Maksimov,
Izv. Akad. Nauk SSSR, Neorg. Mat., 24 (1988) 1158.

59 E.L. Krivovyazov, I.D. Sokolova and N.K. Voskresenskaya,
Izv. Akad. Nauk SSSR, Neorg. Mat., 3 (1967) 530.

60 R.C. Ropp. and M.A. Aia,
Ann. Chem., 34 (1962) 1288.

61 A.G. Nord and K.B. Lindberg,
Acta Chem. Scand., A29 (1975) 1.

62 A.G. Nord,
Cryst. Struct. Comm., 11 (1982) 1467.

63 A.G. Nord,
Acta Chem. Scand., A37 (1983) 539.

64 A.G. Nord,
Mat. Res. Bull., 18 (1983) 765.

65 E.A. Genkina, B.A. Maksimov and O.K. Mel'nikov,
Sov. Phys.Crystallogr., 30 (1985) 513.

66 M.T. Averbuch-Pouchot, A. Durif and M. Bagieu-Beucher,
Acta Cryst., C39 (1983) 25.

67 W. Gunsser, D. Fruehauf, K. Rohwer, A. Zimmermann and A. Wiedenmann,
J. Solid State Chem., 82 (1989) 43.

68 M. Schneider and K.H. Jost,
Z. anorg. allg. Chem., 580 (1990) 175.

69 M.T. Averbuch-Pouchot,
Z. anorg. allg. Chem., 503 (1983) 231.

70 G. Foumakoye,
Thesis, Univ. of Liége, Belgium, 1986.

71 A.V. Lavrov, T.A. Bykanova and N.N. Chudinova,
Izv. Akad. Nauk SSSR, Neorg. Mat., 13 (1977) 3348.

72 M. Schneider, K.H. Jost and H. Fichtner,
Z. anorg. allg. Chem., 500 (1983) 117.

73 U. Schülke,
personal communication in ref. 77.

74 M. Schneider and K.H. Jost,
Z. anorg. allg. Chem., 576 (1989) 267.

75 A. Durif and M.T. Averbuch-Pouchot,
Acta Cryst., 503 (1986) 927.

76 H. Worzala,
Z. anorg. allg. Chem., 421 (1976) 122.

77 B. Klinkert and M. Jansen,
Z. anorg. allg. Chem., 556 (1988) 85.

78 H. Worzala,
Z. anorg. allg. Chem., 445 (1978) 27.

79 M.T. Averbuch-Pouchot and A. Durif,
J. Solid State Chem., 49 (1983) 341.

80 C. Martin,
Thesis, Univ. of Grenoble, France, 1972.

81 M.T. Averbuch-Pouchot and A. Durif,
Acta Cryst., C39 (1983) 811.

82 G.A. Bukhalova and I.A. Tokman,
Izv. Akad. Nauk SSSR, Neorg. Mat., 8 (1972) 528.

83 A. Durif, C. Martin, I. Tordjman and D. Tranqui,
Bull. Soc. fr. Minér. Crist., 89 (1966) 439.

84 I. Tordjman, C. Martin and A. Durif,
Bull. Soc. fr. Minér. Crist., 90 (1967) 293.

85 C. Cavero-Ghersi and A. Durif,
J. Appl. Cryst., 8 (1975) 562.

86 J.B. Gill and R.M. Taylor,
J. Chem. Soc., (1964) 5905.

87 R. Andrieu and R. Diament,
C. R. Acad. Sci., 259 (1964) 4708.

88 I. Mahama, M. Brunel-Laügt and M.T. Averbuch-Pouchot,
C. R. Acad. Sci., 284C (1977) 681.

89 M. Schneider,
Z. anorg. allg. Chem., 503 (1983) 238.

90 I.A. Tokman and G.A. Bukhalova,
Izv. Akad. Nauk SSSR, Neorg. Mat., 11 (1975) 1654.

91 M.T. Averbuch-Pouchot and A. Durif,
Acta Cryst., C41 (1985) 1557.

92 M.T. Averbuch-Pouchot and A. Durif,
Acta Cryst., C44 (1988) 212.

93 A. Durif, M.T. Averbuch-Pouchot and J.C. Guitel,
Acta Cryst., C39 (1983) 812.

94 A. Jouini, M. Dabbabi and A. Durif,
J. Solid State Chem., 60 (1985) 6.

95 A. Jouini, M. Soua and M. Dabbabi,
J. Solid State Chem., 69 (1987) 135.

96 C. Cavero-Ghersi,
Thesis, Univ. of Grenoble, France, 1975.

97 I. Tordjman, R. Masse and J.C. Guitel,
Acta Cryst., B32 (1976) 1643.

98 S.B. Hendricks and R.W.G. Wyckoff,
Amer. J. Sci.,13 (1927) 491.

99 F. D'Yvoire,
Bull. Soc. Chim. Fr., (1962) 1237.

100 P. Rémy and A. Boullé,
C. R. Acad. Sci., 258 (1964) 927.

101 F. Liebau and H.P. Williams, *Angew. Chem.*, 76 (1964) 303.

102 M. Bagieu-Beucher, *J. Appl. Cryst.*, 9 (1976) 368.

103 M. Bagieu-Beucher and J.C. Guitel, *Acta Cryst.*, B34 (1978) 1439.

104 G. Wappler, *Diplom-Arbeit*, Humbolt Univ., Berlin, Germany, 1958.

105 L.P. Mezentseva, A.I. Domanskii and I.A. Bondar, *Russ. J. Inorg. Chem.*, 22 (1977) 43.

106 N.N. Chudinova, *Izv. Akad. Nauk SSSR, Neorg. Mat.*, 15 (1979) 833.

107 N.B.S., Monograph 25, section 2, 1963, p3.

108 M. Beucher, *Les Eléments des Terres Rares*, Intern. Meeting, Paris-Grenoble, 1969.

109 R. Masse, J.C. Guitel and A. Durif, *Acta. Cryst.*, B33 (1977) 630.

110 M.A. Vaivada and Z.A. Konstant, *Izv. Akad. Nauk SSSR, Neorg.Mat.*, 15 (1979) 824.

111 M.A. Vaivada and Z.A. Konstant, *Izv. Akad. Nauk SSSR, Neorg. Mat.*, 16 (1980) 1810.

112 M. Rzaigui and N.K. Ariguib, *J. Solid State Chem.*, 49 (1983) 391.

113 M. Rzaigui, M.T. Averbuch-Pouchot and A. Durif, *Acta Cryst.*, C39 (1983) 1612.

114 N.N. Chudinova, L.P. Shklover, A.E. Balanevaskaya, L.M. Shkol'nikova, A.E. Obodovskaya and G.M. Balagina, *Izv. Akad. Nauk SSSR, Neorg. Mat.*, 14 (1978) 727.

115 O.S. Tarasenkova, G.I. Dorokhova, N.N. Chudinova, B.N. Litvin and N.V. Vinogradova, *Izv. Akad. Nauk SSSR, Neorg. Mat.*, 21 (1985) 452.

116 B.N. Litvin, G.I. Dorokhova and O.S. Filipenko, *Sov. Phys. Dokl.*, 26 (1981) 717.

117 K.K. Palkina, V.G. Kuznetsov, N.N. Chudinova and N.T. Chibiskova, *Izv. Akad. Nauk SSSR, Neorg. Mat.*, 12 (1976) 730.

118 N.N. Chudinova, N.V. Vinogradova, G.M. Balagina and K.K. Palkina, *Izv. Akad. Nauk SSSR, Neorg. Mat.*, 13 (1977) 1494.

119 H. Koizumi and J. Nakano, *Acta Cryst.*, B33 (1977) 2680.

120 G.I. Dorokhova, O.S. Filipenko, L.O. Atovmyan and B.N. Litvin, *Russ. J. Inorg. Chem.*, 33 (1988) 1581.

121 K.K. Palkina, S.I. Maksimova and N.T. Chibiskova, *Izv. Akad. Nauk SSSR, Neorg. Mat.*, 17 (1981) 1248.

122 G.I. Byrappa and B.N. Litvin, *J. Mat. Sc.*, 18 (1983) 2056.

123 K.K. Palkina, S.I. Maksimova and N.T. Chibiskova, *Sov. Phys. Dokl.*, 26 (1981) 254.

124 K. Byrappa, I.I. Plyusnina and G.I. Dorokhova, *J. Mat. Sc.*, 17 (1982) 1847.

125 M.A. Avaliani, N.N. Chudinova and I.V. Tananaev, *Izv. Akad. Nauk SSSR, Neorg. Mat.*, 20 (1984) 282.

126 K.K. Palkina, A.K. Mustaev, S.I. Maksimova, R.Yu. Khusainova and N.T. Chibiskova, *Russ. J. Inorg. Chem.*, 34 (1989) 1533.

127 H. Nariai, A.I. Motooka and M. Tsuhako, *Bull. Chem. Soc. Jpn.*, 61 (1988) 2811.

128 S.A. Linde, Yu.E. Gorbunova and V. Lavrova, *Russ. J. Inorg. Chem.*, 28 (1983) 785.

129 A. Durif, M.T. Averbuch-Pouchot and J.C. Guitel, *J. Solid State Chem.*, 41 (1982) 153.

130 M.T. Averbuch-Pouchot and A. Durif, *Acta.Cryst.*, C43 (1987) 1245.

131 M.T. Averbuch-Pouchot and A. Durif, *C. R. Acad. Sc.*, 304-II (1987) 269.

132 M.T. Averbuch-Pouchot, A. Durif and J.C. Guitel, *Acta Cryst.*, C44 (1988) 888.

133 M.T. Averbuch-Pouchot, A. Durif and J.C. Guitel, *Acta Cryst.*, C44 (1988) 1416.

134 M.T. Averbuch-Pouchot, A. Durif and M. Bagieu-Beucher, *Z. Kristallogr.*, 185 (1988) 506.

135 M.T. Averbuch-Pouchot, A. Durif and J.C. Guitel, *Acta Cryst.*, C45 (1989) 428.

136 M.T. Averbuch-Pouchot and A. Durif, *Acta Cryst.*, C45 (1989) 46.

137 M.T. Averbuch-Pouchot, A. Durif and J.C. Guitel, *Acta Cryst.*, C44 (1988) 1189.

138 M. Bagieu-Beucher, A. Durif and J.C. Guitel, *Acta Cryst.*, C44 (1988) 2063.

139 M. Bdiri and A. Jouini, *C. R. Acad. Sci.*, 308-II (1989) 1345.

140 A. Jouini, *Acta Cryst.*, C45 (1989) 1877.

141 M. Bdiri and A. Jouini, *J. Solid State Chem.*, 83 (1989) 350.

142 M. Bdiri and A. Jouini, *C. R. Acad. Sci.*, 309-II (1989) 881.

143 M. Bdiri and A. Jouini, *Eur. J. Solid State Inorg. Chem.*, 26 (1989) 585.

144 M.T. Averbuch-Pouchot, *Z. anorg. allg. Chem.*, 529 (1985) 143.

145 I. Grunze, K.K. Palkina, S.I. Maksimova and N.T. Chibiskova, *Dokl. Akad. Nauk SSSR*, 275 (1984) 879.

4. CYCLOPENTAPHOSPHATES

Cyclopentaphosphates are very rare compounds and up to now only four of them are clearly characterized.

Thilo and Schülke [1] showed that sodium phosphate glasses with a P/Na ratio close by one contain various cyclophosphates $Na_nP_nO_{3n}$ with the n value varying from 3 to 8. After fractionating solutions of these salts by acetone and hexamminecobalt(III), they have been able to isolate the sodium salt $Na_5P_5O_{15}.4H_2O$ and to prepare the corresponding barium and silver salts $Ba_5(P_5O_{15})_2.10H_2O$ and $Ag_5P_5O_{15}.2.6H_2O$.

The only structural determination of a cyclopentaphosphate was performed by Jost [2] on $Na_4(NH_4)P_5O_{15}.4H_2O$. This salt is triclinic, $P\bar{1}$, with the following bimolecular unit cell :

a = 8.73(2)	b = 15.66(3)	c = 6.81(2) Å
α = 93.9(4)	β = 106.1(4)	γ = 95.1(3)°

Its very simple arrangement is given in Fig.4.1, in projection along the c axis, while Table 4.1 reports the main features of this unique ring anion. From this table, one can notice that if P-P distances (average value : 2.914 Å) and P-O-P angles (130.5°) are similar to what we observed in cyclotri- and cyclotetraphosphates, the P-P-P angles deviate very

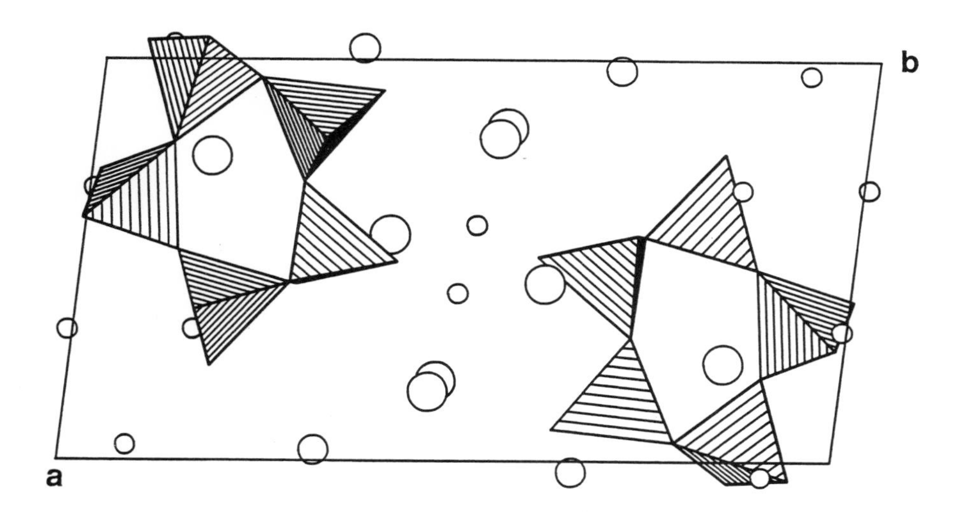

Figure 4.1 Projection, along the c axis of the atomic, arrangement of $Na_4(NH_4)P_5O_{15}.4H_2O$. The smallest empty circles represent the sodium atoms, the intermediate ones the ammonium groups and the largest ones the water molecules.

significantly from their average value in this ring (104.0°) or from the ideal value (108°) for a planar regular ring. As we shall see, the extent of these deviations will increase with larger rings. According to the authors, the accuracies obtained for the description of the ring geometry are of 0.01 Å for distances and 0.3° for angles.

Two of the sodium atoms have distorted octahedral coordinations, one has a fivefold coordination built up by a tetragonal pyramid and the last one an irregular sixfold neighboring. The authors had some difficulties to distinguish between water molecules and ammonium groups, and so, to assign properly their respective locations.

Table 4.1
Main geometrical features of the P_5O_{15} ring anion.

P-P-P (°)	P-O-P (°)	P-P (Å)
95.8	134.4	2.989
113.7	132.7	2.810
94.0	130.1	2.958
114.0	132.0	2.882
102.6	123.4	2.930

REFERENCES

1 E. Thilo and U. Schülke, *Z. anorg. allg. Chem.*, 344 (1965) 293.

2 K.H. Jost, *Acta Cryst.*, B28 (1972) 732.

5. CYCLOHEXAPHOSPHATES

The chemistry of cyclohexaphosphates was slow to develop and even today is relatively poor if compared with that of cyclotri- and cyclotetraphosphates. The lack, for a long time, of reliable processes to produce large amount of a starting material is probably the main reason for this small development. The first evidence for the existence of a P_6O_{18} ring anion was reported by Thilo and Schülke in 1963 [1, 2] and the first structural evidence by Jost in 1965 when describing the crystal structure of the sodium salt $Na_6P_6O_{18}.6H_2O$ [3, 4]. The same year, Griffith and Buxton [5] describe for the first time a reproducible process for the preparation of useful amounts of sodium or lithium cyclohexaphosphates. In spite of this possibility for producing a proper starting material, it is worth noticing that during the twenty years following the publication of this process all the cyclohexaphosphates clearly characterized have been discovered by indirect methods :

(i) $Cu_2Li_2P_6O_{18}$, during the attempts to prepare single crystals of the long-chain polyphosphate, $CuLi(PO_3)_3$, which was characterized during the determination of the $LiPO_3$-$Cu_2P_4O_{12}$ phase-equilibrium diagram [6, 7] ;

(ii) $Cr_2P_6O_{18}$ and $Cs_2(UO_2)_2P_6O_{18}$, during the systematic investigations of the Cr_2O_3-P_2O_5 [8] and Cs_2O-UO_2-P_2O_5 [9] systems by various flux methods.

Recently, Schülke and Kayser [10] reported an important improvement of the Griffith-Buxton process, so that today large amounts of $Li_6P_6O_{18}$ can be prepared in a reproducible way and with almost theoretical yields. This process will be described in details in the "Chemical Preparations" Section. More recently, Averbuch-Pouchot [11] described chemical preparation and accurate structural characterization of $Ag_6P_6O_{18}.H_2O$, so opening the way to metathesis reactions deriving from the Boullé's one [12] for the preparation of water-soluble cyclohexaphosphates.

5.1. Present State of the Cyclohexaphosphate Chemistry

5.1.1. Cyclohexaphosphates of monovalent cations

- *$Li_6P_6O_{18}.nH_2O$*

As said before, Griffith and Buxton [5] were the first to describe a process for the chemical preparation of the anhydrous salt. Later on, Schülke and Kayser [10] improved this process and reported the existence of one monohydrate, two forms for the tetrahydrate and two forms for the anhydrous salt. Crystal structures of a pentahydrate has been performed by Trunov *et al* [13] and that of a more hydrated salt, $Li_6P_6O_{18}.(8+n)H_2O$, (n ~ 0.2) by Bagieu-Beucher and Rzaigui [14]. The solubility of $Li_6P_6O_{18}$ in water and in various aqueous-methanolic mixtures has been investigated by Borodina and Chudinova [15]. The existence of the hexahydrate reported by several authors as the normal hydrate crystallizing at room temperature has not yet been confirmed by a structural investigation.

- *$Na_6P_6O_{18}.6H_2O$*

The hexahydrate $Na_6P_6O_{18}.6H_2O$ is the only well characterized salt. Its crystal structure has been described by Jost [3, 4]. This compound is now easily prepared from the corresponding lithium salt by using ion-exchange resins or the insolubility of lithium fluoride according to :

$$Li_6P_6O_{18} + 6NaF \text{ -----> } Na_6P_6O_{18} + 6LiF$$

- $Ag_6P_6O_{18}.H_2O$ and $Ag_6P_6O_{18}.2.2H_2O$

The monohydrate $Ag_6P_6O_{18}.H_2O$ is up to now the only well characterized silver cyclohexaphosphate. It was first described by Averbuch-Pouchot [11], who reported a detailed process for its chemical preparation and a complete determination of its atomic arrangement. Thilo and Schülke [16] reported in 1965 a 2.2 hydrate. The silver salts are of fundamental interest for the preparation of water soluble cyclohexaphosphates by metathesis reactions deriving from the Boullé's process [12]. See "Chemical Preparations" Section.

- $K_6P_6O_{18}$ and $K_6P_6O_{18}.3H_2O$

Two crystalline potassium cyclohexaphosphates are now well known, $K_6P_6O_{18}$ and $K_6P_6O_{18}.3H_2O$. Chemical preparations and indexed X-ray powder diffraction diagrams of these two salts are reported by Chudinova *et al* [17]. The crystal structure of the anhydrous salt has been performed almost simultaneously by Averbuch-Pouchot [18] and Kholodkovskaya *et al* [19], while that of the trihydrate has been determined by the last authors [20]. The thermal behaviour of the trihydrate has been investigated by Chudinova *et al* [21]. The authors report the following scheme of transformation :

$$K_6P_6O_{18}.3H_2O \xrightarrow{363K} K_6P_6O_{18} \xrightarrow{553K} KPO_3$$

- $(NH_4)_6P_6O_{18}.H_2O$

Chemical preparation and crystal structure of $(NH_4)_6P_6O_{18}.H_2O$ have been reported by Averbuch-Pouchot [22]. The atomic arrangement is close by that reported for $Ag_6P_6O_{18}.H_2O$.

- $Rb_6P_6O_{18}.6H_2O$, $Cs_6P_6O_{18}.6H_2O$, $Rb_6P_6O_{18}$ and $Cs_6P_6O_{18}$

The two hydrates $M_6P_6O_{18}.6H_2O$ (M = Rb, Cs) have been prepared by Averbuch-Pouchot and Durif [23] using the Boullé's reaction ; they are triclinic and isotypic. The authors report a detailed determination of the atomic arrangement using the caesium salt. Crystal structures of these two salts have been also determined almost simultaneously by Kholodkovskaya *et al* [24]. In addition, chemical preparations and indexed X-ray powder diffraction diagrams for $Rb_6P_6O_{18}.6H_2O$ and $Cs_6P_6O_{18}.6H_2O$ are reported in Ref. 17 and their thermal behaviour investigated in Ref. 21. The following schemes of transformation are observed :

$$Rb_6P_6O_{18}.6H_2O \xrightarrow{323K} Rb_6P_6O_{18}.nH_2O \xrightarrow{373K} Rb_6P_6O_{18} \xrightarrow{573K} RbPO_3$$

$$Cs_6P_6O_{18}.6H_2O \xrightarrow{363K} Cs_6P_6O_{18}.nH_2O \xrightarrow{413K} Cs_6P_6O_{18} \xrightarrow{553K} CsPO_3$$

According to the authors [17, 21], the anhydrous salts are cubic, the rubidium one being isotypic with $K_6P_6O_{18}$.

- ***$Li_3Na_3P_6O_{18}.12H_2O$***

The existence of $Li_3Na_3P_6O_{18}.12H_2O$ was first reported by Schülke and Kayser [10]. Later on, Averbuch-Pouchot [25] published a detailed description of its atomic arrangement.

- ***$Li_3K_3P_6O_{18}.H_2O$ and $Ag_3(NH_4)_3P_6O_{18}.H_2O$***

The two trigonal (rhomboedral) salts $Li_3K_3P_6O_{18}.H_2O$ and $Ag_3(NH_4)_3P_6O_{18}.H_2O$ are isotypic. Both have been prepared and investigated by Averbuch-Pouchot [25, 26], who performed their crystal structure determinations. The first one is simply synthesized by mixing in stoichiometric amount aqueous solutions of the two monovalent cyclohexaphosphates, while the second one is prepared by adding an aqueous solution of silver nitrate to an aqueous solution of ammonium cyclohexaphosphate according to the stoichiometry :

Table 5.1
Crystal data for cyclohexaphosphates of monovalent cations.

Formula	a α	b β	c(Å) γ(°)	S. G.	Z	Ref.
$Li_6P_6O_{18}.5H_2O$	9.490(1) 107.16(1)	8.069(1) 113.84(1)	7.810(1) 65.19(1)	$P\bar{1}$	1	13
$Li_6P_6O_{18}.(8+n)H_2O$	15.429(8)	11.794(5) 115.95(5)	14.369(8)	C2/m	4	14
$Na_6P_6O_{18}.6H_2O$	11.58(2)	18.54(4)	10.48(2)	Ccmb	4	3, 4
$Ag_6P_6O_{18}.H_2O$	14.807(10)	14.807(10)	6.597(7)	$R\bar{3}$	3	11
$K_6P_6O_{18}$	15.753(6)	15.753(6)	15.753(6)	Pa3	8	18, 19
$K_6P_6O_{18}.3H_2O$	6.803(5)	17.45(2) 107.18(6)	9.195(8)	$P2_1/m$	2	20
$(NH_4)_6P_6O_{18}.H_2O$	15.445(10)	15.445(10)	7.553(7)	$R\bar{3}$	3	22
$Rb_6P_6O_{18}.6H_2O$	9.662(7) 111.02(5)	9.640(6) 107.83(5)	8.704(3) 60.14(5)	$P\bar{1}$	1	23, 24
$Cs_6P_6O_{18}.6H_2O$	9.904(9) 111.53(2)	9.898(9) 106.53(2)	9.013(8) 60.08(3)	$P\bar{1}$	1	23, 24
$Rb_6P_6O_{18}$	16.23	16.23	16.23	Pa3	8	17, 21
$Cs_6P_6O_{18}$	25.16	25.16	25.16	?	?	17, 21

$$(NH_4)_6P_6O_{18} + 3AgNO_3 \text{------>} Ag_3(NH_4)_3P_6O_{18}.H_2O + 3NH_4NO_3$$

The silver-ammonium salt, less water-soluble, crystallizes first by slow evaporation of the solution at room temperature.

- $Ag_4Li_2P_6O_{18}.2H_2O$

The triclinic salt $Ag_4Li_2P_6O_{18}.2H_2O$ has been prepared by Averbuch-Pouchot and Durif [27] according to the process described for the synthesis of the silver-ammonium salt. The authors performed a detailed determination of the atomic arrangement.

- $Na_2(NH_4)_4P_6O_{18}.2H_2O$ and $Na_2Tl_4P_6O_{18}.2H_2O$

The two isotypic salts $Na_2M_4P_6O_{18}.2H_2O$ (M = NH_4, Tl) were prepared and investigated by Averbuch-Pouchot and Durif [28], who reported the crystal structure of the sodium-ammonium compound. The thallium-sodium salt was prepared by a process similar to that described for the synthesis of $Ag_3(NH_4)_3P_6O_{18}.H_2O$, but using thallium nitrate, while the sodium-ammonium salt was prepared by mixing solutions of the two alkali cyclohexaphosphates.

- $Na_4Rb_2P_6O_{18}.6H_2O$ and $Na_4Cs_2P_6O_{18}.6H_2O$

The two triclinic isotypic cyclohexaphosphates $Na_4M_2P_6O_{18}.6H_2O$ (M = Rb, Cs) have been studied by Averbuch-Pouchot and Durif [28], who

Table 5.2
Crystal data for mixed monovalent cation cyclohexaphosphates.

Formula	a α	b β	c(Å) γ(°)	S. G.	Z	Ref.
$Li_3Na_3P_6O_{18}.12H_2O$	10.474(8)	10.474(8)	41.68(5)	$R\bar{3}c$	6	25
$Li_3K_3P_6O_{18}.H_2O$	15.047(8)	15.047(8)	12.779(8)	$R\bar{3}$	6	25
$Ag_3(NH_4)_3P_6O_{18}.H_2O$	15.172(5)	15.172(5)	13.994(5)	$R\bar{3}$	6	26
$Ag_4Li_2P_6O_{18}.2H_2O$	8.408(2) 107.47(3)	7.602(2) 106.09(3)	7.566(2) 72.64(3)	$P\bar{1}$	1	27
$Na_2(NH_4)_4P_6O_{18}.2H_2O$	13.363(7)	11.580(12) 101.87(5)	6.809(5)	$P2_1/n$	2	28
$Na_2Tl_4P_6O_{18}.2H_2O$	13.215(8)	11.583(8) 101.43(5)	6.874(8)	$P2_1/n$	2	28
$Na_4Rb_2P_6O_{18}.6H_2O$	7.532(3) 113.92(4)	9.752(3) 102.29(4)	8.730(3) 85.00(4)	$P\bar{1}$	1	28
$Na_4Cs_2P_6O_{18}.6H_2O$	7.653(4) 114.64(4)	9.959(4) 102.47(4)	8.740(4) 84.94(4)	$P\bar{1}$	1	28

determined the atomic arrangement of the sodium-rubidium salt.

Main crystallographic data for the monovalent and mixed monovalent cation cyclohexaphosphates are reported in Table 5.1 and Table 5.2.

5.1.2. Cyclohexaphosphates of divalent cations

Eight divalent-cation cyclohexaphosphates have been characterized but only three of them have been structurally investigated

- $Mn_3P_6O_{18}.9H_2O$, $Ni_3P_6O_{18}.17H_2O$, $Cd_3P_6O_{18}.16H_2O$ and $M_3P_6O_{18}.14H_2O$ (M = Co, Cu)

The five cyclohexaphosphate hydrates $M_3P_6O_{18}.14H_2O$ (M = Co, Cu), $Mn_3P_6O_{18}.9H_2O$, $Ni_3P_6O_{18}.17H_2O$ and $Cd_3P_6O_{18}.16H_2O$ have been prepared and characterized by Lazarevski *et al* [29]. They were synthesized by action of cyclohexaphosphoric acid on aqueous solutions of the corresponding perchlorates. The acid was produced through ion-exchange resins from solutions of Li or Na phosphates. According to the X-ray powder diagrams given by the authors, it seems that the cobalt and copper salts, crystallizing with the same degree of hydration, are not isotypic. Thermal behavior of the cadmium, copper, cobalt, nickel and manganese salts has also been investigated by the same authors [29-31].

In the case of $Cu_3P_6O_{18}.14H_2O$, at high temperature, the P_6O_{18} ring is destroyed with formation of the cyclotetraphosphate.

$$2Cu_3P_6O_{18}.14H_2O \dashrightarrow 3Cu_2P_4O_{12} + 14H_2O$$

Single crystals of this salt were later prepared by Averbuch-Pouchot [32] by adding water solutions of $CuCl_2$ (or $Cu(NO_3)_2$) to an aqueous solution of $Li_6P_6O_{18}.5\text{-}6H_2O$, or by adding a water solution of copper nitrate to a solution of guanidinium cyclohexaphosphate in a stoichiometric ratio. After some days of evaporation of these solutions at room temperature, crystals of $Cu_3P_6O_{18}.14H_2O$ appear as large turquoise calcite-like pseudo-rhombohedra. The author reports a complete determination of the atomic arrangement, confirming the hydration state previously observed [29-31].

- $Mn_3P_6O_{18}.6H_2O$ and $Cd_3P_6O_{18}.6H_2O$

Averbuch-Pouchot [33] prepared the cadmium salt $Cd_3P_6O_{18}.6H_2O$ using the Boullé's process [12] and determined its atomic arrangement. The manganese salt $Mn_3P_6O_{18}.6H_2O$ was prepared and characterized by Averbuch-Pouchot and Durif [34] as an isotype of the first one.

- $Ba_3P_6O_{18}.9H_2O$

Thilo and Schülke [16] and Lazarevski *et al* [35] obtained $Ba_3P_6O_{18}.9H_2O$ as a white crystalline substance by reaction between aqueous solutions of $BaCl_2$ and $Na_6P_6O_{18}$. Its thermal behavior has been studied by the latter authors [31].

Table 5.3 gives the main crystal features of the three presently well investigated divalent cation cyclohexaphosphates.

Table 5.3
Crystal data for cyclohexaphosphates of bivalent cations.

Formula	a α	b β	c (Å) γ (°)	S.G.	Z	Ref.
$Mn_3P_6O_{18}.6H_2O$	14.836(8)	14.836(8)	15.781(8)	$R\bar{3}$	6	34
$Cd_3P_6O_{18}.6H_2O$	15.056(8)	15.056(8)	16.080(8)	$R\bar{3}$	6	33
$Cu_3P_6O_{18}.14H_2O$	10.944(8) 110.49(5)	7.539(4) 110.14(5)	8.974(4) 77.82(5)	$P\bar{1}$	1	32

5.1.3. Cyclohexaphosphates of divalent-monovalent cations

With the exception of $Cu_2Li_2P_6O_{18}$ and $Cs_2(UO_2)_2P_6O_{18}$, all the today well characterized divalent-monovalent cation cyclohexaphosphates are hydrates corresponding to the general formula : $M^{II}_2M^{I}_2P_6O_{18}.nH_2O$ belonging to seven different structure types.

- ***$Cu_2Li_2P_6O_{18}$***

$Cu_2Li_2P_6O_{18}$ has been characterized during the elaboration of the $LiPO_3$-$Cu_2P_4O_{12}$ phase-equilibrium diagram, when attempts were made to prepare single crystals of the $CuLi(PO_3)_3$ long-chain polyphosphate. Laügt *et al* [36] observed a low temperature form for this polyphosphate and characterized it by chromatographic analysis as a cyclohexaphosphate. According to the authors, this salt transforms into the polyphosphate at 773K. The atomic arrangement of $Cu_2Li_2P_6O_{18}$ was determined by Laügt and Durif [6].

- ***$Cs_2(UO_2)_2P_6O_{18}$***

$Cs_2(UO_2)_2P_6O_{18}$ has been characterized by Linde *et al* [9] during an investigation on the compounds obtained by heating mixtures of H_3PO_4, $CsNO_3$ and $UO_2(NO_3)_2.6H_2O$ in various conditions. The authors report the crystal structure of this salt.

Most of the hydrates we describe now have been prepared as polycrystalline samples by classical methods of aqueous chemistry according to the general reaction scheme :

$$M^{I}_6P_6O_{18} + 2M^{II}(NO_3)_2 \text{ -----> } M^{I}_2M^{II}_2P_6O_{18}.nH_2O + 4M^{I}NO_3$$

All of them are sparingly water-soluble and in some cases the production of single crystals suitable for structural investigations is tributary of non-orthodox processes.

- $Zn_2Li_2P_6O_{18}.10H_2O$ and $Mn_2Li_2P_6O_{18}.10H_2O$

The two isotypic triclinic compounds $Zn_2Li_2P_6O_{18}.10H_2O$ and $Mn_2Li_2P_6O_{18}.10H_2O$ have been prepared by Averbuch-Pouchot [37], who determined the atomic arrangement with the manganese salt.

- $Ca_2Li_2P_6O_{18}.8H_2O$ and $Ca_2Na_2P_6O_{18}.8H_2O$

The two salts $Ca_2M_2P_6O_{18}.8H_2O$ (M = Li, Na) are also triclinic and have been prepared by adding solid gypsum in an aqueous solution of the alkali cyclohexaphosphate. These two hydrates are isotypic. The crystal structure of the calcium-lithium compound has been performed by Averbuch-Pouchot and Durif [38].

- $Cd_2Na_2P_6O_{18}.14H_2O$

Chemical preparation and crystal structure of the triclinic compound $Cd_2Na_2P_6O_{18}.14H_2O$ have been reported by Averbuch-Pouchot [39].

- $Cu_2(NH_4)_2P_6O_{18}.8H_2O$

$Cu_2(NH_4)_2P_6O_{18}.8H_2O$ can be prepared in a polycrystalline state by classical methods, but the production of single crystals suitable for structural investigations has been difficult to optimize. Averbuch and Durif [40] described the following process. An almost saturated aqueous solution of ammonium cyclohexaphosphate monohydrate $(NH_4)_6P_6O_{18}.H_2O$ is added with the required amount of copper hydroxycarbonate ($CuCO_3.Cu(OH)_2$). In this slurry, drops of concentrated hydrochloric acid are added till the complete disappearance of the copper hydroxycarbonate. Crystal of $Cu_2(NH_4)_2P_6O_{18}.8H_2O$ appear immediately as diamond-like plates. This compound is very sparingly water-soluble.

- $Ca_2(NH_4)_2P_6O_{18}.6H_2O$

$Ca_2(NH_4)_2P_6O_{18}.6H_2O$ is easily prepared as large orthorhombic prisms, after some days of evaporation, at room temperature, of an aqueous solution of ammonium cyclohexaphosphate added with solid gypsum. Averbuch-Pouchot [41] performed the determination of its crystal structure.

- $Zn_2(NH_4)_2P_6O_{18}.8H_2O$ and $Cd_2(NH_4)_2P_6O_{18}.9H_2O$

Chemical preparations and crystal structures of the two triclinic salts $Zn_2(NH_4)_2P_6O_{18}.8H_2O$ and $Cd_2(NH_4)_2P_6O_{18}.9H_2O$ have been reported by Averbuch-Pouchot *et al* [42]. A second crystalline form of the zinc salt, isotypic with the corresponding copper salt has been observed, but not yet investigated.

Table 5.4 gathers the main crystallographic features of the divalent-monovalent cation cyclohexaphosphates.

Table 5.4
Crystal data for divalent-monovalent cation cyclohexaphosphates.

Formula	a α	b β	c(Å) γ(°)	S. G.	Z	Ref.
$Cs_2(UO_2)_2P_6O_{18}$	6.988(2)	10.838(4) 104.25(2)	13.309(3)	$P2_1/n$	4	9
$Cu_2Li_2P_6O_{18}$	9.485(2) 111.73(2)	9.419(2) 106.25(2)	9.379(2) 106.80(2)	$P\bar{1}$	2	7, 36
$Zn_2Li_2P_6O_{18}.10H_2O$	7.160(5) 118.49(5)	9.741(7) 110.57(5)	9.928(6) 86.96(5)	$P\bar{1}$	1	37
$Mn_2Li_2P_6O_{18}.10H_2O$	7.286(5) 118.31(5)	9.761(7) 110.62(5)	10.026(6) 86.27(5)	$P\bar{1}$	1	37
$Ca_2Li_2P_6O_{18}.8H_2O$	7.767(2) 105.17(4)	10.144(3) 102.76(4)	7.225(2) 84.95(4)	$P\bar{1}$	1	38
$Ca_2Na_2P_6O_{18}.8H_2O$	8.031(5) 105.69(5)	10.296(9) 103.27(5)	7.279(5) 85.30(5)	$P\bar{1}$	1	38
$Cd_2Na_2P_6O_{18}.14H_2O$	7.709(1) 108.25(5)	11.028(6) 110.06(5)	9.231(2) 79.77(5)	$P\bar{1}$	1	39
$Cu_2(NH_4)_2P_6O_{18}.8H_2O$	7.413(3) 116.23(5)	9.334(4) 107.98(5)	9.634 83.10(5)	$P\bar{1}$	1	40
$Ca_2(NH_4)_2P_6O_{18}.6H_2O$	12.821(6)	12.537(6)	7.029(2)	$P2_12_12$	2	41
$Zn_2(NH_4)_2P_6O_{18}.8H_2O$	8.717(7) 104.82(2)	10.297(8) 111.03(2)	7.409(7) 70.96(2)	$P\bar{1}$	1	42
$Cd_2(NH_4)_2P_6O_{18}.9H_2O$	9.126(6) 107.95(6)	9.581(8) 112.93(6)	8.993(4) 61.90(5)	$P\bar{1}$	1	42

5.1.4. Cyclohexaphosphates of trivalent cations

- $Cr_2P_6O_{18}$

$Cr_2P_6O_{18}$ has been obtained for the first time by Rémy and Boullé [43, 44] during a systematic investigation of chromium phosphates. Later on, Bagieu-Beucher and Guitel [8] reported the preparation of single crystals and performed the determination of its atomic arrangement.

- $Ga_2P_6O_{18}$

The existence of $Ga_2P_6O_{18}$ has been reported by Chudinova *et al* [45] during a study of the NH_4-P_2O_5-Ga_2O_3-H_2O system. The most stable compound in this system is $GaNH_4HP_3O_{10}$, a triphosphate which, when heated at about 723K, transforms into $Ga_2P_6O_{18}$:

$$2GaNH_4HP_3O_{10} \xrightarrow{723K} Ga_2P_6O_{18} + 2H_2O + 2NH_3$$

This salt is isotypic with the corresponding chromium salt.

- $Al_2P_6O_{18}$

The aluminium cyclohexaphosphate $Al_2P_6O_{18}$ has been obtained as crystalline samples by Kanene *et al* [46] by heating Al_2O_3 and $NH_4H_2PO_4$ at 823K during 4 hours. When heated at 1223K-1273K, it decomposes into aluminium cyclotetraphosphate :

$$2Al_2P_6O_{18} \longrightarrow Al_4(P_4O_{12})_3$$

From crystal data, the authors assumed the isotypy of this salt with the chromium compound.

- $[Ga(OH)_2]_6P_6O_{18}$

During the investigation of the $GaCl_3$-$Na_6P_6O_{18}$ system, Lazarevski *et al* [35] characterized chemically the salt $[Ga(OH)_2]_6P_6O_{18}$. Its thermal behavior was later investigated by the same authors [30].

- $Cr_2P_6O_{18}.21H_2O$

Rzaigui [47] prepared the hydrate $Cr_2P_6O_{18}.21H_2O$ and described its main chemical and physical properties. The crystal structure performed by Bagieu-Beucher *et al* [48] shows that nine of the twenty-one water molecules are of a zeolitic nature in accordance with the thermal behavior reported by Rzaigui [47].

- $Y_2P_6O_{18}.18H_2O$

The hydrate $Y_2P_6O_{18}.18H_2O$ has been prepared by Lazarevski *et al* [29] by action of YCl_3 on $Na_6P_6O_{18}$ in aqueous solution and its thermal behavior investigated by the same authors [30].

- $Ce_2P_6O_{18}.10H_2O$

Rzaigui [49] prepared single crystals of $Ce_2P_6O_{18}.10H_2O$ and described its main chemical properties. Bagieu-Beucher and Rzaigui [50] performed the crystal structure determination. Four of the ten water molecules are not involved in the cerium coordination polyhedra.

- $Nd_2P_6O_{18}.12H_2O$

The crystal structure of $Nd_2P_6O_{18}.12H_2O$ prepared by reaction of $NdCl_3$ with an aqueous solution of $Li_6P_6O_{18}$, has been performed by Trunov *et al* [51].

- $Mn_3Cs_3(P_6O_{18})_2$

The anhydrous trivalent-monovalent cation cyclohexaphosphate, $Mn_3Cs_3(P_6O_{18})_2$, has been characterized by Guzeeva and Tananaev [52] during an investigation of the MnO_2-Cs_2O-P_2O_5-H_2O system between 423 and 637K. No crystal data are reported by the authors.

Crystal data for cyclohexaphosphates of trivalent cations are reported in Table 5.5.

Table 5.5
Crystal data for cyclohexaphosphates of trivalent cations.

Formula	a α	b β	c(Å) γ (°)	S. G.	Z	Ref.
$Cr_2P_6O_{18}$	8.311(4)	15.221(8) 105.85(5)	6.220(3)	$P2_1/a$	2	8
$Ga_2P_6O_{18}$	8.293	15.196 105.89	6.188	$P2_1/a$	2	45
$Al_2P_6O_{18}$	8.201(3)	15.062(3) 105.13(3)	6.101(2)	$P2_1/a$	2	46
$Cr_2P_6O_{18}.21H_2O$	19.052(10)	19.052(10)	19.052(10)	$P\bar{4}3n$	8	48
$Nd_2P_6O_{18}.12H_2O$	9.149(2)	11.693(3) 96.92(1)	11.959(3)	$P2_1/c$	2	51
$Ce_2P_6O_{18}.10H_2O$	13.522(5)	13.105(9)	6.938(3)	$P2_12_12$	2	50

5.1.5. Organic and organometallic cation cyclohexaphosphates

Four organic cation cyclohexaphosphates are up to now well characterized. All of them have been synthesized by the Boullé's process [12] and in all cases the atomic arrangements have been accurately determined.

- $(NH_3OH)_6P_6O_{18}.4H_2O$

Prepared and described by Durif and Averbuch-Pouchot [53] $(NH_3OH)_6P_6O_{18}.4H_2O$, the hydroxylammonium cyclohexaphosphate tetrahydrate, is very stable in normal conditions.

- $(C_2N_2H_{10})_3P_6O_{18}.2H_2O$

Prepared and investigated by Durif and Averbuch-Pouchot [54] $(C_2N_2H_{10})_3P_6O_{18}.2H_2O$, the ethylenediammonium cyclohexaphosphate dihydrate, appears as an interesting starting material for the syntheses of some species of complex cyclohexaphosphates.

- $(C_2H_5NH_3)_6P_6O_{18}.4H_2O$

Prepared and described by Averbuch-Pouchot and Durif [56], the ethylammonium cyclohexaphosphate tetrahydrate $(C_2H_5NH_3)_6P_6O_{18}.4H_2O$ is stable for months at room temperature.

- ***$(N_2H_5)_2(N_2H_6)_2P_6O_{18}$***

The chemical preparation of the dihydrazinium cyclohexaphosphate $(N_2H_6)_3P_6O_{18}$ unexpectedly afforded the monohydrazinium-dihydrazinium salt $(N_2H_5)_2(N_2H_6)_2P_6O_{18}$ as clearly shown by the crystal structure performed by Averbuch-Pouchot and Durif [56].

- ***$M(C_2N_2H_{10})_2P_6O_{18}.6H_2O$*** (M = Cu, Co, Ni, Mg, Zn, Fe)

A series of six isotypic organometallic cation compounds namely, $M(C_2N_2H_{10})_2P_6O_{18}.6H_2O$ (M = Cu, Co, Ni, Mg, Zn, Fe) has been characterized by Durif and Averbuch [55]. These compounds are prepared by adding in a stoichiometric ratio an aqueous solution of the divalent cation nitrate to a water solution of ethylenediammonium cyclohexa-

Table 5.6
Crystal data for organic and organometallic cation cyclohexaphosphates.

Formula	a α	b β	c (Å) γ (°)	S. G.	Z	Ref.
$(NH_3OH)_6P_6O_{18}.4H_2O$	10.365(5) 108.39(5)	9.278(4) 100.30(5)	7.280(3) 96.02(5)	$P\bar{1}$	1	53
$(EDA)_3P_6O_{18}.2H_2O$	11.064(5)	12.317(5) 90.53(5)	9.342(5)	$P2_1/n$	2	54
$(C_2H_5NH_3)_6P_6O_{18}.4H_2O$	16.804(10)	23.883(10) 109.66(2)	10.623(8)	$P2_1/a$	4	56
$(N_2H_5)_2(N_2H_6)_2P_6O_{18}$	8.175(8) 105.05(2)	7.926(8) 102.08(2)	8.457(7) 86.42(2)	$P\bar{1}$	1	56
$Cu(EDA)_2P_6O_{18}.6H_2O$	13.378(8)	11.574(6) 103.15(3)	8.687(3)	$P2_1/a$	2	54
$Co(EDA)_2P_6O_{18}.6H_2O$	13.193(8)	11.604(6) 102.22(5)	8.616(3)	$P2_1/a$	2	55
$Ni(EDA)_2P_6O_{18}.6H_2O$	13.165(8)	11.579(8) 102.61(5)	8.612(4)	$P2_1/a$	2	55
$Mg(EDA)_2P_6O_{18}.6H_2O$	13.234(5)	11.630(5) 102.45(5)	8.674(2)	$P2_1/a$	2	55
$Zn(EDA)_2P_6O_{18}.6H_2O$	13.217(5)	11.617(5) 102.44(5)	8.670(2)	$P2_1/a$	2	55
$Fe(EDA)_2P_6O_{18}.6H_2O$	13.233(5)	11.674(5) 102.06(5)	8.646(2)	$P2_1/a$	2	55

EDA = $[NH_3-(CH_2)_2-NH_3]^{+2}$ (ethylenediammonium)

phosphate. Sulfate has been used in the case of the iron salt. The reaction scheme is :

$$(EDA)_3P_6O_{18} + M(NO_3)_2 \text{ ------> } M(EDA)_2P_6O_{18} + EDA(NO_3)_2$$

If concentrated solutions are used, large stout monoclinic prisms of $M(EDA)_2P_6O_{18}.6H_2O$ appear within some hours in the solution. This structure type has been determined with the copper salt by Durif and Averbuch-Pouchot [54].

Table 5.6 gives the main crystallographic features for organic and organometallic cation cyclohexaphosphates.

5.1.6. Telluric acid-cyclohexaphosphate adducts and cyclohexaphosphate-oxalates

The P_6O_{18} ring anion is also observed in compounds like adducts of alkali cyclohexaphosphates with telluric acid and in mixed-anion compounds like cyclohexaphosphate-oxalates. The property to form adducts with telluric acid has been observed for all kinds of alkali phosphates condensed or not and has been extensively investigated by the reviewers during the past fifteen years, but the cyclohexaphosphate-oxalates are the first example of the coexistence of phosphoric ring anions and C_2O_4 groups in an atomic arrangement.

The adducts with telluric acid are simply prepared by adding an aqueous solution of telluric acid to a solution of the alkali phosphate in the proper stoichiometry. All the compounds reported in Table 5.7 have been prepared by this simple process and for all of them the atomic arrangements have been determined.

- *$(NH_4)_6P_6O_{18}.Te(OH)_6.2H_2O$*

The salt $(NH_4)_6P_6O_{18}.Te(OH)_6.2H_2O$ has been recently prepared by Averbuch-Pouchot and Durif [57]. The crystal structure performed by these authors shows the existence of two statistically distributed water molecules.

- *$K_6P_6O_{18}.2Te(OH)_6.3H_2O$*

The layered arrangement of the trigonal (rhomboedral) adduct $K_6P_6O_{18}.2Te(OH)_6.3H_2O$ has been established by Averbuch-Pouchot and Durif [58].

- *$Rb_6P_6O_{18}.3Te(OH)_6.4H_2O$ and $Cs_6P_6O_{18}.3Te(OH)_6.4H_2O$*

The two isotypic adducts $M_6P_6O_{18}.3Te(OH)_6.4H_2O$ (M = Rb, Cs) have been recently prepared by Averbuch-Pouchot and Durif [58]. The authors determined the atomic arrangement with the rubidium salt.

- *$(EDA)_3P_6O_{18}.2Te(OH)_6.2H_2O$*

The salt $(EDA)_3P_6O_{18}.2Te(OH)_6.2H_2O$ prepared and described by

Averbuch-Pouchot and Durif [59] is one of the rare example of adduct between an organic cation condensed phosphate and telluric acid.

Table 5.7

Crystal data for cyclohexaphosphate adducts with telluric acid and oxalates

Formula	a α	b β	c (Å) γ (°)	S. G.	Z	Ref.
$(NH_4)_6P_6O_{18}.Te(OH)_6.2H_2O$	9.899(4) 109.53(6)	11.042(7) 106.74(6)	7.632(9) 100.91(4)	$P\bar{1}$	1	57
$K_6P_6O_{18}.2Te(OH)_6.3H_2O$	13.084(5)	13.084(5)	34.80(2)	$R\bar{3}c$	6	58
$Rb_6P_6O_{18}.3Te(OH)_6.4H_2O$	11.222(8) 111.11(2)	8.077(6) 104.66(2)	11.731(9) 83.25(2)	$P\bar{1}$	1	58
$Cs_6P_6O_{18}.3Te(OH)_6.4H_2O$	11.549(8) 111.08(5)	8.228(4) 103.13(5)	11.946(6) 82.26(5)	$P\bar{1}$	1	58
$(EDA)_3P_6O_{18}.2Te(OH)_6.2H_2O$	10.945(3) 90.90(5)	11.252(3) 92.97(5)	8.042(7) 116.82(5)	$P\bar{1}$	1	59
$Mg_2(NH_4)_4P_6O_{18}(C_2O_4).6H_2O$	9.656(4) 98.88(5)	9.642(4) 106.21(5)	7.618(3) 100.37(5)	$P\bar{1}$	1	60
$Mn_2(NH_4)_4P_6O_{18}(C_2O_4).6H_2O$	9.747(3) 99.25(5)	9.751(3) 105.88(5)	7.689(3) 100.08(5)	$P\bar{1}$	1	60

EDA = $[NH_3\text{-}(CH_2)_2\text{-}NH_3]^{+2}$ (ethylenediammonium)

- $M_2(NH_4)_4P_6O_{18}(C_2O_4).6H_2O$ (M = Mg, Mn)

The preparation of the phosphate-oxalates is rather difficult and in some cases requires months. A very diluted aqueous solution of the divalent cation oxalate is added drop by drop (some drops a day) to a solution of the alkali phosphate kept at room temperature. After some months, very sparingly water-soluble crystals of the phosphate-oxalates appear in the solution. The two reported examples of phosphate-oxalates, $Mg_2(NH_4)_4P_6O_{18}(C_2O_4).6H_2O$ and $Mn_2(NH_4)_4P_6O_{18}(C_2O_4).6H_2O$, are isotypic compounds. The crystal structure of the manganese salt has been determined by Averbuch-Pouchot and Durif [60].

Table 5.7 reports the main crystal data measured for these new compounds.

5.2. Chemical Preparations of Cyclohexaphosphates

For this very recently developed branch of the cyclophosphate chemistry, it seems useful to report in a way as detailed as possible the main chemical processes used today for the preparation of cyclohexaphosphates.

5.2.1. The Griffith-Buxton process

Griffith and Buxton [5] were the first to optimize the cyclization condensation in the Li_2O-P_2O_5 system leading to the formation of lithium cyclohexaphosphate. The process, as described by the authors, is reported below. An amount of 51.7 g of Li_2CO_3 is added to 115.3 g of H_3PO_4 (85%). When the reaction subsided, the so obtained slurry is heated for one hour at 473K ; the temperature is then increased to 548K and maintained for five hours. The solid so obtained is ground and added to 800 cm^3 of water. After 15 min of stirring, the insoluble solid part (probably long-chain lithium polyphosphate) is removed by filtration. The resulting solution containing both diphosphate and cyclohexaphosphate is passed through a column of strong acid ion-exchange resins in its hydrogen form. The obtained solution is then neutralized with sodium carbonate until the pH is between 5 and 6 and 300 cm^3 of methyl alcohol are added. The first compound to crystallize is $Na_6P_6O_{18}.6H_2O$. If an excess of methyl alcohol is added, the precipitation of the diphosphate occurs. The precipitated $Na_6P_6O_{18}.6H_2O$ is filtered off and air dried. If the sample contains an excessive amount of the diphosphate, the purity can be improved by redissolving and reprecipitating with methyl alcohol. The quantities obtained for such runs vary between 10 and 20 g of sodium cyclohexaphosphate.

5.2.2. The Schülke and Kayser process

Recently, the Griffith-Buxton process has been greatly improved and simplified by Schülke and Kayser [10]. This process is based on the structure-controlled thermal dehydration condensation of LiH_2PO_4 seeded by $Li_6P_6O_{18}$ crystals. The experimental procedure is the following one. A mixture of 104 g of LiH_2PO_4 and 21 g of $Li_6P_6O_{18}.5\text{-}6H_2O$ is finely ground and heated at 623K for 30 min. The resulting product is then ground and dissolved in 500 cm^3 of water. Commonly, three hours of mechanical stirring are necessary to perform this dissolution. The insoluble part, mainly $LiPO_3$ long-chain polyphosphate, is removed by filtration. 500 cm^3 of methanol are then added to the solution leading to a complete precipitation of $Li_6P_6O_{18}.5\text{-}6H_2O$ within one day. The yield is almost the theoretical one. Schematically, the reaction is :

$$6LiH_2PO_4 + 0.1Li_6P_6O_{18} \xrightarrow[30min]{623K} 1.1Li_6P_6O_{18} + 6H_2O$$

5.2.3. The Boullé's process

Since the characterization of the silver salt, $Ag_6P_6O_{18}.H_2O$, by Averbuch-Pouchot [8], a metathesis reaction deriving from the Boullé's process [12] is commonly used for the preparation of water-soluble cyclohexaphosphates. The reaction scheme is :

$$Ag_6P_6O_{18}.H_2O + 6MCl \text{ ------> } M_6P_6O_{18} + 6AgCl$$

Many cyclohexaphosphates have been prepared using this process and among them several alkali cyclohexaphosphates and all the presently known organic cation cyclohexaphosphates.

5.2.4. The use of fluorides

The insolubility of LiF can be used to prepare most of the alkali cyclohexaphosphates according to the following process :

$$6MF + Li_6P_6O_{18} \text{ ------> } M_6P_6O_{18} + 6LiF$$

5.2.5. Ion-exchange resins

Production of cyclohexaphosphoric acid from $Na_6PO_{18}.6H_2O$ or $Li_6PO_{18}.5\text{-}6H_2O$ is a common process for the synthesis of the alkali or other cyclohexaphosphates.

5.2.6. Flux methods

These methods have been successful for the production of some anhydrous cyclohexaphosphates ($Cu_2Li_2P_6O_{18}$, $Cr_2P_6O_{18}$...). As it was already said, the optimization of a flux method for the production of a given compound is a long and tedious work.

5.2.7. Classical methods

Most of the monovalent-divalent cation cyclohexaphosphates have been prepared as polycrystalline samples by classical reactions as :

$$(NH_4)_6P_6O_{18} + 2Cd(NO_3)_2 \text{ ------> } Cd_2(NH_4)_2P_6O_{18} + 4NH_4NO_3$$

Nevertheless, the production of single crystals is in many cases tributary of rather non-orthodox processes we cannot report here in detail.

5.3. Some Atomic Arrangements of Cyclohexaphosphates

We describe now in some details the atomic arrangements of six cyclohexaphosphates we consider as typical of this class of compounds.

5.3.1. Ammonium cyclohexaphosphate monohydrate

$(NH_4)_6P_6O_{18}.H_2O$ [22] is trigonal, $R\overline{3}$, with the following crystal data :

$a = 15.445(10)$, $c = 7.553(7)$ Å, $Z = 3$ (hexagonal setting)
$a = 9.266$ Å, $\alpha = 112.91°$, $Z = 1$ (rhomboedral setting)

The atomic arrangement, given in Fig. 5.1 in projection along the **c** axis, can be simply described by examining what occurs along a threefold axis.

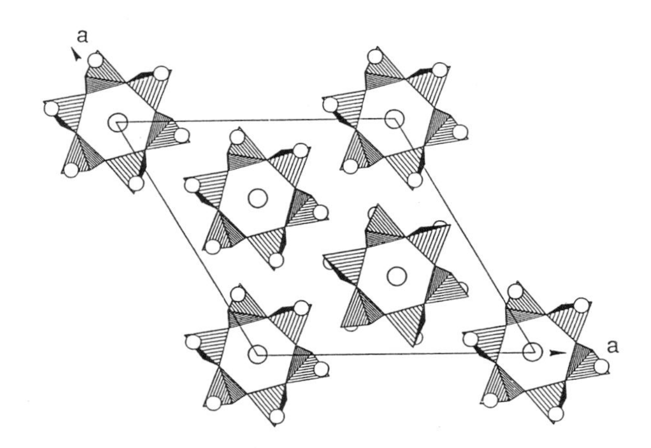

Figure 5.1 Projection, along the **c** axis, of the atomic arrangement of $(NH_4)_6P_6O_{18}.H_2O$. The PO_4 groups are schematized by tetrahedra, the larger circles represent the water molecules and the smaller ones the ammonium groups.

The P_6O_{18} ring anions located around this $\overline{3}$ axis alternate with pseudo-hexagonal rings built up by six ammonium groups. These $(NH_4)_6$ rings are centred by a water molecule located on the threefold axis. This water molecule is not involved in the associated cation coordination. The P_6O_{18} group and the $(NH_4)_6$ ring are separated by a distance of c/2. Located close by z = 0, the phosphoric ring anion has the $\overline{3}$ internal symmetry and is so built by only one independent phosphorus atom.

5.3.2. Silver-ammonium cyclohexaphosphate monohydrate

$Ag_3(NH_4)_3P_6O_{18}.H_2O$ [26] is trigonal, $R\overline{3}$, with the following crystal data :

a = 15.172(5), c = 13.994(5) Å (hexagonal description)
a = 9.924 Å, α = 99.71° (rhomboedral setting)

The hexagonal unit cell contains six formula units. This unit cell is closely related to those of two very similar phosphates, $Ag_6P_6O_{18}.H_2O$ and $(NH_4)_6P_6O_{18}.H_2O$. The observed a value is approximately half-way between the values measured for the ammonium and silver salts and the c value is close by the sum of those measured for these two phosphates. These observations suggest that the atomic arrangement of this salt must be closely related to that determined for $Ag_6P_6O_{18}.H_2O$ and

$(NH_4)_6P_6O_{18}.H_2O$, with in addition an order between the associated cations to explain the doubling of the c axis. The determination of the atomic arrangement confirmed these assumptions. Here, two independent

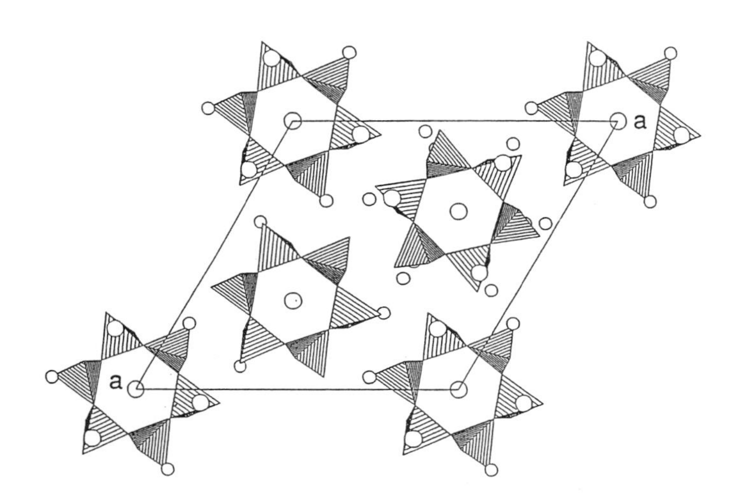

Figure 5.2 Projection of $Ag_3(NH_4)_3P_6O_{18}.H_2O$ along the c axis. The small empty circles represent silver atoms, the larger ones the ammonium groups. The ring anions are centred by water molecules.

P_6O_{18} ring-anions coexist in the structure, both located around the $\overline{3}$ axis, the first one at $z = 0$, the second one at $z = 1/2$, so both of them have a $\overline{3}$ internal symmetry. Half way between these groups, in $z = 1/4$ and $3/4$, are located pseudo-hexagonal rings of associated cations made by three silver atoms and three ammonium groups. These rings of associated cations built around the $\overline{3}$ axis are themselves centred by the water molecules located on this axis. So, the two main differences between the $Ag_6P_6O_{18}.H_2O$ and $(NH_4)_6P_6O_{18}.H_2O$ atomic arrangements and the present one are : the existence of two cristallographic independent P_6O_{18} groups and the order established in the hexagonal ring of associated cations. Fig. 5.2 gives a projection of this structure along the c axis.

5.3.3. Lithium-sodium cyclohexaphosphate duodecahydrate

$Li_3Na_3P_6O_{18}.12H_2O$ [25] is trigonal, $R\overline{3}c$, with $Z = 4$ in the hexagonal description and the following unit-cell dimensions :

$a = 10.474(8)$, $c = 41.68(5)$ Å (hexagonal setting)
$a = 15.152$ Å, $\alpha = 40.44°$ (rhomboedral setting)

In spite of an apparent complexity due to the large size of the unit cell, the

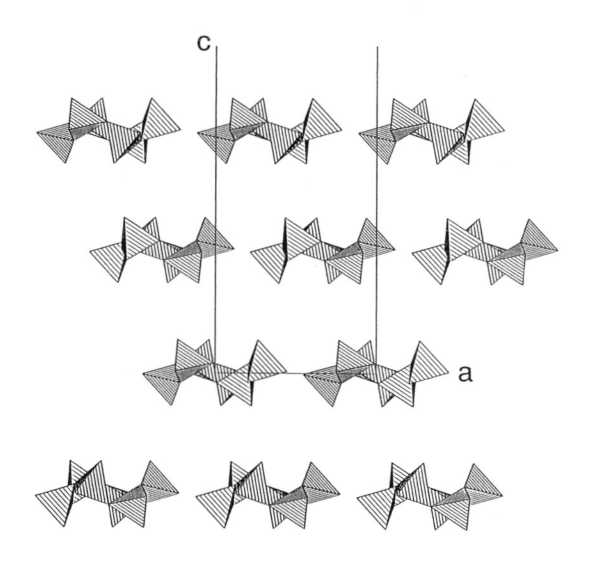

Figure 5.3 Projection, along the **b** axis, of a group of planes of P_6O_{18} ring anions in $Li_3Na_3P_6O_{18}.12H_2O$.

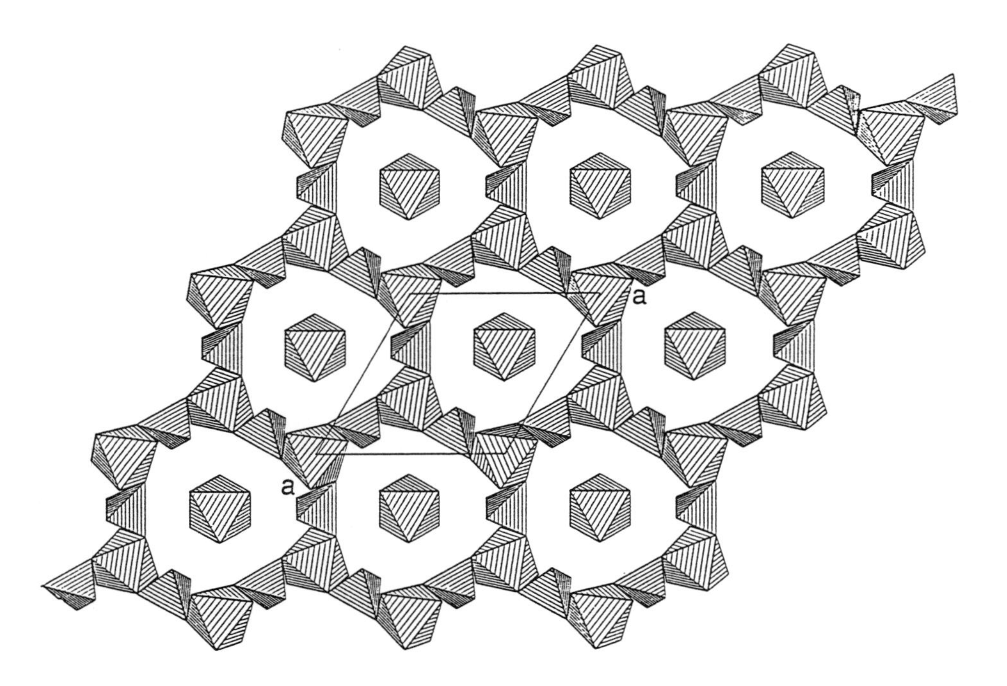

Figure 5.4 Projection, along the **c** axis, of the two-dimensional network of LiO_4 and NaO_6 octahedra situated in z = c/12. Isolated octahedra are the $Na(H_2O)_6$ groups.

atomic arrangement of this salt can be easily described as a succession of planes perpendicular to the threefold axis. A first family of planes contains the P_6O_{18} ring anions. These planes are located approximately in z = n×c/6 and so separated by a distance of about 7 Å. The second family of planes is built up by a two-dimensional network of corner sharing LiO_4 tetrahedra and NaO_6 octahedra creating around the the $\bar{3}$ axes large voids occupied by the isolated $Na(H_2O)_6$ octahedra. These planes located approximately in z = (2n+1)c/12 alternate with those containing the phosphoric anions. As in the precedent structure, there are two independent P_6O_{18} ring anions with a $\bar{3}$ internal symmetry. Fig. 5.3 reports, in projection along the b axis, a group of P_6O_{18} ring anion planes, while Fig. 5.4 gives the representation of the two-dimensional network of LiO_4 and NaO_6 octahedra situated in z = c/12.

5.3.4. Cadmium cyclohexaphosphate hexahydrate

$Cd_3P_6O_{18}.6H_2O$ [33] is trigonal with the following unit-cell dimensions :

a = 15.056(8), c = 16.080(8) Å (hexagonal setting)
a = 10.212 Å, α = 94.98° (rhomboedral description)

The space group is $R\bar{3}$ and the hexagonal unit cell countains six formula units. As depicted by Fig. 5.5, the P_6O_{18} groups are located around the

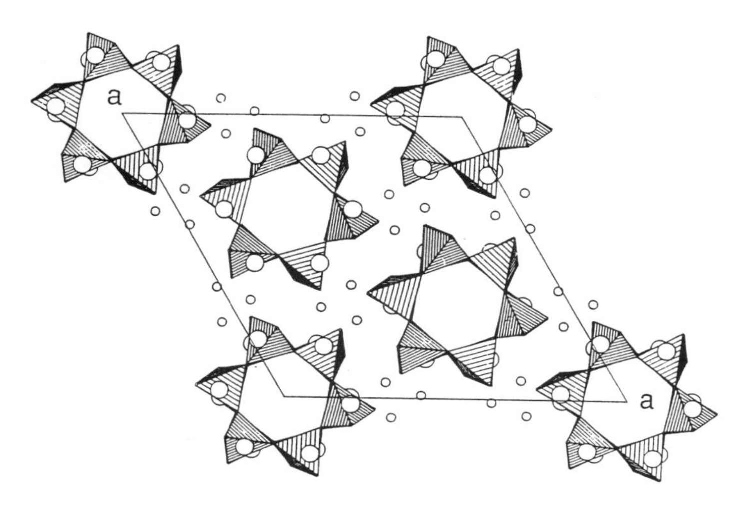

Figure 5.5 Projection of the atomic arrangement of $Cd_3P_6O_{18}.6H_2O$ along the **c** axis. Small circles are cadmium atoms and larger circles represent the water molecules. Hydrogen atoms have been omitted.

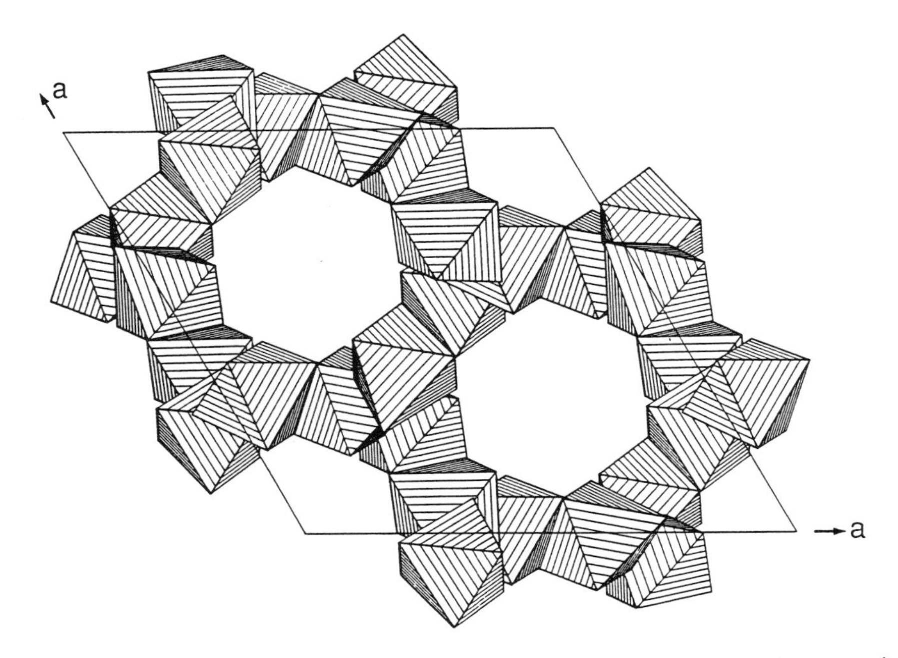

Figure 5.6 Projection along the **c** axis of the three-dimensional network of CdO_6 octahedra. In the large hexagonal channels are located the phosphoric ring anions.

threefold axes. They are built by two independent PO_4 tetrahedra and have a threefold internal symmetry. Along a threefold axis, these rings are separated by a distance of c/2 (8.04 Å) and are almost superimposed in projection. Half way between the phosphoric groups and always around the threefold axes, rings of six water molecules are located. Inside such a ring, each water molecule is connected to its neighbour by an hydrogen-bond, the second hydrogen of the water molecule establishing an H-bridge with an external oxygen atom of the phosphoric group. The cadmium atom coordination is made by four oxygen atoms and two water molecules. These cations being located close by the helical axes, the CdO_6 octahedra form edge-sharing spirals around these axes. These spirals are themselves interconnected as to build the three-dimensional network of octahedra reported in Fig. 5.6, in projection along the c direction. This network creates large hexagonal channels, parallel to the c axis, in which are located the P_6O_{18} anions.

5.3.5. Copper-ammonium cyclohexaphosphate octahydrate

$Cu_2(NH_4)_2P_6O_{18}.8H_2O$ [40] is triclinic, $P\bar{1}$, and crystallizes with the following monomolecular unit cell :

$a = 7.413(3)$, $b = 9.234(4)$, $c = 9.634(4)$ Å
$\alpha = 116.23(5)$, $\beta = 107.98(5)$, $\gamma = 83.10(5)°$

The P_6O_{18} ring anion is centrosymmetric, located around the inversion centre in 0,0,1/2. The two copper atoms are also on inversion centres, the first one in 0,0,0, the second in 1/2,1/2,0. The atomic arrangement,

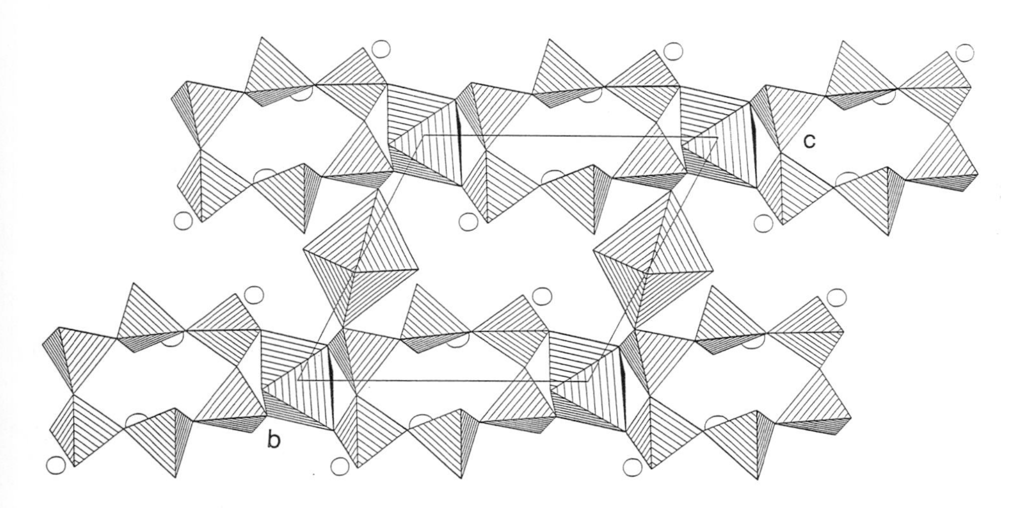

Figure 5.7 Projection of the atomic arrangement of $Cu_2(NH_4)_2P_6O_{18}.8H_2O$ along the **a** axis. Empty circles represent the ammonium groups.

represented in projection along the a axis in Fig. 5.7, can be easily described as a succession of ribbons built up by the P_6O_{18} phosphoric rings and one kind of CuO_6 octahedra spreading parallel to the c direction. These ribbons are themselves interconnected along the b direction by the second kind of CuO_6 octahedra as to form a layer parallel to the (b, c) plane. The connections, so established by the CuO_6 octahedra, are not identical ; one of the CuO_6 octahedron shares four of its oxygen atoms with the two adjacent P_6O_{18} groups, while the other one shares only two. The three-dimensional cohesion between these layers is established by the ammonium polyhedra and the hydrogen-bond network.

5.3.6. Potassium cyclohexaphosphate-telluric acid trihydrate adduct $K_6P_6O_{18}.2Te(OH)_6.3H_2O$ [58] is trigonal with the following unit cell :

$a = 13.084(5)$, $c = 34.80(2)$ Å (hexagonal setting)
$a = 13.843$ Å, $\alpha = 56.40°$ (rhomboedral setting)

The space group is $R\bar{3}c$ and $Z = 6$ for the hexagonal cell unit. The atomic arrangement can be described by alternance of two types of planes perpendicular to the threefold axis and separated by a distance of about c/12. Planes of the first type contain the P_6O_{18} groups, $Te(OH)_6$ octahedra and one kind of potassium atoms. Fig. 5.8 is the representation of such a plane in projection along the c axis. Planes of the second type contain the

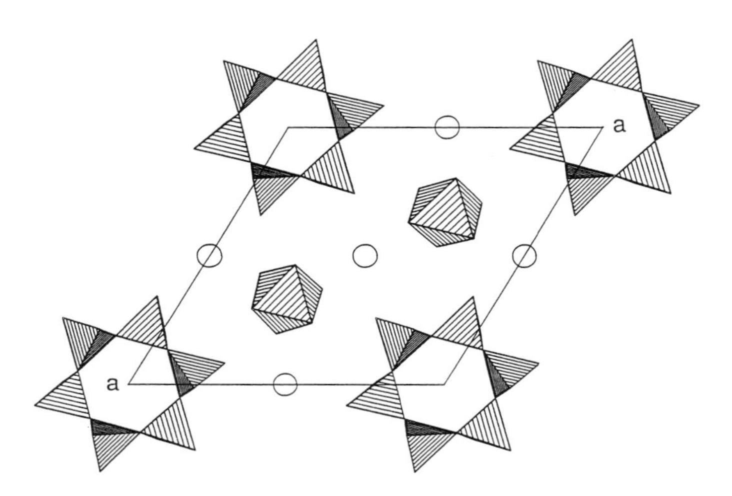

Figure 5.8 One plane of the first type in projection along the **c** axis. Empty circles are potassium atoms and hatched octahedra $Te(OH)_6$ groups.

water molecules and the second category of potassium atoms. Fig. 5.9 gives the respective locations of water molecules and potassium atoms in such a plane. The cohesion between these planes is established through the potassium polyhedra and the hydrogen bonds. The P_6O_{18} groups have a $\overline{3}$ internal symmetry.

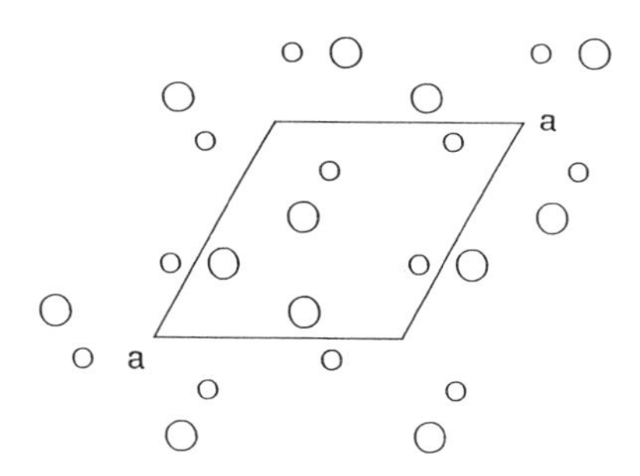

Figure 5.9 Plane of the second type in $K_6P_6O_{18}.2Te(OH)_6.3H_2O$. The smaller empty circles are potassium atoms. The larger ones represent the water molecules.

5.4. The P_6O_{18} Ring Anions

Up to now more than thirty-five accurate determinations of cyclo-hexaphosphate atomic arrangements have been performed, authorizing the comparison of 41 different P_6O_{18} ring anions (several crystallographic independent P_6O_{18} groups can coexist in some atomic arrangements).

Before examining various numerical aspects of bond distances and angles in these rings it seems interesting to give a survey of their re-partition among the various possible internal symmetries. Among the 41 known examples, two have no internal symmetry, one has a m symmetry, two a 2 symmetry, one a 2/m symmetry, three a 3 symmetry, eight a $\overline{3}$ and twenty-four a $\overline{1}$.

In spite of the small number of representatives the examination of bond distances and angles in these rings seems valuable. As for the other types of cyclophosphates we limit this examination to what we consider as

Table 5.8
Main geometrical data for centrosymmetrical P_6O_{18} ring anions (first part).

Formula [Ref.]				Formula [Ref.]			
	P-P (Å)	P-O-P (°)	P-P-P (°)		P-P (Å)	P-O-P (°)	P-P-P (°)
$Li_6P_6O_{18}.5H_2O$ [13]				$Li_6P_6O_{18}.(8+n)H_2O$ [14]			
	2.927	133.0	117.5		2.958	137.2	126.9
	2.927	131.7	106.1		2.935	135.2	111.2
	2.915	133.1	122.6		2.930	134.3	121.8
$Cs_2(UO_2)_2P_6O_{18}$ [9]				$Cs_6P_6O_{18}.6H_2O$ [23]			
	2.866	131.7	107.9		2.960	133.5	93.2
	2.918	129.6	99.1		2.918	127.0	111.5
	2.857	129.4	105.6		2.880	129.9	142.5
$Ag_4Li_2P_6O_{18}.2H_2O$ [27]				$Na_4Rb_2P_6O_{18}.6H_2O$ [28]			
	2.902	145.9	94.9		2.906	130.1	141.7
	3.029	131.1	140.6		2.975	129.1	95.3
	2.940	135.4	98.5		2.885	135.2	109.4
$Na_2(NH_4)_4P_6O_{18}.2H_2O$ [28]				$Cu_3P_6O_{18}.14H_2O$ [32]			
	2.882	127.5	132.6		2.920	132.4	139.9
	2.939	133.1	128.4		2.904	130.3	96.3
	2.927	131.7	98.7		2.883	128.6	114.1
$Mn_2Li_2P_6O_{18}.10H_2O$ [37]				$Ca_2Li_2P_6O_{18}.8H_2O$ [38]			
	2.964	136.5	111.5		2.867	125.9	133.3
	2.960	135.8	90.6		2.906	130.3	97.1
	2.938	132.8	114.1		2.877	127.0	129.3
$Cd_2Na_2P_6O_{18}.14H_2O$ [39]				$Cu_2(NH_4)_2P_6O_{18}.8H_2O$ [40]			
	2.906	132.4	94.2		2.945	134.6	138.9
	2.937	130.5	142.8		2.965	136.4	97.4
	2.896	129.4	110.1		2.922	131.4	95.5

Table 5.9

Main geometrical features for centrosymmetrical P_6O_{18} ring anions (second part).

Formula [Ref.]	P-P (Å)	P-O-P (°)	P-P-P (°)
$Zn_2(NH_4)_2P_6O_{18}.8H_2O$ [42]			
	2.932	133.6	117.8
	2.933	127.9	94.2
	2.874	133.7	135.1
$Cd_2(NH_4)_2P_6O_{18}.9H_2O$ [42]			
	2.977	137.5	109.1
	2.885	128.1	92.2
	2.916	130.8	141.6
$(NH_3OH)_6P_6O_{18}.4H_2O$ [53]			
	2.873	127.7	138.9
	2.932	123.4	112.5
	2.935	133.4	93.5
$Cu(EDA)_2P_6O_{18}.6H_2O$ [54]			
	2.885	127.5	121.1
	2.921	132.8	114.4
	2.919	131.8	106.5
$Te(OH)_6.(NH_4)_6P_6O_{18}.2H_2O$ [57]			
	2.876	127.3	114.3
	2.960	135.3	115.2
	2.933	132.4	96.1
$Mn_2(NH_4)_4(P_6O_{18})(C_2O_4).5H_2O$ [60]			
	2.928	132.0	93.1
	2.942	129.2	103.6
	2.891	134.3	116.1
$Cr_2P_6O_{18}$ [8]			
	2.909	131.9	112.7
	2.970	139.2	121.5
	2.911	133.0	125.5
$Nd_2P_6O_{18}.12H_2O$ [51]			
	2.905	131.3	117.9
	2.948	126.3	127.6
	2.867	132.3	85.9
$(EDA)_3P_6O_{18}.2H_2O$ [54]			
	2.980	126.9	110.8
	2.878	137.8	91.8
	2.930	132.3	111.9
$2Te(OH)_6.(EDA)_3P_6O_{18}.2H_2O$ [59]			
	2.929	132.1	101.1
	2.935	131.4	139.5
	2.973	137.2	99.5
$3Te(OH)_6.Rb_6P_6O_{18}.4H_2O$ [58]			
	2.940	133.3	96.7
	2.911	127.3	144.9
	2.884	131.1	102.2
$(N_2H_5)_2(N_2H_6)_2P_6O_{18}$ [56]			
	2.857	124.7	108.5
	2.929	133.4	110.3
	2.934	133.1	94.1

EDA = $[NH_3\text{-}(CH_2)_2\text{-}NH_3]^{+2}$ (ethylenediammonium)

Table 5.10

Main geometrical features for P_6O_{18} ring anions with no symmetry.

Formula [Ref.]	P-P (Å)	P-O-P (°)	P-P-P (°)
$Cu_2Li_2P_6O_{18}$ [7]			
	2.892	131.3	113.5
	2.931	135.0	134.3
	2.992	141.4	111.1
	2.926	134.4	113.9
	2.950	136.3	133.6
	2.989	141.1	112.4
$(C_2H_5NH_3)_6P_6O_{18}.4H_2O$ [56]			
	2.925	131.2	104.5
	2.963	129.3	105.5
	2.902	134.0	115.7
	2.954	131.4	99.6
	2.917	136.0	111.1
	2.974	135.7	108.5

Table 5.11
Main geometrical features for P_6O_{18} ring anions with higher symmetries.

Formula [Ref.]	P-P (Å)	P-O-P (°)	P-P-P (°)	Formula [Ref.]	P-P (Å)	P-O-P (°)	P-P-P (°)
Rings with a 2 symmetry							
$Ca_2(NH_4)_2P_6O_{18}.6H_2O$ [41]				$Ce_2P_6O_{18}.10H_2O$ [50]			
	2.900	130.3	116.1		2.906	131.1	87.5
	2.955	134.3	112.9		2.922	133.8	114.4
	2.945	134.3	87.8		2.944	134.4	115.6
Rings with a 3 symmetry							
$K_6P_6O_{18}$ [18]				$Cd_3P_6O_{18}.6H_2O$ [33]			
	2.957	130.0	105.8		2.934	133.8	116.4
	2.917	134.9	103.4		2.933	133.7	115.8
$Cr_2P_6O_{18}.21H_2O$ [48]							
	2.926	133.6	104.4				
	2.929	135.2	109.5				
Rings with a $\overline{3}$ symmetry							
$(NH_4)_6P_6O_{18}.H_2O$ [22]				$Ag_6P_6O_{18}.H_2O$ [11]			
	2.954	134.3	108.5		2.909	130.3	112.2
$Li_3K_3P_6O_{18}.12H_2O$, two independent rings in the unit cell [25]							
	2.921	130.6	111.9		2.937	133.9	113.9
$Ag_3(NH_4)_3P_6O_{18}.H_2O$, two independent rings in the unit cell [26]							
	2.916	131.0	110.4		2.907	131.0	110.3
$Li_3Na_3P_6O_{18}.H_2O$ [25]				$2Te(OH)_6.K_6P_6O_{18}.3H_2O$ [58]			
	2.931	133.2	106.5		2.934	134.8	107.8
Rings with a m symmetry				*Rings with a 2/m symmetry*			
$K_6P_6O_{18}.3H_2O$ [13]				$Na_6P_6O_{18}.6H_2O$ [4]			
	3.004	141.9	106.1		2.870	131.3	131.7
	2.965	136.2	106.5		2.935	124.9	96.6
	2.865	125.6	107.1				
	2.962	131.1					

the main framework of a ring : the P-P distances and the P-O-P and P-P-P angles. These numerical values are reported in Tables 5.8-5.11.

The observed *P-P distances* are quite comparable with what has been previously measured in all condensed phosphoric anions. Here, the general average is P-P = 2.927 Å, with a small dispersion, the extreme values being 2.857 and 3.029 Å. The same observations can be formulated for the *P-O-P angles* with an average P-O-P = 132.4 and extrema of 141.9 and 123.4°.

Things are very different when one starts the examination of the *P-P-P angles*, mainly in comparison with the other types of well-represented phosphoric rings as cyclotri- and cyclotetraphosphates. For these two classes of cyclophosphates P-P-P angles never depart significantly from their ideal values, $60 \pm 2°$ for cyclotriphosphates and $90 \pm 4°$ for cyclotetraphosphates. For the P_6O_{18} ring anions, if we examine first the most common ones ($\bar{1}$ internal symmetry), we observe very large deviations from the ideal value (120°). The average angle is 111.9° with extrema of 85.9 and 142.8°. For higher symmetry rings and so with more geometrical stresses one could expect a more regular distribution of the P-P-P angles, but here also the dispersion is large since the extreme values are 87.5 and 131.7°.

Most of the structural data concerning cyclohexaphosphates are recent ones, so for the numerical data reported in Tables 5.8 to 5.11 the estimated standard deviations can be evaluated as better than 0.001 Å for the P-P distances, between 0.01 and 0.05° for the P-P-P angles and always lower than 0.1° for the P-O-P angles.

REFERENCES

1 E. Thilo and U. Schülke, *Angew. Chem., Intern. Edit.*, 2 (1963) 742.

2 E. Thilo and U. Schülke, *Angew. Chem.*, 75 (1963) 1175.

3 K.H. Jost, *Acta Cryst.*, 19 (1965) 555.

4 A. McAdam, K.H. Jost, B. Beagley, *Acta Cryst.*, B28 (1972) 2740.

5 E.J. Griffith and R.L. Buxton, *Inorg. Chem.*, 4 (1965) 549.

6 M. Laügt, *C. R. Acad. Sci.*, 269C (1969) 1122.

7 M. Laügt and A. Durif, *Acta Cryst.*, B30 (1974) 2118.

8 M. Bagieu-Beucher and J.C. Guitel, *Acta Cryst.*, B33 (1977) 2529.

9 S.A. Linde, Yu.E. Gorbunova, A.V. Lavrov and V.G. Kuznetsov, *Dokl. Akad. Nauk SSSR*, 241 (1978) 1083.

10 U. Schülke and R. Kayser, *Z. anorg. allg. Chem.*, 531 (1985) 167.

11 M.T. Averbuch-Pouchot, *Z. Kristallogr.*, 189 (1990) 17.

12 A. Boullé, *C. R. Acad. Sci.*, 206 (1938) 517.

13 W.K. Trunov, L.N. Kholodkovskaya, L.A. Borodina and N.N. Chudinova, *Kristallografyia*, 34 (1989) 748.

14 M. Bagieu-Beucher and M. Rzaigui, in preparation.

15 L.A. Borodina and N.N. Chudinova, *Izv. Akad. Nauk SSSR, Neorg. Mat.*, 24 (1988) 345.

16 E. Thilo and U. Schülke, *Z. anorg. allg. Chem.*, 341 (1965) 293.

17 N.N. Chudinova, L.A. Borodina and U. Schülke, *Izv. Akad. Nauk SSSR, Neorg. Mat.*, 25 (1989) 303.

18 M.T. Averbuch-Pouchot, *Acta Cryst.*, C45 (1989) 1273.

19 L.N. Kholodkovskaya, L.A. Borodina, W.K. Trunov and N.N. Chudinova, *Izv. Akad. Nauk SSSR, Neorg. Mat.*, 25 (1989) 466.

20 L.N. Kholodkovskaya, L.A. Borodina, W.K. Trunov and N.N. Chudinova, *Izv. Akad. Nauk SSSR, Neorg. Mat,* 25 (1989) 454.

21 N.N. Chudinova, L.A. Borodina, U. Schülke and K.H. Jost, *Izv. Akad. Nauk SSSR, Neorg. Mat.*, 25 (1989) 459.

22 M.T. Averbuch-Pouchot, *Acta Cryst.*, C45 (1989) 539.

23 M.T. Averbuch-Pouchot and A. Durif, *C. R. Acad. Sci.*, 308-II (1989) 1699.

24 L.N. Kholodkovskaya, L.A. Borodina, W.K. Trunov and N.N. Chudinova, *Izv. Akad. Nauk SSSR, Neorg. Mat.*, 25 (1989) 470.

25 M.T. Averbuch-Pouchot, *Z. anorg. allg. Chem.*, 574 (1989) 225.

26 M.T. Averbuch-Pouchot, *Acta Cryst.*, C47 (1991) 930.

27 M.T. Averbuch-Pouchot and A. Durif, Acta Cryst., C47 (1991) 1150.

28 M.T. Averbuch-Pouchot and A. Durif, Acta Cryst., C47 (1991) 932.

29 E.V. Lazarevski, L.V. Kubasova, N.N. Chudinova and I.V. Tananaev, *Izv. Akad. Nauk SSSR, Neorg. Mat.*, 16 (1980) 120.

30 E.V. Lazarevski, L.V. Kubasova, N.N. Chudinova and I.V. Tananaev, *Izv. Akad. Nauk SSSR, Neorg. Mat.*, 18 (1982) 1550.

31 E.V. Lazarevski, L.V. Kubasova, N.N. Chudinova and I.V. Tananaev, Izv. *Akad. Nauk SSSR, Neorg. Mat.*, 18 (1982) 1544.

32 M.T. Averbuch-Pouchot, *Acta Cryst.*, C45 (1989) 1275.

33 M.T. Averbuch-Pouchot, *Z. anorg. allg. Chem.*, 570 (1989) 138.

34 M.T. Averbuch-Pouchot and A. Durif, *C. R. Acad. Sci.*, 309-II (1989) 535.

35 E.V. Lazarevski, L.V. Kubasova, N.N. Chudinova and I.V. Tananaev, *Izv. Akad. Nauk SSSR, Neorg. Mat.*, 17 (1981) 486.

36 M. Laügt, A. Durif and C. Martin, *J. Appl. Cryst.*, 7 (1974) 448.

37 M.T. Averbuch-Pouchot, *Acta Cryst.*, C45 (1989) 1856.

38 M.T. Averbuch-Pouchot and A. Durif, *Acta Cryst.*, C46 (1990) 968.

39 M.T. Averbuch-Pouchot, *Acta Cryst.*, C46 (1990) 10.

40 M.T. Averbuch-Pouchot and A. Durif, *Acta Cryst.*, C 47 (1991) 1148.

41 M.T. Averbuch-Pouchot, *Acta Cryst.*, C46 (1990) 2005.

42 M.T. Averbuch-Pouchot, I. Tordjman and A. Durif, *Acta Cryst.*, in press.

43 P. Rémy and A. Boullé, *C. R. Acad. Sci.*, 258 (1964) 927.

44 P. Rémy and A. Boullé, *Bull. Soc. Chim. Fr.*, (1972) 2213.

45 N.N. Chudinova, I. Grunze and L.S. Guzeeva, *Izv. Akad. Nauk SSSR, Neorg. Mat.*, 23 (1987) 616.

46 Z.Ya. Kanene, Z.A. Konstant and V.V. Krasnikov, *Izv. Akad. Nauk SSSR, Neorg. Mat,* 21 (1985) 1552.

47 M. Rzaigui, *Solid State Chem.*, 89 (1990) 340.

48 M. Bagieu-Beucher, M.T. Averbuch-Pouchot and M. Rzaigui, *Acta Cryst.*, C47.(1991) 1364

49 M. Rzaigui, *Solid State Chem.*, in press.

50 M. Bagieu-Beucher and M. Rzaigui, *Acta Cryst.*, C47 (1991) 1789.

51 V.K. Trunov, N.N. Chudinova and L.A. Borodina, *Dokl. Akad. Nauk SSSR,* 300 (1988), 1375.

52 L.S. Guzeeva and I.V. Tananaev, *Izv. Akad. Nauk SSSR, Neorg. Mat.*, 24 (1988) 651.

53 A. Durif and M.T. Averbuch-Pouchot, *Acta Cryst.*, C46 (1990) 2026.

54 A. Durif and M.T. Averbuch-Pouchot, *Acta Cryst.*, C45 (1989) 1884.

55 M.T. Averbuch-Pouchot and A. Durif, Eur. J. Solid State Inorg. Chem.,28 (1991) 9.

56 M.T. Averbuch-Pouchot and A. Durif, *Acta Cryst.*, C46 (1990) 2028.

57 M.T. Averbuch-Pouchot and A. Durif, *Acta Cryst.*, C46 (1990) 179.

58 M.T. Averbuch-Pouchot and A. Durif, *Acta Cryst.*, C47 (1991) 1576.

59 M.T. Averbuch-Pouchot and A. Durif, *Acta Cryst.*, C46 (1990) 2236.

60 M.T. Averbuch-Pouchot and A. Durif, *Acta Cryst.*, C46 (1990) 965.

6. CYCLOOCTAPHOSPHATES

As it occurs for other large phosphoric rings, $[P_8O_{24}]^{8-}$ anions have been clearly characterized during paper-chromatography experiments as early as 1956 by Van Wazer and Karl-Kroupa [1]. However, one must wait till 1968 for the first characterization, by Schülke [2, 3], of a crystalline cyclooctaphosphate, $Pb_4P_8O_{24}$. The author describes the chemical preparation of this salt and used it for producing the corresponding sodium salt [2, 3]. The Schülke's process will be described in more details in the next Section. By using ion-exchange resin techniques, the other alkali cycloctaphosphates have been prepared by Schülke and Chudinova [4]. In spite of this possibility to produce good starting materials to investigate this class of compounds, it must be noticed that the first structural characterization of a P_8O_{24} ring anion did not occur before 1975, when Laügt and Guitel [5] determined the crystal structure of $Cu_3(NH_4)_2P_8O_{24}$. It must be also noticed that till 1992, all the cyclooctaphosphates well investigated at a structural point of view were characterized during investigations of various phase-equilibrium diagrams and prepared by flux methods at relatively high temperatures. Since 1992, a relatively good number of new cyclooctaphosphates have been clearly characterized. Most of them were prepared by classical chemical methods involving the sodium salt as starting material.

6.1. Present State of the Cyclooctaphosphate Chemistry

During a very careful investigation of the thermal behavior of $Pb_2P_4O_{12}.4H_2O$, Schülke [2, 3] observed that this salt can be converted into anhydrous lead cyclooctaphosphate under well defined conditions. According to the author, the best yields are obtained when the tetrahydrate salt is first heated at 383K for 30 min and then at 623K again during 30 min. This transformation can be schematized by the following two steps :

$$2Pb_2P_4O_{12}.4H_2O \dashrightarrow 2Pb_2H_4(PO_4)(P_3O_{10}) + 4H_2O \dashrightarrow Pb_4P_8O_{24} + 8H_2O$$

In fact, the process is not so simple and must be separated into four distinct steps :

(i) formation of crystalline $Pb_2P_4O_{12}.2H_2O$ by partial dehydration ;

(ii) hydrolysis of the cyclotetraphosphate anion by the residual crystal water leading to a mixture of mono-, di- and triphosphates ;
(iii) condensation of these various anions into 30% of lead polyphosphate, $Pb(PO_3)_2$, and 70% of anhydrous lead cyclotetraphosphate, $Pb_2P_4O_{12}$;
(iv) conversion of $Pb_2P_4O_{12}$ into $Pb_4P_8O_{24}$.
According to Schülke, the steps (iii) and (iv) occur simultaneously.

Then, sodium cyclooctaphosphate hexahydrate can be obtained from the lead salt by action of sodium sulphide :

$$Pb_4P_8O_{24} + 4Na_2S \text{ -----> } Na_8P_8O_{24} + 4PbS$$

During the same study, the existence of $Ag_8P_8O_{24}.AgNO_3.(5\text{-}6)H_2O$, $K_8P_8O_{24}.6H_2O$ and $Ca_4P_8O_{24}.16H_2O$ has been reported by the author.

Schülke and Chudinova [4] described chemical preparation and investigated the solubilities of five new hydrates $M_8P_8O_{24}.6H_2O$ (M = Li, K, Rb, Cs, NH_4) prepared through ion-exchange resins from the sodium salt. According to these authors, the solubility of the sodium salt is relatively small in comparison with that of other alkali cyclooctaphosphates.

Lavrov *et al* [6] prepared and characterized three hydrates : $Ni_4P_8O_{24}.19H_2O$, $Mn_4P_8O_{24}.17H_2O$ and $Cd_4P_8O_{24}.12H_2O$ obtained as crystalline compounds by reaction of $H_8P_8O_{24}$ on the corresponding divalent cation perchlorates.

The chemical preparation of $Sn_2P_8O_{24}.5H_2O$ has been described by Nariai *et al* [7]. The authors investigated the effect of water vapor on the thermal decomposition of this salt.

A series of four isotypic cyclooctaphosphates of general formula $Cu_3M_2P_8O_{24}$ (M = Rb, Cs, Tl, NH_4) was characterized by Laügt [8] during a systematic study of the $Cu_2P_4O_{12}$-MPO_3 (M = Rb, Cs, Tl) phase-equilibrium diagrams and a careful investigation of the P_2O_5-$(NH_4)_2O$-Cu_2O-H_2O system. Two of these compounds, $Cu_3Rb_2P_8O_{24}$ and $Cu_3Tl_2P_8O_{24}$, were not observed in the first versions of the $Cu_2P_4O_{12}$-$RbPO_3$ and $Cu_2P_4O_{12}$-$TlPO_3$ phase-equilibrium diagrams reported by Laügt *et al* [9], but appear as non-congruent melting compounds in the revised versions elaborated by Laügt [10, 11]. Fig 6.1 and 6.2 report the new versions of these two phase-equilibrium diagrams. $Cu_3Rb_2P_8O_{24}$ and $Cu_3Tl_2P_8O_{24}$ decompose at 893 and 909K, respectively. The $Cu_2P_4O_{12}$-$CsPO_3$ phase-equilibrium diagram elaborated by Laügt and Martin [12] shows the existence of $Cu_3Cs_2P_8O_{24}$ as a non-congruent melting compound decomposing at 939K and isotypic with the corresponding

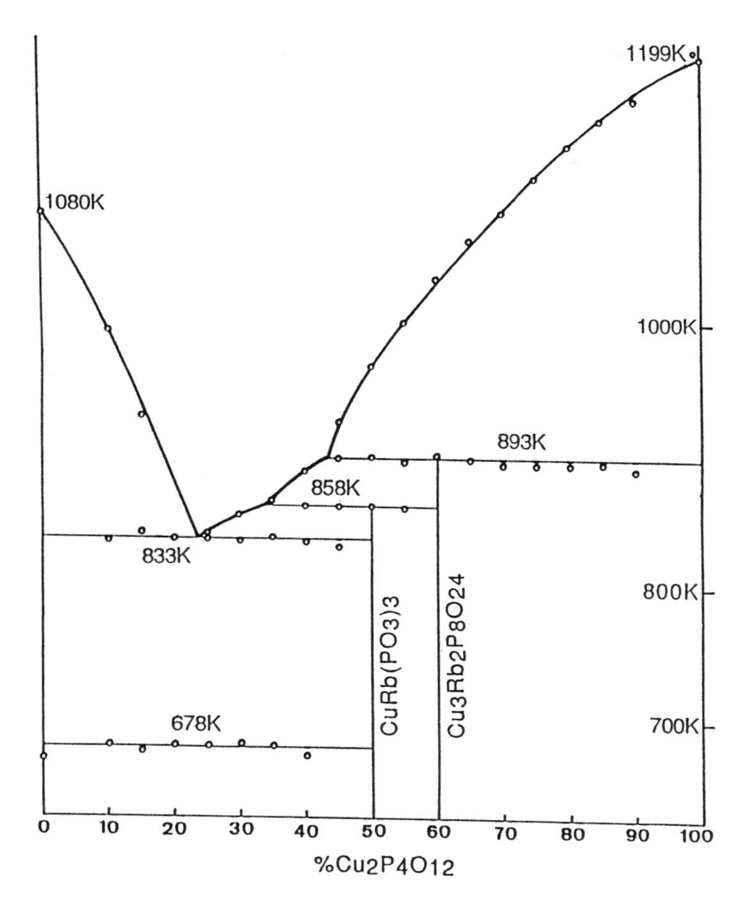

Figure 6.1 The revised $Cu_2P_4O_{12}$-$RbPO_3$ phase-equilibrium diagram.

rubidium and thallium salts. $Cu_3(NH_4)_2P_8O_{24}$ has been prepared by Laügt and Guitel [5] and recognized as an isotype of the previous ones. Its accurate structure determination, performed by these authors, will be described in the next Section.

Nine compounds of general formula $A_2B_2P_8O_{24}$ (A = Al, Ga, Fe, V ; B = K, Rb, NH_4) have been described in chemical literature. In spite of some deficiencies in the crystallographic characterizations of some of them, it can be assumed that they are isotypic. All these compounds have been characterized during investigations of various P_2O_5-A_2O-B_2O_3-H_2O systems :

- $Ga_2K_2P_8O_{24}$ during the study of the P_2O_5-K_2O-Ga_2O_3-H_2O system between 423 and 773K by Chudinova *et al* [13, 14] ;

- $Ga_2Rb_2P_8O_{24}$ by Chudinova *et al* [15] within the same range of temperatures ;

- $Cr_2K_2P_8O_{24}$, $Cr_2(NH_4)_2P_8O_{24}$ and $Cr_2Rb_2P_8O_{24}$ by Grunze and Chudinova [16] during the investigations of the same type of systems between 473 and 673K ;

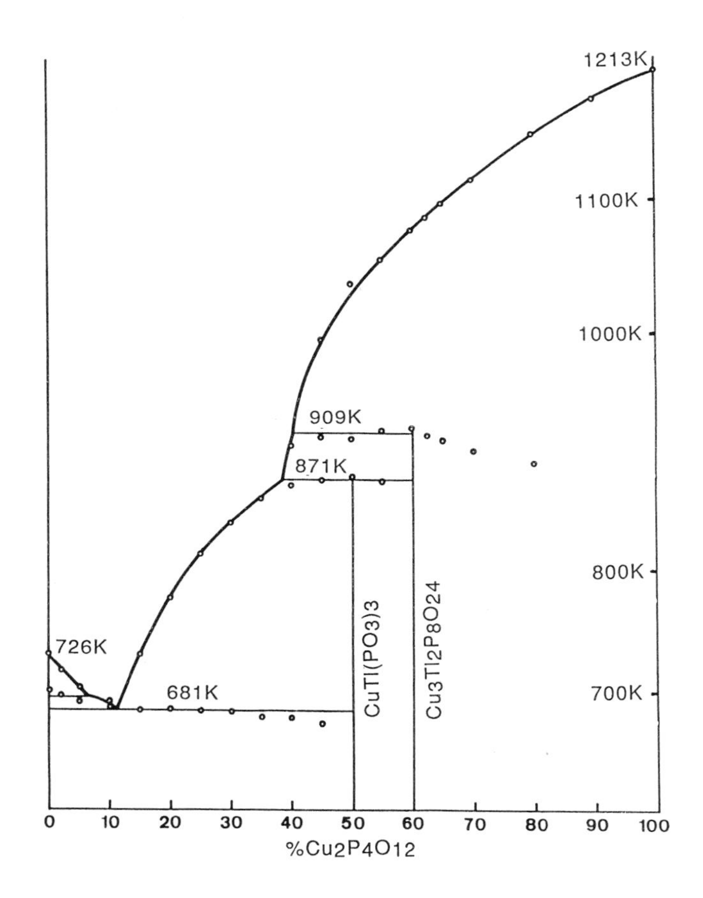

Figure 6.2 The revised $Cu_2P_4O_{12}$-$TlPO_3$ phase-equilibrium diagram.

- $V_2K_2P_8O_{24}$ by Lavrov *et al* [17] during an investigation of several V_2O_3-M_2O-P_2O_5-H_2O systems. This compound is stable under argon up to 933K. Crystals are pale yellow acicular prisms. According to the authors, the process described for the preparation gives very high yields.

- $Al_2K_2P_8O_{24}$ and $Ga_2K_2P_8O_{24}$ obtained by Grunze *et al* [18] by thermal decomposition of $MK(H_2P_2O_7)_2$ at 663K for M = Al and at 648K for M = Ga. In the case of the aluminium compound the reaction is :

$$2AlK(H_2P_2O_7)_2 \dashrightarrow Al_2K_2P_8O_{24} + 4H_2O$$

- $Al_2K_2P_8O_{24}$, $Fe_2K_2P_8O_{24}$ and $Al_2Rb_2P_8O_{24}$ prepared by Grunze *et al* [19, 20].

The atomic arrangement common to these nine compounds was determined by Palkina *et al* [21] with the gallium-potassium salt : $Ga_2K_2P_8O_{24}$. It will be described in the next part of this Chapter devoted to the structural aspect of cyclooctaphosphates.

As we said above, since 1992 a good number of cyclooctaphosphates have been characterized. We report below some comments on these new compounds and on the structural investigations performed on previously known salts.

A detailed structure determination of the sodium salt $Na_8P_8O_{24}.6H_2O$ has been performed by Schülke *et al* [22]. The atomic arrangement cannot explain the relatively low water solubility of this salt.

Single crystals of the ammonium cyclooctaphosphate trihydrate $(NH_4)_8P_8O_{24}.3H_2O$ have been obtained and the crystal structure determination performed by Schülke *et al* [23].

Two adducts with telluric acid have also been described by Averbuch-Pouchot and Durif : $Te(OH)_6.K_8P_8O_{24}.2H_2O$ [24] and the corresponding ammonium derivative $Te(OH)_6.(NH_4)_8P_8O_{24}.2H_2O$ [25]. These two adducts are not isotypic. The two atomic arrangements are reported by the authors.

During attempts to prepare the silver salt, in order to generalize for the cyclooctaphosphates the Boullé's process, Averbuch-Pouchot and Durif [26] characterized a new type of adduct, a cyclooctaphosphate-nitrate $Ag_9NaP_8O_{24}(NO_3)_2.4H_2O$, whose crystal structure is described.

Two organic cation derivatives have been recently characterized by Averbuch-Pouchot and Durif, the guanidinium cyclooctaphosphate dihydrate $[C(NH_2)_3]_8P_8O_{24}.2H_2O$ (27) and $[NH_3(CH_2)_2NH_3]_4P_8O_{24}.6H_2O$ the ethylenediammonium cyclooctaphosphate hexahydrate (28).

6.2. Structural Chemistry of Cyclooctaphosphates

Till very recently, our knowledge in the domain of structural chemistry of cyclooctaphosphates was restricted to two structure types we describe below. The first one is represented by a series of four salts of general formula : $Cu_3M_2P_8O_{24}$ (M = NH_4, Rb, Cs, Tl). Crystal structure has been determined by using the ammonium salt [5].

The second type of structure is represented by nine compounds of general formula $A_2B_2P_8O_{24}$ (A = Ga, Al, V, Fe ; B = K, Rb, NH_4). For this last family of compounds, crystal structure has been performed with the gallium salt [21]. We report in Tables 6.1 and 6.2 the main crystallographic features for these two classes of compounds whose atomic arrangements will be described below and in Table 6.3 those corresponding to the compounds characterized during very recent investigations.

6.2.1. The $Cu_3(NH_4)_2P_8O_{24}$ structure type

The common space group for these four salts is $P\overline{1}$ and the unit cell contains one formula unit. Fig. 6.3 gives a representation of this atomic arrangement in projection along the b axis. The centrosymmetric P_8O_{24} ring anion is located around the inversion centre in (1/2,1/2,1/2). Its main

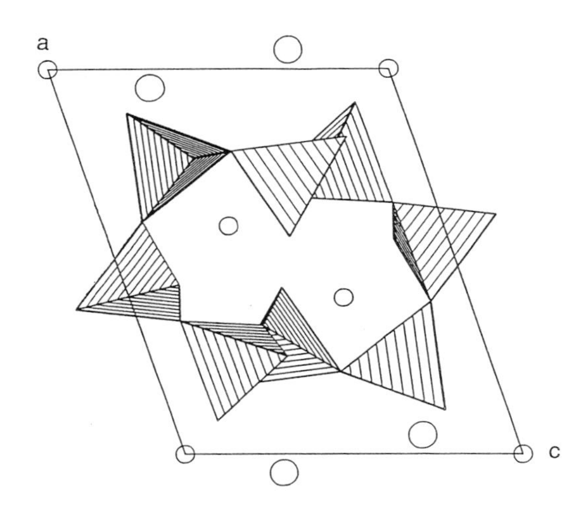

Figure 6.3 Projection of the atomic arrangement of $Cu_3(NH_4)_2P_8O_{24}$ along the b axis. The smaller empty circles represent the copper atoms, the larger ones the ammonium groups.

geometrical features, P-P distances and P-O-P and P-P-P angles are reported in Table 6.4. One of the copper atom occupies a special position on the inversion centre in (0,1/2,0), the second one is located on a general position ; both have a distorted octahedral coordination with Cu-O distances spreading between 1.927 and 2.873 Å.

The ammonium group has a ninefold coordination. The associated cation polyhedra build a three-dimensional network in which chains of corner-sharing CuO_6 octahedra connect planes built by CuO_6 and NO_9 polyhedra.

The determination of this atomic arrangement provided the first structural evidence for the existence of a P_8O_{24} ring anion.

Table 6.1
Unit-cell dimensions of the four salts isotypic with $Cu_3(NH_4)_2P_8O_{24}$.

Formula	a α	b β	c Å) γ (°)	Ref.
$Cu_3(NH_4)_2P_8O_{24}$	9.846(2) 80.98(3)	7.962(2) 110.79(3)	7.261(2) 110.61(3)	5, 10
$Cu_3Rb_2P_8O_{24}$	9.797(4) 80.93(3)	8.035(3) 110.35(3)	7.256(3) 110.48	10, 11
$Cu_3Cs_2P_8O_{24}$	9.913(2) 81.55(2)	7.998(1) 109.20(2)	7.298(1) 109.14	10, 12
$Cu_3Tl_2P_8O_{24}$	9.862(2) 81.62(3)	7.922(2) 110.43(3)	7.273(2) 109.88	10

6.2.2. The $Ga_2K_2P_8O_{24}$ structure type

Among the nine compounds belonging to this structure type only four of them have been investigated at a crystallographic point of view. The common space group is A2/m and the unit cell contains two formula units. Fig. 6.4 reports a projection of the atomic arrangement of $Ga_2K_2P_8O_{24}$ along the a axis. As it can be seen from this figure, the respective locations of the P_8O_{24} rings and of the associated cations is such that

Table 6.2
Unit-cell dimensions of the four salts isotypic with $Ga_2K_2P_8O_{24}$.

Formula	a	b	c (Å)	β (°)	Ref.
$K_2Ga_2P_8O_{24}$	5.138(3)	12.290(5)	16.802(13)	101.04(5)	21
$K_2V_2P_8O_{24}$	5.223(2)	12.277(4)	16.867(4)	101.19(5)	17
$K_2Al_2P_8O_{24}$	5.070	12.266	16.510	97.5	20
$K_2Fe_2P_8O_{24}$	5.351	12.290	17.735	110.97	20

wide infinite empty channels (5.2 x 5.4 Å), parallel to the a axis, pass through the rings. The P_8O_{24} groups are interconnected by slightly distorted GaO_6 octahedra and KO_6 polyhedra. The KO_6 polyhedra are distorted trigonal prisms arranged in space as isolated pairs with a common edge. The P_8O_{24} ring anion has a 2/m internal symmetry.

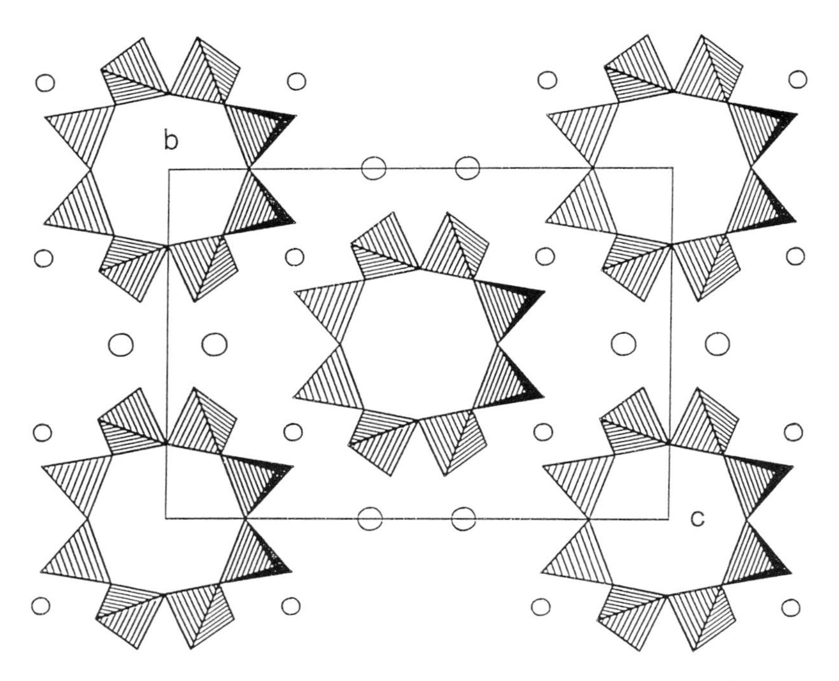

Figure 6.4 Projection of the atomic arrangement of $Ga_2K_2P_8O_{24}$ along the a axis. Small empty circles represent gallium atoms, while the larger ones represent potassium atoms.

6.2.3. Other structure types

Since we started writing this review a significant progress has been made in our knowledge of crystal chemistry of cyclooctaphosphate chemistry.

Seven new structure types have been clearly characterized providing the opportunity to examine eight new geometries of P_8O_{24} ring anions. The main geometrical features of these new examples, P-P distances, P-O-P and P-P-P angles have been included in Table 6.4 and are briefly discussed in the next Section.

6.3. The P_8O_{24} Ring Anions

Till now, nine structural studies of cyclooctaphosphates have been reported. The main geometrical features for the ten rings observed during these studies are reported in Table 6.4. In front of such a restricted number of examples, any kind of discussion similar to what was possible for other smaller rings seems here fruitless. It is nevertheless interesting to notice that among the ten P_8O_{24} ring anions presently known, seven of them are centrosymmetrical, two have a 2/m internal symmetry and one has no internal symmetry. One can also observe that the values obtained has no internal symmetry. One can also observe that the values obtained in these studies for the P-P-P, the P-O-P angles and the P-P distances are within the range commonly observed in the closest type of rings, the P_6O_{18} groups.

The observed *P-P distances* spread between 2.818 and 3.018 Å while the range of values in cyclohexaphosphates is 2.857-3.029 Å.

The *P-P-P angles* spreading between 85.9 and 142.8° in the P_6O_{18} ring anions are in cyclooctaphosphates observed in the range 92.1-146.7°.

The *P-O-P angles* observed in a range of 126.3 to 146.3° are within the limits commonly encountered in all condensed phosphoric anions.

Most of the structural data dealing with cyclooctaphosphates are very recent ones and rather accurate, so that for all the numerical data reported in Table 6.4 the estimated standard deviations can be evaluated as better than 0.001 Å for the P-P distances, between 0.01 and 0.03 ° for the P-P-P angles and in all cases lower than 0.1° for the P-O-P angles.

Table 6.3
Crystal data for the cyclooctaphosphates recently characterized

Formula	a α	b β	c Å) γ (°)	S.G.	Z	Ref.
$Na_8P_8O_{24}.6H_2O$	6.6222) 104.06(5)	10.031(4) 101.21(5)	11.250(4) 90.88(5)	$P\bar{1}$	1	22
$(NH_4)_8P_8O_{24}.3H_2O$	24.27(1)	6.700(3) 112.06(6)	20.59(1)	Cc	4	23
$Ag_9NaP_8O_{24}(NO_3)_2.4H_2O$	17.254(5)	7.543(1)	23.465(5)	Cmcm	4	26
$K_8P_8O_{24}.Te(OH)_6.2H_2O$	11.315(9) 108.72(5)	10.67(1) 100.30(2)	7.547(3) 66.80(5)	$P\bar{1}$	1	24
$(NH_4)_8P_8O_{24}.Te(OH)_6.2H_2O$	15.146(6) 117.15(4)	11.049(6) 109.72(4)	12.189(6) 90.54(4)	$P\bar{1}$	2	25
$(EDA)_4P_8O_{24}.6H_2O$	11.833(2)	21.844(3) 98.49(3)	7.467(2)	$P2_1/n$	2	28
$(Gua)_8P_8O_{24}.2H_2O$	12.621(4)	20.41(1) 110.45(4)	9.365(6)	$P2_1/n$	2	27

EDA = $[NH_3(CH_2)_2NH_3]^{2+}$ (ethylenediammonium)
Gua = $[C(NH_2)_3]^+$ (guanidinium)

Table 6.4
Main geometrical features in the P_8O_{24} ring anions.

Formula [Ref.] P-P (Å)	P-O-P (°)	P-P-P (°)	Formula [Ref.] P-P (Å)	P-O-P (°)	P-P-P (°)
Rings with a 2/m symmetry					
$K_2Ga_2P_8O_{24}$ [20]			$Ag_9NaP_8O_{24}(NO_3)_2.4H_2O$ [26]		
131.4	123.1	2.818	102.5	133.9	2.958
138.0	134.4	2.933	108.7	129.5	2.889
	136.0	2.947		128.8	2.900
Ring with $\overline{1}$ symmetry					
$Na_8P_8O_{24}.6H_2O$ [22]			$Cu_3(NH_4)_2P_8O_{24}$ [5]		
123.9	138.4	3.011	119.9	129.1	2.888
121.0	126.3	2.880	92.1	134.8	2.928
147.8	127.3	2.891	112.2	146.3	2.930
146.7	128.5	2.902	123.3	134.9	3.018
$(NH_4)_8P_8O_{24}.Te(OH)_6.2H_2O$, two independent rings in the unit cell [25]					
106.7	136.5	2.960	110.3	130.0	2.884
103.7	134.9	2.952	118.7	137.0	2.936
114.5	126.3	2.868	105.6	134.6	2.946
119.9	131.0	2.908	106.8	136.4	2.965
$K_8P_8O_{24}.Te(OH)_6.2H_2O$ [24]			$(EDA)_4P_8O_{24}.6H_2O$ [27]		
107.4	132.4	2.946	103.4	133.8	2.896
105.1	132.6	2.939	123.0	129.1	2.912
117.5	132.2	2.925	122.7	130.9	2.884
101.1	133.3	2.953	123.0	127.6	2.942
$(Gua)_4P_8O_{24}.2H_2O$ [28]					
127.6	127.9	2.886			
112.5	133.0	2.924			
100.7	132.2	2.938			
114.5	128.8	2.881			
Ring with no symmetry					
$(NH_4)_8P_8O_{24}.3H_2O$ [23]					
128.5	126.7	2.887	123.1	128.2	2.887
118.2	130.5	2.915	121.7	128.7	2.890
100.6	129.7	2.910	100.4	131.5	2.930
109.9	127.8	2.876	112.1	126.2	2.874

EDA = $[NH_3(CH_2)_2NH_3]^{+2}$ (ethylenediammonium)
Gua = $[C(NH2)3]^+$ (guanidinium)

REFERENCES

1 J.R. Van Wazer and E. Karl-Kroupa, *J. Am. Chem. Soc.*, 78 (1956) 1772.

2 U. Schülke,
Z. anorg. allg. Chem., 360 (1968) 231.

3 U. Schülke,
Angew. Chem. (International Edition), 7 (1968) 71.

4 U. Schülke and N.N. Chudinova,
Izv. Akad. Nauk SSSR, Neorg. Mat., 10 (1974) 1697.

5 M. Laügt and J.C. Guitel,
Z. Kristallogr., 141 (1975) 203.

6 A.V. Lavrov, T.A. Bykanova and N.N. Chudinova,
Izv. Akad. Nauk SSSR, Neorg. Mat., 13 (1977) 334.

7 H. Nariai, I. Motooka and M. Tsuhako,
Bull. Chem. Soc. Jpn., 61 (1988) 2811.

8 M. Laügt,
Thesis, Univ. of Grenoble, France, 1974.

9 M. Laügt, M. Scory and A. Durif,
Mat. Res. Bull., 3 (1968) 963.

10 M. Laügt,
C. R. Acad. Sci., 278C (1974) 1497.

11 M. Laügt,
C. R. Acad. Sci., 278C (1974) 1197.

12 M. Laügt and C. Martin,
Mat. Res. Bull., 7 (1972) 1525.

13 N.N. Chudinoval, I.V. Tananaev and M.A. Avaliani,
Izv. Akad. NaukSSSR, Neorg. Mat., 13 (1977) 2234.

14 N.N. Chudinova, M.A. Avaliani, L.S. Guzeeva and I.V. Tananaev,
Izv. Akad. Nauk SSSR, Neorg. Mat., 14 (1978) 2054.

15 N.N. Chudinova, M.A. Avaliani, L.S. Guzeeva and I.V. Tananaev,
Izv. Akad. Nauk SSSR, Neorg. Mat., 15 (1979) 2176.

16 I. Grunze and N.N. Chudinova,
Izv. Akad. Nauk SSSR, Neorg. Mat.,24 (1988) 988.

17 A.V. Lavrov, M.Ya. Voitenkov and E.G. Tselebrovskaya,
Izv. Akad.Nauk SSSR, Neorg. Mater., 17 (1981) 99.

18 I. Grunze, W. Hilmer, N.N. Chudinova and H. Grunze,
Izv. Akad.Nauk SSSR, Neorg. Mat., 20 (1984) 287.

19 I. Grunze and H. Grunze,
Z. anorg. allg. Chem., 512 (1984) 39.

20 I. Grunze, N.N. Chudinova and K.K. Palkina,
Izv. Akad. Nauk SSSR, Neorg. Mater., 19 (1983) 1943.

21 K.K. Palkina, S.I. Maksimova, V. G. Kusznetsov and N.N. Chudinova,
Dokl. Akad. Nauk SSSR, 245 (1979) 1386.

22 U Schülke, M.T. Averbuch-Pouchot and A. Durif,
J. Solid State Chem., 98, (1992) 213.

23 U Schülke, M.T. Averbuch-Pouchot and A. Durif,
Z. anorg allg Chem., in press.

24 M.T. Averbuch-Pouchot and A. Durif,
Z. Kristallogr., in press.

25 M.T. Averbuch-Pouchot and A. Durif, *Acta Cryst.*, C49 (1993) 361.

26 M.T. Averbuch-Pouchot and A. Durif, *Acta Cryst.*, C48 (1992) 1173.

27 M.T. Averbuch-Pouchot, A. Durif and U Schülke *Eur. J. Solid State Inorg. Chem.*, in press.

28 M.T. Averbuch-Pouchot, A. Durif and U Schülke *Eur. J. Solid State Inorg. Chem.*, 30 (1993) 557.

7. CYCLODECAPHOSPHATES

$P_{10}O_{30}$ anions have been observed more than thirty years ago during thin-layer chromatography experiments, but for a long time the only example of a crystalline cyclodecaphosphate was given by $Ba_2Zn_3P_{10}O_{30}$.

While investigating the $Zn(PO_3)_2$-$Ba(PO_3)_2$ system, Bagieu-Beucher and El-Horr [1] tried to optimize a flux method for the production of single crystals of $Ba_2Zn(PO_3)_6$, a long-chain polyphosphate appearing in this phase-equilibrium diagram. During these experiments, they produced crystals corresponding to the formula $Ba_2Zn_3P_{10}O_{30}$, a compound not observed in the diagram, and suspected this salt to be a cyclodecaphosphate

The crystal structure determination of this salt, performed by Bagieu-Beucher *et al* [2, 3] confirmed this assumption and provided the first structural evidence for the existence of a $P_{10}O_{30}$ ring anion.

Infrared and Raman spectra of polycrystalline $Ba_2Zn_3P_{10}O_{30}$ have been recorded and discussed by Cabello and Baran [4].

We describe now with some details this first example of cyclodecaphosphate. $Ba_2Zn_3P_{10}O_{30}$ is monoclinic, P2/n, with the following bimolecular unit-cell dimensions :

$$a = 21.738(15),\ b = 5.356(5),\ c = 10.748(8)\ \text{Å},\ \beta = 99.65(3)°$$

Such an atomic arrangement can be simply described as a three-dimensional framework of ZnO_6 octahedra and BaO_9 polyhedra organized as to build large channels in which the phosphoric groups and the ZnO_4 tetrahedra are located. Fig. 7.1, which represents a projection of the structure along the b axis, shows the respective locations of the phosphoric ring anions and of associated cations. The phosphoric anions form rows parallel to the b axis. In these rows, the $P_{10}O_{30}$ groups are interconnected by the ZnO_4 tetrahedra. Fig. 7.2 represents such a row in projection along the c axis. BaO_9 polyhedra and ZnO_6 octahedra are interconnected in a three-dimensional way as to build wide channels

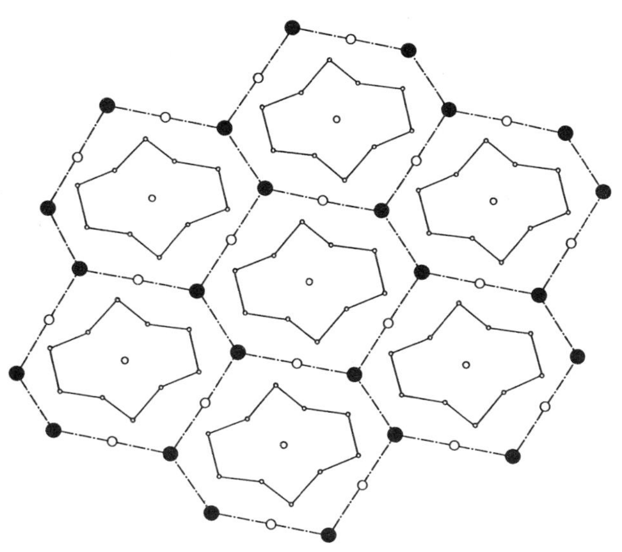

Figure 7.1 Projection along the b axis of the atomic arrangement of $Ba_2Zn_3P_{10}O_{30}$. The black circles are barium atoms, the largest empty circles represent the zinc atoms in octahedral coordination, the intermediate empty circles represent the zinc atoms in tetrahedral coordination and the smallest ones are the phosphorus atoms.

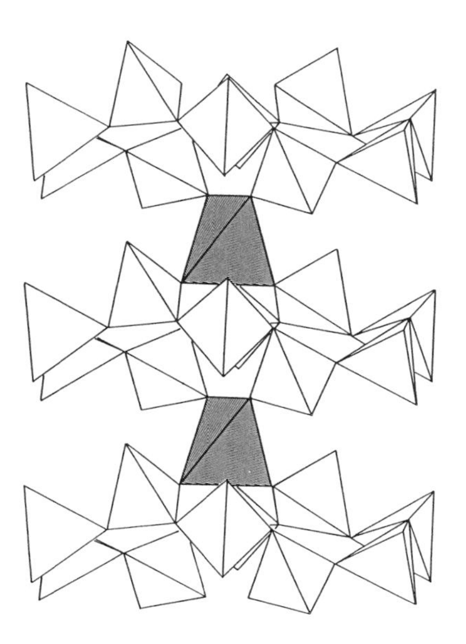

Figure 7.2 Representation of a row of $P_{10}O_{30}$ ring anions interconnected by ZnO_4 tetrahedra (grey tetrahedra).

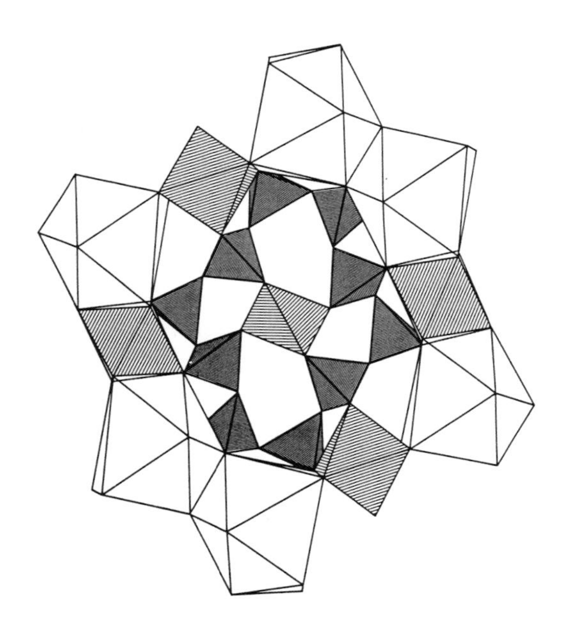

Figure 7.3 Details of the organization of a channel of ZnO_6 and BaO_9 polyhedra around a phosphoric ring seen along the b axis. The central ZnO_4 tetrahedron interconnecting the $P_{10}O_{30}$ groups along the b direction is also figurated.

parallel to the b axis. Inside these channels are located the $P_{10}O_{30}$ - ZnO_4 - $P_{10}O_{30}$... rows. Fig. 7.3 depicts, in projection along the b axis, the organization of such a channel. The phosphoric ring anion has a twofold internal symmetry.

Very recently, Schülke *et al* [5] using $Ba_2Zn_3P_{10}O_{30}$ as starting material reported a process for the production of alkali cyclodecaphosphates. Following this discovery, systematic investigations involving the use of the potassium salt as starting material were undertaken and several new cyclodecaphosphates clearly characterized.

It seems of interest to report first in detail the process described by Schülke [5] for obtaining the potassium salt, since this salt is at the present time the parent compound for the synthesis of all cyclodecaphosphates.

Before giving experimental details, let us say first that this process can be schematized by three steps corresponding to the following reactions :

$$Ba_2Zn_3P_{10}O_{30} + 2K_2SO_4 \dashrightarrow 2BaSO_4 + K_4Zn_3P_{10}O_{30} \text{ (amorphous)}$$

$$K_4Zn_3P_{10}O_{30} + 3Na_2S \text{ ---> } 3ZnS + K_4Na_6P_{10}O_{30} \text{ (solution)}$$

$$K_4Na_6P_{10}O_{30} + K^+\text{cation exchanger ---> } K_{10}P_{10}O_{30} \text{ (solution)}$$

The following experimental conditions are used : 18.9 g of $Ba_2Zn_3P_{10}O_{30}$, 5.2 g of K_2SO_4 and 40 cm^3 of H_2O are intensively grinded in an agate swing mill during 8 hours, then under vigorous stirring a solution of 10.5 g of $Na_2S.9H_2O$ in 400 cm^3 of H_2O is added to the suspension. The ZnS and $BaSO_4$, formed during the first two steps, are then eliminated by filtration and an aqueous solution of $K_4Na_6P_{10}O_{30}$ so obtained. This last salt can be precipitated by addition of 400 cm^3 of methanol.

The pure potassium salt, $K_{10}P_{10}O_{30}.4H_2O$, is then prepared from $K_4Na_6P_{10}O_{30}$.aq. by ion exchange with a strong acidic cationic exchanger in the K^+ form or by precipitation of the insoluble silver salt, $Ag_{10}P_{10}O_{30}$, and reaction of this salt with KCl in aqueous solution. The yield is about 70%.

Crystal structure of the potassium salt $K_{10}P_{10}O_{30}.4H_2O$ was then performed by Schülke *et al* [5] and some new derivatives clearly characterized :

- $Ag_4K_6P_{10}O_{30}.10H_2O$ obtained during various attempts to prepare the silver salt was investigated by Averbuch-Pouchot *et al* [6] ;
- $Mn_4K_2P_{10}O_{30}.18H_2O$ prepared by Schülke and Averbuch-Pouchot [7], who performed its crystal structure determination ;
- till more recently, the first organic derivative, guanidinium cyclodecaphosphate tetrahydrate $[C(NH_2)_3]_{10}P_{10}O_{30}.4H_2O$, was prepared

Table 7.1
Main crystal data for recently described cyclodecaphosphates.

Formula	a α	b β	c (Å) γ (°)	S. G.	Z	Ref.
$K_{10}P_{10}O_{30}.4H_2O$	15.342(5)	11.846(5) 91.27(5)	19.264(5)	C2/c	4	5
$Ag_4K_6P_{10}O_{30}.10H_2O$	14.267(7) 105.38(5)	7.305(1) 101.03(5)	10.319(4) 87.51(5)	$P\bar{1}$	1	6
$Mn_4K_2P_{10}O_{30}.18H_2O$	14.546(10)	15.211(10) 105.12(4)	9.860(6)	$P2_1/a$	2	7
$(Gua)_{10}P_{10}O_{30}.4H_2O$	12.192(8) 91.25(3)	14.083(9) 103.61(3)	9.317(6) 71.22(2)	$P\bar{1}$	1	8

Gua = $[C(NH_2)_3]^+$ (guanidinium)

by Averbuch-Pouchot and Schülke [8], who reported a detailed description of its atomic arrangement.

Table 7.1 reports the main crystal data for these recently characterized compounds, while Table 7.2 gives the geometrical features of the five $P_{10}O_{30}$ ring anions.

Table 7.2
Main geometrical features for $P_{10}O_{30}$ ring anions.

Formula [Ref.]	P-P (Å)	P-O-P (°)	P-P-P (°)
Rings with 2 symmetry			
$Ba_2Zn_3P_{10}O_{30}$ [2]			
	2.915	130.0	112.9
	2.915	133.8	102.1
	3.002	144.3	131.9
	2.838	123.3	99.7
	2.870	128.6	126.6
$K_{10}P_{10}O_{30}.4H_2O$ [5]			
	3.011	137.1	86.6
	2.950	133.1	111.5
	2.911	128.9	133.3
	2.936	131.3	107.1
	2.963	134.9	90.5
	3.006	138.2	
Ring with $\bar{1}$ symmetry			
$Ag_4K_6P_{10}O_{30}.10H_2O$ [6]			
	2.986	131.5	99.8
	2.925	130.1	96.0
	2.911	127.9	136.9
	2.891	134.2	116.3
	2.945	137.1	99.4
$Mn_4K_2P_{10}O_{30}.18H_2O$ [7]			
	2.938	135.9	114.6
	2.928	129.4	138.8
	2.869	133.5	85.5
	2.890	134.3	87.9
	2.944	127.1	98.0
$(Gua)_{10}P_{10}O_{30}.4H_2O$ [8]			
	2.925	130.9	138.7
	2.867	125.6	144.1
	2.958	135.5	138.8
	2.881	127.8	121.4
	2.904	129.2	107.6

Gua = $[C(NH_2)_3]^+$ (guanidinium)

REFERENCES

1 M. Bagieu-Beucher and N. El Horr, *personal communication.*

2 M. Bagieu-Beucher, A. Durif and J.C. Guitel, *J. Solid State Chem.,* 45 (1982) 159.

3 M. Bagieu-Beucher, A. Durif. and J.C. Guitel, *J. Solid State Chem.,* 40 (1981) 248.

4 C.I. Cabello and E. Baran,
J. Spectrochim. Acta, 41A (1985) 1359.
5 U. Schülke, M.T. Averbuch-Pouchot and A. Durif,
Z. anorg. allg., 612 (1992) 107.
6 M.T. Averbuch-Pouchot, A. Durif and U. Schülke,
J. Solid State Chem., 97 (1992) 299.
7 U. Schülke and M.T. Averbuch-Pouchot
Z. anorg. allg. Chem., in press
8 M.T. Averbuch-Pouchot and U. Schülke,
J. Solid State Chem., in press

8. CYCLODUODECAPHOSPHATES

8.1. *Present State of the Cycloduodecaphosphate Chemistry*

Cycloduodecaphosphates are very rare compounds. One of them, $Cs_3V_3P_{12}O_{36}$, was recently characterized by Lavrov *et al* [1] during an investigation of the V_2O_3-M_2O-P_2O_5-H_2O systems. The crystalline compound appears as greenish-yellow pentagonal dodecahedra belonging to the cubic system. This salt is stable under argon up to 1173K. At this temperature, it decomposes into $V(PO_3)_3$ (C form) and $CsPO_3$:

$$Cs_3V_3P_{12}O_{36} \text{ -----> } 3CsPO_3 + 3V(PO_3)_3$$

The corresponding iron-caesium compound is isotypic. The crystal structure determination of $Cs_3V_3P_{12}O_{36}$ has been performed by Lavrov *et al* [2].

Grunze *et al* [3] investigated the thermal behavior of various gallium-caesium phosphates. They obtained $Cs_3Ga_3P_{12}O_{36}$ by thermal decomposition of $CsGaHP_3O_{10}$ according to the following scheme :

$$6CsGaHP_3O_{10} \text{ -----> } Cs_3Ga_3P_{12}O_{36} + 3CsGaP_2O_7 + 3H_2O$$

$CsGaHP_3O_{10}$ is trimorphic ; the decomposition temperature is a function of the form used for the preparation and it varies from 823 to 853K. The same authors carried out the crystal structure determination of $Cs_3Ga_3P_{12}O_{36}$, which appears as isotypic with the corresponding vanadium compound.

During an investigation of the P_2O_5-Cr_2O_3-M_2O-H_2O systems between 473 and 673K, Grunze and Chudinova [4] characterized three compounds : $(NH_4)_3Cr_3P_{12}O_{36}$, $Rb_3Cr_3P_{12}O_{36}$ and $Cs_3Cr_3P_{12}O_{36}$, but they did not report their unit-cell dimensions. Nevertheless, from the non-indexed X-ray patterns given by the authors, one can assume they are probably isotypic with the caesium-vanadium salt.

8.2. Crystal Chemistry

All the cycloduodecaphosphate compounds are isotypic and crystallize with cubic unit cells close by 14.5 Å. Only two of these unit cells have been accurately measured and are reported below :

$V_3Cs_3P_{12}O_{36}$, a = 14.543Å [1, 4]
$Ga_3Cs_3P_{12}O_{36}$, a = 14.374 [3]

The common space group is Pa3 with Z = 4. We describe the structure of the vanadium salt. This highly symmetrical atomic arrangement is rather difficult to explain and to represent. There are two crystallographic independent vanadium atoms, both located on ternary axes and in almost regular octahedral coordinations. The two independent caesium atoms have nine- and twelvefold coordinations. The $P_{12}O_{36}$ ring anions built

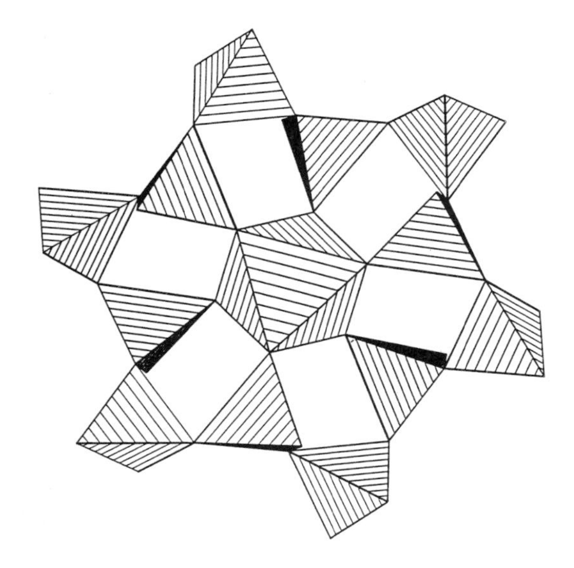

Figure 8.1 Projection along the ternary axis of the $P_{12}O_{36}$ ring anion centred by a VO_6 octahedron.

around the ternary axes have in fact a $\overline{3}$ internal symmetry and are centred by a vanadium atom building its octahedral coordination with six of the external atoms of the ring. This organization, represented in projection along a threefold axis in Fig. 8.1, is the unique example of a phosphoric ring centred by an associated cation polyhedron. Along the threefold axis, the CsO_9 polyhedra and the other type of VO_6 octahedra form columns by sharing faces. The geometrical features of the phosphoric ring as observed in the vanadium salt are reported below :

P-P-P(°)	P-O-P(°)	P-P(Å)
85.8	133.8	2.945
111.3	136.3	2.952

They are not significantly different from what has been observed in other large phosphoric rings.

REFERENCES

1 A.V. Lavrov, M.Ya. Voitenko and E.G. Tselebrovskaya, *Izv. Akad. Nauk SSSR, Neorg. Mat.,* 17 (1981) 99.
2 A.V. Lavrov, V.P. Nikolaev, G.G. Sadikov and M.Ya. Voitenko, *Dokl. Akad. Nauk SSSR,* 259 (1981) 103.
3 I. Grunze, K.K. Palkina, N.N. Chudinova, L.S. Guzeeva, M.A. Avaliani and S.I. Maksimova, *Izv. Akad. Nauk SSSR, Neorg. Mat.,* 23 (1987) 610.
4 I. Grunze and N.N. Chudinova, *Izv. Akad. Nauk SSSR, Neorg. Mat.,* 24 (1988) 988.

9. RING ANIONS IN MIXED ANION PHOSPHATES

More than twenty phosphates crystallize with two kinds of phosphoric anions having different degrees of condensation. There is no rule for the chemical preparation of such phosphates. With very few exceptions, the great majority of them are the unexpected fruit of various flux-method experiments and their real constitutions have been in fact proved once their crystal structures have been solved. Some of them contain cyclotetraphosphoric anions. Other types of ring anions have not yet been observed in this class of compounds.

9.1. Present State Chemistry of the Mixed Phosphoric Anion **Compounds Including a Ring Anion**

- $MN_2(H_2P_3O_{10})(P_4O_{12})$

The existence of eight compounds, corresponding to the general formula $MN_2(H_2P_3O_{10})(P_4O_{12})$ (M = K, NH_4, Rb, Cs ; N = Al, Fe) has been reported by Grunze and Grunze [1]. They are prepared by heating various mixtures of MH_2PO_4, $Al(OH)_3$ or $Fe(NO_3)_3.9H_2O$ and H_3PO_4 at temperatures ranging from 443 to 573K. Grunze *et al* [2] showed that the four aluminium salts are isotypic. The atomic arrangement of this kind of compounds was determined using the aluminium-potassium salt by Grunze *et al* [3], who produced single crystals through a new investigation of the crystallization products obtained by heating various mixtures of KH_2PO_4, $Al(OH)_3$ and H_3PO_4. The authors report a detailed chemical preparation of this salt and an accurate crystal structure showing the coexistence of both P_4O_{12} ring anions and P_3O_{10} groups in the atomic arrangement.

- *$Pb_2Cs_3(P_4O_{12})(PO_3)_3$ and $Sr_2Cs_3(P_4O_{12})(PO_3)_3$*

The two salts $M_2Cs_3(P_4O_{12})(PO_3)_3$ (M = Pb, Sr) were identified by Averbuch-Pouchot [4] as two isotypic compounds during investigations of the $PbO-Cs_2O-P_2O_5$ and $SrO-Cs_2O-P_2O_5$ systems. A detailed crystal structure determination carried out by the author, with the lead salt, shows the coexistence of both P_4O_{12} ring anions and $(PO_3)_n$ infinite phosphoric chains in the atomic arrangement.

9.2. Atomic Arrangement of $Pb_2Cs_3(P_4O_{12})(PO_3)_3$

$Pb_2Cs_3(P_4O_{12})(PO_3)_3$ is triclinic, $P\bar{1}$ with Z = 2 and the unit-cell dimensions [4] :

$$a = 6.808(5),\ b = 7.875(6),\ c = 22.12(1)\ \text{Å}$$
$$\alpha = 86.23(1),\ \beta = 96.96(1),\ \gamma = 113.98(1)°$$

If one considers only the stacking of the anions, this atomic arrangement can be considered as a layer organization. As it can be seen in Fig. 9.1, all the P_4O_{12} anions are located in planes z = 0 and 1/2, while the infinite $(PO_3)_n$ chains spread in planes z = 1/4 and 3/4. Two crystallo-

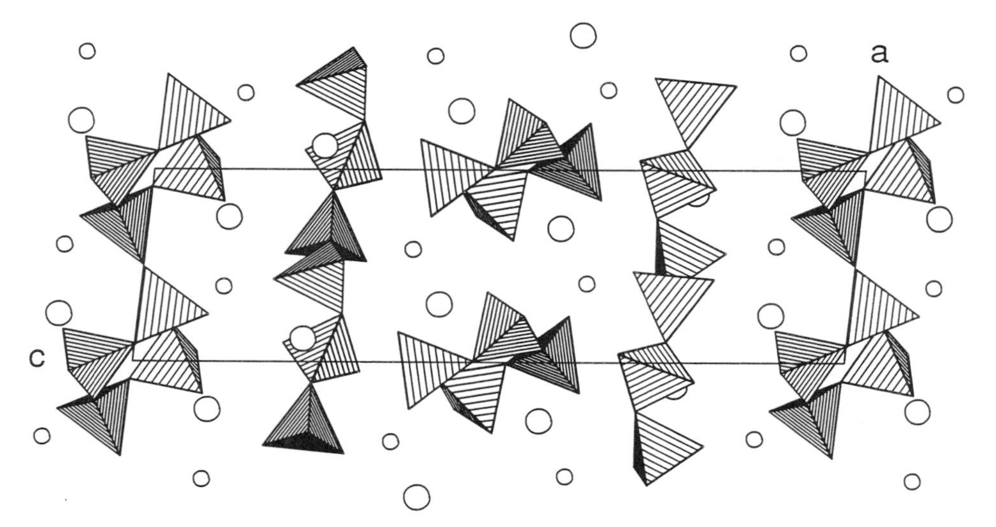

Figure 9.1 Projection of the strucure of $Pb_2Cs_3(P_4O_{12})(PO_3)_3$ along the b axis. Larger circles are caesium atoms. The smaller ones represent the lead atoms.

graphic independent ring anions, both centrosymmetrical, coexist in the structure. Their main geometrical features have already been reported in Table 3.12. The infinite $(PO_3)_n$ anion has a period of three tetrahedra. The general aspect of such a chain as observed in a layer centred on the plane z = 0.25 is given in Fig.9.2. The two independent lead atoms both have sevenfold coordinations, while within a range of 3.5 Å two of the caesium atoms have eight neighbors and the third one nine.

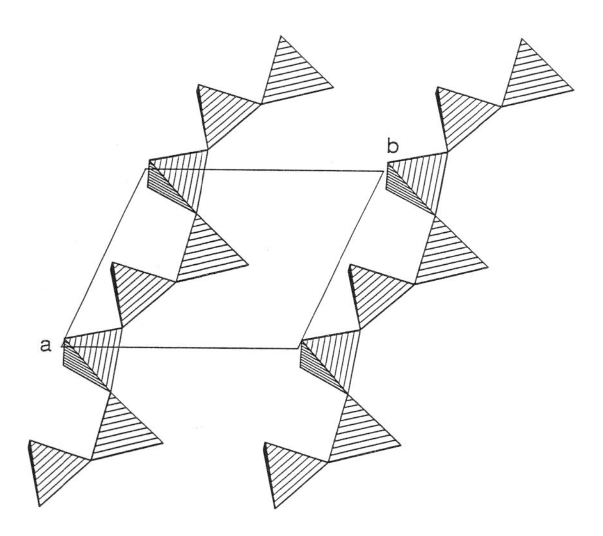

Figure 9.2 The infinite phosphoric chain as observed in projection along the c axis in a layer centred in z = 0.25.

The strontium salt, $Sr_2Cs_3(P_4O_{12})(PO_3)_3$, is isotypic with the following unit-cell dimensions :

$$a = 6.922(5), \ b = 8.055(5), \ c = 21.97(5) \text{ Å}$$
$$\alpha = 86.91(5), \ \beta = 97.99(5), \ \gamma = 115.15(5)°$$

9.3. Atomic Arrangement of $KAl_2(H_2P_3O_{10})(P_4O_{12})$

The salt $KAl_2(H_2P_3O_{10})(P_4O_{12})$ [3] is monoclinic, C2/c, with Z = 4 and the following unit-cell dimensions :

$$a = 11.864(3), \ b = 8.332(3), \ c = 17.317(4) \text{ Å}, \ \beta = 99.67(2)°$$

The atomic arrangement of this salt, as shown in projection along the b axis in Fig. 9.3, is a typical layer organization. Planes z = 0 and 1/2 contain the P_4O_{12} ring anions, while planes z = 1/4 and 3/4 contain the potassium atoms and the P_3O_{10} triphosphate groups. Between these two kinds of layers, the aluminium atoms are intercalated in planes z = (2n+1)/8. The P_4O_{12} ring anions are centrosymmetrical, while the P_3O_{10} groups have a twofold internal symmetry. According to the authors, the hydrogen atoms belonging to terminal oxygen atoms of the P_3O_{10} group establish a set of strong hydrogen-bonds connecting the phosphoric anions along the c direction.

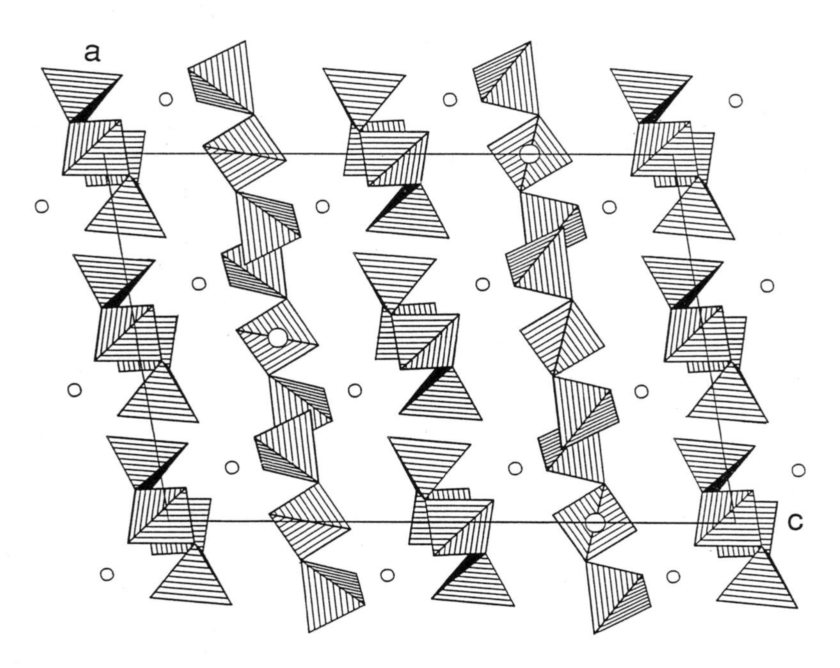

Figure 9.3 Projection, along the b axis, of the atomic arrangement of $KAl_2(H_2P_3O_{10})(P_4O_{12})$. The larger empty circles represent potassium atoms, the smaller ones represent the aluminium atoms.

REFERENCES

1 I. Grunze and H. Grunze, *Z. anorg. allg. Chem.*, 512 (1984) 39.

2 I. Grunze, N.N. Chudinova and K.K. Palkina, *Izv. Akad. Nauk SSSR, Neorg. Mat.*, 20 (1984) 1053.

3 I. Grunze, K.K. Palkina, S.I. Maksimova and N.T. Chibiskova, *Phys. Dokl.*, 29 (1984) 255.

4 M.T. Averbuch-Pouchot, *Z. anorg. allg. Chem.*, 529 (1985) 143.

10. THIOCYCLOPHOSPHATES

Wolf and Meisel [1] described the synthesis of several phosphorus oxide-sulfides $P_4O_{(10-n)}S_n$ by reaction of P_4S_{10} with P_4O_{10}. According to the ratio P_4S_{10} / P_4O_{10} in the starting mixture, various phosphorus oxide-sulfides are formed. Besides the well known $P_4O_6S_4$, the following compounds have been obtained for the first time by the authors : $P_4S_5O_5$, $P_4O_4S_6$, $P_4O_3S_7$, $P_4O_2S_8$ and P_4OS_9. They are separated by fractional distillation or crystallization. A detailed study of these phosphorus oxide-sulfides has been reported by Meisel [2]. Some of these compounds have been used for the preparation of thiocyclophosphates by a process similar to that described for the production of cyclotetraphosphoric acid by hydrolysis of P_4O_{10} at low temperature. Here, the reaction is :

$$P_4O_6S_4 + 2H_2O \longrightarrow H_4P_4O_8S_4$$

In the so obtained compounds, when compared with cyclophosphates, one of the two external oxygen atoms of each tetrahedron is replaced by a sulfur atom, except for $(NH_4)_3P_3S_9$, in which all the oxygen atoms seem substituted.

10.1. Present State of the Thiocyclophosphate Chemistry

10.1.1. Thiocyclotriphosphates

- *$(NH_4)_3P_3S_9$*

Reaction of P_4S_{10} or P_4S_9 in liquid ammonia below 240K, as investigated by Wolf and Meisel [3], leads to ammonium nonathiocyclotriphosphate, $(NH_4)_3P_3S_9$, which crystallizes on cooling to 195K. The reaction schemes are the following ones :

$$P_4S_9 + 6NH_3 \longrightarrow (NH_4)_3P_3S_9 + P(NH_2)_3$$
$$P_4S_{10} + 6NH_3 \longrightarrow (NH_4)_3P_3S_9 + SP(NH_2)_3$$

$(NH_4)_3P_3S_9$ reacts with PCl_3 rebuilding the adamantine-like structure of P_4S_9 :

$$(NH_4)_3P_3S_9 + PCl_3 \longrightarrow P_4S_9 + 3NH_4Cl$$

Heating $(NH_4)_3P_3S_9$ at 513K for 100 hours, in vacuum, leads to the formation of PNS :

$$(NH_4)_3P_3S_9 \longrightarrow 3PNS + 6H_2S$$

No structural study ascertained the ring nature of the anion suggested by the chemical formula of this salt.

- $(NH_4)_3P_3O_6S_3$

The salt $(NH_4)_3P_3O_6S_3$ was obtained by fluoridrolysis of $P_4O_6S_4$ in glacial acetic acid [4]. Finely grinded $P_4O_6S_4$ is slowly added to an ice-cooled saturated solution of ammonium fluoride in acetic acid. Crystals of $(NH_4)_3P_3O_6S_3$ precipitate after two hours of strirring. The reaction scheme is :

$$3NH_4F + P_4O_6S_4 + 2AcOH \text{ -----> } (NH_4)_3P_3O_6S_3 + H(POSF_2) + Ac_2O + HF$$

The crystal structure determination has been performed by Meisel *et al* [4].

10.1.2. Thiocyclotetraphosphates

The chemical preparations and the main properties of salts of tetrathiocyclotetraphosphoric acid with all monovalent cations and some organic cations have been described by Kuvshinova *et al* [5]. Some of them have been investigated from a structural point of view and they are discussed below.

- $Na_4P_4O_8S_4.6H_2O$

$Na_4P_4O_8S_4.6H_2O$ has been obtained by Ilyukhin *et al* [6] by adding slowly ground $P_4O_6S_4$ into a cooled aqueous solution of $NaHCO_3$ (an excess of about 20% of carbonate with respect to the stoichiometry of the reaction is used). At the end of the reaction, the excess of carbonate is destroyed by acetic acid and ethanol added to precipitate the sodium salt. $Na_4P_4O_8S_4.6H_2O$ is very difficult to crystallize and not very stable in air. Nevertheless, the authors could perform the crystal structure determination.

- $Cs_4P_4O_8S_4$

The caesium salt $Cs_4P_4O_8S_4$ was prepared by Ilyukhin *et al* [7] by adding slowly powdered $P_4O_6S_4$ in a cold aqueous solution of caesium carbonate. This solution must contain an excess of about 15% of carbonate. The excess of carbonate is removed by acetic acid and $Cs_4P_4O_8S_4$ is precipitated by ethanol. The so obtained crystals are lamellar. The authors performed the determination of the atomic arrangement.

Later on, Ilyukhin *et al* [8] reported a comparative study of the atomic arrangements of $Cs_4P_4O_8S_4$ and $Na_4P_4O_8S_4.6H_2O$.

- $Ba_2P_4O_8S_4.10H_2O$

Nikolaev and Kuvshinova [9] reported the chemical preparation and crystal structure of $Ba_2P_4O_8S_4.10H_2O$. It was synthesized by adding $P_4O_6S_4$ to a suspension of $BaCO_3$ in water. Large prismatic colorless crystals are obtained by recrystallization in acetic acid. They are stable for years in the mother liquor, but in air decompose into the hexahydrate within some minutes.

- $[C(NH_2)_3]_4P_4O_8S_4$

Guanidinium tetrathiocyclotetraphosphate $[C(NH_2)_3]_4P_4O_8S_4$ was prepared by adding slowly guanidinium carbonate to a cooled solution of tetrathiocyclotetraphosphoric acid, which is obtained by careful dissolution of the phosphorus oxide-sulfide, $P_4O_6S_4$, in water at 273-278K, as reported in Ref. 5. The reactions are :

$$P_4O_6S_4 + 2H_2O \text{ -----> } H_4P_4O_8S_4$$
$$H_4P_4O_8S_4 + 2(Gua)_2CO_3 \text{ ------> } (Gua)_4P_4O_8S_4 + 2CO_2 + 2H_2O$$

The salt is relatively little water soluble and is stable at room temperature. Crystals, up to 1mm long, have been obtained as colorless prisms by evaporation at room temperature. Its atomic arrangement has been determined by Meisel *et al* [10].

The main crystal data for thiocyclophosphates are given in Table 10.1.

Table 10.1
Main crystallographic features for thiocyclophosphates.

Formula	a α	b β	c (Å) γ (°)	S. G.	Z	Ref.
$(NH_4)_3P_3O_6S_3$	12.450(8)	12.755(8)	8.154(6)	Pnma	4	4
$Na_4P_4O_8S_4.6H_2O$	8.815(1) 78.33(4)	9.313(4) 95.07(2)	14.259(3) 119.26(2)	$P\bar{1}$	2	6, 8
$Cs_4P_4O_8S_4$	7.784(3) 95.55(3)	7.889(2) 114.67(4)	8.610(5) 75.46(3)	$P\bar{1}$	1	7, 8
$Ba_2P_4O_8S_4.10H_2O$	8.600(2) 88.69(1)	8.904(2) 73.20(1)	9.234(1) 61.77(1)	$P\bar{1}$	1	9
$(Gua)_4P_4O_8S_4$	10.894(8)	13.126(8) 105.74(5)	8.919(6)	$P2_1/n$	2	10

Gua = $[C(NH_2)_3]^+$ (guanidinium)

10.2. Some Atomic Arrangements of Thiocyclophosphates

10.2.1. Ammonium trithiocyclotriphosphate

$(NH_4)_3P_3O_6S_3$ [4] is orthorhombic, Pnma, Z = 4, with the following unit-cell dimensions :

$$a = 12.450(8),\ b = 12.755(8),\ c = 8.154(6)\ \text{Å}$$

Fig. 10.1 reports the projection along the a axis of this atomic arrangement. The ring anions and one kind of ammonium are located in the mirror planes (y = 1/4 and 3/4), while the second type of ammonium group, in general position, interconnects these planes through hydrogen bonds. The $P_3O_6S_3$ ring anion has a mirror symmetry and is built by two independent, strongly distorted tetrahedra including one P-S bond length larger than 1.9 Å and three normal P-O distances. One of the ammonium group, located in the mirror plane, has a sixfold coordination of oxygen atoms, while the second one, in a general position, has a sevenfold including two sulfur atoms.

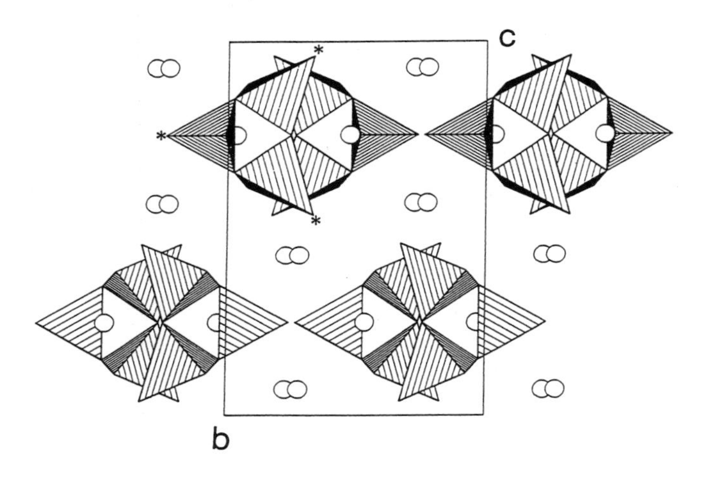

Figure 10.1 Projection, along the a axis, of the atomic arrangement of $(NH_4)_3P_3O_6S_3$. Starred tetrahedron corners correspond to sulfur atoms. The empty circles represent the ammonium groups.

10.2.2. Caesium tetrathiocyclotetraphosphate

$Cs_4P_4O_8S_4$ [7] is triclinic, $P\bar{1}$, Z = 1, with the following unit cell :

$$a = 7.784(3),\quad b = 7.889(2),\quad c = 8.610(5)\ \text{Å}$$
$$\alpha = 95.55(3),\quad \beta = 114.67(4),\quad \gamma = 75.46(3)^\circ$$

The centrosymmetric $P_4O_8S_4$ ring anion is located around the centre of inversion in 1/2,0,0. As it can be expected the two independent PO_3S tetrahedra are very distorted with three normal P-O distances and P-S distances of 1.926 Å in the first tetrahedron and 1.968 Å for the second one. Fig.10.2 shows clearly the layer organization of this arrangement built by planes of ring anions and caesium atoms parallel to the $[\bar{1}01]$ direction. Within a range of 3.80 Å, the two independent caesium atoms have an eightfold coordination made by four oxygen and four sulfur atoms.

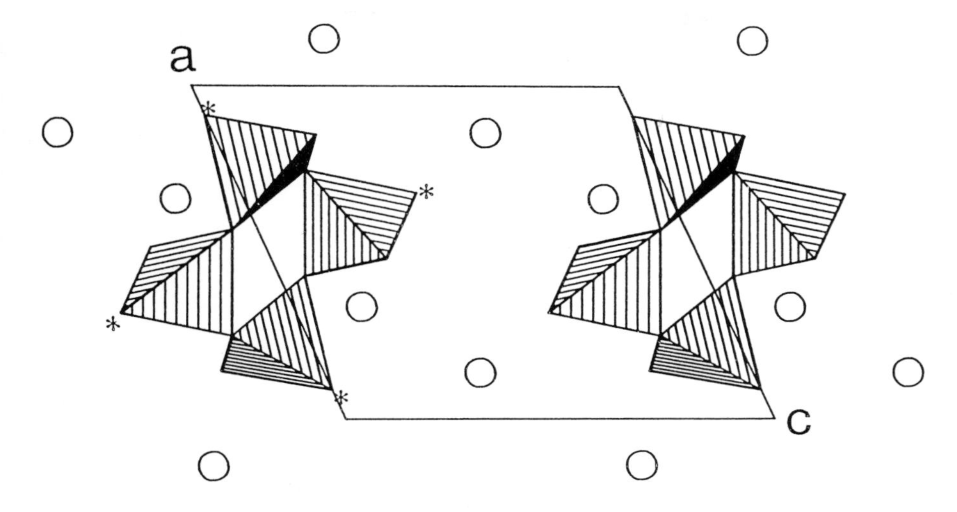

Figure 10.2 Projection, along the b axis, of the atomic arrangement of $Cs_4P_4O_8S_4$. Empty circles are caesium atoms. Starred corners of the tetrahedra indicate the locations of the sulfur atoms.

10.3. The Thiocyclophosphate Rings

Today, structural examples of thiocyclophosphate rings are not enough numerous to authorize any valuable comparisons. Nevertheless, one can notice from Table 10.2 that the five $P_4O_8S_4$ rings, all centrosymmetrical, have P-P-P angles departing more significantly from the ideal value than in the P_4O_{12} rings. Thus, for these five $P_4O_8S_4$ rings the P-P-P angles spread from 97.7 to 95.6°.

The most important difference between P_3O_9 or P_4O_{12} phosphoric rings and the present thio-rings lies in the constitution of the tetrahedron, since the presence of a sulfur atom leading to an average P-S distance larger than 1.93 Å induces a strong deformation of this polyhedron. Table 10.3 reports the geometry of such a PO_3S tetrahedron as observed in $[C(NH_2)_3]_4P_4O_8S_4$.

If compared with the normal PO_4 tetrahedra observed in other types of condensed phosphoric anions, the main differences arise from the P-S bond, significantly longer than the three P-O distances. Due to this last feature, three edges of the tetrahedron, corresponding to the S-O distances, are much longer ($2.876 < S\text{-}O < 2.976$ Å) than the three remaining ones corresponding to the O-O distances ($2.478 < O\text{-}O < 2.560$ Å). On the contrary, the O-P-O(S) angles are in a remarkable accordance with all those measured in P_nO_{3n} ring anions. The value of 119.37°

Table 10.2

Main geometrical features for $P_3O_6S_3$ and $P_4O_8S_4$ anions.

Formula [Ref.]			Formula [Ref.]		
P-P (Å)	P-O-P (°)	P-P-P (°)	P-P (Å)	P-O-P (°)	P-P-P (°)
Ring with a m symmetry					
$(NH_4)_3P_3O_6S_3$ [4]					
2.938	131.3	2x60.4			
2x2.977	2x136.0	59.1			
Ring with a $\overline{1}$ symmetry					
$Na_4P_4O_8S_4.6H_2O$, two independent rings in the unit cell [6, 8]					
2.951	132.0	97.7	2.957	133.0	83.7
2.993	133.0	82.3	3.015	141.5	96.3
$Cs_4P_4O_8S_4$ [7, 8]			$Ba_2P_4O_8S_4.10H_2O$ [9]		
2.973	133.7	82.8	2.935	131.1	82.6
2.981	134.2	97.2	2.953	132.0	97.4
$(Gua)_4P_4O_8S_4$ [10]					
2.981	135.3	95.6			
2.977	135.1	84.4			

Gua = $[C(NH_2)_3]^+$ (guanidinium)

measured for the S-P-O(E) is quite comparable with the O(E)-P-O(E) values observed in other phosphoric rings and the averaged value O-P-O(S) in this tetrahedron (109.23°) is similar to what is commonly measured in any kind of condensed phosphoric anions.

Table 10.3

Main geometrical features of a PSO_3 tetrahedron in the $P_4S_4O_8$ anion observed in $[C(NH_2)_3]_4P_4O_8S_4$. The underlined values reported in the diagonal correspond to P-O or P-S distances. The O-P-O(S) angles are given in the lower left triangle and the O-O(S) distances in the upper right triangle. O(E) is an external oxygen of the ring, while O(L) atoms belong to the ring.

P	S	O(E)	O(L1)	O(L2)
S	$\underline{1.9427}$(8)	2.976(2)	2.876(2)	2.926(2)
O(E)	119.37(8)	$\underline{1.495}$(2)	2.560(2)	2.478(2)
O(L1)	107.84(6)	111.17(9)	$\underline{1.608}$(2)	2.504(2)
O(L2)	110.04(7)	105.29(8)	101.70(8)	$\underline{1.621}$(1)

REFERENCES

1 G.U. Wolf and M. Meisel, *Z. anorg. allg. Chem.*, 509 (1984) 101.

2 M. Meisel, *Z. Chem.*, 23 (1983) 117-125.

3 G.U. Wolf and M. Meisel, *Z. anorg. allg. Chem.*, 494 (1982) 49.

4 M. Meisel, G.U. Wolf and M.T. Averbuch-Pouchot, *Acta Cryst.*, in preparation.

5 T.B. Kuvshinova, G.U. Wolf and M. Meisel, *Izv. Akad. Nauk SSSR, Neorg. Mat.*, 20 (1984) 1056.

6 V.V. Ilyukhin, V.R. Kalinin, T.B. Kuvshinova and I.V.Tananaev, Dokl. *Akad. Nauk SSSR,* 266 (1981) 1387.

7 V.V. Ilyukhin, V.R. Kalinin, T.B. Kuvshinova and I.V. Tananaev, *Dokl. Akad. Nauk SSSR,* 267 (1982) 85.

8 V.V. Ilyukhin, V.R. Kalinin and T.B. Kuvshinova, *Izv. Akad. Nauk SSSR, Neorg. Mat.*, 24 (1988) 828.

9 V.P. Nikolaev and T.B. Kuvshinova, *Izv. Akad. Nauk SSSR, Neorg. Mat.*, 23 (1987) 622.

10 M. Meisel, G.U. Wolf and M.T. Averbuch-Pouchot, *Acta Cryst.*, C46 (1990) 2239.

11. THE PO_4 TETRAHEDRON IN CYCLOPHOSPHATES

Along the various sections of this survey, we reported the main characteristics of ring anions, but nothing was said concerning the geometrical aspects of the basic unit of such rings, the PO_4 tetrahedron. A common feature to all PO_4 tetrahedra involved in the constitution of a phosphoric ring is to share two of their oxygen atoms with two adjacent tetrahedra. So, in a first step, we can distinguish two kinds of oxygen atoms inside the PO_4 tetrahedron of a ring. Two of them are involved in P-O-P bonds and are generally denoted O(L), the two remaining ones are usually called external oxygens and denoted O(E). In the great majority of condensed phosphates only the O(E) oxygen atoms are involved in the associated cation coordinations. In some rare cases, O(L) atoms are given as belonging to some associated cation polyhedra, but this assuption is difficult to verify for, in the case of large cations, the limit of the coordination polyhedron is never well determined.

The two P-O(E) distances are the shortest inside the tetrahedron. They vary normally from 1.470 to 1.502 Å, with an average value of 1.481 Å, while the two P-O(L) bonds are significantly longer and spread within a range of 1.595-1.628 Å, with an average value of 1.609 Å. The most remarkable feature for the P-O distances is the almost constant value of their average in a given tetrahedron. This value never departs significantly from 1.545 Å, with some small, but signifiant variations in tetrahedra having an internal symmetry (m, 2, 3). The rather large difference between P-O(E) and P-O(L) distances has been discussed and explained by Cruickshank [1] in terms of dπ-sπ orbitals and later on by Pakhomov [2].

Among the six O-P-O angles of the tetrahedron, one can distinguish three categories : the O(E)-P-O(E), the O(L)-P-O(L) and the O(E)-P-O(L) angles. The angles of the first category never depart significantly from their average value (119.6°) spreading between 118.4 and 121.3°. The O(L)-P-O(L) angles vary between 99.3 and 102.7°, with an average value of 101.3°, and the O(E)-P-O(L) between 104.8 and 112.0°, with an average of 108.7°. The averaged value for the O-P-O angles is of 109.3°, close by the value of a regular tetrahedron.

The numerical values reported here have been extracted from forty tetrahedra choosen among very accurately determined atomic arrangements of cyclophosphates. A more extended review is planned covering, this time, all types of condensed phosphoric anions.

REFERENCES

1 D.W. Cruickshank, *J. Chem. Soc.*, (1961) 5486.

2 V.I. Pakhomov, *Izv. Akad. Nauk SSSR, Neorg. Mat.*, 13 (1977) 1341.

STEREOCHEMICAL ASPECTS OF THE REDOX PROPENSITY OF HOMOMETAL CARBONYL CLUSTERS

P. Zanello

STEREOCHEMICAL ASPECTS OF THE REDOX PROPENSITY OF HOMOMETAL CARBONYL CLUSTERS

Piero Zanello

1. INTRODUCTION

During the last two decades, we have seen interest in metal clusters to grow exponentially, and the different aspects of their chemistry (structure |1-13|, bonding |1-4,6,7,9,12,13|, chemical reactivity |12,14-17|, and electron counting rules |9,13,18-22|) have been extensively reviewed and continuously updated. The well-recognized ability of clusters to undergo multiple, non-destructive electron-transfer processes has also received a good deal of attention thanks to some stimulating review papers |23-27| which surveyed their electrochemical properties.

Our aim is to give a systematic picture of the intrinsic ability to add or to lose electrons of the most important classes of cluster compounds, while pointing out the stereochemical consequences accompanying such electron additions/removals. In this connection, we recently dealt with metal-sulfur clusters |28,29|, as well as with heterometal-carbonyl clusters |30|; in particular, the present review is devoted to homometal-carbonyl clusters.

Structural reorganizations induced by electron transfers and electrochemical parameters have been clearly correlated by W.E.Geiger |24|. They are intimately connected to the kinetic aspects of electron exchange between an electrode and the metallic complex, in the sense that: fast electron-transfer (defined as *electrochemically reversible*) preludes no significant geometrical reorganization; a relatively slow electron-transfer (*electrochemically quasireversible*) is a prelude to some significant stereochemical rearrangement, leaving intact the starting molecular assembly; finally, an extremely slow electron-transfer (*electrochemically irreversible*) induces so deep structural modifications as to destroy the starting frame.

Further, knowledge of the thermodynamic potential values of the redox changes is not the less important from both practical and theoretical viewpoints. In fact, concerning the synthetic aspect, it can help in a skilful choice of the most appropriate redox agents to obtain the desired (and only those) congeners (if judged stable through a brief glance at their electrochemical fingerprint). On the other hand, bonding descriptions derived through theoretical molecular orbital schemes can receive invaluable experimental support from such knowledge, as far as the variation of the electronic configuration of the frontier orbitals is concerned.

Unless otherwise specified, throughout this paper, potential values will be referred to the aqueous, saturated calomel electrode (*S.C.E.*).

2. IRON CLUSTERS

2.1. *Trinuclear Binary Complexes*

2.1.1. $Fe_3(CO)_{12}$. The molecular structure of $Fe_3(CO)_{12}$ is shown in Figure 1 |31-33|.

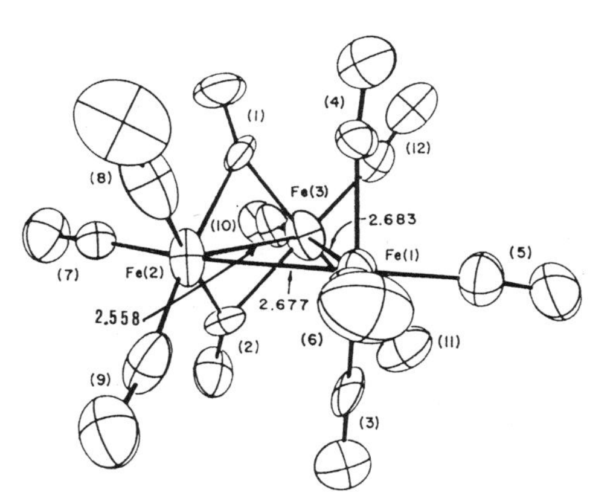

Figure 1. *Molecular structure of $Fe_3(CO)_{12}$ with relevant Fe-Fe bond lengths (from Ref. 33).*

It is constituted by a bonding triiron triangle, having ten terminal and two asymmetrically bridging CO groups. The CO-bridged Fe-Fe distance is significantly shorter (0.1 Å) than the unbridged ones.

The redox propensity of the 48-cluster-valence-electron $Fe_3(CO)_{12}$ has been widely studied by electrochemical methods |34-37|.

Figure 2 shows the cyclic voltammetric response exhibited by $Fe_3(CO)_{12}$ in dichloromethane solution |35|.

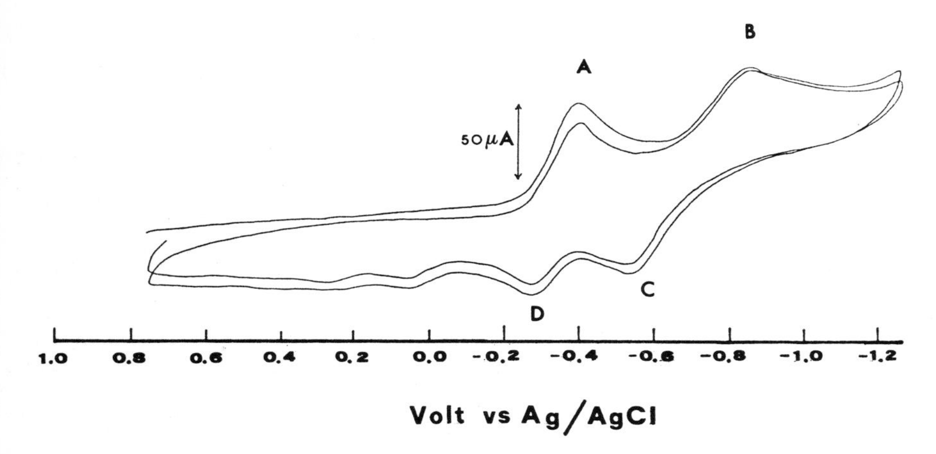

Figure 2. *Cyclic voltammogram obtained at a platinum electrode on a deaerated dichloromethane solution of $Fe_3(CO)_{12}$. Scan rate 0.2 Vs^{-1}. Room temperature (from Ref. 35).*

The less cathodic peak-system A/D is due to the chemically reversible, one-electron reduction $Fe_3(CO)_{12}/[Fe_3(CO)_{12}]^-$. The most cathodic reduction (peak-system B/C) is attributed to the subsequent one-electron reduction $[Fe_3(CO)_{12}]^-/[Fe_3(CO)_{12}]^{2-}$, complicated by irreversible decomposition in the long times of electrolysis, which makes the dianion a transient species, particularly in tetrahydrofuran and acetone solutions |35,38|. Only in the shorter timescale of cyclic voltammetry, it has been proved that reduction to dianion is mainly followed by the conproportionation reaction |36,37|:

$$[Fe_3(CO)_{12}]^{2-} + Fe_3(CO)_{12} \longrightarrow 2\ [Fe_3(CO)_{12}]^-$$

Table 1 summarizes the redox potentials for the two successive one-electron reductions of $Fe_3(CO)_{12}$.

As suggested by the ΔEp values, the chemically reversible redox change $Fe_3(CO)_{12}/[Fe_3(CO)_{12}]^-$ is electrochemically quasi-reversible, departure from the value of 59 mV for a purely reversible electrochemical one-electron transfer being significant. This would mean that, upon one-electron addition, the geometry of $Fe_3(CO)_{12}$ undergoes some important reorganization. This rearrangement is not unlikely if one takes into account that, in an ideally symmetric D_{3h} geometry, the electron added to the *saturated* electron precise, $M_3(CO)_{12}$ (M = Fe, Ru, Os) enters a metal-metal antibonding LUMO level |39,40| which, as ESR evidence |38| shows, receives some contribution from carbonyl groups.

TABLE 1. *Formal electrode potentials for the cathodic processes exhibited by $Fe_3(CO)_{12}$ in different nonaqueous solvents*

redox step	E°' (V)	ΔEp[a] (mV)	solvent	reference
$Fe_3(CO)_{12}/[Fe_3(CO)_{12}]^-$	-0.35	110	CH_2Cl_2	35
	-0.44		CH_2Cl_2	36
	-0.25	105	Me_2CO	35
$[Fe_3(CO)_{12}]^-/[Fe_3(CO)_{12}]^{2-}$	-0.70	320	CH_2Cl_2	35
	-1.05[b]		CH_2Cl_2	36
	-1.0[b]		Me_2CO	35

[a]*Measured at 0.2* Vs^{-1}; [b]*peak potential value for irreversible process*

2.1.2. $[Fe_3(CO)_{11}]^{2-}$. Figure 3 shows the molecular structure of the dianion $[Fe_3(CO)_{11}]^{2-}$ |41|. It is constituted by a bonding triiron, almost equilateral, triangle (Fe-Fe, 2.60 Å), capped by a triply-bridging carbonyl group. One edge (Fe2-Fe2') bears one doubly-bridging carbonyl too.

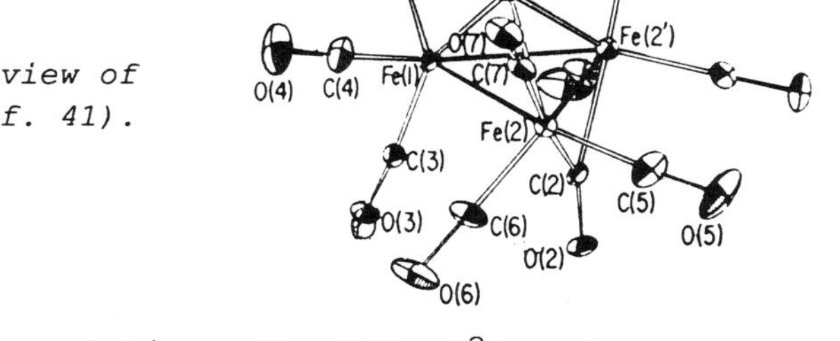

Figure 3. *Perspective view of $[Fe_3(CO)_{11}]^{2-}$ (from Ref. 41).*

In tetrahydrofuran solution, $[Fe_3(CO)_{11}]^{2-}$ undergoes an easy (E°' = -0.76 V), chemically reversible, one-electron oxidation to the stable monoanion $[Fe_3(CO)_{11}]^-$ |42|. Based on its ESR features, $[Fe_3(CO)_{11}]^-$ has been assigned a molecular geometry qualitatively similar to that of the parent dianion |43|.

2.2. *Trinuclear Mixed-Ligand Complexes*

2.2.1. $[HFe_3(CO)_{11}]^-$. The structure of the hydride anion $[HFe_3(CO)_{11}]^-$ is schematized in Figure 4 |44|.

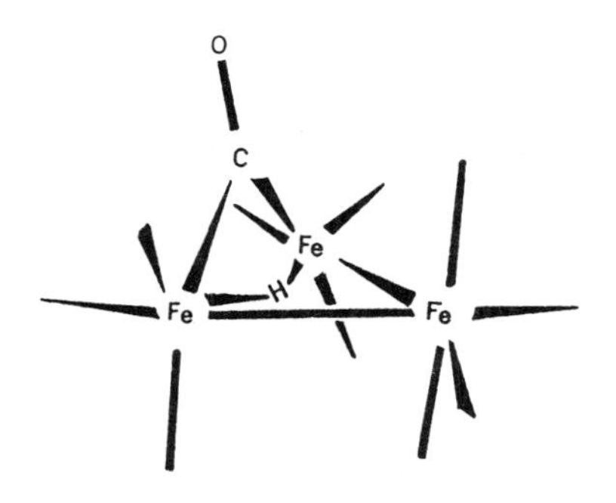

Figure 4.
Schematic representation of the molecular structure of $[HFe_3(CO)_{11}]^-$.

The trimetallic core has the geometry of an isosceles triangle, the minor edge of which is doubly bridged by both a carbonyl and the hydride hydrogen. The hydrogen atom occupies the position of the second carbonyl bridging group in $[Fe_3(CO)_{12}]^{2-}$. The mean Fe-Fe bond length is 2.69 Å, whereas the Fe-Fe distance in the two-electron three-center bond Fe-H-Fe is 2.58 Å. In acetone, at -70°C, the 48-electron $[HFe_3(CO)_{11}]^-$ is able to undergo reversibly both, a one-electron oxidation and a one-electron reduction according to the sequence |35|:

$$\begin{array}{ccccccc} & E^{\circ\prime}=+0.43\text{ V} & & E^{\circ\prime}=-1.18\text{ V} & & Ep=-1.95\text{ V} & \\ HFe_3(CO)_{11} & \underset{-e}{\overset{+e}{\rightleftarrows}} & [HFe_3(CO)_{11}]^- & \underset{-e}{\overset{+e}{\rightleftarrows}} & [HFe_3(CO)_{11}]^{2-} & \overset{+e}{\longrightarrow} & [HFe_3(CO)_{11}]^{3-} \\ & & & & & & \downarrow \\ & & & & & & \text{fast decomposition} \end{array}$$

It has been postulated that the paramagnetic species $HFe_3(CO)_{11}$ largely maintains the structure of the monoanion precursor |45|.

2.2.2. $Fe_3(CO)_{12-n}(L)_n$ (L = PR_3, n = 1-3). Sequential substitution of CO groups in $Fe_3(CO)_{12}$ by phosphines or phosphites leads to a series of complexes of formula $Fe_3(CO)_{12-n}(L)_n$, with L = PR_3, n = 1-3 (see references cited in Ref. 35).

Figure 5 shows the crystal structure of two isomeric forms of $Fe_3(CO)_{11}(PPh_3)$ |46|.

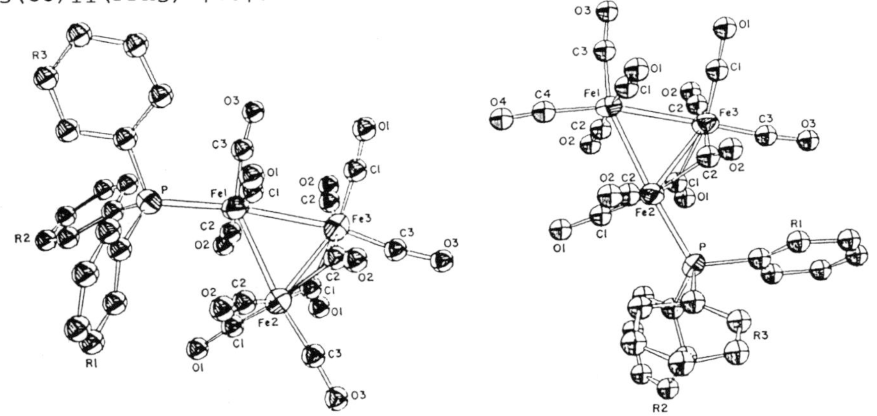

Figure 5. *Perspective view of two structural isomers of $Fe_3(CO)_{11}(PPh_3)$ (from Ref. 46).*

Their overall geometry is quite similar to that of the isoelectronic $Fe_3(CO)_{12}$. One molecule contains the phosphine group bonded to one Fe atom of the doubly-CO-bridged Fe-Fe edge of the Fe_3 isosceles triangle. The other molecule has the phosphine bonded to the unbridged Fe atom. In both cases, the phosphine occupies an equatorial position. The Fe-Fe bond lengths are substantially similar to those found in $Fe_3(CO)_{12}$, except for a slight elongation ($\sim$0.02 Å) of the ones bonding the phosphine, likely because of the electron withdrawing ability of the phenyl groups. The existence of the two isomers is likely a consequence of the dynamic interchange between bridging and terminal CO groups occurring in solution |47|.

The successive substitution of CO groups for phosphorus containing ligands enters, gradually, one equatorial position of the different Fe atoms, as proved by the X-ray structure of $Fe_3(CO)_9(PMe_2Ph)_3$, Figure 6 |48|. The Fe-Fe bond lengths are quite consistent with those of the parent $Fe_3(CO)_{12}$.

Like the unsubstituted precursor, these substituted derivatives undergo a first, one-electron reduction, which is chemically reversible only at very low temperature, followed by a second one-electron addition. The redox potentials of these steps, which are summarized in Table 2, are significantly more negative than those exhibited by $Fe_3(CO)_{12}$, because of the σ-donor ability of PR_3 ligands. The trisubstituted $Fe_3(CO)_9[P(OMe)_3]_3$ is also able to undergo, reversibly, a one-electron removal.

As an example, Figure 7 illustrates the low-temperature cyclic voltammogram of $Fe_3(CO)_{11}[P(OPh)_3]$ in acetone solution |35|.

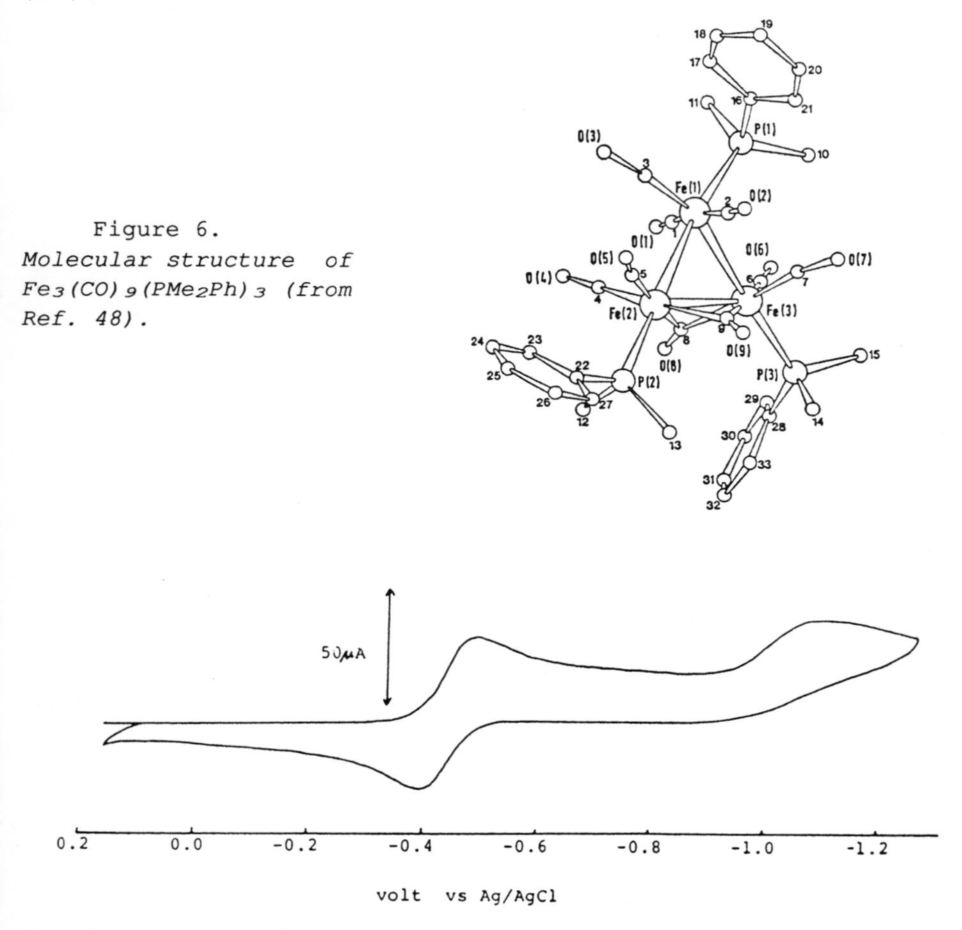

Figure 6.
Molecular structure of $Fe_3(CO)_9(PMe_2Ph)_3$ (from Ref. 48).

Figure 7. *Cyclic voltammogram recorded at a platinum electrode on an acetone solution of $Fe_3(CO)_{11}[P(OPh)_3]$. T = 228 K. Scan rate 0.2 Vs^{-1} (from Ref. 35).*

It has been also confirmed by ESR analysis that the radical monoanions $[Fe_3(CO)_{12-n}(L)_n]^-$ are shorter lived than $[Fe_3(CO)_{12}]^-$, and, within the substituted species, the more substituted is the complex, the faster is its decomposition, which mainly follows a disproportionation pathway to the corresponding neutral and dianionic congeners |38|.

TABLE 2. *Redox potentials (in Volt) for the two one -electron reduction steps exhibited by $Fe_3(CO)_{12-n}(L)_n$ in acetone solution |35|*

complex	E°' (0/1-)	Ep[a] (1-/2-)	E°' (1+/0)
$Fe_3(CO)_{11}(PPh_3)$	-0.45	-1.00	
$Fe_3(CO)_{11}[P(OMe)_3]$	-0.49	-1.20	
$Fe_3(CO)_{11}[P(OPh)_3]$	-0.46	-1.1	
$Fe_3(CO)_{10}[P(OMe)_3]_2$	-0.81	-1.22	
$Fe_3(CO)_{10}[P(OPh)_3]_2$	-0.64	-1.15	
$Fe_3(CO)_9[P(OMe)_3]_3$	-1.08	-1.36	+0.67
$Fe_3(CO)_9[P(OPh)_3]_3$	-0.96	-1.0	

[a]*Measured at 0.2 Vs^{-1}*

2.2.3. $Fe_3(CO)_9(RC_2R)$. The alkyne-carbonyl series $Fe_3(CO)_9(RC_2R)$ possesses a structure in which the acetylene moiety is placed above the metallic triangle, with the triple bond perpendicular to one Fe-Fe edge. Figure 8 illustrates just such a feature for $Fe_3(CO)_9(PhC_2Ph)$ |49|.

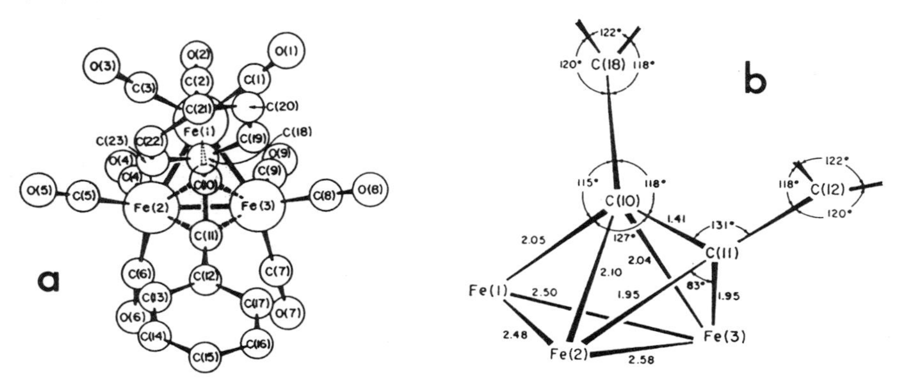

Figure 8. *Perspective view (a) of $Fe_3(CO)_9(PhC_2Ph)$, together with (b) selected structural parameters (bond distances (in Å) and angles) (from Ref. 49).*

These derivatives are considered formally unsaturated 46-electron species, so that it is not surprising that they undergo two successive one-electron additions |50|. As shown in Figure 9 for $Fe_3(CO)_9(PhC_2Ph)$, the two one-electron steps (electrochemically near reversible) have features of chemical reversibility. As a matter of fact, controlled potential electrolysis produces the stable dianions $[Fe_3(CO)_9(RC_2R)]^{2-}$.

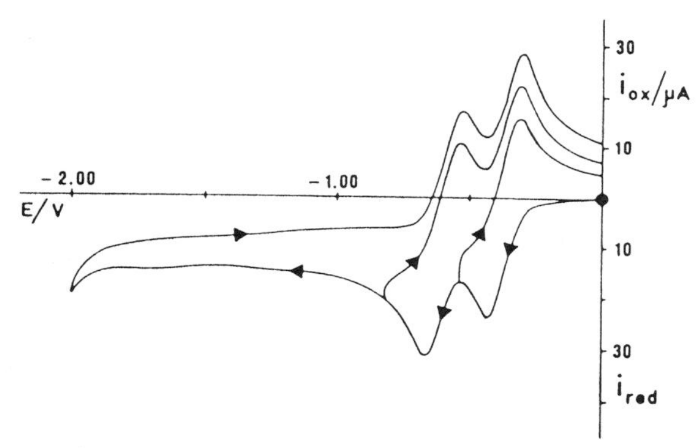

Figure 9. *Cyclic voltammetric response obtained at a platinum electrode on a CH_2Cl_2 solution of $Fe_3(CO)_9(PhC_2Ph)$. Scan rate 0.2 Vs^{-1} (from Ref. 50).*

Table 3 summarizes the redox potentials of these cathodic processes.

TABLE 3. *Redox potentials (in Volt) for the two distinct one-electron reductions exhibited by $Fe_3(CO)_9(RC_2R)$ |50|*

complex	$E^{\circ\prime}_{(0/1-)}$	$E^{\circ\prime}_{(1-/2-)}$	solvent
$Fe_3(CO)_9(EtC_2Et)$	-0.49	-0.67	CH_2Cl_2
	-0.39	-0.76	Me_2CO
$Fe_3(CO)_9(PhC_2Ph)$	-0.36	-0.60	CH_2Cl_2
	-0.33	-0.65	Me_2CO
$Fe_3(CO)_9(MeC_2Me)$	-0.47	-0.65	CH_2Cl_2
	-0.40	-0.75	Me_2CO
$Fe_3(CO)_9(MeC_2Ph)$ [a]	-0.43	-0.62	CH_2Cl_2
	-0.33	-0.70	Me_2CO

[a] *Mixture of two isomers*

R.Hoffman has theoretically foreseen that addition of two electrons to the model compound $Fe_3(CO)_9(HC_2H)$ should cause reorientation of the acetylene molecule from perpendicular to parallel to one Fe-Fe edge |40|. Actually, we found good spectroscopic evidence that, in the electrogenerated dianions $[Fe_3(CO)_9(RC_2R)]^{2-}$, the alkyne orients parallel to one Fe-Fe bond |50|. Furthermore, the quasireversibility of the two electrochemical steps lends legitimacy to the speculation that the alkyne rotation starts simultaneously with the electron addition.

2.2.4. $Fe_3(CO)_9(RC_2R)(PR')$. As illustrated in Figure 10, which shows the crystal structure of $Fe_3(CO)_9(PhC_2Ph)(P\text{-}C_6H_4\text{-}p\text{-}OMe)$, insertion of a phosphinidene fragment into the $Fe_3(CO)_9(RC_2R)$ frame cause opening of the metallic triangle |51|.

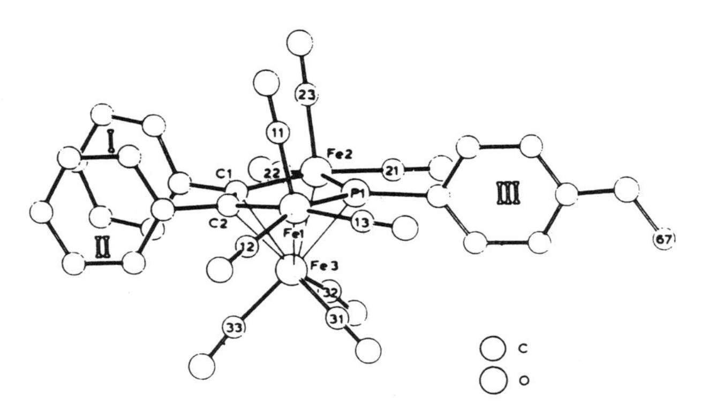

Figure 10. *Crystal structure of $Fe_3(CO)_9(PhC_2Ph)(P\text{-}C_6H_4\text{-}p\text{-}OMe)$. C1-C2, 1.42 Å; Fe-Fe, averaged 2.66 Å; Fe-C, 1.99 Å; Fe1-P = Fe2-P, 2.17 Å; Fe3-P, 2.29 Å (from Ref. 51).*

The insertion of the phosphinidene group also affects significantly the electron addition process, making it notably more difficult than that previously discussed for the $Fe_3(CO)_9(RC_2R)$ derivatives. As a matter of fact, the $Fe_3(CO)_9(RC_2R)(P\text{-}C_6H_4\text{-}p\text{-}OMe)$ complexes allow access only to the first reduction step |52|. As summarized in Table 4, the nature of the alkyne substituents has a marginal effect on the redox potential.

TABLE 4. *Redox potentials (in Volt) for the one-electron reduction exhibited in dichloromethane solution by $Fe_3(CO)_9(RC_2R')(P-C_6H_4-p-OMe)$. Glassy-carbon working electrode |52|*

R	R'	E°' (0/1-)
Ph	Ph	-1.01
H	Bu^n	-1.08
H	Ph	-1.04

2.2.5. $[Fe_3(CO)_9(E)]^{2-}$ (E = O,S). The triiron dianions $[Fe_3(CO)_9(\mu_3-O)]^{2-}$ and $[Fe_3(CO)_9(\mu_3-S)]^{2-}$ are isostructural, and their structure is shown in Figure 11. They are formed by an equilateral metallic triangle capped by the triply bridging chalcogen atom |53,54|.

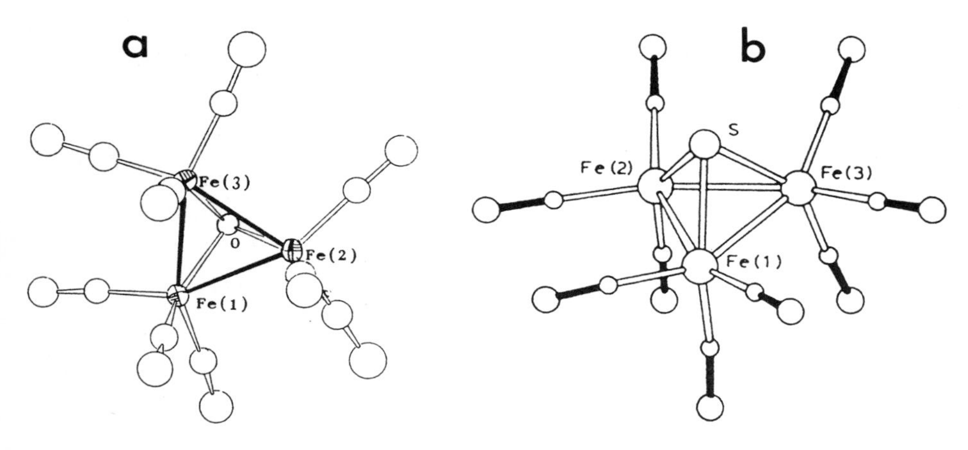

Figure 11. *Molecular structure of: (a) $[Fe_3(CO)_9(\mu_3-O)]^{2-}$. Fe-Fe mean length, 2.48 Å; Fe-O, 1.89 Å (from Ref. 53); (b) $[Fe_3(CO)_9(\mu_3-S)]^{2-}$. Fe-Fe mean length, 2.57 Å; averaged Fe-S distance, 2.19 Å (from Ref. 54).*

The 48-electron $[Fe_3(CO)_9(O)]^{2-}$ undergoes, in acetonitrile solution, two distinct, framework-destroying one-electron removals (Ep = -0.26 V, +0.13 V, respectively) |55|, thus indicating it to be the only stable member among the potential oxo-capped congeners.

The isoelectronic $[Fe_3(CO)_9(S)]^{2-}$ undergoes, in dichloroethane solution, a quasireversible one-electron oxidation at E°′ = -0.40 V. In contrast with $[Fe_3(CO)_9(O)]^-$, the monoanion $[Fe_3(CO)_9(S)]^-$, even if apparently not fully stable (i_{pa}/i_{pc} = 0.95), seems relatively long-lived |56|.

2.2.6. $Fe_3(CO)_9(E)_2$ (E = S,Se). Figure 12 shows the molecular structure of $Fe_3(CO)_9(\mu_3\text{-}S)_2$ |57,58|. The Fe_3S_2 core consists of an open triiron triangle, capped above and below by a triply bridging sulfur atom. The nonbonding Fe···Fe distance is 3.37 Å.

This 50-electron molecule undergoes two subsequent one-electron reductions, only the first one having the features of chemical reversibility, Figure 13 |59|.

Figure 12. *Perspective view of $Fe_3(CO)_9(S)_2$. Bond lengths in Å (from Ref. 57).*

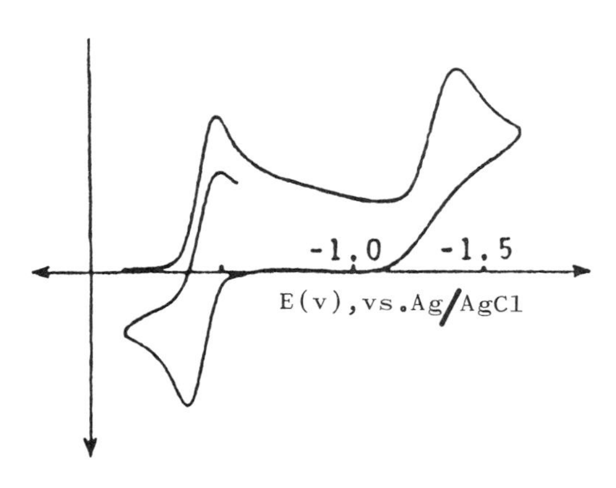

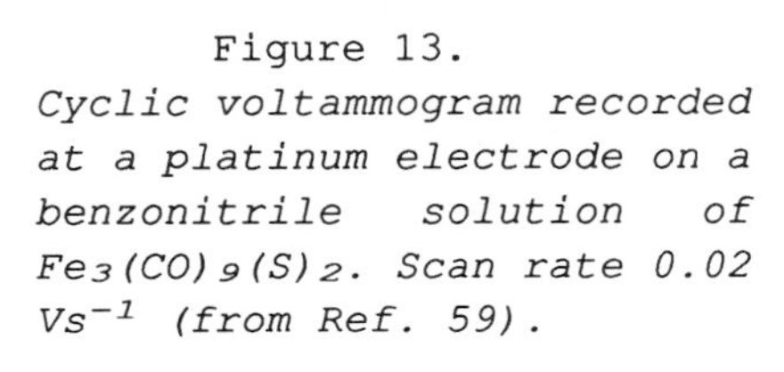

Figure 13.
Cyclic voltammogram recorded at a platinum electrode on a benzonitrile solution of $Fe_3(CO)_9(S)_2$. Scan rate 0.02 Vs^{-1} (from Ref. 59).

The redox potentials for such electron transfers are reported in Table 5, together with those of some related species which will be discussed later on.

Similar neutral/monoanion reductions have been observed for the isoelectronic and isostructural derivatives $Fe_3(CO)_9(Se)_2$ |62| and $Fe_3(CO)_8(C_3H_2S_2)(S)_2$ |63|. In this latter case, the presence of the strong σ-donor carbenic substituent makes the one-electron addition notably more difficult.

The quasireversibility of the 0/1- electron transfer, preluding significant structural reorganization, agrees either with an elongation of the Fe-Fe bonds, predicted on the basis that the extra-electron enters a metal-metal antibonding orbital |64|, or with some bond-breaking at the capping sulfur atoms, hypothesized as an intimate mechanism of the Electron-Transfer-Chain catalytic substitution of CO groups for phosphite ligands |60,61|.

TABLE 5. *Formal electrode potentials for the electron transfers exhibited by $Fe_3(CO)_9(S)_2$ and related compounds*

complex	$E°'_{(0/1-)}$ (V)	ΔEp (mV)	$Ep_{(1-/2-)}$ (V)	solvent	reference
$Fe_3(CO)_9(S)_2$	-0.48	76[a]	-1.43[a]	C_6H_5CN	59
	-0.72	120[b]			52
	-0.61	130[c]		CH_2Cl_2	60
	-0.40			DMF	61
$Fe_3(CO)_8(C_3H_2S_2)(S)_2$	-0.73	60[d]		DMF	61
$Fe_3(CO)_9(Se)_2$	-0.56	100[c]		CH_2Cl_2	60
$Fe_3(CO)_9(S)(CO)$	-0.31[e]			C_6H_5CN	59
$Fe_3(CO)_9(S)(SO)$	-0.36[e]			C_6H_5CN	59

[a]*Measured at 0.02 Vs^{-1};* [b]*measured at 0.2 Vs^{-1};* [c]*measured at 0.5 Vs^{-1};* [d]*measured at 0.1 Vs^{-1};* [e]*peak potential value for an irreversible process*

2.2.7. $Fe_3(CO)_9(S)(X)$ (X = CO,SO). The analogies between $Fe_3(CO)_9(\mu_3\text{-}S)(\mu_3\text{-}CO)$ and $Fe_3(CO)_9(\mu_3\text{-}S)(\mu_3\text{-}SO)$ are simply limited to their formulation. In fact, as shown in Figure 14, the 48-electron

$Fe_3(CO)_9(S)(CO)$ is constituted by a bicapped, total bonding iron triangle (Fe-Fe, 2.61 Å) |65|, whereas the 50-electron $Fe_3(CO)_9(S)(SO)$ is constituted by a bicapped, open iron triangle (Fe-Fe, 2.62 Å) |66|.

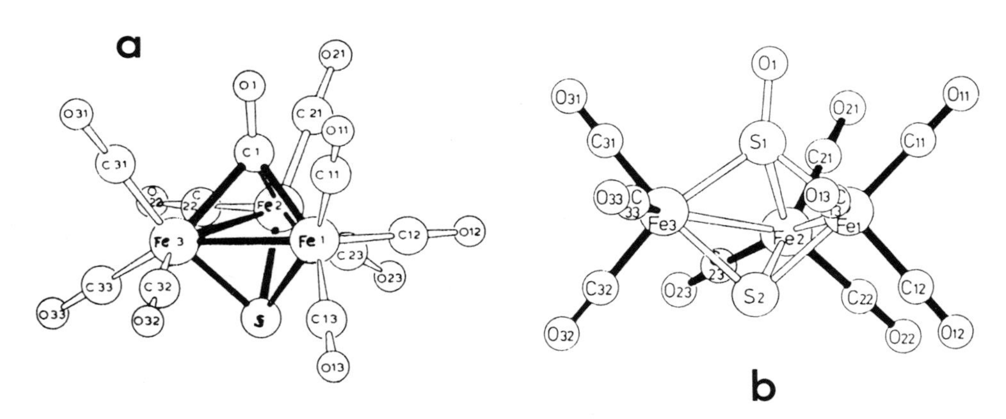

Figure 14. *Molecular structures of: (a) $Fe_3(CO)_9(S)(CO)$ (from Ref. 65); (b) $Fe_3(CO)_9(S)(SO)$ (from Ref. 66).*

In contrast with the related species $Fe_3(CO)_9(E)_2$ (E = S, Se), the present derivatives display an irreversible reduction to the corresponding monoanion (Table 5), supporting the view that the stability of bicapped triiron-monoanions is remarkably dependent upon the nature of the capping units |60|.

Even if not structurally characterized, it is not unlikely that the derivatives $(C_5H_5)_3Fe_3(CO)_2(S)(SR)$ belong to the family of complexes presented in this Section. They seem however to undergo reversibly two one-electron removals |67|.

2.2.8. $Fe_3(CO)_9(E)(PR)$ (E = S,Se,Te). Figure 15 shows the molecular structure of the two complexes $Fe_3(CO)_9(S)(P\text{-}Ph)$ |68,69| and $Fe_3(CO)_9(Te)(P\text{-}Pr^i)$ |70|. In both cases the metallic triangle is open.

As illustrated in Figure 16, which refers to $Fe_3(CO)_9(S)(P\text{-}Ph)$, the derivatives $Fe_3(CO)_9(E)(PR)$ (E = S, Se, Te) display two separated one-electron reduction steps, with features of chemical reversibility at least in the short times of cyclic voltammetry |71|. The relevant redox potentials are reported in Table 6.

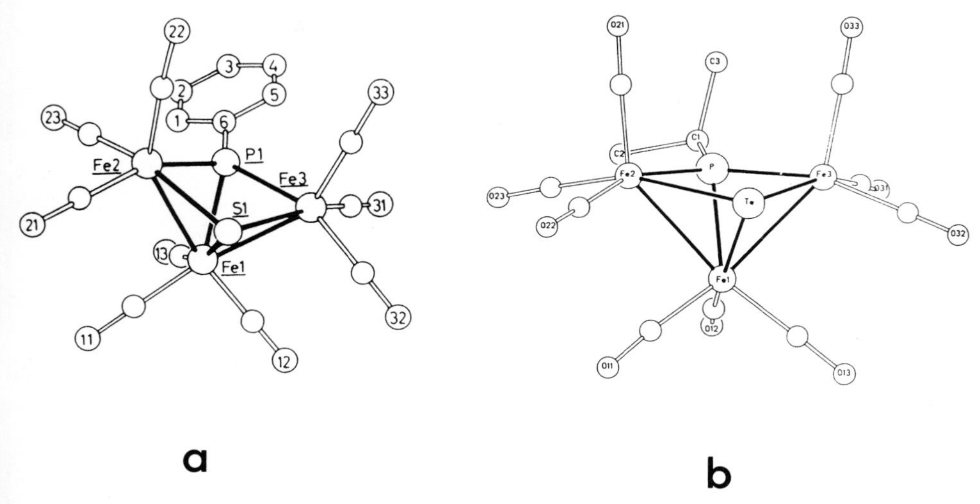

Figure 15. *X-Ray structures of: (a) $Fe_3(CO)_9(S)(P\text{-}Ph)$. Averaged Fe-Fe distance, 2.65 Å; averaged Fe-S distance, 2.26 Å; Fe2-P = Fe3-P = 2.19 Å; Fe1-P = 2.26 Å (from Ref. 68); (b) $Fe_3(CO)_9(Te)(P\text{-}Pr^i)$. Averaged Fe-Fe distance, 2.70 Å; averaged Fe-Te distance, 2.55 Å; Fe2-P = Fe3-P = 2.19 Å; Fe1-P = 2.24 Å (from Ref. 70).*

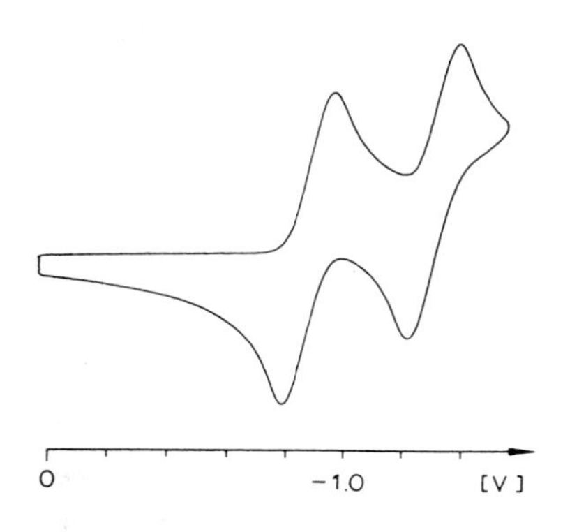

Figure 16. *Cyclic voltammogram exhibited by $Fe_3(CO)_9(S)(PPh)$ in dichloromethane solution. Glassy-carbon electrode. Scan rate 0.2 Vs^{-1} (from Ref. 71).*

TABLE 6. *Formal electrode potentials for the two electron transfers exhibited by $Fe_3(CO)_9(E)(PR)$ in dichloromethane solution*

E	R	$E^{\circ\prime}_{(0/1-)}$ (V)	$E^{\circ\prime}_{(1-/2-)}$ (V)	reference
S	Ph	-0.92	-1.35	71
	Pr^i	-0.91	-1.34	71
	Bu^t	-0.82	-1.29	52
Te	Pr^i	-0.76[a]	-1.13[a]	52
Se	Pr^i	-0.79[a]	-1.23[a]	52

[a]*Converted to the common value of +0.45 V for the Fc/Fc^+ reference couple*

2.2.9. $Fe_3(CO)_8(S)(CR_2)(PR')$. Figure 17 shows that the triron triangle in the carbene substituted complex $Fe_3(CO)_8(S)[C(NH_2)_2](PPr^i)$ is once again open |72|.

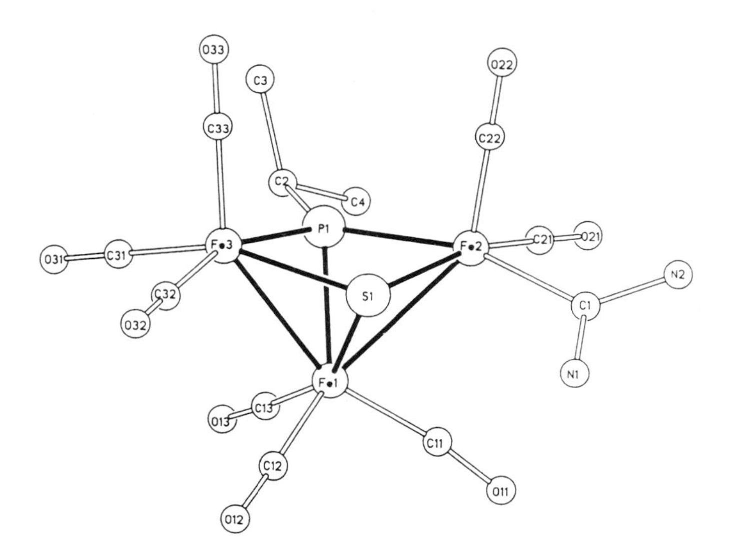

Figure 17. *Perspective view of $Fe_3(CO)_8(S)[C(NH_2)_2](PPr^i)$. Averaged Fe-Fe distance, 2.66 Å; averaged Fe-S distance, 2.25 Å; Fe2-P = Fe3-P, 2.20 Å; Fe1-P. 2.25 Å (from Ref. 72).*

These carbene complexes also undergo two separated one-electron reductions, the extent of chemical reversibility of which depends upon the carbene substituents. The relevant redox potentials are summarized in Table 7.

TABLE 7. *Formal electrode potentials for the electron transfers exhibited by $Fe_3(CO)_8(CRR')(S)(PR'')$ in dichloromethane solution |71|*

R	R′	R″	$E^{\circ\prime}_{(0/1-)}$ (V)	$E^{\circ\prime}_{(1-/2-)}$ (V)
NH_2	NH_2	Ph	-1.27	-1.50
Me	NH_2	Pr^i	-1.19[a]	-1.64[a]

[a]*Peak potential value for irreversible steps*

2.2.10. $Fe_3(CO)_{9-n}(L)_n(PPh)_2$ (L = PR_3, n = 0-3). The parent molecule of the complexes of general formula $Fe_3(CO)_{9-n}(L)_n(PPh)_2$ is $Fe_3(CO)_9(\mu_3\text{-}PPh)_2$, the structure of which is shown in Figure 18 |73|. The assembly is built around an open triiron triangle having an averaged Fe-Fe bond length of 2.72 Å [0.1 Å longer than the corresponding distance in $Fe_3(CO)_9(S)_2$]. The nonbonding Fe···Fe distance is 3.54 Å. The two phenylphosphinidene groups cap, on opposite sides, the trimetallic fragment.

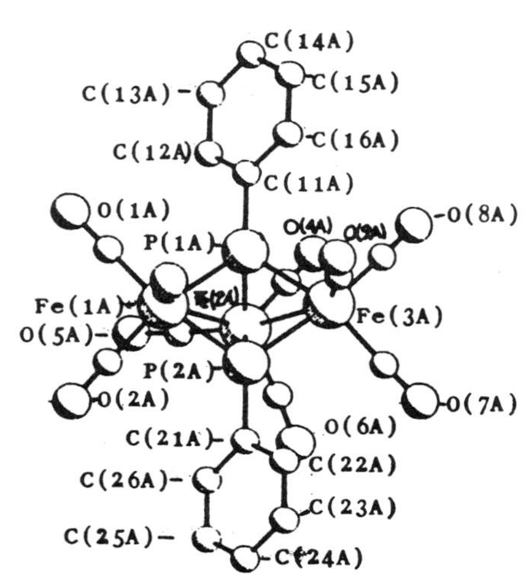

Figure 18. *Perspective view of $Fe_3(CO)_9(PPh)_2$ (from Ref. 73).*

As illustrated in Figure 19, $Fe_3(CO)_9(PPh)_2$ undergoes two subsequent one-electron reductions, with marked features of chemical and electrochemical reversibility |74,75|. The relevant potential values are summarized in Table 8.

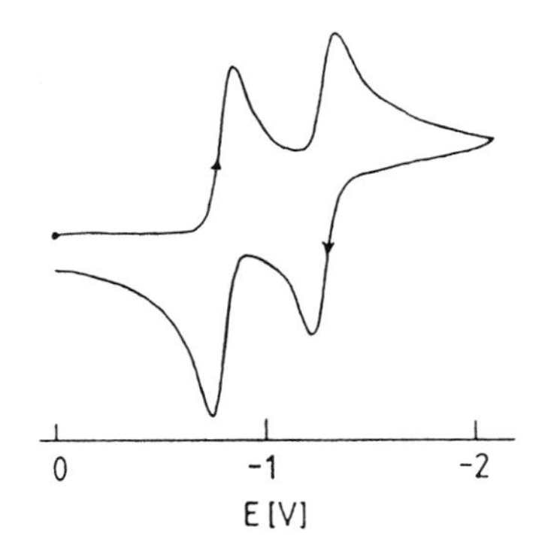

Figure 19. *Cyclic voltammetric response recorded at a platinum electrode on a THF solution of $Fe_3(CO)_9(PPh)_2$. Scan rate 0.5 Vs^{-1} (from Ref. 74).*

In agreement with the substantial reversibility of the first one-electron addition, the instantaneously generated monoanion $[Fe_3(CO)_9(PPh)_2]^-$ was assigned, on the basis of its ESR spectrum, a geometry essentially similar to that of the neutral parent. Nevertheless, on standing, the monoanion likely undergoes the skeletal-opening rearrangement shown in Scheme 1, which, exposing a coordinatively unsaturated 17-electron iron center, favours substitution of CO groups for phosphorus donor ligands by electron-trasfer catalysis |74,75|.

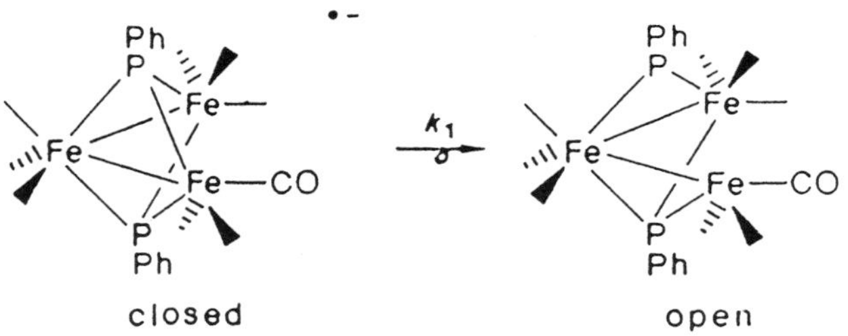

closed $\xrightarrow{k_1}$ open

SCHEME 1

Similar reorganizations are also hypothesized for the second reduction step. In fact, in the timescale of cyclic voltammetry, it is presumed that $[Fe_3(CO)_9(PPh)_2]^{2-}$ maintains the geometry of its precursors. In longer times, hydridic attack at the phosphinidene capping unit occurs, with formation of the species schematized in Figure 20 |74,75|.

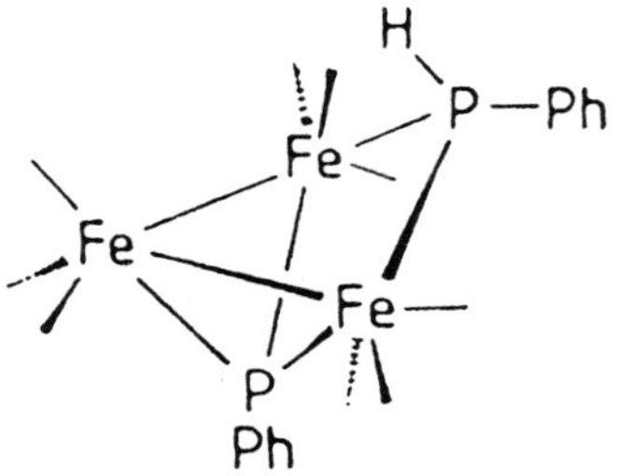

Figure 20.
Schematic representation of the species formed by hydridic attack to the dianion $[Fe_3(CO)_9(PPh)_2]^{2-}$.

As briefly above outlined, $Fe_3(CO)_9(PPh)_2$ undergoes stepwise substitution of CO ligands by phosphines or phosphites at the three different iron sites. There is, in fact, spectral evidence that the first substitution occurs at one of the two equivalent iron centres, whereas the second substitution takes place at the apical iron site, Scheme 2 |74,75|.

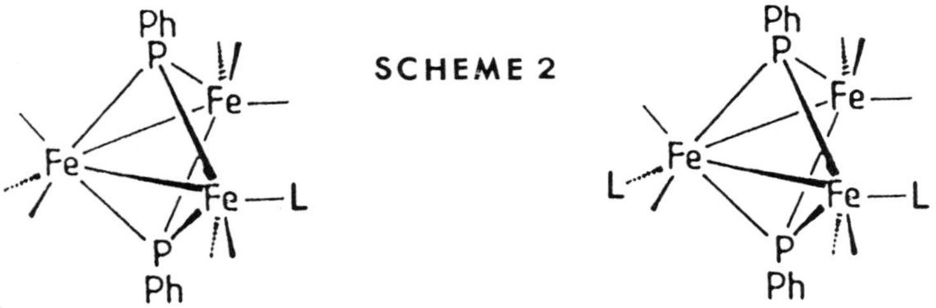

SCHEME 2

On the other hand, the crystal structure of the trisubstituted species is known through that of $Fe_3(CO)_6(PPh)_2[P(OMe)_3]_3$ |75|, which is illustrated in Figure 21.

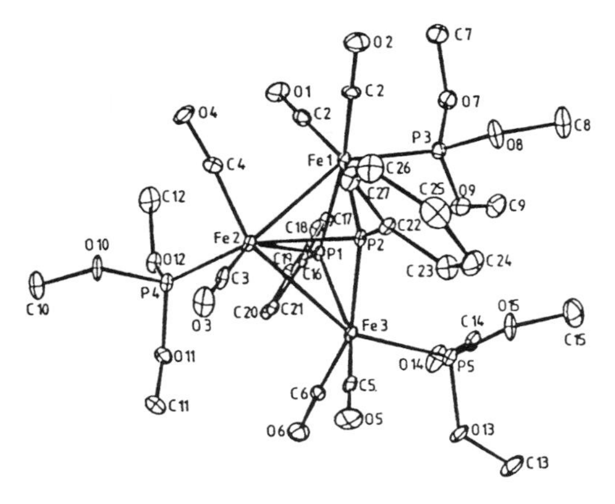

Figure 21. *Perspective view of $Fe_3(CO)_6(PPh)_2[P(OMe)_3]_3$ (from Ref. 75).*

The stereochemistry of the Fe_3P_2 core is similar to that of the unsubstituted precursor. The mean Fe-Fe bond length (Fe1-Fe2, 2.68 Å; Fe2-Fe3, 2.74 Å) is also similar. The nonbonding edge is 3.62 Å.

As deducible from Table 8, the two reversible electron reductions, typical for the unsubstituted precursor, also characterize the substituted congeners, but such electron additions become more and more difficult with the increasing of the number of substituents, in agreement with the σ-donor ability of the phosphorus-based ligands, which raises in energy the LUMO level of the Fe_3P_2 molecule. Concomitantly, the oxidation process becomes easier and easier; as a matter of fact, the one-electron removal from either the unsubstituted or the monosubstituted species is framework-destroying, while it is chemically reversible at -50 °C for $Fe_3(CO)_7(PPh)_2[P(OMe)_3]_2$, and becomes reversible, even at room temperature, for $Fe_3(CO)_6(PPh)_2[P(OMe)_3]_3$.

Concerning the stereochemistry of these redox changes, they are assumed, on the basis of ESR analysis, to be equivalent to those described for the unsubstituted derivative |74|.

TABLE 8. *Redox potentials for the electron-transfer sequences exhibited by $Fe_3(CO)_{9-n}(PPh)_2(L)_n$ in tetrahydrofuran solution*

complex	E°' (0/-) (V)	E°' (-/2-) (V)	E°' (+/0) (V)	reference
$Fe_3(CO)_9(PPh)_2$	-0.79	-1.30	+1.37[a]	74, 75
$Fe_3(CO)_8(PPh)_2(PPh_3)$	-1.01	-1.62	+1.09[a]	74
$Fe_3(CO)_8(PPh)_2(PEt_3)$	-1.16	-1.64	+1.05[a]	75
$Fe_3(CO)_8(PPh)_2[P(OMe)_3]$	-1.09	-1.54	+1.18[a]	75
$Fe_3(CO)_7(PPh)_2(PEt_3)_2$	-1.44	-1.85	+0.71	75
$Fe_3(CO)_7(PPh)_2[P(OMe)_3]_2$	-1.32	-1.68	+0.72[b]	75
$Fe_3(CO)_6(PPh)_2[P(OMe)_3]_3$	-1.52	-1.88	+0.49	75

[a]*Peak potential value for irreversible step;* [b]*at -50 °C*

2.2.11. $Fe_3(CO)_{9-n}(L)_n(NPh)_2$ [L = $P(OMe)_3$, n = 0-2]. The phenylnitrene derivatives $Fe_3(CO)_{9-n}[P(OMe)_3]_n(NPh)_2$ are homologues of the phenylphosphinidene-capped family discussed in Section 2.2.10.

The available crystallographic data concern $Fe_3(CO)_9(\mu_3\text{-}NPh)_2$ |76| and $Fe_3(CO)_8(\mu_3\text{-}NPh)_2[P(OMe)_3]$ |60|; their structural features are illustrated in Figure 22. The bond lengths in the basal, open metallic triangle are almost unaffected by the absence or the presence of the phosphite substituent. In fact, they are 2.48 Å *vs.* 2.42 Å, and 2.49 Å *vs.* 2.48 Å, respectively.

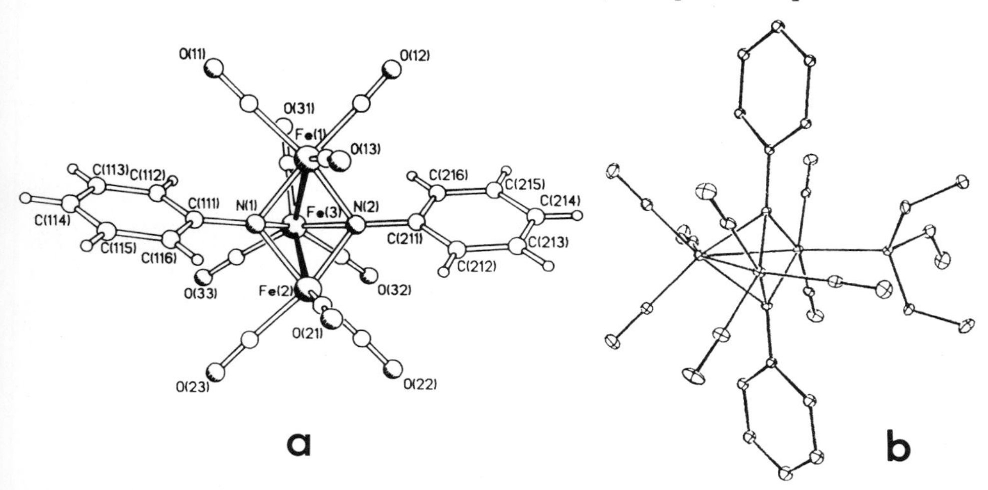

Figure 22. *Perspective view of: (a) $Fe_3(CO)_9(NPh)_2$ (from Ref. 76); (b) $Fe_3(CO)_8(NPh)_2[P(OMe)_3]$ (from Ref. 60).*

In contrast with the relative stability of the monoanion species $[Fe_3(CO)_{9-n}(L)_n(PPh)_2]^-$, the monoanion derivatives of the phenylnitrene family are notably short-lived. For instance, a lifetime of about 0.01 sec has been estimated for $[Fe_3(CO)_9(NPh)_2]^-$ |60|. In this connection, Figure 23 shows the cyclic voltammetric response exhibited by $Fe_3(CO)_9(NPh)_2$, in dichloromethane solution, both under argon and carbon monoxide atmosphere |60|. The appearance of peak N in Figure 23a, under argon atmosphere, is likely due to the reduction of minor amounts of $[Fe_3(CO)_8(NPh)_2]^-$ arising from fast decarbonylation of the electrogenerated monoanion $[Fe_3(CO)_9(NPh)_2]^-$. Such a decomposition is, on the other hand, retarded under CO atmosphere, Figure 23b.

Table 9 summarizes the redox potentials for the neutral/ monoanion reduction of the present series. It is once again confirmed that the more substituted is the cluster by phosphorus-containing substituents, the more difficult is the electron addition.

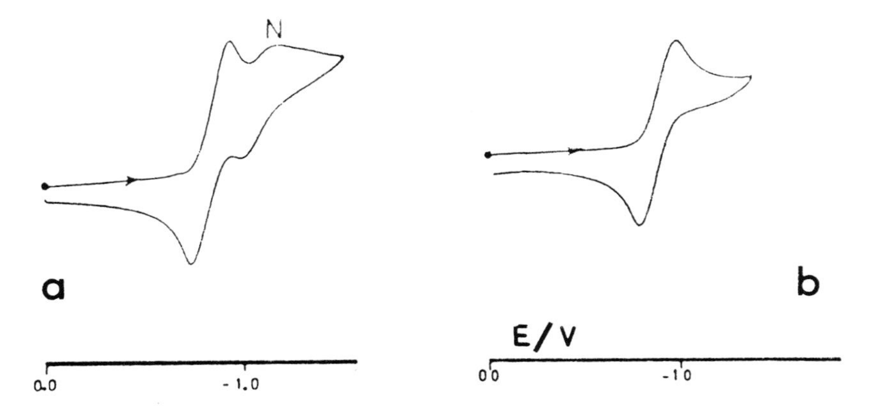

Figure 23. *Cyclic voltammograms recorded at a platinum electrode on a deaerated CH_2Cl_2 solution of $Fe_3(CO)_9(NPh)_2$. (a) Under argon atmosphere; (b) under CO atmosphere. $[NBu_4][ClO_4]$ supporting electrolyte. Scan rate 0.2 Vs^{-1} (from Ref. 60).*

TABLE 9. *Redox potentials for the one-electron reduction of $Fe_3(CO)_{9-n}(L)_n(NPh)_2$ |60|*

complex	E°' (0/-) (V)	ΔEp[a] (mV)	solvent
$Fe_3(CO)_9(NPh)_2$	-0.86	220	CH_2Cl_2
	-0.71	110	THF
$Fe_3(CO)_8(NPh)_2[P(OMe)_3]$	-0.96	330	CH_2Cl_2
$Fe_3(CO)_7(NPh)_2[P(OMe)_3]_2$	-1.43[b]		CH_2Cl_2

[a]*Measured at 0.5 Vs^{-1};* [b]*peak potential value for an irreversible process*

At variance with the substantial chemical and electrochemical reversibility of the one-electron reduction of $Fe_3(CO)_{9-n}(L)_n(PPh)_2$, the same redox change in $Fe_3(CO)_{9-n}(L)_n(NPh)_2$ not only is kinetically complicated by decarbonylation, but it is also decidedly quasireversible in character. This means that some significant structural rearrangement occurs in the framework of $[Fe_3(CO)_{9-n}(L)_n(NPh)_2]^-$ with respect to the parent precursor.

The important question concerning the extremely fast loss of CO by $[Fe_3(CO)_9(NPh)_2]^-$ with respect to the extremely slow loss of CO by $[Fe_3(CO)_9(PPh)_2]^-$ still remains theoretically unresolved.

2.2.12. $Fe_3(CO)_{8-n}(L)_n(RC_2R)_2$ [L = $P(OMe)_3$, n = 0,1]. As illustrated in Figure 24, the *black* form of the ferracyclopentadienyl cluster $Fe_3(CO)_6(\mu\text{-}CO)_2(PhC_2Ph)_2$, which is the most stable precursor of the family $Fe_3(CO)_6(\mu\text{-}CO)_2(RC_2R)_2$, has the peculiar structural feature that the open trimetallic triangle is perpendicularly intersected by a butadiene chain forming a five-membered ring with the apical Fe atom |77|. Two edge bridging carbonyls are located at the two Fe-Fe bonds (2.43 Å). The two iron atoms of the nonbonding edge (Fe···Fe, 3.36 Å) form nearly symmetrical π-bonds with each carbon atom of the bisecting chain.

In contrast, in the less stable *violet* form, the different components of the complex still maintain their original shapes, in that the closed, isosceles, metallic triangle is capped, at opposite sides, by the two acetylene fragments |77|.

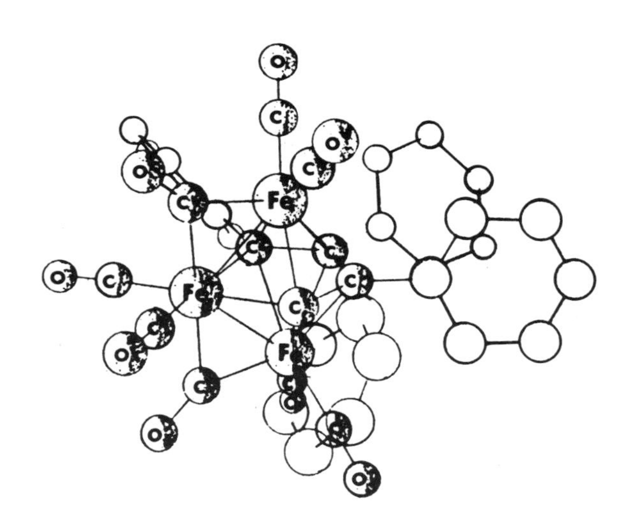

Figure 24. *Perspective view of the black isomer of $Fe_3(CO)_8(PhC_2Ph)_2$ (from Ref. 77).*

The complex $Fe_3(CO)_6(\mu\text{-}CO)_2[(Ph)CC_2(NEt_2)_2C(C_3H_5)]$ also possesses a geometry in which the open metallic triangle is perpendicularly intersected by the ferrocyclopentadiene plane, Figure 25 |78|.

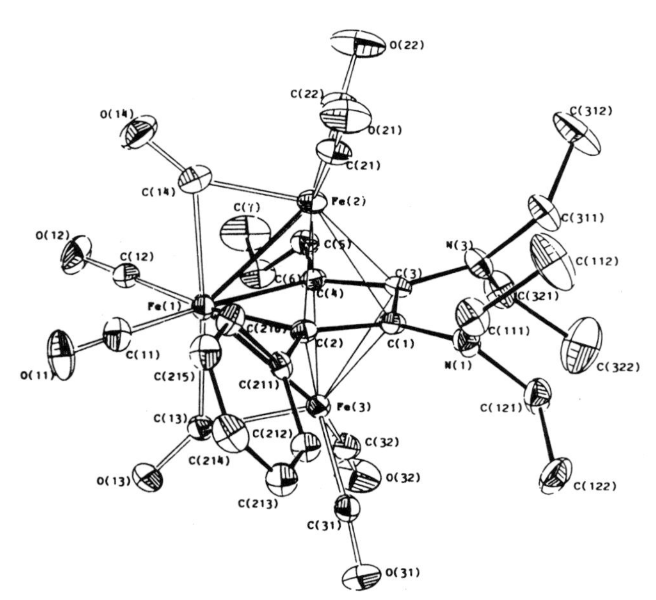

Figure 25. *X-Ray structure of* $Fe_3(CO)_8[(Ph)CC_2(NEt_2)_2C(C_3H_5)]$. *Averaged Fe-Fe distance, 2.42 Å (from Ref. 78).*

Typically, the present 46-cluster-valence-electron derivatives undergo sequentially two one-electron additions as well as two one-electron removals |78-80|. While the two electron removals are always declustering steps, the two successive one-electron additions, as shown in Figure 26 for $Fe_3(CO)_8(PhC_2Ph)_2$, have features of chemical reversibility, apparently in agreement with the formal unsaturation of the assembly.

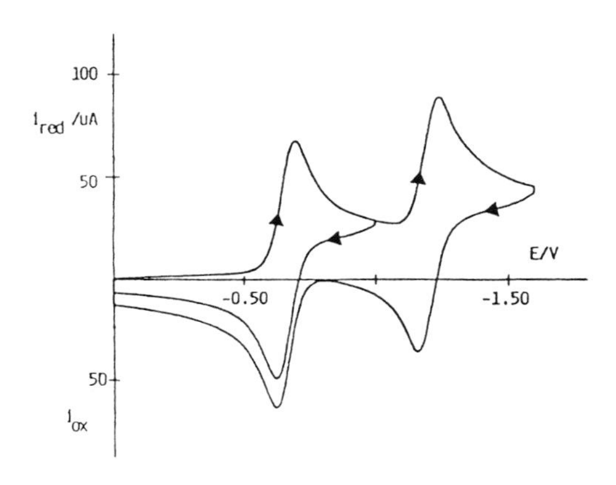

Figure 26.
Cyclic voltammogram exhibited at a mercury electrode by an acetonitrile solution of $Fe_3(CO)_8(PhC_2Ph)_2$. *Scan rate 0.2* Vs^{-1} *(from Ref. 80).*

Indeed, only the monoanions $[Fe_3(CO)_8(RC_2R)_2]^-$ are stable in the longer times of electrolysis, whereas the corresponding dianions mainly revert, through a conproportionation reaction, to their monoanions.

It must be noted that the near electrochemical reversibility of the neutral/monoanion reduction testifies to a substantial preservation of the starting geometry. This is in keeping with the fact that the first added electron enters a complex LUMO level, which is antibonding with respect to the Fe-Fe vectors, but bonding with respect to the iron-butadiene ferrocyclopentadienyl fragment |70,81|, so that the increased electron density is in part diverted towards the organic chain.

Table 10 reports the redox potentials of the electron transfers exhibited by the present series of triiron clusters.

As expected on the basis of the LUMO's character, the electron-donating ability of the acetylene substituents significantly affects the potentials of the reduction processes.

The phosphite monosubstituted species exists in two isomeric forms, which, on the basis of spectral data, have assigned the structure seen in Scheme 3 |82|.

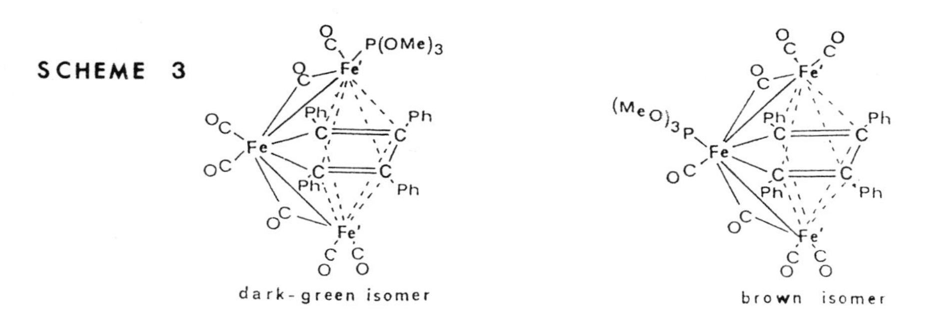

The presence of the phosphite must affect significantly the redox orbitals; in fact, it not only makes the addition of electrons more difficult (see Table 10), but the monoanion $[Fe_3(CO)_7\{P(OMe)_3\}(PhC_2Ph)_2]^-$ becomes no more stable ($t_{1/2} \cong 5$ sec).

2.2.13. $Fe_3(CO)_7(L)_2$ [L = α-C_6H_4(S)(NN)]. The complex $Fe_3(CO)_7[\alpha$-$C_6H_4(S)(NN)]_2$, the structure of which is illustrated in Figure 27, represents the only cluster having a linear triiron assembly |83|.

TABLE 10. *Formal electrode potentials (in Volt) for the redox changes exhibited by the complexes* $Fe_3(CO)_{8-n}|P(OMe)_3|_n(RC_2R)_2$ |70,78|

complex	$Ep_{(2+/1+)}$ [a]	$Ep_{(1+/0)}$ [a]	$E^{\circ\prime}_{(0/1-)}$	$E^{\circ\prime}_{(1-/2-)}$	solvent
$Fe_3(CO)_8(PhC_2Ph)_2$	+1.36	+1.21	-0.66	-1.20	MeCN
	+1.45	+1.31	-0.75	-1.31	CH_2Cl_2
$Fe_3(CO)_8(EtC_2Et)_2$	+1.08	+0.97	-0.81	-1.45	MeCN
	+1.43	+1.21	-0.94	-1.61	CH_2Cl_2
$Fe_3(CO)_8(PhC_2Ph)(EtC_2Et)$	+1.12	+0.98	-0.76	-1.38	MeCN
	+1.38	+1.21	-0.83	-1.44	CH_2Cl_2
$Fe_3(CO)_8(PhC_2Ph)(HC_2H)$	+1.13	+0.94	-0.70	-1.28	MeCN
	+1.42	+1.13	-0.74	-1.33	CH_2Cl_2
$Fe_3(CO)_8(EtC_2Et)(HC_2H)$	+1.17	+0.99	-0.81	-1.45	MeCN
	+1.56	+1.15	-0.86	-1.54	CH_2Cl_2
$Fe_3(CO)_8(MeC_2Ph)(HC_2H)$	+1.23	+1.06	-0.74	-1.33	MeCN
	+1.56	+1.23	-0.83	-1.49	CH_2Cl_2
$Fe_3(CO)_8(MeC_2Ph)(PhC_2Ph)$	+1.30	+1.18	-0.73	-1.29	MeCN
	+1.39	+1.27	-0.80	-1.35	CH_2Cl_2
$Fe_3(CO)_8|(Ph)CC_2(NEt_2)_2C(C_3H_5)|$		+1.20	-0.97	-1.30	CH_2Cl_2
$Fe_3(CO)_8|(C_3H_5)CC_2(NEt_2)_2C(C_3H_5)|$		+1.13	-1.02	-1.35	CH_2Cl_2
$Fe_3(CO)_7|P(OMe)_3|(PhC_2Ph)_2$ [b]			-0.98	-1.24	CH_2Cl_2
$Fe_3(CO)_7|P(OMe)_3|(PhC_2Ph)_2$ [c]			-1.03	-1.24	CH_2Cl_2

[a] *Peak potential value for irreversible process;* [b] *dark-green isomer (see text);* [c] *brown isomer (see text)*

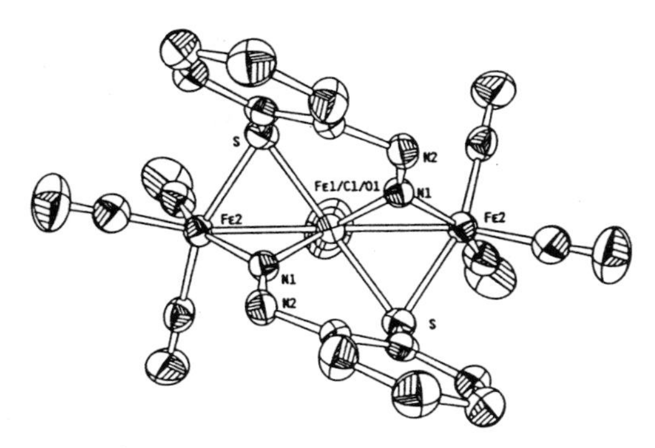

Figure 27. *Perspective view of $Fe_3(CO)_7[C_6H_4(S)(NN)]_2$ (from Ref. 83).*

The linearity of the trimetallic core is imposed by the two (unusual) α-diazo thioketone ligands, which lie at the same side of the Fe_3 unit, with a dihedral angle of 80°, imparting an overall butterfly shape to the cluster. Each one of the two outer iron atoms bears three terminal CO groups, whereas the central iron atom binds one terminal carbonyl. The mean Fe-Fe bond length is 2.49 Å.

As shown in Figure 28, the present cluster undergoes, in dichloromethane solution, both a one-electron reduction ($E°' = -0.88$ V) and a one electron oxidation ($E°' = +0.81$ V).

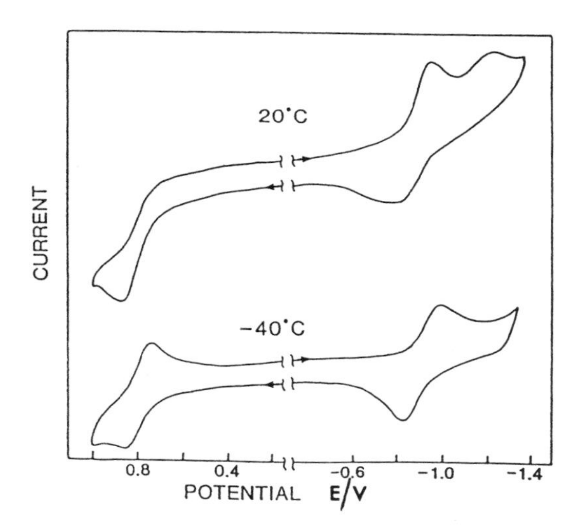

Figure 28. *Cyclic voltammograms recorded at a platinum electrode on a CH_2Cl_2 solution of $Fe_3(CO)_7[C_6H_4(S)(NN)]_2$ (from Ref. 84).*

Such redox changes are complicated, at room temperature, by decomposition of the primarily electrogenerated monoanion and monocation, respectively. Only at -40 °C such species are stable, and they have been characterized by EPR spectroscopy |84|.

2.3. Tetranuclear Binary Complexes

2.3.1. $[Fe_4(CO)_{13}]^{2-}$. The tetrametal core of the dianion $[Fe_4(CO)_{13}]^{2-}$ possesses a nearly regular tetrahedral geometry, but some slight structural differences may occur depending upon the nature of the counterion. In fact: (*i*) in $[Fe(C_5H_5N)_6][Fe_4(CO)_{13}]$ the apical iron atom bears three terminal carbonyls, whereas the three basal iron atoms are bonded to two terminal and one edge-bridging carbonyls; finally the thirteenth carbonyl is triply bridging with respect to the basal trimetallic plane, Figure 29a |85|; (*ii*) in $[\{(C_6H_5)_3P\}_2N]_2[Fe_4(CO)_{13}]$ all four metal atoms are bonded to three terminal carbonyl groups, and the thirteenth carbonyl maintains its triply-bridging capping nature with respect to the basal plane, Figure 29b |86|.

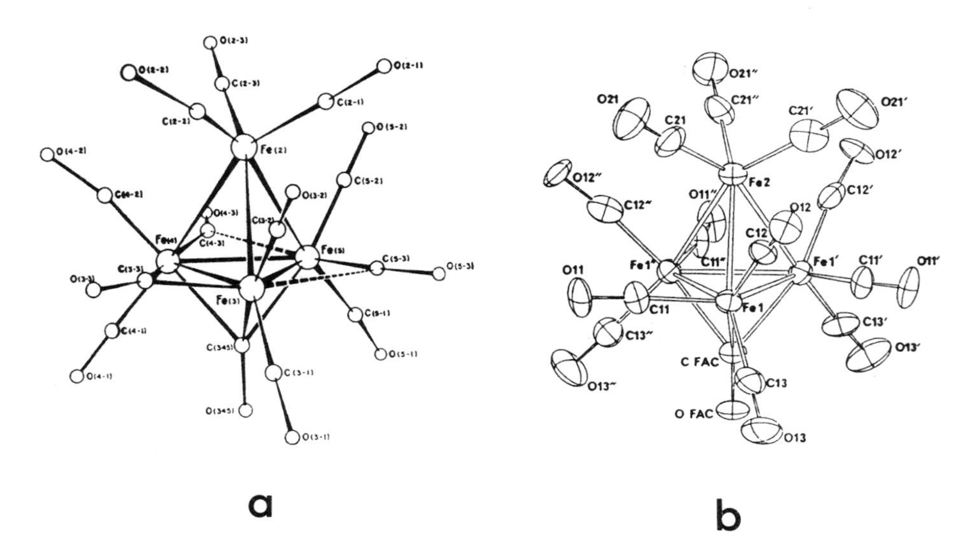

Figure 29. *Perspective views of $[Fe_4(CO)_{13}]^{2-}$. (a) $[Fe(C_5H_5N)_6]^{2+}$ counterion (from Ref. 85); (b) $[\{(C_6H_5)_3P\}_2N]^+$ counterion (from Ref. 86).*

Such variation in the disposition of the carbonyl groups induces minor differences in the metal-metal bonding distances. So, while the $Fe_{(apical)}$-$Fe_{(basal)}$ bond lengths are likely not significantly different (2.59 Å *vs.* 2.61 Å in the clusters of Figure 29a and 29b, respectively), the averaged $Fe_{(basal)}$-$Fe_{(basal)}$ distance widens from 2.50 Å to 2.54 Å, concomitantly with the transition of the CO groups from semibridging to terminal.

In tetrahydrofuran solution, the present 60-electron dianion undergoes an easy electrochemical one-electron oxidation (E°′ = -0.17 V), leading to the stable monoanion $[Fe_4(CO)_{13}]^-$ |42|; which, on the basis of its EPR spectrum, was assigned a structure closely related to that of the parent dianion |43|.

2.4. Tetranuclear Carbido Complexes

2.4.1. $[Fe_4(CO)_{12}C]^{2-}$. Figure 30 shows the crystal structure of $[Fe_4(CO)_{12}(\mu_4\text{-}C)]^{2-}$ in its benzyltrimethylammonium salt |87|, which is, on the other hand, coincident with that of its tetraammoniumzinc salt |88|.

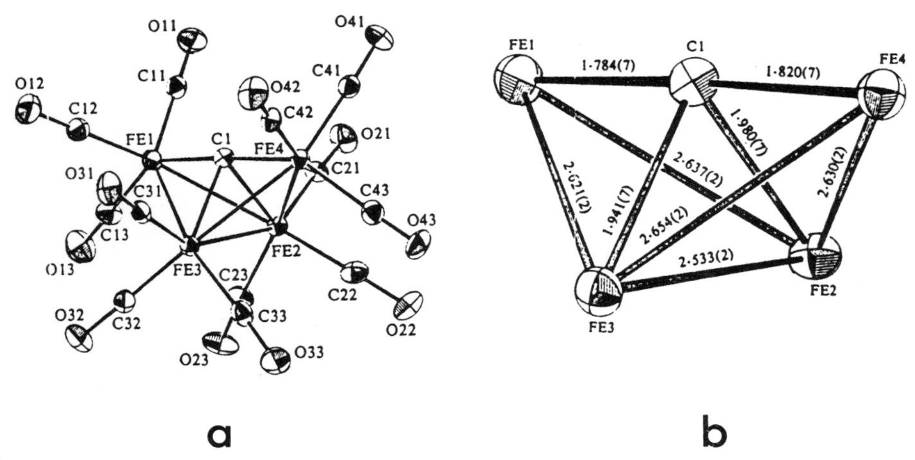

Figure 30. *Perspective view (a) and bond lengths (in Å) (b) of the dianion $[Fe_4(CO)_{12}C]^{2-}$. Dihedral angle between the Fe(1)Fe(2)Fe(3) and Fe(4)Fe(2)Fe(3) planes = 102.4° (from Ref. 87).*

The cluster, which can be obtained by reductive cleavage of the triply bridging CO of $[Fe_4(CO)_{13}]^{2-}$ |16|, has a butterfly arrangement of iron atoms with a carbon atom lying halfway in between the two wingtip ones.

It has been briefly reported |89| that such 62-electron dianion undergoes, in nonaqueous solutions (MeCN or CH_2Cl_2), four subsequent anodic steps, the first two likely generating the congeners $[Fe_4(CO)_{12}C]^-$ and $Fe_4(CO)_{12}C$, stable only in the short times of cyclic voltammetry (some seconds), but unstable in the longer times of electrolysis, because of their coordinative unsaturation |90|. As a matter of fact, molecular orbital calculations |91-94| suggest that the highest energy occupied orbitals are iron-iron bonding; in particular the HOMO should be essentially localized across the backbone iron atoms |92,93| [Fe(2)-Fe(3) in Figure 30]. This datum makes plausible deep molecular reorganizations upon electron removals, unless further attack by nucleophiles stabilizes the new species, as it happens by oxidation of $[Fe_4(CO)_{12}C]^{2-}$ in the presence of CO |88|. In this case, removal of the two electrons from the backbone-localized HOMO, is followed by addition of the two-electron donor CO across the backbone irons. The resulting 62-electron species $Fe_4(CO)_{13}C$ really maintains the starting butterfly-shaped Fe_4C core, with the thirteenth carbonyl bridging the backbone iron atoms, Figure 31 |95|.

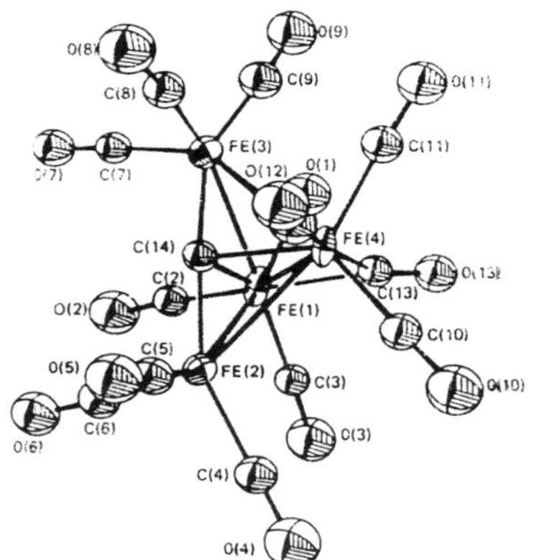

Figure 31. *Perspective view of $Fe_4(CO)_{13}C$. Fe(1)-Fe(4), 2.54 Å; Fe(1)-Fe(3) ≈ Fe(1)-Fe(2) ≈ Fe(3)-Fe(4) ≈ Fe(2)-Fe(4), 2.64 Å; Fe(2)-C(14) ≈ Fe(3)-C(14), 1.80 Å; dihedral angle between the Fe(1)Fe(2)Fe(4) and Fe(1)Fe(3)Fe(4) planes = 101° (from Ref. 95).*

2.4.2. $[Fe_4(CO)_{12}CC(O)R]^-$. One of the possible pathways in alkylation reactions of $[Fe_4(CO)_{12}C]^{2-}$ leads to μ_4-methylidyne clusters $[Fe_4(CO)_{12}CC(O)R]^-$ |89,96|, according to Scheme 4.

$$[Fe_4C]^{2-} \xrightarrow[CO]{R^+} [Fe_4C\text{–}C(O)R]^-$$

R = Me
R = CH_2Ph

SCHEME 4

These compounds have the usual tetrairon butterfly array, Figure 32, but significant differences exist with respect to the previously described 62-electron carbido derivatives {namely, $[Fe_4(CO)_{12}C]^{2-}$ and $Fe_4(CO)_{13}C$}. In fact, the actual bond length between the backbone and the wingtip irons of 2.51 Å is significantly shorter (≈0.13 Å); also, the wingtip iron-carbide carbon length of 2.01 Å is notably longer (≈0.21 Å). In addition, the present butterfly geometry is more opened, the dihedral angle between the two wings being 128° (in comparison with the value of 101°-102° for the preceding carbido clusters). The origin of this change in geometry must be attributed to the fact that the present acetyl-derivatized clusters have 60 valence electrons.

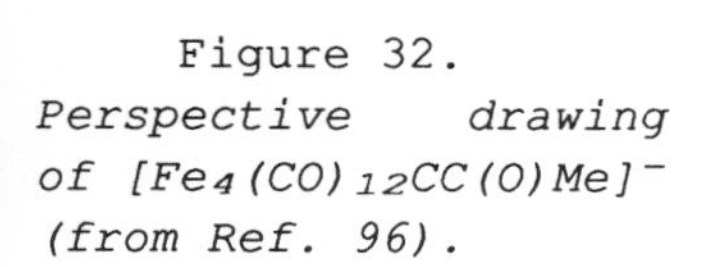

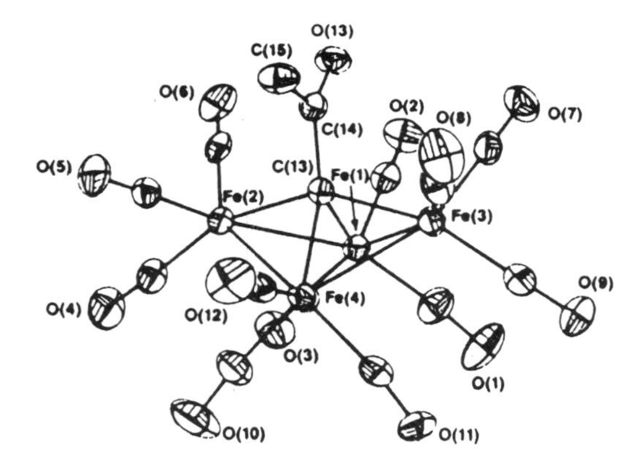

Figure 32. *Perspective drawing of* $[Fe_4(CO)_{12}CC(O)Me]^-$ *(from Ref. 96).*

Thus, Shriver and coworkers |96| have shown that these species can reversibly add electrons. In this connection, Figure 33 illustrates the cyclic voltammetric response exhibited by $[Fe_4(CO)_{12}CC(O)CH_2Ph]^-$ in acetonitrile solution |97|.

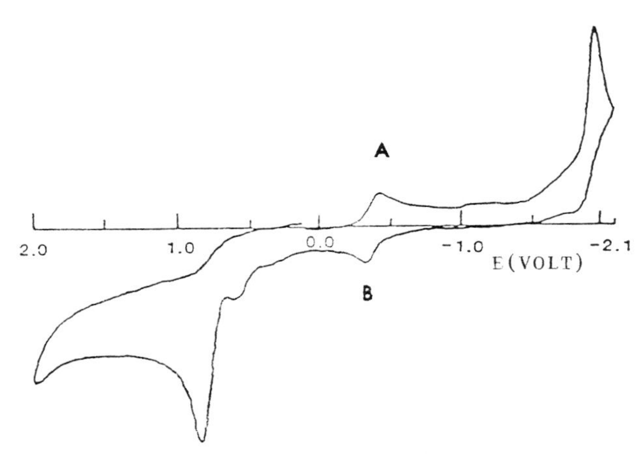

Figure 33. *Cyclic voltammogram recorded at a platinum electrode on a MeCN solution of [PPN][$Fe_4(CO)_{12}CC(O)CH_2Ph$]. Scan rate 0.1 Vs^{-1}. Ag/AgCl reference electrode (from Ref. 97).*

Apart from an irreversible multielectron process at very negative potential values [due to the reduction of bis (triphenyl phosphine)iminium cation] and an irreversible multielectron oxidation (attributable to declustering electron removals), a well defined peak-system A/B is present at $E°' = -0.43$ V (*vs.* S.C.E.) (both for the benzyl and methyl derivatives), due to a quasi-reversible, but chemically reversible, one-electron reduction.

The corresponding dianions have been obtained by chemical reduction (Na/Hg), and $[Fe_4(CO)_{12}CC(O)Me]^{2-}$ has been X-ray characterized, Figure 34 |97|.

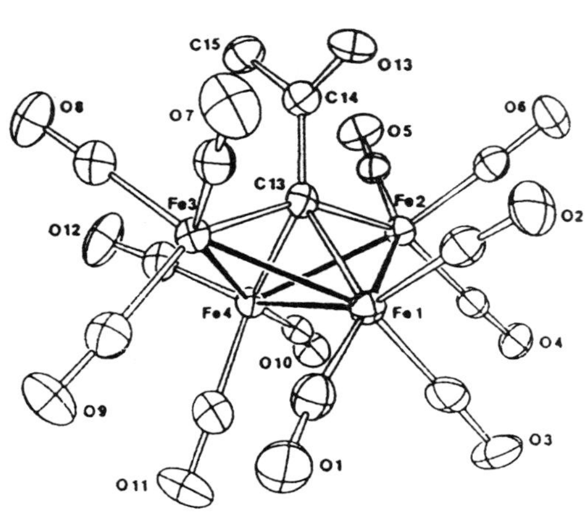

Figure 34. *ORTEP drawing of $[Fe_4(CO)_{12}CC(O)Me]^{2-}$ (from Ref. 97).*

It is interesting to note that the 60-electron/61-electron redox change is accompanied by an increase of the wingtip-backbone iron-iron bond (from 2.51 Å to 2.56 Å), as well as by a decrease of the butterfly dihedral angle (from 128° to 124°). These data point out to a tendency for restoration of the 62 cluster-valence-electron butterfly symmetry.

Further, it must be noted that the structure of the methylidyne group remains essentially unchanged. This datum, together with the insensitivity of both the redox potential and the ESR parameters towards methylidyne substituents, gives evidence for the fact that the added electron enters the $Fe_4(CO)_{12}$ moiety, rather than the $CC(O)CH_3$ group |97|.

Molecular orbital calculations can account for the easy addition of one-electron to the monoanions $[Fe_4(CO)_{12}CC(O)R]^-$. In fact, in contrast to $[Fe_4(CO)_{12}C]^{2-}$, a low-energy, essentially metal-based LUMO is available (the HOMO-LUMO gap is around 1.2-1.3 eV), which is antibonding between hinge and wingtip irons |97|.

2.5. Tetranuclear Nitrido Complexes

2.5.1. $[Fe_4(CO)_{12}N]^-$. Figure 35 shows the molecular structure of the nitrido monoanion $[Fe_4(CO)_{12}N]^-$ |98|. The Fe_4N butterfly core is quite reminiscent of the Fe_4C assembly of the isoelectronic dianion $[Fe_4(CO)_{12}C]^{2-}$ previously discussed; in the present case the two iron wings are significantly less opened wide.

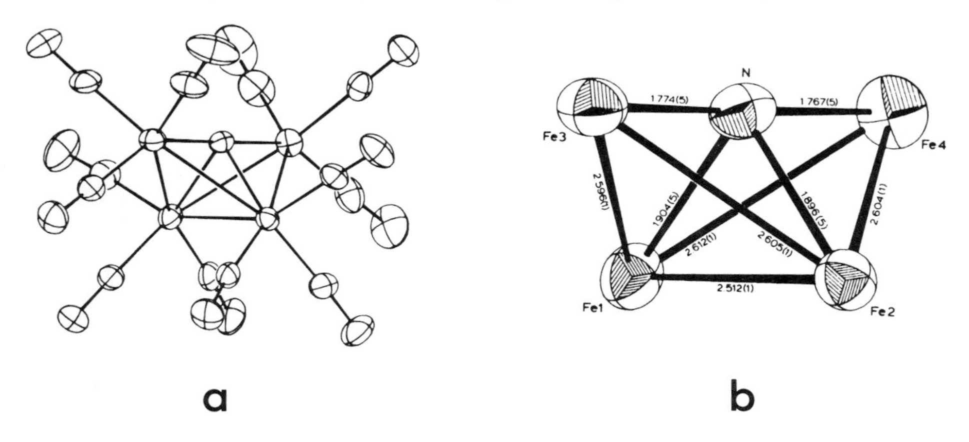

Figure 35. *Perspective view (a) and bond lengths (in Å) (b) of the monoanion $[Fe_4(CO)_{12}N]^-$. Dihedral angle between the Fe(1)Fe(2)Fe(3) and Fe(4)Fe(1)Fe(3) planes = 78.2° (from Ref. 98).*

As illustrated in Figure 36, in acetonitrile solution, the 62-electron complex $[Fe_4(CO)_{12}N]^-$ is able to add reversibly and stepwisely two more electrons according to the sequence |99|:

$$[Fe_4(CO)_{12}N]^- \underset{-e}{\overset{+e}{\rightleftharpoons}} [Fe_4(CO)_{12}N]^{2-} \underset{-e}{\overset{+e}{\rightleftharpoons}} [Fe_4(CO)_{12}N]^{3-}$$

$E°' = -1.24$ V (first step); $E°' = -1.58$ V (second step)

Indeed, only the dianion congener is indefinitely stable, and, based on the substantial electrochemical reversibility of the $[Fe_4N(CO)_{12}]^{-/2-}$ redox change, it likely maintains the geometry of the monoanion precursor.

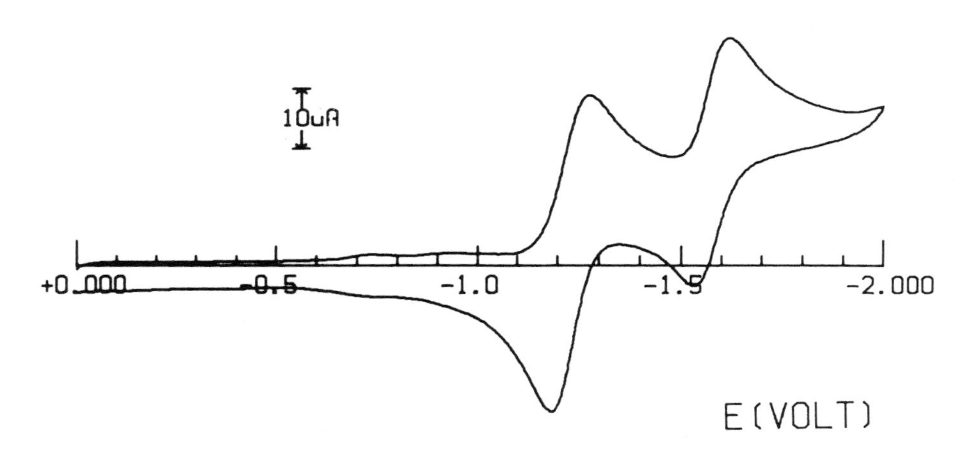

Figure 36. *Cyclic voltammogram recorded at a mercury electrode on an acetonitrile solution of $[Fe_4(CO)_{12}N]^-$. Scan rate 0.2 Vs^{-1}.*

We have gained preliminary evidence that, in the presence of phosphines, the cathodic reduction of $[Fe_4N(CO)_{12}]^-$ may easily afford the monosubstituted species $[Fe_4N(CO)_{11}(PR_3)]^-$ according to an Electron-Transfer-Chain catalytic process of the type:

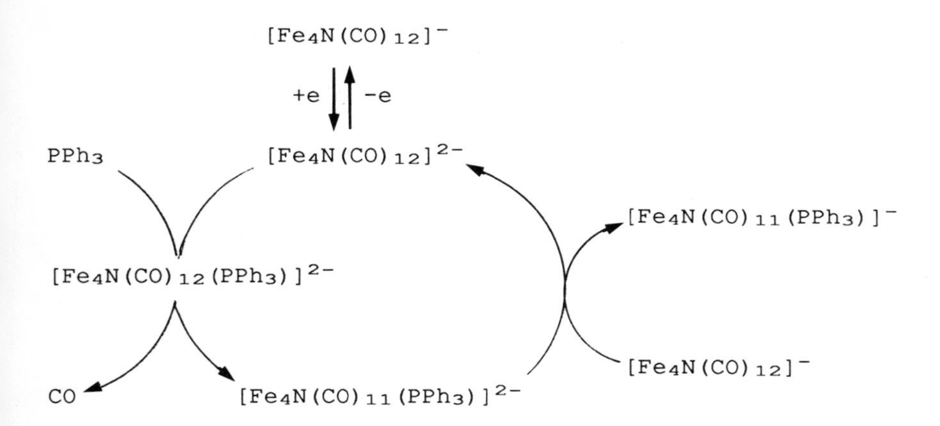

We are investigating if such an electrocatalytic path might favourably compete with the recently proposed phoshine substitution by thermal activation, which has led to the X-ray characterization of $[Fe_4N(CO)_{11}(PMe_2Ph)]^-$ |100|.

2.6. Tetranuclear Mixed-Ligand Complexes

2.6.1. $(\eta^5\text{-}C_5H_5)_4Fe_4(CO)_4$. The tetranuclear species $(\eta^5\text{-}C_5H_5)_4Fe_4(CO)_4$ is an air-stable product able to undergo multiple, reversible, one-electron redox changes |101,102|. Such redox propensity is here illustrated in Figure 37, which refers to $[(\eta^5\text{-}C_5H_5)_4Fe_4(CO)_4]^+$ |102|.

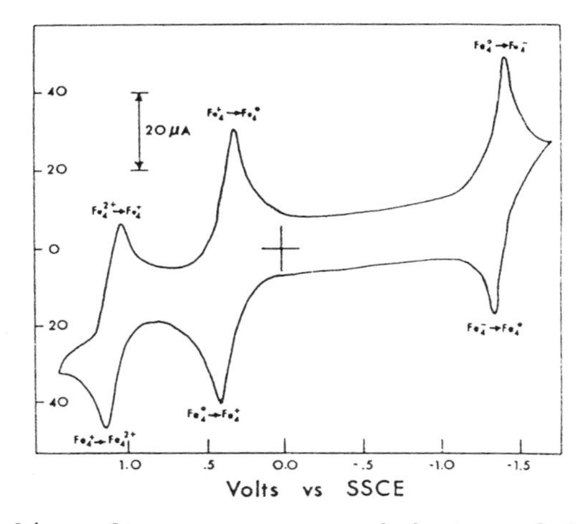

Figure 37. *Cyclic voltammogram recorded at a platinum electrode on a MeCN solution of $[(\eta^5\text{-}C_5H_5)_4Fe_4(CO)_4][PF_6]$. Scan rate 0.1 Vs^{-1} (from Ref. 102).*

At least in the time-window of cyclic voltammetry, the following electron transfer sequence holds:

$$[(C_5H_5)_4Fe_4(CO)_4]^{2+} \underset{-e}{\overset{+e}{\rightleftarrows}} [(C_5H_5)_4Fe_4(CO)_4]^{+} \underset{-e}{\overset{+e}{\rightleftarrows}} \quad E^{\circ\prime}=+0.84\ V; \quad E^{\circ\prime}=+0.08\ V$$

$$(C_5H_5)_4Fe_4(CO)_4 \underset{-e}{\overset{+e}{\rightleftarrows}} [(C_5H_5)_4Fe_4(CO)_4]^{-} \quad E^{\circ\prime}=-1.54\ V$$

Two terms of this four-membered series have been structurally characterized: $(C_5H_5)_4Fe_4(CO)_4$ and $[(C_5H_5)_4Fe_4(CO)_4]^+$ |103,104|. Both the clusters have basically the tetrahedral Fe_4 assembly illustrated in Figure 38 for the neutral derivative |103|.

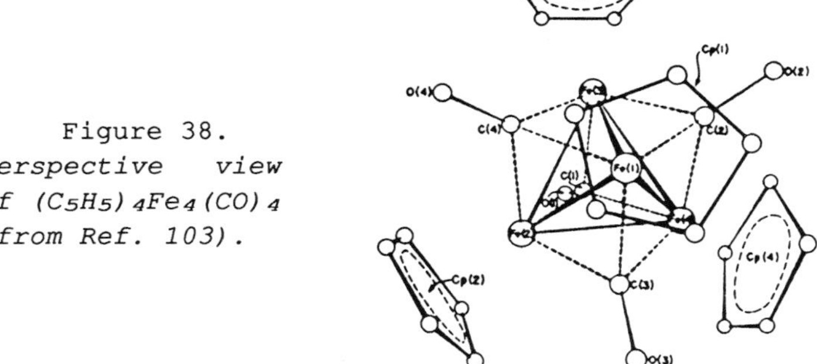

Figure 38. *Perspective view of $(C_5H_5)_4Fe_4(CO)_4$ (from Ref. 103).*

Each face of the tetrahedron is triply bridged by one carbonyl ligand; each cyclopentadienyl ring is symmetrically bonded to each iron atom.

The neutral species has assigned a regular tetrahedral geometry (with a mean Fe-Fe bond length of 2.52 Å), very reminiscent of the geometry of $[Fe_4(CO)_{13}]^{2-}$, described previously. The monocation is tetragonally distorted in that the two opposite Fe-Fe edges [Fe(1)-Fe(3) and Fe(2)-Fe(4)], normal to the S_4 axis, are slightly longer than the remaining four Fe-Fe distances; the iron-iron mean length decreases to 2.48 Å. Conversely, the $Fe-C_{(CO)}$ distance remains unvaried.

According to the MO diagram developed by Dahl |104|, in the neutral/monocation oxidation, the electron is removed by a 8-MO's level essentially nonbonding with respect to the tetrametal core, the highest one being however somewhat antibonding, thus accounting for the slight enforcement of the metal-metal bond. It can be foreseen that further oxidation to the dication would not cause significant geometrical changes in that the six, strongly bonding tetrametal orbitals lie at lower energy. In contrast, one-electron reduction to $[(C_5H_5)_4Fe_4(CO)_4]^-$ should involve significant increase of the Fe-Fe bond length, since the extra electron enters a 6-MO's level strongly tetrairon antibonding.

Finally, photochemically induced oxidation of the neutral species to the corresponding monocation has been reported |105|.

2.6.2. $(\eta^5\text{-}C_5H_4Me)_4Fe_4(CO)_6(P)_8$. In spite of the vague assonance with the preceding tetrairon derivative, the crystal structure of the polynuclear species $(\eta^5\text{-}C_5H_4Me)_4Fe_4(CO)_6(P)_8$ reveals that no Fe-Fe bonding exists. As a matter of fact, it is constituted by a central wedge-shaped P_8 cage bearing two chelating $(\eta^5\text{-}C_5H_4Me)Fe(CO)$ and two terminal $(\eta^5\text{-}C_5H_4Me)Fe(CO)_2$ fragments, Figure 39 |106|.

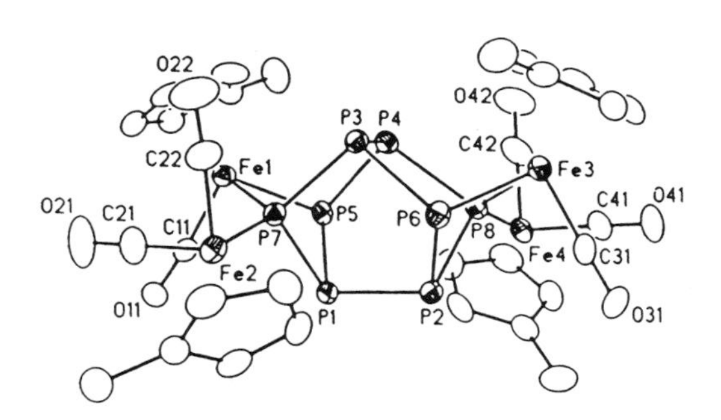

Figure 39. *X-Ray structure of $(\eta^5\text{-}C_5H_4Me)_4Fe_4(CO)_6(P)_8$ (from Ref. 106).*

The most prominent redox feature of the actual complex is its oxidation to the corresponding transient monocation (E°' = +0.34 V, in dichloromethane solution) |106|.

2.6.3. $[Fe_4(CO)_{12}(S)_4]^{2-}$. Two apparently congeners of formula $[Fe_4(CO)_{12}(S)_4]^{2-}$ and $Fe_4(CO)_{12}(S)_4$ have been prepared |107,108|. Really, as shown in Figure 40, they display marked structural differences.

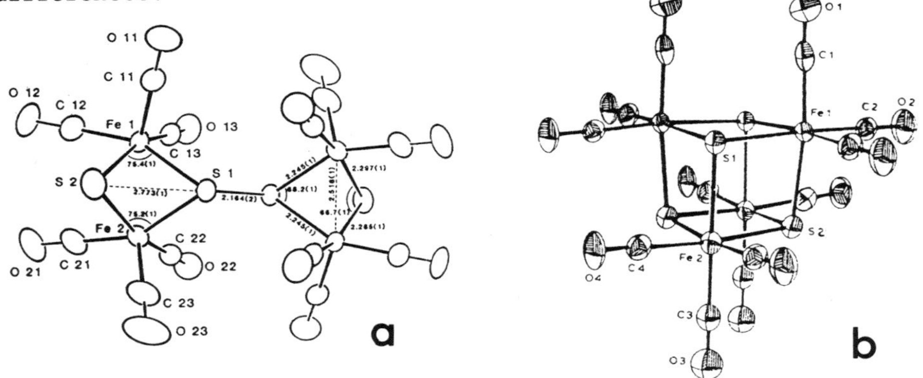

Figure 40. *View of: (a) $[Fe_4(CO)_{12}(S)_4]^{2-}$ (from Ref. 107); (b) $Fe_4(CO)_{12}(S)_4$ (from Ref. 108).*

In fact, the dianion is formed by two S-S linked, puckered Fe_2S_2 units with a Fe/Fe distance within the bonding value (2.52 Å), whereas the neutral derivative has a cubane-like Fe_4S_4 geometry with nonbonding Fe/Fe distances (3.47 Å).

The likely lack of mutual relations between these compounds can be speculatively proved by the redox behaviour of the dianion |109|. In fact, in tetrahydrofuran solution, it undergoes both an irreversible two-electron reduction (Ep = -1.90 V) to $[Fe_2(CO)_6(S)_2]^{2-}$, and an irreversible two-electron oxidation (Ep = -0.07 V) to $Fe_2(CO)_6(S)_2$. In the course of such redox changes, no evidence to the electrogeneration of the insoluble $Fe_4(CO)_{12}(S)_4$ was gained |109|.

Finally, it has been reported that chemical oxidation of $Fe_4(CO)_{12}(S)_4$ (Br_2 under CO atmosphere) gives the monocation $[Fe_4(CO)_{12}(S)_4]^+$, not yet crystallographycally characterized. According to Dahl's bonding description, this monocation should possess a distorted (*via* Jahn-Teller effect) cubic geometry with reduced Fe···Fe nonbonding distances, in that the electron is removed from the completely filled tetrairon antibonding 6-MO's level |108|.

2.6.4. $Fe_4(CO)_{12-n}(PR)_2$ (n = 0-2). A tetrairon planar (or essentially planar) array is present in the phosphorus-bicapped clusters presented in Scheme 5.

SCHEME 5

$Fe_4(CO)_{12}(PPh)_2$

$Fe_4(CO)_{11}(PPh)_2$

$Fe_4(CO)_{11}(PBu^t)_2$

$Fe_4(CO)_{10}(P(OMe)_3)(PPh)_2$

$Fe_4(CO)_{10}(p\text{-tol-NC})(PPh)_2$

$Fe_4(CO)_{10}(PPh)_2\text{-}(OMe)_2P\text{-}C_6H_4\text{-}P(OMe)_3\text{-}Fe_4(CO)_{10}(PPh)_2$

$Fe_4(CO)_{10}(PPh)_2\text{-}CN\text{-}C_6H_4\text{-}CN\text{-}Fe_4(CO)_{10}(PPh)_2$

In this series, $Fe_4(CO)_{12}(PPh)_2$ obeys the 18-electron rule and is considered "saturated" (having 64 cluster-valence-electrons or 8 skeletal electron pairs) |110|, whereas the $Fe_4(CO)_{11}$- or $Fe_4(CO)_{10}L$- analogues are "unsaturated" by two missing electrons. The stability of both the classes of compounds offers the opportunity to test their aptitude to lose or to add electrons.

The molecular structure of the saturated $Fe_4(CO)_{12}(PPh)_2$ can be deduced by that of the monosubstituted analog $Fe_4(CO)_{11}[P(OMe)_3](P\text{-}p\text{-}Tol)_2$ |111,112|, which is reported in Figure 41. The tetrairon fragment of the octahedral assembly is slightly bent with respect to the Fe(1)Fe(4) vector; the dihedral angle between the Fe(1)Fe(2)Fe(4)/Fe(1)Fe(3)Fe(4) planes is 163°. This geometry causes a significant spreading of the Fe-P lengths (ranging from 2.21 Å to 2.40 Å), but a limited variation in the Fe-Fe bond distances (from 2.67 Å to 2.71 Å; mean length 2.68 Å). In addition, the strain produced by the slight folding of the tetrametal moiety causes the 12 equatorial ligands to dispose alternatively one above and two below the virtual plane.

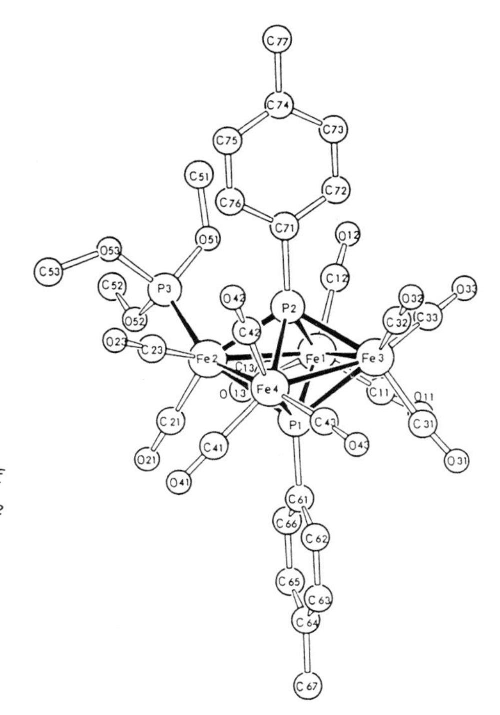

Figure 41.
Molecular structure of $Fe_4(CO)_{11}[P(OMe)_3](P\text{-}p\text{-}Tol)_2$ (from Ref. 112).

As expected on the basis of its saturation, $Fe_4(CO)_{12}(PPh)_2$ undergoes oxidation processes, Figure 42 |113|.

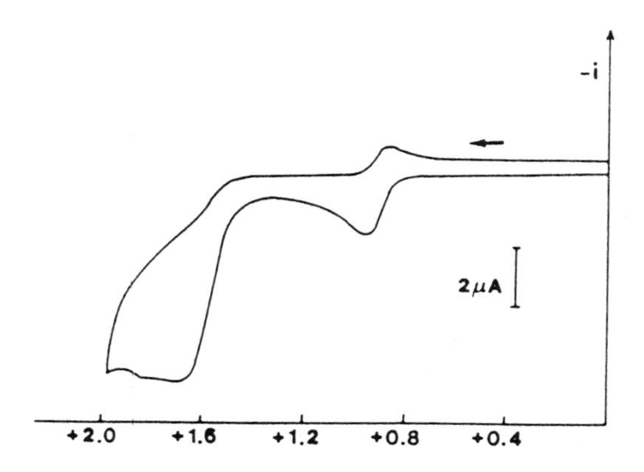

Figure 42. *Cyclic voltammetric response recorded at a platinum electrode on a benzonitrile solution of $Fe_4(CO)_{12}(PPh)_2$. Scan rate 0.2 Vs^{-1}. Potential values vs. Ag/AgCl (from Ref. 113).*

The first anodic step (E°'= +0.86 V) involves a quasi-reversible one-electron removal to the stable $[Fe_4(CO)_{12}(PPh)_2]^+$, which has not been isolated, but ESR characterized |113|. The stability of the monocation can be easily accounted for by considering that MO calculations on the model compound $Fe_4(CO)_{12}(PH)_2$ foresee that the HOMO level is essentially nonbonding (its weak Fe-Fe antibonding character is counterbalanced by some Fe-P bonding character) |110|, thus electron removal should not affect the cluster stability. Indeed, further one-electron removal should be theoretically allowed, affording the corresponding 62-electron species, but this does not seem the case since an irreversible multielectron process takes place at +1.6 V.

As also expected, the neutral compound undergoes an irreversible declustering reduction process (Ep = -0.69 V). In fact, the added electron populates an antibonding LUMO level |110|.

As far as the 62-electron unsaturated single clusters in Scheme 5 are concerned, their structure is representatively illustrated in Figure 43 by that of $Fe_4(CO)_{11}$(P-*p*-Tol)$_2$ and $Fe_4(CO)_{10}[P(OMe)_3]$(P-*p*-Tol)$_2$ |111,112,114,115|.

At variance with the saturated $Fe_4(CO)_{11}[P(OMe)_3]$(P-*p*-Tol)$_2$, the tetrairon fragment is here practically planar, and the three substituents of each iron atom regularly lie one in the tetrametal plane, one above and one below it. The mean iron-iron bond distance is 2.63 Å (0.05 Å shorter than that in the saturated compound), but one of the iron-iron distances is shortened (by the CO bridge) (2.45 Å) with respect to the nearly equivalent other ones (mean = 2.69 Å).

The shortening of the mean metal-metal length can be accounted for by considering that in the saturated compound the populated HOMO level has some antibonding character.

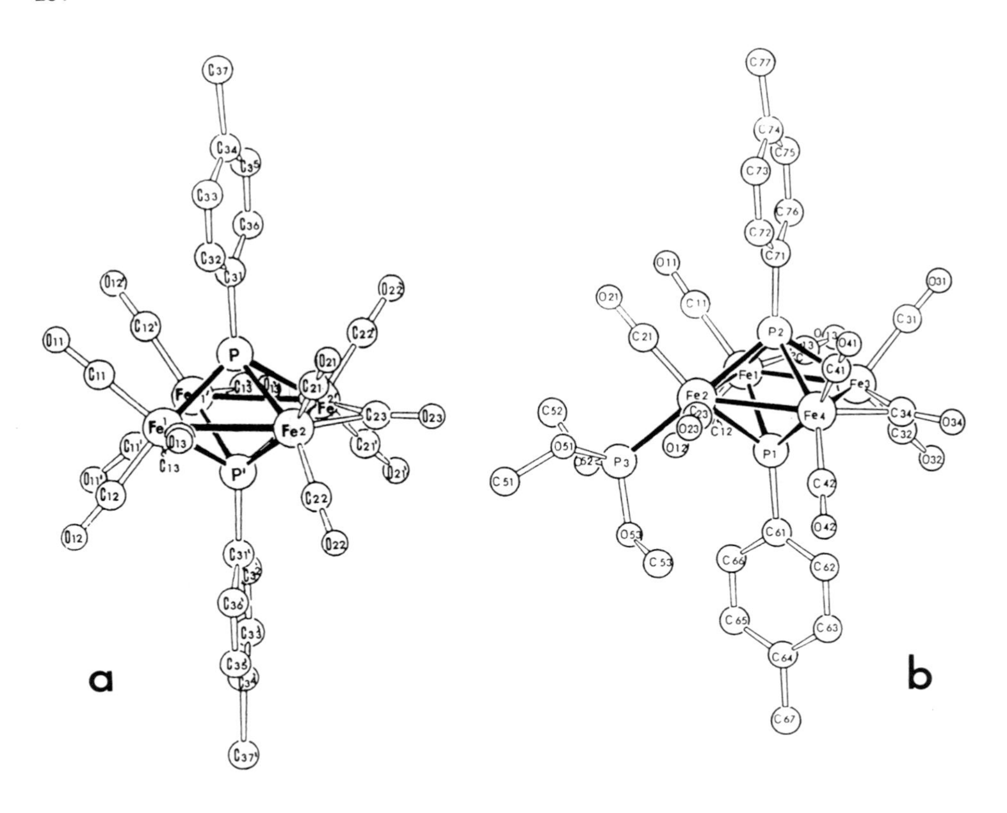

Figure 43. *Molecular structure of: (a)* $Fe_4(CO)_{11}(P\text{-}p\text{-}Tol)_2$ *(from Ref. 115); (b)* $Fe_4(CO)_{10}[P(OMe)_3](P\text{-}p\text{-}Tol)_2$ *(from Ref. 112).*

The electrochemical fingerprint of these saturated clusters is representatively illustrated in Figure 44, which refers to $Fe_4(CO)_{11}(PPh)_2$ |113|. As it can be seen, it undergoes two successive reduction steps, each one involving a quasireversible one-electron addition, at relatively high potential values (Table 8). The stability of the one-electron added species in the long times of macroelectrolysis, as well as of the two-electron added species in cyclic voltammetry, confirms the existence of both $[Fe_4(CO)_{11}(PPh)_2]^-$ and $[Fe_4(CO)_{11}(PPh)_2]^{2-}$ and lends support to the fact that actually the nonbonding LUMO level (the HOMO level of the saturated 64-electron derivative) is becoming populated.

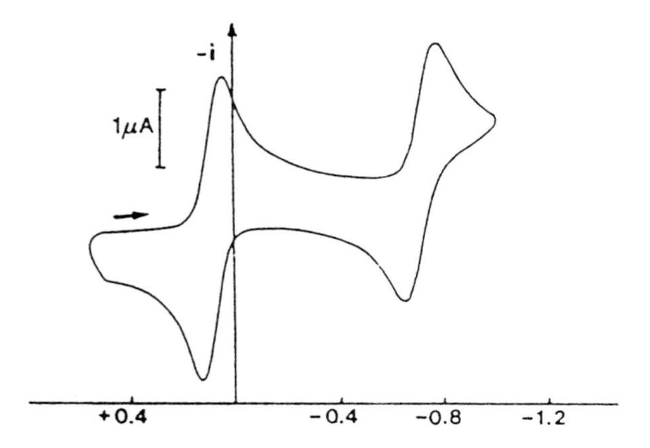

Figure 44. *Cyclic voltammogram exhibited by $Fe_4(CO)_{11}(PPh)_2$ under the experimental conditions reported in Figure 43 (from Ref. 113).*

It is interesting to note that ESR spectra of $[Fe_4(CO)_{11}(PPh)_2]^-$ indicate a substantial preservation of the starting geometry, whereas $[Fe_4(CO)_{11}(PPh)_2]^{2-}$ should have a theoretically assigned *nido* structure, which can be thought as a pentagonal bipyramid having an unoccupied vertex |116|.

In addition, even if electron removal to stable 60-electron species is theoretically allowed, irreversible oxidation occurs.

Such voltammetric behaviour is common to all other single clusters, and the relevant characteristics are summarized in Table 11. The presence of electron-donating substituents either in the equatorial planar fragment or in the bicapping units makes slightly more difficult the addition of electrons.

Finally, concerning the double clusters in Scheme 5, their molecular structure is exemplified in Figure 45, which refers to $Fe_4(CO)_{10}(PPh)_2$-$(MeO)_2P$-C_6H_4-$P(OMe)_2$-$Fe_4(CO)_{10}(PPh)_2$ |117|. Apart from some minor differences induced by the *p*-phenylene-bis(phosphite) bridging unit, each unsaturated octahedral cluster is structurally similar to those just now discussed.

TABLE 11. *Redox potentials for the most significant redox changes exhibited in benzonitrile solution by the tetrairon compounds schematized in Scheme 5 |113|*

complex	redox changes					
	1+/0		0/1-		1-/2-	
	E°' (V)	ΔEp[a] (mV)	E°' (V)	ΔEp[a] (mV)	E°' (V)	ΔEp[a] (mV)
$Fe_4(CO)_{12}(PPh)_2$	+0.86	100	-0.69[a,b]			
$Fe_4(CO)_{11}(PPh)_2$	+0.71[a,b]		+0.05	80	-0.77	120
$Fe_4(CO)_{11}(PBu^t)_2$			-0.06	80	-1.12	100
$Fe_4(CO)_{10}[P(OMe)_3](PPh)_2$			-0.17	80	-0.97	100
$Fe_4(CO)_{10}$(*p*-Tol-NC)$(PPh)_2$			-0.14	80	-0.89	200
$Fe_4(CO)_{10}(PPh)_2$-$(OMe)_2P$-C_6H_4-P$(OMe)_2$-$Fe_4(CO)_{10}(PPh)_2$			-0.17[c]	140	-0.17[c]	140
$Fe_4(CO)_{10}(PPh)_2$-CN-C_6H_4-CN-$Fe_4(CO)_{10}(PPh)_2$			-0.04	100	-0.13	100

[a]*Measured at 0.2 Vs^{-1};* [b]*peak potential value for irreversible process;* [c]*two overlapping one-electron steps*

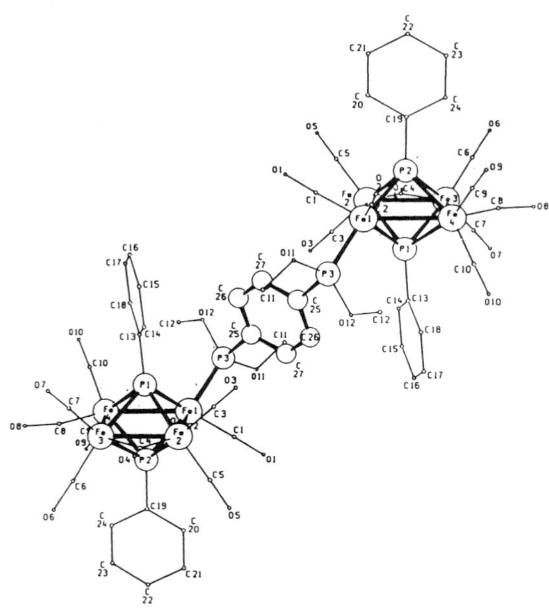

Figure 45. *Molecular structure of $Fe_4(CO)_{10}(PPh)_2$-$(MeO)_2PC_6H_4P(OMe)_2$-$Fe_4(CO)_{10}(PPh)_2$ (from Ref. 117).*

As such, it should be expected that these double clusters are able to add four electrons (two for each cluster). Indeed, as shown in Figure 46, they undergo two main distinct reduction steps |113|. The first one involves two, overlapping [in the case of bis(phosphite) bridged species] or very close-spaced [in the case of bis(isonitrile) bridged species], one-electron additions, chemically reversible (Table 11); the second one, always irreversible in character (in agreement with the unstability of organometallic tri- or tetra-anions), is difficult to analyze.

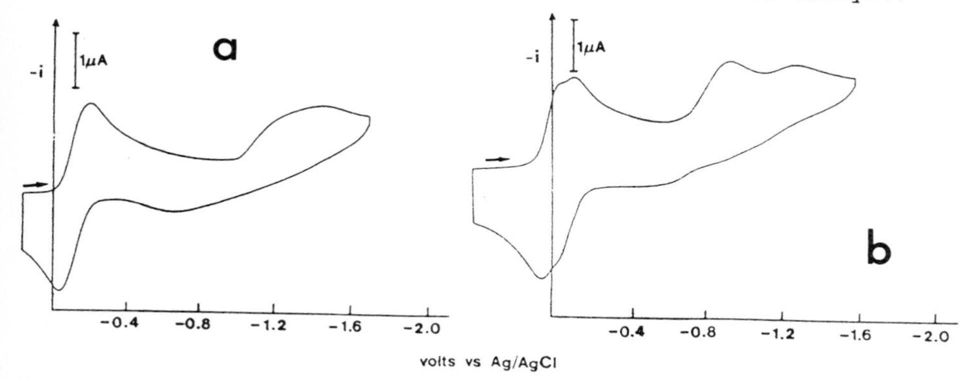

Figure 46. *Cyclic voltammograms recorded on benzonitrile solutions of: (a)* $Fe_4(CO)_{10}(PPh)_2-(MeO)_2PC_6H_4P(MeO)_2-Fe_4(CO)_{10}(PPh)_2$*; (b)* $Fe_4(CO)_{10}(PPh)_2-CNC_6H_4NC-Fe_4(CO)_{10}(PPh)_2$*. Same experimental conditions than those indicated in Figure 42 (from Ref. 113).*

It is interesting to examine briefly the first two-electron reduction step. It is well known |118| that the separate addition of two electrons in two non-interacting sites of a molecule causes the two redox changes to occur at electrode potentials separated by 36 mV, *i.e.* practically overlapping in the cyclic voltammetric response. In the likely hypothesis that in the present cases the first two electrons enter separately the two octahedral clusters, it seems that the bis(phosphite) bridge is more effective in keeping electronically separated the two Fe_4P_2 clusters than the bis(isonitrile) bridge is able to do. In fact, in the latter compound the two one-electron charge transfers are slightly, but definitely, separated ($\Delta E°' = 90$ mV), thus suggesting some extent of interaction between the two Fe_4P_2 units. This situation is quite reminiscent of what happens in double-cubane assemblies of iron-molybdenum (or tungsten) sulfur clusters |29|.

2.6.5. $Fe_4(CO)_{10}(\mu_2\text{-}COEt)(PR)_2$. Chemical reduction of $Fe_4(CO)_{11}(PR)_2$ by PhLi affords, as expected, the corresponding monoanions $[Fe_4(CO)_{11}(PR)_2]^-$, which, by treatment with Et_3OBF_4, unexpectedely undergo alkylation at one carbonyl ligand to give the paramagnetic species $Fe_4(CO)_{10}(\mu_2\text{-}COEt)(PR)_2$ |119|.

The molecular structure of $Fe_4(CO)_{10}(\mu_2\text{-}COEt)(PBu^t)_2$ is shown in Figure 47 |119|.

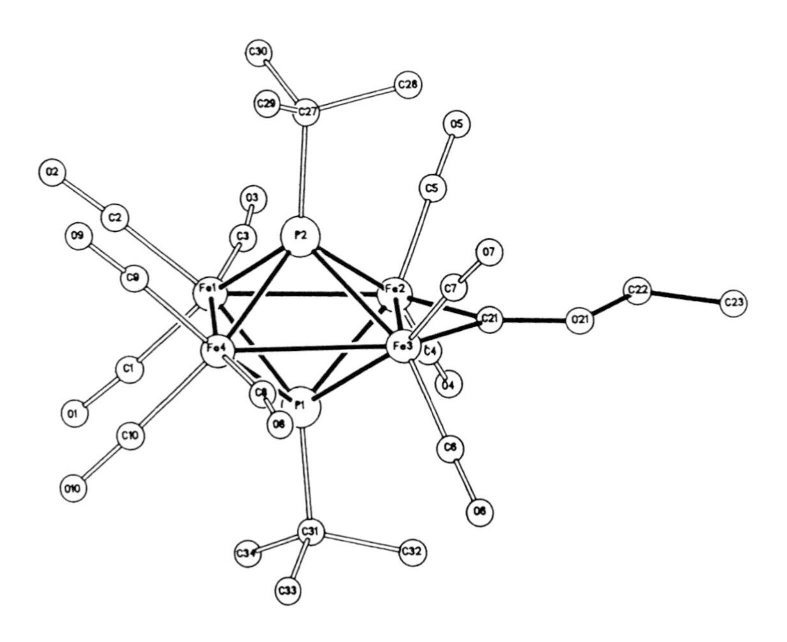

Figure 47. *Perspective view of $Fe_4(CO)_{10}(\mu_2\text{-}COEt)(PBu^t)_2$ (from Ref. 119).*

The overall assembly is quite reminiscent of that of the precursors $Fe_4(CO)_{11}(PR)_2$, with the COEt-bridged Fe-Fe distance (2.48 Å) significantly shorter (0.22 Å) than the averaged value of the remaining ones.

In confirmation that $Fe_4(CO)_{10}(\mu_2\text{-}COEt)(PR)_2$ possess one more electron than $Fe_4(CO)_{11}(PR)_2$, they undergo reversibly both a one-electron oxidation and a one-electron reduction |119|. The relevant redox potentials are summarized in Table 12.

TABLE 12. *Redox potentials (in Volt) for the electron-transfer steps exhibited by $Fe_4(CO)_{10}(\mu_2\text{-}COEt)(PR)_2$ in dichloromethane solution* |119|

R	oxidation	reduction
Bu^t	+0.31	-0.69
Ph	+0.38	-0.57

2.6.6. $Fe_4(CO)_{11}(E)(PR)$ (E = S,Se,Te). The crystal structure of the two calchogena-phosphinidene species $Fe_4(CO)_{11}(S)(PPh)$ |71| and $Fe_4(CO)_{11}(Te)(PPr^i)$ |52| is shown in Figure 48.

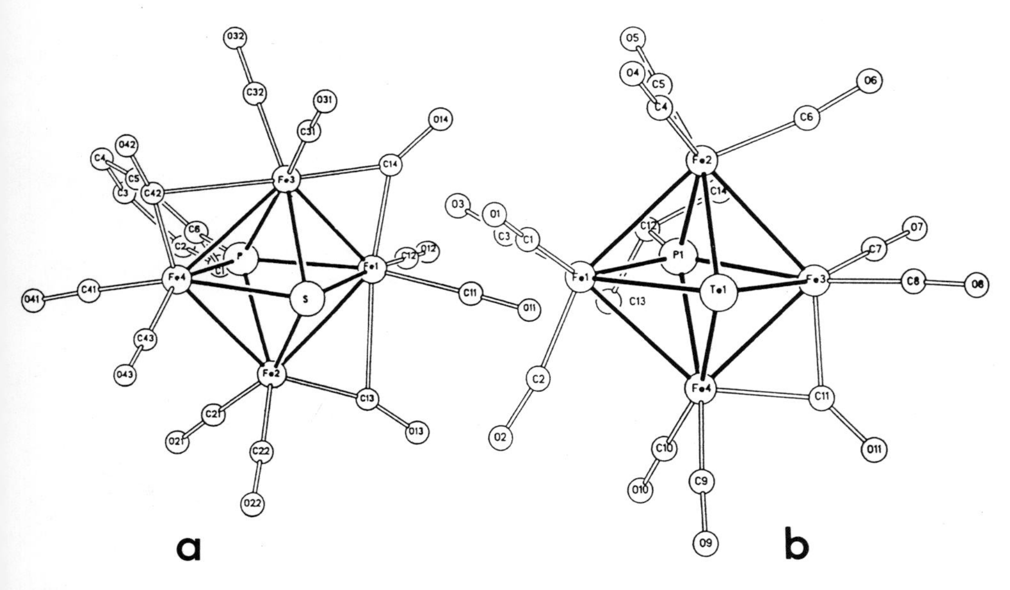

Figure 48. *X-Ray structure of: (a) $Fe_4(CO)_{11}(S)(PPh)$ (from Ref. 71); (b) $Fe_4(CO)_{11}(Te)(PPr^i)$ (from Ref. 52).*

The Fe_4PE core has the usual octahedral geometry. The most evident difference lies on the coordination mode of some carbonyl ligands: three *vs.* one doubly bridging dispositions. This causes some spreading in the bonding lengths, the averaged values of which are: (*i*) in the sulfur-capped complex, Fe-Fe = 2.59 Å, Fe-S 2.32 Å, Fe-P 2.27 Å; (*ii*) in the tellurium-capped complex, Fe-Fe = 2.68 Å, Fe-Te 2.61 Å, Fe-P 2.28 Å.

As illustrated in Figure 49 for $Fe_4(CO)_{11}(S)(PPh)$, these complexes undergo, in dichloromethane solution, two one-electron reduction steps with features of marked chemical reversibility |71|.

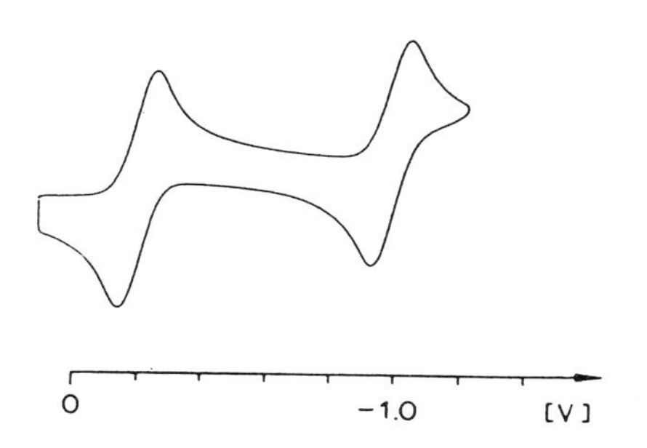

Figure 49.
Cyclic voltammetric response exhibited by $Fe_4(CO)_{11}(S)(PPh)$ in CH_2Cl_2 solution. Platinum working electrode. Scan rate 0.2 Vs^{-1} (from Ref. 71).

The redox potentials for the series $Fe_4(CO)_{11}(E)(PR)$ are reported in Table 13.

TABLE 13. *Redox potentials (in Volt) for the two successive one-electron reductions exhibited by $Fe_4(CO)_{11}(E)(PR)$ in dichloromethane solution*

E	R	$E°'_{0/-}$	$E°'_{-/2-}$
S	Ph	-0.21	-1.01
	Pr^i	-0.24	-1.05
Se	Pr^i	-0.23	-1.05
Te	Pr^i	-0.23	-1.20

2.6.7. $Fe_4(CO)_{11}(PR)(R'C_2R'')$. Substitution of a phosphinidene group in $Fe_4(CO)_{11}(PR)_2$ for an alkyne molecule leads to the series $Fe_4(CO)_{11}(R'C_2R')(PR)$, the structure of which can be represented according to Figure 50 |120|.

Like the diphosphinidene complexes, the present derivatives undergo two sequential one-electron reductions, which in cyclic voltammetry look as chemically reversible processes |52|. As deducible from Table 14, the addition of the second electron to the alkyne-substituted molecules is significantly easier than that to the unsubstituted precursor.

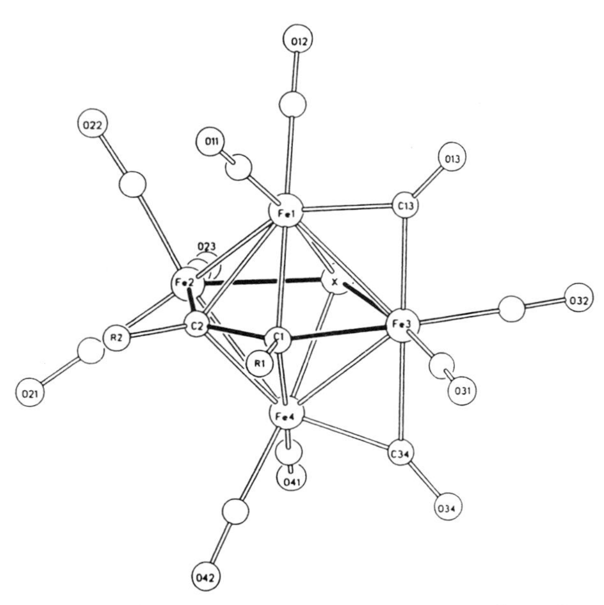

Figure 50. *Perspective view of $Fe_4(CO)_{11}(R^1C_2R^2)(X)$ (R^1=R^2=Ph; X=PBu^t) (from Ref. 120).*

TABLE 14. *Comparison between the redox potentials (in Volt) of the two successive one-electron reductions exhibited by $Fe_4(CO)_{11}(R'C_2R'')(PR)$ and $Fe_4(CO)_{11}(PR)_2$, in dichloromethane solution |52|*

complex	$E°'_{0/-}$	$E°'_{-/2-}$
$Fe_4(CO)_{11}(HC_2Bu^t)(P\text{-}C_6H_4\text{-}p\text{-}OMe)$	-0.55	-0.70
$Fe_4(CO)_{11}(HC_2Ph)(P\text{-}C_6H_4\text{-}p\text{-}OMe)$	-0.53	-0.65
$Fe_4(CO)_{11}(P\text{-}Ph)_2$	-0.20	-1.20

2.6.8. $Fe_4(CO)_{11}(S)(RC_2R)$. The complex $Fe_4(CO)_{11}(S)(HC_2Ph)$ possesses a geometry similar to that illustrated in Figure 50 for the species $Fe_4(CO)_{11}(R'C_2R')(PR)$ |120|. Like these latter complexes, it undergoes, in dichloromethane solution, two relatively close-spaced, one-electron reductions ($E°'_{0/-}$ = -0.48 V; $E°'_{-/2-}$ = -0.65 V) |52|.

2.6.9. $Fe_4(CO)_{10}(S)(PR)(CR'_2)$. The geometry of the carbene complexes $Fe_4(CO)_{10}(S)(CR'_2)(PR)$ can be represented by that of $Fe_4(CO)_{10}(S)[C(NH_2)_2](PPh)$, which is shown in Figure 51 |71|.

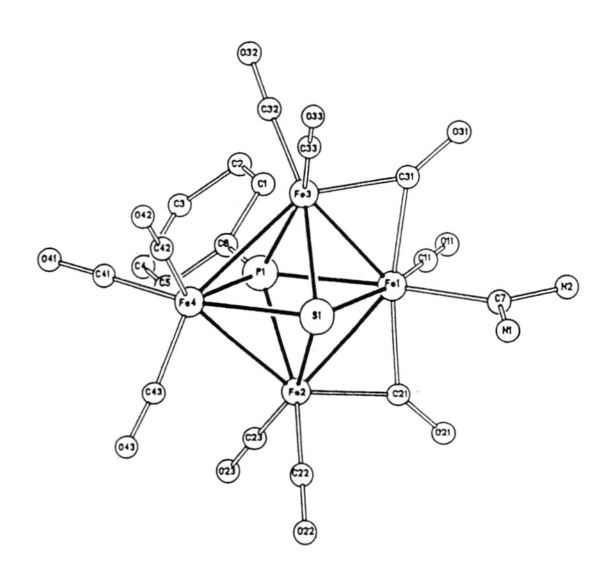

Figure 51. *Perspective view of $Fe_4(CO)_{10}(S)[C(NH_2)_2](PPh)$ (from Ref. 71).*

These Fe-Fe bonds which do not bear bridging carbonyl ligands (averaged value, 2.68 Å) are significantly longer than the edge-bridged ones (2.50 Å). The Fe1-S (2.48 Å) and the Fe1-P (2.33 Å) distances are also significantly longer than the remaining ones (averaged values, Fe-S, 2.30 Å; Fe-P, 2.25 Å).

As illustrated in Figure 52 for $Fe_4(CO)_{10}(S)[C(NH_2)_2](PPh)$, these complexes undergo reversibly two significantly separated one electron reductions {$Fe_4(CO)_{10}(S)[C(NH_2)_2](PPh)$: $E°'_{0/-} = -0.64$ V, $E°'_{-/2-} = -1.35$ V; $Fe_4(CO)_{10}(S)[C(NH_2)(Me)](PPr^i)$: $E°'_{0/-} = -0.63$ V, $E°'_{-/2-} = -1.34$ V} |71|.

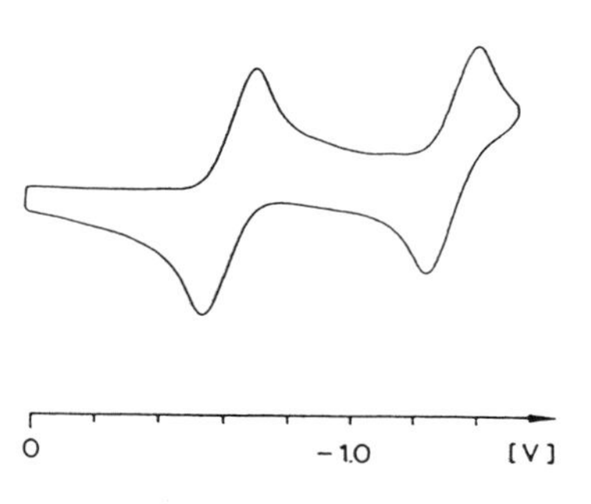

Figure 52.
Cyclic voltammogram recorded on a dichloromethane solution of $Fe_4(CO)_{10}(S)[C(NH_2)_2](PPh)$. Scan rate 0.2 Vs^{-1} (from Ref. 71).

2.7. Pentanuclear Carbido Complexes

2.7.1. $[Fe_5(CO)_{14}C]^{2-}$. The molecular geometry of the 74-electron pentairon carbido cluster $[Fe_5(CO)_{14}C]^{2-}$ is shown in Figure 53 |121|.

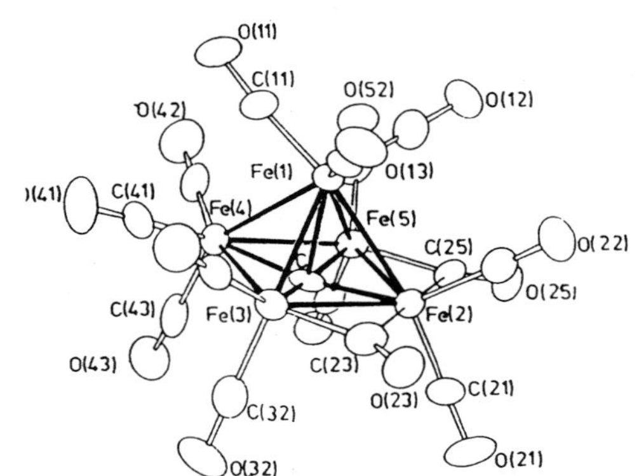

Figure 53. *Perspective view of the dianion $[Fe_5(CO)_{14}C]^{2-}$ (from Ref. 121).*

The dianion possesses an octahedrally distorted square-based pyramidal structure, the exposed carbide atom being displaced out of the basal plane (opposite to the apical iron) by 0.18 Å. The iron-iron bond length of the base has a mean value of 2.62 Å, but the presence of two semibridging carbonyl groups (C23, C25), shortening the relevant metal-metal distances by about 0.1 Å with respect to the other ones, makes irregular the tetrairon plane. The metal-metal bond length from the apical iron is 2.60 Å. The averaged Fe-C length from the basal iron atoms is 1.89 Å, obviously shorter than the Fe-C distance from the apex (1.95 Å).

The redox ability of such a dianion is illustrated in Figure 54 |121|.

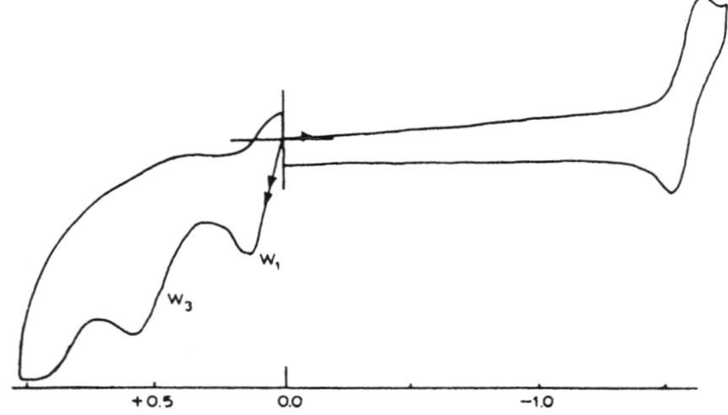

Figure 54. *Cyclic voltammogram recorded at a carbon electrode on a CH_2Cl_2 solution of $[Fe_5(CO)_{14}C][NEt_4]_2$. Scan rate 0.1 Vs^{-1} (from Ref. 121).*

It undergoes a single-stepped quasireversible two electron reduction at E°'= -1.58 V, generating a somewhat stable tetraanion $[Fe_5(CO)_{14}C]^{4-}$, as well as a single-stepped two-electron oxidation to the corresponding neutral species, with some extent of chemical reversibility, at E°'= +0.07 V, followed by further irreversible anodic steps.

Chemical activation of $[Fe_5(CO)_{14}C]^{2-}$ towards multielectron donor ligands has been recently achieved by the use of chemical oxidants selected on the basis of electrochemical investigation |122|.

2.7.2. $Fe_5(CO)_{15}C$. The 74-electron $Fe_5(CO)_{15}C$ is the first carbido-carbonyl cluster synthesized and structurally characterized |123|, and it is the founder of a rich cluster chemistry |8|. It has the square pyramidal geometry illustrated in Figure 55 |121,122|.

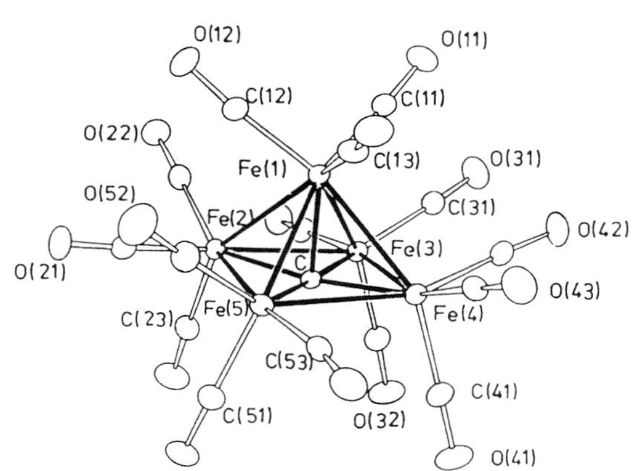

Figure 55.
Molecular structure of $Fe_5(CO)_{15}C$ (from Ref. 121).

The mean metal-metal distance in the basal plane (2.67 Å) is slightly longer (0.05 Å) than that in the isoelectronic $[Fe_5(CO)_{14}C]^{2-}$, likely because of the lacking of (semi)bridging CO groups. The Fe-Fe mean length from the apical iron atom is 2.65 Å, once again slightly longer (0.05 Å) than that in the cited dianion The carbide atom is pushed down from the tetrametal plane by only 0.09 Å (it has been noted that the displacement increases with the increase of the cluster charge). The iron-carbide bond lengths are essentially coincident with those of the dianion |121|.

The redox behaviour of $Fe_5(CO)_{15}C$ can be inferred by the cyclic voltammograms shown in Figure 56 |121|.

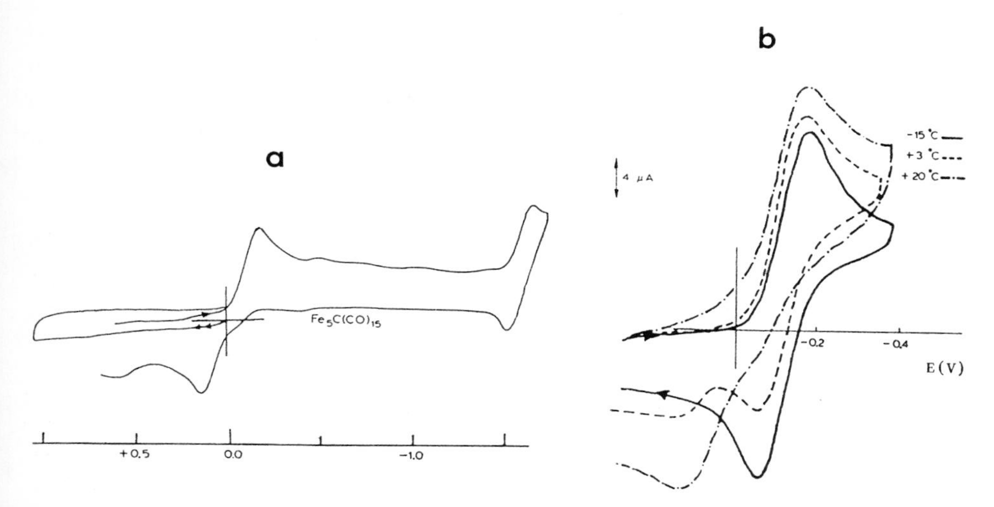

Figure 56. *Cyclic voltammograms of $Fe_5(CO)_{14}C$ recorded at a carbon electrode, in CH_2Cl_2 solution: (a) at 20°C; (b) at different temperatures. Scan rate 0.1 Vs^{-1} (from Ref. 121).*

The pentairon cluster undergoes a first two-electron reduction process at Ep = -0.17 V, which, as shown in Figure 56b, is chemically irreversible at room temperature, but chemically reversible at -15°C. The cathoanodic system at -1.58 V is attributed to the previously discussed two-electron reduction of $[Fe_5(CO)_{14}C]^{2-}$, chemically generated in the decomposition of the unstable $[Fe_5(CO)_{15}C]^{2-}$, primarily electrogenerated. As a consequence, in the reverse scan, the previously discussed oxidation of $[Fe_5(CO)_{14}C]^{2-}$ occurs |121|. The fact that, upon reduction, $Fe_5(CO)_{15}C$ affords $[Fe_5(CO)_{14}C]^{2-}$ has been also chemically proved |124|.

The stated stability of $[Fe_5(CO)_{15}C]^{2-}$ at low temperature deserves some comments. According to an extended Hückel calculation |125|, the two added electrons should enter an antibonding iron-iron orbital, thus suggesting that in the elusive dianion $[Fe_5(CO)_{15}C]^{2-}$ some significant lengthening of the mean Fe-Fe distance must occur.

2.8. Pentanuclear Nitrido Complexes

2.8.1. $[Fe_5(CO)_{14}N]^-$. The nitrido-carbonyl anion $[Fe_5(CO)_{14}N]^-$ is isoelectronic with the previously discussed carbido-carbonyl clusters $[Fe_5(CO)_{14}C]^{2-}$ and $Fe_5(CO)_{15}C$, and like them {as well as

$HFe_5(CO)_{14}N$ |126|}, it possesses the square-based pyramidal geometry shown in Figure 57. It must however taken into account that, in the solid state, some minor structural isomerizations can occur (mainly in the basal sites) as a function of the counterion |121,127|.

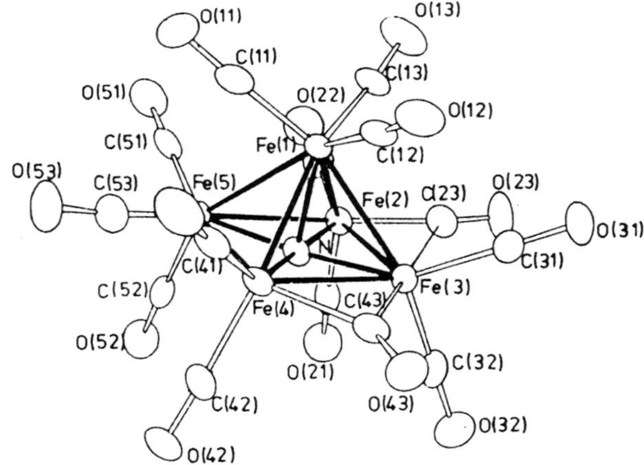

Figure 57. *Perspective view of $[Fe_5(CO)_{14}N]^-$, as $[\{(C_6H_5)_3P\}_2N]^+$ salt. Averaged Fe-Fe bond length in the basal plane, 2.61 Å; averaged $Fe_{(apical)}$-$Fe_{(basal)}$, 2.58 Å; averaged $Fe_{(basal)}$-N, 1.84 Å; $Fe_{(apical)}$-N, 1.92 Å; the nitride atom is displaced below the tetrametal base by 0.11 Å (from Ref. 121).*

It has been reported that $[Fe_5(CO)_{14}N]^-$ undergoes, in dichloromethane solution, a single-stepped two-electron reduction to the somewhat stable $[Fe_5(CO)_{14}N]^{3-}$ ($E°'$= -1.24 V), as well as a two-electron oxidation to the monocation $[Fe_5(CO)_{14}N]^+$ ($E°'$= +0.56 V) |121|. Really, as shown in Figure 58 |127|, in nonaqueous solutions the nitrido monoanion undergoes two separated one-electron additions, the first of which leads to the quite stable dianion $[Fe_5(CO)_{14}N]^{2-}$.

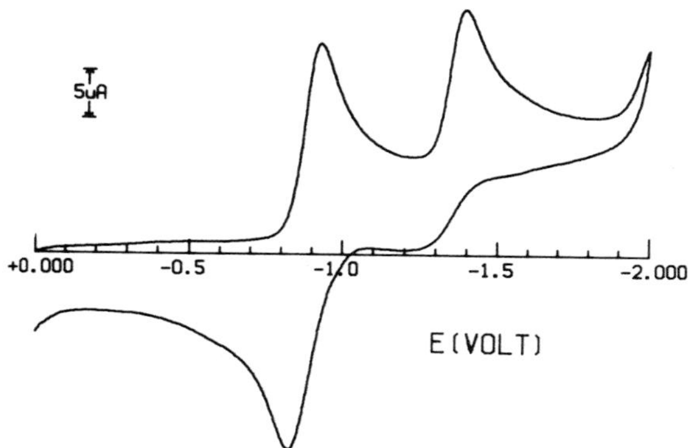

Figure 58. *Cyclic voltammogram recorded at a mercury electrode on a MeCN solution of $[Fe_5(CO)_{14}N][NEt_4]$. Scan rate 0.2 Vs^{-1}.*

We have also evidence |99| that in the presence of phosphines the cathodic reduction affords the monosubstituted species $[Fe_5N(CO)_{13}(PR_3)]^-$ according to the following Electron Transfer Chain catalytic process:

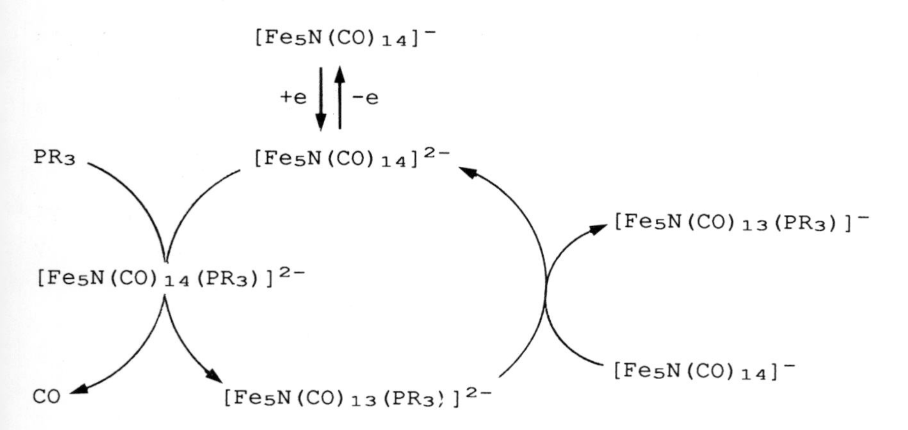

2.9. Pentanuclear Mixed-Ligand Complexes

2.9.1. $[Fe_5(CO)_{13}C(SO_2)]^{2-}$. Substitution of one SO_2 molecule for one CO group in $[Fe_5(CO)_{14}C]^{2-}$ affords the dianion $[Fe_5(CO)_{13}C(SO_2)]^{2-}$ shown in Figure 59 |128|.

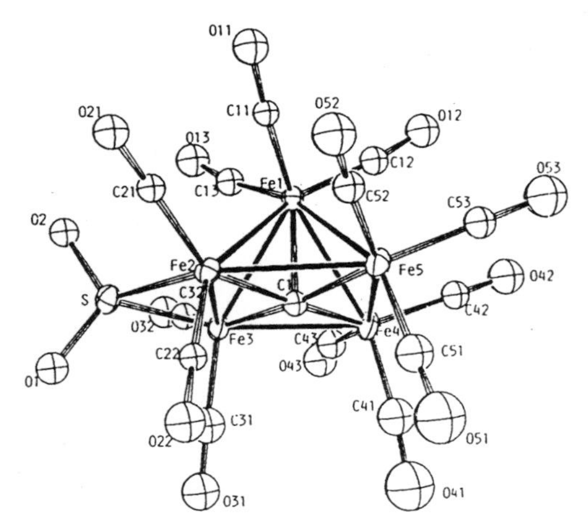

Figure 59. Perspective view of $[Fe_5(CO)_{13}C(SO_2)]^{2-}$ (from Ref. 128).

The square-based pyramidal assembly is substantially similar to that of the parent $[Fe_5(CO)_{14}C]^{2-}$. The averaged bond lengths are: $Fe_{(basal)}$-$Fe_{(basal)}$, 2.64 Å; $Fe_{(apical)}$-$Fe_{(basal)}$, 2.63 Å; $Fe_{(basal)}$-$C_{(carbide)}$, 1.87 Å; $Fe_{(apical)}$-$C_{(carbide)}$, 1.95 Å. The carbide atom is displaced below the tetrametallic base by 0.10 Å.

The presence of a SO_2 ligand makes the previously discussed redox changes more difficult in oxidation ($Ep_{2-/0}$ = +0.4 V*, in MeCN) and easier in reduction ($E°'_{2-/4-}$ = -1.4 V*) |128|, thus indicating that SO_2 is more electron withdrawing than CO.

2.9.2. $Fe_5(CO)_{12}C(PMe_2Ph)_3$. Replacement of three terminal carbonyl groups of $Fe_5(CO)_{15}C$, one for each basal iron atom, by dimethylphenyl-phosphine molecules affords $Fe_5(CO)_{12}C(PMe_2Ph)_3$, the structure of which is illustrated in Figure 60 |129|.

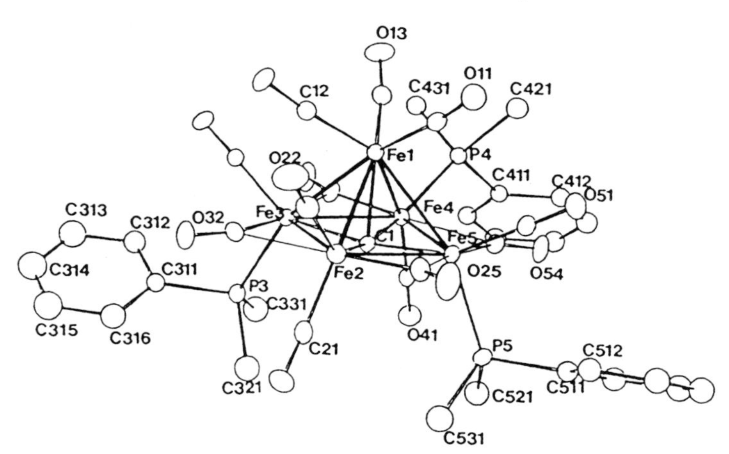

Figure 60. *Perspective view of $Fe_5(CO)_{12}C(PMe_2Ph)_3$ (from Ref. 129).*

Two phosphine ligands (P3 and P5) lie in axial positions with respect to the basal plane, whereas the third one (P4) occupies an equatorial position. The asymmetry induced by substitution at three of the four basal sites causes spreading of all the Fe-Fe

---

*The potential values here reported are only indicative, in that Ref. 128 refers to an unquotable pseudoreference Ag electrode.

distances, which assume averaged lengths of 2.63 Å in the basal plane and 2.60 Å in the apical-to-basal vectors. The displacement of the carbide atom from the basal plane is 0.20 Å, much higher than that in the unsubstituted precursor (0.09 Å). This complete the previous observation that such a displacement seems to increase not only with the charge of the cluster, but also with the electron density poured in the core by the peripheral ligands (actually, the phosphine ligands).

At variance with the fully carbonylated precursor, which was able to undergo a two-electron reduction to the corresponding dianion, stable at low temperature, here, the primarily electro-generated dianion $[Fe_5(CO)_{12}C(PMe_2Ph)_3]^{2-}$ ($E°'_{0/2-}$ = -0.72 V) is quite unstable |129|.

In contrast, the bidentate-phosphine substituted species $Fe_5(CO)_{13}C(Me_2PCH_2CH_2PMe_2)$ resembles more $Fe_5(CO)_{15}C$, in that it undergoes a two-electron reduction ($E°'_{0/2-}$ = -0.71 V) to $[Fe_5(CO)_{13}C(Me_2PCH_2CH_2PMe_2)]^{2-}$. Such a dianion is relatively stable at low temperature, but it decarbonylates to $[Fe_5(CO)_{12}C(Me_2PCH_2CH_2PMe_2)]^{2-}$, at room temperature |129|.

2.10. Hexanuclear Carbido Complexes

2.10.1. $[Fe_6(CO)_{16}C]^{2-}$. The hexanuclear dianion $[Fe_6(CO)_{16}C]^{2-}$ is obtained by reacting $[Fe_5(CO)_{14}C]^{2-}$ with $Fe_2(CO)_9$. As illustrated in Figure 61, it possesses an octahedral geometry, which completely encapsulates the carbide carbon atom |130,131|.

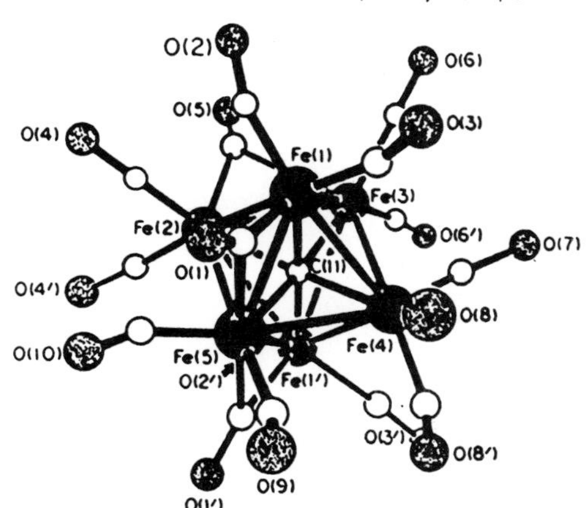

Figure 61.
Molecular structure of $[Fe_6(CO)_{16}C]^{2-}$ (from Ref. 130).

The presence of three bridging carbonyls [two between the $Fe_{(axial)}$-$Fe_{(equatorial)}$ bonds and one between the $Fe_{(equatorial)}$-$Fe_{(equatorial)}$ linkage], shortening the relevant metal-metal distances, makes slightly irregular the octahedral core. Nevertheless, the so-induced spreading of the metal-metal bond lengths causes no significant variation between the equatorial and the axial ones, so that an averaged Fe-Fe of 2.67 Å can be reliably assigned. The carbide atom is almost equidistant from the six iron atoms at a mean distance of 1.89 Å.

The electrochemical behaviour of the 86-electron $[Fe_6(CO)_{16}C]^{2-}$, in dichloroethane, is illustrated in Figure 62 |132|.

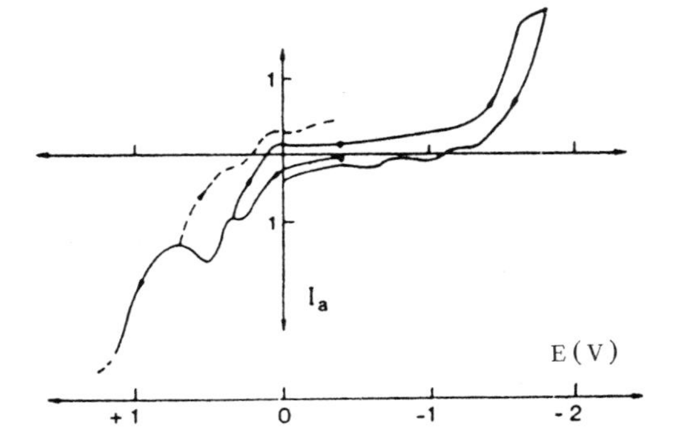

Figure 62.
Cyclic voltammetric response exhibited by $[Fe_6(CO)_{16}C]2-$, in $C_2H_4Cl_2$ solution, at a platinum electrode. Scan rate 0.1 Vs^{-1} (from Ref. 132).

As it can be seen, it undergoes two distinct (one-electron) irreversible oxidations, which, on the basis of the macro-electrolysis products, can be assigned to the main process |132|:

$$[Fe_6(CO)_{16}C]^{2-} \xrightarrow[-e]{Ep=+0.20\ V} [Fe_6(CO)_{16}C]^{-} \xrightarrow[-e]{Ep=+0.50\ V} Fe_6(CO)_{16}C \xrightarrow{-CO} Fe_5(CO)_{15}$$

This pathway follows the chemically induced oxidation process |133|. Indeed, according to a computed energy level diagram |125|, one-electron removal should not lead to fragmentation, the metal-centred bonding HOMO remaining half-filled. One could speculate that, in reality, in the cyclic voltammogram a residual, directly associated, reduction peak occurs after traversing the first anodic step, which should indicate a definite, even if short-lived, existence of $[Fe_6(CO)_{16}C]^-$. On the contrary, two-electron addition should cause filling of the metal-metal antibonding LUMO, thus giving rise to declustering processes. Accordingly, Figure 62 displays an irreversible two-electron reduction, which proved to be due to |133|:

$$[Fe_6(CO)_{16}C]^{2-} \xrightarrow[+2e]{Ep=-1.50\ V} [Fe_6(CO)_{16}C]^{4-} \rightarrow [Fe_5(CO)_{14}C]^{2-}$$

2.11. *Hexanuclear Mixed-Ligand Complexes*

2.11.1. $[Fe_6(CO)_{15}C(SO_2)]^{2-}$. In Section 2.9.1. we saw that substitution of one carbonyl group for one SO_2 group in $[Fe_5(CO)_{14}C]^{2-}$ does not involve significant structural rearrangements. The same happens in the case of $[Fe_6(CO)_{16}C]^{2-}$. As shown in Figure 63, $[Fe_6(CO)_{15}C(SO_2)]^{2-}$ possesses on octahedral geometry similar to that of $[Fe_6(CO)_{16}C]^{2-}$ |128|.

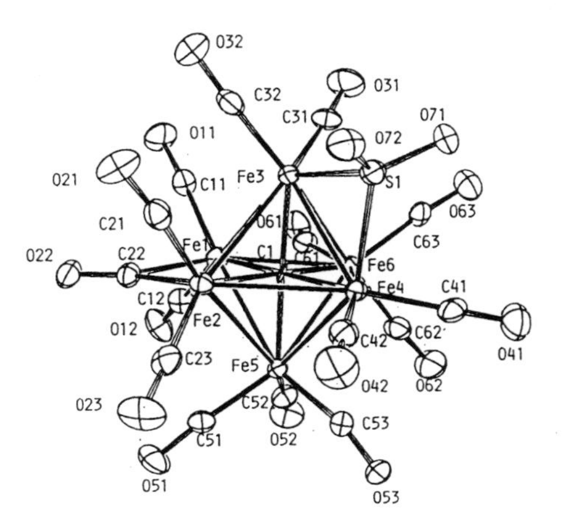

Figure 63. *Perspective view of $[Fe_6(CO)_{15}C(SO_2)]^{2-}$ (from Ref. 128).*

Some reorganization occurs as far as the disposition of the non-terminal carbonyls is concerned (the SO_2 group assuming an edge-bridging linkage), but the averaged Fe-Fe bond length remains fixed at 2.67 Å, as well as the mean bond length $Fe-C_{(carbide)}$ remains fixed at 1.89 Å.

The previously discussed electron withdrawing ability of the SO_2 group makes $[Fe_6(CO)_{15}C(SO_2)]^{2-}$ more difficult to oxidize and easier to reduce by about 0.3 V with respect to $[Fe_6(CO)_{16}C]^{2-}$ |128|.

2.11.2. $[Fe_6(CO)_{12}(S)_6]^{2-}$. As illustrated in Figure 64, the dianion $[Fe_6(CO)_{12}(S)_6]^{2-}$ is really constituted by an aggregation of three Fe_2S_2 fragments |134|.

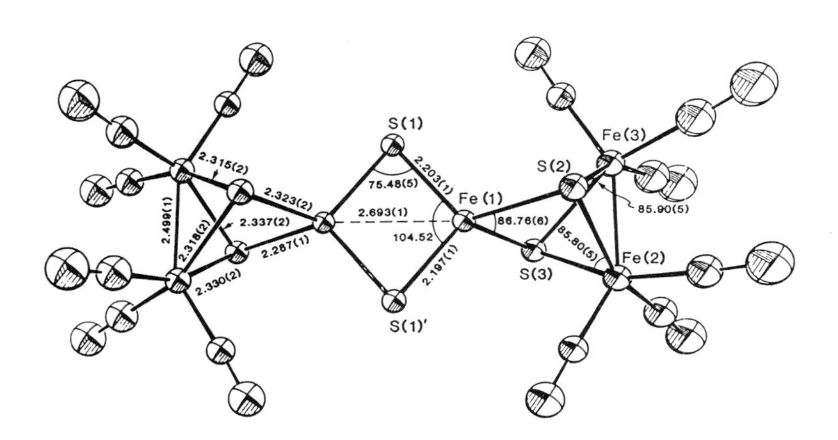

Figure 64. *Molecular structure of $[Fe_6(CO)_{12}(S)_6]^{2-}$, with selected distances (Å) and angles (deg) (from Ref. 134).*

The central planar $[Fe_2S_2]^{2+}$ unit is linked to two butterfly-like $[Fe_2S_2(CO)_6]^{2-}$ groups (Fe-Fe hinge).

This iron-sulfur-carbonyl derivative undergoes, in acetonitrile solution, two distinct, one-electron, quasireversible reduction steps ($E°'_{2-/3-}$= -1.13 V; $E°'_{3-/4-}$= -1.73 V), and an irreversible, multielectron oxidation (Ep= -0.08 V) |134|, thus suggesting a chemical stability for the corresponding tri- and tetra-anions.

3. RUTHENIUM CLUSTERS

3.1. Trinuclear Binary Complexes

3.1.1. $Ru_3(CO)_{12}$. Figure 65 shows the crystal structure of $Ru_3(CO)_{12}$ |135,136|.

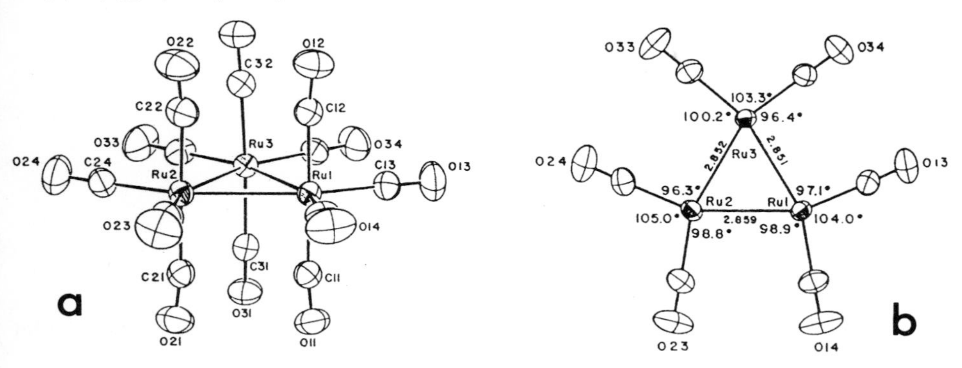

Figure 65. *Perspective view (a) and selected structural parameters (b) of $Ru_3(CO)_{12}$ (from Ref. 136).*

In contrast with the isoelectronic $Fe_3(CO)_{12}$, which possesses two edge-bridging carbonyls, all the carbonyl groups are terminal.

Electrochemistry of $Ru_3(CO)_{12}$ is not straightforward, since it depends upon the nature of the supporting electrolyte as well as the concentration of the cluster itself, and, besides that, it is extremely sensitive to traces of water, which affect the decomposition pathway of transient electrogenerated species |35, 137-139|. The voltammetric response consists of a single-stepped two-electron reduction, irreversible even at -80°C, which seems to involve a rate determining E.C.E. mechanism of the type |138,139|:

$$Ru_3(CO)_{12} \xrightarrow{+e} [Ru_3(CO)_{12}]^- \xrightarrow{k_1} [Ru_3(CO)_{12}^*]^- \xrightarrow{+e} [Ru_3(CO)_{12}^*]^{2-}$$

or, alternatively, an E.E. mechanism of the type |140|:

$$Ru_3(CO)_{12} \xrightarrow{+e} [Ru_3(CO)_{12}^*]^- \xrightarrow{+e} [Ru_3(CO)_{12}^*]^{2-}$$

where the asterisk denotes a supposed open trimetallic intermediate. The two mechanisms tend to discriminate between a *triangulo*-Ru_3/*open*-Ru_3 reorganization occurring *after* or *concomitantly with* the first electron addition. Such initial two-electron addition is followed in turn by fast (dissociative |138| or H_2O-induced |139|) decarbonylation to $[Ru_3(CO)_{11}{}^*]^{2-}$, which again puts in motion a sequence of the type:

$$[Ru_3(CO)_{11}{}^*]^{2-} \xrightarrow{k_2} [Ru_3(CO)_{11}]^{2-}$$

$$[Ru_3(CO)_{11}]^{2-} + Ru_3(CO)_{12} \longrightarrow [Ru_6(CO)_{18}]^{2-} + 5\ CO$$

Such mechanism could explain the predominant formation of $[Ru_6(CO)_{18}]^{2-}$ in the long times of macroelectrolysis.

Table 15 summarizes the electrode potential for the irreversible reduction of $Ru_3(CO)_{12}$ in different solvents.

TABLE 15. *Peak potential values (in Volt) for the irreversible reduction of* $Ru_3(CO)_{12}$

solvent	Ep[a]	reference
CH_2Cl_2	-1.05	139
Me_2CO	-1.10	138,139
THF	-1.03	139
MeCN	-0.87	139
MeCN	-1.05	137

[a]*Slight variations occur depending upon the supporting electrolyte*

The fact that electron addition to $Ru_3(CO)_{12}$ leads to deep structural reorganization is not unexpected, since the added electron enters an orbital which is antibonding with respect to both metal-metal and metal-ligand interactions |141|.

3.1.2. $[Ru_3(CO)_{11}]^{2-}$. The structure assigned to the dianion $[Ru_3(CO)_{11}]^{2-}$, on the basis of spectral data |142,143|, is schematized in Figure 66.

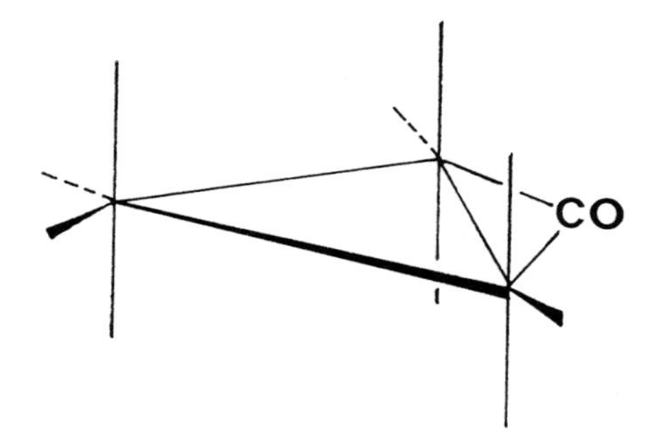

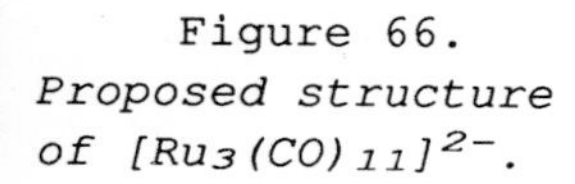

Figure 66.
Proposed structure of $[Ru_3(CO)_{11}]^{2-}$.

It has been briefly reported that, in acetone solution, $[Ru_3(CO)_{11}]^{2-}$ undergoes a complex oxidation process, a possible pathway being the following one |140,142,143|:

$$[Ru_3(CO)_{11}]^{2-} \xrightarrow{-e} [Ru_3(CO)_{11}]^{-} \xrightarrow{+CO} [Ru_3(CO)_{12}]^{-} \xrightarrow{-e} Ru_3(CO)_{12}$$

E_p=-0.27 V (first step); $[Ru_3(CO)_{11}]^{-}$ $\downarrow$ Ar: fast decomposition [partially to $Ru(CO)_5$]

3.2. Trinuclear Mixed-Ligand Complexes

3.2.1. $[HRu_3(CO)_{11}]^-$. The solid state structure of the hydrido anion $[HRu_3(CO)_{11}]^-$ parallels that of the isoelectronic $[HFe_3(CO)_{11}]^-$ (Section 2.2.1.). As shown in Figure 67, the ruthenium atoms define an isosceles triangle, the shorter edge of which bears, on opposite sides, both a bridging carbonyl group and a bridging hydride. The two longer Fe-Fe bonds are 2.84 Å, whereas the doubly bridged Fe-Fe distance is 2.81 Å |144|.

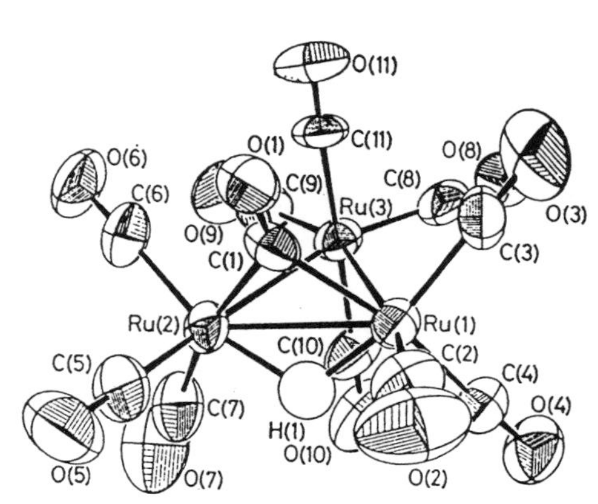

Figure 67.
Molecular structure of $[HRu_3(CO)_{11}]^-$ (from Ref. 144).

In dichloromethane solution, $[HRu_3(CO)_{11}]^-$ undergoes an irreversible one-electron oxidation, which, under CO atmosphere, ultimately forms $Ru_3(CO)_{12}$ |139|, according to the overall sequence:

$$[HRu_3(CO)_{11}]^- \xrightarrow[Ep= +0.07\ V]{-e} HRu_3(CO)_{11} \xrightarrow[+CO]{-H^+} Ru_3(CO)_{12}$$

3.2.2. $Ru_3(CO)_{12-n}(PPh_3)_n$ (n = 1-3). Preparation and reactivity of the family $Ru_3(CO)_{12-n}(PPh_3)_n$ (n = 1-3) have been widely reported |145,146|.

Progressive substitution of one carbonyl at each Ru centre generates the present family. Figure 68 shows the X-ray structure of the monosubstituted derivative |147|.

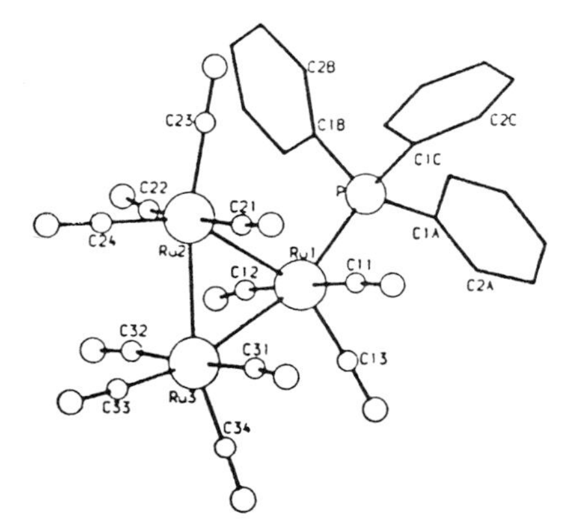

Figure 68. *Stereoscopic view of $Ru_3(CO)_{11}(PPh_3)$ (from Ref. 147).*

The phosphine group has substituted one equatorial carbonyl. All the carbonyls are terminal. The trimetallic triangle has two slightly shorter edges (2.88 Å) than the third one (2.91 Å). The averaged Ru-Ru distance is 0.04 Å longer than that in $Ru_3(CO)_{12}$; this has been attributed to the increased electronic density controlled by the phosphine, which mainly affects the Ru1-Ru2 bond *cis* to the substituent |147|.

The progressive substitution of one equatorial carbonyl at the other Ru centres is definitively proved by X-ray structures of molecules having phosphine or phosphite substituents |148|.

Figure 69 shows the cyclic voltammetric response exhibited by the series $Ru_3(CO)_{12-n}(PPh_3)_n$ (n = 1-3) in dichloromethane solution |139|.

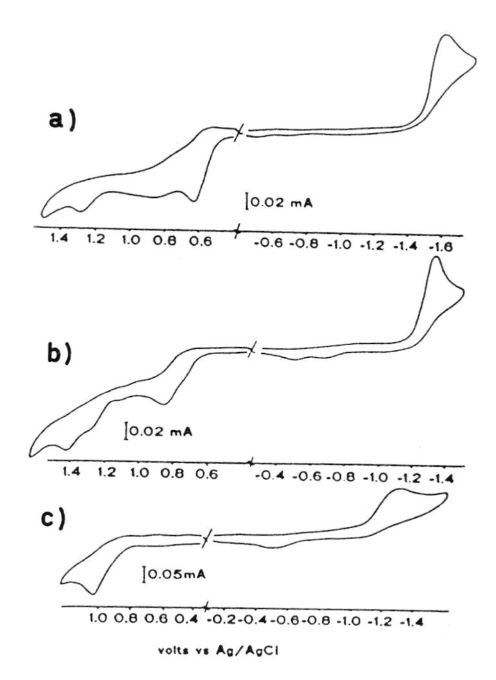

Figure 69. *Cyclic voltammograms recorded at a platinum electrode on a dichloromethane solution of: (a) $Ru_3(CO)_{11}(PPh_3)$; (b) $Ru_3(CO)_{10}(PPh_3)_2$; (c) $Ru_3(CO)_9(PPh_3)_3$. $[NBu_4][ClO_4]$ supporting electrolyte. Scan rate 0.1 Vs^{-1} (from Ref. 139).*

The main feature consists in the complete irreversibility of both oxidation and reduction processes. This is in sharp contrast with the iron analogues, which were able to add one electron reversibly (at least at low temperature). Speculatively, one can think that the lack of bridging carbonyls in Ru_3 complexes (substituted or unsubstituted) really confers a pure metal-metal antibonding character to their LUMO level, while in Fe_3 complexes the bridging carbonyls relieve in part such antibonding property.

Table 16 summarizes the redox potentials for the series $Ru_3(CO)_{12-n}(PPh_3)_n$. As expected, the higher the number of electron-donating phosphine groups, the more difficult is the addition of electrons as well as the easier their removal.

TABLE 16. *Peak potentials values (in Volt) for the irreversible electron-transfer processes exhibited by* $Ru_3(CO)_{12-n}(PPh_3)_n$ *in different solvents |139|*

complex	$Ep_{(0/2-)}$	$Ep_{(+/0)}$	solvent
$Ru_3(CO)_{11}(PPh_3)$	-1.29	+1.00	CH_2Cl_2
	-1.25	+1.22	THF
	-1.11	+1.02	Me_2CO
$Ru_3(CO)_{10}(PPh_3)_2$	-1.35	+0.80	CH_2Cl_2
	-1.32	+1.02	THF
	-1.25	+0.91	Me_2CO
$Ru_3(CO)_9(PPh_3)_3$	-1.65	+0.58	CH_2Cl_2
	-1.70	+0.68	THF
	-1.54	+0.62	Me_2CO

3.2.3. $H_3Ru_3(CO)_{9-n}(L)_n(C-X)$ (L = phosphines,arsines). The triangulated Ru_3 clusters capped by a methylidyne fragment $H_3Ru_3(CO)_{9-n}(L)_n(C-X)$ have been prepared, also to compare and contrast their properties with those of the series $Co_3(CO)_{9-n}(L)_n$ (C-X), which will be examined in Section 5.1.4.

Figure 70, which refers to $H_3Ru_3(CO)_9$(C-*p*-Tol), shows the molecular structure typical for the unsubstituted precursors |149|.

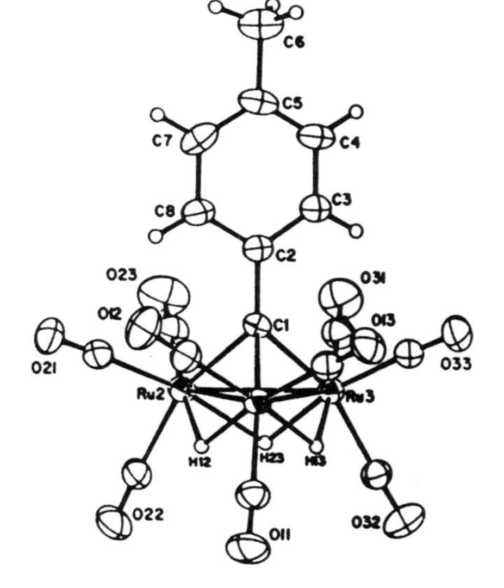

Figure 70. *Perspective view of* $(\mu-H)_3Ru_3(CO)_9[\mu_3-C-p-Tol]$ *(from Ref. 149).*

The triruthenium triangle is almost equilateral (averaged Ru-Ru distance, 2.95 Å) and symmetrically capped by the methylydine carbon (Ru-C, 2.11 Å). Each ruthenium atom bears two equatorial carbonyls, which are *trans* to the bridging hydrides, and one axial carbonyl, which is *trans* to the capping methylydine carbon |149|.

It has been demonstrated that progressive substitution by large ligands occurs at the axial carbonyl groups. This is clearly put in evidence in Figure 71, which shows the crystal structure of $H_3Ru_3(CO)_7(AsPh_3)_2(C\text{-}Ph)$ |150| and $H_3Ru_3(CO)_6[(PPh_2CH_2)_3CMe](C\text{-}OMe)$ |151|, respectively.

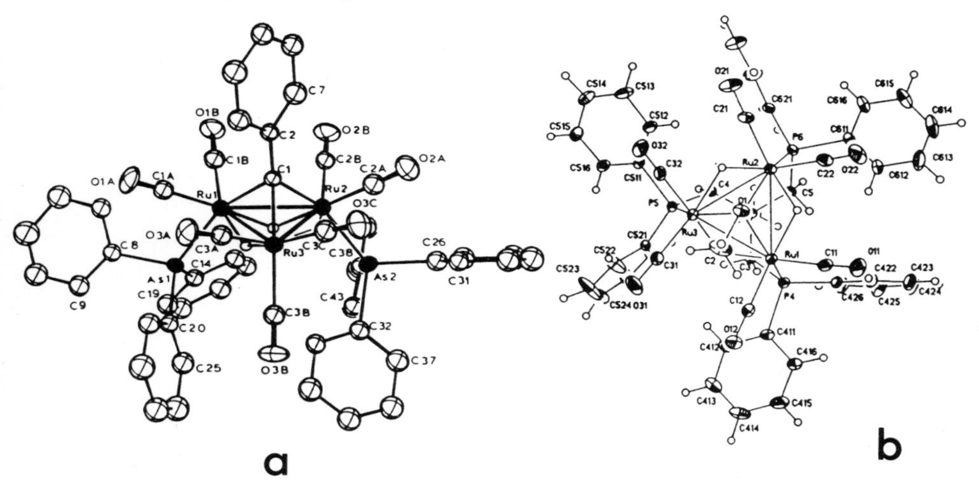

Figure 71. *X-Ray crystal structures of: (a)* $(\mu\text{-}H)_3Ru_3(\mu_3\text{-}CPh)(CO)_7(AsPh_3)_2$ *(from Ref. 150); (b)* $(\mu\text{-}H)_3Ru_3(\mu_3\text{-}COMe)(CO)_6[\mu_3\text{-}(PPh_2CH_2)_3CMe]$ *(from Ref. 151).*

As far as the redox properties of such complexes are concerned, the unsubstituted precursors as well as the momosubstituted species commonly undergo a single irreversible anodic step |152|. In contrast, the disubstituted and trisubstituted species undergo a first one-electron removal, with features of chemical reversibility, followed by a second one-electron irreversible process. Figure 72 shows the typical cyclic voltammetric profile |152|.

The kinetic stability of, as well as the thermodynamic redox potential of the access to, the 47-electron monocations $[H_3Ru_3(CO)_{9-n}(L)_n(C\text{-}X)]^+$ (n = 2,3) depend upon the substituents of the capping carbon atom. It has been argued from spectroscopic evidence that, upon one-electron removal, cluster reorganization occurs through the exchange of the fosphine or arsine substituents from axial to equatorial positions |151,152|.

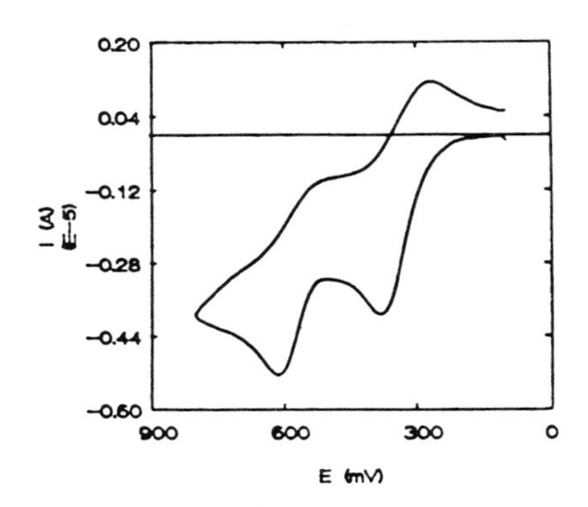

Figure 72.
Cyclic voltammogram recorded on a dichloromethane solution of $H_3Ru_3(CO)_6(PPh_3)_3(C{-}OMe)$. Scan rate 0.1 Vs^{-1} (from Ref. 152).

Table 17 reports the redox potentials of the electron-transfers exhibited by the complexes $H_3Ru_3(CO)_{9-n}(L)_n(C{-}X)$ (n = 2,3).

TABLE 17. *Formal electrode potential values (in Volt) for the electron-transfer processes exhibited by $H_3Ru_3(CO)_{9-n}(L)_n(C{-}X)$ (n = 2,3) in CH_2Cl_2 solution*

complex	$E°'_{(0/+)}$	$Ep_{(+/2+)}$	reference
$H_3Ru_3(CO)_7(PPh_3)_2(COMe)$	+0.51	+0.77	152
$H_3Ru_3(CO)_6(PPh_3)_3(COMe)$	+0.31	+0.60	152
$H_3Ru_3(CO)_6(AsPh_3)_3(COMe)$	+0.45	+0.68	152
$H_3Ru_3(CO)_6[(PPh_2CH_2)_3CMe](C{-}OMe)$	+0.42	+0.65	151
$H_3Ru_3(CO)_6(PPh_3)_3(CSEt)$	+0.35	+0.81	152
$H_3Ru_3(CO)_6(PPh_3)_3(CNMeCH_2Ph)$	-0.06	+0.35	152

It is interesting to note preliminarily that the structurally similar complexes $Co_3(CO)_{9-n}(L)_n(C{-}X)$ undergo reduction processes.

3.2.4. $H_2Ru_3(CO)_9(EtC_2Et)$. The dihydrido complex $H_2Ru_3(CO)_9(EtC_2Et)$ is a formally saturated 48-electron alkyno-carbonyl cluster, which, together with the osmium-analog (Section 4.2.5.), has assigned the structure schematized in Figure 73, with the alkyne parallel to one Ru-Ru edge of the underlying metallic triangle (μ_3- η^2-|| coordination) |153|.

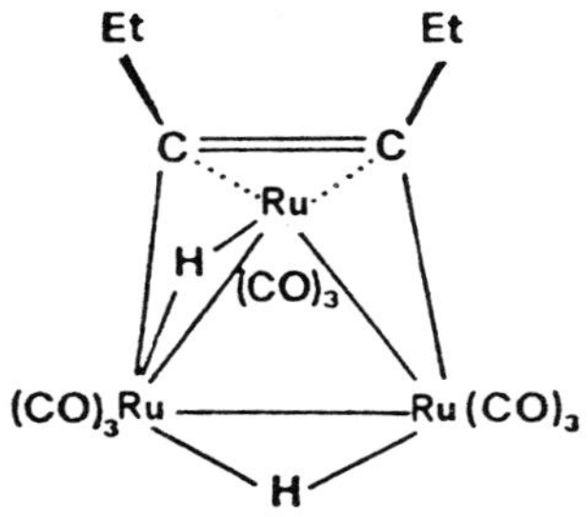

Figure 73.
Schematic structure assigned to $H_2Ru_3(CO)_9(EtC_2Et)$.

Figure 74 shows that, in agreement with its electron-saturation, it is very difficult to add electrons to it, and the irreversibility of the single-stepped two-electron reduction (Ep = -1.74 V) testifies to the consequent fragmentation |50|.

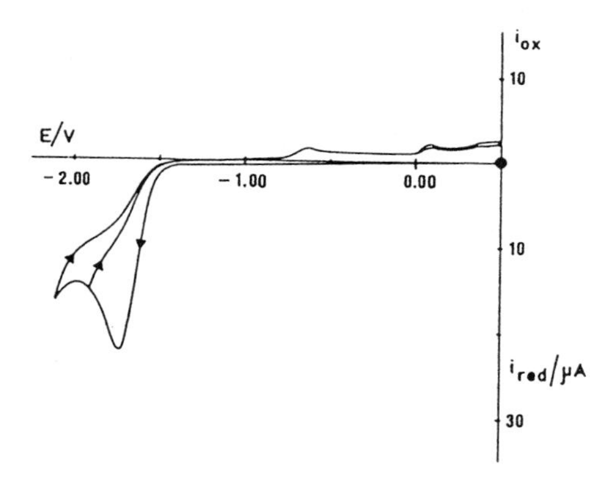

Figure 74. *Cyclic voltammogram recorded at a platinum electrode on a CH_2Cl_2 solution of $H_2Ru_3(CO)_9(EtC_2Et)$. Scan rate 0.2 Vs^{-1} (from Ref. 50).*

3.2.5. $[Ru_3(CO)_{10}](HC\equiv C-C_6H_4-X)$ (X = H, $C\equiv CH[Ru_3(CO)_{10}]$). Figure 75 schematically shows the molecular structure assigned to the two clearly related cluster compounds $[Ru_3(CO)_{10}](HC\equiv C-C_6H_5)$ and $[Ru_3(CO)_{10}](HC\equiv C-C_6H_4-C\equiv CH)[Ru_3(CO)_{10}])$ |154,155|.

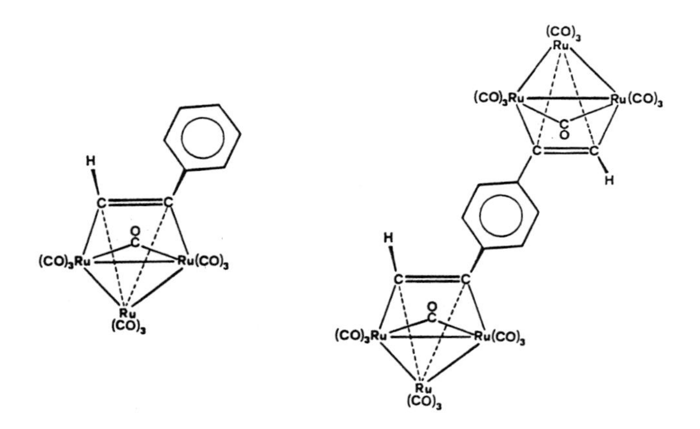

Figure 75. *Schematic representation of the molecular structure of: (a) $[Ru_3(CO)_9(\mu\text{-}CO)](HC{\equiv}C\text{-}C_6H_5)$; (b) $[Ru_3(CO)_9(\mu\text{-}CO)]_2(HC{\equiv}C\text{-}C_6H_4\text{-}C{\equiv}CH)$.*

As illustrated in Figure 76, the cyclic voltammetric profiles exhibited by the two complexes display a single reduction peak for the mononucluster as well as for the dicluster assemblies |155|.

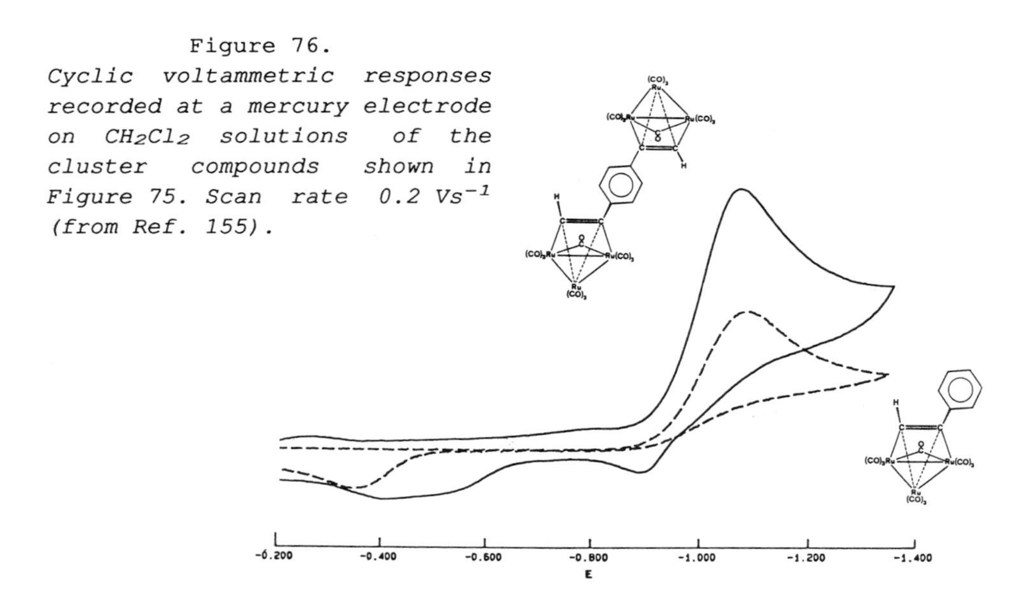

Figure 76. *Cyclic voltammetric responses recorded at a mercury electrode on CH_2Cl_2 solutions of the cluster compounds shown in Figure 75. Scan rate 0.2 Vs^{-1} (from Ref. 155).*

As expected, the saturated 48-electron species $[Ru_3(CO)_{10}](HCC-C_6H_5)$ undergoes an irreversible two-electron reduction (Ep = -1.05 V). The fact that also the dicluster species $[Ru_3(CO)_{10}]_2(HCC-C_6H_4-CCH)$ undergoes an irreversible, single-stepped four-electron reduction points out that no electronic communication exists between the two Ru_3 fragments |155|.

3.2.6. $HRu_3(CO)_9[RC{=}C{=}C(R')R'']$. The most kinetically stable product from reaction of $Ru_3(CO)_{12}$ with substituted alkynes, having a hydrogen atom adjacent to the triple bond, is the *allenic* family $HRu_3(CO)_9[RC{=}C{=}(R')R'']$ |22|. As an example, Figure 77 shows the X-ray structure of $HRu_3(CO)_9[EtC{=}C{=}C(H)Me]$ |156|.

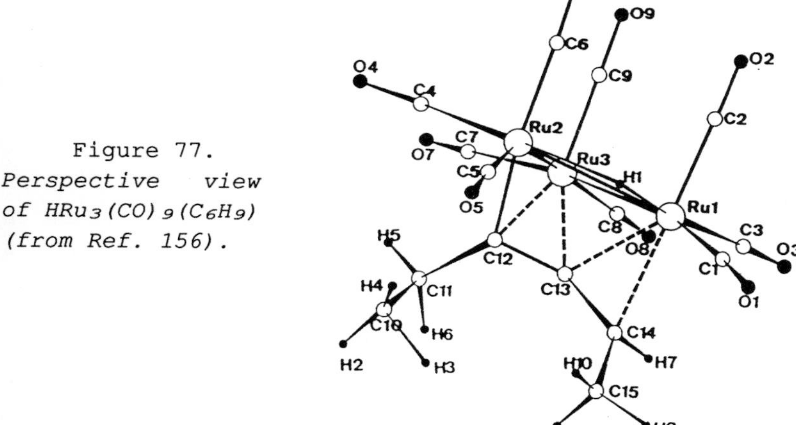

Figure 77. *Perspective view of $HRu_3(CO)_9(C_6H_9)$ (from Ref. 156).*

The cluster consists of an almost isosceles ruthenium triangle (Ru1-Ru2, 2.99 Å; Ru1-Ru3, 2.78 Å; Ru2-Ru3, 2.74 Å), in which each ruthenium atom bears three terminal CO groups. The hydride atom lies on the trimetallic plane and bridges the longer Ru1-Ru2 bond. Three carbon atoms of the organic moiety interact with the metallic fragment such as to make the hexacarbon chain an overall five-electron donor.

Figure 78, which refers to $HRu_3(CO)_9(MeC{=}C{=}CMe_2)$, illustrates the typical electrochemical response exhibited by the present derivatives |137|. Apparently, the cluster, in agreement with its saturated $Ru_3(CO)_{12}$-like electronic configuration, undergoes irreversible reduction steps. Nevertheless, the increase of scan rate (broken curve), preventing in part the decomposition of the instantaneously generated monoanion, allows us to compute a half-life of few tens of milliseconds for $[HRu_3(CO)_9\{RC{=}C{=}C(R')R''\}]^-$. This datum indicates that capping the triruthenium cluster by selected organic fragments can improve the lifetime of the reduced congeners, likely subtracting electron density from the metallic frame.

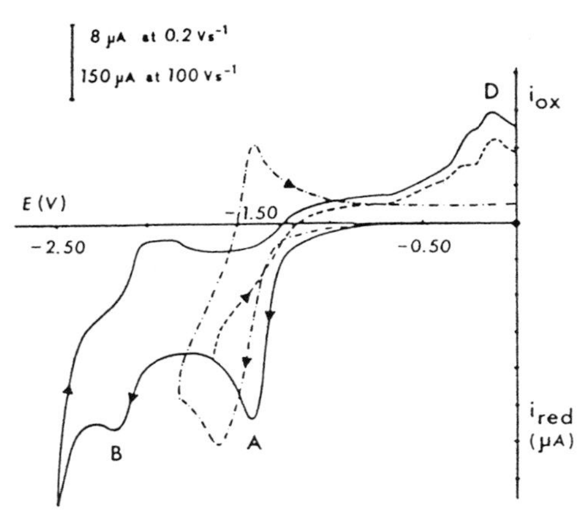

Figure 78. *Cyclic voltammograms recorded at a mercury electrode on a MeCN solution of $HRu_3(CO)_9(MeC{=}C{=}CMe_2)$. Scan rates: (----) 0.2 Vs^{-1}; (-·-·-) 100 Vs^{-1} (from Ref. 137).*

Table 18 reports the redox potentials for the electron transfers exhibited by the present complexes.

TABLE 18. *Formal electrode potentials (in Volt) for the redox changes exhibited by $HRu_3(CO)_9[RC{=}C{=}C(R')R'']$ in acetonitrile solution |137|*

complex	$E°'_{0/-}$	$Ep^a{}_{-/2-}$
$HRu_3(CO)_9[EtC{=}C{=}C(H)Me]$	-1.56	-1.98
$HRu_3(CO)_9[MeC{=}C{=}CMe_2]$	-1.54	-2.18
$HRu_3(CO)_9[MeC{=}C{=}C(H)NMe_2]$	-1.46[a]	

[a]*Peak potential value for totally irreversible steps*

3.2.7. $HRu_3(CO)_9[RC\overset{\cdot\cdot\cdot}{—}C(R')\overset{\cdot\cdot\cdot}{—}CR'']$. The *allenic* derivatives $HRu_3(CO)_9[RC{=}C{=}C(R')R'']$ undergo thermal rearrangement to the *allylic* forms $HRu_3(CO)_9[RC\overset{\cdot\cdot\cdot}{—}C(R')\overset{\cdot\cdot\cdot}{—}CR'']$ |157|. Figure 79 shows the structure of $HRu_3(CO)_9[EtC\overset{\cdot\cdot\cdot}{—}C(H)\overset{\cdot\cdot\cdot}{—}CMe]$ |157|, which is the *allylic* isomer of the *allenic* molecule illustrated in Figure 77.

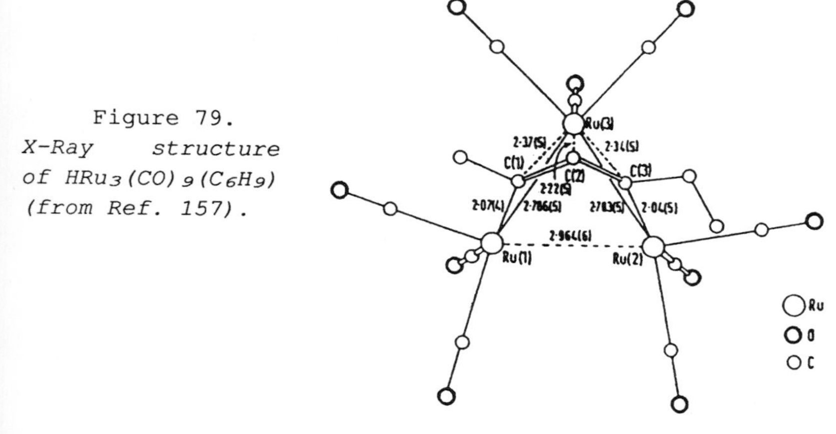

Figure 79.
X-Ray structure of $HRu_3(CO)_9(C_6H_9)$ (from Ref. 157).

The hexacarbon chain is π-bonded to Ru_3 (through the allylic C1, C2, C3 atoms) and σ-bonded to Ru1 and Ru2 through C1 and C3 atoms. The hydride atom is thought to bridge the Ru1-Ru2 edge.

Like the *allenic* family, the present 48-electron *allylic* derivatives undergo irreversible reductions. However, the capping by the actual organic system seems less effective in stabilizing the corresponding monoanions ($t_{1/2}$ = some milliseconds) |137|. The relevant reduction potentials are given in Table 19.

TABLE 19. *Redox potentials (in Volt) for the reduction steps exhibited by $HRu_3(CO)_9[RC\overset{\cdot\cdot\cdot}{—}C(R')\overset{\cdot\cdot\cdot}{—}CR'']$ in acetonitrile solution |137|*

complex	$Ep^a{}_{0/-}$	$Ep^a{}_{-/2-}$
$HRu_3(CO)_9(EtC\overset{\cdot\cdot\cdot}{—}CH\overset{\cdot\cdot\cdot}{—}CMe)$	-1.43	-1.80
$HRu_3(CO)_9(MeC\overset{\cdot\cdot\cdot}{—}CH\overset{\cdot\cdot\cdot}{—}CMe)$	-1.40	-1.83
$HRu_3(CO)_9(MeC\overset{\cdot\cdot\cdot}{—}CH\overset{\cdot\cdot\cdot}{—}CNMe_2)$	-1.59	-1.80

[a]*Peak potential value for irreversible process*

3.2.8. $HRu_3(CO)_9[C\equiv C\text{-}C(Me)_3]$. 3,3-Dimethyl-1-butyne reacts with $Ru_3(CO)_{12}$ to afford $HRu_3(CO)_9[C\equiv C\text{-}C(Me)_3]$, the structure of which is shown in Figure 80 |158|.

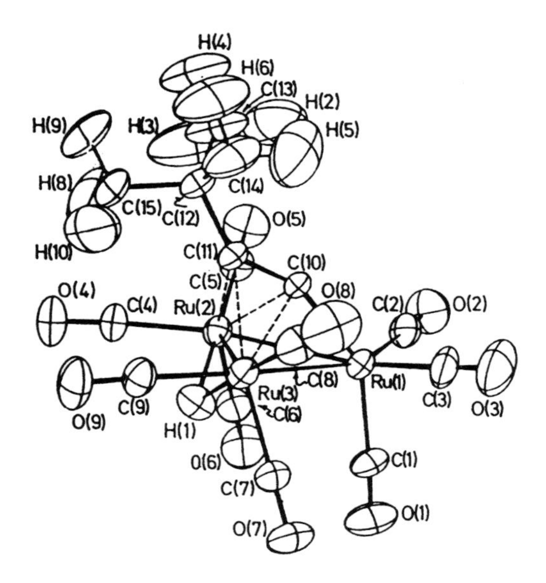

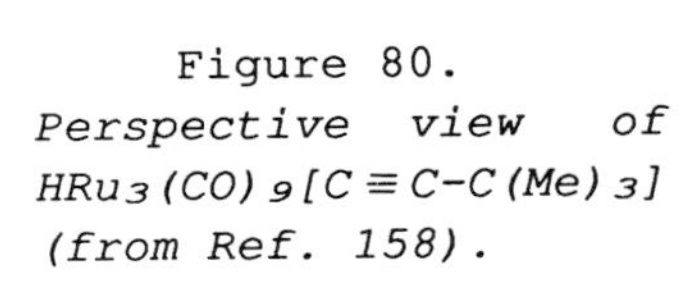
Figure 80. *Perspective view of $HRu_3(CO)_9[C\equiv C\text{-}C(Me)_3]$ (from Ref. 158).*

The alkyne caps the equilateral ruthenium triangle (Ru-Ru averaged distance, 2.79 Å) orienting itself along the median. The hydride atom, which was located by neutron diffraction, bridges the Ru2-Ru3 edge.

Once again, this 48-electron cluster undergoes, in acetonitrile solution, a first one-electron reduction ($E°'_{0/-}$ = -1.60 V) complicated by relatively fast decomposition of the corresponding monoanion, which has assigned an half-life of about 10 msec (intermediate between that of the *allenic* and *allylic* capped clusters) |137|.

3.2.9. $Ru_3(CO)_8(RC_2R)_2$. The 46-electron clusters $Ru_3(CO)_6(\mu\text{-}CO)_2(RC_2R)_2$ are isostructural with the $Fe_3(CO)_8(RC_2R)_2$ series discussed in Section 2.2.12. |159|. Their structure is schematized in Figure 81.

Figure 82 shows the cyclic voltammetric responses exhibited by $Ru_3(CO)_8(PhC_2Ph)_2$ in dichloromethane solution, at different temperatures |81|.

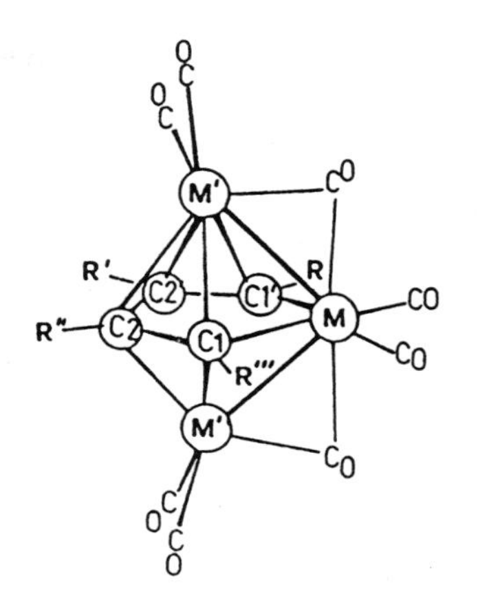

Figure 81. *Idealized structure of $Ru_3(CO)_8(RC_2R)_2$.*

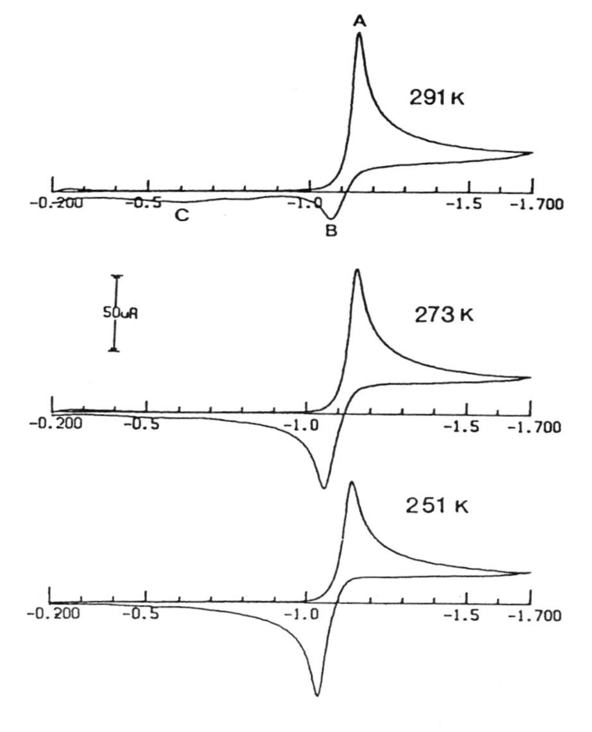

Figure 82. *Cyclic voltammograms recorded on a CH_2Cl_2 solution of $Ru_3(CO)_8(PhC_2Ph)_2$, at different temperatures. Mercury working electrode. Scan rate 0.2 Vs^{-1} (from Ref. 81).*

In contrast with the Fe_3-analogs, which gave quite stable monoanions and relatively long-lived dianions, the present Ru_3 species undergo a complex, single-stepped reduction, which initially generates the corresponding monoanions. Such monoanions, in turn, decompose to electroreducible derivatives [namely, $Ru_2(CO)_6(RC_2R)_2$ and a higly unsaturated "$Ru(CO)_2$" species], through a chemical reaction, which, as deducible from Figure 82, is significantly quenched by decreasing the temperature.

Table 20 summarizes the redox potentials and the half-life of the different monoanions for the complexes studied.

TABLE 20. *Redox features for the reduction process of $Ru_3(CO)_8(RC_2R)_2$ in dichloromethane solution, at 18°C |81|*

complex	E°' (V)	$t_{1/2}$(monoanion) (sec)
$Ru_3(CO)_8(PhC_2Ph)_2$	-1.16	0.7
$Ru_3(CO)_8(EtC_2Et)_2$	-1.42	0.1
$Ru_3(CO)_8(MeC_2Et)(PhC_2Ph)$	-1.28	0.3

While the occurrence of the reduction process at potentials more negative (≈ 0.4 V) than those for the corresponding iron clusters is easily accounted for by the location, at higher energy, of the LUMO of the Ru_3 complex, the instability of $[Ru_3(CO)_8(RC_2R)]^-$ with respect to $[Fe_3(CO)_8(RC_2R)]^-$ has no easy explanations. This remains so inspite the fact that stabilization of these types of monoanions involves a delicate balance between the metal-metal antibonding and metal-ligand bonding characters of their LUMO levels |80,81|.

Recently, in the attempt to prepare another member of the present series having R=R'=Ph, R"=Me, R"'=CH_2NMe_2 (see Figure 81), we discovered a quite different structural geometry |160|. Figure 83 just shows the geometry of $Ru_3(CO)_8(PhC_2Ph)(MeC_2CH_2NMe_2)$. The molecule consists of a closed isosceles ruthenium triangle (two Ru-Ru bonds of 2.88 Å, and one Ru-Ru bond of 2.77 Å), which, through Ru3, forms an outer ruthenacyclopentadiene ring. Coordination is completed by bonds (C9, C16, C23, C25) with Ru2 and (N27) with Ru1. Consequently, the occurrence of significant structural differences induced by the presence of electron-donating NMe_2 groups within a class of compounds is not new |161|.

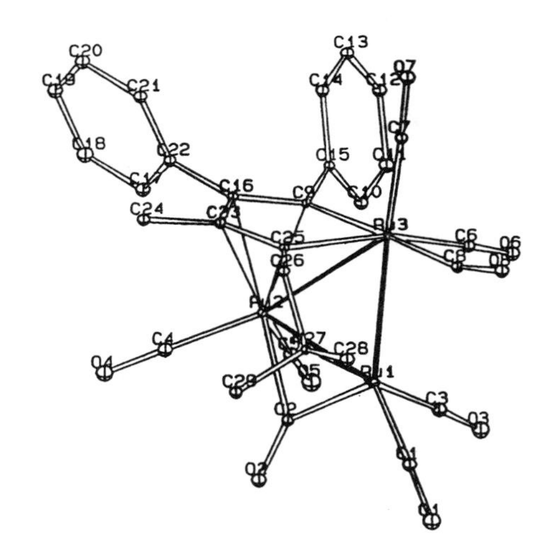

Figure 83.
Perspective view of $Ru_3(CO)_8(PhC_2Ph)(MeC_2CH_2NMe_2)$ (from Ref. 160).

In accordance with the different molecular structure, also the electrochemical response is different, in that, upon reduction (in MeCN, Ep = -1.45 V), the corresponding monoanion is never detectable, either at high scan rates or at low temperature. This behaviour is more reminescent of that exhibited by total bonding triangular, saturated 48-electron species, than that of the actual, open triangular 46-electron clusters.

3.3. Tetranuclear Mixed-Ligand Complexes

3.3.1. $Ru_4(CO)_{11-n}(PPh)_2[P(OMe)_3]_n$ (n = 0,1). The phenyl-phosphinidene-capped ruthenium derivatives $Ru_4(CO)_{11}(PPh)_2$ and $Ru_4(CO)_{10}[P(OMe)_3](PPh)_2$ are the unsaturated (62-electron, or 7-skeletal electron pairs) analogs of the iron complexes discussed in Section 2.6.4. Their molecular structure is typified by that of $Ru_4(CO)_{11}(PPh)_2$ shown in Figure 84 |162,163|.

As in the case of the iron compound, the tetrametal fragment is essentially planar. The mean Ru-Ru bond length is 2.83 Å, notably longer than that of the iron cluster (2.63 Å), but the presence of the bridging carbonyl between Ru2 and Ru3, as usual, shortens this distance to 2.72 Å with respect to the other three equivalent lengths of 2.87 Å.

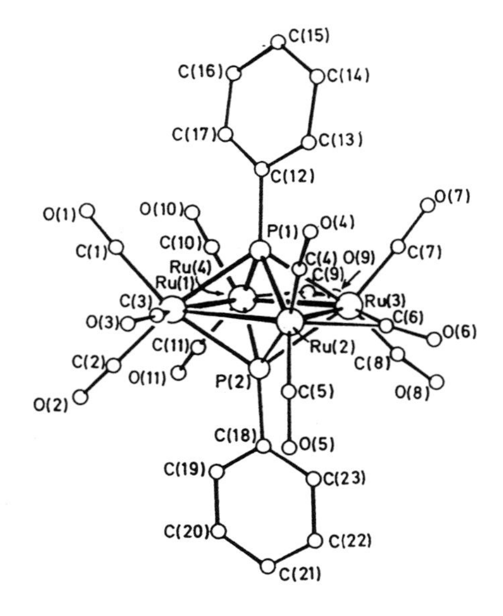

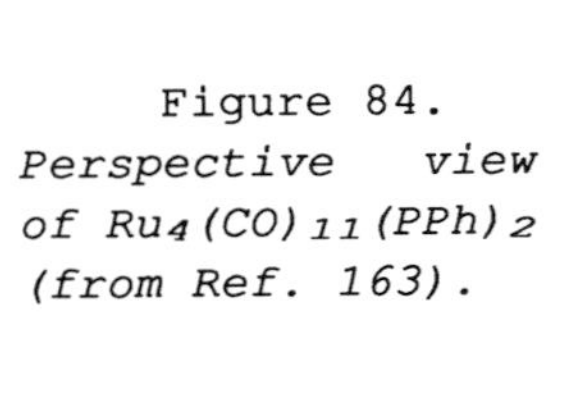
Figure 84. *Perspective view of $Ru_4(CO)_{11}(PPh)_2$ (from Ref. 163).*

As previously discussed, these unsaturated derivatives should be predisposed to add two electrons in the nonbonding level interposed between the bonding and antibonding orbitals so to reach a stable 64-electron configuration |110|. Thus, as deducible from Figure 85, $Ru_4(CO)_{11}(PPh)_2$ undergoes a first chemically reversible one-electron reduction, while the second reduction step leads to a dianion stable only on the short timescale of cyclic voltammetry. Then, further electron addition becomes chemically irreversible, because antibonding orbitals become populated |113|.

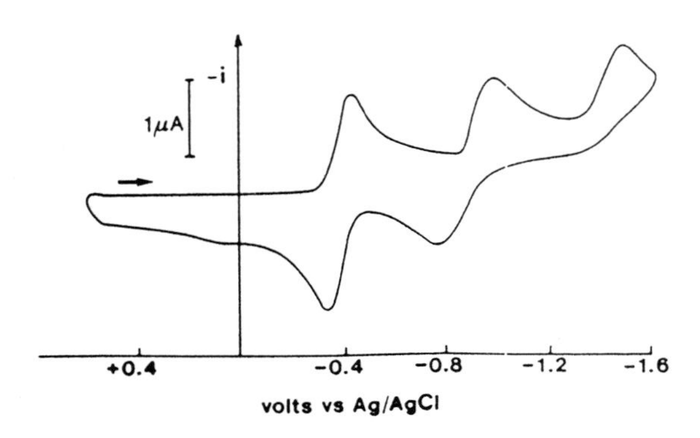

Figure 85. *Cyclic voltammogram exhibited by $Ru_4(CO)_{11}(PPh)_2$ under the experimental conditions of Figure 42 (from Ref. 113).*

Table 21 compares the redox potentials of the two-electron addition for the tetraruthenium clusters with those of the corresponding iron analogues.

TABLE 21. *Redox potentials (in Volt) and peak-to-peak separation (in mV) for the reduction steps of the phenylphosphinidene-capped ruthenium and iron unsaturated clusters, in benzonitrile |113|*

cluster	$E^{\circ\prime}_{0/1-}$	ΔEp^{a}	$E^{\circ\prime}_{1-/2-}$	ΔEp^{a}
$Ru_4(CO)_{11}(PPh)_2$	-0.43	100	-0.95	200
$Ru_4(CO)_{10}[P(OMe)_3](PPh)_2$	-0.65	80	-1.20	300
$Fe_4(CO)_{11}(PPh)_2$	+0.05	80	-0.77	120
$Fe_4(CO)_{10}[P(OMe)_3](PPh)_2$	-0.17	80	-0.97	100

[a]*Measured at 0.2 Vs^{-1}*

It is easy to deduce that the ruthenium derivatives are harder to reduce than the corresponding iron derivatives (in particular, the difference for the first step is about 0.5 V). This has been attributed to the fact that, with reference to the model $Fe_4(CO)_{12}(PH)_2$ |110|, the lower electronegativity of ruthenium makes higher in energy, and hence less easily accessible, the oft-cited nonbonding orbital responsible for the stability of the 8-skeletal electron pair configuration |113|.

3.3.2. $[Ru_4(CO)_{12-n}(PhC_2Ph)]^{2n-}$ (n = 0,1). Reaction between $Ru_3(CO)_{12}$ and alkynes affords different products, depending upon the reaction conditions as well as the nature of the alkyne |164|. One class of such compounds is constituted by the tetraruthenium derivatives $Ru_4(CO)_{12}(RC_2R)$ |165-169|.

Figure 86 shows the molecular structure of $Ru_4(CO)_{12}(PhC_2Ph)$ |166|. The ruthenium atoms form a butterfly arrangement capped by the alkyne, which disposes the unsaturated C-C chain parallel to the Ru1-Ru2 hinge bond. The dihedral angle between the two wings is 115.5°. Interestingly, in $Ru_4(CO)_{12}(MeC_2Me)$ the corresponding dihedral angle is 63.1° |167|, which gives evidence to the effect of less or more bulky substituents in the alkyne moiety on the geometrical assembly of the tetrametal butterfly.

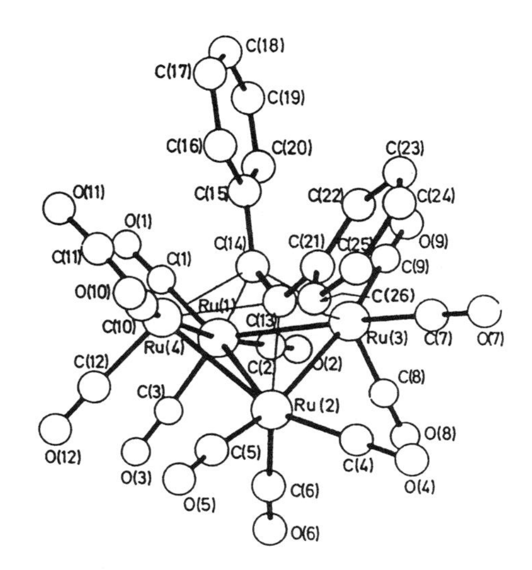

Figure 86. *X-Ray structure of $Ru_4(CO)_{12}(PhC_2Ph)$. Ru1-Ru2, 2.85 Å; averaged $Ru_{(wing)}$-$Ru_{(hinge)}$, 2.73 Å; averaged C13-Ru2, C13-Ru3, C13-Ru4, 2.22 Å; C13-C14, 1.46 Å (from Ref. 166).*

The redox propensity of $Ru_4(CO)_{12}(C_2Ph_2)$ in acetonitrile solution is shown in Figure 87 |168|.

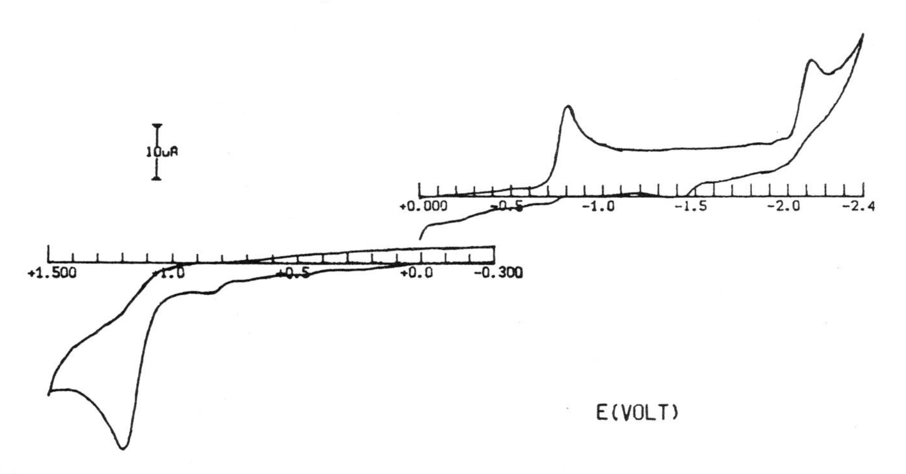

Figure 87. *Cyclic voltammogram recorded at a platinum electrode on a deaerated MeCN solution of $Ru_4(CO)_{12}(C_2Ph_2)$. $|NEt_4||ClO_4|$ supporting electrolyte. Scan rate 0.2 Vs^{-1}.*

It displays two irreversible two-electron reduction steps at -0.81 V and -2.12 V, respectively, as well as a single-stepped four-electron oxidation at + 1.18 V.

A detailed study of the first reduction process showed that it involves the following overall path |170|:

$$Ru_4(CO)_{12}(C_2Ph_2) \xrightarrow{+2e} [Ru_4(CO)_{12}(C_2Ph_2)]^{2-} \xrightarrow[-CO]{fast} [Ru_4(CO)_{11}(C_2Ph_2)]^{2-}$$

Chemical reduction by sodium amalgam afforded the decarbonylated dianion, the X-ray structure of which is shown in Figure 88 |170|.

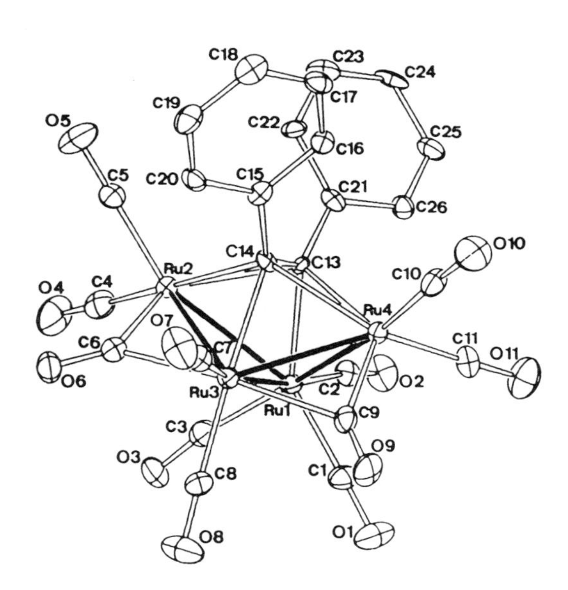

Figure 88. *Perspective view of* $[Ru_4(CO)_{11}(C_2Ph_2)]^{2-}$. *Ru1-Ru3, 2.82 Å; averaged* $Ru_{(wing)}$-$Ru_{(hinge)}$, *2.74 Å; averaged C13-Ru1, C13-Ru2, C13-Ru4, 2.26 Å; C13-C14, 1.42 Å (from Ref. 170).*

With respect to the neutral precursor $Ru_4(CO)_{12}(C_2Ph_2)$, the molecular assembly does not undergo significant geometrical reorganization. The loss of one CO group induces two terminal CO ligands to assume edge-bridging positions, but the bonding lengths in the cluster core remain substantially unvaried. The dihedral angle between the two wings is unaffected too (115.8°) |170|.

The dianion $[Ru_4(CO)_{11}(PhC_2Ph)]^{2-}$ exhibits, in acetonitrile solution, an irreversible two-electron oxidation (Ep = +0.33 V), which, under CO, regenerates the neutral precursor $Ru_4(CO)_{12}(PhC_2Ph)$ |170|.

It is useful to note that $Ru_4(CO)_{12}(PhC_2Ph)$ is considered a 60-electron cluster compound, *i.e.* it possesses two fewer electrons than that required for a butterfly arrangement. Unfortunately, the loss of one CO molecule following the two-electron addition, leaving unvaried the electron counting, does not allow the complex to obey the classical 62-electron pattern.

3.4. Pentanuclear Carbido Complexes

3.4.1. $[Ru_5(CO)_{14}C]^{2-}$. On the basis of spectroscopic evidence, the dianion $[Ru_5(CO)_{14}C]^{2-}$ has been assigned a square-based pyramidal geometry, just similar to that described in Section 2.7.1. for $[Fe_5(CO)_{14}C]^{2-}$ |169|.

This dianion undergoes, in dichloromethane solution, irreversible oxidation steps (the second of which at +0.2 V |122|), which lead, among different decomposition products, to $Fe_5(CO)_{15}C$ |169|.

3.4.2. $Ru_5(CO)_{15}C$. Like the corresponding iron analogue (Section 2.7.2.), the 74-electron $Ru_5(CO)_{15}C$ has the square pyramidal geometry shown in Figure 89, with the carbide atom slightly displaced below the basal plane (by 0.11 Å) |171,172|.

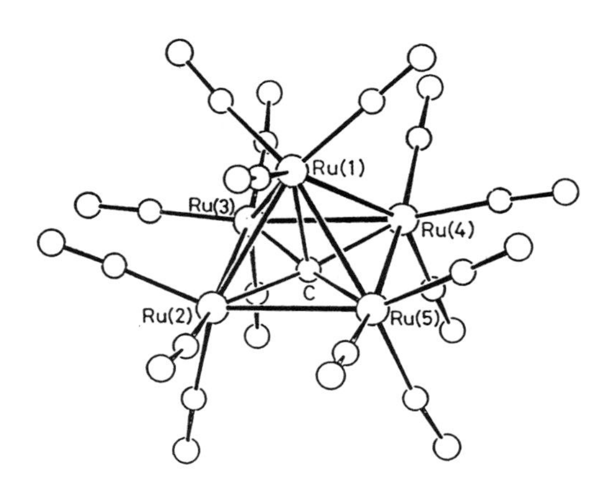

Figure 89. *Perspective view of $Ru_5(CO)_{15}C$. Mean bond lengths: $Ru_{(basal)}$-$Ru_{(basal)}$, 2.85 Å; $Ru_{(basal)}$ - $Ru_{(apical)}$, 2.83 Å; Ru-$C_{(carbide)}$, 2.04 Å (from Ref. 171).*

In dichloromethane solution, $Ru_5(CO)_{15}C$ undergoes an irreversible single-stepped two-electron reduction (in the range from -1.4 V to -1.5 V), affording, in accordance with the chemical pathway |173|, the previously cited dianion $[Ru_5(CO)_{14}C]^{2-}$ |169|.

It is worth noticing, how, although the geometry of the Ru_5C core remains likely unchanged upon two-electron addition, expulsion of one carbonyl group followed by the reorganization of some of the remaining ones (one bridging or two semibridging) makes totally irreversible the electron transfer step. It seems hence likely to hypothesize that the LUMO level of $Ru_5(CO)_{15}C$ has mainly an antibonding metal-carbon$_{(carbonyl)}$ character. In spite of this observation, $Ru_5(CO)_{15}C$ in the presence of one (two-electron donor) acetonitrile molecule is by no means fragmented, but undergoes a deep structural reorganization to a butterfly assembly "$Ru_4(CO)_2C$", which is flyover-bridged by a $Ru(CO)_3{\cdot}NCMe$ fragment, Figure 90 |172-174|.

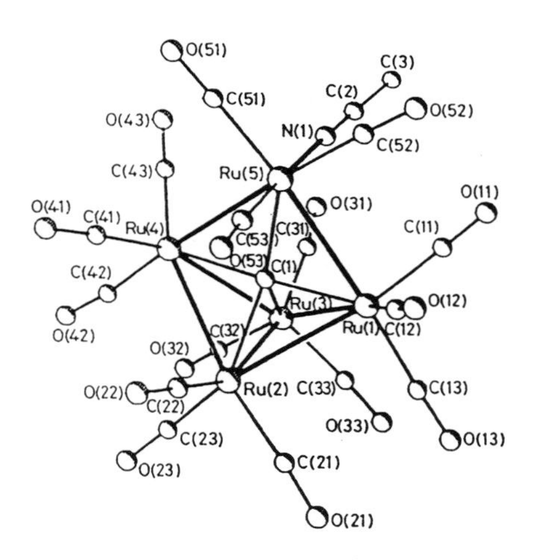

Figure 90.
Molecular structure of $Ru_5(CO)_{15}C{\cdot}MeCN$ (from Ref. 174).

The Ru-Ru hinge bond length (2.72 Å) is shorter than the remaining metal-metal bonds (2.88 Å). In the 76-electron $Ru_5(CO)_{15}C{\cdot}MeCN$ the averaged Ru-Ru distance is 2.86 Å, as contrasted with 2.84 Å in $Ru_5(CO)_{15}C$; analogously, the Ru-C$_{(carbido)}$ mean distance (2.04 Å) remains substantially unaltered.

3.5. Pentanuclear Mixed-Ligand Complexes

3.5.1. $Ru_5(CO)_{13}(C_2PPh_2)(PPh_2)$. By reacting $Ru_3(CO)_{12}$ with the alkyno-diphosphine $Ph_2P\text{-}C{\equiv}C\text{-}PPh_2$, the bis-triangled ruthenium complex $[Ru_3(CO)_{11}]_2(\mu\text{-}Ph_2PCCPPh_2)$ forms |175|. Thermal rearrangement of this latter species affords $Ru_5(CO)_{13}(\eta^5\text{-}C_2PPh_2)(\mu\text{-}PPh_2)$, the X-ray structure of which is shown in Figure 91 |175|.

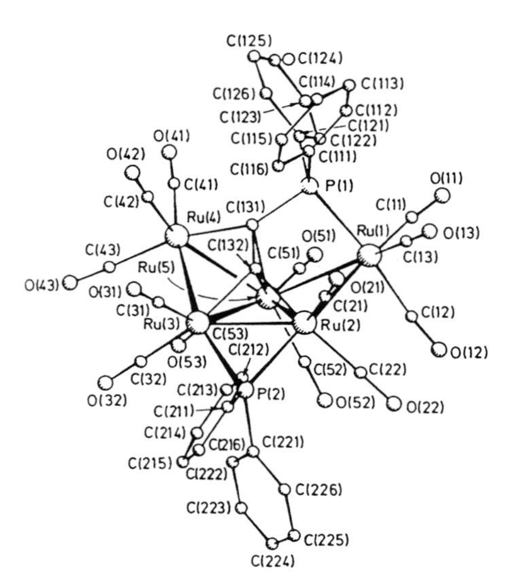

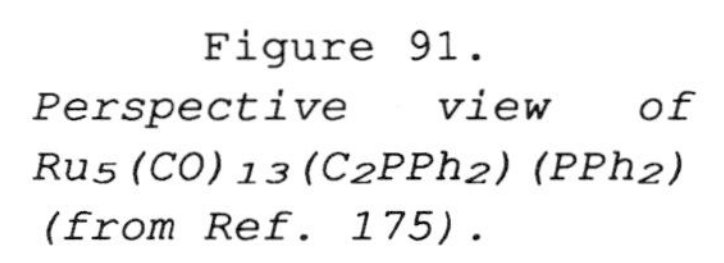
Figure 91. *Perspective view of $Ru_5(CO)_{13}(C_2PPh_2)(PPh_2)$ (from Ref. 175).*

The five ruthenium atoms are arranged according to a basal Ru_3 triangle, two edges of which bear two fused triangular wings. The Ru2-Ru3 vector is bridged by a phosphinidene group. The C_2PPh_2 fragment is placed above the swallow-like metallic core. The Ru-Ru distances vary from 2.73 Å of the shortest one (Ru2-Ru3) to about 2.93 Å of the longest ones (Ru1-Ru2, Ru1-Ru5). The C C distance is 1.38 Å.

As illustrated in Figure 92, $Ru_5(CO)_{13}(C_2PPh_2)(PPh_2)$ undergoes, in dichloromethane solution, a quasireversible reduction step ($E°' = -0.78$ V; $\Delta Ep_{(0.2\ Vs^{-1})} = 210$ mV) to the corresponding monoanion $[Ru_5(CO)_{13}(C_2PPh_2)(PPh_2)]^-$ |176|.

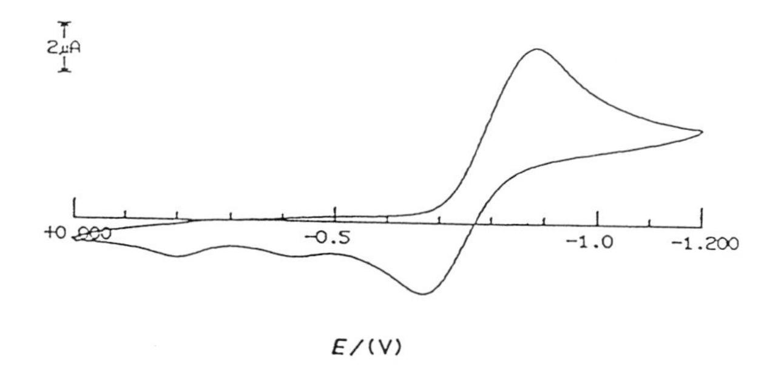

Figure 92. *Cyclic voltammogram recorded on a CH_2Cl_2 solution of $Ru_5(CO)_{13}(C_2PPh_2)(PPh_2)$. Scan rate 0.2 Vs^{-1} (from Ref. 176).*

The minor reoxidation peaks at -0.43 V and -0.20 V support the chemical evidence that the monoanion is not completely stable.

3.6. Hexanuclear Binary Complexes

3.6.1. $[Ru_6(CO)_{18}]^{2-}$. Different routes to the preparation of the dianion $[Ru_6(CO)_{18}]^{2-}$ have been worked out |177-180|.

The geometrical assembly of $[Ru_6(CO)_{18}]^{2-}$ is shown in Figure 93 |178|.

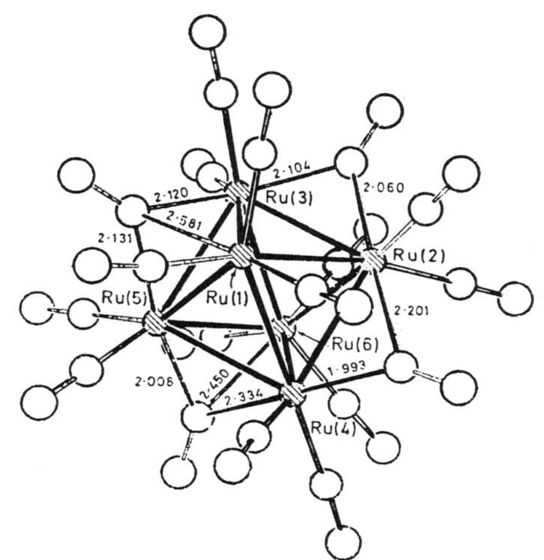

Figure 93. *Perspective view of $[Ru_6(CO)_{18}]^{2-}$ (from Ref. 178).*

The Ru_6 core forms a substantially regular octahedron with the Ru-Ru distances varying from 2.80 Å to 2.89 Å. Of the eighteen carbonyl groups, two (symmetrically disposed) are doubly bridging, while two are triply bridging, giving hence rise to a significant spreading of Ru-C distances.

It is widely known that $[Ru_6(CO)_{18}]^{2-}$ undergoes an irreversible two-electron oxidation |138,139,181|, the potential of which depends upon the solvent employed (two apparently distinct one-electron steps at Ep = +0.2 V and + 0.5 V in acetone |138,140|; a single-stepped two-electron process at Ep = +0.17 V in tetrahydrofuran |181|, or at Ep = -0.40 V in dichloromethane |139|). It is however interesting to note that, just in CH_2Cl_2, a transient oxidized species ($t_{1/2} \cong 0.001$ sec) {either $[Ru_6(CO)_{18}]^-$ or $Ru_6(CO)_{18}$} has been put in evidence |139|. The lability of the one- or two-electron oxidized species may be accounted for taking into account that the added electrons enter antibonding orbitals |182|.

3.7. Hexanuclear Carbido Complexes

3.7.1. $[Ru_6(CO)_{16}C]^{2-}$. The dianion $[Ru_6(CO)_{16}C]^{2-}$ possesses the octahedral geometry shown in Figure 94 |183|.

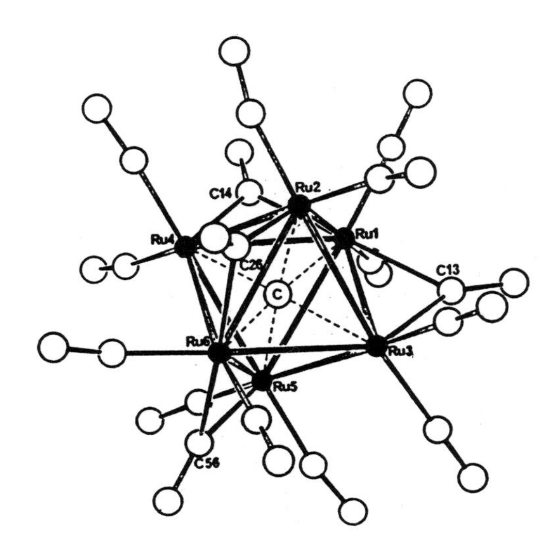

Figure 94. *Perspective view of the dianion $[Ru_6(CO)_{16}C]^{2-}$. $[Ph_4As]^+$ counteranion (from Ref. 183).*

In it, the carbido atom lies in the centre of the octahedral Ru_6 core. Although the presence of four edge-bridging and twelve terminal CO groups, the octahedral geometry is substantially regular, with a mean Ru-Ru bond distance of 2.89 Å, and a mean Ru-$C_{(carbido)}$ distance of 2.05 Å.

A slightly different disposition of the CO groups has been found in the dianion having $[NMe_4]^+$ counteranion |184,185|; nevertheless, the presence of three edge-bridging CO groups and a fourth one likely asymmetrically edge-bridging, together with the fact that the Ru-Ru and Ru-C distances are coincident with the above cited ones, make crystal packing anomalies responsible for this slight discrepancy.

In dichloromethane solution the 86-electron $[Ru_6(CO)_{16}C]^{2-}$ undergoes an irreversible two-electron oxidation at Ep = +0.43 V, Figure 95 |122,186|.

The exhaustive two-electron oxidation (particularly under CO atmosphere) leads to the formation of $Ru_6(CO)_{17}C$, which will be discussed in the next Section.

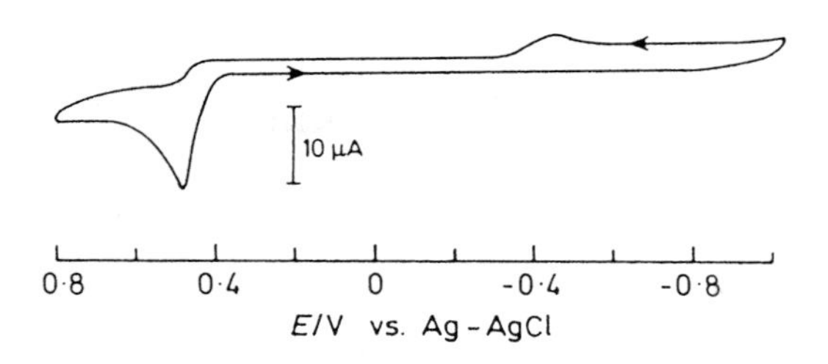

Figure 95. *Cyclic voltammogram recorded at a platinum electrode on a CH_2Cl_2 solution of $[Ru_6(CO)_{16}C]^{2-}$ ($[NBu_4][BF_4]$ supporting electrolyte). Scan rate 0.1 Vs^{-1} (from Ref. 186).*

Interestingly, chemical (by ferrocenium ions) or electrochemical oxidation of $[Ru_6(CO)_{16}C]^{2-}$ in the presence of phosphines or phosphites leads to the main formation of monosubstituted species $Ru_6(CO)_{16}C(PR_3)$ [R = OMe,Ph,Ph_2Me] in high yields. In these species, based on the X-ray structure of the related $Ru_6(CO)_{16}C(PPh_2Et)$ |187|, the octahedral core Ru_6C remains substantially unaltered. The same oxidative process performed in the presence of alkynes leads to the formation of the species $Ru_6(CO)_{15}C(RCCR')$, in which the alkyne ligand binds the octahedral Ru_6C core in a μ_3- η^2 mode |8,188|.

3.7.2. $Ru_6(CO)_{17}C$. $Ru_6(CO)_{17}C$ is isoelectronic with $[Ru_6(CO)_{16}C]^{2-}$ and, like it, assumes a substantially regular octahedral geometry with the carbide atom buried at the centre, Figure 96 |189|.

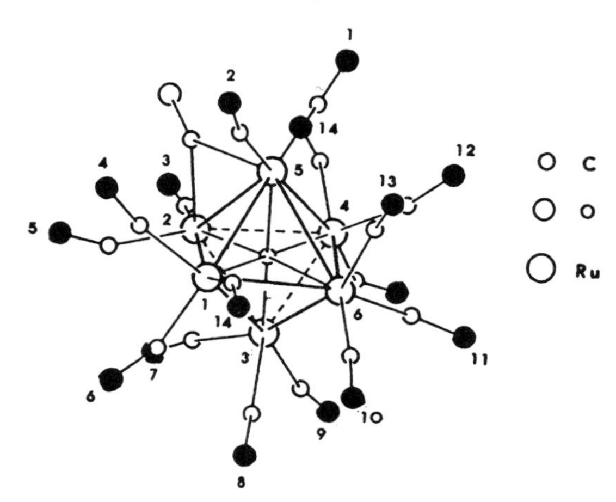

Figure 96. *Crystal structure of $Ru_6(CO)_{17}C$. Ru-Ru mean bond length, 2.90 Å; Ru-$C_{(carbide)}$, 2.05 Å (from Ref. 189).*

Except for a somewhat different disposition of CO groups (one edge-bridging and sixteen terminal), the Ru_6C core is substantially identical to that of $[Ru_6(CO)_{16}C]^{2-}$.

In dichloromethane solution, $Ru_6(CO)_{17}C$ undergoes a single-stepped, irreversible, two-electron reduction (Ep = -0.50 V), which proved to be consistent with the reaction:

$$Ru_6(CO)_{17}C \xrightarrow[-CO]{+2e} [Ru_6(CO)_{16}C]^{2-}$$

In confirmation, the relevant cyclic voltammogram, Figure 97, is quite complementary to that displayed by $[Ru_6(CO)_{16}C]^{2-}$ shown in Figure 95 |186|.

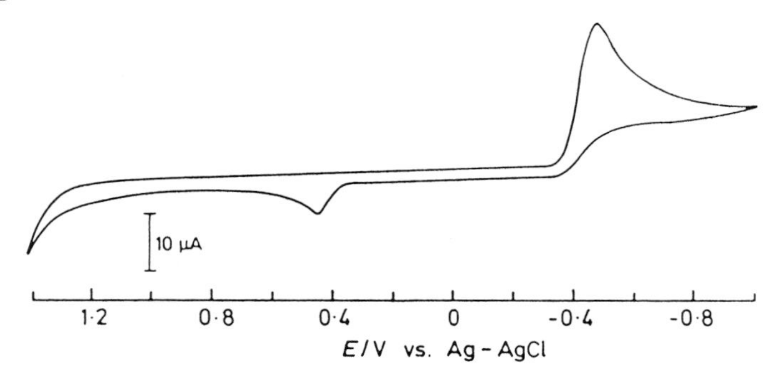

Figure 97. *Cyclic voltammogram recorded at a platinum electrode on a CH_2Cl_2 solution of $Ru_6(CO)_{17}C$ ($[NBu_4][BF_4]$ supporting electrolyte). Scan rate 0.1 Vs^{-1} (from Ref. 186).*

In this picture, the observation of the ESR signal of the one-electron reduced species $[Ru_6(CO)_{17}C]^-$ is not easily explainable |190|.

3.8. Hexanuclear Mixed-Ligand Complexes

3.8.1. $Ru_6(CO)_{12}(PPh)_5$. The molecular structure of $Ru_6(CO)_{12}(\mu_4\text{-}PPh)_3(\mu_3\text{-}PPh)_2$ is shown in Figure 98 |162|. The Ru_6P_5 core is constituted by a distorted trigonal prismatic Ru_6 array, with each face capped by a phenyl-phosphinidene group. Both the basal Ru_3 triangles have to be considered "quasi-open"; in fact two edges are 2.88 Å, while the third one is nearly 3.4 Å. The prism height is of about 3.0 Å.

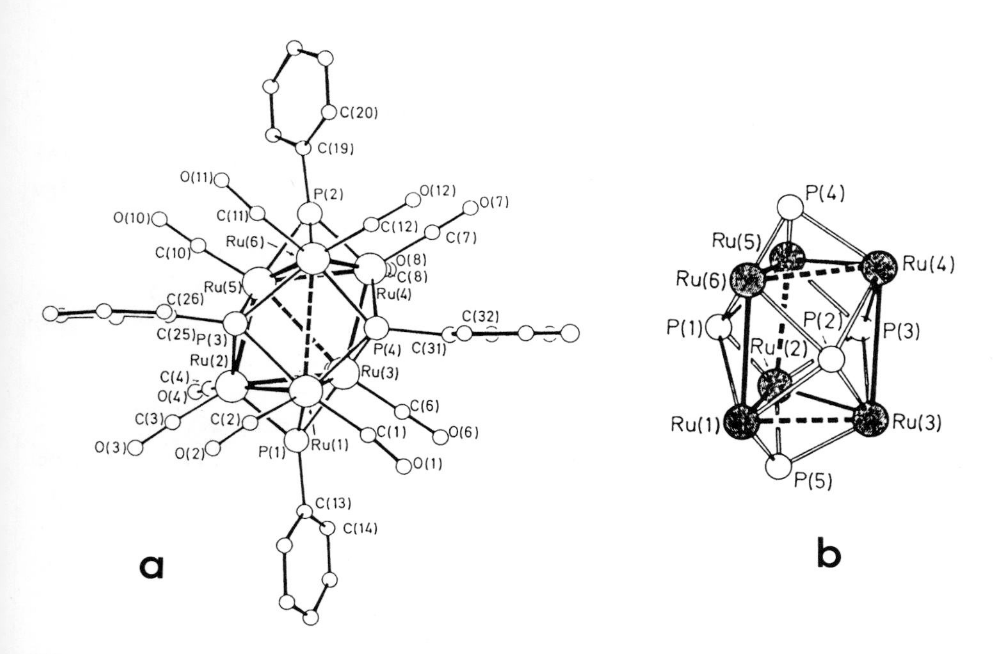

Figure 98. *Molecular structure of $Ru_6(CO)_{12}(PPh)_5$. (a) Perspective view; (b) core geometry (from Ref. 162).*

In dichloromethane solution, the 92-valence-electron $Ru_6(CO)_{12}(PPh)_5$ undergoes a one-electron removal to the stable, structurally uncharacterized, $[Ru_6(CO)_{12}(PPh)_5]^+$ ($E^{\circ\prime}_{0/+}$ = +0.36 V), and an irreversible reduction at -1.08 V |162|.

4. OSMIUM CLUSTERS

4.1. Trinuclear Binary Complexes

4.1.1. $Os_3(CO)_{12}$. The crystal structure of $Os_3(CO)_{12}$ is shown in Figure 99 |191,192|. The metallic triangle is nearly equilateral (averaged Os-Os distance, 2.88 Å). All the carbonyl groups are terminal.

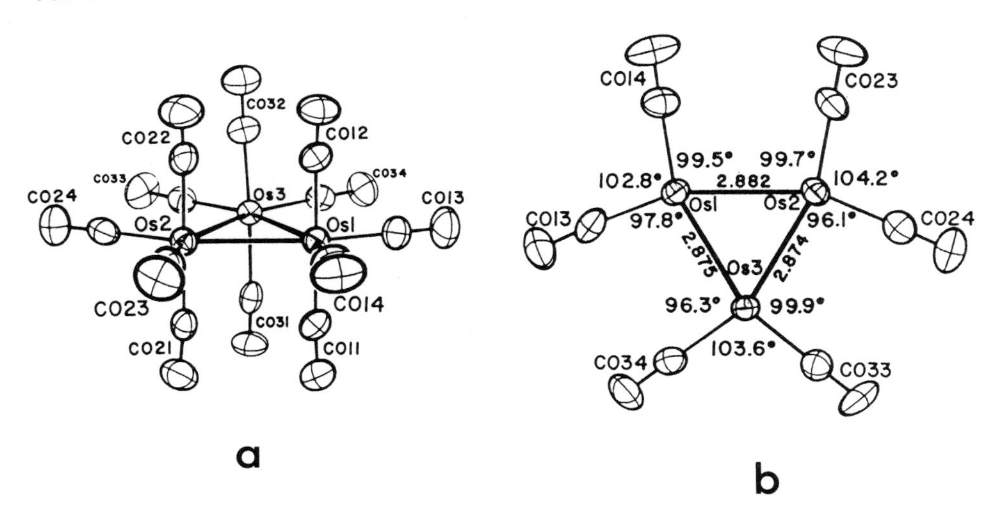

Figure 99. *Molecular structure (a) and selected structural parameters (b) of $Os_3(CO)_{12}$ (from Ref. 192).*

Like the isoelectronic $Ru_3(CO)_{12}$, the saturated 48-electron $Os_3(CO)_{12}$ undergoes a very complex reduction process, quite sensitive to the experimental conditions. It is however firmly established that the primarily electrogenerated $[Os_3(CO)_{12}]^-$ is completely unstable. The relevant redox potentials are reported in Table 22.

TABLE 22. *Electrode potentials (in Volt) for the irreversible reduction process of $Os_3(CO)_{12}$ |139|*

solvent	Ep
CH_2Cl_2	-1.59
THF	-1.33

It results evident that the addition of electrons is more and more difficult on passing from $Fe_3(CO)_{12}$ to $Ru_3(CO)_{12}$ to $Os_3(CO)_{12}$, thus reflecting the relative increase in energy of the relevant metal-metal antibonding LUMO levels |141|.

4.1.2. $[Os_3(CO)_{11}]^{2-}$. Like $[Ru_3(CO)_{11}]^{2-}$, $[Os_3(CO)_{11}]^{2-}$ has assigned the triangular structure schematized in Figure 100 |142|.

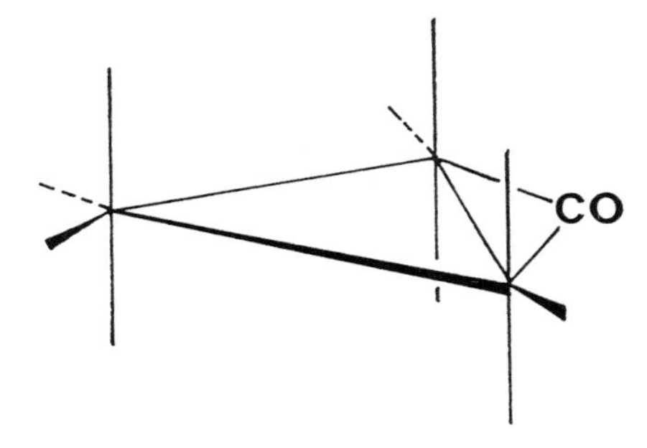

Figure 100.
Schematic representation of the structure proposed for $[Os_3(CO)_{11}]^{2-}$.

It has been briefly reported that, in dichloromethane solution, $[Os_3(CO)_{11}]^{2-}$ undergoes an irreversible oxidation at Ep = -0.1 V |139|.

4.2. Trinuclear Mixed-Ligand Complexes

4.2.1. $H_2Os_3(CO)_{10}$. A number of studies have dealt with solving the structure of $(\mu\text{-}H)_2Os_3(CO)_{10}$ |193-196|; finally, combined X-ray and neutron diffraction analysis has definitively localized the hydride atoms |195,196|. Figure 101 shows the relevant structure |196|. Both the two hydride atoms bridge the shorter edge of the isosceles metallic triangle. All the carbonyls are terminal.

This unsaturated 46-electron cluster easily reacts with Lewis bases to reach the saturated 48-electron configuration. We have recently found that $H_2Os_3(CO)_{10}$ also undergoes electrochemical reduction through a first one-electron addition, exhibiting remarkable extent of chemical reversibility, and, apparently, a second one-electron addition irreversible in character |197|. Figure 102 shows the relevant cyclic voltammogram.

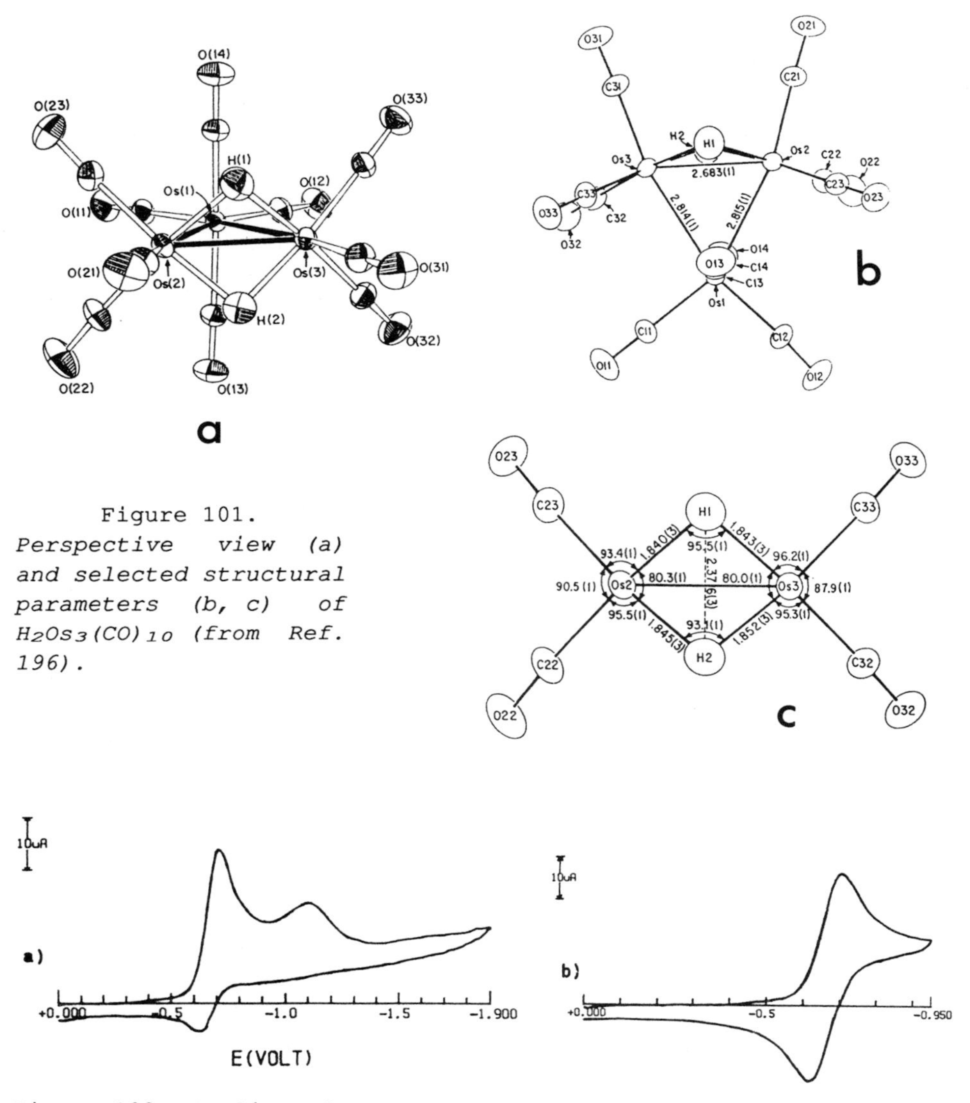

Figure 101. *Perspective view (a) and selected structural parameters (b, c) of $H_2Os_3(CO)_{10}$ (from Ref. 196).*

Figure 102. *Cyclic voltammograms recorded at a platinum electrode on a Me_2CO solution of $H_2Os_3(CO)_{10}$. T = 20 °C. Scan rate 0.2 Vs^{-1}.*

The 47-electron monoanion, though not fully stable (even at the temperature of -20 °C), has a lifetime sufficiently long to allow its detection by EPR spectroscopy. In contrast, the 47/48-electron redox change proceeds through a complicated electrode mechanism. In view of the quasi-reversibility of the neutral/ monoanion transition (in acetone, $E°' = -0.66$ V, $\Delta E_{p(0.2\ Vs^{-1})} = 80$ mV), a significant structural reorganization seems likely. By way of speculation, we note that, on the basis of the nonbonding or antibonding Os-H character of the LUMO level, it has been hypothesized that at least one hydrogen atom may become terminal, upon reduction |196|.

4.2.2. $H_2Os_3(CO)_{10}(L)$ (L = CO, PPh_3). As above mentioned, the unsaturated 46-electron $(\mu\text{-H})_2Os_3(CO)_{10}$ easily reacts with Lewis bases to afford a series of saturated 48-electron species of formula $(\mu\text{-H})(H)Os_3(CO)_{10}(L)$. Two of these species are $H_2Os_3(CO)_{11}$ and $H_2Os_3(CO)_{10}(PPh_3)$, the molecular structure of which is illustrated in Figures 103 and 104, respectively |192,198,199|.

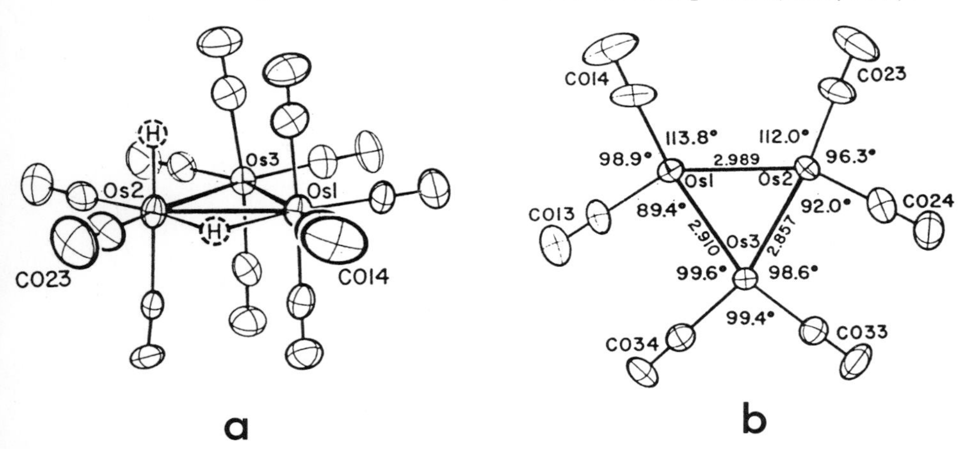

Figure 103. *Perspective view (a) and structural parameters (b) of $H_2Os_3(CO)_{11}$ (from Ref. 198).*

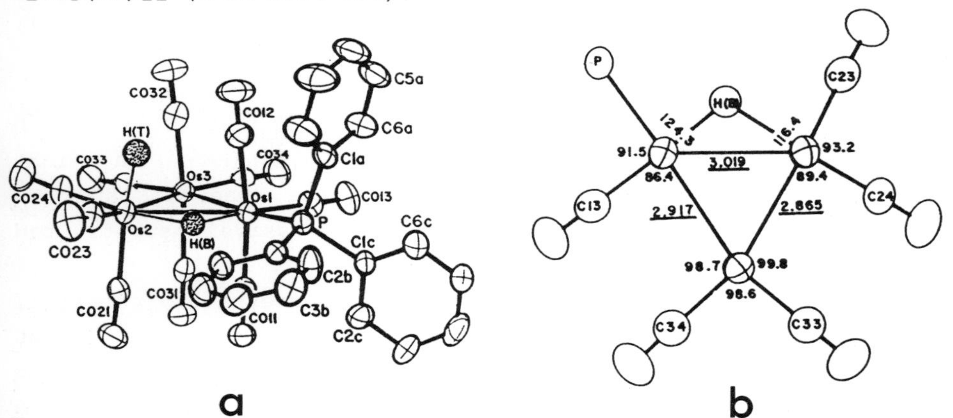

Figure 104. *Perspective view (a) and structural parameters (b) of $H_2Os_3(CO)_{10}(PPh_3)$ (from Ref. 199).*

Even if not directly located, in both cases the two hydride ligands have reliably assigned one equatorially bridging and one axially terminal position with respect to the triangular triosmium core. The $_2$-hydrido-bridged Os-Os distance is, in both cases, significantly longer (meanly, by 0.11-0.12 Å) than the unbridged ones. All the CO groups assume terminal positions. The phosphine group in $H_2Os_3(CO)_{10}(PPh_3)$ occupies a less sterically hindered equatorial site.

In confirmation of their electronic saturation, these species undergo, in acetone solution, a fully irreversible, and hence framework-destroying, two-electron reduction [Ep = -1.63 V for $H_2Os_3(CO)_{11}$; Ep = -1.71 V for $H_2Os_3(CO)_{10}(PPh_3)$], Figure 105 |197|. As expected, the electron-donating ability of PPh_3 makes the electron addition more difficult.

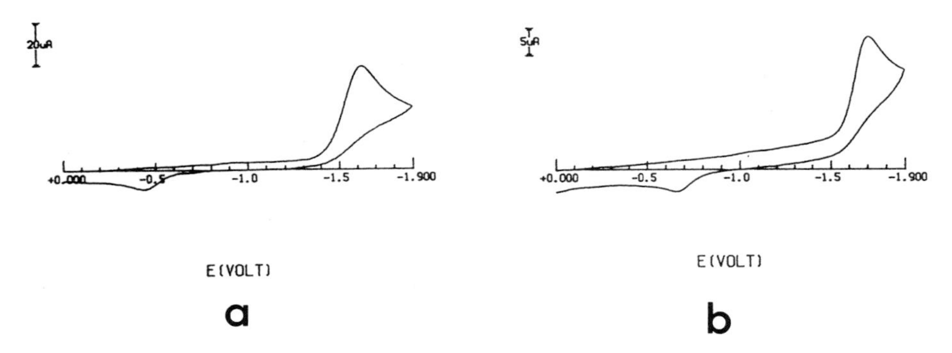

Figure 105. *Cyclic voltammograms recorded at a platinum electrode on Me_2CO solutions of: (a) $H_2Os_3(CO)_{11}$; (b) $H_2Os_3(CO)_{10}(PPh_3)$. Scan rate 0.2 Vs^{-1}.*

4.2.3. $Os_3(CO)_{11}(PPh_3)$. Thermal reaction of $Os_3(CO)_{12}$ with tertiary phosphines affords the series $Os_3(CO)_{12-n}(PR_3)_n$ (n = 1-3) |200|, but under more drastic conditions (thermal and UV irradiation) up to six CO groups can be progressively substituted |201|.

In all cases, the substitution for CO progressively occurs at the equatorial positions; in fact, it is presumed that substitution by bulky ligands in axial position would lead to severe steric interactions with the remaining axial CO groups on the same side of the triosmium plane.

Since redox changes have been observed only for the monosubstituted species $Os_3(CO)_{11}(PPh_3)$ |139|, we report in Figure 106, as related example, the structure of $Os_3(CO)_{11}[P(OMe)_3]$ |202|.

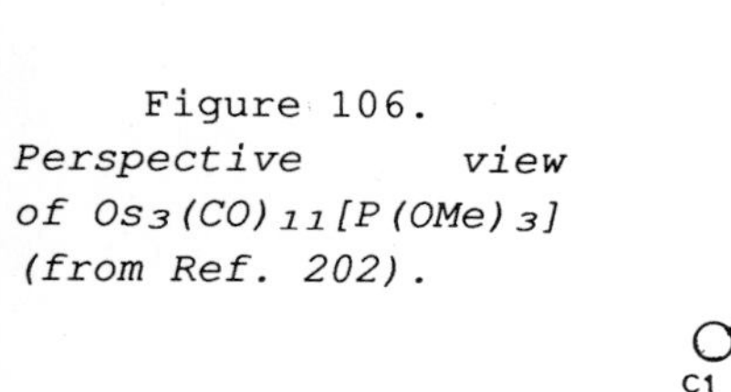
Figure 106.
Perspective view of $Os_3(CO)_{11}[P(OMe)_3]$ (from Ref. 202).

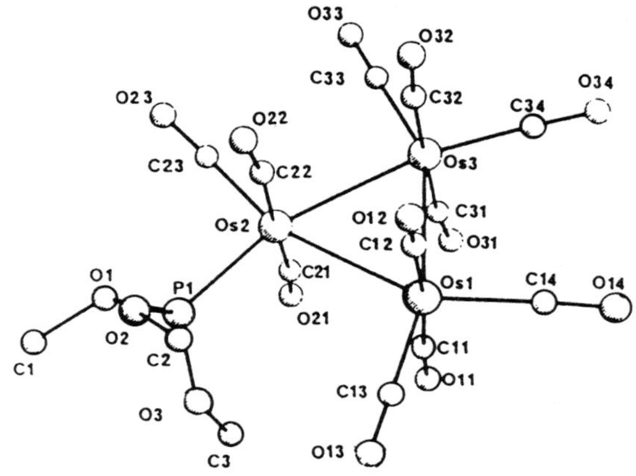

The edges of the trimetallic, nearly-equilateral triangle (averaged Os-Os distance, 2.90 Å) are slightly longer (by 0.02 Å) than those of $Os_3(CO)_{12}$. This effect is rather common in the presence of phosphorus-donor substituents, which pour electron density into the metallic fragment. All the carbonyl groups are terminal in the solid state, but there is spectral evidence that, in solution, dynamic terminal-to-bridging processes take place |203|.

It has been briefly reported that the saturated 48-electron $Os_3(CO)_{11}(PPh_3)$ undergoes, in dichloromethane solution, one two-electron reduction step, with features of chemical reversibility at low temperature ($E°' = -1.71$ V) |139|. Unfortunately, the reduction processes for more substituted species are obscured by the solvent discharge.

4.2.4. $H_nOs_3(CO)_{10-m}[(C_6H_5)_2PCH_2P(C_6H_5)(C_6H_{5-n})]$ ($n = 0$, $m = 0$; $n = 1$, $m = 1,2$). As illustrated in Scheme 6, the diphenyl phosphinomethane-triosmium cluster species $HOs_3(CO)_8[Ph_2PCH_2PPh(C_6H_4)]$ (**1**), $HOs_3(CO)_9[Ph_2PCH_2PPh(C_6H_4)]$ (**2**), $Os_3(CO)_{10}(Ph_2PCH_2PPh_2)$ (**3**) are each other related by simple carbonylation/ decarbonylation reactions |204|.

SCHEME 6

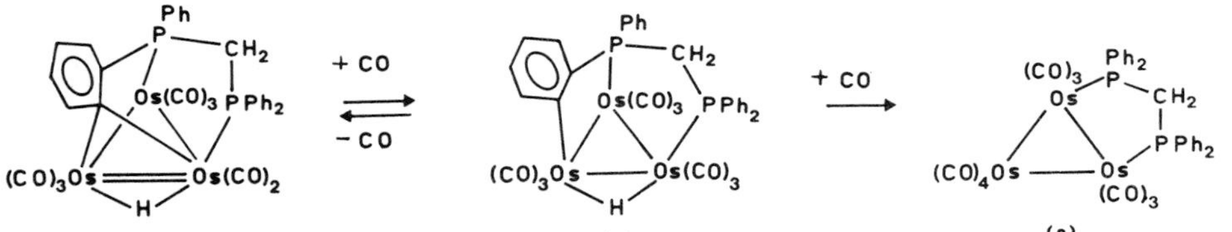

(1) $\underset{-CO}{\overset{+CO}{\rightleftharpoons}}$ (2) $\xrightarrow{+CO}$ (3)

The saturated cluster **3** has assigned a molecular structure similar to that of $Ru_3(CO)_{10}(Ph_2PCH_2PPh_2)$, with the bidentate phosphine bridging two osmium atoms and occupying equatorial positions |205|.

The structure of the unsaturated (46-electron) compound **1** is shown in Figure 107 |206|.

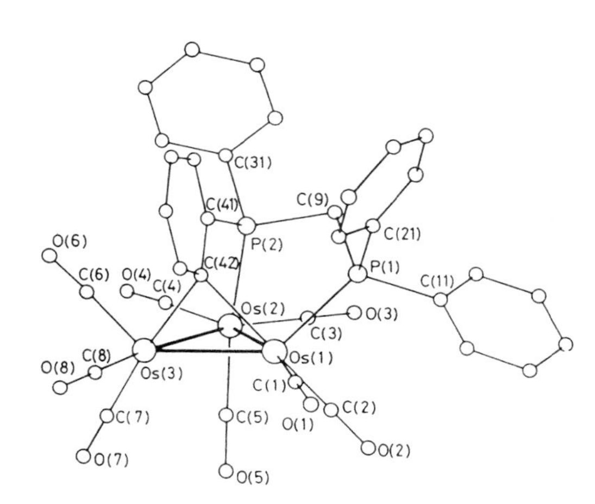

Figure 107.
Perspective view of $HOs_3(CO)_8[Ph_2PCH_2PPh(C_6H_4)]$. Os1-Os2 ~ Os2-Os3, 2.84 Å; Os1-Os3, 2.75 Å (from Ref. 206).

It is well evident that the removal of two CO molecules from **3** leads to metallation of one phenyl group of the diphosphine ligand. The hydride ligand is thought to bridge the shorter Os1-Os3 edge |206|.

Figure 108 gives an overall picture of the redox propensity of the whole family **1-3** |204|. The unsaturated species **1** is able to add one electron, at the peak-system A/B ($E°'$ = -1.14 V), to form the paramagnetic 47-electron monoanion $[HOs_3(CO)_8\{Ph_2PCH_2PPh(C_6H_4)\}]^-$, which is only relatively stable ($t_{1/2}$ ~ 10 s). In contrast, both the saturated clusters **2** and **3** undergo fast framework-destroying one-electron reduction at peaks D (Ep = -1.60 V) and E (Ep = -1.71 V), respectively.

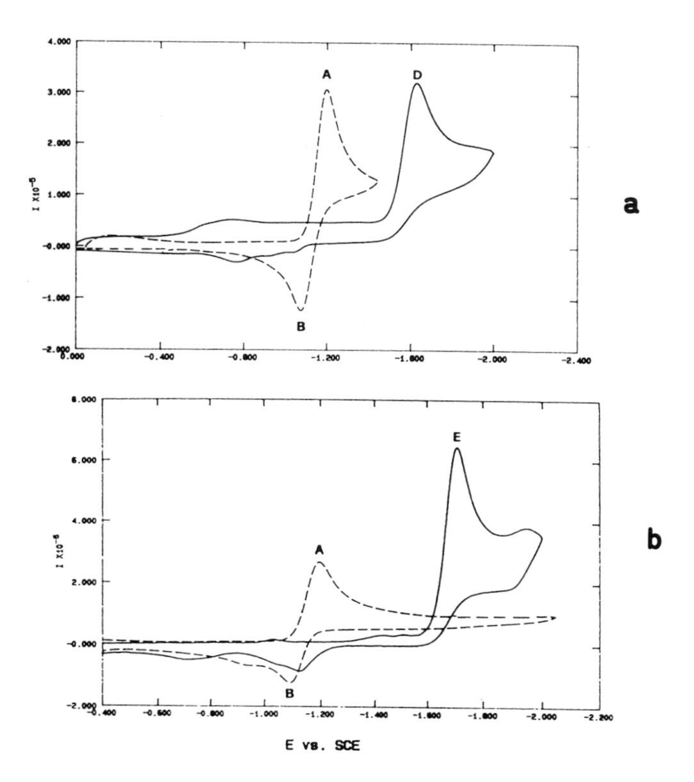

Figure 108. *Cyclic voltammograms recorded at a mercury electrode on MeCN solutions of: (a)* ***1*** *(dotted line) and* ***2*** *(bold line); (b)* ***1*** *(dotted line) and* ***3*** *(bold line). Scan rate 0.2* Vs^{-1} *(from Ref. 204).*

4.2.5. $\mathbf{H_2Os_3(CO)_9(EtC_2Et)}$**.** As briefly outlined in Section 3.2.4., $H_2Os_3(CO)_9(EtC_2Et)$ has assigned a structure in which the alkyne moiety is placed above the metallic triangle, with the unsaturated carbon-carbon bond parallel to one Os-Os edge |148|, Figure 109.

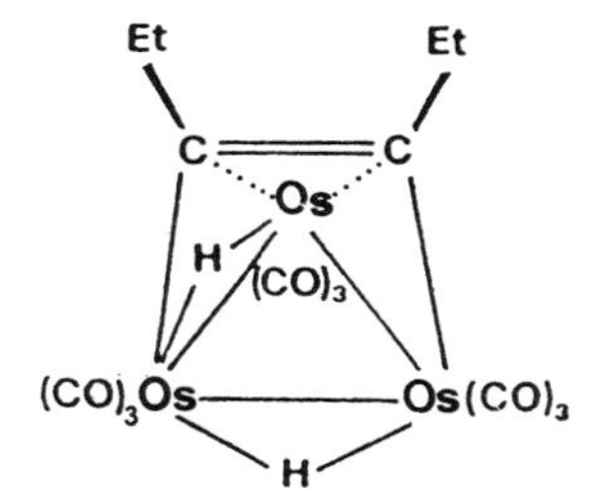

Figure 109. *Schematic representation of the structure of* $H_2Os_3(CO)_9(EtC_2Et)$.

Like the isoelectronic ruthenium congener, this saturated 48-electron cluster adds electrons with great difficulty (one two-electron step at Ep = -2.02 V, in dichloromethane solution), with immediate decomposition of the primarily electrogenerated dianion |50|.

4.2.6. $Os_3(CO)_{10}(RC_2R)$. The saturated 48-electron clusters $Os_3(CO)_{10}(RC_2R)$ can assume one of the two isomeric structures schematized in Figure 110 |6,207|.

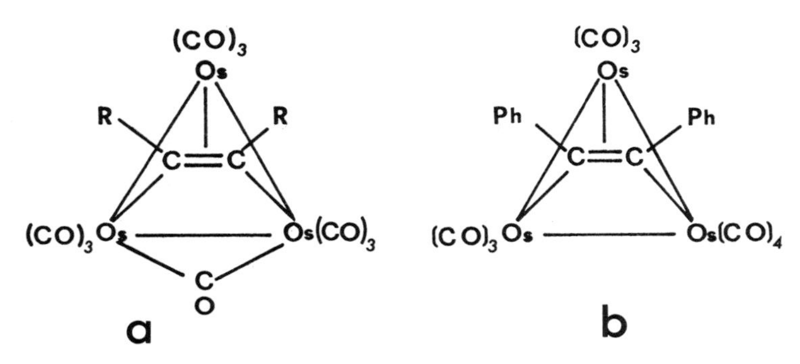

Figure 110. *Schematic representation of the isomeric forms of* $Os_3(CO)_{10}(RC_2R)$.

The structure (b), having only terminal carbonyls, has been proved only for R = Ph |207|. The species $Os_3(CO)_{10}(EtC_2Et)$ was assigned a structure of type (a) |50,207|.

$Os_3(CO)_9(\mu\text{-}CO)(EtC_2Et)$ undergoes, in dichloromethane solution, an irreversible one-electron reduction (Ep = -1.46 V), which indicates complete instability for the corresponding 49-electron monoanion |50|.

4.2.7. $Os_3(CO)_{7+n}(RC_2R)[(C_6H_5)_2PCH_2P(C_6H_5)_2]$ (n = 0,1). Treatment of the unsaturated 46-electron cluster $HOs_3(CO)_8[Ph_2PCH_2PPh(C_6H_4)]$ (previously discussed in Section 4.2.4.) with alkynes afford different products, depending on the bulkiness of the alkyne substituents. Diphenylacetylene as well as di-*p*-tolylacetylene afford the unsaturated 46-electron species $Os_3(CO)_7(RC_2R)[Ph_2PCH_2PPh_2]$; but-2-yne as well as hexafluorobut-2-yne afford the saturated 48-electron species $Os_3(CO)_8(RC_2R)[Ph_2PCH_2PPh_2]$ |208|.

As expected, in the saturated species the alkyne is placed above the trimetal triangle and assumes a coordination parallel to one metal-metal bond, while in the unsaturated species it disposes perpendicular to one metal-metal bond. Figure 111 just shows the two different coordination modes in the complexes $Os_3(CO)_7(PhC_2Ph)[Ph_2PCH_2PPh_2]$ |209| and $Os_3(CO)_8(MeC_2Me)[Ph_2PCH_2PPh_2]$ |208|.

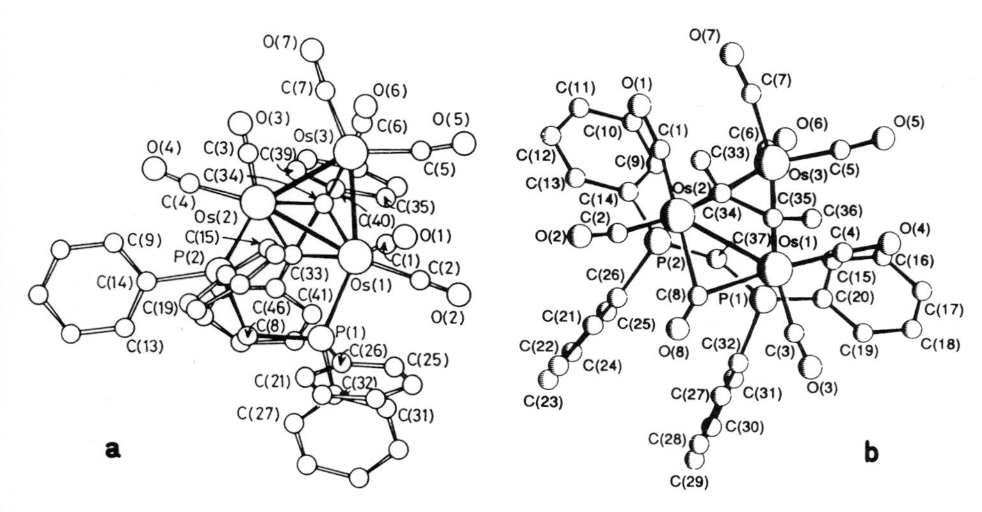

Figure 111. *Molecular structures of: (a) $Os_3(CO)_7(PhC_2Ph)[Ph_2PCH_2PPh_2]$. Os1-Os2, 2.85 Å; Os1-Os3 ~ Os2-Os3, 2.68 Å (from Ref. 209); (b) $Os_3(CO)_8(MeC_2Me)[Ph_2PCH_2PPh_2]$. Os1-Os2, 2.81 Å; Os1-Os3 ~ Os2-Os3, 2.76 Å (from Ref. 208).*

Bubbling CO through solutions of $Os_3(CO)_7(PhC_2Ph)[Ph_2PCH_2PPh_2]$ leads to the saturated 48-electron species $Os_3(CO)_8(PhC_2Ph)[Ph_2PCH_2PPh_2]$. It has been spectroscopically proved that such a conversion is accompanied by rotation of the alkyne from perpendicular to parallel coordination. The same 46/48-electron change can be electrochemically accomplished. In fact, as illustrated in Figure 112a, $Os_3(CO)_7(PhC_2Ph)[Ph_2PCH_2PPh_2]$ undergoes, in acetone solution, one chemically reversible two-electron reduction |208|. It has been proved that the electrogenerated 48-electron dianion $[Os_3(CO)_7(PhC_2Ph)\{Ph_2PCH_2PPh_2\}]^{2-}$ also has the alkyne parallel to one osmium-osmium bond. In contrast, as illustrated in Figure 112b for $Os_3(CO)_8(PhC_2Ph)[Ph_2PCH_2PPh_2]$, the saturated clusters $Os_3(CO)_8(RC_2R)[Ph_2PCH_2PPh_2]$ undergo a two-electron reduction which causes immediate decomposition to the stable dianions $[Os_3(CO)_7(RC_2R)\{Ph_2PCH_2PPh_2\}]^{2-}$.

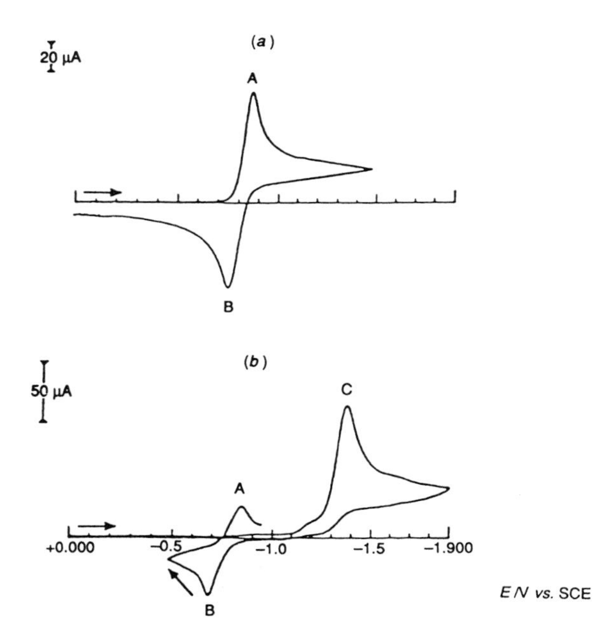

Figure 112. *Cyclic voltammograms recorded at a mercury electrode on Me_2CO solutions of: (a) $Os_3(CO)_7(PhC_2Ph)(dppm)$; (b) $Os_3(CO)_8(PhC_2Ph)(dppm)$ [dppm = $Ph_2PCH_2PPh_2$]. Scan rate 0.2 Vs^{-1} (from Ref. 208).*

The redox potentials for such redox changes are summarized in Table 23.

TABLE 23. *Formal electrode potentials (in Volt) for the electron transfers exhibited by the complexes $Os_3(CO)_{7+n}(RC_2R)[(C_6H_5)_2PCH_2P(C_6H_5)_2]$, in acetone solution |208|*

complex	$E^{\circ\prime}$ (0/2−)
$Os_3(CO)_7(PhC_2Ph)[Ph_2PCH_2PPh_2]$	−0.80
$Os_3(CO)_7(TolC_2Tol)[Ph_2PCH_2PPh_2]$	−0.78
$Os_3(CO)_8(PhC_2Ph)[Ph_2PCH_2PPh_2]$	−1.40[a]
$Os_3(CO)_8(TolC_2Tol)[Ph_2PCH_2PPh_2]$	−1.42[a]
$Os_3(CO)_8(MeC_2Me)[Ph_2PCH_2PPh_2]$	−1.53[a]
$Os_3(CO)_8(CF_3C_2CF_3)[Ph_2PCH_2PPh_2]$	−1.18[a]

[a]*Peak-potential value for irreversible process*

4.3. Tetranuclear Hydrido Complexes

4.3.1. $[H_3Os_4(CO)_{12}]^-$. Tetraosmium hydrido-carbonyl complexes have been prepared and structurally characterized in the deprotonated sequence: $H_4Os_4(CO)_{12}$ |210|, $[H_3Os_4(CO)_{12}]^-$ |211|, and $[H_2Os_4(CO)_{12}]^{2-}$ |212|. All the complexes are characterized by a tetrahedral Os_4 assembly, distorted by the presence of the edge-bridging hydrido atoms, which significantly elongate the involved Os-Os bonds with respect to the unbridged ones.

Commonly such complexes display redox activity towards chemical oxidants |213-216|, but only for $[H_3Os_4(CO)_{12}]^-$ the intrinsic ability to add or lose electrons has been studied by electrochemical techniques |217|.

A perspective view of one of the two isomers |214| of the monoanion $[H_3Os_4(CO)_{12}]^-$ is shown in Figure 113 |211|.

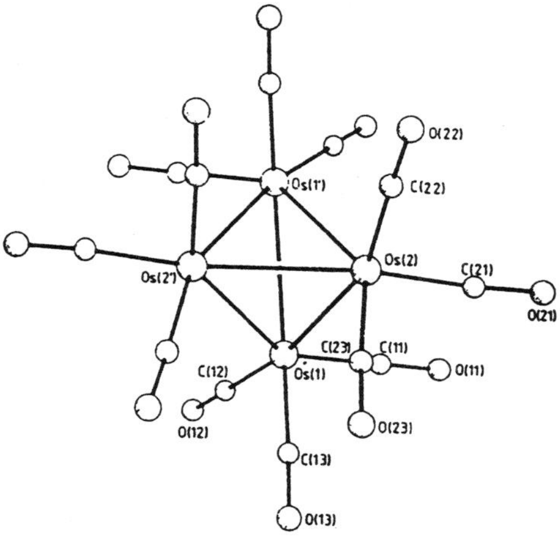

Figure 113. *Perspective view of the tetrahedral geometry of $[H_3Os_4(CO)_{12}]^-$ (from Ref. 211).*

The three hydrido atoms are thought to bridge the Os(1)-Os(1′), Os(1)-Os(2′), Os(2)-Os(1′) vectors, giving rise to three-centre two-electron bonds of mean length 2.93 Å. The unbridged Os-Os bond lengths average 2.80 Å.

In acetonitrile solution the 60-electron $[H_3Os_4(CO)_{12}]^-$ undergoes a single-stepped two-electron oxidation at Ep = +0.78 V, which is apparently irreversible because being coupled to the following chemical reaction |217|:

$$[H_3Os_4(CO)_{12}]^- \xrightarrow{-2e} [H_3Os_4(CO)_{12}]^+ \xrightarrow{+2MeCN} [H_3Os_4(CO)_{12}(NCMe)_2]^+$$

It is conceivable that the primarily electrogenerated, unsaturated 58-electron monocation $[H_3Os_4(CO)_{12}]^+$ is, as such, highly unstable, and tends to coordinate solvent molecules, giving rise to the 62-electron $[H_3Os_4(CO)_{12}(NCMe)_2]^+$. This 60-62 electron conversion, electrochemically promoted, has the effect to open the Os_4 tetrahedron to a butterfly arrangement, Figure 114 |217|.

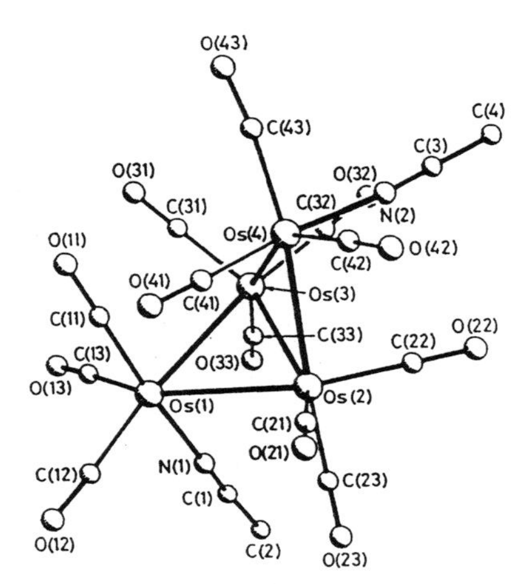

Figure 114. *Perspective view of* $|H_3Os_4(CO)_{12}(NCMe)_2|^+$ *(from Ref. 217).*

As expected, three of the Os-Os bonds are longer (mean, 3.07 Å) than the remaining two (mean, 2.84 Å), likely because hydrido edge-bridged.

4.4. Pentanuclear Binary Complexes

4.4.1. $[Os_5(CO)_{15}]^{2-}$. Spectroscopic properties of $[Os_5(CO)_{15}]^{2-}$ suggest a trigonal bipyramidal assembly of metal atoms, each one bearing three terminal carbonyl groups, just like what happens in $[HOs_5(CO)_{15}]^-$, Figure 115 |218,219|.

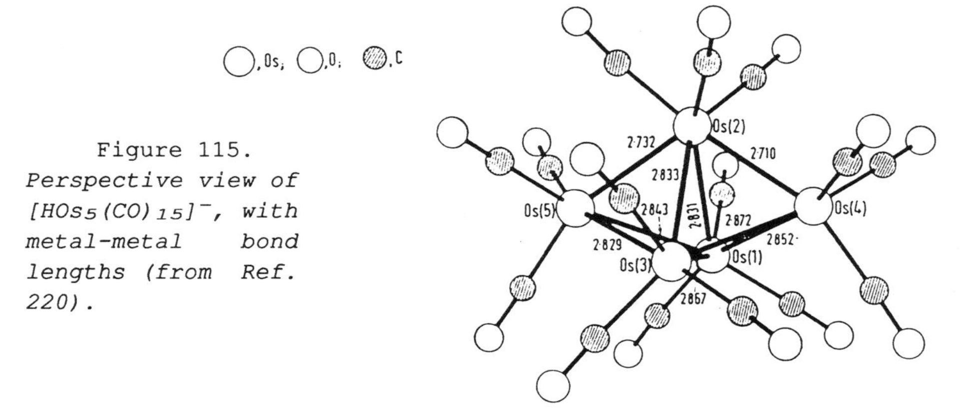

Figure 115. *Perspective view of* $[HOs_5(CO)_{15}]^-$, *with metal-metal bond lengths (from Ref. 220).*

It has been briefly reported that the 72-electron $[Os_5(CO)_{15}]^{2-}$ undergoes, in dichloromethane solution, two-electron removal through irreversible anodic steps (the second of which at Ep = +0.43 V) |122|.

The use of appropriate chemical oxidants in the presence of two-electron donor ligands affords, by oxidative addition, the 72-electron systems $Os_5(CO)_{16}$ and $Os_5(CO)_{15}(PR_3)$ (R = Ph, OPh) |122|. These 72-electron species maintain, as proved by $Os_5(CO)_{16}$, Figure 116 |220,221|, a bipyramidal trigonal geometry.

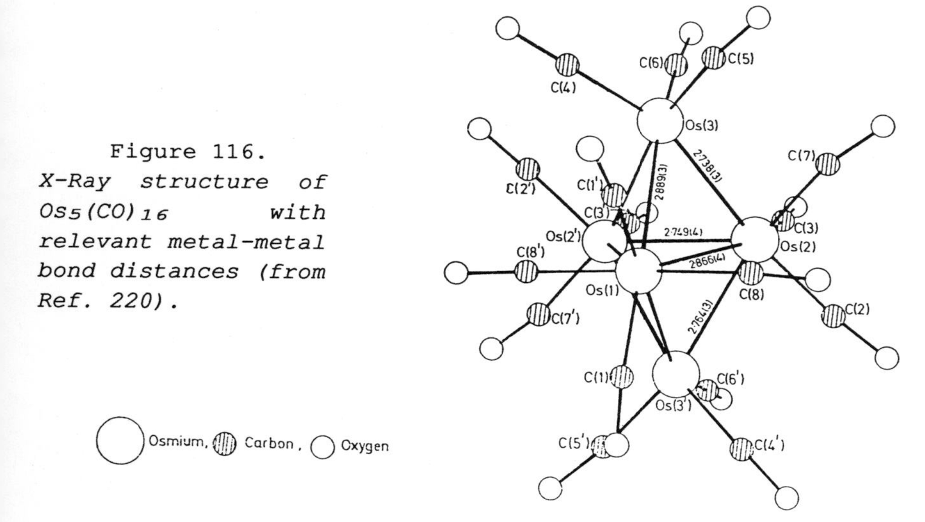

Figure 116.
X-Ray structure of $Os_5(CO)_{16}$ with relevant metal-metal bond distances (from Ref. 220).

At variance with $[Os_5(CO)_{15}]^{2-}$, one of the equatorial Os atoms |Os(1)| coordinates four CO groups.

4.5. Pentanuclear Carbido Complexes

4.5.1. $[Os_5(CO)_{14}C]^{2-}$. The dianion $[Os_5(CO)_{14}C]^{2-}$, which forms by chemical reduction of $Os_5(CO)_{15}C$ (see next Section 4.5.2.), has the square based pyramidal geometry shown in Figure 117 |169|. The Os-Os bond distances are spread out by the presence of one bridging CO group in the basal plane, which makes the Os(1)-Os(2) length (2.77 Å) notably shorter than the remaining basal ones (mean, 2.91 Å). The same spreading obviously occurs as far as the $Os_{(basal)}$-$Os_{(apical)}$ distances are concerned. Thus the two bonds connecting the CO-bridged Os atoms are longer (2.90 Å) than the other two (2.81 Å). Finally the carbide atom lies 0.21 Å below the basal plane, so that the $Os_{(apical)}$-C distance is 2.22 Å, whereas the $Os_{(basal)}$-C mean distance is 2.05 Å.

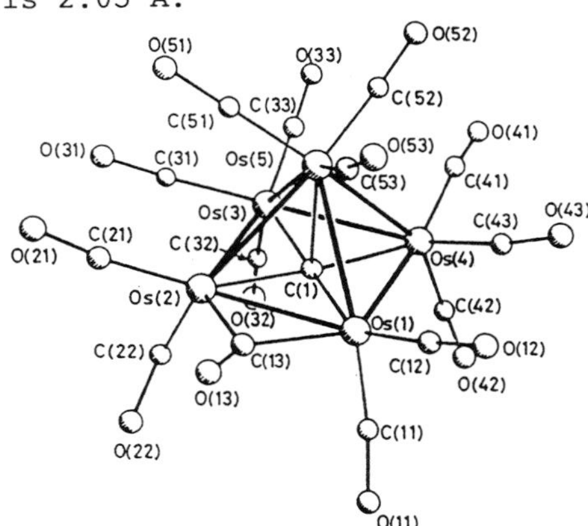

Figure 117.
Perspective view of the dianion $[Os_5(CO)_{14}C]^{2-}$ *(from Ref. 171).*

This 74-electron dianion undergoes, in dichloromethane solution, an irreversible two-electron oxidation (Ep = +0.69 V), complicated by subsequent chemical reactions, which produce, among different decomposition products, also the 74-electron $Os_5(CO)_{15}C$ |169|, whose molecular structure will be discussed in the next Section.

4.5.2. $Os_5(CO)_{15}C$. The 74-electron carbido system $Os_5(CO)_{15}C$ has the regular square-based pyramidal geometry illustrated in Figure 118 |222|.

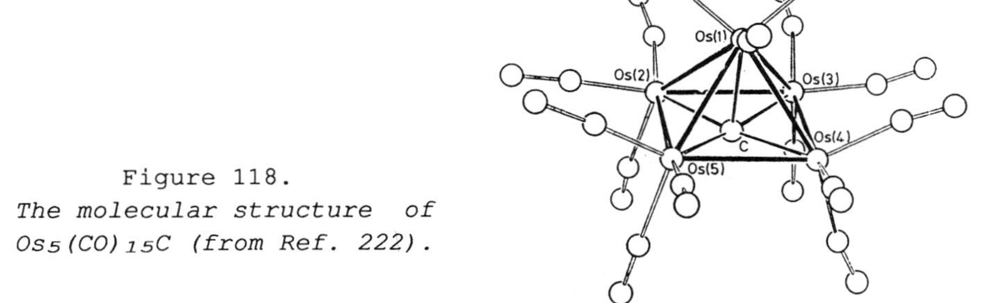

Figure 118.
The molecular structure of $Os_5(CO)_{15}C$ *(from Ref. 222).*

Each Os atom bears three terminal CO groups. The $Os_{(basal)}$-$Os_{(basal)}$ distances are 2.88 Å; the $Os_{(basal)}$-$Os_{(apical)}$ ones are 2.85 Å. The carbide atom lies 0.12 Å below the basal plane, with the Os-C distances being 2.06 Å, on average.

$Os_5(CO)_{15}C$ undergoes, in dichloromethane solution, a two-electron reduction at -1.16 V, which is followed by CO loss, thus generating the dianion $[Os_5(CO)_{14}C]^{2-}$ |169|.

In summary, the here discussed pentanuclear carbido derivatives are mutually correlated by:

$$Os_5(CO)_{15}C \underset{-2e,+CO}{\overset{+2e,-CO}{\rightleftarrows}} [Os_5(CO)_{14}C]^{2-}$$

These redox processes, while maintaining the square-based geometry typical of the 74-electron systems, are accompanied by geometrical reorganizations of the CO groups, which, also affecting the Os-Os bond distances, cause a stereochemical (regular square pyramidal/distorted square pyramidal) rearrangement.

4.6. Hexanuclear Binary Complexes

4.6.1. $Os_6(CO)_{18}$. The hexaosmium complex $Os_6(CO)_{18}$ has the bicapped tetrahedral geometry with three terminal CO groups per osmium atom illustrated in Figure 119 |223|.

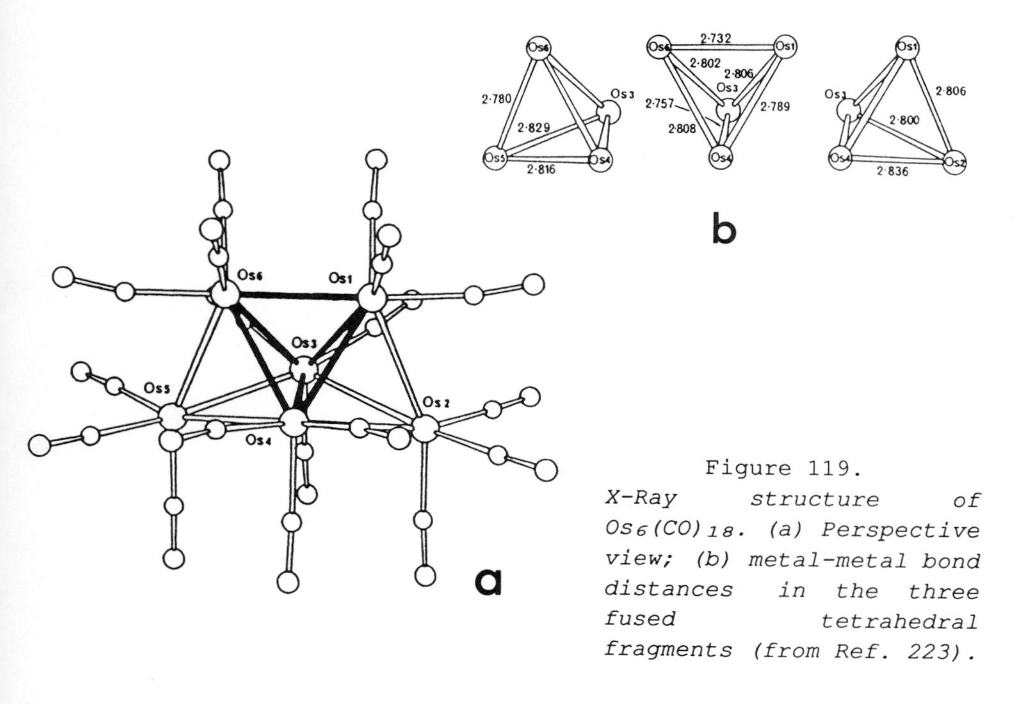

Figure 119.
X-Ray structure of $Os_6(CO)_{18}$. (a) Perspective view; (b) metal-metal bond distances in the three fused tetrahedral fragments (from Ref. 223).

Slight variations in metal-metal bond lengths occur, since each of the Os(2) and Os(5) atoms form three metal-metal bonds, whereas Os(1) and Os(6) form four metal-metal bonds. A mean Os-Os distance of 2.80 Å can be computed.

This 84-electron system undergoes, both chemically |181,224| and electrochemically |181|, a two-electron reduction to the stable dianion $[Os_6(CO)_{18}]^{2-}$. In this connection, Figure 120 shows the cyclic voltammetric profile of such redox change in tetrahydrofuran solution |181|.

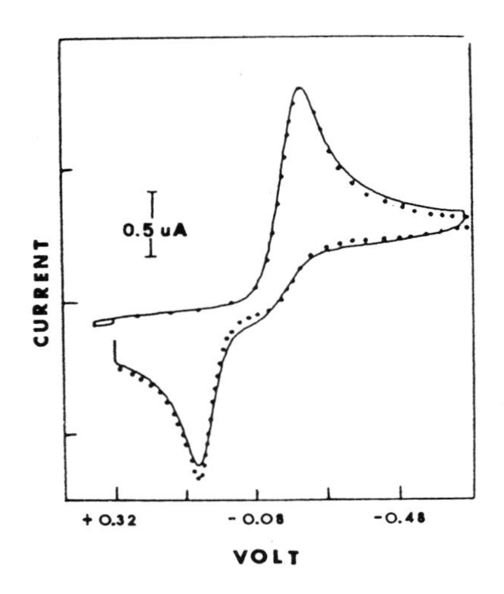

Figure 120.
Cyclic voltammogram (solid line) exhibited by $Os_6(CO)_{18}$ at a platinum electrode in THF solution. Scan rate 0.04 Vs^{-1} (from Ref. 181).

The marked quasireversibility of the cathodic step, as put in evidence by a ΔEp value increasing from 255 mV at 0.01 Vs^{-1} to 429 mV at 0.609 Vs^{-1}, has been the basis for a seminal study tending to correlate departure from pure electrochemical reversibility with significant structural reorganization accompanying the redox change |181|. In fact, the 86-electron dianion $[Os_6(CO)_{18}]^{2-}$ assumes a substantially regular octahedral geometry, Figure 121, with an averaged Os-Os bond length of 2.86 Å |225|.

It has been concluded that the bicapped-tetrahedron/ octahedron reorganization entirely occurs during the $Os_6(CO)_{18}$/ $[Os_6(CO)_{18}]^-$ rate-determining step |181|. A formal electrode potential of + 0.04 V in tetrahydrofuran |181|, and of -0.21 V in dichloromethane |139| is assigned to the two-electron reduction.

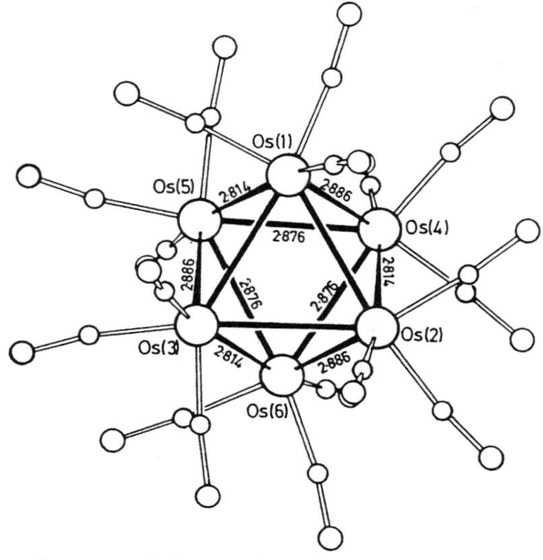

Figure 121. *Perspective view of $[Os_6(CO)_{18}]^{2-}$ together with Os-Os bond lengths (from Ref. 225).*

Finally, a further quasireversible reduction to the transient $[Os_6(CO)_{18}]^{4-}$ has been recorded at Ep = -2.1 V in tetrahydrofuran solution |181|.

4.7. Hexanuclear Phosphido Complexes

4.7.1. $[Os_6(CO)_{18}P]^-$. The geometry of the phosphido monoanion $[Os_6(CO)_{18}P]^-$ is shown in Figure 122 |226,227|.

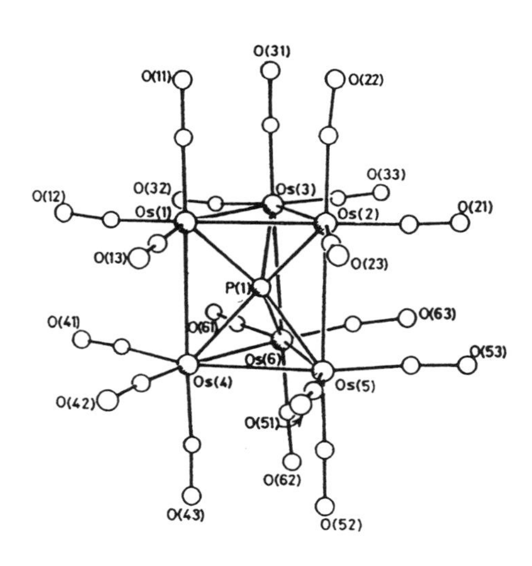

Figure 122. *Perspective view of the anion $[Os_6(CO)_{18}P]^-$ (from Ref. 227).*

It contains an interstitial phosphorus atom encapsulated in a Os_6 trigonal prism. Each osmium atom bears three terminal carbonyl groups. Within the basal triangles, the Os-Os bond length averages 2.93 Å, whereas the averaged interbasal Os-Os distance elongates to 3.14 Å. The Os-P mean bond length is 2.31 Å.

It has been briefly reported that the 90-electron $[Os_6(CO)_{18}P]^-$ undergoes, in dichloromethane solution, a single-stepped two-electron reduction to the corresponding trianion $[Os_6(CO)_{18}P]^{3-}$ |228|. In spite of the low extent of electrochemical reversibility for such redox change, based on the infrared characterization, it is thought that the two added electrons delocalize over the hexametal core, leaving unchanged the trigonal-prismatic geometry.

4.8. Hexanuclear Mixed-Ligand Complexes

4.8.1. $Os_6(CO)_{21-n}[PR_3]_n$ (n = 1-6). The complex $Os_6(CO)_{21}$, so far not fully characterized |229|, can be thought to undergo stepwise substitution of CO groups for phosphite or phosphine units to afford a series of compounds of formula $Os_6(CO)_{21-n}[PR_3]_n$ (R = OMe, OPh, Ph; n = 1-6) |229,230|.

Either the binary precursor or the substituted derivatives have assigned a planar Os_6 geometry in which an almost equilateral triangle is formed by four fused triangles, as illustrated in Figure 123 for $Os_6(CO)_{17}[P(OMe)_3]_4$ |230|.

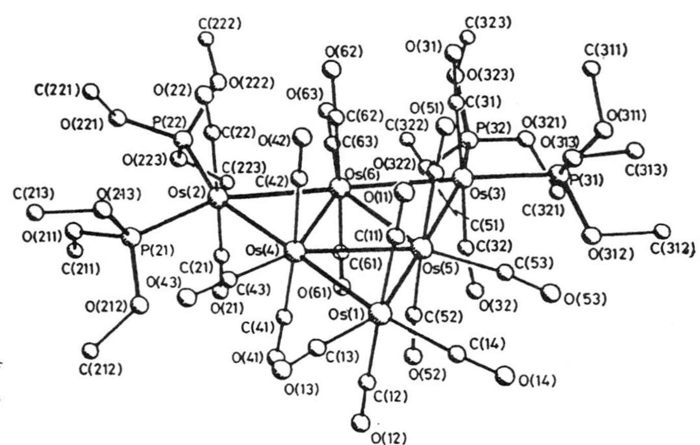

Figure 123. *Perspective view of $Os_6(CO)_{17}[P(OMe)_3]_4$ (from Ref. 232).*

In this raft-like hexa-metal assembly the mean Os-Os bond length is 2.87 Å; it must however be taken into account that Os-Os distances arising from Os atoms bound to $P(OMe)_3$ groups are, on average, longer by about 0.05 Å than those arising from Os atoms bound only to CO ligands.

The Os-Os mean length is relatively longer than that in the bicapped tetrahedral arrays of $Os_6(CO)_{18}$ |223| and its substituted derivatives |231,232|. In particular, in $Os_6(CO)_{16}[PPh_3]_2$ it is 2.82 Å, where lengthening of Os-Os bonds arising from Os atoms bound to phosphines also occurs, attributed to the poorer π-accepting ability of phosphorus-containing ligands with respect to carbonyls |226|.

In the present complex, each of the Os(2), Os(3) atoms have the two phosphite groups equatorial and the two carbonyl groups axial with respect to the metal plane.

This class of 90-electron derivatives undergo, as exemplified in Figure 124 for $Os_6(CO)_{20}[P(OMe)_3]$, two distinct one-electron reduction steps, generating the corresponding mono- and dianions, stable at least in the short times of cyclic voltammetry |229|.

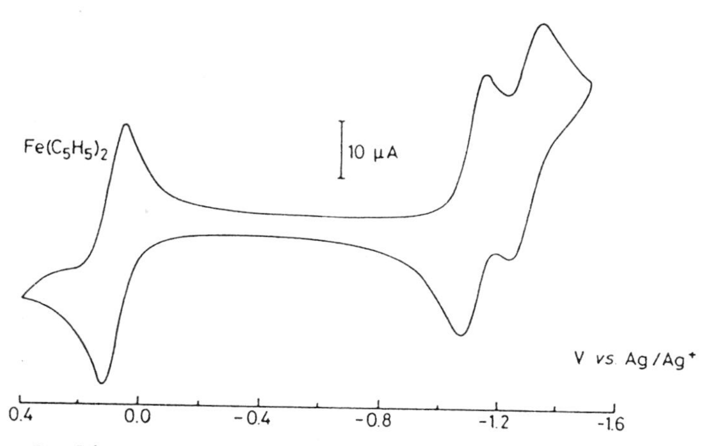

Figure 124. *Cyclic voltammogram recorded at a platinum electrode on a MeCN solution of* $Os_6(CO)_{20}[P(OMe)_3]$ *(phosphite group equatorial to the metal plane) (from Ref. 229).*

Table 24 summarizes the redox potentials of such reduction steps, together with the relevant peak-to-peak separations, which, when compared with the theoretical value of 59 mV for a purely reversible electron-transfer, suggest that, at least in the first electron addition, no significant geometrical reorganization should occur.

It is interesting to note how redox potentials are significantly affected by the position of the phosphorus-containing substituents equatorial or axial with respect to the hexaosmium plane, so allowing an identification of the two isomeric forms, the axial species being generally more easily reducible than the equatorial ones.

TABLE 24. *Redox potentials and peak-to-peak separations for the two consecutive electron additions displayed by the derivatives* $Os_6(CO)_{21-n}(PR_3)_n$ *in MeCN solution |229|*

complex	$E°'_{0/1-}$ (V)	ΔEp^a (mV)	$E°'_{1-/2-}$ (V)	ΔEp^a (mV)
$Os_6(CO)_{20}[P(OMe)_3]^b$	-0.63	65	-0.79	90
$Os_6(CO)_{20}[P(OMe)_3]^c$	-0.76	67	-0.95	85
$Os_6(CO)_{20}[P(OPh)_3]^b$	-0.68	62	-0.94	57
$Os_6(CO)_{20}[P(OPh)_3]^c$	-0.82	64	-1.05	72
$Os_6(CO)_{20}[P(Ph)_3]^c$	-0.69	67	-0.97	62
$Os_6(CO)_{19}[P(OMe)_3]_2{}^b$	-0.91	60	-1.10	75
$Os_6(CO)_{19}[P(OMe)_3]_2{}^c$	-0.91	80	-1.13	85
$Os_6(CO)_{18}[P(OMe)_3]_3{}^b$	-0.95	75	-1.16	140
$Os_6(CO)_{18}[P(OMe)_3]_3{}^c$	-1.09	68	-1.29	125
$Os_6(CO)_{17}[P(OMe)_3]_4{}^c$	-1.10		-1.27	
$Os_6(CO)_{16}[P(OMe)_3]_5$	-1.25	64	-1.45	85
$Os_6(CO)_{15}[P(OMe)_3]_6$	-1.51	62	-1.71	75

[a]*Measured at 0.2* Vs^{-1}*;* [b]*phosphorus-containing ligands axial with respect to the* Os_6 *plane;* [c]*phosphorus-containing ligands equatorial with respect to the* Os_6 *plane*

The separation between the redox potentials of the first electron addition and the the second electron addition ranges from 160 to 280 mV. This means that the two added electrons interact electrostatically each other.

The ability of these 90-electron systems to add as many as two electrons has been theoretically foreseen by Mingos and coworkers |233|.

4.8.2. $Os_6(CO)_{21-n}(MeCN)_n$ (n = 1-3). The derivatives $Os_6(CO)_{21-n}(MeCN)_n$ (n = 1-3) belong to the family of 90-electron hexanuclear raft-like planar clusters discussed in the preceding Section. Indeed, they are the starting material for the preparation of such $Os_6(CO)_{21-n}[PR_3]_n$ (n = 1-6) derivatives |229|. Like these, $Os_6(CO)_{20}(MeCN)$ and $Os_6(CO)_{19}(MeCN)_2$ undergo subsequent one-electron reductions, in acetonitrile solution, according to |229|:

$$Os_6(CO)_{20}(MeCN) \underset{-e}{\overset{+e}{\rightleftarrows}} [Os_6(CO)_{20}(MeCN)]^- \underset{-e}{\overset{+e}{\rightleftarrows}} [Os_6(CO)_{20}(MeCN)]^{2-}$$

$E°' = -0.65$ V; $E°' = -0.89$ V

$$Os_6(CO)_{19}(MeCN)_2 \underset{-e}{\overset{+e}{\rightleftarrows}} [Os_6(CO)_{19}(MeCN)_2]^-$$

$E°' = -0.87$ V

The redox behaviour of $Os_6(CO)_{18}(MeCN)_3$ has not yet been reported.

4.8.3. $Os_6(CO)_{20}[C{=}C(H)Ph]$. Once again, $Os_6(CO)_{20}(MeCN)$ is the starting material for the synthesis of the vinylidene cluster $Os_6(CO)_{20}[C{=}C(H)Ph]$ |234|. Interestingly, this 92-electron system maintains, to a large extent, a planar geometry, but with a somewhat different Os_6 array with respect to the previously described raft species. In fact, as shown in Figure 125, the trapezoidal fragment Os(1)Os(2)Os(4)Os(5), formed by three fused triangles, has the same shape of the preceding derivatives (even if with a significantly longer mean Os-Os bond length [3.01 Å vs. 2.82 Å]), but, the fourth triangle, having at the apex the Os(6) atom and capped by the organic ligand, is displaced out from forming a unique big triangle, as well as not being coplanar with the other ones (forming in fact a dihedral angle of 34°) |201|.

In confirmation of its 92-electron nature, it has been briefly reported that it undergoes, in cyclic voltammetry, two irreversible (and hence likely framework destroying) one-electron removals |234|.

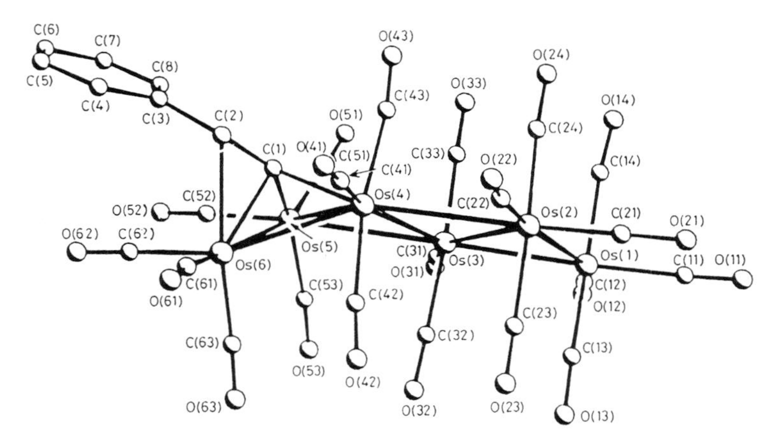

Figure 125. *The molecular structure of $Os_6(CO)_{20}[C{=}C(H)Ph]$ (from Ref. 234).*

4.9. Heptanuclear Binary Complexes

4.9.1. $[Os_7(CO)_{20}]^{2-}$. The heptaosmium cluster dianion $[Os_7(CO)_{20}]^{2-}$, obtained by treating $Os_8(CO)_{23}$ with bases |6|, undergoes, in dichloromethane solution, an irreversible two-electron oxidation at Ep = +0.46 V. By performing such an oxidation by chemical oxidants in the presence of the two-electron donors CO or $P(OMe)_3$, $Os_7(CO)_{21}$ and $Os_7(CO)_{20}[P(OMe)_3]$ can be obtained in high yields |122|.

The 98-electron dianion $[Os_7(CO)_{20}]^{2-}$ possesses likely a molecular structure similar to that of the isoelectronic $Os_7(CO)_{21}$, Figure 126 |235|.

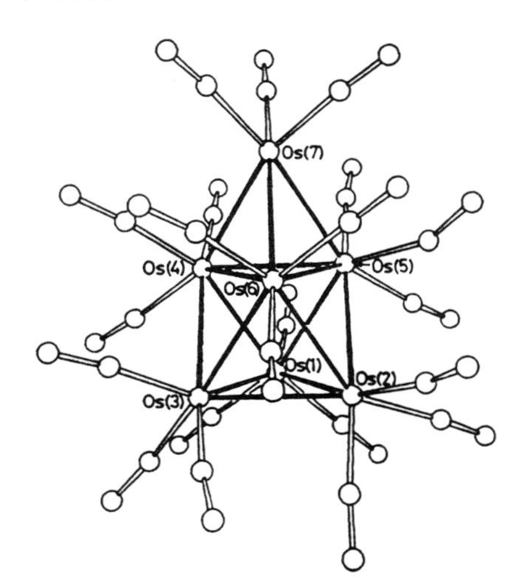

Figure 126. *Perspective view of $Os_7(CO)_{21}$ (from Ref. 235).*

In it, the seven Os atoms adopt an octahedral geometry having one face capped by the triply bridged seventh Os atom. Within the octahedron the mean Os-Os bond length is 2.85 Å, whereas in the capping fragment it is 2.82 Å.

4.10. Decanuclear Carbido Complexes

4.10.1. $[Os_{10}(CO)_{24}C]^{2-}$. Different routes to the synthesis of the decaosmium carbido dianion $[Os_{10}(CO)_{24}C]^{2-}$ have been set up |180, 236,237|. Figure 127 illustrates the molecular structure of such an high nuclearity cluster |236,237|.

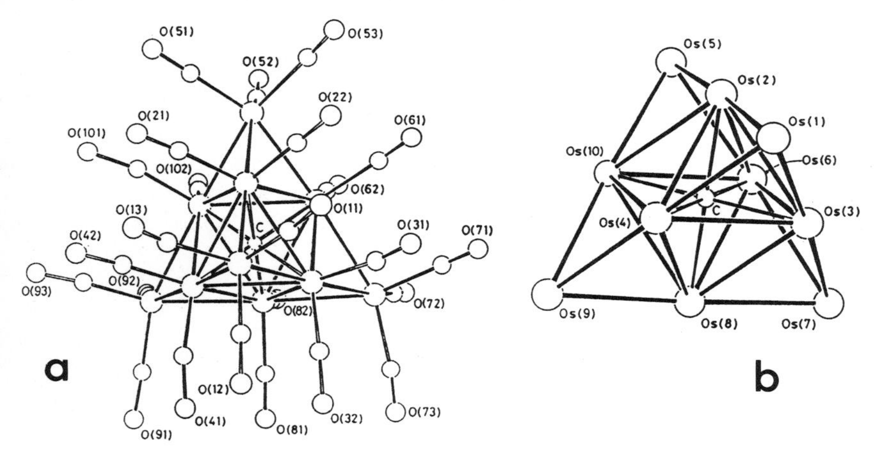

Figure 127. *Perspective view of the dianion $[Os_{10}(CO)_{24}C]^{2-}$ (a), and its $Os_{10}C$ core (b) (from Ref. 237).*

The geometry can be described as a carbido-encapsulated hexaosmium octahedron, four faces of which are capped by an $Os(CO)_3$ unit, so to form an overall tetrahedral assembly. The Os-Os bond length within the central octahedron are significantly longer (mean, 2.88 Å) than those in the four capping units (mean, 2.79 Å), more likely to accomodate the carbido carbon atom. The averaged $Os-C_{(carbido)}$ distance in the octahedral moiety is 2.04 Å.

The redox propensity of this 134-electron cluster, preliminarily pointed out by its reactivity towards the couple I_2/I^- |238|, has been extensively examined by electrochemical techniques |122,228,239-241|.

Figure 128 shows the relevant anodic behaviour in dichloromethane solution at a platinum electrode |239|.

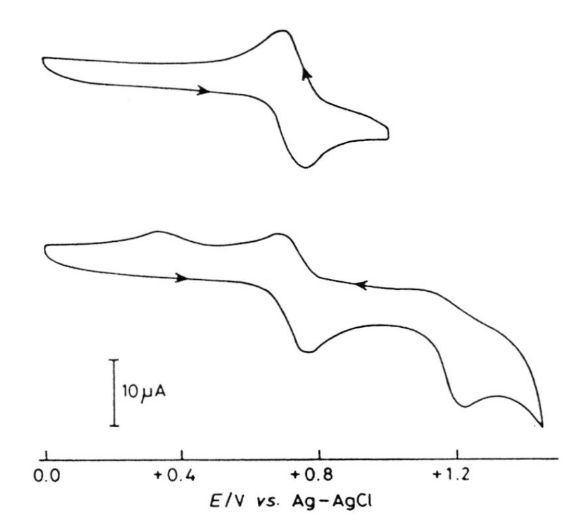

Figure 128. *Anodic cyclic voltammograms recorded at a platinum electrode on a CH_2Cl_2 solution of $[Os_{10}(CO)_{24}C]^{2-}$ ($[NBu_4][BF_4]$ supporting electrolyte). Scan rate 0.1 Vs^{-1}. Room temperature (from Ref. 239).*

The dianion undergoes two successive one-electron oxidations, only the first one being chemically and electrochemically reversible, according to the scheme |239|:

$$[Os_{10}(CO)_{24}C]^{2-} \underset{+e}{\overset{-e}{\rightleftharpoons}} [Os_{10}(CO)_{24}C]^{-} \xrightarrow{-e} Os_{10}(CO)_{24}C$$

(stable only in the solid state)

This oxidation pattern becomes more complicated when mercury working electrodes are employed. In fact, in this case the mercury electrode surface partakes in the oxidation process generating the heteronuclear cluster $[Os_{20}Hg_2(CO)_{48}C_2]^{2-}$ |241|, which has been discussed elsewhere |30|.

Concerning the reduction behaviour of $[Os_{10}(CO)_{24}C]^{2-}$, the relevant cyclic voltammogram in tetrahydrofuran solution, at different temperatures, is illustrated in Figure 129 |240,241|. At relatively high temperature, the dianion apparently undergoes a single-stepped two-electron addition, complicated by successive chemical reactions. In fact, the $i_{p(B)}/i_{p(A)}$ ratio is notably lower than 1 and one reoxidation peack C, very far from peak A, is present in the reverse scan. Nevertheless, decreasing the temperature makes peak A to split in two peaks, D and E, so that peak B is directly associated to peak D, while peak C is associated with peak E.

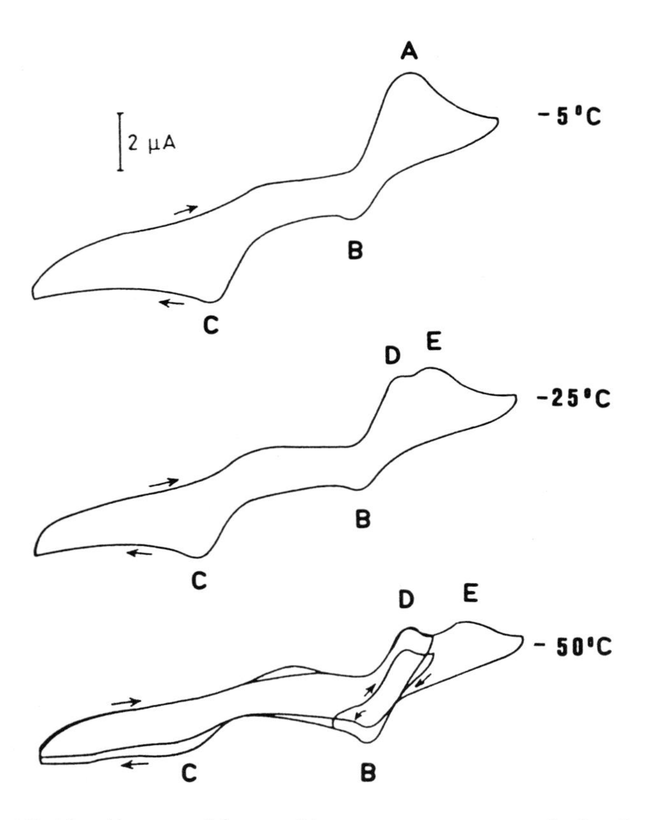

Figure 129. *Cathodic cyclic voltammograms recorded at a platinum electrode on a THF solution of* $[Os_{10}(CO)_{24}C]^{2-}$, *at different temperatures. Scan rate: 0.1* Vs^{-1} *(from Ref. 240).*

In summary, based also on chemical evidence, the two-electron reduction of the decaosmium dianion proceeds according to the pathway |240,241|:

$$[Os_{10}(CO)_{24}C]^{2-} \underset{-e \text{ (peak B)}}{\overset{+e \text{ (peak D)}}{\rightleftarrows}} [Os_{10}(CO)_{24}C]^{3-} \underset{-e \text{ (peak C)}}{\overset{+e \text{ (peak E)}}{\rightleftarrows}} [Os_{10}(CO)_{24}C]^{4-}$$

The large separation in potential values between peak E and peak C (about 800-900 mV) is hardly conducible to a chemically reversible redox change (like $[Os_{10}(CO)_{24}C]^{3-/4-}$), unless geometrically important, even if not framework destroying, reorganizations accompany such electron addition/removal. This indeed seems the case on the basis of spectroscopic evidence.

In conclusion, the tetracapped octahedron-tetrahedral dianion is able to support both a one-electron addition and a one-electron removal without any substantial change in geometry. By contrast, the two-electron addition leads to the corresponding tetraanion, which likely has a somewhat different geometry. Analogously, the two-electron removal leads to the corresponding neutral species, which also significantly differs in geometry with respect to the starting precursor.

Table 25 summarizes the electrode potentials of these redox changes.

TABLE 25. *Redox potentials for the electron transfers exhibited by* $[Os_{10}(CO)_{24}C]^{2-}$ *in different nonaqueous solutions |239-242|*

	Redox changes					
	0/-	-/2-		2-/3-		3-/4-
solvent	Ep (V)	E°' (V)	ΔEp^a (mV)	E°' (V)	ΔEp^a (mV)	Ep (V)
CH_2Cl_2	+1.16	+0.74	56	-1.39	120	-1.41
THF				-1.29	113	-1.35
THF				-1.23^b	110	-1.47^b
MeCN				-1.15	75	-1.20

^{a}*Measured at 0.1 Vs^{-1}*; ^{b}*measured at -50°C*

In the likely assumption that, with respect to the starting dianion, the first one-electron is removed from the HOMO level as well as the first one-electron is added to the LUMO level, an energy gap of 2.18 eV can be computed in dichloromethane; such value is in sharp contrast with the 0.5 eV value theoretically computed |243|.

4.11. Decanuclear Hydrido Complexes

4.11.1. $[H_4Os_{10}(CO)_{24}]^{2-}$. The tetrahydrido-decaosmium dianion $[H_4Os_{10}(CO)_{24}]^{2-}$, isoelectronic with the carbido dianion $[Os_{10}(CO)_{24}C]^{2-}$, has an overall structure very similar to this latter, Figure 130 |244|.

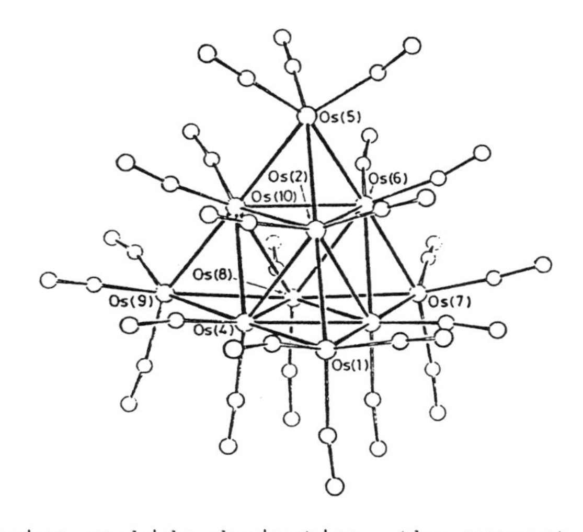

Figure 130. *Perspective view of $[H_4Os_{10}(CO)_{24}]^{2-}$ (from Ref. 244).*

Like the decaosmium carbido derivative, the present hydrido cluster has a tetracapped octahedral assembly forming an overall tetrahedral geometry. Within the octahedron there is a wide spreading of Os-Os distances (from 2.77 Å to 2.92 Å), with a mean bond length of 2.86 Å, not very different from the mean value of 2.83 Å in the tetrahedral capping units. Location of the hydrido ligands is not univocal, however there is a good chance that they are interstitial to the decaosmium core, possibly three within the tetrahedral capping fragments (perhaps not casually three of them are larger than the remaining one) and the fourth one within the central octahedron |244|.

Like $[Os_{10}(CO)_{24}C]^{2-}$, the 134-valence-electron $[H_4Os_{10}(CO)_{24}]^{2-}$ undergoes, in dichloromethane solution, at a platinum electrode, a first, chemically and electrochemically, reversible one-electron oxidation ($E°' = +0.35$ V), followed by a second irreversible one-electron oxidation at $E_p = +0.57$ V |239|.

As in the case of the related carbido cluster, it is likely that in the $[H_4Os_{10}(CO)_{24}]^{2-/-}$ redox change no significant structural change occurs, whereas further oxidation to $H_4Os_{10}(CO)_{24}$ involves deep structural reorganizations.

5. COBALT CLUSTERS

5.1. Trinuclear Mixed-Ligand Complexes

5.1.1. $Co_3(CO)_9(X)$ (X = S,Se). The basic molecular architecture of the tricobalt clusters $Co_3(CO)_9(\mu_3\text{-}X)$, (X = S,Se), is shown in Figure 131 |245,246|.

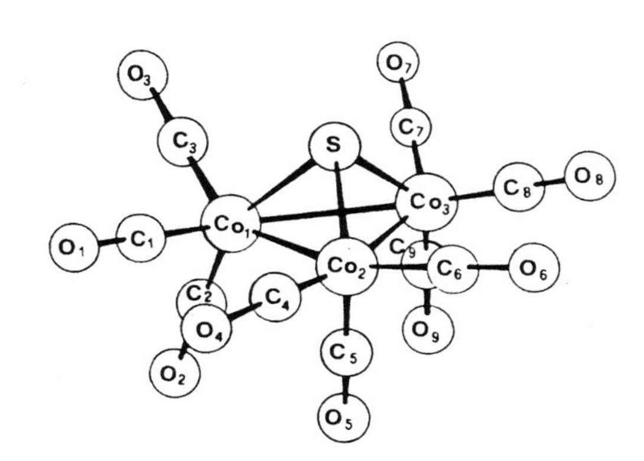

Figure 131.
Schematic representation of the molecular structure of $Co_3(CO)_9(X)$ (X = S, Se).

The tetrahedral core Co_3X is constituted by an essentially metallic equilateral triangle (Co-Co averaged distances: 2.64 Å, X = S; 2.62 Å, X = Se) capped by the triply bridging chalcogen atom (Co-S, 2.14 Å; Co-Se, 2.28 Å).

These 49-electron clusters undergo, in 1,2-dichloroethane solution, a reversible one-electron oxidation ($E°'$ = +0.44 V, X = S; $E°'$ = +0.42 V, X = Se) as well as an irreversible two-electron reduction (Ep = -0.70 V, X = S; Ep = -0.77 V, X = Se) |56|.

The chemically reversible oxidation step is electrochemically quasireversible in character (ΔEp = 97 mV, X = S; ΔEp = 86 mV, X = Se). This likely preludes some significant structural change upon one-electron removal. The monocation congeners have not been X-ray characterized. Nevertheless, structural comparisons have been drawn on the basis of the structures of the 48-electron species $Co_2Fe(CO)_9(\mu_3\text{-}S)$ and $Co_2Fe(CO)_2(\mu_3\text{-}Se)$, respectively |246|. Since the covalent radii for Co and Fe atoms are virtually identical, the fact that in $Co_2Fe(CO)_9(X)$ the metal-metal distances significantly shorten (by 0.09 Å, X = S; by 0.04 Å, X = Se), whereas the metal-chalcogen distances remain substantially unaltered, provides good evidence that the electron in excess, present in both $Co_3(CO)_9(X)$ clusters, resides in a metal-metal antibonding HOMO level |40,246|.

5.1.2. $Co_3(CO)_7(L)(S)$ (L = bidentate ligands). A few series of homonuclear 48-electron Co_3S clusters have been prepared by substituting two equatorial CO groups of $Co_3(CO)_9(S)$ for three-electron bidentate ligands |247-249|. They have general formula: $Co_3(CO)_7[\mu\text{-}1,2\text{-}C(R)N(C_6H_{11})](S)$, R = Me, Ph; $Co_3(CO)_7[\mu\text{-}1,2\text{-}\eta^2\text{-}SC(NMe_2)](S)$; $Co_3(CO)_7[\mu\text{-}1,3\text{-}\eta^2\text{-}NHC(R)S](S)$, R = Me, Ph.

Figure 132 shows the molecular geometry of such class of clusters. The constrains imposed by the bidentate substituents make isosceles the basal tricobalt triangles, shortening the bridged Co-Co bond length; as a consequence, the Co-S distances from the bridged Co atoms slightly shorten also. Some structural parameters are reported in Table 26 (together with the relevant electrochemical parameters).

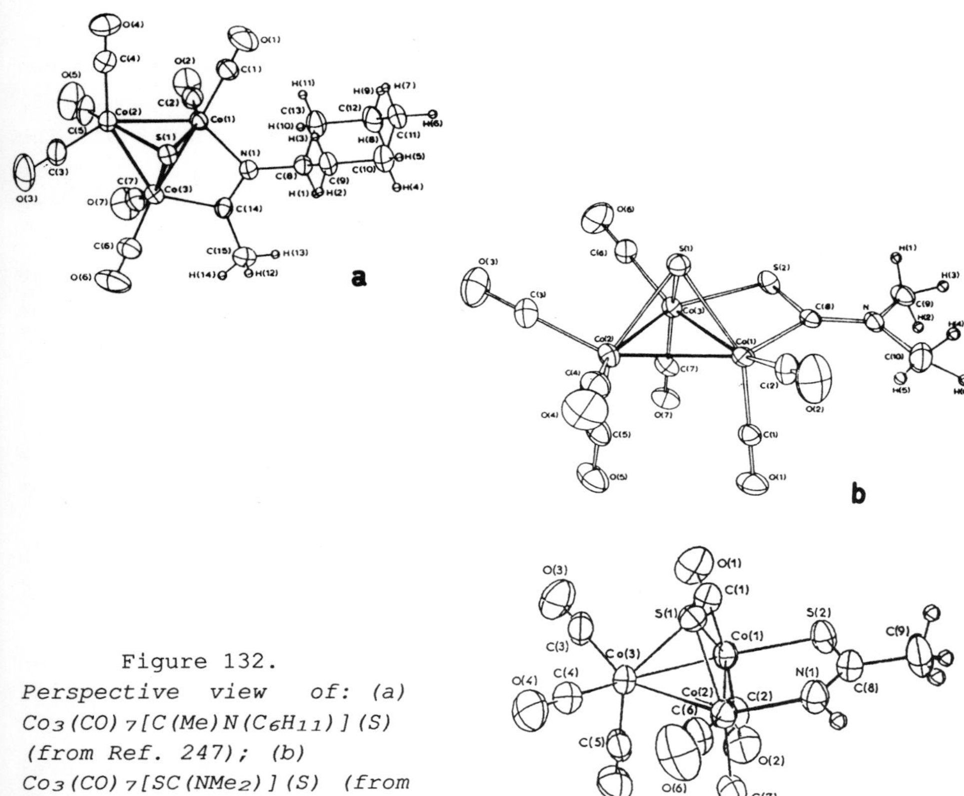

Figure 132.
Perspective view of: (a) $Co_3(CO)_7[C(Me)N(C_6H_{11})](S)$ *(from Ref. 247); (b)* $Co_3(CO)_7[SC(NMe_2)](S)$ *(from Ref. 248); (c)* $Co_3(CO)_7[NHC(Me)S](S)$ *(from Ref. 249).*

TABLE 26. *Structural and electrochemical parameters for the clusters $Co_3(CO)_7(L)(S)$ |247-249|*

L	Co-Co (Å)		Co-S (Å)		$E^{\circ\prime}_{0/-}$ (V)	
$C(Me)-N(C_6H_{11})$	2.55[a]	2.42[b]	2.15[c]	2.17[d]	-0.67[e]	
$C(Ph)-N(C_6H_{11})$					-0.63[e]	-0.80[f]
$S-C(NMe_2)$	2.54[a]	2.44[b]	2.15[c]	2.17[d]	-0.67[e]	-0.82[f]
S-C(Me)-NH	2.52[a]	2.48[b]	2.15[c]	2.17[d]	-0.44[e]	-0.52[f]
S-C(Ph)-NH						-0.49[f]

[a]*Unbridged Co-Co edges;* [b]*bridged Co-Co edge;* [c]*from bridged Co atoms;* [d]*from the unbridged Co atom;* [e]*in DMF solution;* [f]*in CH_2Cl_2 solution*

In accordance with the expectation that these 48-electron clusters can reach the 49-electron configuration of $Co_3(CO)_9S$, they undergo a relatively easy, one-electron reduction |249|. The corresponding monoanions seem relatively stable under anaerobic conditions, confirming that, also in this case, population of the antibonding LUMO does not lead to declustering processes.

The redox potentials of such one-electron additions are reported in Table 26. They are, obviously, affected by the electronic effects of the bidentate substituents.

5.1.3. $\{[Co_3(CO)_9]C\}_nR$ (n = 1,2). Methylidyne tricobalt clusters of formula $Co_3(CO)_9(C-X)$ are one of the most popular class of cobalt clusters |250-252|.

Some structural examples of early studied derivatives, in particular three species examined also from the electrochemical viewpoint, are illustrated in Figure 133 |253-255|.

More recently, alkyne-capped Co_3 cluster compounds have been also prepared and their structure is exemplified in Figure 134 |256|.

In all cases, the Co_3C assembly is formed by an equilateral trimetallic base (Co-Co bond lengths, constantly equal to 2.47 Å) capped by a triply bridging carbon atom (mean Co-C bond distances around 1.90 Å).

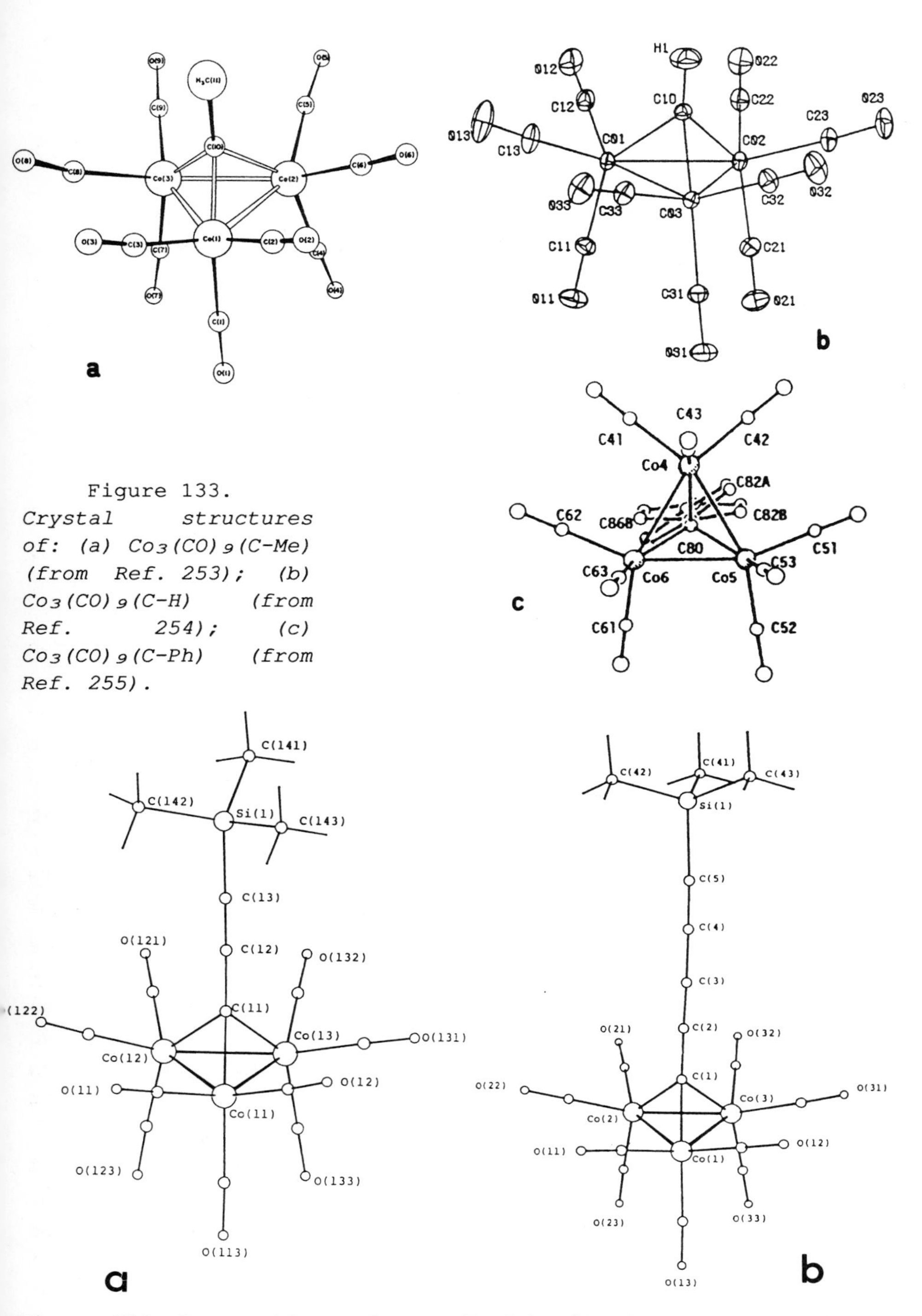

Figure 133. *Crystal structures of: (a)* $Co_3(CO)_9(C\text{-}Me)$ *(from Ref. 253); (b)* $Co_3(CO)_9(C\text{-}H)$ *(from Ref. 254); (c)* $Co_3(CO)_9(C\text{-}Ph)$ *(from Ref. 255).*

Figure 134. *Perspective views of: (a)* $[Co_3(CO)_9](C\text{-}C{\equiv}C\text{-}SiMe_3)$*; (b)* $[Co_3(CO)_9](C\text{-}C{\equiv}C\text{-}C{\equiv}C\text{-}SiMe_3)$ *(from Ref. 256).*

The most significant redox change exhibited by these 48-electron clusters is their one-electron reduction. In this connection, Figure 135 shows the cyclic voltammogram exhibited by $Co_3(CO)_9(C\text{-}Me)$, in acetone solution |257|.

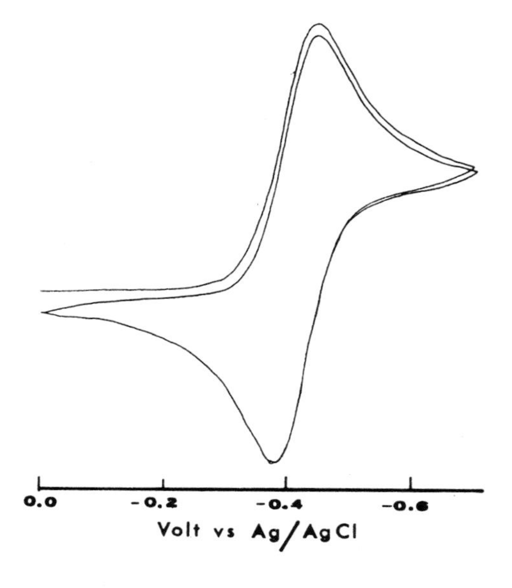

Figure 135.
Cyclic voltammetric response recorded at a platinum electrode on an acetone solution of $Co_3(CO)_9(C\text{-}Me)$. Scan rate 0.05 Vs^{-1} (from Ref. 257).

Although the voltammetric picture appears due to a chemically reversible one-electron transfer, it has been recently showed that the monoanions $[Co_3(CO)_9(C\text{-}X)]^-$ really undergo, during long times, slow fragmentation |258,259|, unless electrogeneration is performed under CO atmosphere |258|.

Table 27 summarizes the formal electrode potentials of the neutral/monoanion redox change for the present series of tricobalt clusters. It is well conceivable that the inductive electronic effects from the apical substituents may affect the electrode potentials, in the sense that electron-withdrawing groups make easier the electron addition.

TABLE 27. *Redox potentials for the one-electron reduction of $Co_3(CO)_9(C{-}X)$ in different nonaqueous solutions*

X	$E^{\circ\prime a}$ (V)	ΔEp^{b} (mV)	solvent	reference
H	-0.71	170	CH_2Cl_2	260
	-0.45	120	Me_2CO	257
F	-0.35	98	Me_2CO	257
Cl	-0.59	170	CH_2Cl_2	260
	-0.36	100	Me_2CO	257
Br	-0.60	140	CH_2Cl_2	260
	-0.34	108	Me_2CO	257
I	-0.59	180	CH_2Cl_2	260
OMe	-0.72	130	CH_2Cl_2	260
Me	-0.49	102	Me_2CO	257
	-0.63	70	CH_2Cl_2	261
	-0.79	120	MeCN	258
CF_3	-0.29	95	Me_2CO	257
Et	-0.82	70	CH_2Cl_2	260
$CH{=}CH_2$	-0.72	80	CH_2Cl_2	260
Ph	-0.72	140	CH_2Cl_2	260
	-0.63	65	CH_2Cl_2	261
	-0.69	60	CH_2Cl_2	262
	-0.45	80	Me_2CO	257
$SiMe_3$	-0.47	82	Me_2CO	257
$C{\equiv}C{-}Ph$	-0.62	180	CH_2Cl_2	262
$C{\equiv}C{-}SiMe_3$	-0.55	140	CH_2Cl_2	262
$C{\equiv}C{-}C{\equiv}CSiMe_3$	-0.52	100	CH_2Cl_2	262
CH(OH)Me	-0.70	120	CH_2Cl_2	260
CH(OH)Ph	-0.71	110	CH_2Cl_2	260
$C(OH)Me_2$	-0.73	110	CH_2Cl_2	260
$C(O)SC(Me)_3$	-0.60	150	CH_2Cl_2	260

[a]*Some significant discrepancies among different solvents likely arise from the use of different reference couples;* [b]*measured at 0.2 Vs^{-1}*

Theoretical |40| and spectral results |263| indicate that, upon reduction, the added electron enters an antibonding LUMO level essentially localized on the trimetallic triangle. This intuitively leads to some elongation of the Co-Co distances, with concomitant structural rearrangement of the Co_3C unity. Such geometrical reorganization could explain the electrochemical quasireversibility of the electron transfer step.

Speaking about $Co_3(CO)_9C$ assemblies, it has to be taken into account that clusters containing more than one tetrahedral Co_3C fragment are long since known |250,252|. In this connection, we wish here to examine a series of di-cluster compounds in which the two Co_3 units are each other connected by different carbon chains.

Figure 136 shows the X-ray structure of four molecules of known redox activity.

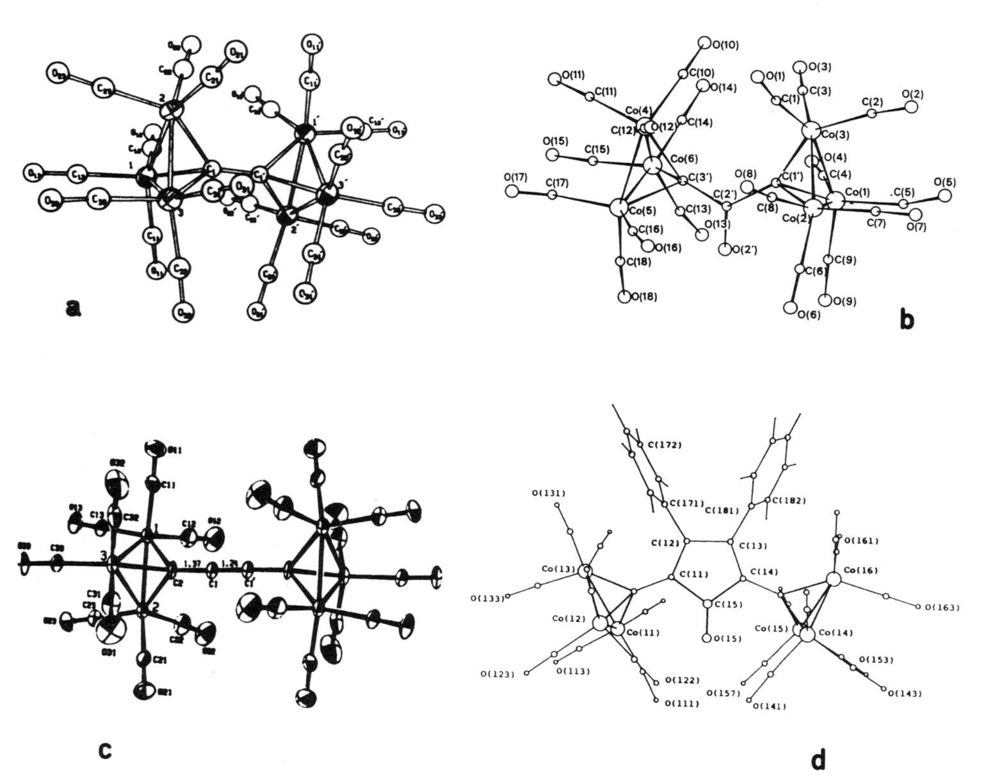

Figure 136. *Perspective views of: (a)* $[Co_3(CO)_9C]_2$ *(from Ref. 264); (b)* $[Co_3(CO)_9C]_2C(O)$ *(from Ref. 265); (c)* $[Co_3(CO)_9C]_2(C{\equiv}C)$ *(from Ref. 266); (d)* $[Co_3(CO)_9C]_2(Ph_2C_5O)$ *(from Ref. 267).*

In the two $Co_3(CO)_9C$ tetrahedral subunits, the metallic triangle remains unaltered with respect to the momocluster precursors, while the averaged Co-$C_{(apical)}$ distance tends to elongate slightly (by about 0.02-0.03 Å in $[Co_3(CO)_9C]_2C(O)$, $[Co_3(CO)_9C]_2(C\equiv C)$ and $[Co_3(CO)_9C]_2(Ph_2C_5O)$; by about 0.06 Å in $[Co_3(CO)_9C]_2$), likely because of electron delocalization along the bridging carbon chain.

Usually such di-cluster compounds undergo two separated one-electron reductions, which testify electronic communication between the two tricobalt fragments. Indeed, the two electron additions have features of chemical reversibility only at low temperature; at room temperature, chemical complications follows the reduction processes |262|. As an example, Figure 137 shows the cyclic voltammetric profiles exhibited by $[Co_3(CO)_9C]C(O)$ $[Co_3(CO)_9C]$ at different temperatures |262a|.

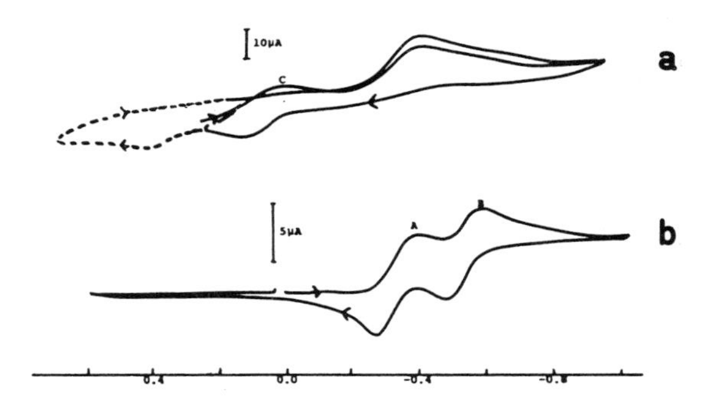

Figure 137. *Cyclic voltammetric responses recorded at a platinum electrode on a dichloromethane solution of $[Co_3(CO)_9C]_2C(O)$. (a) T = 20 °C, scan rate 0.2 Vs^{-1}; (b) T = -50 °C, scan rate 0.1 Vs^{-1} (from Ref. 262a).*

Only in the case of $[Co_3(CO)_9C]_2(Ph_2C_5O)$ a different pathway holds. The low-temperature cyclic voltammogram shown in Figure 138 indicates that the cluster undergoes: (*i*) a first two-electron reduction at peak A, attributed to the simultaneous addition of one electron into the two, likely non-interacting, Co_3C subunits; (*ii*) a second one-electron addition at peak B, centred on the cyclopentadienone ring; (*iii*) a third two-electron reduction at peak D, corresponding to the irreversible addition of a second electron to the two tetrahedral Co_3 units |267|.

Table 28 summarizes the redox potentials of the two electron addition steps in the present di-cluster species |262|.

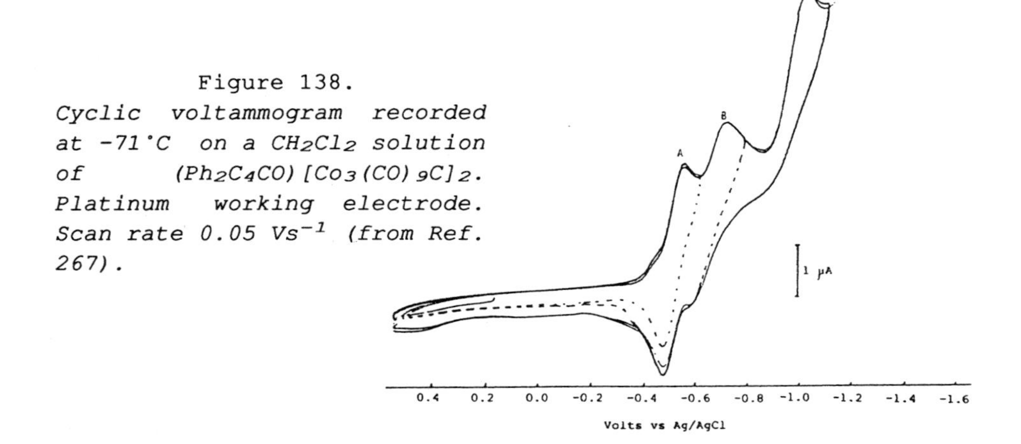

Figure 138.
Cyclic voltammogram recorded at -71°C on a CH_2Cl_2 solution of $(Ph_2C_4CO)[Co_3(CO)_9C]_2$. Platinum working electrode. Scan rate 0.05 Vs^{-1} (from Ref. 267).

TABLE 28. *Formal electrode potentials for the sequential one-electron additions to the two Co_3 units of bis-tricobalt clusters. Low temperature experiments (see text)*

complex	$E^{\circ\prime}_{(0/-)}$ (V)	$E^{\circ\prime}_{(-/2-)}$ (V)	solvent	reference
$(CO)_9Co_3C\text{-}CCo_3(CO)_9$	-0.54	-0.90	CH_2Cl_2	262a
	-0.45	-0.92	THF	262b
	-0.38	-0.74	Me_2CO	262b
$(CO)_9Co_3C\text{-}C(O)\text{-}CCo_3(CO)_9$	-0.45	-0.70	CH_2Cl_2	262a
$(CO)_9Co_3C(C\equiv C)CCo_3(CO)_9$	-0.52	-0.73	CH_2Cl_2	262a
	-0.36[a]	-0.66[a]	THF	262b
	-0.32	-0.53	Me_2CO	262b
$(CO)_9Co_3C(C\equiv C\text{-}C\equiv C)CCo_3(CO)_9$	-0.52	-0.63	CH_2Cl_2	262a
$(CO)_9Co_3C(Ph_2C_5O)CCo_3(CO)_9$	-0.64	-0.64	CH_2Cl_2	267

[a]*Room temperature*

5.1.4. $Co_3(CO)_{9-n}(L)_n(C\text{-}X)$ (L = phosphines or phosphites; n=1-3). A remarkably high number of substituted derivatives of formula $Co_3(CO)_{9-n}(L)_n(C\text{-}X)$ have been synthesized by varying X and/or L. The main class belong to phosphine or phosphite derivatives. Some interesting structural features within the monosubstituted species are illustrated in Figure 139.

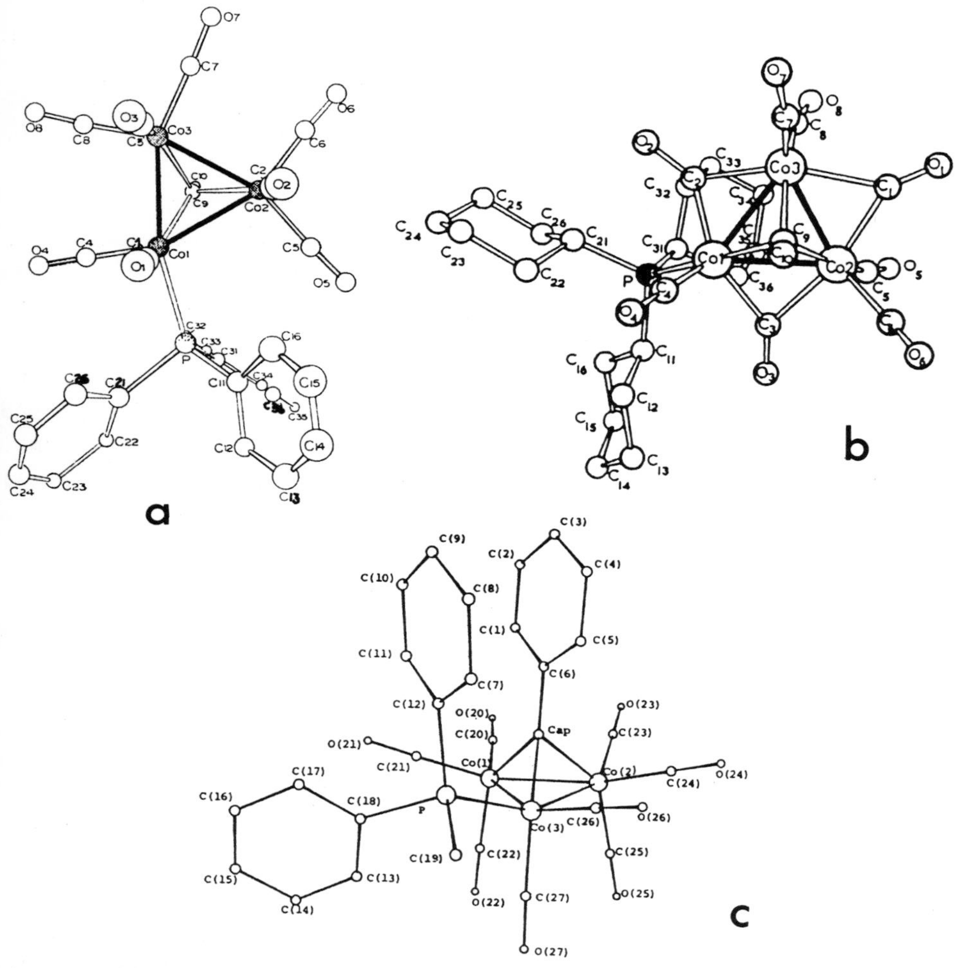

Figure 139. *Perspective view of: (a) $Co_3(CO)_8(PPh_3)(C\text{-}Me)$ (from Ref. 268); (b) $Co_3(CO)_8[P(C_6H_{11})_3](C\text{-}Me)$ (from Ref. 269); half of the molecule of $[Co_3(CO)_8(C\text{-}Ph)]_2(Ph_2PCH_2CH_2PPh_2)$ (from Ref. 270).*

In $Co_3(CO)_8(PPh_3)(CMe)$ and $[Co_3(CO)_8(CPh)]_2(Ph_2PCH_2CH_2PPh_2)$, the phosphine or half of the bidentate diphosphine (which acts, here, as monodentate towards each Co_3C cluster core) substitute one equatorial carbonyl, leaving the remaining CO groups terminal. In $Co_3(CO)_8[P(C_6H_{11})_3](CMe)$, the phosphine not only substitutes one axial carbonyl, but, in addition, induces three CO groups to assume an edge-bridging position.

Commonly, the presence of the electron-donating phosphorus-containing substituents induces the metallic triangle, which in the unsubstituted parent is equilateral (Co-Co, meanly 2.47 Å), to become isosceles, with the Co-Co distance opposite to the Co atom linked to the substituent being the minor edge (PPh_3: Co-Co, 2.50 Å and 2.49 Å, respectively; $P(C_6H_{11})_3$: Co-Co, 2.52 Å and 2.38 Å, respectively; $Ph_2PCH_2CH_2PPh_2$: Co-Co, 2.48 Å and 2.47 Å, respectively).

As far as the disubstituted species are concerned, the solved X-ray structures concern the diphosphine-substituted complexes $Co_3(CO)_7(Ph_2PCH_2PPh_2)(C\text{-}X)$ (X = Cl |271a|, Me |270, 271b|, Ph |271c|) and $Co_3(CO)_7(Me_2PCH_2CH_2PMe_2)(C\text{-}Ph)$ |271d|. As exemplified in Figure 140 for $Co_3(CO)_7(Ph_2PCH_2PPh_2)(CMe)$, the bidentate phosphine substitutes two equatorial CO groups, leaving the remaining carbonyls terminal. The phosphino-bridged Co-Co bond (2.49 Å) becomes slightly longer than the other two (2.47 Å). Important distortions arise further with respect to the metal triangle-$C_{(apical)}$ interactions; in fact, the Co-C distances assume three significantly different values (1.86 Å, 1.89 Å, 1.95 Å).

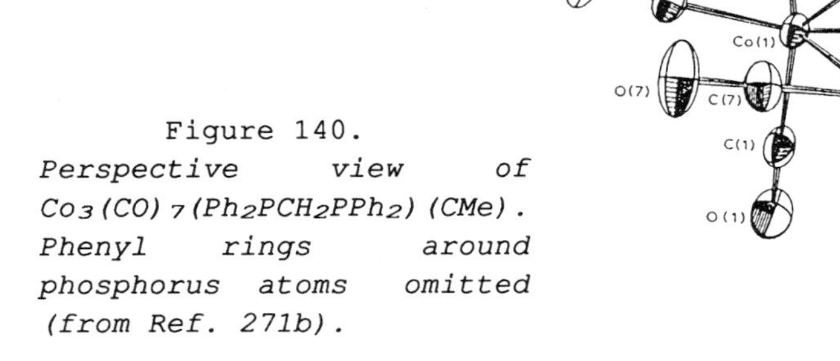

Figure 140. *Perspective view of $Co_3(CO)_7(Ph_2PCH_2PPh_2)(CMe)$. Phenyl rings around phosphorus atoms omitted (from Ref. 271b).*

Finally, Figure 141 shows the crystal structure of the trisubstituted species $Co_3(CO)_6[P(OMe)_3](C\text{-}Me)$ |272|. The three phosphite ligands substitute one equatorial carbonyl on each of the three cobalt atoms; the remaining CO groups maintain their terminal coordination. The increased electron density elongates slightly the Co-Co averaged bond length (2.48 Å).

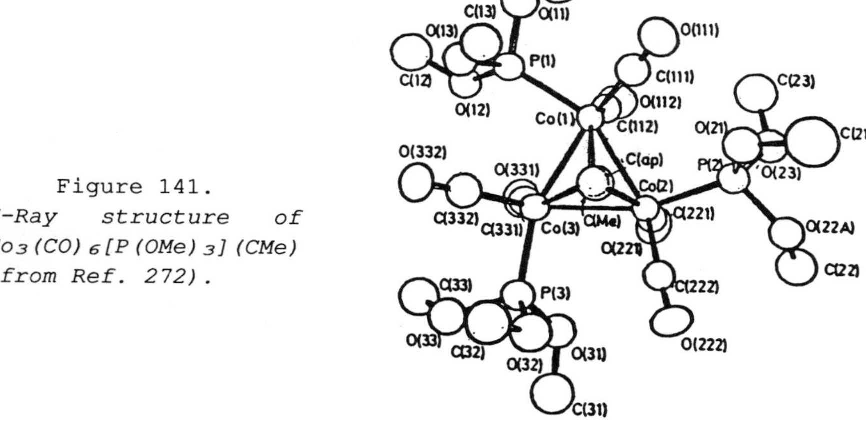

Figure 141.
X-Ray structure of $Co_3(CO)_6[P(OMe)_3](CMe)$ *(from Ref. 272).*

In spite of the precise locations of both L substituents and CO groups now discussed in the solid state, spectral measurements in solution sometimes indicate the occurrence of isomerization processes (L axial/equatorial; CO, terminal/bridging) |250,272, 273| so that the electrochemical responses below discussed, in some cases, are not correlatable with an ascertained structural geometry.

Electrochemistry of $Co_3(CO)_{9-n}(L)_n(C\text{-}X)$ is not straightforward and only recently it has been elucidated in detail |258|. An useful example of these difficulties is that represented by the electrode behaviour of $Co_3(CO)_8(PPh_3)(C\text{-}Me)$, studied by many authors |251,274-276|. As shown in Figure 142a, the cluster seems to undergo, at room temperature, a reversible one-electron reduction. However, by decreasing the temperature, it can be seen, Figure 142b, that the one-electron reduction is accompanied by fast decay of the monoanion. Only at low temperature and extremely fast scan rates, Figure 142c, a directly associated response C can be observed |258|. The reoxidation peak B (which at low temperature and high scan rate is shifted towards more positive potential values) is attributed to the reoxidation of the intermediate $[Co_3(CO)_8(CMe)]^{2-}$, produced through the following ECE pathway:

$$Co_3(CO)_8(PPh_3)(CMe) \underset{-e}{\overset{+e}{\rightleftharpoons}} [Co_3(CO)_8(PPh_3)(CMe)]^-$$

$$\downarrow -PPh_3 \text{ (fast)}$$

$$[Co_3(CO)_8(CMe)]^- \underset{-e}{\overset{+e}{\rightleftharpoons}} [Co_3(CO)_8(CMe)]^{2-}$$

The apparent reversibility of the redox changes is only due to a superposition of a series of redox processes; this is proved by the impossibility to obtain an ESR spectrum of $[Co_3(CO)_8(PPh_3)(CMe)]^-$ |274|.

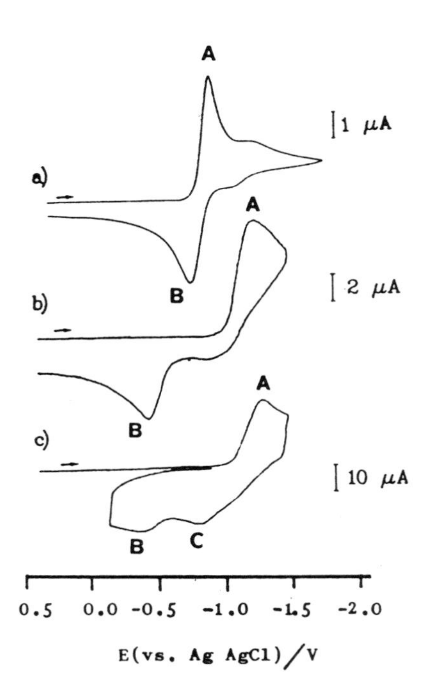

Figure 142. *Cyclic voltammograms recorded on a MeCN solution of $Co_3(CO)_8(PPh_3)(CMe)$. (a) $T = 24°C$, $v = 0.3\ Vs^{-1}$; (b) $T = -43°C$, $v = 10\ Vs^{-1}$; (c) $T = -43°C$, $v = 150\ Vs^{-1}$ (from Ref. 258).*

The reduction pathways of the substituted Co_3C derivatives are additionally complicated by the nature of the phosphine, which alternatively or furtherly can induce decarbonylation of the intermediates, as well as by reassociation of the released ligands.

The following scheme may be assumed as a general representation of the redox propensity of the present 48-electron family of tricobalt clusters |258|.

$$[Co_3(CO)_{8-n}(L)_n(CX)]^- \xrightarrow{+L} [Co_3(CO)_{8-n}(L)_{n+1}(CX)]^-$$

$-CO \uparrow \downarrow +CO \qquad +Co_3(CO)_{9-n}(L)_n(CX) \downarrow \uparrow -Co_3(CO)_{8-n}(L)_{n+1}(CX)$

$$Co_3(CO)_{9-n}(L)_n(CX) \underset{-e}{\overset{+e}{\rightleftharpoons}} [Co_3(CO)_{9-n}(L)_n(CX)]^-$$

$-L \downarrow\uparrow +L$

$$[Co_3(CO)_{9-n}(L)_{n-1}(CX)]^- \underset{-CO}{\overset{+CO}{\rightleftharpoons}} [Co_3(CO)_{10-n}(L)_{n-1}(CX)]^-$$

$+e \downarrow\uparrow -e \qquad\qquad \downarrow -L$

$$[Co_3(CO)_{9-n}(L)_{n-1}(CX)]^{2-} \qquad [Co_3(CO)_{10-n}(L)_{n-2}(CX)]^-$$

Such scheme reconciles the apparently conflicting electrochemical findings concerning these substituted species; in fact, it has been found that reduction processes can range from single one-electron to single two-electron transfer steps. The dominant key role is exerted by the instability of the primarily electrogenerated 49-electron monoanion $[Co_3(CO)_{9-n}(L)_n(CX)]^-$. Particularly with monosubstituted species, the monoanion decay can follows either the chain initiated by CO elimination, or that initiated by phosphine elimination. In this latter case the single-stepped two-electron reduction can take place. The competitive pathway is selected by the nature of the phosphine. In fact, electronic factors should stabilize kinetically the monoanions containing the less basic phosphines towards phosphine releasing, while destabilizing them towards CO elimination; however, the steric factors are also important, so that elimination of the bulkier phosphines is sometime favoured.

These arguments make clear the difficulties connected to oversimplifications of the electrochemistry of monosubstituted species.

In contrast, in the monoanions of bis- and tris-substituted species the CO elimination is less important, so that only the chain initiated by phosphine elimination becomes important.

In this picture, we think appropriate to summarize in Table 29, as a qualitative indication, only the potentials of the first cathodic step for the wide number of substituted tricobalt clusters studied, irrespective of both the reversibility/irreversibility of the redox change and the number of electrons globally involved.

TABLE 29. *Peak potential values (in Volt) for the first reduction step exhibited by $Co_3(CO)_{9-n}(L)_n(C-R)$*

n	L	R	Ep	solvent	reference
1	PPh_3	Me	-0.99	MeCN	258
			-0.77	Me_2CO	274
			-0.84	CH_2Cl_2	276
1	PPh_3	Ph	-0.70	Me_2CO	274
			-0.80	CH_2Cl_2	259
1	PPh_3	F	-0.62	Me_2CO	274
1	PPh_3	Cl	-0.86	CH_2Cl_2	252
1	$PMePh_2$	Me	-1.07	MeCN	258
1	$PMePh_2$	H	-0.78	Me_2CO	274
1	PMe_2Ph	Me	-1.09	MeCN	258
1	PEt_2Ph	Me	-0.86	Me_2CO	274
1	PMe_3	Me	-1.09	MeCN	258
1	$P(n\text{-}Bu)_3$	Me	-0.90	Me_2CO	274
1	$P(C_6H_{11})_3$	Me	-0.6	Me_2CO	274
1	$P(C_6H_{11})_3$	F	-0.68	Me_2CO	274
1	$P(OPh)_3$	F	-0.57	Me_2CO	274
1	$P(OMe)_3$	Me	-0.79	Me_2CO	274
1	$P(OMe)_3$	F	-0.67	Me_2CO	274
1	$P(OEt)_3$	F	-0.63	Me_2CO	274
1	$Ph_2PCH_2CH_2PPh_2$[a]	Ph	-0.60	CH_2Cl_2	276
1	$Ph_2PCH_2CH_2PPh_2$[a]	Me	-0.67	CH_2Cl_2	276
2	$PMePh_2$	Me	-1.34	MeCN	258
2	PMe_2Ph	Me	-1.40	MeCN	258
2	PMe_3	Me	-1.52	MeCN	258
2	$P(OPh)_3$	Me	-0.87	Me_2CO	274
2	$P(OPh)_3$	F	-0.78	Me_2CO	274
2	$P(OMe)_3$	Me	-1.05	Me_2CO	274
2	$P(OMe)_3$	F	-0.94	Me_2CO	274
2	$Ph_2PCH_2CH_2PPh_2$	Me	-1.10	CH_2Cl_2	276

[a]*Bidentate ligand acting as monodentate*

TABLE 29. *(continued)*

n	L	R	Ep	solvent	reference
2	$Ph_2PCH_2CH_2PPh_2$	Ph	-1.09	CH_2Cl_2	276
2	$Ph_2PCH_2PPh_2$	Me	-1.05	CH_2Cl_2	276
2	$Ph_2PCH_2PPh_2$	Ph	-1.00	CH_2Cl_2	276
3	PEt_2Ph	Me	-1.82	CH_2Cl_2	258
			-1.55	Me_2CO	274
3	PMe_2Ph	Me	-1.91	CH_2Cl_2	258
3	PMe_3	Me	-1.95	CH_2Cl_2	258
3	$P(OPh)_3$	Me	-1.08	Me_2CO	274
3	$P(OPh)_3$	F	-0.95	Me_2CO	274
3	$P(OMe)_3$	Me	-1.35	Me_2CO	274
3	$P(OMe)_3$	F	-1.23	Me_2CO	274
3	$MeC(CH_2PPh_2)_3$	Ph	-1.48	CH_2Cl_2	276

The trend in redox potentials is substantially governed by two factors both involving the metal antibonting character of the LUMO: (*i*) the higher the number of σ-donor phosphorus-containing ligands substituting the π-acceptor carbonyl groups, the more difficult the addition of electrons; (*ii*) the inductive effect of the substituents present either at the apical carbon atom or at the phosphorus-containing molecules operates so that electron-withdrawing groups favour, as well as electron-donating groups disfavour, the reduction processes.

As above said, more subtle information about the effects of the stereochemical assembly of the tricobalt clusters (CO terminal/bridging, L equatorial/axial) upon the redox potentials seem hardly deducible, not only because of their possible dynamic interconversions, but also for the non-thermodynamic significance, in many cases, of such redox potentials due to the instability of the monoanions $[Co_3(CO)_{9-n}(L)_n(C-X)]^-$.

5.1.5. $Co_3(CO)_9(P-Ph)$. The X-ray structure of $Co_3(CO)_9(P-Ph)$ is shown in Figure 143 |277|. The almost equilateral tricobalt triangle (Co-Co averaged, 2.71 Å) is tetrahedrically capped by the phosphorus atom (Co-P averaged, 2.13 Å), with all the carbonyl groups staying in terminal positions.

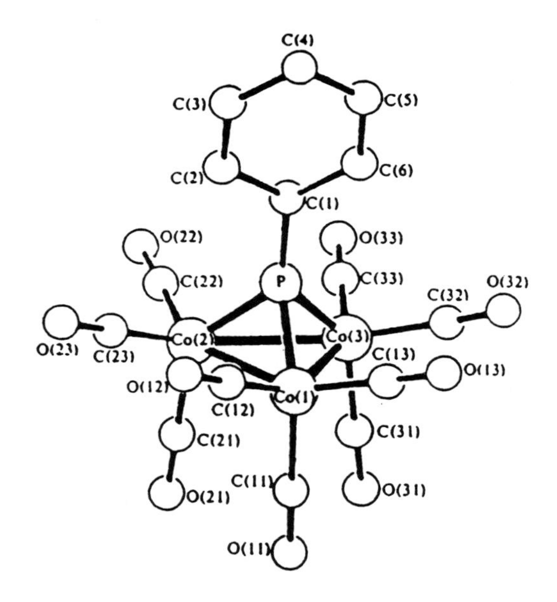

Figure 143. *Crystal structure of $Co_3(CO)_9(PPh)$ (from Ref. 277).*

The core dimensions of the present 49-electron cluster are roughly similar to those of the isoelectronic $Co_3(CO)_9S$, and significantly higher than those of the 48-electron species $Co_3(CO)_9(CX)$ and, more interestingly, $Co_2Fe(CO)_9(PPh)$ |278|. This datum, together with ESR evidence that the unpaired electron does not couple with the apical phosphorus atom, suggested the electron in excess to reside in a metal-metal antibonding level.

In agreement with such electronic picture, $Co_3(CO)_9(PPh)$ undergoes in nonaqueous solutions a chemically reversible one-electron removal. In contrast, reduction processes are framework-destroying |56,261|. Table 30 reports the redox potentials for the more interesting 49/48-electron redox change, also in comparison with the analogous redox change in the monocapped tricobalt clusters previously discussed. It is clearly evident that, starting from 48-electron clusters, the LUMO level of $[Co_3(CO)_9(PPh)]^+$ and $[Co_3(CO)_9(S)]^+$ would seem significantly more easily accessible (and hence lower in energy of about 1 eV) than that of $Co_3(CO)_9(CPh)$. This datum sounds however unrealistic, so that we attribute this difference simply to the favourable coulombic effect of adding one electron to monocationic species, in the case of phosphane and sulfur clusters, with respect to the neutral methylidine species.

TABLE 30. *Formal electrode potentials (in Volt) for the 49/48-electron redox change in monocapped tricobalt clusters*

complex	$E^{\circ\prime}_{49/48-e}$	solvent	reference
$Co_3(CO)_9(PPh)$	$+0.46^{a}$	$C_2H_4Cl_2$	56
$Co_3(CO)_9(PPh)$	$+0.43^{a}$	CH_2Cl_2	261
$Co_3(CO)_9(S)$	$+0.44^{a}$	$C_2H_4Cl_2$	56
$Co_3(CO)_9(CPh)$	-0.63^{b}	CH_2Cl_2	261

[a]*Oxidation step;* [b]*reduction step*

It has once again to be noted that the oxidation $[Co_3(CO)_9(PPh)]^{0/+}$ is electrochemically quasi-reversible. Although the 48-electron monocation $[Co_3(CO)_9(PPh)]^{+}$ has not been isolated, the quasireversibility of the electron transfer allows us to foresee that some significant structural reorganization must take place, likely deeper than the obvious shortening of the Co-Co bond lengths.

A double tetrahedral cluster, of formula and structure depicted in Scheme 7, has been characterized |279|.

SCHEME 7

The connecting benzene ring must exert an effective electronic separation between the two 49-electron Co_3P cores, if this species undergoes, in dichloroethane solution, a chemically reversible, single-stepped two electron oxidation at $E^{\circ\prime} = +0.40$ V |279|.

5.1.6. $Co_3(C_5Me_5)_3(CO)_2$. The first member of a series of bicapped tricobalt cluster is $Co_3(\eta^5\text{-}C_5Me_5)_3(\mu_3\text{-}CO)_2$, the structure of which is shown in Figure 144 |280|.

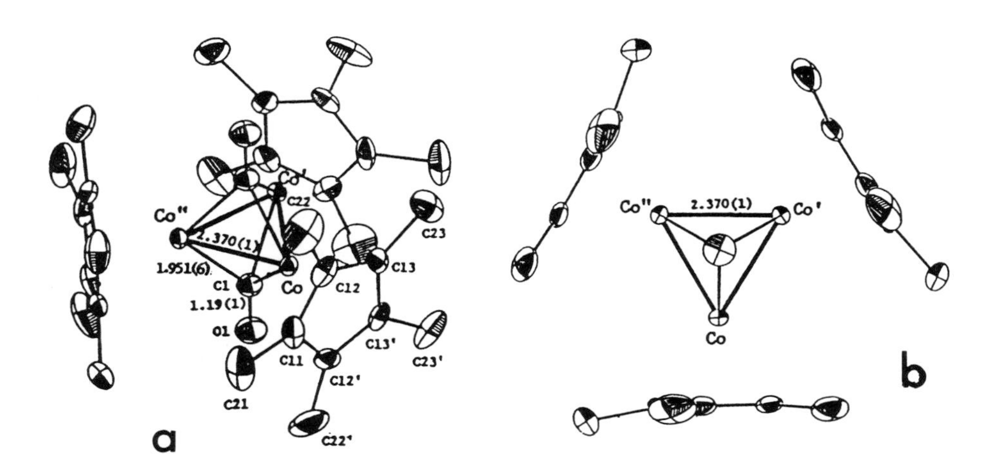

Figure 144. *Perspective view (a) and view down the 3-fold rotational axis (b) of* $Co_3(C_5Me_5)_3(CO)_2$ *(from Ref. 280).*

The equilateral Co_3 triangle is capped, on opposite sides, by a triply bridging carbonyl group. The unmethylated analogue has been also recently structurally characterized |281|.

Such an unsaturated 46-electron cluster undergoes, in tetrahydrofuran solution, a chemically reversible, but electrochemically quasireversible, one-electron reduction at $E°' = -1.13$ V, Figure 145. No further reduction step occurs before the solvent discharge |280|.

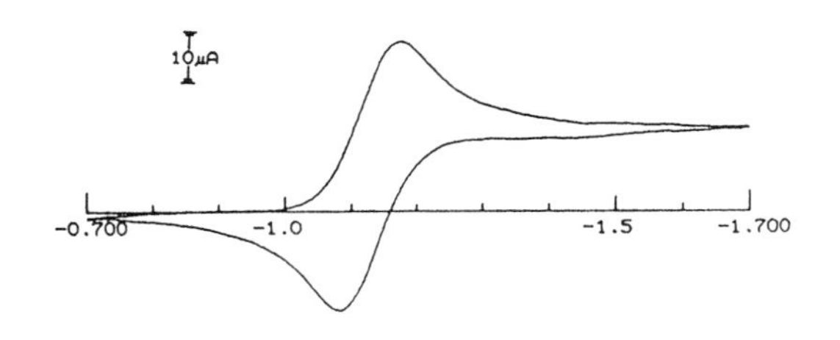

Figure 145. *Cyclic voltammogram recorded at a platinum electrode on a THF solution of* $Co_3(C_5Me_5)_3(CO)_2$. *Scan rate 0.01* Vs^{-1} *(from Ref. 280).*

Theoretical analysis shows that $Co_3(C_5H_5)_3(CO)_2$ possesses two unpaired electrons in two half-filled, doubly degenerated, HOMOs, essentially Co-Co antibonding, but with some bonding contribution towards Co-CO interactions |40,280,282|. It is hence foreseeable that pairing of the added electron causes some significant structural reorganization in the apparently stable monoanion $[Co_3(C_5Me_5)_3(CO)_2]^-$.

5.1.7. $[Co_3(C_6H_6)_3(CO)_2]^+$. The structure of the 48-electron cluster $[Co_3(\eta^6-C_6H_6)_3(\mu_3-CO)_2]^+$ is shown in Figure 146 |282|.

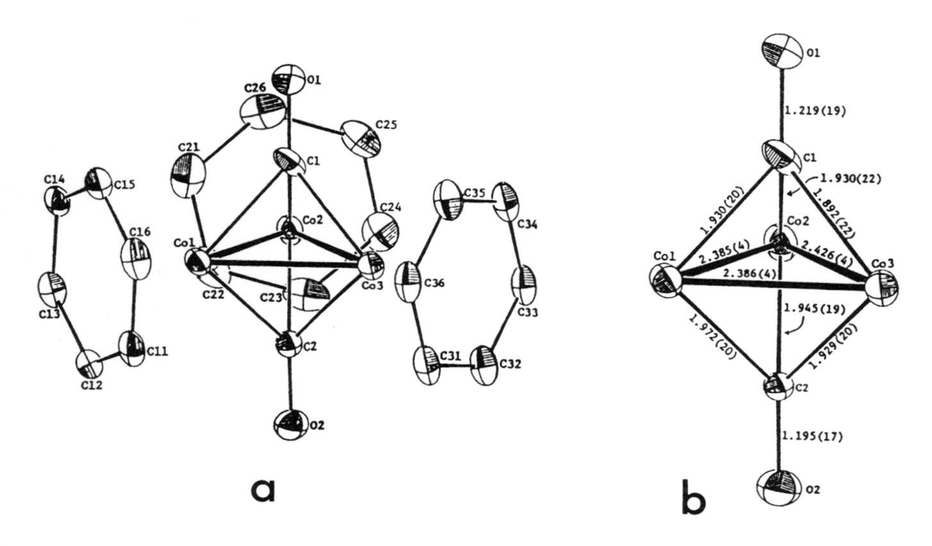

Figure 146. *Perspective view (a) and selected bond lengths (b) in* $[Co_3(C_6H_6)_3(CO)_2]^+$ *(from Ref. 282).*

In comparison with the strictly correlated 46-electron $Co_3(C_5Me_5)_3(CO)_2$, the core dimensions increase; taking into account the different steric effects of C_5Me_5 and C_6H_6 ligands, it has been estimated an averaged elongation of the Co-Co bonds by about 0.05-0.06 Å |282|. This result agrees with the assumption that on going from the 46-electron Co_3C_2 core to the 48-electron Co_3C_2 core, the two electrons fill the Co-Co antibonding HOMO levels, as previously discussed.

$[Co_3(C_6H_6)_3(CO)_2]^+$ undergoes, in acetonitrile solution an apparently reversible, both chemically and electrochemically, one-electron reduction at $E°' = -0.87$ V, Figure 147 |282|.

Even if attempts to obtain the neutral 49-electron $Co_3(C_6H_6)_3(CO)_2$ failed, its finite lifetime is not surprising in view of the stability of the isoelectronic and structurally similar $Ni_3(\eta^5-C_5H_5)_3(\mu_3-CO)_2$, which will be discussed later.

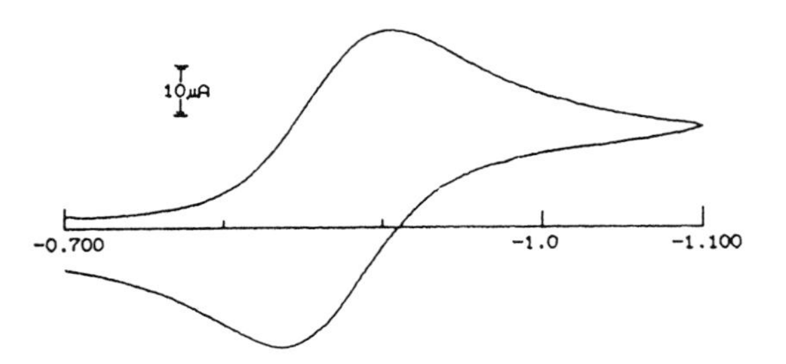

Figure 147. *Cyclic voltammogram recorded at a platinum electrode on a MeCN solution of $[Co_3(C_6H_6)_3(CO)_2]^+$. Scan rate 0.1 Vs^{-1} (from Ref. 282).*

It has been concluded that, in nice agreement with the substantial electrochemical reversibility of the 48/49-electron addition, the neutral $Co_3(C_6H_6)_3(CO)_2$ should display an elongation (by about 0.03 Å) of the tricobalt triangle, without significant geometrical distortions |282|.

5.1.8. $Co_3(C_5R_5)_3(CO)(X)$ (X = O,NR'). One of the two capping atoms of the previously discussed series $Co_3(C_5R_5)_3(CO)_2$ can be substituted by different ligands, such as O or NR'.

Figure 148 illustrates the molecular structure of the 48-electron $Co_3(\eta^5\text{-}C_5H_5)_3(\mu_3\text{-}CO)(\mu_3\text{-}O)$ |283|, isostructural with the isoelectronic $Co_3(\eta^5\text{-}C_5H_5)_3(\mu_3\text{-}CO)(\mu_3\text{-}S)$ |284|.

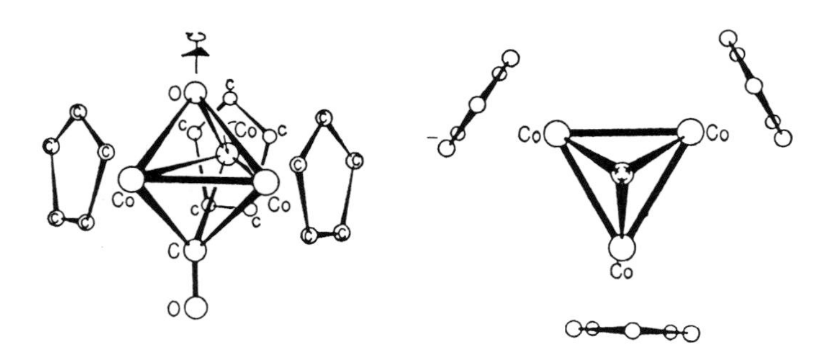

Figure 148. *X-Ray structure of $Co_3(C_5H_5)_3(CO)(O)$ (from Ref. 283).*

The equilateral metallic triangle (Co-Co, 2.36 Å) is symmetrically triply bridged by one carbonyl group and one oxygen atom, respectively [Co-O, 2.00 Å; $Co\text{-}C_{(carbonyl)}$, 1.78 Å].

It has been briefly reported that the methylated analogue $Co_3(C_5Me_5)_3(CO)(O)$ undergoes, in dichloromethane solution, a substantially chemically reversible oxidation at $E°' = -0.09$ V. The number of electrons involved in such electron transfer has not been determined |280|.

More precise redox informations have been presented for a number of 48-electron nitrene derivatives of formula $Co_3(\eta^5\text{-}C_5H_{5-n}Me_n)_3(\mu_3\text{-}CO)(\mu_3\text{-}NR)$. Figure 149 shows the usual bicapped structure of two members of this series |285|.

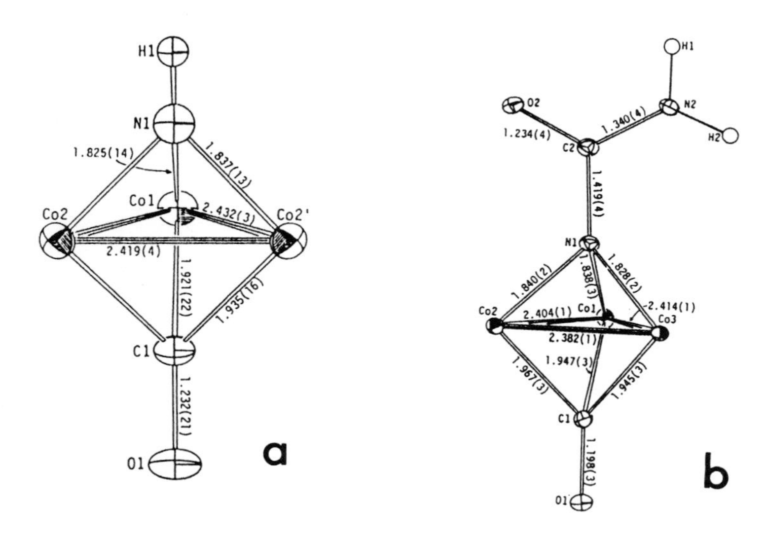

Figure 149. *Perspective views and bond lengths of the (N)Co_3(C) core in: (a) $Co_3(C_5Me_5)_3(CO)(NH)$; (b) $Co_3(C_5H_4Me)_3(CO)[NC(O)NH_2]$ (from Ref. 285).*

Most of the present clusters [(n = 0,1; R = $SiMe_3$), (n = 0,1; R = H), (n = 1; R = $C(O)NH_2$)] exhibit both simultaneous two-electron oxidation and simultaneous two-electron reduction. In contrast, $Co_3(C_5Me_5)_3(CO)(NH)$ displays two distinct one-electron oxidations and a one-electron reduction |286|. This redox behaviour is representatively illustrated in Figure 150.

Table 31 reports the formal electrode potentials for such redox changes.

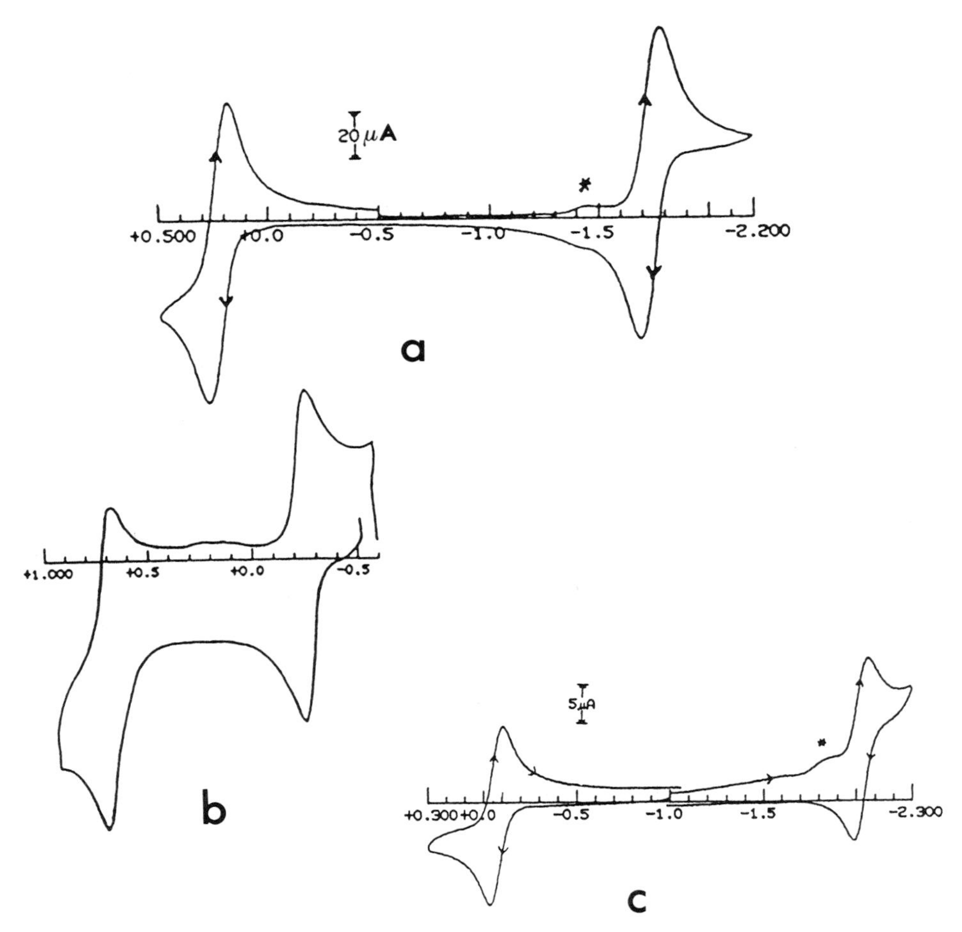

Figure 150. *Cyclic voltammograms recorded at a platinum electrode on: (a) THF solution of $Co_3(C_5H_4Me)_3(CO)[NC(O)NH_2]$; (b) CH_2Cl_2 solution of $Co_3(C_5Me_5)_3(CO)(NH)$; (c) THF solution of $Co_3(C_5Me_5)_3(CO)(NH)$. Scan rate 0.1 Vs^{-1} (from Ref. 286).*

These redox changes appear generally chemically reversible, but electrochemically quasireversible. Since single-stepped two-electron transfers are usually found in those redox changes in which the second-electron transfer is thermodynamically favoured with respect to the first one, the present redox propensity is attributed to the fact that removal of two electrons from the HOMO, or addition of two electrons to the LUMO, leads to a greater stabilization, and hence minor geometrical distortions, than that operated by single one-electron transfers. On the other hand, increased steric effects operated by C_5Me_5 rings likely change this picture, favouring lower geometrical distortions through one-electron redox changes. In this connection, the 47-electron [$Co_3(\eta^5-C_5Me_5)_3(\mu_3-CO)(\mu_3-NH)]^+$ has been X-ray characterized |287|. With respect to the parent $Co_3(\eta^5-C_5Me_5)(\mu_3-CO)(\mu_3-NH)$, removal of one electron from the two filled (somewhat Co-Co antibonding) HOMOs causes a shortening of about 0.02 Å of the metal-metal mean length (from 2.43 Å to 2.41 Å) in the tricobalt triangle (now, slightly less equilateral; in fact, the edges vary from 2.42 Å, 2.43 Å, 2.43 Å in the neutral species to 2.40 Å, 2.40 Å, 2.42 Å in the monocation). Concomitantly, in accordance with the somewhat Co-CO bonding character of the HOMOs, the Co-CO mean length elongates from 1.93 Å to 1.98 Å. Finally, the Co-NH mean length shortens from 1.83 Å to 1.81 Å.

TABLE 31. *Redox potentials (in Volt) for the redox changes exhibited by* $Co_3(C_5H_{5-n}Me_n)_3(CO)(NR)$ *|286|*

n	R	E°' (46/47-e)	E°' (47/48-e)	E°' (48/49-e)	E°' (49/50-e)	solvent
0	$SiMe_3$	+0.40	+0.40	-1.60	-1.60	THF
1	$SiMe_3$	+0.33	+0.33	-1.68	-1.68	THF
0	H	+0.31	+0.31	-1.63	-1.63	THF
1	H	+0.25	+0.25	-1.76	-1.76	THF
5	H		-0.07	-2.03		THF
5	H	+0.75	-0.21			CH_2Cl_2
1	$C(O)NH_2$	+0.22	+0.22	-1.73	-1.73	THF

5.1.9. $Co_3(C_5H_5)_3(CO)_3$. The Co_3-monocapped structure of $(\eta^5-C_5H_5)_3Co_3(\mu_3-CO)(\mu-CO)_2$ is shown in Figure 151 |288|. The two CO-bridged edges of the metallic triangle (averaged Co-Co length, 2.44 Å) are shorter than the unbridged one (Co-Co, 2.52 Å). The $Co-C_{(apical)}$ distance is 1.96 Å in average.

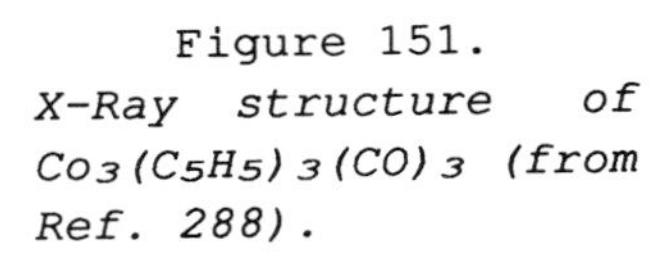

Figure 151.
X-Ray structure of $Co_3(C_5H_5)_3(CO)_3$ (from Ref. 288).

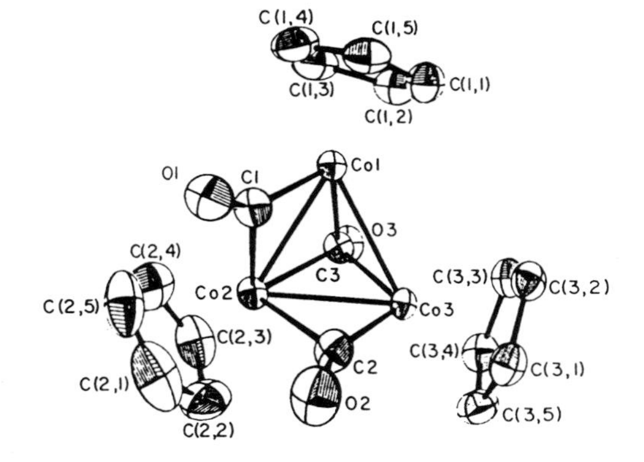

As illustrated in Figure 152, $Co_3(C_5H_5)_3(CO)_3$ undergoes two subsequent one-electron reductions, only the first one having features of chemical reversibility ($E°'_{0/-}$ = -1.03 V; $Ep_{-/2-}$ = -1.9 V). Really, even the first one-electron addition, as deducible from the inset, is complicated at ambient temperature by following chemical reactions, which proved to afford $[(C_5H_5)_2Co_2(CO)_2]^-$. The lifetime of the primarily electrogenerated monoanion $[(C_5H_5)_3Co_3(CO)_3]^-$ is estimated to be around 2-3 sec, at 25 °C |289|.

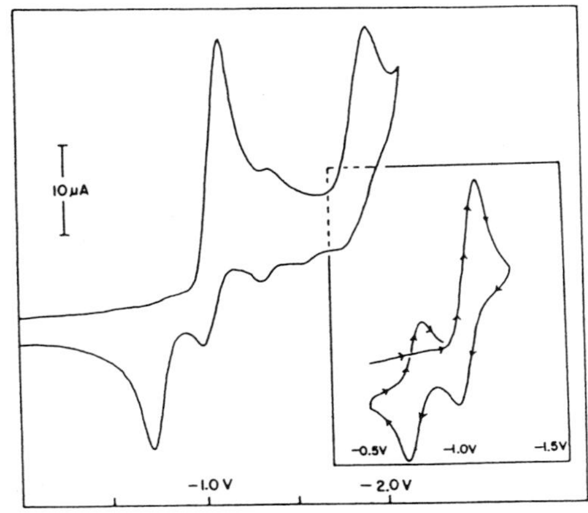

Figure 152. *Cyclic voltammograms recorded at a platinum electrode on a THF solution of $Co_3(C_5H_5)_3(CO)_3$. Scan rate 0.05 Vs^{-1}. Ambient temperature (from Ref. 289).*

5.2. Tetranuclear Binary Complexes

5.2.1. $Co_4(CO)_{12}$. Despite the difficulties imposed by disorder, the crystal structure of $Co_4(CO)_{12}$ reveals a tetrahedral tetrametal assembly, with nine terminal and three bridging CO groups, these latter disposed in the basal plane, Figure 153 |290-294|.

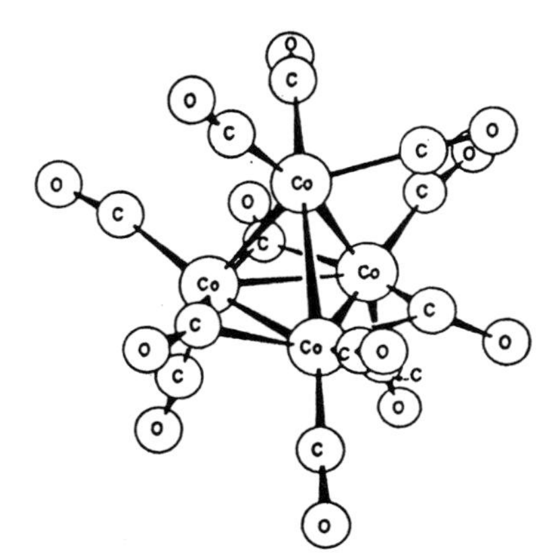

Figure 153.
Molecular structure of $Co_4(CO)_{12}$ (from Ref. 291).

Of the nine terminal carbonyls, three belong to the apical cobalt, the remaining three couples are bound to the basal cobalt atoms, each one both in axial and equatorial positions. The mean Co-Co bond distance is 2.49 Å.

Even if some controversy exists, it is accepted that such a geometry holds also in solution |294|.

As shown in Figure 154, the 60-electron system $Co_4(CO)_{12}$ undergoes in dichloroethane solution: (*i*) a first quasireversible one-electron reduction at $E°' = -0.30$ V ($\Delta Ep = 100$ mV at 0.1 Vs^{-1}) complicated by slow decomposition of the primarily electrogenerated monoanion $[Co_4(CO)_{12}]^-$ to $[Co(CO)_4]^-$, as put in evidence by the spurius reoxidation peak at +0.20 V present in the reverse scan; (*ii*) a second irreversible one-electron reduction at Ep = -1.38 V; (*iii*) an irreversible four-electron oxidation at Ep = +1.40 V |295,296|.

A similar behaviour holds in dichloromethane solution ($E°'_{0/-} = -0.36$ V, $\Delta Ep_{0.1\ Vs^{-1}} = 104$ mV; $Ep_{-/2-} = -1.00$ V). A rough evaluation of the lifetime of $[Co_4(CO)_{12}]^-$ at ambient temperature gives $t_{1/2} \cong 7$ s |297|. By contrast, in tetrahydrofuran solution, $Co_4(CO)_{12}$ undergoes: (*i*) an irreversible one-electron reduction at

Ep = -0.10 V; (*ii*) an irreversible oxidation at Ep = +1.7 V |294| (see footnote b in Table 33). This means that in such solvent decomposition of $[Co_4(CO)_{12}]^-$ is very fast.

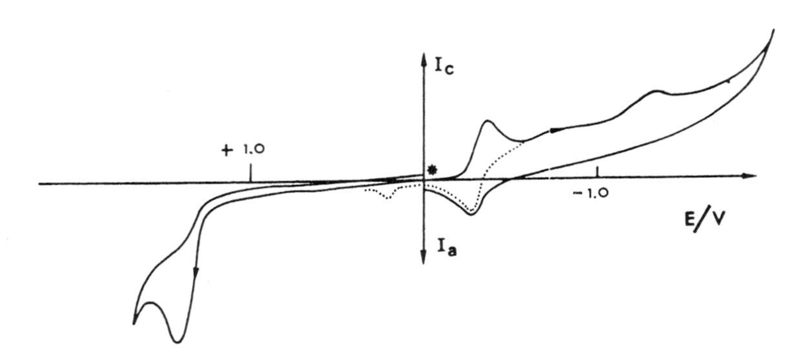

Figure 154. *Cyclic voltammogram recorded at a platinum electrode on a $C_2H_4Cl_2$ solution of $Co_4(CO)_{12}$ (tetra-n-hexylammonium perchlorate supporting electrolyte). Scan rate 0.1 Vs^{-1} (from Ref. 296).*

The most interesting redox change hence results in $Co_4(CO)_{12}/[Co_4(CO)_{12}]^-$. Neglecting the behaviour in tetrahydrofuran, in which the solvent itself can cause fragmentation |298|, the relative stability of $[Co_4(CO)_{12}]^-$ must be correlated to the fact that the added electron enters an antibonding orbital, mainly centred along the tetrahedral faces |298,299|. Unfortunately, ESR data on electrogenerated $[Co_4(CO)_{12}]^-$ do not give informations on the relevant structural reorganization |295|.

5.3. Tetranuclear Mixed-Ligand Complexes

5.3.1. $Co_4(CO)_{12-2n}(Ph_2PCH_2PPh_2)_n$ (n = 1,2). As we shall see, a number of ligands are able to substitute progressively CO groups in $Co_4(CO)_{12}$; one of these is bis(diphenylphosphino)methane, which affords the clusters $Co_4(CO)_{10}(Ph_2PCH_2PPh_2)$ and $Co_4(CO)_8(Ph_2PCH_2PPh_2)_2$, respectively |295|. On the basis of spectroscopic evidence, such two clusters have been assigned a tetrahedral geometry similar to that of $Co_4(CO)_{12}$, with the diphosphine ligand substituting terminal CO groups, while leaving intact the bridging basal ones, Figure 155 |295|. By analogy with $Co_4(CO)_{10}[PMe_2(CH_2CHCH_2)]_2$ |300| in the former and $Rh_4(CO)_8(Ph_2PCH_2PPh_2)_2$

|294| in the latter case, it is likely that in both the species, the unsubstituted carbonyl groups in the basal plane are the equatorial ones.

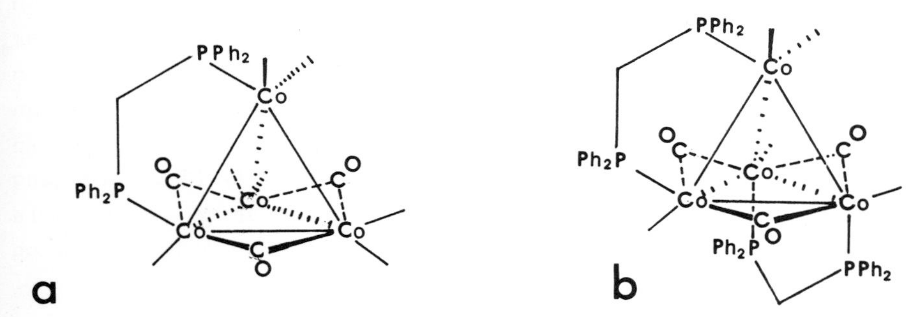

Figure 155. *Schematic representation of the molecular structure assigned to: (a) $Co_4(CO)_{10}(Ph_2PCH_2PPh_2)$; (b) $Co_4(CO)_8(Ph_2PCH_2PPh_2)_2$.*

Figure 156 compares the redox ability of these two substituted derivatives with that of the parent $Co_4(CO)_{12}$ (previously shown in Figure 154) |295|.

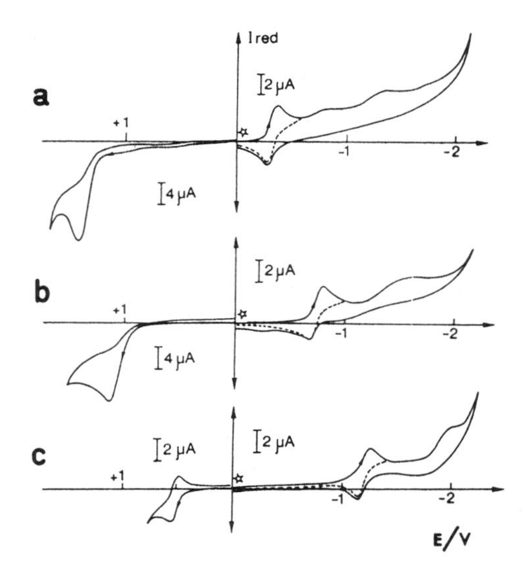

Figure 156. *Cyclic voltammograms recorded at a platinum electrode on a $C_2H_4Cl_2$ solution of: (a) $Co_4(CO)_{12}$; (b) $Co_4(CO)_{10}(Ph_2PCH_2PPh_2)$; (c) $Co_4(CO)_8(Ph_2PCH_2PPh_2)_2$. Scan rate: 0.1 Vs^{-1} (from Ref. 295).*

Like $Co_4(CO)_{12}$, these 60-electron substituted clusters undergo a first quasireversible one-electron reduction step, complicated by subsequent slow decomposition of the electrogenerated monoanion; a second irreversible one-electron reduction is also present. A significant difference holds as far as the anodic behaviour of $Co_4(CO)_8(Ph_2PCH_2PPh_2)_2$ is concerned. In fact, while $Co_4(CO)_{12}$ and $Co_4(CO)_{10}(Ph_2PCH_2PPh_2)$ undergo an irreversible four-electron oxidation, $Co_4(CO)_8(Ph_2PCH_2PPh_2)_2$ undergoes a quasireversible one-electron oxidation, leading however to the not fully stable monocation $[Co_4(CO)_8(Ph_2PCH_2PPh_2)_2]^+$. Table 32 summarizes the redox potentials of such electrochemical steps.

TABLE 32. *Redox potentials for the electrode activity exhibited by the series $Co_4(CO)_{12-2n}(Ph_2PCH_2PPh_2)_n$ in 1,2-dichloroethane solution |295|*

complex	oxidation		reductions		
	Ep (V)	ΔEp^a (mV)	$E°'_{0/-}$ (V)	ΔEp^a (mV)	$Ep_{-/2-}$ (V)
$Co_4(CO)_{12}$	+1.40[b]		-0.30	100	-1.38
$Co_4(CO)_{10}(Ph_2PCH_2PPh_2)$	+1.03[b]		-0.73	100	-1.58
$Co_4(CO)_8(Ph_2PCH_2PPh_2)_2$	+0.51[c]	65	-1.22	110	-2.00

[a]*Measured at 0.1 Vs^{-1};* [b]*peak potential value for an irreversible four-electron process;* [c]*formal electrode potential for a quasi-reversible one-electron process*

As matter of fact, the substitution of CO groups for diphosphinomethane makes the reduction progressively more difficult (*i.e.*, the energy of the LUMO more and more increases), without appreciably stabilizing the relative monoanions; correspondingly, the oxidation becomes easier and easier. Interestingly, introduction of two diphosphino molecules stabilizes to a good extent the monocation $[Co_4(CO)_8(Ph_2PCH_2PPh_2)]^+$. Either for the monoanions, or for the monocation, there are not sufficient data to judge the structural reorganizations accompanying the involved redox changes

5.3.2. $Co_4(CO)_9(\eta\text{-}C_6H_5Me)$. Substitution of the three terminal CO groups of the apical cobalt in $Co_4(CO)_{12}$ with benzenoid molecules affords a series of η-arene-tetracobalt complexes |298,301,302|, the structure of which can be exemplified by that of $Co_4(CO)_9(\eta\text{-}C_6H_4Me_2)$ shown in Figure 157 |301|.

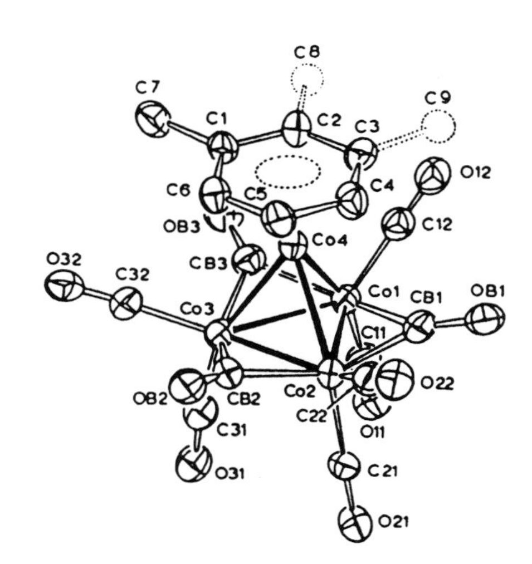

Figure 157. *Perspective view of $Co_4(CO)_9(\eta\text{-}C_6H_4Me_2)$ (from Ref. 301).*

Also in this case, the basal bridging CO groups remain untouched. As we shall discuss, the fact that the averaged Co-Co bond length of 2.47 Å is apparently only slightly shorter than that of $Co_4(CO)_{12}$ (2.49 Å) is indeed of some relevancy. The distance of the apical cobalt to the centre of the capping xylene ring is 1.60 Å (not very different from 1.64 Å for the benzene-capped molecule |301|), notably shorter than that of the capping terminal CO groups (1.88 Å) in $Co_4(CO)_{12}$.

The only electrochemical report concerns $Co_4(CO)_9(\eta\text{-}C_6H_5Me)$ |298|. As shown in Figure 158, in dichloromethane solution, it undergoes a first quasireversible one-electron reduction ($E°'$ = -0.57 V (see footnote b in Table 33), ΔEp = 95 mV at 0.2 Vs^{-1}), generating the likely stable monoanion $[Co_4(CO)_9(\eta\text{-}C_6H_5Me)]^-$ (data at the longer times of electrolysis are lacking), as well as a second irreversible two-electron reduction (Ep = -1.18 V). An irreversible oxidation at Ep = +1.2 V is also present.

Such redox behaviour, which is quite reminiscent of that of $Co_4(CO)_{12}$, agrees with a rather similar electronic configuration of their HOMO-LUMO orbitals |298|. It seems that the toluene capping, while making the reduction more difficult by about 0.2 V, slightly improves the kinetic stability of the monoanion. This has been ascribed |298| to the fact that M(η-arene) fragments favour metal-metal bonding to a higher extent than the isolobal $M(CO)_3$ fragment |303|. This higher tetrametal cohesion may be also supported by the fact that the mean $Co_{(apical)}$-$Co_{(basal)}$ distance is 2.50 Å in $Co_4(CO)_{12}$ *vs.* 2.48 Å in $Co_4(CO)_9(\eta$-arene), as well as that the mean basal Co-Co distance is 2.48 Å in $Co_4(CO)_{12}$ *vs.* 2.46 Å in $Co_4(CO)_9(\eta$-arene).

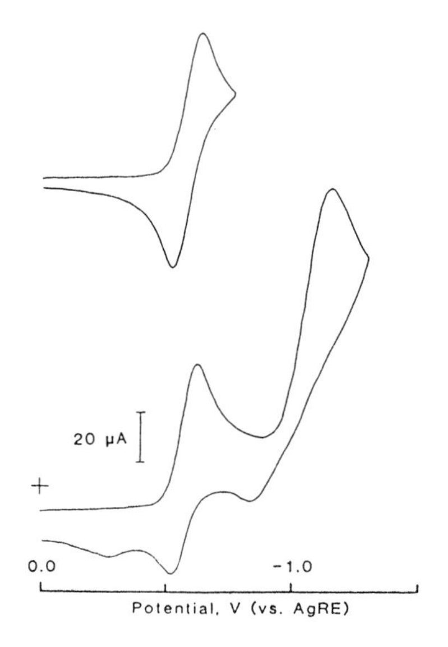

Figure 158.
Cyclic voltammograms recorded at a platinum electrode on a CH_2Cl_2 solution of $Co_4(CO)_9$ (η-C_6H_5Me). Scan rate: 0.2 Vs^{-1} (from Ref. 298).

5.3.3. $Co_4(CO)_{9-n}[HC(PPh_2)_3](L)_n$ (L = PR_3; n = 0,1). The tripodal ligand 1,1,1-tris(diphenylphosphino)methane, $HC(PPh_2)_3$, is able to substitute three terminal CO groups in the basal plane (in particular the axial ones) of $Co_4(CO)_{12}$, one for each cobalt atom, giving rise to $Co_4(CO)_9[HC(PPh_2)_3]$ |304,305|.

Figure 159 shows the molecular structure of $Co_4(CO)_9[HC(PPh_2)_3]$ |306|. It can be seen that the tripodal ligand caps one face of the tetrahedral Co_4 assembly. In the tetrametal core, the mean Co-Co bond length is 2.50 Å, near coincident to that of $Co_4(CO)_{12}$, but the difference between the mean bond lengths of $Co_{(apical)}$-$Co_{(basal)}$ (2.54 Å) and $Co_{(basal)}$-$Co_{(basal)}$ (2.46 Å) is notably more marked than in the fully carbonylated precursor (0.08 Å vs. 0.02 Å).

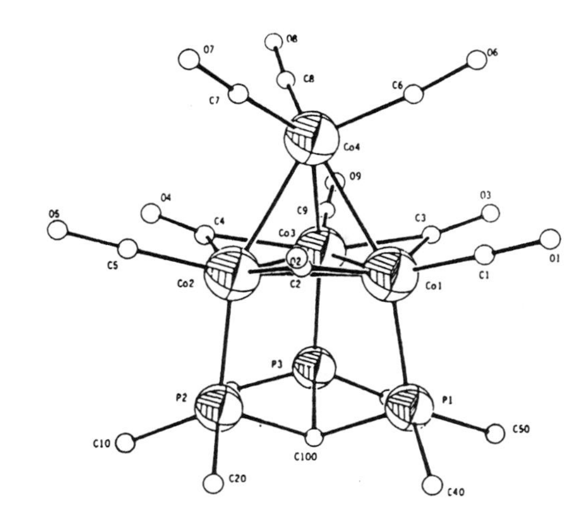

Figure 159. *Molecular structure of $Co_4(CO)_9[HC(PPh_2)_3]$ (from Ref. 306).*

Figure 160 illustrates the redox behaviour of $Co_4(CO)_9[HC(PPh_2)_3]$ in dichloromethane solution |298|.

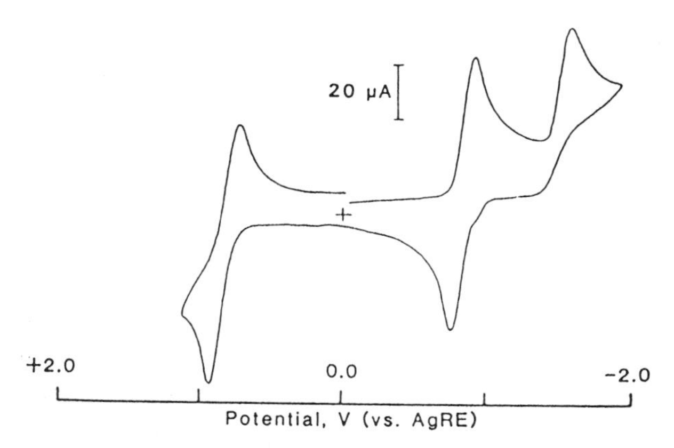

Figure 160. *Cyclic voltammogram recorded at a platinum electrode on a CH_2Cl_2 solution of $Co_4(CO)_9[HC(PPh_2)_3]$. Scan rate: 0.2 Vs^{-1} (from Ref. 298).*

It undergoes a quasireversible one-electron reduction, chemically reversible, followed by a second irreversible one-electron reduction; a quasireversible one-electron oxidation, complicated by slow decomposition of the electrogenerated monocation, is also present. Table 33 summarizes the redox potentials for these electrochemical steps, both in dichloromethane |298| and dichloroethane |296| solution. As expected, the presence of the tripodal phosphorus ligand, which donates electron density to the tetrametal moiety, makes the reduction more difficult as well as the oxidation easier with respect to the fully carbonylated $Co_4(CO)_{12}$, both the HOMO and LUMO orbital being essentially centred on the Co_4 core |298|.

The difference between the formal electrode potentials of the redox changes $[Co_4(CO)_9\{HC(PPh_2)_3\}]^+/Co_4(CO)_9\{HC(PPh_2)_3\}$ and $Co_4(CO)_9\{HC(PPh_2)_3\}/[Co_4(CO)_9\{HC(PPh_2)_3\}]^-$ should reflect just the HOMO-LUMO gap. Such an experimental value (1.71 eV) is somewhat different from the value of 1.46 eV theoretically computed for the model compound $Co_4(CO)_9[HC(PH_2)_3]$ |298|, but it is very close to the value of 1.63 eV obtained from the electronic absorption spectrum |298|.

Once again, it appears that the capping by the tridentate phosphine ligand, like in the arene-capping, increases the tetrametal cohesion in these 60-electron tetrahedral clusters so to stabilize the corresponding monoanion, *i.e.*, to buffer the addition of one electron in an antibonding orbital. The likely occurring geometrical effects of this one-electron addition are so far unknown; however the stability of the monoanion is a good omen for its X-ray structural determination.

$Co_4(CO)_9[HC(PPh_2)_3]$ undergoes both thermal |307,308| and redox-induced |296| carbonyl substitution by phosphorus donor ligands affording the series $Co_4(CO)_8[HC(PPh_2)_3](PR_3)$. The spectroscopic inference that the phosphorus donor ligands substitute one terminal CO group of the apical cobalt atom is supported by the X-ray structure of $Co_4(CO)_8[HC(PPh_2)_3](PMe_3)$, Figure 161 |307|. Within the tetrahedral metallic core, the Co-Co distance averages 2.51 Å, the $Co_{(apical)}$-$Co_{(basal)}$ and $Co_{(basal)}$-$Co_{(basal)}$ mean lengths being of 2.53 Å and 2.48 Å, respectively. This means that the tetrametal cohesion is substantially similar to that of $Co_4(CO)_9[HC(PPh_2)_3]$.

Concerning the redox behaviour of the family $Co_4(CO)_8[HC(PPh_2)_3](PR_3)$, it has been briefly reported that the presence of the phosphine ligand, furtherly adding electron density to the metal core, makes the reduction more difficult by about 0.15-0.30 V (depending upon the electron-donating ability of

TABLE 33. *Redox potentials for the redox changes exhibited by the series* $Co_4(CO)_{9-n}|HC(PPh_2)_3|(L)_n$

complex	$E°'_{+/0}$ (V)	ΔEp (mV)	$E°'_{0/-}$ (V)	ΔEp (mV)	$Ep_{-/2-}$ (V)	solvent	reference
$Co_4(CO)_9|HC(PPh_2)_3|$	+0.82	97[a]	-0.89	100[a]	-1.62	$1,2-C_2H_4Cl_2$	296
	+0.96[b]	100[c]	-0.76[b]	100[c]	-1.47[b]	CH_2Cl_2	298
$Co_4(CO)_8|HC(PPh_2)_3|(PMe_3)$			-1.22[d]			$1,2-C_2H_4Cl_2$	296
$Co_4(CO)_8|HC(PPh_2)_3||P(Me_2Ph)|$			-1.18[d]			THF	296
$Co_4(CO)_8|HC(PPh_2)_3||P(MePh_2)|$			-1.12[d]			THF	296
$Co_4(CO)_8|HC(PPh_2)_3||(PPh_3)$			-1.08[d]			$1,2-C_2H_4Cl_2$	296

[a] *Measured at 0.1* Vs^{-1}; [b] *potential values arbitrarily referenced to S.C.E.. In Ref. 298 potential values are quoted vs. a pseudoreference Ag electrode having a potential of +0.31 V vs. S.C.E.. Indeed, the fact that ferrocene is reported to oxidize at +0.47 V makes doubted this assignment, since we find that ferrocene is oxidized at +0.49 V vs. S.C.E. in* CH_2Cl_2 *solution,* [c] *measured at 0.2* Vs^{-1}; [d] *peak potential value for processes of unknown degree of reversibility*

the phosphine substituent) (see Table 33) |296|. No informations are available on the kinetic stability of the corresponding monoanions.

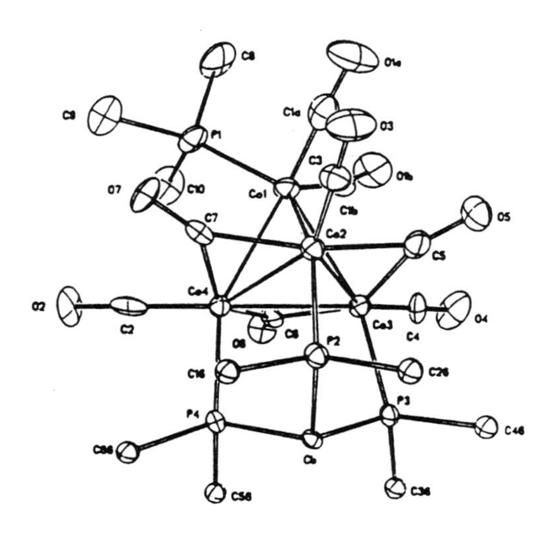

Figure 161.
Perspective view of $Co_4(CO)_8[HC(PPh_2)_3](PMe_3)$ *(from Ref. 307).*

5.3.4. $Co_4(CO)_6[HC(PPh_2)_3](\eta\text{-}C_6X_5Me)$ **(X = H,Me).** The last class of tetrahedral 60-electron tetracobalt clusters is $Co_4(CO)_6[HC(PPh_2)_3](\eta\text{-}C_6X_5Me)$ (X = H,Me), in which the three basal cobalt atoms are capped by the tripodal ligand and the apical cobalt is capped by an arene molecule. In this connection, Figure 162 shows the molecular structure of $Co_4(CO)_6[HC(PPh_2)_3](\eta\text{-}C_6H_5Me)$ |305|.

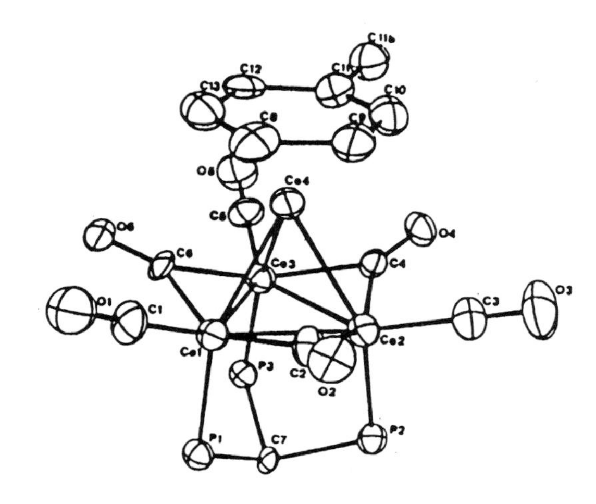

Figure 162.
The molecular structure of $Co_4(CO)_6[HC(PPh_2)_3](C_6H_5Me)$ *(from Ref. 305).*

The averaged Co-Co bond length is 2.46 Å, with the $Co_{(apical)}-Co_{(basal)}$ and $Co_{(basal)}-Co_{(basal)}$ mean distances of 2.47 Å and 2.45 Å, respectively. These data confirm that, with respect to $Co_4(CO)_{12}$, capping of cobalt atoms reinforces the tetrametal cohesion; in particular, in the present case, in which all the cobalt atoms are capped, the shortest metal-metal bonds are observed.

As a consequence of the high electron density present in the tetrametal core, it is expected that the actual clusters add electrons with great difficulty. As shown in Figure 163, this is just the case |296|. In comparison with $Co_4(CO)_9[HC(PPh_2)_3]$, the one-electron reduction occurs at potentials more negative by 0.44 V; analogously, the one-electron oxidation is made much easier. Table 34 summarizes the electrode potentials of such redox changes.

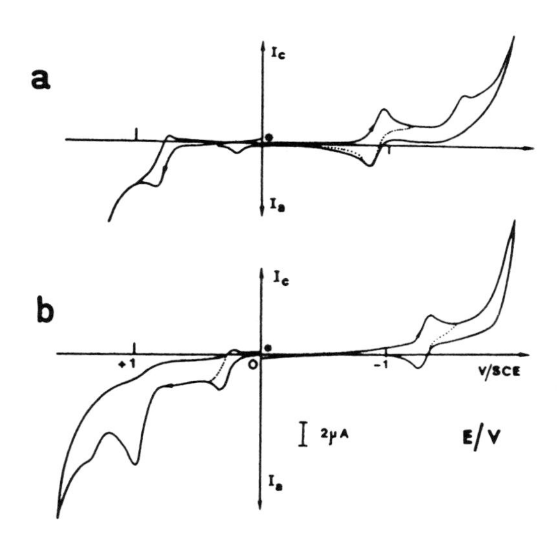

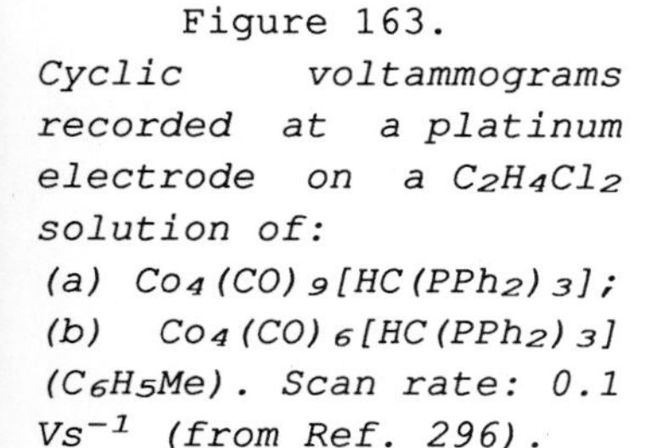
Figure 163. *Cyclic voltammograms recorded at a platinum electrode on a $C_2H_4Cl_2$ solution of: (a) $Co_4(CO)_9[HC(PPh_2)_3]$; (b) $Co_4(CO)_6[HC(PPh_2)_3](C_6H_5Me)$. Scan rate: 0.1 Vs^{-1} (from Ref. 296).*

As in the case of the related clusters previously discussed, the quasireversible one-electron reduction leads to a stable (at least in the cyclic voltammetric timescale) monoanion $[Co_4(CO)_6\{HC(PPh_2)_3\}(\eta\text{-arene})]^-$. In contrast, the corresponding monocation $[Co_4(CO)_6\{HC(PPh_2)_3\}(\eta\text{-arene})]^+$ is short-lived.

The HOMO-LUMO gap computable by electrochemistry (1.6 eV) is close enough to that obtainable from electronic absorption

TABLE 34. *Formal electrode potentials for the redox changes exhibited by the complexes* $Co_4(CO)_6|HC(PPh_2)_3|(\eta\text{-arene})$

complex	$E°'_{+/0}$ (V)	ΔEp (mV)	$E°'_{0/-}$ (V)	ΔEp (mV)	solvent	reference
$Co_4(CO)_6|HC(PPh_2)_3|(C_6H_5Me)$	+0.29	105[a]	-1.33	100[a]	$1,2\text{-}C_2H_4Cl_2$	296
	+0.41[b]	100[c]	-1.19	100[c]	CH_2Cl_2	298
$Co_4(CO)_6|HC(PPh_2)_3|(C_6Me_6)$	+0.47[b,d]		-1.16[b]	120[c]	CH_2Cl_2	298

[a] *Measured at 0.1* Vs^{-1}; [b] *see footnote b in Table 33*; [c] *measured at 0.2* Vs^{-1}; [d] *peak potential value for an irreversible process*

transition energies (1.4 eV), but significantly higher than that theoretically computed from the model $Co_4(CO)_6[HC(PH_2)_3](\eta-C_6H_6)$ (1.1 eV) |298|.

5.3.5. $Co_4(CO)_{10-n}(PPh)_2(L)_n$ (L = monophosphorus donor ligands; n = 0-4). Tetracobalt clusters having a planar Co_4 array are quite common. The first characterized derivatives were $Co_4(CO)_8(\mu_2-CO)_2(\mu_4-E)_2$, E = S,Te |309-311|, in which the two chalcogenide atoms cap above and below, by a quadruple bridge, the planar cobalt rectangle (arising from the presence of bridging and terminal carbonyl groups, which affect the Co-Co bond distances). The insolubility of these compounds has likely made unknown their redox chemistry.

The two capping sulfur atoms can be replaced partially |312| or entirely |311,313-315| by phosphinidene groups, maintaining the planar disposition of the four cobalt atoms. In this connection, Figure 164 shows the X-ray structure of $Co_4(CO)_8(\mu_2-CO)_2(\mu_4-PPh)_2$ |311|.

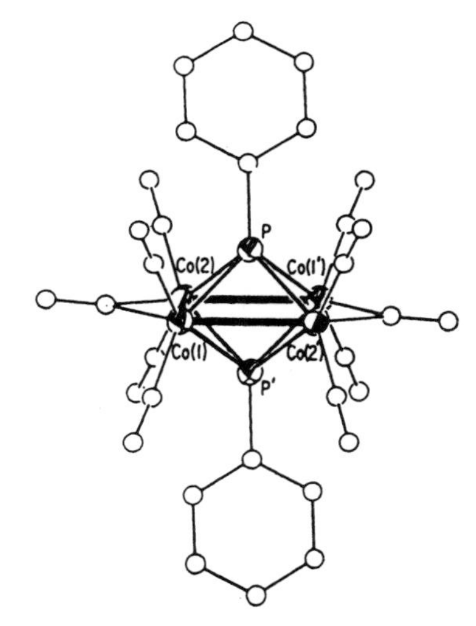

Figure 164. *Molecular structure of $Co_4(CO)_8(CO)_2(PC_6H_5)_2$ (triclinic form) (from Ref. 311).*

The octahedral Co_4P_2 core is based on a tetracobalt rectangle bearing eight terminal and two symmetrical bridging carbonyl groups. The CO-bridged Co-Co bond lengths are significantly shorter (2.52 Å) than the unbridged ones (2.70 Å). The averaged Co-P distance is 2.24 Å. The nonbonding P···P distance is 2.54 Å.

This 64-electron system is redox active, and its redox ability is illustrated in Figure 165 |315|.

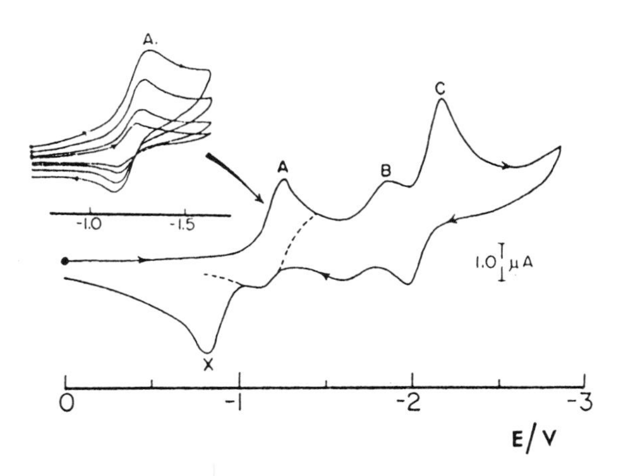

Figure 165.
Cyclic voltammogram of $Co_4(CO)_{10}(PPh)_2$ at a platinum electrode in THF solution. Scan rate 0.2 Vs^{-1} (from Ref. 315).

Three subsequent one-electron reduction processes (A, B, C) are displayed. The first one, quasireversible in character, is assigned to the chemically reversible redox change $Co_4(CO)_{10}(PPh)_2/[Co_4(CO)_{10}(PPh)_2]^-$. The subsequent ones, attributed to the -/2- and 2-/3- steps, respectively, are complicated by relatively slow decomposition of the dianion and trianion. The relevant redox potentials are reported in Table 35 (together with those of the related ligand-substituted complexes).

The significant departure from electrochemical reversibility (ΔEp = 390 mV) likely implies an important geometrical change upon forming the monoanion. However the lack of hyperfine splitting in its ESR spectrum, the only structural measure so far available |315|, does not allow a reliable hypothesis to be advanced. It is however interesting to note that according to the previously cited MO description |110,116| (Section 2.6.4.), these added electrons enter an antibonding metal-metal orbital, so if the kinetic lability of the -/2- step should be conceivable (note that one possible structure for the two-added electron species is a bicapped, one-side open, rectangular array), the finite lifetime of the monoanion is likely unexpected and pleasanty surprising (a possible structure will be discussed in Section 5.3.7.).

$Co_4(CO)_{10}(PPh)_2$ undergoes easily the gradual substitution of up to four CO groups by monophosphorus donor ligands affording the relevant substituted species |313,315-318|, namely: (*i*) *monosubstituted*, $Co_4(CO)_9(PPh)_2(L)$, L = PPh_3, PEt_3, $P(OMe)_3$; (*ii*) *bis-substituted*, $Co_4(CO)_8(PPh)_2(L)_2$, L = PPh_3, $P(OMe)_3$; (*iii*) *tris-substituted*, $Co_4(CO)_7(PPh)_2(L)_3$, L = $P(OMe)_3$; (*iv*) *tetrakis-*

substituted, $Co_4(CO)_6(PPh)_2(L)_4$, L = $P(OMe)_3$. Part of these clusters have been characterized also by X-ray crystallography, and their molecular structures are illustrated in Figure 166.

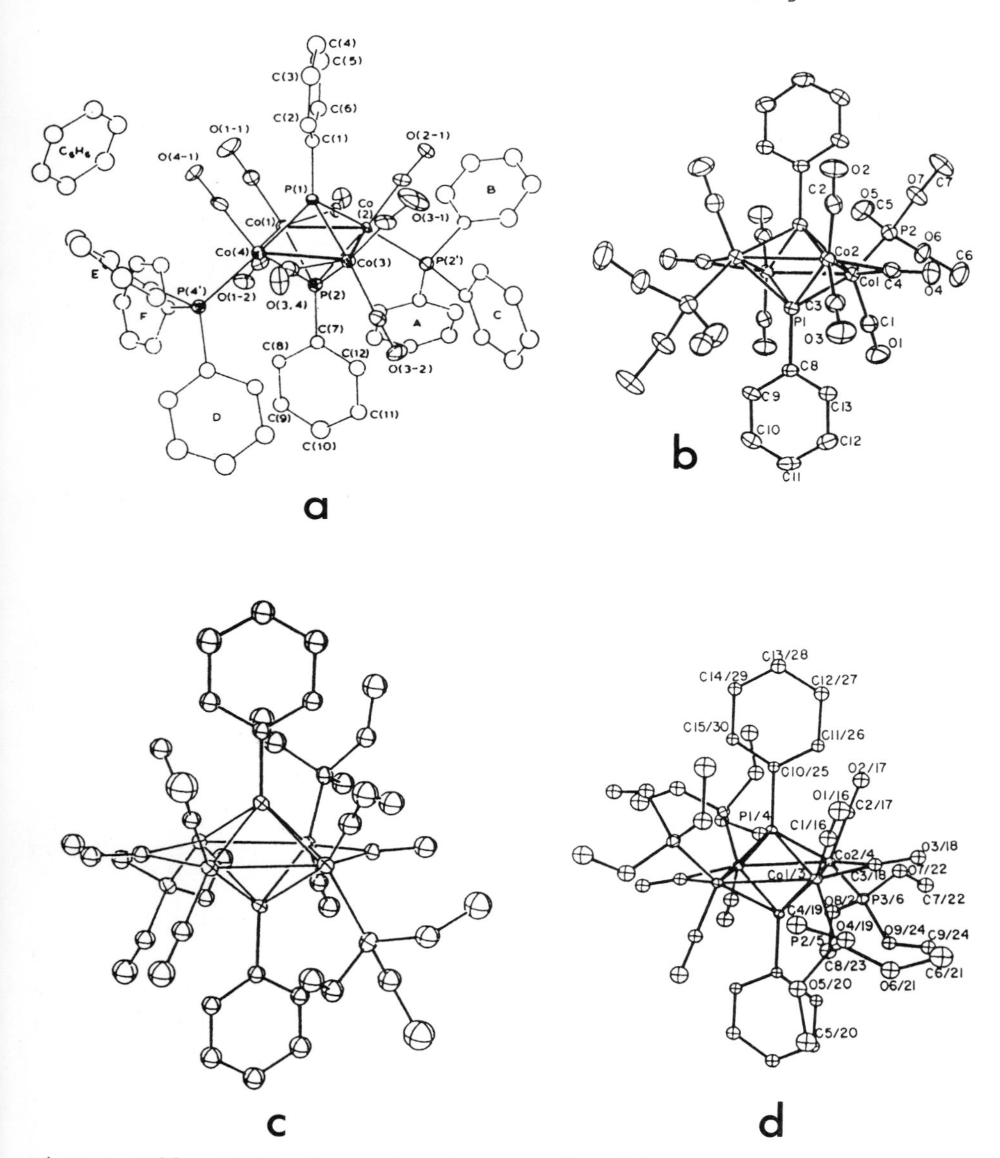

Figure 166. *Molecular structures of: (a)* $Co_4(CO)_8(PPh)_2(PPh_3)_2$ *(from Ref. 313); (b)* $Co_4(CO)_8(PPh)_2[P(OMe)_3]_2$ *(from Ref. 317); (c)* $Co_4(CO)_7(PPh)_2[P(OMe)_3]_3$ *(from Ref. 318); (d)* $Co_4(CO)_6(PPh)_2[P(OMe)_3]_4$ *(from Ref. 317).*

TABLE 35. *Electrochemical parameters for the most significant redox changes exhibited by the series* $Co_4(CO)_{10-n}(PPh)_2(L)_n$ *in tetrahydrofuran solution*

complex	$E°'_{0/-}$ (V)	ΔEp (mV)	$E°_{+/0}$ (V)	ΔEp (mV)	reference
$Co_4(CO)_{10}(PPh)_2$	-1.22	390[a]			315,316
$Co_4(CO)_9(PPh)_2(PPh_3)$	-1.45[b]	145[b,c]	+0.64[b,d]		315
$Co_4(CO)_9(PPh)_2(PEt_3)$	-1.53[b]	120[e]	+0.55[b,d]		315
$Co_4(CO)_9(PPh)_2|P(OMe)_3|$	-1.45[b]	120[e]	+0.64[b,d]		315
$Co_4(CO)_8(PPh)_2(PPh_3)_2$	-1.84[b]	280[b,c]	+0.43[b]	160[b,c]	315
$Co_4(CO)_8(PPh)_2|P(OMe)_3|_2$	-1.80[b]	175[e]	+0.41[b]	55[e]	315
$Co_4(CO)_7(PPh)_2|P(OMe)_3|_3$	-2.00[b]	407[b,c]	-0.03[b]	185[b,c]	315
$Co_4(CO)_6(PPh)_2|P(OMe)_3|_4$	-2.17[f]	185[e]	-0.14[f]	60[e]	315

[a] *Measured at 0.5* Vs^{-1}; [b] *measured at -40°C;* [c] *measured at 0.2* Vs^{-1}; [d] *peak potential value;* [e] $E_{3/4}-E_{1/4}$ *value from the analysis of the wave obtained by cyclic voltammetry at a microelectrode;* [f] *measured at -70°C*

It can be seen that in all cases, the substitution occurs at the terminal carbonyl groups of different cobalt atoms so that the graded four substitution follows the order: 1; 1,3; 1,3,3; 1,3,2,4.

In addition, the phosphorus donor ligands assume a configuration such as to keep the maximal interligand separation |317|. Table 36 summarizes the most significant bond lengths within the Co_4P_2 cores. Except for the bis-substituted triphenylphosphine derivative, no gross variation occurs with respect to the fully carbonylated procursor. In the case of the triphenylphosphine substituted complex, the remarkable elongation of both the unbridged Co-Co bonds and the Co-P$_{(phosphinidene)}$ distance (as well as a slight puckering of the tetrametal plane) is attributed to the intramolecular steric repulsion between the bulky phenyl rings of phosphine ligands and likely the phenyl rings of the capping phosphinidene units. It is hence evident that in the presence of the less steric encumbering tris(phosphite) substituents such repulsions are notably attenuated. Please note that in the bis(phosphite) complexes, the two phosphorus ligands assume a 1,3-*trans* configuration, whereas in the bis(phosphine) complex they assume a 1,3-*cis* position.

Let us now examine the electrochemical behaviour of these substituted species.

As shown in Figure 167, the monosubstituted $Co_4(CO)_9(PPh)_2(PPh_3)$ undergoes a first quasireversible one-electron reduction step (peak A), complicated by decomposition of the monoanion at room temperature (Figure 167a), but uncomplicated at -40°C (Figure 167b), followed by a second irreversible reduction (peak B in Figure 167c). In addition, an anodic process, peak C, complicated by decomposition of the monocation even at low temperature, is also present |315|. Such a behaviour is typical for all the studied monosubstituted species.

Concerning the bis-substituted derivatives, their redox propensity is exemplified by the cyclic voltammograms reported in Figure 168, which refer to $Co_4(CO)_8(PPh)_2(PPh_3)_2$. Also in this case, both the 0/- and 0/+ redox changes are complicated by decomposition of the primarily electrogenerated species at room temperature. Decreasing the temperature more and more stabilizes both the monoanion and the monocation.

TABLE 36. *Significant bond lengths in the* Co_4P_2 *octahedral core of the family* $Co_4(\mu_2\text{-}CO)_2(CO)_{8-n}(PPh)_2(L)_n$

complex	Co-Co (CO bridged)	Co-Co (unbridged)	Co-P	P...P	reference
$Co_4(CO)_{10}(PPh)_2$	2.52	2.70	2.24	2.54	311, 313
$Co_4(CO)_8(PPh)_2(PPh_3)_2$	2.55	2.78	2.29[a]; 2.25[b]	2.54	313
$Co_4(CO)_8(PPh)_2|P(OMe)_3|_2$	2.52	2.71	2.25[a]; 2.25[b]	2.53	317
$Co_4(CO)_7(PPh)_2|P(OMe)_3|_3$	2.52	2.73[c]	2.26[d]	2.56[d]	318
$Co_4(CO)_6(PPh)_2|P(OMe)_3|_4$	2.52	2.73	2.25	2.55	317

[a] *Distance from Co atoms flanked by P-donor ligand;* [b] *distance from Co atoms bearing only carbonyl groups;* [c] *averaged distance (from two isomeric molecules) between* $|P(OMe)_3|Co\text{-}Co|P(OMe)_3|$ *and* $|P(OMe)_3|Co\text{-}Co(CO)$ *bond lengths;* [d] *averaged distance from two isomeric molecules*

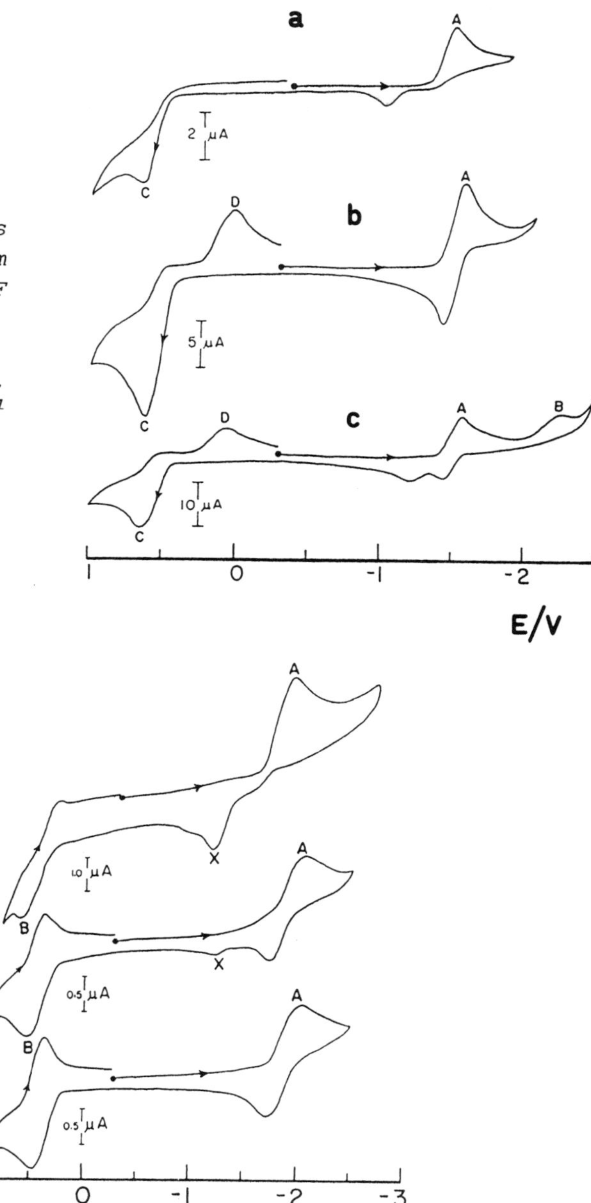

Figure 167. *Cyclic voltammograms recorded at a platinum electrode on a THF solution of $Co_4(CO)_9(PPh)_2(PPh_3)$. (a) 25°C; (b,c) -40°C. Scan rate 0.2 Vs^{-1} (from Ref. 315).*

Figure 168. *Cyclic voltammograms recorded at a platinum electrode on a THF solution of $Co_4(CO)_8(PPh)_2(PPh_3)_2$. (a) 25°C; (b) -40°C; (c) -65°C (from Ref. 315).*

A qualitatively similar trend holds for both the tris- and tetrakis-substituted compounds. The relevant redox potentials are summarized in Table 36.

It can be seen that the gradual substitution of CO groups for P-donor ligands in $Co_4(CO)_{10}(PPh)_2$ progressively makes the one-electron reduction more difficult as well as the one-electron oxidation easier; not only that, but the kinetic stability of the monoanions decreases more and more as the electron density on the Co_4P_2 core is increased by increasing the number of P-donor substituents. This is just opposite to what happened in the previously discussed Co_4 tetrahedral assemblies.

Also in this case, we have no information at hand to foresee the geometrical reorganizations accompanying the 0/+ and the 0/- redox changes, which are likely significant in view of the electrochemical quasireversibility of such steps. In fact, once again, ESR spectra on the electrogenerated paramagnetic species failed in exhibiting resolvable hyperfine absorptions |315|.

We note that the electrochemically computed HOMO-LUMO gap for all the substituted species is around 2.1 eV, somewhat higher (by about 0.5 eV) than that theoretically computed for the unsubstituted model compound $Co_4(CO)_{10}(PH)_2$ |116|.

5.3.6. $Co_4(CO)_{10-2n}(PPh)_2(L)_n$ (L = diphosphorus donor ligands; n = 0,1). The tetracobalt cluster $Co_4(CO)_{10}(PPh)_2$ reacts with bidentate phosphines affording the disubstituted species $Co_4(CO)_8(PPh)_2$(diphosphine), in which the diphosphine either bridges a pair of adjacent cobalt atoms or chelates a single cobalt atom, Figure 169 |316,319,320|.

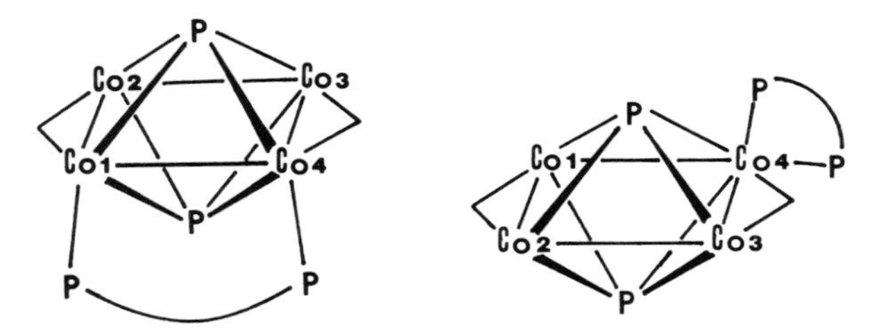

Figure 169. *Schematic representation of the binding mode of diphosphines to the tetracobalt core in $Co_4(CO)_8(PPh)_2$ (diphosphine).*

Two of these species have been studied from the electrochemical viewpoint, namely $Co_4(CO)_8(PPh)_2$(1,2-diphenylphosphinoethane) and $Co_4(CO)_8(PPh)_2$[bis(difluorophosphino)methylamine] (here referred as $Co_4(CO)_8(PPh)_2$(dppe) and $Co_4(CO)_8(PPh)_2[(F_2P)_2NMe]$, respectively).

Figure 170 shows the X-ray structure of $Co_4(CO)_8(PPh)_2$(dppe) |319|.

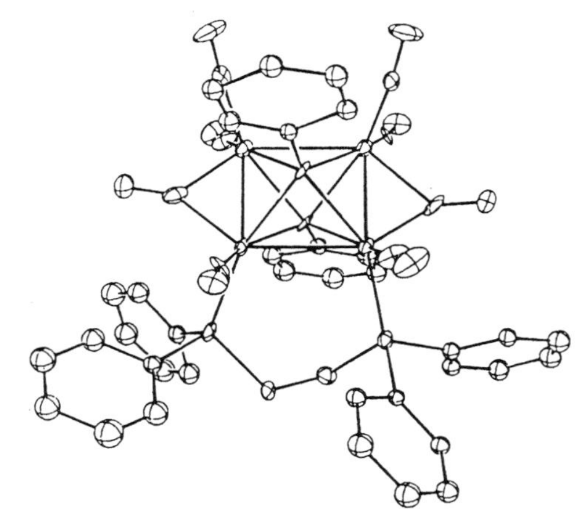

Figure 170. *Perspective view of $Co_4(CO)_8(PPh)_2$(dppe) (from Ref. 319).*

It can be seen that, with respect to the unsubstituted tetracobalt parent $Co_4(CO)_{10}(PPh)_2$, the bridging diphosphine has substituted two terminal CO groups from two adjacent non-carbonyl-bridged cobalt atoms. The Co_4P_2 core is virtually identical to that of $Co_4(CO)_{10}(PPh)_2$; in fact, the carbonyl-bridged Co-Co bond length is of 2.53 Å, the unbridged Co-Co bond length is of 2.69 Å, the averaged Co-P distance is 2.24 Å, the nonbonding P···P distance is 2.55 Å. This suggests that the diphosphine does not exert significant perturbations on the cluster framework |319|.

A similar structure, Scheme 8, is assigned to $Co_4(CO)_8(PPh)_2[(F_2P)NMe]$ |316|.

SCHEME 8

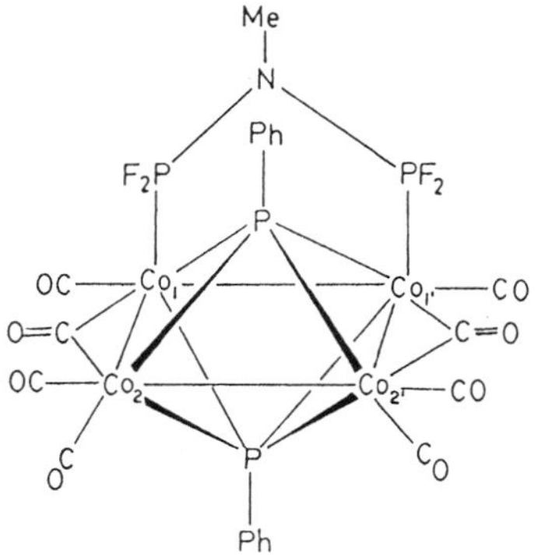

It is useful to note that the synthesis of this complex is rather unusual. In fact, it is prepared by electrogenerating the anion $[Co_4(CO)_{10}(PPh)_2]^-$ in the presence of $(F_2P)_2NMe$, followed by anodic reoxidation of the so-formed substituted anion $[Co_4(CO)_8(PPh)_2\{(F_2P)_2NMe\}]^-$. This electrochemical procedure is required in view of the fact that the thermal reaction of $Co_4(CO)_{10}(PPh)_2$ with $(F_2P)_2NMe$ affords directly the tetrakis derivative $Co_4(CO)_3(PPh)_2[(F_2P)_2NMe]_4$ |316|, which will be treated in the next Section.

The electrochemical behaviour of these diphosphine derivatives is strictly related to that of the parent $Co_4(CO)_{10}(PPh)_2$ |315,316|. In fact, the most significant redox change is their one-electron reduction, quasireversible in character, which gives a stable monoanion in the case of the fluorodiphosphine derivative; in the case of the diphenyl phosphine, the monoanion undergoes decomposition reactions, which can be prevented by decreasing the temperature to -40°C. Under such experimental conditions, the latter complex also gives a stable monocation through a quasireversible electrochemical step.

Table 37 summarizes the relevant redox potentials.

TABLE 37. *Redox potentials for the cathodic and anodic steps exhibited by the series* $Co_4(CO)_{10-2n}(PPh)_2(diphosphine)_n$, *in tetrahydrofuran solution*

complex	$E°'_{0/-}$ (V)	ΔEp (mV)	$E°'_{+/0}$ (V)	ΔEp (mV)	reference
$Co_4(CO)_{10}(PPh)_2$	-1.22	390[a]			315, 316
$Co_4(CO)_8(PPh)_2[(F_2P)_2NMe]$	-1.25	330[a]	+0.70[b]		316
$Co_4(CO)_8(PPh)_2(dppe)$	-1.83[c]	125[c,d]	+0.26[c]	70[d]	315

[a]*Measured at 0.5* Vs^{-1}; [b]*peak potential for an irreversible process;* [c]*measured at - 40°C;* [d]$E_{3/4}-E_{1/4}$ *values from cyclic voltammetry at microelectrodes*

It is expected that phosphorus donor ligands push electron density into the Co_4P_2 core, so making difficult the electron addition as well as easy the electron removal. This is really the case for the dppe derivative. In contrast, in the case of the fluorodiphosphine complex it is thought that such ligand, being able to promote extensive electron delocalization, can act as an "electron reservoir" accepting or donating electron density via low-lying π and π^* orbitals, so to leave the tetracobalt cluster substantially unperturbed |316|.

5.3.7. $Co_4(CO)_{3-2n}(PPh)_2[(F_2P)_2NMe]_{4+n}$ (n = 0,1). In the preceding Section, we have briefly reported that reaction of $Co_4(CO)_{10}(PPh)_2$ with bis(difluorophosphino)methylamine affords the tetrasubstituted cluster $Co_4(CO)_3(PPh)_2[(F_2P)_2NMe]_4$ |316,321|.

This process, which globally involves a formal two-electron addition, is quite interesting in that it implies not only the substitution of six terminal CO groups, but also that of one CO-bridged group, so causing the cleavage of one Co-Co bond in the tetrametal plane, Figure 171 |321|.

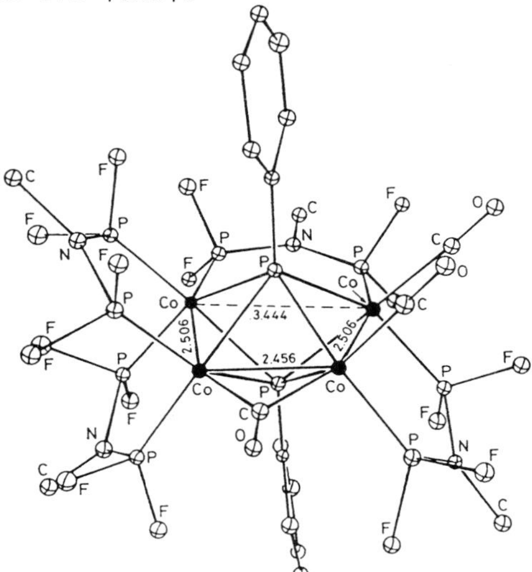

Figure 171. *X-Ray structure of $Co_4(CO)_3(PPh)_2[(F_2P)_2NMe]_4$ (from Ref. 321).*

With respect to $Co_4(CO)_{10}(PPh)_2$, the scission of the Co-Co bond significantly affects all the other bond lengths in the Co_4P_2 core. The CO-unbridged Co-Co distances (2.51 Å) shorten by about 0.2 Å; the CO-bridged Co-Co bond length (2.46 Å) decreases by about 0.06 Å; the averaged Co-P bond length (2.31 Å) increases by about 0.07 Å; the nonbonding P···P distance (2.44 Å) shortens by 0.1 Å.

Such an open-rectangular Co_4 array is assigned, on the basis of spectroscopic evidence, also to the pentakis derivative $Co_4(CO)(PPh)_2[(F_2P)_2NMe]_5$, Scheme 9 |316|.

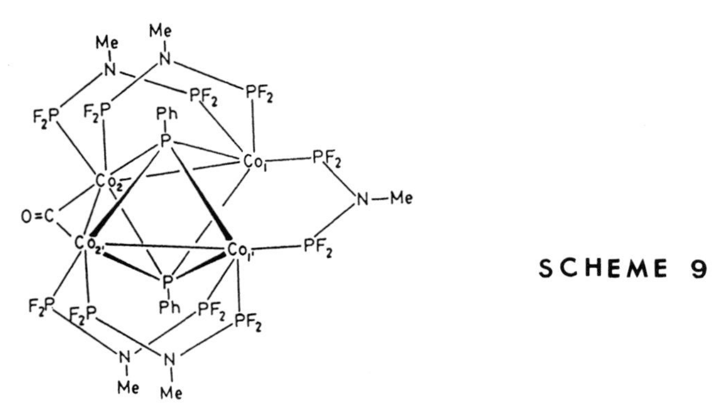

SCHEME 9

Figure 172 shows the cyclic voltammetric behaviour of both the 66-electron tetrakis and pentakis difluorophosphine derivatives |316|.

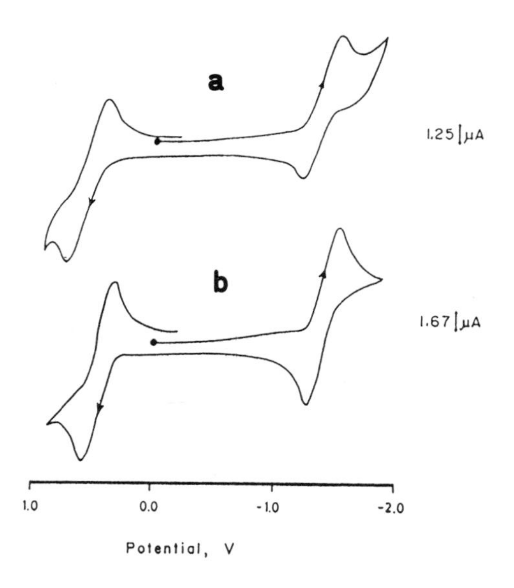

Figure 172.
Cyclic voltammograms recorded at a platinum electrode on THF solutions of: (a) $Co_4(CO)_3(PPh)_2[(F_2P)_2NMe]_4$*; (b)* $Co_4(CO)(PPh)_2[(F_2P)_2NMe]_5$*. Scan rate: 0.5* Vs^{-1} *(from Ref. 316).*

As it happens for the 64-electron substituted Co_4P_2 derivatives, they undergo quasireversibly both a one-electron oxidation and a one-electron reduction. The corresponding anions and cations are stable at least in the short times of cyclic voltammetry (macroelectrolysis tests in fact lead to EPR silent species, which should indicate decomposition of the primarily electrogenerated species). Table 38 reports the relevant redox potentials.

TABLE 38. *Formal electrode potentials for the redox changes exhibited by $Co_4(CO)_{3-2n}(PPh)_2[(F_2P)_2NMe]_{4+n}$ in tetrahydrofuran solution |316|*

complex	$E^{\circ\prime}_{0/-}$ (V)	ΔEp^a (mV)	$E^{\circ\prime}_{+/0}$ (V)	ΔEp^a (mV)
$Co_4(CO)_3(PPh)_2[(F_2P)_2NMe]_4$	-1.31	290	+0.58	300
$Co_4(CO)(PPh)_2[(F_2P)_2NMe]_5$	-1.36	260	+0.49	260

[a]*Measured at 0.5 Vs^{-1}*

Under the assumption that the MO description for M_4E_2 clusters may be still valid |110,116|, since the actual monocations are formally isoelectronic with the monoanions of the series $Co_4(CO)_{10-n}(PPh)_2(L)_n$ (Section 5.3.5.), a structure intermediate between a bis-capped closed-rectangular and a bis-capped open-rectangular Co_4 assembly could be proposed for both the series. In contrast, the relative stability of the 67-electron monoanions is unexpected, since the electron should enter one further metal-metal antibonding orbital (indeed the full occupancy of the antibonding orbitals needs one more electron).

5.3.8. $Co_4(CO)_8(L)_2(RC_2R)$ (L = CO,PPh$_3$; R = H,Et,Ph). $Co_4(CO)_{12}$ reacts with alkynes to afford a series of 60-electron tetracobalt complexes of formula $Co_4(CO)_8(\mu\text{-}CO)_2$(alkyne), having a tetrametal butterfly arrangement |322-324|. In this connection, Figure 173 illustrates the molecular structure of $Co_4(CO)_{10}(HC_2H)$ |324|. In the Co_4 butterfly core, the hinge Co-Co bond length is longer (2.56 Å) than the wing-bridged Co-Co distances (averaged 2.45 Å). The dihedral angle between the two wings is 119°. The acetylene is disposed nearly parallel to the Co2-Co4 hinge bond, through a μ_4-η^2 coordination consisting of two σ bonds (C1-Co2 and C2-Co4) and a delocalized four-center π-bonding system between the two acetylenic carbon atoms and the two wing-tip cobalt atoms. Accordingly, the $Co_{(hinge)}$-C distances (1.98 Å) are shorter than the $Co_{(wing)}$-C distances (2.07 Å). Indeed the presence of two bridging carbonyls in the wings, shortening the involved Co-Co bonds, causes an overall small distortion. Finally, the C-C bond distance of 1.40 Å confirm the olefinic character of the coordinating alkyne.

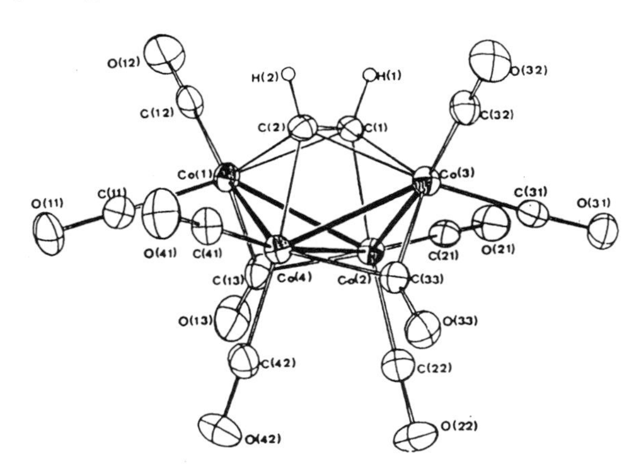

Figure 173. *Perspective view of $Co_4(CO)_{10}(HCCH)$ (from Ref. 324).*

Substitution of two carbonyl groups for two triphenyl-phosphine molecules affords $Co_4(CO)_8(PPh_3)_2(HC_2H)$, the structure of which is shown in Figure 174 |325|.

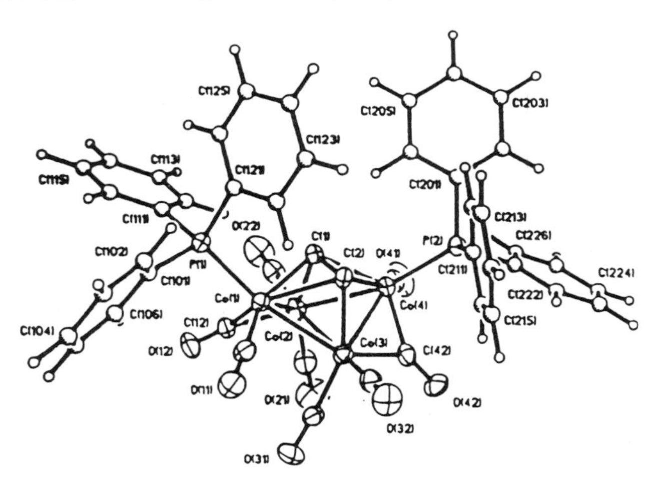

Figure 174. *Perspective view of $Co_4(CO)_8(PPh_3)_2(HCCH)$ (from Ref. 325).*

It can be seen that substitution of the CO groups has occurred at the wing-tip cobalt atoms. The geometrical assembly of the tetracobalt core remains unvaried with respect to the unsubstituted precursor.

$Co_4(CO)_{10}(C_2H_2)$ is relatively unstable in nonaqueous solvents and tends to decompose to $Co_2(CO)_6(C_2H_2)$. In dichloromethane solution, at -20°, such decomposition is however substantially blocked. As shown in Figure 175, under these experimental conditions, the tetracobalt cluster undergoes two subsequent one-electron reduction steps |325|.

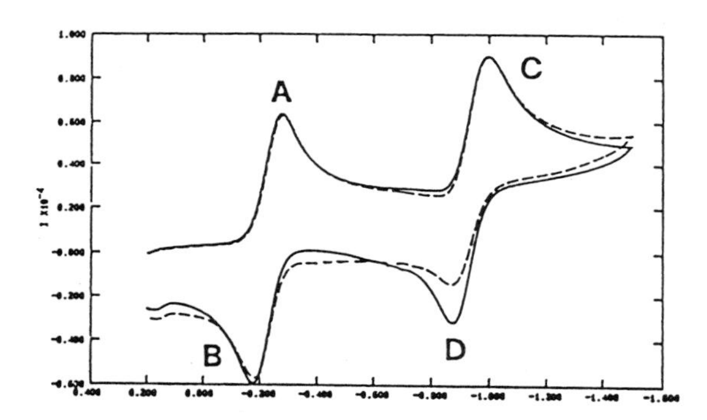

Figure 175. *Cyclic voltammograms recorded at a mercury electrode on a CH_2Cl_2 solution of $Co_4(CO)_{10}(HCCH)$. Scan rate: 0.2 Vs^{-1}. Dotted line, under Ar; bold line, under CO (from Ref. 325).*

The first reduction is chemically reversible, and the electrogenerated monoanion $[Co_4(CO)_{10}(HCCH)]^-$ has been spectroscopically characterized. The second one, which primarily generates the short-lived dianion $[Co_4(CO)_{10}(HCCH)]^{2-}$, improves its chemical reversibility under CO atmosphere, in that the framework-destroying decarbonylation process is retarded.

Table 39 summarizes the redox potentials of the two reduction processes in the cluster compounds so far studied.

It is interesting to note how the introduction of -donor PPh_3 ligands mekes notably more difficult the reduction processes.

Since theoretical calculations show the LUMO of the actual complexes to be Co_4-centred, it is expected that the addition of electrons leads to a diffuse lengthening of the Co-Co distances |325|.

TABLE 39. *Formal electrode potentials for the two sequential reductions exhibited by $Co_4(CO)_8(L)_2(RC_2R)$ in dichloromethane solution |325|*

complex	E°′ 0/- (V)	E°′ -/2- (V)
$Co_4(CO)_{10}(HC_2H)$	-0.33	-1.03
$Co_4(CO)_8(PPh_3)_2(HC_2H)$	-0.81	-1.42
$Co_4(CO)_{10}(EtC_2Et)$	-0.36	-1.05
$Co_4(CO)_{10}(PhC_2Ph)$	-0.20	-0.91

5.3.9. $Co_4(\eta^5\text{-}C_5H_5)_2(CO)_{5-n}(L)_n(CF_3C_2CF_3)$ (L = $P(OMe)_3$, t-BuNC; n = 0-2). Like the preceding case, in $Co_4(\eta^5\text{-}C_5H_5)_2(CO)_4(\mu\text{-}CO)(CF_3C_2CF_3)$, a butterfly arrangement of a tetracobalt fragment is connected to an alkyne molecule through a μ_4- η^2 coordination |326 -328|. Figure 176 illustrates the relevant X-ray structure |327|.

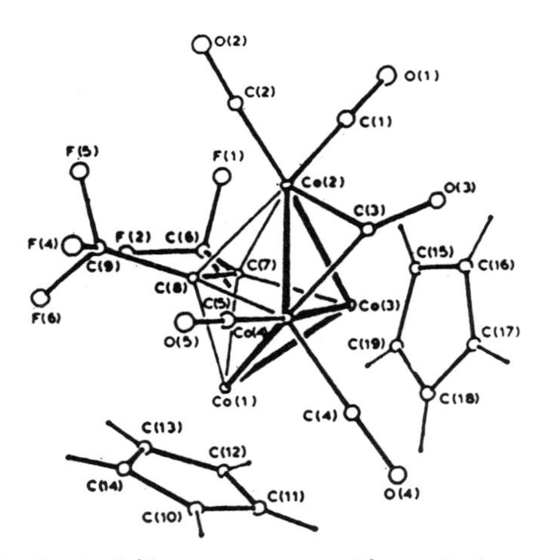

Figure 176. *Perspective view of $Co_4(C_5H_5)_2(CO)_5(CF_3C_2CF_3)$ (from Ref. 327).*

In this overall octahedral-like core, the tetracobalt assembly is constituted by an hinge Co-Co bond of length 2.47 Å, longer than the $Co_{(hinge)}$-$Co_{(wing)}$ distances (averaged 2.40 Å). The dihedral angle between the two wings is 117°. The σ-bonds $Co_{(hinge)}$-$C_{(alkyne)}$ (1.96 Å) are, as expected, shorter than the π-bonds $Co_{(wing)}$-$C_{(alkyne)}$ (averaged 2.01 Å). The olefinic character assumed by the coordinated hexafluorobutyne is confirmed by a C7-C8 distance of 1.44 Å. With respect to $Co_4(CO)_{10}(RCCR)$ (R = C_2H_5

|323|, R = H |324|), the two cyclopentadienyl rings substitute two terminal carbonyl groups per Co atom [in the two cobalt wing and cobalt hinge atoms (Co1, Co3), respectively] as well as the bridging carbonyl group connecting them.

Figure 177 shows the cyclic voltammogram exhibited by the present 60-electron (the authors, indeed, proposing the alkyne as a 6-electron donor, assign a 62-electron count) tetracobalt cluster in tetrahydrofuran solution |328|.

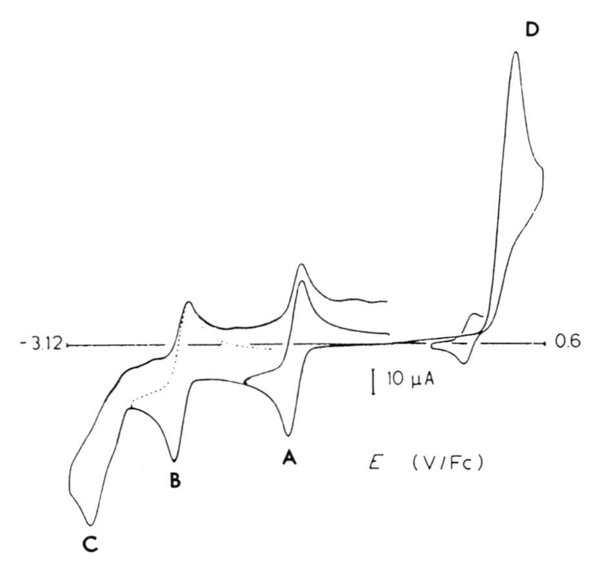

Figure 177. *Cyclic voltammogram recorded at a vitreous carbon electrode on a THF solution of* $Co_4(C_5H_5)_2(CO)_5(CF_3C_2CF_3)$. *Scan rate 0.2* Vs^{-1} *(from Ref. 328).*

Apart from the anodic (peak D) and the most cathodic (peak C) multielectron irreversible processes, $Co_4(C_5H_5)_2(CO)_5(CF_3C_2CF_3)$ undergoes two distinct quasireversible one-electron reductions (peaks A and B, respectively), which generate the corresponding monoanion and dianion, stable in the cyclic voltammetric time-scale. The relevant redox potentials are reported in Table 40.

Concerning the structural changes accompanying such one-electron additions, it is likely that, with respect to the parent 60-electron Co_4 butterfly derivative, the 61-electron monoanion and 62-electron dianion do not vary appreciably their structure, in view of the near reversibility of the relevant electrochemical

TABLE 40. *Redox potentials for the two one-electron additions exhibited by the series* $Co_4(\eta^5\text{-}C_5H_5)_2(CO)_{4-n}(\mu\text{-}CO)(L)_n(CF_3C_2CF_3)$ *in different nonaqueous solutions* |328|

complex	$E°'_{0/-}$ (V)	ΔEp[a] (mV)	$E°'_{-/2-}$ (V)	ΔEp[a] (mV)	solvent
$Co_4(\eta^5\text{-}C_5H_5)_2(CO)_4(\mu\text{-}CO)(CF_3C_2CF_3)$	-0.57	80	-1.47	80	THF
	-0.73	60	-1.43	70	MeCN
	-0.71	70	-1.41	80	CH_2Cl_2
$Co_4(\eta^5\text{-}C_5H_5)_2(CO)_3(\mu\text{-}CO)\vert P(OMe)_3\vert(CF_3C_2CF_3)$	-0.86	80	-1.66	90	THF
	-0.96	60	-1.63	90	MeCN
$Co_4(\eta^5\text{-}C_5H_5)_2(CO)_3(\mu\text{-}CO)(t\text{-}BuNC)(CF_3C_2CF_3)$	-0.88	70	-1.68	90	THF
$Co_4(\eta^5\text{-}C_5H_5)_2(CO)_2(\mu\text{-}CO)\vert P(OMe)_3\vert_2(CF_2C_2CF_3)$	-1.10	90	-1.90		THF
$Co_4(\eta^5\text{-}C_5H_5)_2(CO)_2(\mu\text{-}CO)(t\text{-}BuNC)_2(CF_3C_2CF_3)$	-1.17	75	-1.92		THF

[a] *Measured at 0.2 Vs^{-1}*

steps. Unfortunately no structural information is available, also because the electrogenerated monoanion slowly tends to decompose, as well as the fact that the dianion quickly reverts to the monoanion |328|.

$Co_4(\eta^5-C_5H_5)_2(CO)_4(\mu-CO)(CF_3C_2CF_3)$ undergoes thermal substitution by phosphite or isocyanide affording both the mono- and di-substituted species $Co_4(\eta^5-C_5H_5)_2(CO)_3(\mu-CO)(L)(CF_3C_2CF_3)$ and $Co_4(\eta^5-C_5H_5)_2(CO)_2(\mu-CO)(L)_2(CF_3C_2CF_3)$, (L = $P(OMe)_3$, *t*-BuNC) |328|. Lacking X-ray structures, it is thought from spectroscopic inferences that, in the monosubstituted species, the substitution of one terminal CO may occur preferentially at the carbon C5 (bound to the hinge Co4) as well as at the carbon C8 (bound to the wing Co2) of the retained framework of the parent compound. Indeed a mixture of two monosubstituted isomers is obtained.

In a similar manner, it is argued that in the disubstituted complexes two isomers are obtained with the preferential substitution of the terminal carbonyls at C2, C5 and C2, C4, respectively.

Electrochemistry seems not apt to distinguish between the isomeric derivatives both in the mono- and di-substituted mixtures, so we shall speak in general of monosubstituted or disubstituted species.

Like the parent cluster, the mono- and di-substituted species undergo two quasireversible one-electron reductions, generating the corresponding monoanions and dianions, somewhat stable in the short times of cyclic voltammetry. The relevant redox potentials are summarized in Table 40. It can be noted that each addition of one substituent molecule shifts the redox potentials towards negative values by about 250-300 mV. In addition, although the different donor/acceptor ability of $P(OMe)_3$ and *t*-BuNC, the relevant redox potentials are practically coincident. This datum served the authors to support the tetrametal-centred character of the LUMO of the parent cluster.

5.3.10. $Co_4(\eta^5-C_5H_5)_2(CO)_{5-2n}(L)_n(CF_3C_2CF_3)$ (L = diphosphines; n = 0,1). The tetracobalt derivative discussed in the preceding Section, $Co_4(\eta^5-C_5H_5)_2(CO)_4(\mu-CO)(CF_3C_2CF_3)$, is able to undergo carbonyl substitution by diphosphines such as bis (diphenylphosphino)methane (dppm) and 1,2-bis(diphenylphosphino) ethane (dppe) to give $Co_4(\eta^5-C_5H_5)_2(CO)_2(\mu-CO)(dppm)(CF_3C_2CF_3)$, and $Co_4(\eta^5-C_5H_5)_2(CO)_2(\mu-CO)(dppe)(CF_3C_2CF_3)$, respectively |328|.

On the basis of spectroscopic information, the most probable structure for these species is schematized in Figure 178.

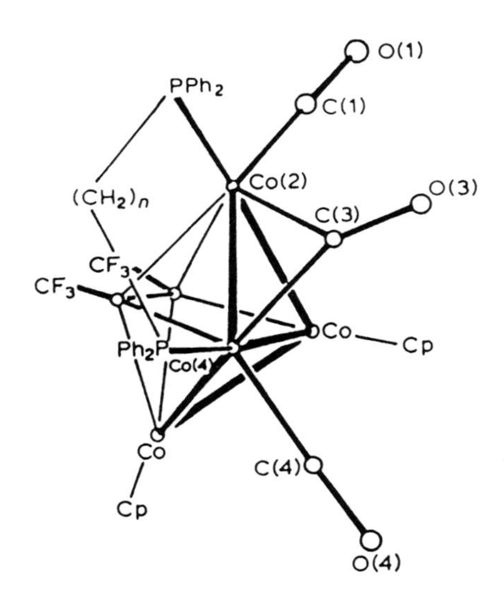

Figure 178.
Probable structure for $Co_4(C_5H_5)_2(CO)_3(L)(CF_3C_2CF_3)$, L = dppm, dppe (from Ref. 328).

Figure 179 illustrates the cyclic voltammetric behaviour exhibited by $Co_4(\eta^5-C_5H_5)_2(CO)_2(\mu-CO)(dppm)(CF_3C_2F_3)$ |328|.

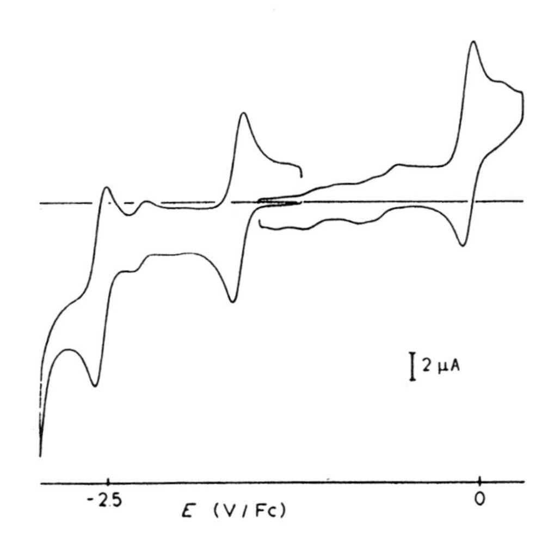

Figure 179.
Cyclic voltammogram recorded at a vitreous carbon electrode on a THF solution of $Co_4(C_5H_5)_2(CO)_3(dppm)(CF_3C_2CF_3)$. Scan rate: 0.2 Vs^{-1} (from Ref. 328).

As in the case of both the parent and its monosubstituted clusters, Section 5.3.9., the present derivative undergoes two successive quasireversible one-electron reductions to the corresponding, somewhat stable, monoanion and dianion, respectively; but, at variance with these former, it also undergoes a quasireversible one-electron oxidation to the relatively stable monocation.

The same behaviour is displayed by $Co_4(\eta^5\text{-}C_5H_5)_2(CO)_2(\mu\text{-}CO)$ (dppe)$(CF_3C_2CF_3)$.

The redox potentials for such electron-transfers are reported in Table 41.

The thermodynamic effect of the ligation by a diphosphine (as expressed by formal electrode potentials) is not substantially different from that obtained by substituting two CO groups for two phosphite molecules (see Table 40). In contrast, the kinetic stability of the cluster framework seems enhanced by bridged coordination to a diphosphine, in that these disubstituted clusters not only support two subsequent one-electron additions without fast decomposition, but also a one-electron removal, so that the cluster valence electrons can range from 59 to 62.

5.4. Hexanuclear Binary Complexes

5.4.1. $[Co_6(CO)_{15}]^{2-}$. The molecular structure of the hexacobalt cluster anion $[Co_6(CO)_{15}]^{2-}$ is shown in Figure 180 |329,330|.

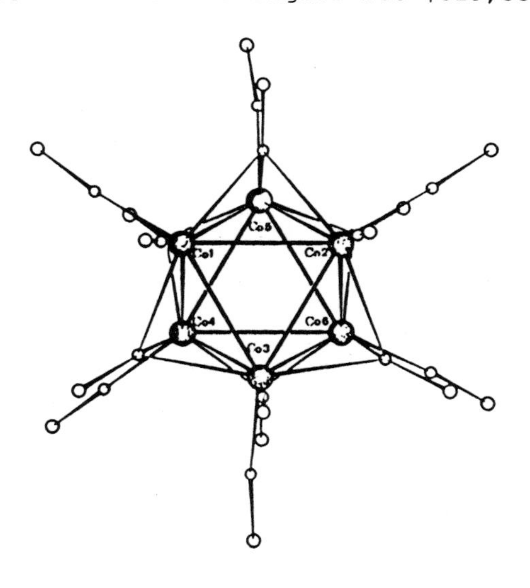

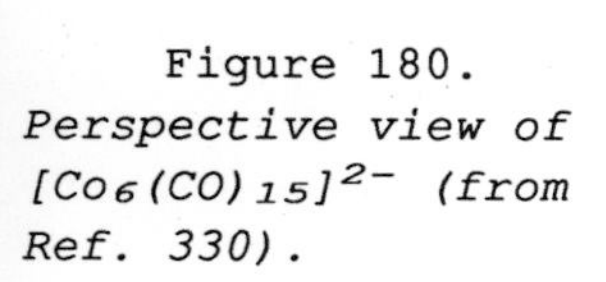
Figure 180. *Perspective view of $[Co_6(CO)_{15}]^{2-}$ (from Ref. 330).*

TABLE 41. *Formal electrode potentials for the redox changes exhibited by the series* $Co_4(\eta^5\text{-}C_5H_5)_2(CO)_{4-2n}(diphosphine)_n(\mu\text{-}CO)(CF_3C_2CF_3)$ *in tetrahydrofuran solution* |328|

complex	$E°'_{+/0}$ (V)	ΔEp[a] (mV)	$E°'_{0/-}$ (V)	ΔEp[a] (mV)	$E°'_{-/2-}$ (V)	ΔEp[a] (mV)
$Co_4(\eta^5\text{-}C_5H_5)_2(CO)_4(\mu\text{-}CO)(CF_3C_2CF_3)$			-0.57	80	-1.47	80
$Co_4(\eta^5\text{-}C_5H_5)_2(CO)_2(\mu\text{-}CO)(dppm)(CF_3C_2CF_3)$	+0.48	70	-1.09	70	-2.02	90
$Co_4(\eta^5\text{-}C_5H_5)_2(CO)_2(\mu\text{-}CO)(dppe)(CF_3C_2CF_3)$	+0.49	b	-1.14	80	-2.09	

[a] *Measured at 0.2* Vs^{-1}; [b] *value not reported in the original reference*

The cobalt atoms assume an octahedral geometry with an averaged Co-Co bond length of 2.51 Å. However, the presence of three triply bridging, three doubly bridging, and nine terminal carbonyl groups induces some distortions. In particular, the Co-Co distances involved in edge bridging by carbonyls [namely Co4-Co5, Co5-Co6, and Co4-Co6 (mean 2.47 Å)] are 0.05 Å shorter than the remaining nine (mean 2.52 Å).

It is known |329| that such an 86-electron cluster undergoes redox reactions by chemical agents which do not change the number of cluster valence electrons because of the involvement of carbon monoxide:

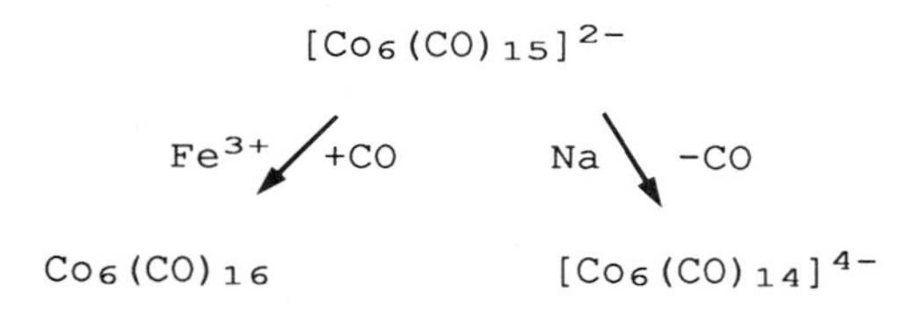

Nevertheless, electrochemistry puts in evidence that, even if short-lived, its one-electron oxidized congener $[Co_6(CO)_{15}]^-$ does exist. Figure 181 shows the cyclic voltammogram exhibited by $[Co_6(CO)_{15}]^{2-}$ in dichloromethane solution |297|.

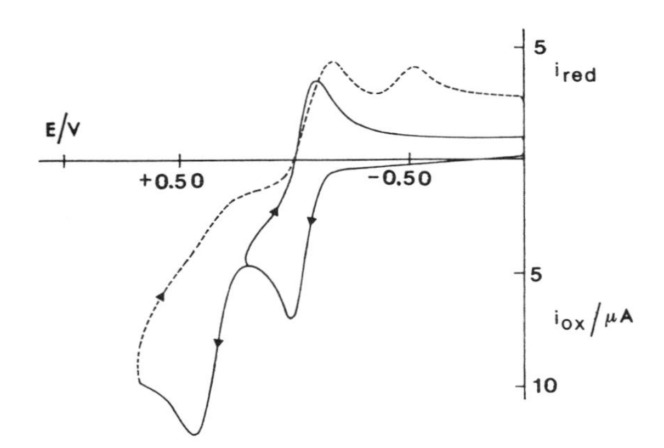

Figure 181. *Cyclic voltammogram recorded at a platinum electrode on a CH_2Cl_2 solution of $[Co_6(CO)_{15}]^{2-}$. Scan rate 0.2 Vs^{-1}.*

A first quasireversible one-electron oxidation ($E^{\circ\prime}$ = -0.05 V; $\Delta Ep_{(0.2\ Vs^{-1})}$ = 118 mV) leads to $[Co_6(CO)_{15}]^-$, which undergoes slow decomposition ($t_{1/2} \approx 3s$) mainly to $Co_4(CO)_{12}$; a second one-electron irreversible oxidation step (Ep = +0.37 V) generates the fast decomposing neutral species $Co_6(CO)_{15}$.

5.5. Hexanuclear Carbido Complexes

5.5.1. $[Co_6(CO)_{15}C]^{2-}$. The hexacobalt carbide dianion $[Co_6(CO)_{15}C]^{2-}$ possesses a trigonal prismatic geometry encapsulating one carbide atom, Figure 182 |331|.

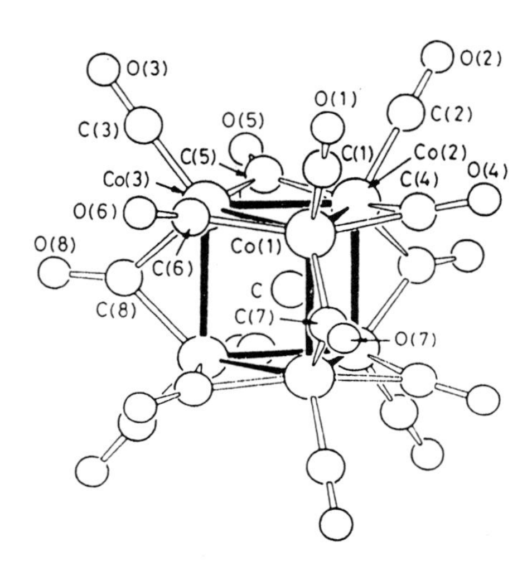

Figure 182. *X-Ray structure of $[Co_6(CO)_{15}C]^{2-}$ (from Ref. 331).*

The basal and inter-basal Co-Co bond lengths are 2.54 Å and 2.58 Å, respectively, with an averaged distance of 2.56 Å. The mean $Co\text{-}C_{(carbide)}$ length is 1.95 Å.

Such a 90-electron cluster dianion undergoes in dichloroethane solution two close-spaced one-electron oxidation steps leading to the quite unstable monoanion and neutral congeners, respectively ($Ep_{2-/-}$ = 0.00 V ; $Ep_{-/0}$ = +0.15 V) |132|. In accordance with the chemical oxidation |331|, exhaustive electrolysis in correspondence to the most anodic process affords, among different oxidation products, the partially decarbonylated hexacobalt carbide $[Co_6(CO)_{14}C]^-$, which will be discussed in the next Section.

Finally, $[Co_6(CO)_{15}C]^{2-}$ also undergoes a two-electron reduction at Ep = -1.75 V, irreversible in character, which, as such, deeply changes its molecular framework |132|.

5.5.2. $[Co_6(CO)_{14}C]^-$. Chemical or electrochemical oxidation of the 90-electron dianion $[Co_6(CO)_{15}C]^{2-}$ affords the paramagnetic 87-electron decarbonylated monoanion $[Co_6(CO)_{14}C]^-$ |132,331|. The crystal structure of this monoanion is illustrated in Figure 183 |332,333|.

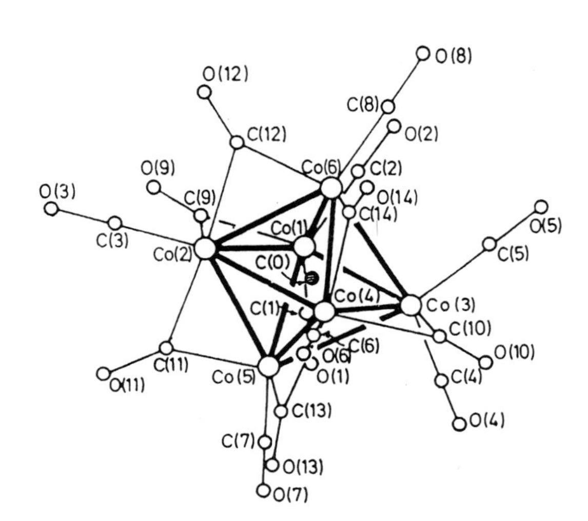

Figure 183. *Perspective view of the anion $[Co_6(CO)_{14}C]^-$ (from Ref. 333).*

The Co_6C core has a distorted octahedral geometry with the carbide atom accomodated inside. Of the fourteen carbonyls, eight are terminal and six are edge-bridging. The averaged Co-Co bond length of 2.65 Å is significantly longer than that of $[Co_6(CO)_{15}C]^{2-}$.

The disposition of the CO groups, as well as the need to encapsulate the carbide atom, cause a notable spreading of the metal-metal distances. The six CO-bridged edges have a mean length of 2.53 Å, the four unbridged edges have a mean length of 2.77 Å; the distance between the two-terminal carbonyl bearing cobalt atoms (Co1-Co3) is 2.66 Å, the Co2-Co4 distance is 2.92 Å. The Co-$C_{(carbide)}$ averaged is 1.88 Å, the distance from the Co2, Co4 atoms being longer (0.08 Å) than all other ones.

These structural data support the view that the HOMO level of the monoanion is localized around the Co2-Co4 bond, which has an antibonding character. Indeed, EPR measurements do not exclude the possibility that the 87th, unpaired, electron is delocalized over the ligand molecules |334|.

It has been briefly reported that this paramagnetic monoanion undergoes, in dichloroethane solution, an irreversible reduction step at Ep = -0.72 V |132|, which likely regenerates $[Co_6(CO)_{15}C]^{2-}$ |331|.

5.6. Octanuclear Carbido Complexes

5.6.1. $[Co_8(CO)_{18}C]^{2-}$. The octanuclear carbide dianion $[Co_8(CO)_{18}C]^{2-}$ can be prepared by reacting $[Co_6(CO)_{15}C]^{2-}$ with $Co_4(CO)_{12}$ |331,332,335|. Its molecular structure is shown in Figure 184 |332,335|.

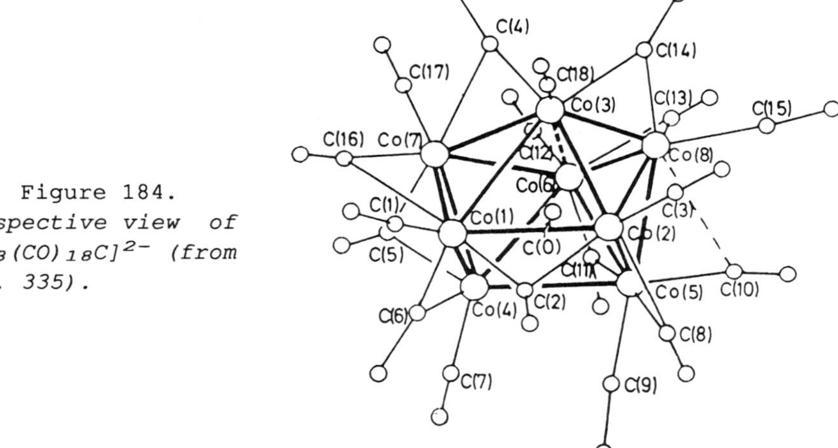

Figure 184. *Perspective view of $[Co_8(CO)_{18}C]^{2-}$ (from Ref. 335).*

The tetragonal antiprismatic arrangement of the eight cobalt atoms, formed by the two parallel planes Co(1,2,4,5) and Co(3,6,7,8), can be more simply viewed as a prismatic Co_6 array (from the structure of the precursor $[Co_6(CO)_{15}C]^{2-}$), two rectangular faces of which are capped by an extra cobalt atom (Co7, Co8). Ten carbonyl groups are edge bridging and eight are terminal. This causes a spreading of the Co-Co bond lengths from 2.46 Å to 2.60 Å, with a mean distance of 2.52 Å. Since the interstitial cavity of a regular antiprism would be larger than that required for an optimal Co-C contact, a slight structural deformation occurs, in which four cobalt-carbide distances become shorter (mean 1.99 Å) than the remaining four (mean 2.15 Å).

The present 114-electron dianion displays a rich redox propensity. In fact, as illustrated in Figure 185, it undergoes, in dichloroethane solution, both two subsequent one-electron reductions and two subsequent oxidations, the first one involving a one-electron step, the second one involving a single two-electron step |132|.

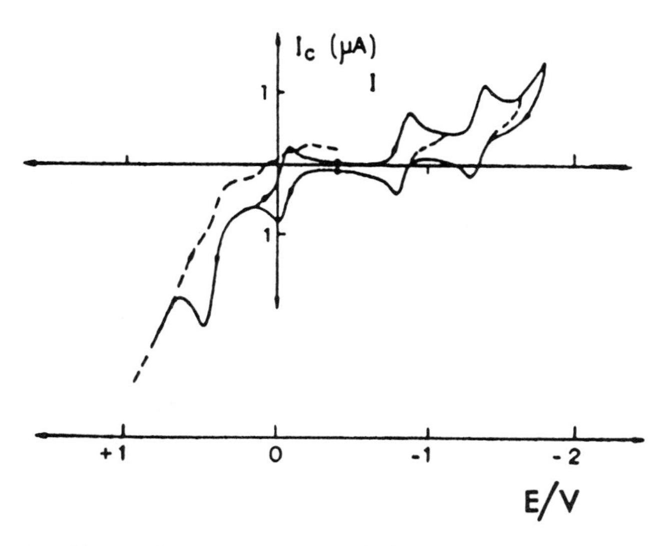

Figure 185. *Cyclic voltammogram exhibited at a platinum electrode by* $[Co_8(CO)_{18}C]^{2-}$ *in* 1,2-$C_2H_4Cl_2$. *Scan rate 0.1* Vs^{-1} *(from Ref. 132).*

The two electrochemically quasireversible reduction steps $[Co_8(CO)_{18}C]^{2-/3-}$ (E°' = -0.81 V; $\Delta Ep_{(0.1\ Vs^{-1})}$ = 80 mV and $[Co_8(CO)_{18}C]^{3-/4-}$ (E°' = -1.35 V; $\Delta Ep_{(0.1\ Vs^{-1})}$ = 80 mV), which are chemically reversible in the cyclic voltammetric timescale, are unfortunately complicated by slow decomposition of the electrogenerated tri- and tetra-anions in the longer times of macroelectrolysis, so to prevent their recovery, and hence the ascertainment of the relevant structural features.

The same happens for the $[Co_8(CO)_{18}C]^{2-/-}$ redox change (E°' = -0.01 V; $\Delta Ep_{(0.1\ Vs^{-1})}$= 80 mV). Finally the $[Co_8(CO)_{18}C]^{-/+}$ step (Ep = +0.50 V) is irreversible in character, thereby indicating a fast decomposition of the primarily electrogenerated monocation.

5.7. *Tridecanuclear Carbido Complexes*

5.7.1. $[Co_{13}(CO)_{24}(C)_2]^{4-}$. Pyrolysis of $[Co_6(CO)_{15}C]^{2-}$ affords the tridecacobaltate species $[Co_{13}(CO)_{24}(C)_2]^{4-}$ |336,337|. Figure 186 shows a perspective view of such a tetraanion |337|.

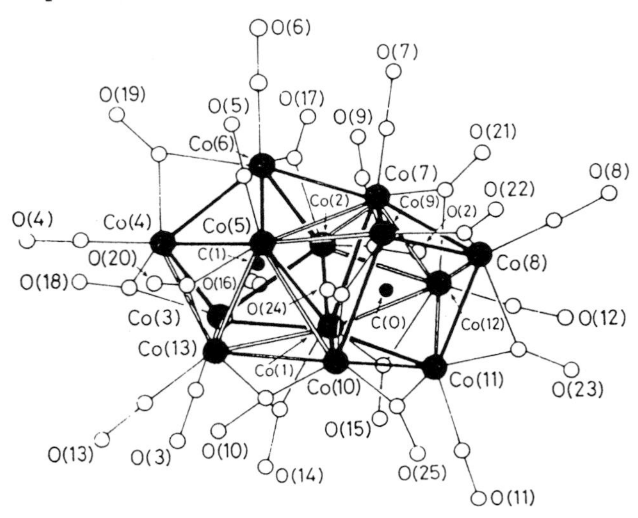

Figure 186. *Molecular structure of the tetraanion $[Co_{13}(CO)_{24}(C)_2]^{4-}$ (from Ref. 337).*

This complex geometry has been described as a three-layer stack containing two squares in the outer layers and three fused triangles in the central one, or, more intuitively, as two, mutually rotated, prismatic Co_6C units (from the precursor dianion cluster) sharing the vertex Co1, cappped at opposite sides, by two five-bridging cobalt atoms (Co12 and Co13), respectively] |337|. Twelve carbonyls are terminal and twelve are edge-bridging. The Co-Co bond lengths are spread from 2.39 Å to 2.80 Å, with an averaged value of 2.57 Å. In particular the two distances Co1-Co5 and Co1-Co7 are the longest ones (mean, 2.76 Å). The $Co-C_{(carbide)}$ distance in the two prismatic subunits is meanly 1.98 Å.

Such a paramagnetic 177-electron cluster |338| acts as an effective electron sponge, in that, as shown in Figure 187, it undergoes, in acetonitrile solution, through chemically reversible redox changes, both a one-electron oxidation ($E°'_{3-/4-} = -0.54$ V, $\Delta Ep_{(0.2\ Vs^{-1})} = 66$ mV) and two subsequent one-electron reductions ($E°'_{4-/5-} = -1.06$ V, $\Delta Ep_{(0.2\ Vs^{-1})} = 60$ mV; $E°'_{5-/6-} = -1.68$ V, $\Delta Ep_{(0.2\ Vs^{-1})} = 66$ mV) |338|.

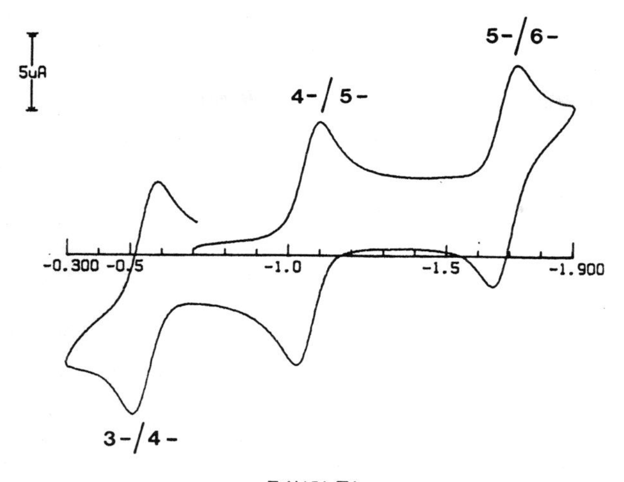

Figure 187. *Cyclic voltammogram recorded at a platinum electrode on a deaerated MeCN solution of* $[Co_{13}(CO)_{24}(C)_2]^{4-}$. *Scan rate 0.2* Vs^{-1}.

The substantial electrochemical reversibility of these electron-transfer sequences preludes a substantial maintaining of the starting molecular framework.

Indeed, up to now, only the congener $[Co_{13}(CO)_{24}(C)_2]^{3-}$ has been structurally characterized |339|. The trianion is, as expected, strictly isostructural with the parent tetraanion. The averaged value of the Co-Co bond length remains 2.57 Å. The two longest Co-Co distances (Co1-Co5, Co1-Co7) slightly shorten (by 0.02 Å), as on the other hand the averaged $Co-C_{(carbide)}$ does (by 0.01 Å).

Attempts are now in progress to characterize structurally the remaining members of the cited redox sequence, in order to evaluate further the concomitant structural rearrangements.

6. RHODIUM CLUSTERS

6.1. *Trinuclear Mixed-Ligand Complexes*

6.1.1. $(C_5H_5)_3Rh_3(CO)_3$. Two isomeric species of formula $(C_5H_5)_3Rh_3(CO)_3$ have been X-ray charaterized, namely $(C_5H_5)_3Rh_3(\mu\text{-}CO)_2(CO)$ |340| and $(C_5H_5)_3Rh_3(\mu\text{-}CO)_3$ |341,342|, Figure 188.

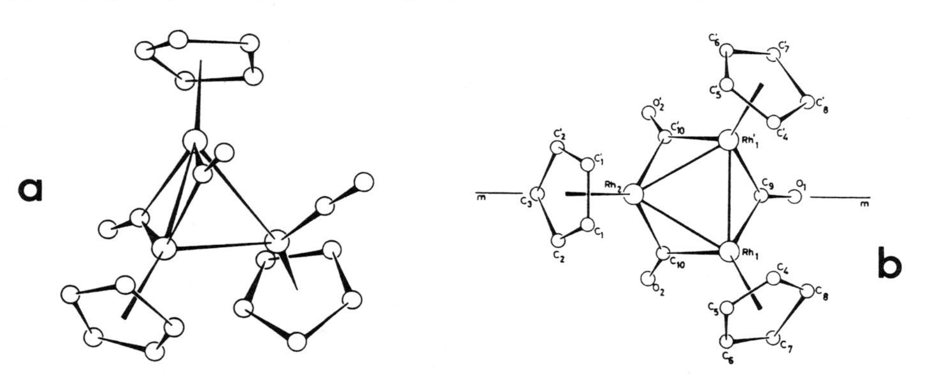

Figure 188. *Perspective view of. (a) $(\eta^5\text{-}C_5H_5)_3Rh_3(\mu\text{-}CO)_2(CO)$, doubly bridged Rh-Rh distance, 2.62 Å; unbridged Rh-Rh distances, 2.69 Å, in average (from Ref. 340); (b) $(\eta^5\text{-}C_5H_5)_3Rh_3(\mu\text{-}CO)_3$, averaged Rh-Rh distance, 2.62 Å (from Ref. 341).*

In spite of the usual assumption that the bridging/terminal conversion of CO groups has a low energy barrier, the two Rh_3 isomers display a markedly different redox propensity |289|. As illustrated in Figure 189, the all-bridged trirhodium complex exhibits, in tetrahydrofuran solution, two reversible one-electron reductions ($E°'_{0/-}$ = -1.01 V; $E°'_{-/2-}$ = -1.75 V), whereas the doubly bridged complex undergoes reversibly a first one-electron reduction ($E°'_{0/-}$ = -1.22 V) followed by an irreversible second step (Ep = -1.73 V), which in turn leads to decomposition products reducible at very negative potential values (Ep = -2.13 V).

As a matter of fact, $[(C_5H_5)_3Rh_3(\mu\text{-}CO)_3]^-$ is indefinitely stable under inert atmosphere, whereas $[(C_5H_5)_3Rh_3(\mu\text{-}CO)_2(CO)]^-$ fragments within few minutes. It is thought that the symmetrical distribution of three bridging CO groups in the former complex may relieve more electron density from the trirhodium core than that removed by the two asymmetrical CO bridges in the latter complex. This should prevent a too strong weakening of the Rh-Rh bonds upon electron addition |289|.

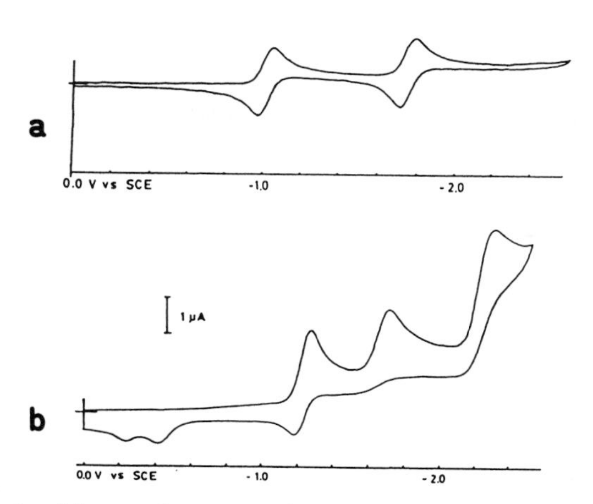

Figure 189. *Cyclic voltammetric responses recorded at a platinum electrode on THF solutions of: (a) $(\eta^5\text{-}C_5H_5)_3Rh_3(\mu\text{-}CO)_3$; (b) $(\eta^5\text{-}C_5H_5)_3Rh_3(\mu\text{-}CO)_2(CO)$. Scan rate 0.2 Vs^{-1} (from Ref. 289).*

6.2. Tetranuclear Binary Complexes

6.2.1. $Rh_4(CO)_{12}$. $Rh_4(CO)_{12}$ is isostructural with $Co_4(CO)_{12}$ (Section 5.2.1.) |293,294|, and its structure is briefly schematized in Scheme 10.

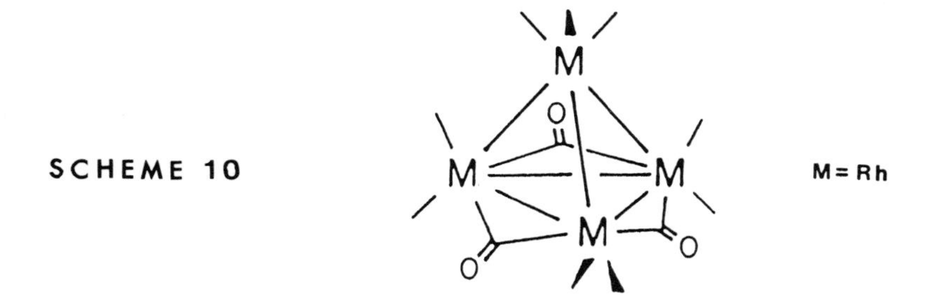

As expected on the basis of the metallic radius of rhodium (larger than that of cobalt), the mean Rh-Rh bond length, 2.73 Å, is higher than the averaged Co-Co distance in $Co_4(CO)_{12}$, 2.49 Å |293|. As in the case of $Co_4(CO)_{12}$, it is thought that such structure persists also in solution.

As shown in Figure 190, such 60-electron species undergoes, in dichloromethane solution, an irreversible one-electron reduction (Ep = -0.69 V), as well as a four-electron irreversible oxidation (Ep = +1.44 V) |296|.

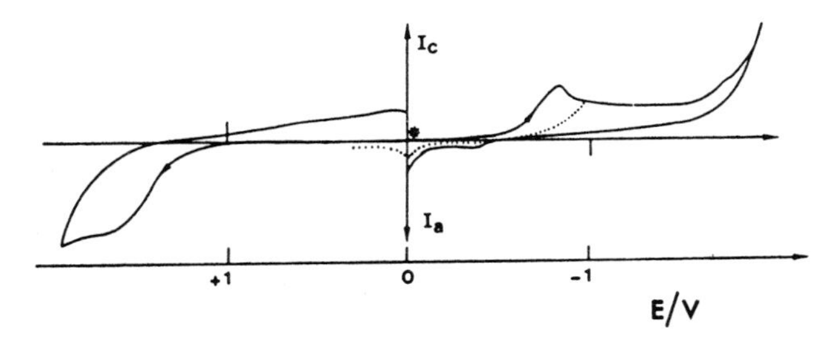

Figure 190. *Cyclic voltammogram recorded at a platinum electrode on a 1,2-$C_2H_4Cl_2$ solution of $Rh_4(CO)_{12}$. Scan rate 0.1 Vs^{-1} (from Ref. 296).*

It is noteworthy that, at variance with $[Co_4(CO)_{12}]^-$, which has a finite lifetime, the primarily electrogenerated $[Rh_4(CO)_{12}]^-$ undergoes fast decomposition (to $[Rh(CO)_4]^-$). The same happens in tetrahydrofuran solution |299|. In addition, reduction of $Rh_4(CO)_{12}$ is notably more difficult than that of $Co_4(CO)_{12}$ (E = 0.4 V). This last result agrees with an higher destabilization of the antibonding LUMO orbital (having an high metal-metal content) of $Rh_4(CO)_{12}$ with respect to the corresponding level of $Co_4(CO)_{12}$ |299,343|.

6.3. Tetranuclear Mixed-Ligand Complexes

6.3.1. $Rh_4(CO)_9[HC(PPh_2)_3]$. The tetranuclear complex $Rh_4(CO)_9[HC(PPh_2)_3]$ is thought to be isostructural with the isoelectronic, tetrahedral tetracobalt analogue (Section 5.3.3.) |299,304|, Scheme 11.

SCHEME 11

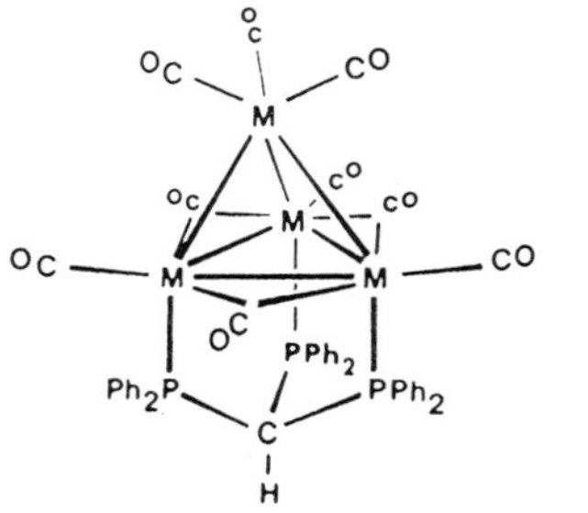

M=Co, Rh

Such 60-electron tetrarhodium species undergoes, in dichloroethane solution, an irreversible one-electron oxidation as well as an irreversible one-electron reduction, Figure 191a. In the presence of CO atmosphere, however, the electrogenerated $[Rh_4(CO)_9\{HC(PPh_2)_3\}]^-$ is able to exist for a finite lifetime ($t_{1/2} \cong$ 2-3 sec), Figure 191b |296|.

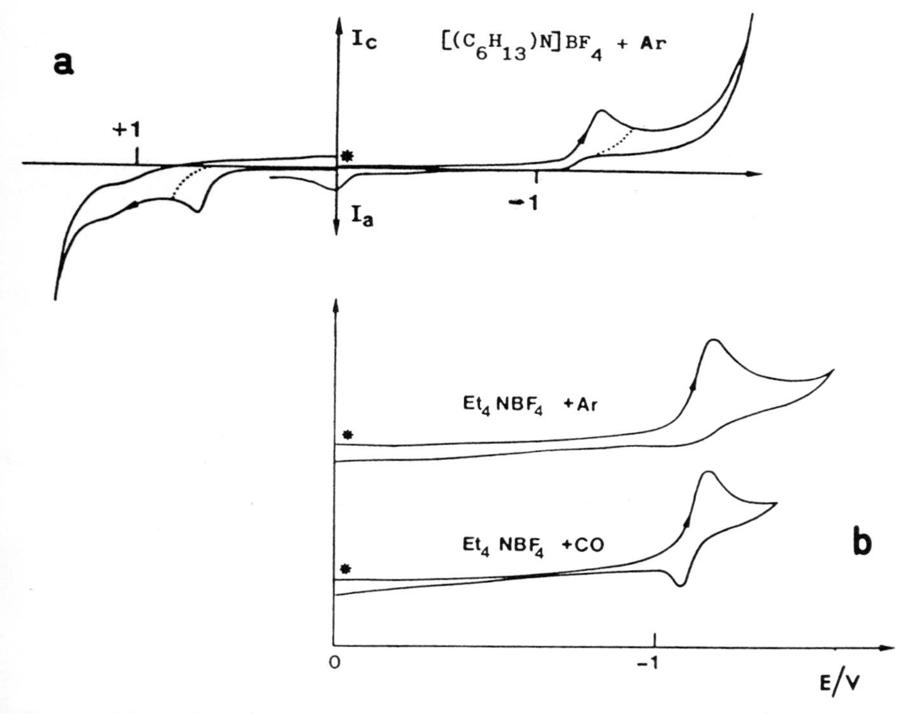

Figure 191. *Cyclic voltammograms recorded at a platinum electrode on 1,2-*$C_2H_4Cl_2$ *solutions of* $Rh_4(CO)_9[HC(PPh_2)_3]$ *under the indicated experimental conditions. Scan rate 0.1* Vs^{-1} *(from Ref. 296).*

In contrast with what happens for the cobalt analogue, the presence of the capping tripodal phosphine is not able to stabilize the entering of one-electron. Decomposition of the monoanion seems to be induced by a decarbonylation step, rather than by metal-metal breakage.

As expected on the basis of the electron-donating ability of the poliphosphine, with respect to the parent $Rh_4(CO)_{12}$, the cited redox changes are more difficult in reduction and easier in oxidation. The relevant potential values are summarized in Table 42.

TABLE 42. *Redox potentials for the electron transfers exhibited by $Rh_4(CO)_9[HC(PPh_2)_3]$, with comparison to those exhibited by the parent $Rh_4(CO)_{12}$*

complex	oxidation Ep (V)	n^a	reduction E°'$_{0/-}$ (V)	ΔEp (mV)	solvent	reference
$Rh_4(CO)_{12}$	+1.44	4	-0.69^b		$C_2H_4Cl_2$	296
			-0.80^b		THF	299
$Rh_4(CO)_9[HC(PPh_2)_3]$	+0.68	1	-1.15	70	$C_2H_4Cl_2$	296
			-1.12^c		CH_2Cl_2	298

[a]*Number of electrons involved in the electron-transfer;* [b]*peak potential value for irreversible process;* [c]*see footnote b in Table 33*

6.4. Pentanuclear Binary Complexes

6.4.1. $[Rh_5(CO)_{15}]^-$. The pentarhodium anion $[Rh_5(CO)_{15}]^-$ has the trigonal-bipyramidal geometry schematized in Figure 192 |344|.

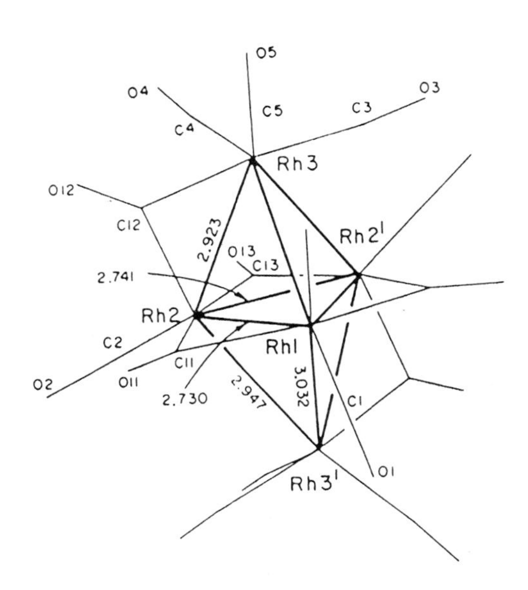

Figure 192.
Schematic drawing of the molecular structure of $[Rh_5(CO)_{15}]^-$, together with metal-metal bond distances (from Ref. 344).

The averaged Rh-Rh distance in the equatorial plane is 2.73 Å, whereas the bond lengths involving the axial metal atoms span from 2.92 Å, for those CO-edge bridged, to 3.03 Å for the unbridged ones. Ten carbonyl groups are terminal and five edge bridging.

The actual cluster is one of the members of the main equilibrium reaction among binary rhodium carbonyl polynuclear species, which makes such derivatives relatively unstable in solution |344|:

$$6\ [Rh_5(CO)_{15}]^- \underset{CO}{\overset{N_2}{\rightleftarrows}} 2\ [Rh_{12}(CO)_{30}]^{2-} + [Rh_6(CO)_{15}]^{2-} + 15\ CO$$

This reaction also demonstrates that these clusters have a molecular framework which rearranges very easily, so allowing to foresee that electron additions/removals are unlikely to leave unchanged the starting molecular assembly. As a matter of fact, $[Rh_5(CO)_{15}]^-$ undergoes two irreversible reduction processes, Figure 193 |345|.

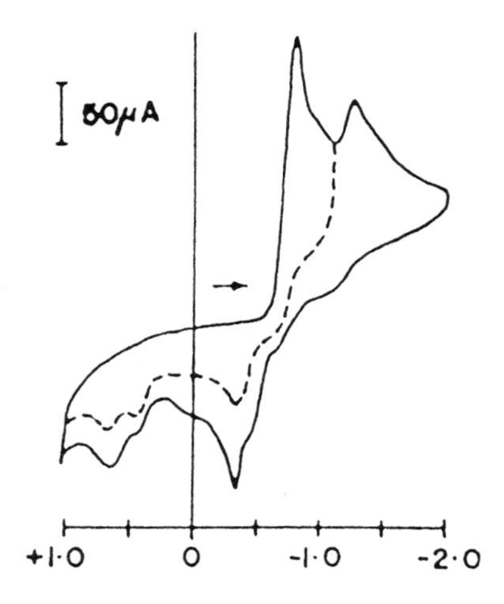

Figure 193.
Cyclic voltammetric response exhibited by $[Rh_5(CO)_{15}]^-$ in THF solution, at -50 °C. Scan rate 2 Vs^{-1} (from Ref. 345).

The most significant redox change is the first two-electron reduction step, (Ep = -1.05 V*), attributed to the generation of $[Rh_5(CO)_{15}]^{3-}$, which undergoes immediate rearrangement to unidentified species even at -50°C and at the high scan rate of 2 Vs^{-1}.

---

*Arbitrarily referenced to S.C.E. In Ref. 345 potential values refer to an unquoted pseudoreference Ag electrode.

6.5. Hexanuclear Binary Complexes

6.5.1. $[Rh_6(CO)_{15}]^{2-}$. Based on spectroscopic evidence, the hexanuclear dianion $[Rh_6(CO)_{15}]^{2-}$ was assigned an octahedral geometry similar to that of $[Co_6(CO)_{15}]^{2-}$ (Section 5.4.1.) |346|.

As illustrated in Figure 194, and in agreement with previous investigations |345|, such an 86-electron system undergoes, in tetrahydrofuran solution, an irreversible, close-spaced two-electron oxidation ($Ep_{2-/-}$ = -0.11 V; $Ep_{-/0}$ = -0.05 V) as well as a quasireversible one-electron reduction ($E°'$ = -1.18 V; $\Delta Ep_{(0.2 Vs^{-1})}$ = 165 mV) complicated by relatively fast decomposition of the primarily electrogenerated trianion $[Rh_6(CO)_{15}]^{3-}$ ($t_{1/2}$ ≅ 1 s), thus indicating it to be the only stable member of the present family of hexarhodium clusters.

Figure 194.
Cyclic voltammogram recorded at a platinum electrode on a THF solution of $[Rh_6(CO)_{15}]^{2-}$. Scan rate 0.2 Vs^{-1}.

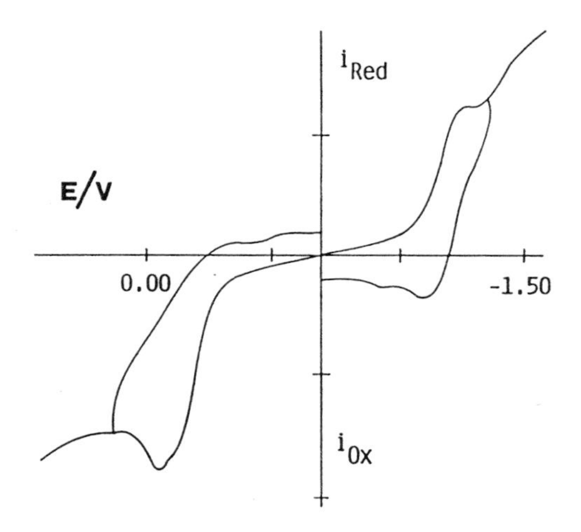

The full instability of the first one-electron oxidation product $[Rh_6(CO)_{15}]^-$ deserves some attention, since its dimeric species $[Rh_{12}(CO)_{30}]^{2-}$ is, by contrast, completely stable (Section 6.7.1.).

6.6. Hexanuclear Carbido Complexes

6.6.1. $[Rh_6(CO)_{15}C]^{2-}$. The X-ray structure of the hexarhodium carbido-carbonyl dianion $[Rh_6(CO)_{15}C]^{2-}$ is shown in Figure 195 |347|.

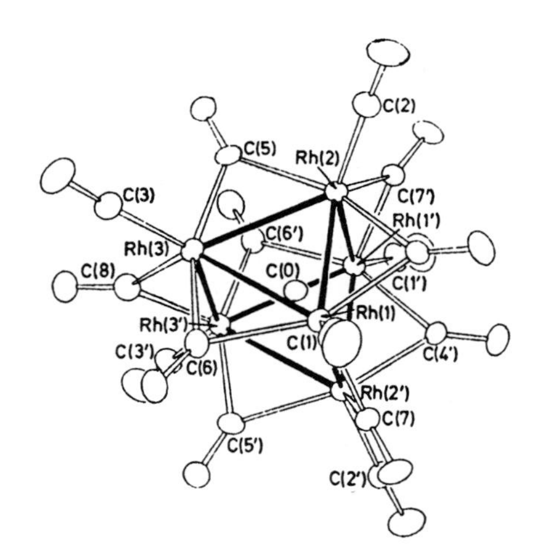

Figure 195. *Molecular structure of* $[Rh_6(CO)_{15}C]^{2-}$ *(from Ref. 347).*

Like the cobalt analogue, the rhodium cluster displays a trigonal prismatic geometry encapsulating one carbide atom. The mean basal Rh-Rh bond length is 2.78 Å, whereas the interbasal Rh-Rh distance is 2.82 Å. The averaged $Rh-C_{(carbide)}$ distance is 2.13 Å. All the nine edges of the metal prism are bridged by CO groups. The remaining six carbonyls are terminal.

The redox pattern of this 90-electron dianion, Figure 196, indicates it undergoes, in dichloroethane solution, a single stepped irreversible two-electron removal (Ep = +0.34 V); controlled potential electrolysis under CO atmosphere leads to $[Rh_{12}(CO)_{24}(C)_2]^{2-}$ |132|.

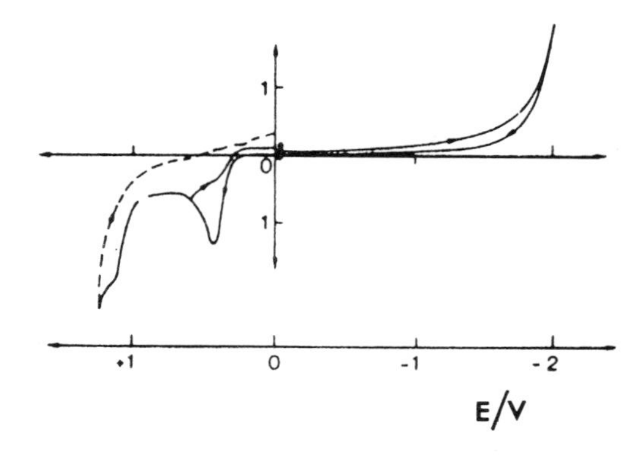

Figure 196. *Cyclic voltammogram recorded at a platinum electrode on a 1,2-* $C_2H_4Cl_2$ *solution of* $[Rh_6(CO)_{15}C]^{2-}$*. Scan rate 0.1* Vs^{-1} *(from Ref. 132).*

This result confirms once again the scarce flexibility of the hexarhodium clusters to add/lose electrons.

6.7. Dodecanuclear Binary Complexes

6.7.1. $[Rh_{12}(CO)_{30}]^{2-}$. The molecular geometry of the dodecarhodium dianion $[Rh_{12}(CO)_{30}]^{2-}$ is shown in Figure 197 |348,349|. It is constituted by two hexarhodium octahedral subunits connected by a Rh-Rh bond (2.82 Å), edge-bridged by two CO groups. Although the presence of significant spreading in Rh-Rh bond lengths, the mean Rh-Rh distance of 2.77 Å is very close to that of $Rh_6(CO)_{16}$ (2.78 Å) |350|.

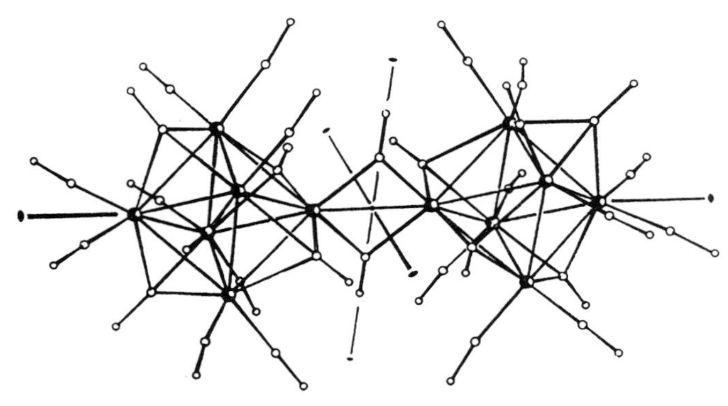

Figure 197. *Perspective view of the dianion $[Rh_{12}(CO)_{30}]^{2-}$ (from Ref. 349).*

Figure 198 shows the redox pattern exhibited by $[Rh_{12}(CO)_{30}]^{2-}$ in tetrahydrofuran solution |345|. Also in this case, there is no flexibility in adding or removing electrons. In fact, the first irreversible reduction process (Ep = -1.25 V, see footnote in Section 6.4.1.) is attributed to the two-electron addition $[Rh_{12}(CO)_{30}]^{2-}/[Rh_{12}(CO)_{30}]^{4-}$, with the tetraanion fast decomposing ($t_{1/2} \cong$ 50 ms, at -40°C) to $[Rh_6(CO)_{15}]^{2-}$, which in turn should undergo reduction in correspondence to the second cathodic peak (Ep = -1.7 V)*.

---

*This latter conclusion, really, does not agree with the redox behaviour of $[Rh_6(CO)_{15}]^{2-}$, as experienced by this author (see Section 6.5.1.)

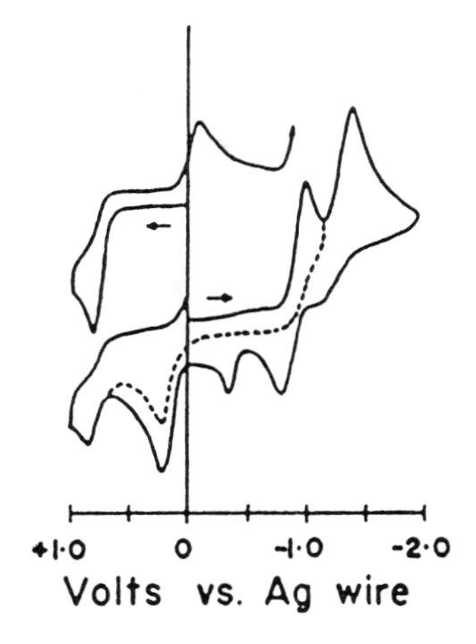

Figure 198.
Cyclic voltammograms recorded on a THF solution of $[Rh_{12}(CO)_{30}]^{2-}$ (from Ref. 345).

Finally, $[Rh_{12}(CO)_{30}]^{2-}$ also undergoes a two-electron removal (Ep = +0.55 V), leading to the corresponding, unstable neutral species ($t_{1/2} \cong 60$ ms, at -40°C).

6.8. Dodecanuclear Carbido Complexes

6.8.1. $[Rh_{12}(CO)_{24}(C)_2]^{2-}$. The dodecarhodate carbido dianion $[Rh_{12}(CO)_{24}(C)_2]^{2-}$ has a polyhedral geometry which can be described in terms of a stacked sequence of square-rhomb-square layers or, alternatively, as two prismatic Rh_6C cores sharing one edge, capped, at opposite sides, by two extra rhodium atoms, Figure 199 |351|.

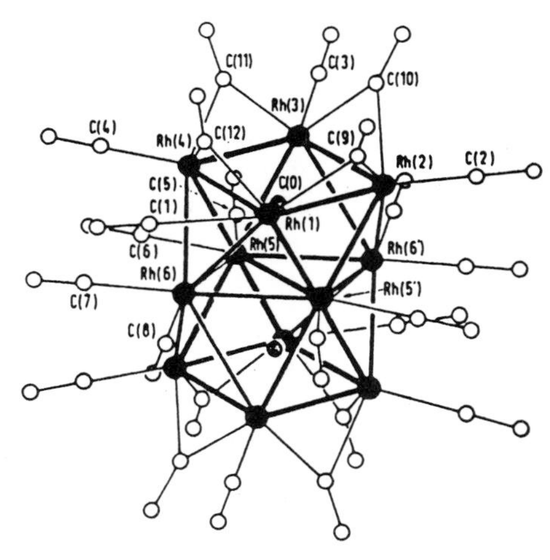

Figure 199.
Molecular structure of the dianion $[Rh_{12}(CO)_{24}(C)_2]^{2-}$ (from Ref. 351).

Each Rh_6 prismatic unit encapsulates one carbide atom. Eight CO groups are edge-bridging with respect to the outer squares; eight of the remaining sixteen terminal carbonyls are in the plane of the central tetrametal rhomb. The geometrical complexity of the metal core causes spreading of the Rh-Rh bond lengths from 2.75 Å to 2.92 Å, with an averaged value of 2.81 Å. The mean Rh-$C_{(carbide)}$ distance within the two fused prismatic units is 2.12 Å, quite close to that found in the Rh_6C core of $[Rh_6(CO)_{15}C]^{2-}$ (2.13 Å).

Figure 200 illustrates the cyclic voltammetric response exhibited by $[Rh_{12}(CO)_{24}(C)_2]^{2-}$ in acetonitrile solution |352|.

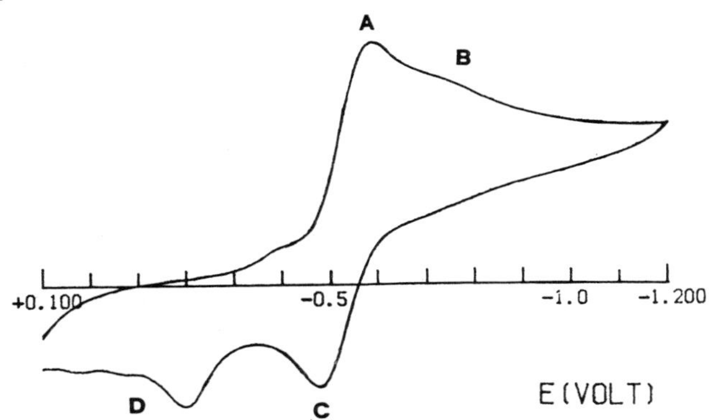

Figure 200. *Cyclic voltammogram recorded at a platinum electrode on a deaerated MeCN solution of* $[Rh_{12}(CO)_{24}(C)_2][N(PPh_3)_2]_2$ *(*$[NEt_4][ClO_4]$*) supporting electrolyte). Scan rate 0.02* Vs^{-1}.

The reduction process occurring at peak A is difficult to analyse because of the presence of adsorption phenomena at the electrode surface in correspondence to peak B. As a matter of fact, controlled potential electrolysis affords the tetraanion $[Rh_{12}(CO)_{23}(C)_2]^{4-}$, but consuming a number of Faradays/mole varying from 0.9 to 1.1, in the temperature range from +20 °C to -20°C. This datum has been interpreted in terms of electrogeneration of an intermediate species able to react with the starting dianion according to an ECE scheme of the type:

$$[Rh_{12}(CO)_{24}(C)_2]^{2-} \underset{-e}{\overset{+e}{\rightleftharpoons}} [Rh_{12}(CO)_{24}(C)_2]^{3-} \quad (E°' = -0.54\ V) \quad \underset{-CO}{\overset{*}{\longrightarrow}} [Rh_{12}(CO)_{23}(C)_2]^{3-}$$

$$[Rh_{12}(CO)_{23}(C)_2]^{3-} \underset{-e}{\overset{+e}{\rightleftharpoons}} [Rh_{12}(CO)_{23}(C)_2]^{4-} \quad E°' = -0.46\ V$$

*partial decomposition to species able to react with $[Rh_{12}(CO)_{24}(C)_2]^{2-}$

The redox change $[Rh_{12}(CO)_{23}(C)_2]^{3-/4-}$ will be discussed in the next Section.

6.8.2. $[Rh_{12}(CO)_{23}(C)_2]^{4-}$. Chemical reduction of the dianion $[Rh_{12}(CO)_{24}(C)_2]^{2-}$ affords the isoelectronic carbido-tetraanion $[Rh_{12}(CO)_{23}(C)_2]^{4-}$, the crystal structure of which is shown in Figure 201 |353|.

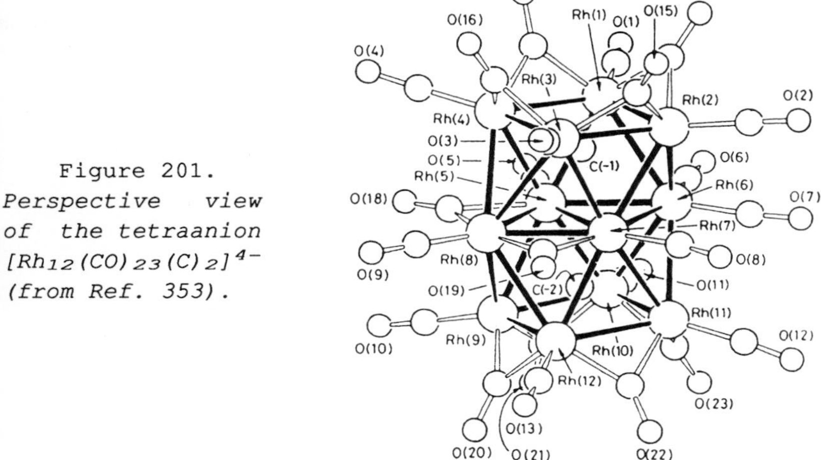

Figure 201. *Perspective view of the tetraanion $[Rh_{12}(CO)_{23}(C)_2]^{4-}$ (from Ref. 353).*

It is easily observed that the metal polyhedron is quite similar to that of the precursor $[Rh_{12}(CO)_{24}(C)_2]^{2-}$, the only macroscopic difference being that in the central tetrametal rhomb of the tetraanion three of the seven CO groups are edge bridging, whereas in the dianion all the eight carbonyls are terminal. Also, with respect to the dianion, a slightly wider spreading of the Rh-Rh distances holds (from 2.71 Å to 3.09 Å), with a slightly longer (by 0.02 Å) averaged value of 2.83 Å. The $Rh\text{-}C_{(carbide)}$ distance within the two Rh_6C subunits slightly elongates to 2.13 Å.

Figure 202 shows the redox pattern exhibited by the present tetraanion in acetonitrile solution |352|. Two distinct one-electron oxidations are displayed, only the first one being chemically, and almost electrochemically too, reversible. In addition, a single-stepped multielectron reduction is present, in turn complicated by following chemical reactions.

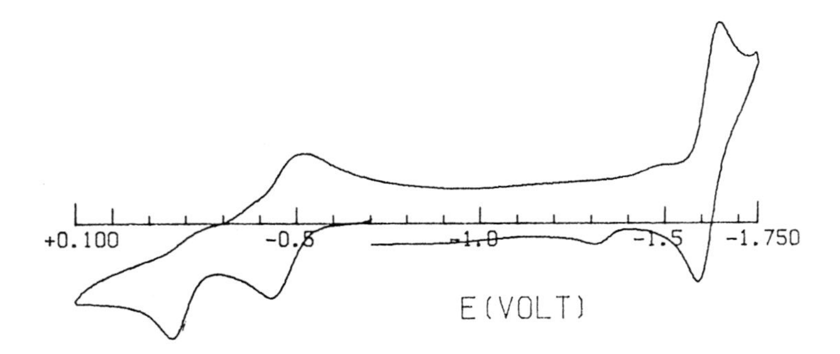

Figure 202. *Cyclic voltammogram exhibited, at a platinum electrode, by* $[Rh_{12}(CO)_{23}(C)_2]^{4-}$ *in acetonitrile solution. Scan rate 0.2* Vs^{-1}.

In summary, the redox propensity of $[Rh_{12}(CO)_{23}(C)_2]^{4-}$ can be so schematized:

$$
\begin{array}{lcl}
 & E^{\circ\prime} = -1.62\ V & \\
[Rh_{12}(CO)_{23}(C)_2]^{4-} & \underset{-2e}{\overset{+2e}{\rightleftharpoons}} & [Rh_{12}(CO)_{23}(C)_2]^{6-} \\
E^{\circ\prime} = -0.46\ V \quad -e \downarrow\uparrow +e & & \downarrow t_{1/2} = 1s \\
[Rh_{12}(CO)_{23}(C)_2]^{3-} & & \text{decomposition to furtherly electroreducible species} \\
Ep = -0.16\ V \quad -e \downarrow & & \\
[Rh_{12}(CO)_{23}(C)_2]^{2-} & & \\
\downarrow \text{fast} & & \\
\text{decomposition products} & &
\end{array}
$$

The geometry of the only stable redox congeners $[Rh_{12}(CO)_{23}(C)_2]^{3-}$ has been X-ray determined |354|. In agreement with the electrochemical reversibility of the relevant redox change, this paramagnetic 165-electron trianion |334| is strictly isostructural with the parent tetraanion. Even if the Rh-Rh bond distances have a slightly larger range (from 2.66 Å to 3.16 Å), the mean metal-metal bond length remains 2.83 Å, and the mean Rh-$C_{(carbide)}$ distance is still 2.13 Å.

This may be interpreted assuming that the one-electron addition/removal is really delocalized over the whole cluster framework.

7. IRIDIUM CLUSTERS

7.1. *Trinuclear Mixed-Ligand Complexes*

7.1.1. $Ir_3(SBu^t)_3(RCCR)(CO)_6$. A limited series of trinuclear, alkyno-carbonyl iridium complexes of general formula $Ir_3(SBu^t)_3(RCCR)(CO)_6$ (R = CF_3, CO_2Me) have been characterized. Figure 203 shows the molecular geometry of $Ir_3(\mu\text{-}SBu^t)_3(CF_3C_2CF_3)(CO)_6$ |355|.

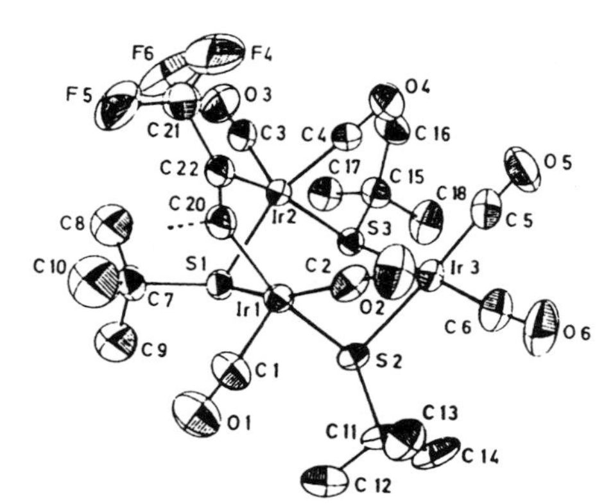

Figure 203. *Perspective view of $Ir_3(SBu^t)_3[C_2(CF_3)_2](CO)_6$ (from Ref. 355).*

In this triiridium species, only the alkyne-bridged metal-metal vector lies at a bonding distance (Ir1-Ir2, 2.69 Å). The core can be described as a six-membered ring with Ir and S atoms alternating along a chair conformation. All the CO groups are in terminal positions.

As illustrated in Figure 204 for $Ir_3(SBu^t)_3(CF_3C_2CF_3)(CO)_6$, these complexes undergo, in acetonitrile solution, both an irreversible one-electron oxidation and an irreversible one-electron reduction |356|.

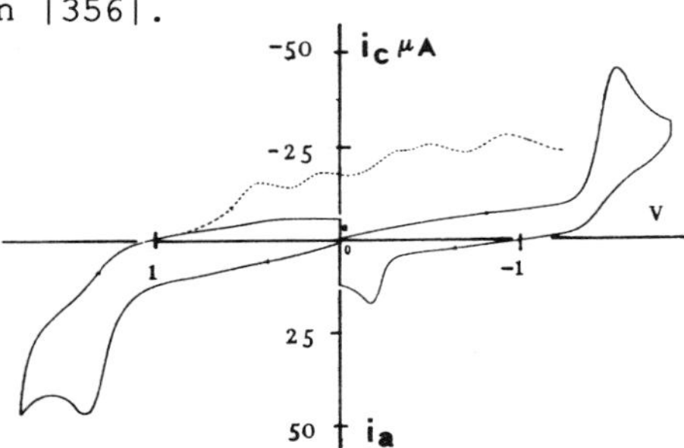

Figure 204. *Cyclic voltammetric response exhibited by $Ir_3(SBu^t)_3(CF_3C_2CF_3)(CO)_6$ in MeCN solution. Scan rate 0.1 Vs^{-1} (from Ref. 356).*

Both the redox changes (R = CF_3, $E_{p+/o}$ = +1.45 V, $E_{po/-}$ = -1.50 V; R = CO_2Me, $E_{p+/o}$ = +1.33 V, $E_{po/-}$ = -1.60 V) lead to fragmentation. The difficulty to add or to remove electrons testifies to the high redox stability of these species.

7.2. Tetranuclear Binary Complexes

7.2.1. $Ir_4(CO)_{12}$. At variance with the isoelectronic complexes $Co_4(CO)_{12}$ and $Rh_4(CO)_{12}$, which bear three edge-bridging carbonyl groups in the basal trimetallic plane of the tetrahedral assembly (see Sections 5.2.1., 6.2.1.), the structure of $Ir_4(CO)_{12}$ reveals a tetrahedral Ir_4 geometry displaying only terminal CO groups, Figure 205 |357|.

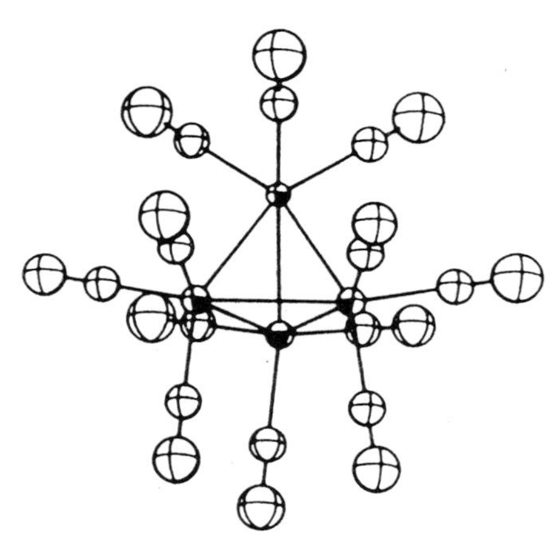

Figure 205. *Perspective view of $Ir_4(CO)_{12}$. Averaged Ir-Ir bond length, 2.69 Å (from Ref. 357).*

It seems interesting to point out that in the substituted derivatives $[Ir_4(CO)_{11}X]^-$ (X = halide or pseudohalide |358,359|, COOR |360|, H |361|), $[Ir_4(CO)_{10}(X)_2]^{n-}$ (X = PPh_3, n = 0 |362|; X = 1/2 diars [diars = 1,2-bis(dimethylarsino)benzene], n = 0 |363|; X = 1/2 1,5-cyclooctadiene, n = 0 |364|; X = H, n = 2 |365|), $Ir_4(CO)_9(X)_3$ (X = PPh_3 |362|), $Ir_4(CO)_8(X)_4$ (X = 1/2 dppm [dppm = bis(diphenylphosphino)methane] |366|, PMe_3 |367|) the tetrametal tetrahedral assembly bears a bridging-carbonyl disposition similar to that of $M_4(CO)_{12}$, M = Co, Rh. Only for $Ir_4(CO)_{11}[CNBu^t]$ a nonbridged tetrahedral assembly has been found |368,369|.

Indeed, confirmation of the spectroscopic evidence that the energy barrier to the CO-bridged/CO-unbridged interconversion is small |370| comes from the X-ray structure of $[Ir_4(CO)_{11}(SCN)]^-$, which shows all terminal CO groups when it crystallyzes with

$[NMe_2(CH_2Ph)_2]^+$ counteranion |371|, while displaying the cited CO-bridged configuration when it crystallyzes with $[N(PPh_3)_2]^+$ counteranion |372|.

The redox propensity of $Ir_4(CO)_{12}$ is difficult to be ascertained because of its insolubility in organic solvents. In dichloromethane solution, at ambient temperature, we recorded, at a mercury electrode, an irreversible reduction peak at -1.39 V. No further investigation could be performed because of the notably low current intensity |373|. This result contrasts with the obtainment, by chemical reduction, of the stable paramagnetic monoanion $[Ir_4(CO)_{12}]^-$ |190|. Indeed, the instability of $[Ir_4(CO)_{12}]^-$, as well as its generation at potential values significantly more negative than $[Rh_4(CO)_{12}]^-$, can be accounted for by both the antibonding character of the LUMO level of $Ir_4(CO)_{12}$ and its destabilization to higher energies (by about 0.5 eV) |343|.

7.3. Tetranuclear Mixed-Ligand Complexes

7.3.1. $Ir_4(CO)_9[HC(PPh_2)_3]$. Just like in $Co_4(CO)_9[HC(PPh_2)_3]$, the poliphosphine $HC(PPh_2)_3$ maintains the basal carbonyl-bridged tetrahedral geometry of the precursor $Co_4(CO)_{12}$ (Section 5.3.3.), in the substitution of the CO groups in (all-terminal carbonyl structured) $Ir_4(CO)_{12}$ it gives rise to $Ir_4(CO)_9[HC(PPh_2)_3]$, in which all the carbonyl groups are terminal, Figure 206 |374|.

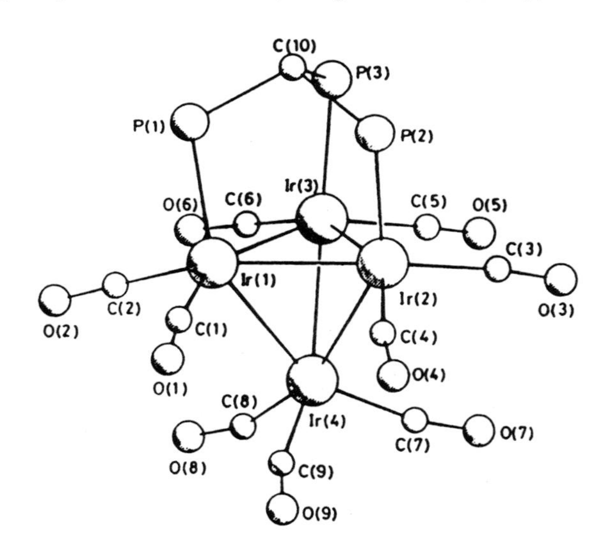

Figure 206. *Perspective view of $Ir_4(CO)_9[HC(PPh_2)_3]$ (from Ref. 374).*

At variance with $Co_4(CO)_9[HC(PPh_2)_3]$, in which the apical to basal metal-metal distances are significantly longer than the basal-basal ones, here they are almost equal each other (basal-basal, 2.69 Å; basal-apical, 2.68 Å), and coincident with those of the fully carbonylated precursor.

Unlike $Co_4(CO)_9[HC(PPh_2)_3]$, but like $Rh_4(CO)_9[HC(PPh_2)_3]$, $Ir_4(CO)_9[HC(PPh_2)_3]$ undergoes, both in dichloromethane and tetrahydrofuran solution, a one-electron reduction (Ep = -1.80 V (see-footnote b in Table 33), which affords the unstable monoanion $[Ir_4(CO)_9\{HC(PPh_2)_3\}]^-$ |299|.

7.4. Hexanuclear Binary Complexes

7.4.1. $[Ir_6(CO)_{15}]^{2-}$. The hexairidium dianion $[Ir_6(CO)_{15}]^{2-}$ has the octahedral geometry shown in Figure 207 |375|.

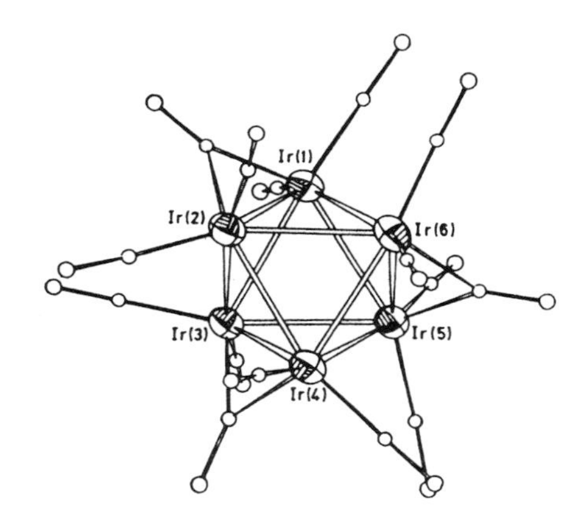

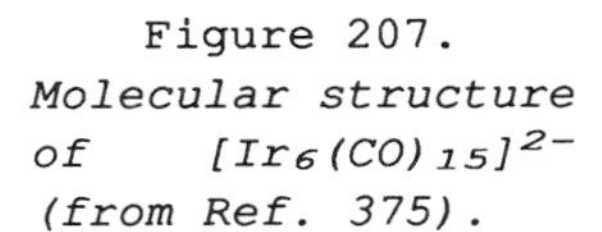

Figure 207. *Molecular structure of $[Ir_6(CO)_{15}]^{2-}$ (from Ref. 375).*

In confirmation of the lower tendency of iridium atoms to assume carbonyl-bridged assemblies with respect to cobalt (and rhodium), twelve CO groups are terminal and three edge-bridging, whereas in the isoelectronic $[Co_6(CO)_{15}]^{2-}$ there are six terminal, three edge-bridging, and three face-bridging carbonyls (see Section 5.4.1.). The averaged Ir-Ir bond length is 2.77 Å, but the CO-bridged metal-metal distances (mean 2.74 Å) are, as usual, significantly shorter than the unbridged one (2.79 Å).

This 86-electron dianion undergoes, in tetrahydrofuran solution, two subsequent one-electron oxidations, both showing features of chemical reversibility, Figure 208 |297|.

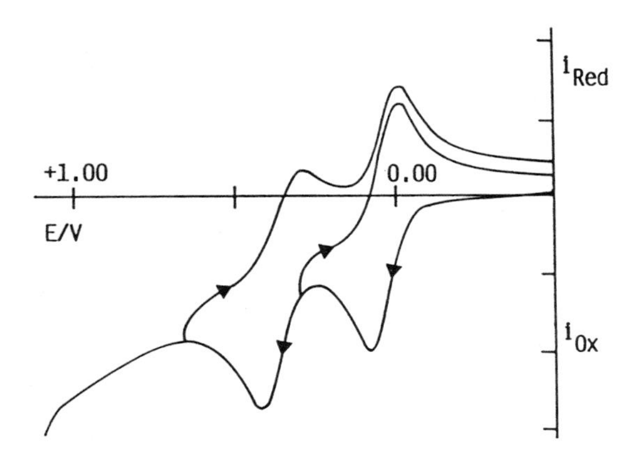

Figure 208. *Cyclic voltammogram recorded at a platinum electrode on a THF solution of $[Ir_6(CO)_{15}]^{2-}$. Scan rate 0.2 Vs^{-1}.*

It is interesting to note that, at variance with the redox behaviour of $[Co_6(CO)_{15}]^{2-}$ and $[Rh_6(CO)_{15}]^{2-}$, which do not afford stable congeners, not only the $[Ir_6(CO)_{15}]^{2-/-}$ redox change ($E°'$ = +0.04 V, $\Delta Ep_{(0.2\ Vs^{-1})}$ = 96 mV) is fully chemically reversible, but also the further $[Ir_6(CO)_{15}]^{-/0}$ oxidation ($E°'$ = +0.36 V) leads to the neutral $Ir_6(CO)_{15}$, which, even if transient, possesses a finite lifetime ($t_{1/2} \cong 2$ sec).

The near reversibility of the $[Ir_6(CO)_{15}]^{2-/-}$ redox change allows one to foresee that no significant stereochemical variation occurs in the structure of the monoanion with respect to that of the parent dianion.

7.5. Hexanuclear Mixed-Ligand Complexes

7.5.1. $[Ir_6(CO)_{15}Cl]^-$. The monoanion $[Ir_6(CO)_{15}Cl]^-$ is obtained by reaction of $[Ir_6(CO)_{15}]^{2-}$ with $FeCl_3$. Its molecular structure is shown in Figure 209 |376|. The anion is formed by an octahedron of iridium atoms surrounded by eleven terminal, one edge-bridging, and three face-bridging carbonyl ligands. The chlorine atom occupies a terminal position. Although the spreading of Ir-Ir distances imposed by the different dispositions of the CO ligands, the averaged value (2.77 Å) remains unvaried with respect to that of $[Ir_6(CO)_{15}]^{2-}$. The distance Ir1-Cl is 2.40 Å, notably longer than the averaged value of 1.85 Å for the distance $Ir\text{-}C_{(CO\ terminal)}$ |376|.

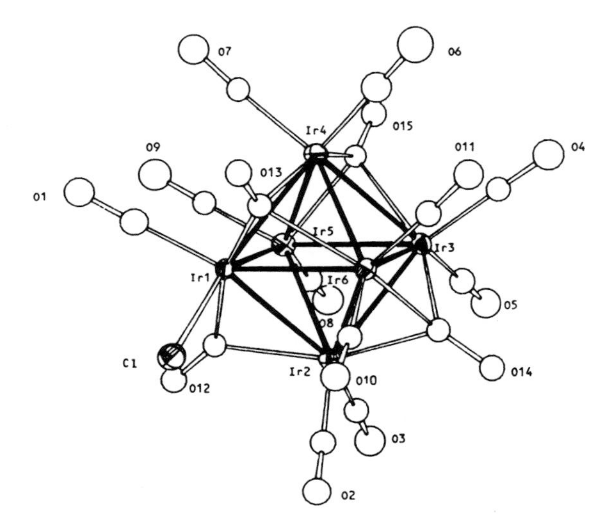

Figure 209.
Crystal structure of $[Ir_6(CO)_{15}Cl]^-$ (from Ref. 376).

In tetrahydrofuran solution, $[Ir_6(CO)_{15}Cl]^-$ undergoes an irreversible two-electron reduction (Ep = -0.93 V) which generates the precursor dianion $[Ir_6(CO)_{15}]^{2-}$ |297|.

7.6. Dodecanuclear Binary Complexes

7.6.1. $[Ir_{12}(CO)_{24}]^{2-}$. Oxidation of $[Ir_6(CO)_{15}]^{2-}$ with $[Cu(NCCH_3)_4]^+$ affords the dodecanuclear dianion $[Ir_{12}(\mu\text{-}CO)_5(CO)_{19}]^{2-}$, the molecular structure of which is illustrated in Figure 210 |377|.

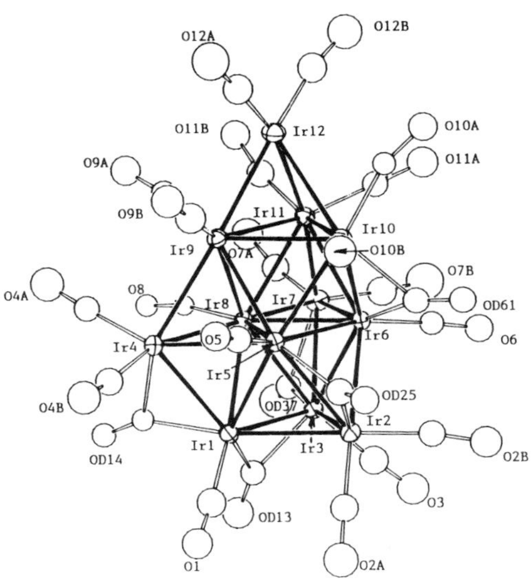

Figure 210.
Perspective view of the dianion $[Ir_{12}(CO)_{24}]^{2-}$ (from Ref. 377).

The close packed metallic assembly can be viewed as a large trigonal bipyramid lacking one apical and one equatorial vertex. The averaged Ir-Ir distance is 2.77 Å. Nineteen carbonyls are terminal, the remaining five are edge-bridging.

As shown in Figure 211a, in acetonitrile solution, $[Ir_{12}(CO)_{24}]^{2-}$ undergoes a series of single-stepped two-electron redox changes [namely, two reductions, ($[Ir_{12}(CO)_{24}]^{2-/4-}$, Ep = -0.85 V; $[Ir_{12}(CO)_{24}]^{4-/6-}$, Ep = -1.50 V) and two oxidations ($[Ir_{12}(CO)_{24}]^{2-/0}$, Ep = +0.23 V; $[Ir_{12}(CO)_{24}]^{0/2+}$, Ep = +0.83 V), all of which look like complicated by subsequent decomposition processes. Really, the first reduction process, Figure 211b, proved to be chemically reversible. In view of the large peak-to-peak separation Ep_H-Ep_A = 520 mV, it can be foreseen that a large structural reorganization of the original metallic frame takes place upon two-electron addition |377|.

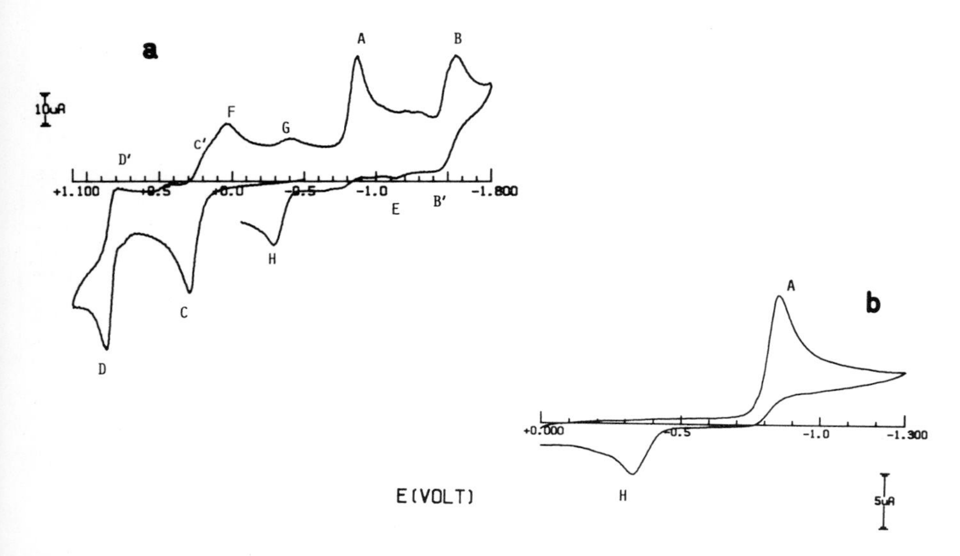

Figure 211. *Cyclic voltammetric responses recorded at a platinum electrode on a MeCN solution of $[Ir_{12}(CO)_{24}]^{2-}$. Scan rates: (a) 0.2 Vs^{-1}; (b) 0.02 Vs^{-1} (from Ref. 377).*

7.7. *Tetradecanuclear Binary Complexes*

7.7.1. $[Ir_{14}(CO)_{27}]^-$. Oxidation of $[Ir_6(CO)_{15}]^{2-}$ with $[(C_5H_5)_2Fe]^+$ affords the paramagnetic tetradecanuclear anion $[Ir_{14}(CO)_{27}]^-$, the molecular structure of which is illustrated in Figure 212 |378|.

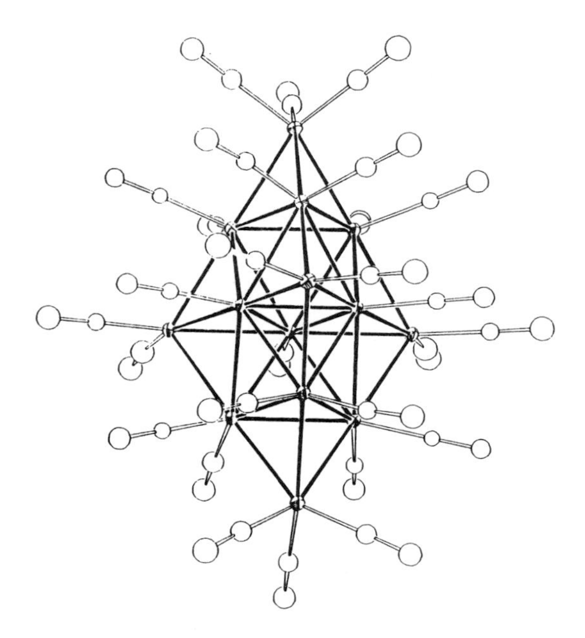

Figure 212. *Perspective view of the monoanion $[Ir_{12}(CO)_{24}]^{2-}$ (from Ref. 378).*

The large metallic frame possessing two more iridium atoms than $[Ir_{12}(CO)_{24}]^{2-}$ now completes the trigonal bipyramidal assembly. The averaged Ir-Ir distance is 2.72 Å. All the CO groups are terminal.

As deducible from Figure 213, such a monoanion is sufficiently flexible to support both a one-electron removal ($E°'$ = +0.39 V) and a one-electron addition ($E°'$ = 0.00 V) without cluster degradation |378|.

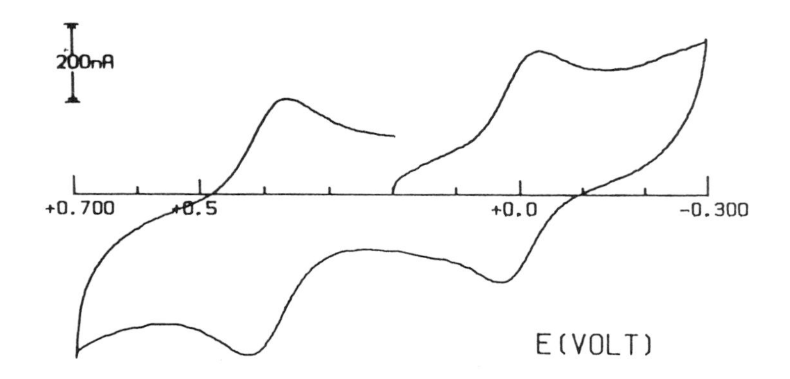

Figure 213. *Cyclic voltammogram recorded at a platinum electrode on a CH_2Cl_2 solution of $[Ir_{14}(CO)_{27}]^-$. Scan rate 0.05 Vs^{-1} (from Ref. 378).*

8. NICKEL CLUSTERS

8.1. *Trinuclear Mixed-Ligand Complexes*

8.1.1. $Ni_3(C_5H_{5-n}Me_n)_3(CO)_2$ (n = 0,1,5). The Fischer-Palm cluster $Ni_3(\eta^5-C_5H_5)_3(\mu_3-CO)_2$ was the first derivative displaying a geometry in which a trimetallic triangle is capped above and below by triply bridging carbonyl groups |379,380|. This paramagnetic 49-electron molecule can be chemically reduced to the corresponding diamagnetic 50-electron monoanion $[Ni_3(\eta^5-C_5H_5)_3(\mu_3-CO)_2]^-$, the molecular structure of which is shown in Figure 214 |381|.

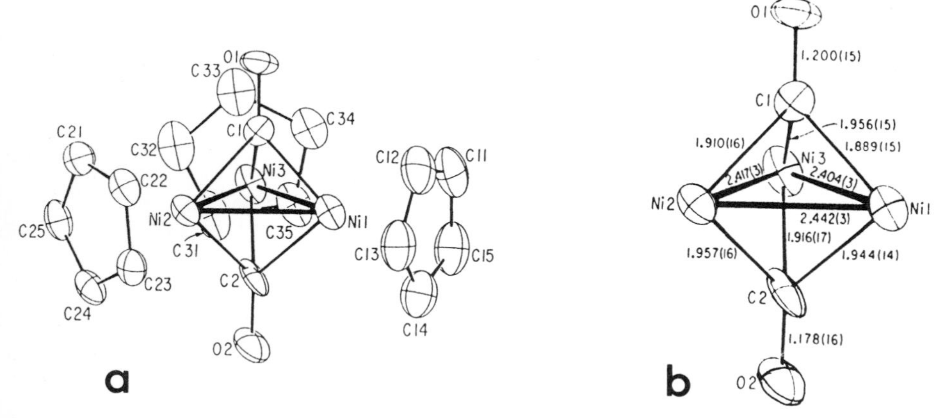

Figure 214. *Perspective view (a) and selected bond lengths (b) of $[Ni_3(C_5H_5)_3(CO)_2]^-$ (from Ref. 381).*

The neutral and monoanion species can be considered substantially isostructural. The more interesting difference lies in the Ni-Ni distances, which are significantly larger in the monoanion derivative, Table 43.

TABLE 43. *Selected interatomic distances (in Å) for the couple $[Ni_3(C_5H_5)_3(CO)_2]^{0/-}$ |381|*

distance	$Ni_3(C_5H_5)_3(CO)_2$	$[Ni_3(C_5H_5)_3(CO)_2]^-$
Ni-Ni	2.39	2.42
$Ni-C_{(CO)}$	1.93	1.93
$Ni-C_{(ring)}$	2.13	2.16
OC···CO	2.71	2.66

This metal-metal elongation, upon one-electron addition, agrees with the assumption that the metal-metal antibonding HOMO, which is half-filled in the neutral molecule (in the 48-electron analogue $CoNi_2(C_5H_5)_3(CO)_2$, the Ni-Ni distance is somewhat shorter, 2.36 Å |382|), can be filled without destruction of the architectural assembly |381,382|. Nevertheless, the redox flexibility of the Fischer-Palm cluster and its methylated analogues is, really, more extended. Figure 215 illustrates the cyclic voltammetric patterns exhibited by $Ni_3(C_5H_{5-n}Me_n)_3(CO)_2$ (n = 0,1,5) |286,381|.

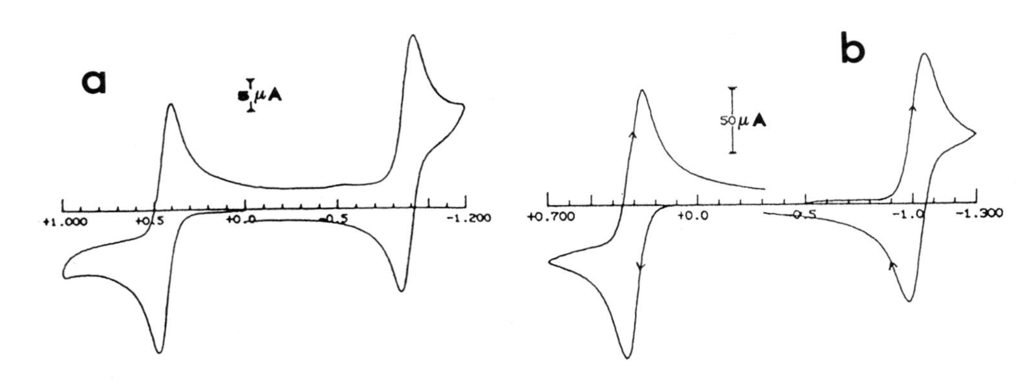

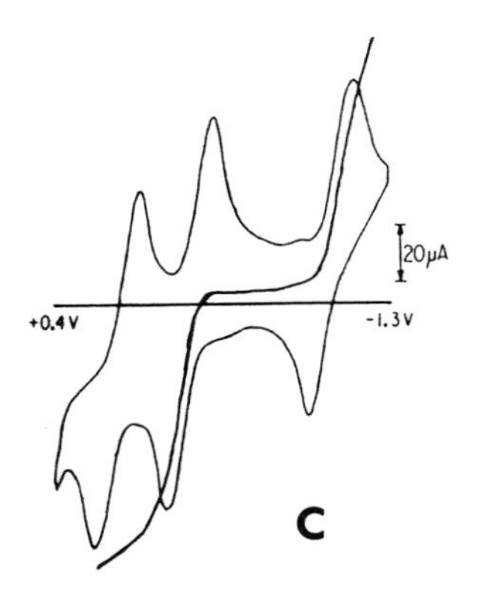

Figure 215. *Cyclic voltammograms exhibited by: (a) $Ni_3(C_5H_5)_3(CO)_2$ in CH_2Cl_2 solution at a platinum electrode; (b) $Ni_3(C_5H_4Me)_3(CO)_2$ in CH_2Cl_2 solution at a platinum electrode; (c) $Ni_3(C_5Me_5)_3(CO)_2$ in THF solution at a gold electrode (from Refs. 286,381).*

It can be seen that, besides the common one-electron reduction, whose stereodynamics have been discussed above, all the derivatives exhibit an apparently chemically reversible one-electron oxidation. In addition the permethylated compound displays a further one-electron removal. Table 44 summarizes the relevant electrochemical characteristics.

TABLE 44. *Electrochemical parameters of the redox changes exhibited by the series $Ni_3(C_5H_{5-n}Me_n)_3(CO)_2$ (n = 0,1,5)*

n	E°' 50/49-e	ΔEp	E°' 49/48-e	ΔEp	E°' 48/47-e	ΔEp	solvent	reference
0	-0.88	90[a]	+0.44	100[a]			CH_2Cl_2	286
0	-0.85	62[b]	+0.37[c]				C_6H_5CN	59
1	-1.02	65[a]	+0.30	65[a]			CH_2Cl_2	286
5	-1.17	245[d]	-0.40	270[d]	+0.03	260[d]	THF	381

[a]*Measured at 0.2 Vs^{-1};* [b]*measured at 0.02 Vs^{-1};* [c]*peak potential value for irreversible process;* [d]*measured at 0.5 Vs^{-1}*

As expected, the presence of methyl substituents in the cyclopentadienyl ligand favours electron removal, while disfavouring electron addition.

Even if cationic species failed to be isolated, it is not surprising that removal of the 49-th electron from the metal-metal antibonding orbital leads to stable congeners.

8.1.2. $Ni_3(C_5H_4Me)_3(CO)(CS)$. The crystal structure of the thiocarbonyl complex $Ni_3(C_5H_4Me)_3(CO)(CS)$, analogue of the Fischer-Palm molecule, is illustrated in Figure 216 |383|.

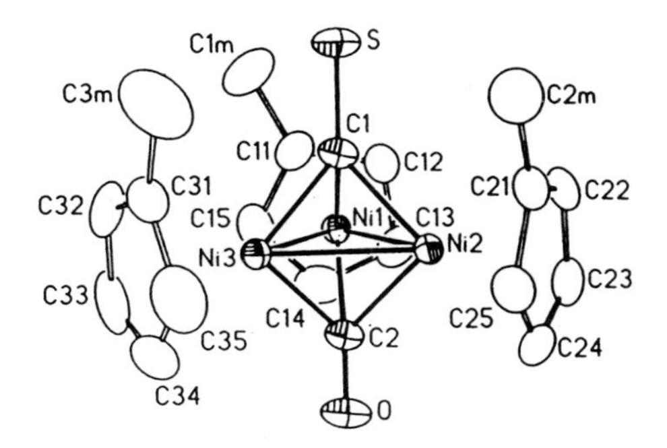

Figure 216. *Perspective view of $Ni_3(\eta^5\text{-}C_5H_4Me)_3(\mu_3\text{-}CO)(\mu_3\text{-}CS)$. Ni-Ni, 2.39 Å; $Ni\text{-}C_{(CO)}$, 1.95 Å; $Ni\text{-}C_{(CS)}$, 1.92 Å (from Ref. 383).*

With respect to $Ni_3(C_5H_5)_3(CO)_2$, the most significant difference lies on the $Ni\text{-}C_{(CS)}$ distance, which is appreciably shorter (by 0.03 Å) than the $Ni\text{-}C_{(CO)}$ distance.

Figure 217 shows the cyclic voltammetric behaviour of $Ni_3(C_5H_4Me)_3(CO)(CS)$ in dichloromethane solution |383|.

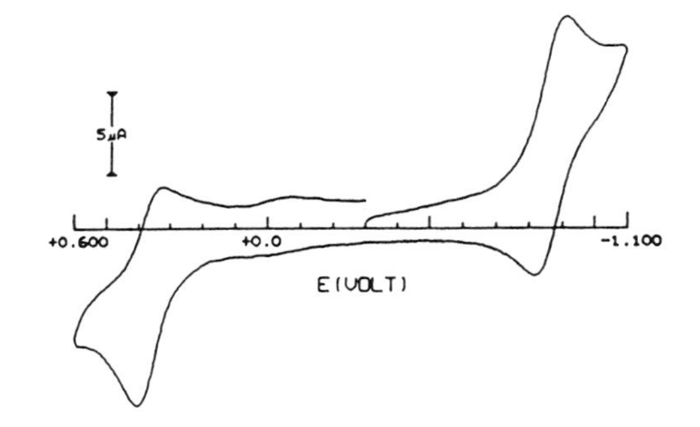

Figure 217.
Cyclic voltammogram exhibited by $Ni_3(C_5H_4Me)_3(CO)(CS)$ in CH_2Cl_2 solution at a platinum electrode. Scan rate 0.2 Vs^{-1} (from Ref. 383).

With respect to the redox aptitude of $Ni_3(C_5H_4Me)_3(CO)_2$ illustrated in Figure 215b, it is well evident from the $ip_{(backward)}/ip_{(foreward)}$ ratios minor than 1, that both the redox congeners $[Ni_3(C_5H_4Me)_3(CO)(CS)]^+$ and $[Ni_3(C_5H_4Me)_3(CO)(CS)]^-$ are decidedly less kinetically stable. In addition, the one-electron removal ($E°'$ = +0.37 V) is made more difficult, whereas the one-electron addition ($E°'$ = -0.86 V) becomes easier. This result confirms that the thiocarbonyl ligand is a better π-acceptor than the carbonyl ligand, and it may account for the increase of Ni_3 backbonding to the π^*(CS) orbitals, which should cause the Ni-$C_{(CS)}$ distance to be shorter than the Ni-$C_{(CO)}$ distance |383|.

8.2. Tetranuclear Mixed-Ligand Complexes

8.2.1. $Ni_4(CO)_6(PR_3)_4$. A series of tetranickel phosphino-carbonyl complexes have been prepared, of formula $Ni_4(CO)_6(PR_3)_4$ (R = C_2H_4CN |384|, Me |385,386|, *n*-Bu |386|).

As shown in Figure 218 for $Ni_4(CO)_6(PMe_3)_4$ |386|, they all possess an almost regular tetrahedral geometry, with each Ni-Ni bond edge-bridged by CO, and in which each Ni atom is bound to a phosphine group. Differences in phosphino groups cause slight variations in Ni-Ni bond distances (averaged, 2.48 Å for R = *n*-Bu; 2.49 Å for R = Me; 2.51 Å for R = CH_2CH_2CN).

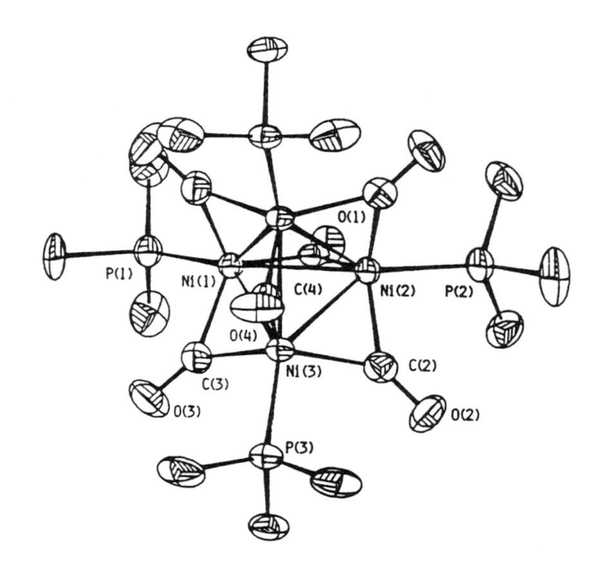

Figure 218.
X-Ray structure of $Ni_4(CO)_6(PMe_3)_4$ (from Ref. 386).

These 60-electron systems undergo, in dichloromethane solution, a first reversible one-electron oxidation followed by a second irreversible oxidation step |386|, according to:

R = *n*-Bu $E°' = +0.29$ V Ep = ?

R = Me $E°' = +0.22$ V Ep = +0.72 V

$$Ni_4(CO)_6(PR_3)_4 \underset{+e}{\overset{-e}{\rightleftharpoons}} [Ni_4(CO)_6(PR_3)]^+ \xrightarrow{-e} [Ni_4(CO)_6(PR_3)]^{2+}$$

(unstable)

No structural information about the monocation species is available, except for indirect spectroscopic evidence that the positive charge is delocalized over the Ni_4 core |386|.

8.3. Octanuclear Mixed-Ligand Complexes

8.3.1. $Ni_8(CO)_8(PPh)_6$. The octanickel cluster $Ni_8(CO)_8(\ {}_4\text{-}PPh)_6$ possesses the metal cubic assembly shown in Figure 219 |387|. All the carbonyl groups are terminal. Each phosphinidene ligand caps each square face of the Ni_8 cube. The Ni-Ni bond distances range from 2.64 Å to 2.68 Å, with an averaged value of 2.65 Å. The Ni-P averaged value is 2.18 Å, and the P atom is displaced out of the Ni_4 plane by about 1.1 Å.

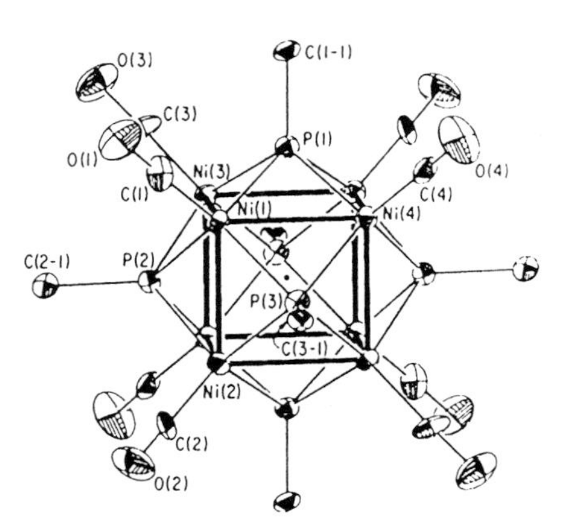

Figure 219. *Perspective view of $Ni_8(CO)_8(PPh)_6$ (from Ref. 387).*

Such a cluster undergoes a chemically reversible one-electron reduction to the corresponding monoanion $[Ni_8(CO)_8(PPh)_6]^-$, which has been X-ray characterized but not yet reported |388|.

8.3.2. $[Ni_8(CO)_{12}(PMe)_4]^{2-}$. The dianion $[Ni_8(CO)_{12}(PMe)_4]^{2-}$ is the low-nuclearity member of a series of noncentered icosahedral cages of general formula $Ni_{12-x}P_x$ which will be here discussed. Figure 220 shows its molecular structure |389|.

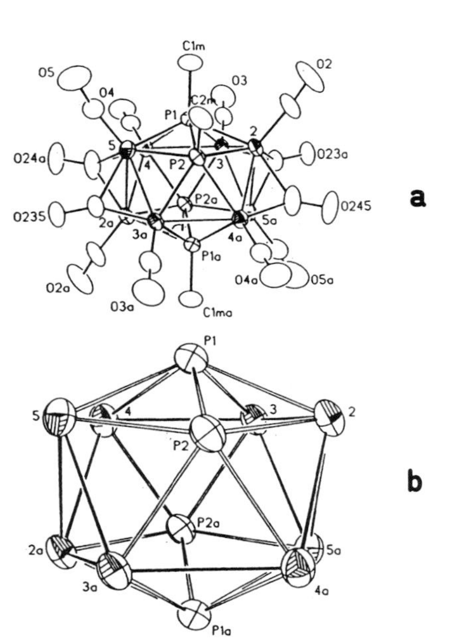

Figure 220. *(a) Perspective view of $[Ni_8(CO)_{12}(PMe)_4]^{2-}$, in the $[PPh_3Me]^+$ salt; (b) its icosahedral Ni_8P_4 core. Averaged bonding distances: Ni-Ni, 2.53 Å; Ni-P, 2.31 Å (from Ref. 389).*

Eight nickel and two phoshorus atoms partake in the formation of the pentagonal antiprism, which is in turn capped at opposite sides by two methylphosphinidene fragments. Each nickel atom bears one terminal carbonyl; the remaining four CO groups are asymmetrically triply bridging |389|.

In tetrahydrofuran solution, the present dianion does not display significant redox flexibility, in that it undergoes either an oxidation step at E°' = -0.34 V, complicated by subsequent chemical reactions, or an irreversible reduction at -1.0 V |389|.

8.4. Enneanuclear Carbido Complexes

8.4.1. $[Ni_9(CO)_{17}C]^{2-}$. Figure 221 shows the crystal structure of the enneanickel dianion $[Ni_9(CO)_{17}C]^{2-}$ |390|.

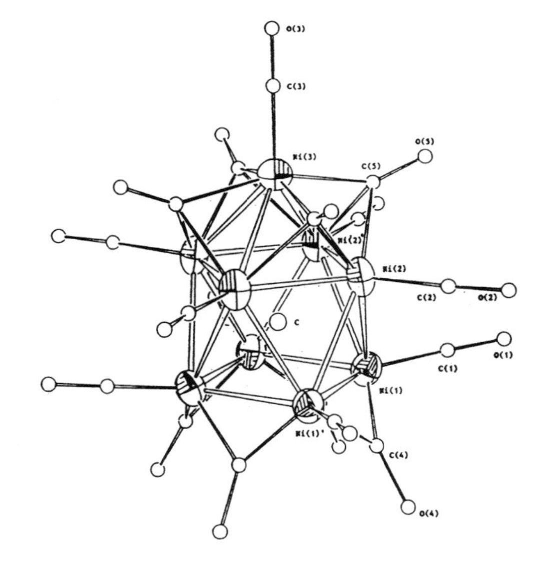

Figure 221. *Molecular structure of $[Ni_9(CO)_{17}C]^{2-}$ (from Ref. 390).*

It is constituted by a square-antiprismatic Ni_8 core, accomodating an interstitial carbide carbon atom, capped over one squared face by a ninth nickel atom. The two Ni_4 planar fragments have a mean Ni-Ni distance of about 2.49 Å; the interlayer Ni-Ni distance is about 2.61 Å; the apical Ni is at 2.53 Å from the basal nickel atoms; the inner carbide atom is 2.1 Å from the two Ni_4 layers. The upper square pyramidal Ni_5 fragment bears four triply-bridging carbonyl groups, whereas the lower Ni_4 base has four edge-bridging carbonyls.

In acetonitrile solution, $[Ni_9(CO)_{17}C]^{2-}$ undergoes a quasi-reversible one-electron oxidation, complicated by subsequent chemical reactions, at E°′ = - 0.41 V as well as an irreversible oxidation at +0.67 V |391|. This means that the dianion is the only stable member of the present family of enneanickel complexes.

8.5. Enneanuclear Mixed-Ligand Complexes

8.5.1. $[Ni_9(CO)_{15}(PMe)_3]^{2-}$. Substitution of one PMe group for one $Ni(CO)_3$ fragment in the preceding dianion $[Ni_8(CO)_{12}(PMe)_4]^{2-}$ affords the dianion $[Ni_9(CO)_{15}(PMe)_3]^{2-}$, the structure of which is shown in Figure 222 |389|.

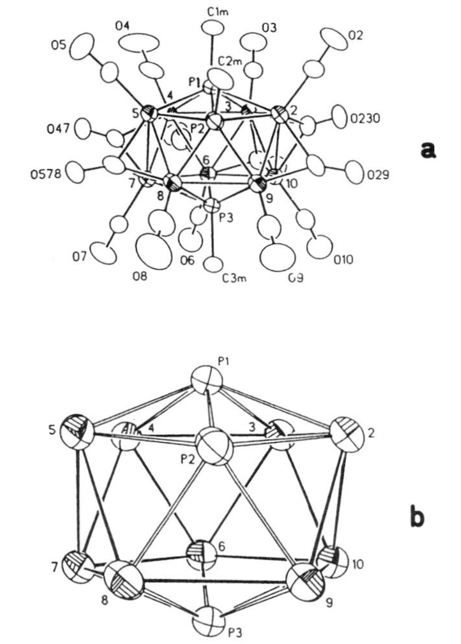

Figure 222.
(a) Perspective view of $[Ni_9(CO)_{15}(PMe)_3]^{2-}$, in the $[PPh_3Me]^+$ salt; (b) its icosahedral Ni_9P_3 core. Averaged bonding distances: Ni-Ni, 2.55 Å; Ni-P, 2.33 Å (from Ref. 389).

The central pentagonal antiprism is now constituted by nine nickel and one phosphorus atoms.

Like the octanuclear precursor, this enneanuclear dianion does not exhibit interesting redox properties. In tetrahydrofuran solution, it undergoes two closely spaced oxidation processes centred at +0.3 V |389|.

8.6. Decanuclear Mixed-Ligand Complexes

8.6.1. $[Ni_{10}(CO)_{18}(PMe)_2]^{2-}$. Substitution of a second PMe group for a second $Ni(CO)_3$ fragment in the preceding dianion $[Ni_8(CO)_{12}(PMe)_4]^{2-}$ affords the dianion $[Ni_{10}(CO)_{18}(PMe)_2]^{2-}$, the structure of which is shown in Figure 223 |389|.

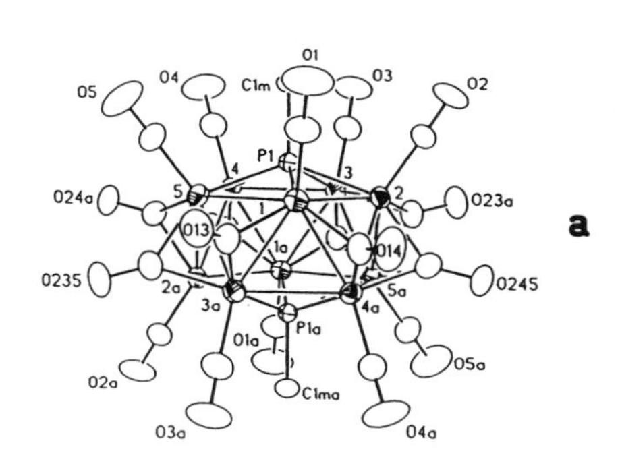

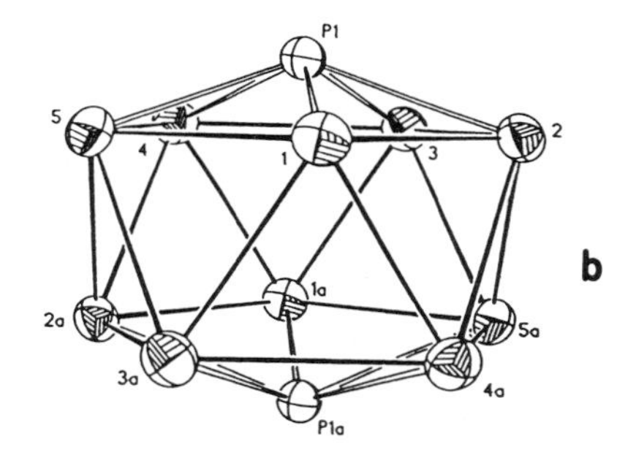

Figure 223.
(a) Perspective view of $[Ni_{10}(CO)_{18}(PMe)_2]^{2-}$, in the $[PPh_3Me]^+$ salt; (b) its icosahedral $Ni_{10}P_2$ core. Averaged bonding distances: Ni-Ni, 2.55 Å; Ni-P, 2.35 Å (from Ref. 389).

In the icosahedral assembly, the ten sites of the central pentagonal antiprism are now occupied only by nickel atoms.

Confirming that in the series $[Ni_{12-x}(CO)_{24-3x}(PMe)_x]^{2-}$, the increase of $Ni(CO)_3$ fragments over PMe groups makes more amd more difficult the removal of electrons, the decanickel dianion undergoes, in tetrahydrofuran solution, an irreversible oxidation at +0.75 V |389|.

9. PLATINUM CLUSTERS

9.1. Enneadecanuclear Binary Complexes

9.1.1. $[Pt_{19}(CO)_{22}]^{4-}$. The structural geometry of the tetraanion $[Pt_{19}(CO)_{12}(\mu_2\text{-}CO)_{10}]^{4-}$ is shown in Figure 224 |392|.

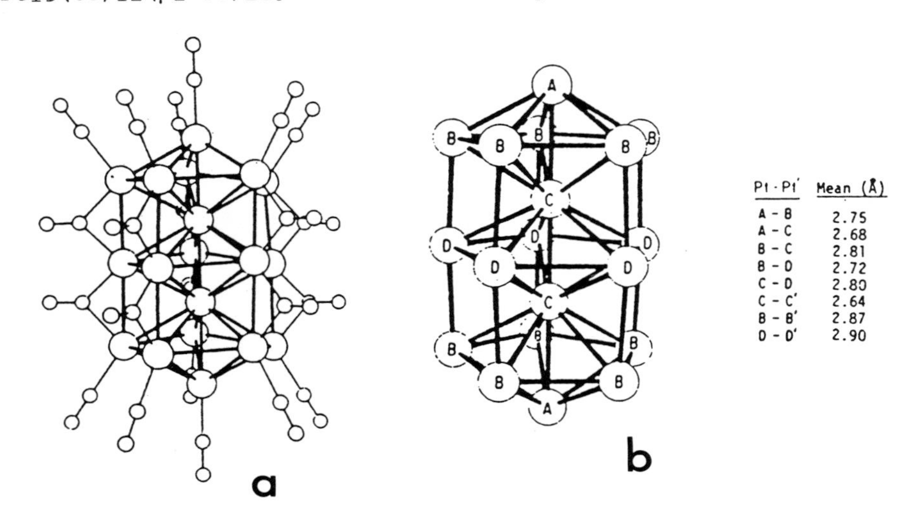

a
b
Pt · Pt' Mean (Å)
A - B 2.75
A - C 2.68
B - C 2.81
B - D 2.72
C - D 2.80
C - C' 2.64
B - B' 2.87
D - D' 2.90

Figure 224. *(a) Perspective view of $[Pt_{19}(CO)_{22}]^{4-}$; (b) interatomic distances in the $[PPh_4]^+$ salt (from Ref. 392).*

The Pt_{19} core consists of a stacked sequence of three pentagonal layers, with two internally encapsulated and two externally capping platinum atoms. Each platinum atom of the two outer Pt_6 fragments bear one terminal carbonyl group, whereas the ten interlayer edges bear one bridging carbonyl group. The 58 Pt-Pt bond lengths range representatively from 2.64 Å of the shortest one, relative to the two interstitial atoms, to 2.90 Å of the longest ones, in the central pentagon (in the $[PPh_4]^+$ salt).

This 238-electron tetraanion displays a rich redox chemistry. In fact, as shown in Figure 225, it is able to undergo two consecutive pairs of close-spaced one-electron reductions, as well as two close-spaced one-electron oxidations |393|. The first three-electron additions and the first one-electron removal exhibit features of chemical reversibility in the short times of cyclic voltammetry. In the longer times of electrolysis, such electron transfers are coupled to slow decomposition of the primarily electrogenerated species, unless exhaustive reductions/oxidations are performed at low temperature (-20 °C).

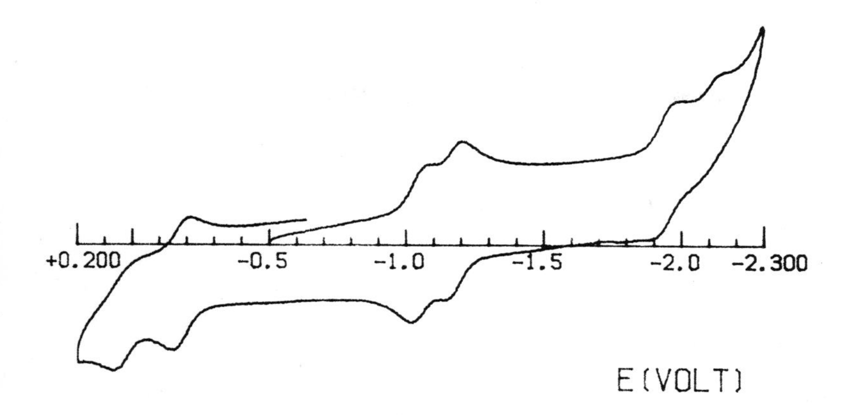

Figure 225. *Cyclic voltammetric response recorded at a platinum electrode on a deaerated MeCN solution of* $[Pt_{19}(CO)_{22}][NBu_4]_4$*. Scan rate 0.2* Vs^{-1}*.*

The redox potentials of these sequential electron transfers are summarized in Table 45, together with those of higher nuclearity platinum carbonyl cluster compounds.

9.2. Tetraduodecanuclear Binary Complexes

9.2.1. $\mathbf{[Pt_{24}(CO)_{30}]^{2-}}$**.** The crystal structure of the dianion $[Pt_{24}(CO)_{22}(\mu_2\text{-}CO)_8]^{2-}$ is shown in Figure 226 |394|.

Figure 226. *Perspective view of* $[Pt_{24}(CO)_{22}(\mu_2\text{-}CO)_8]^{2-}$ *(from Ref. 394).*

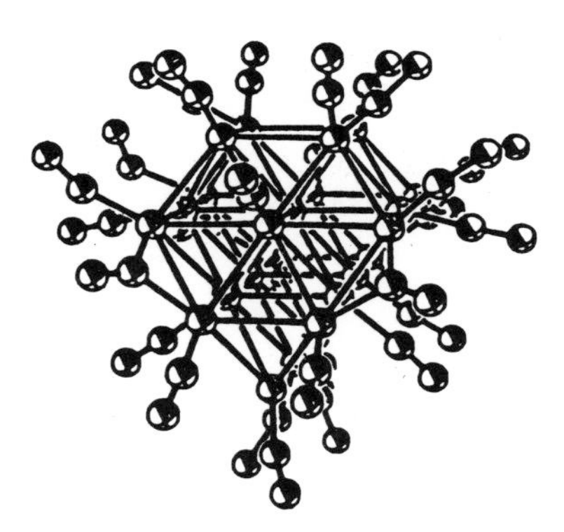

The wedge-shaped Pt_{24} core has a cubic closest packing of platinum atoms.

As shown in Figure 227, the dianion undergoes, in dichloromethane solution, both two consecutive pairs of close-spaced one-electron reductions and one pair of one-electron oxidations |394,395|.

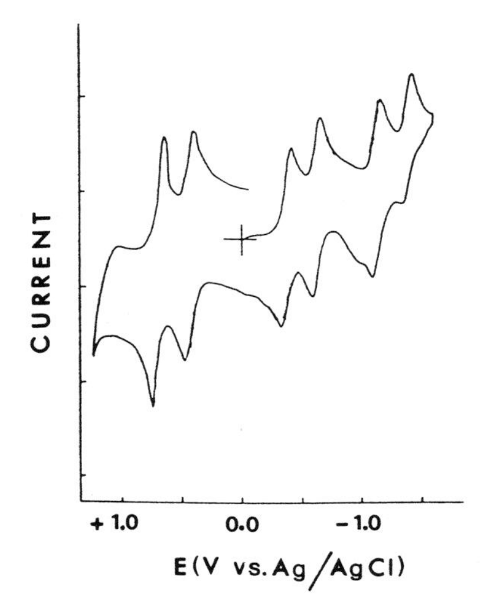

Figure 227.
Cyclic voltammogram exhibited at a platinum electrode by a CH_2Cl_2 solution of $[Pt_{24}(CO)_{30}]^{2-}$ (from Ref. 394).

These sequential redox changes are described as substantially electrochemically reversible. Such a redox pathway is quite reminescent of that previously illustrated for $[Pt_{19}(CO)_{22}]^{4-}$, and confirms the aptitude of these high nuclearity platinum clusters to add or lose electrons in a pairwise fashion, according to their known tendency to maintain an even electron count |394|.

The electrode potentials of the redox changes are reported in Table 45.

9.3. Hexaduodecanuclear Binary Complexes

9.3.1. $[Pt_{26}(CO)_{32}]^{2-}$. As illustrated in Figure 228, the Pt_{26} core of the dianion $[Pt_{26}(CO)_{23}(\mu_2\text{-}CO)_9]^{2-}$ is constituted by a stacked sequence of Pt_7:Pt_{10}:Pt_7 layers.

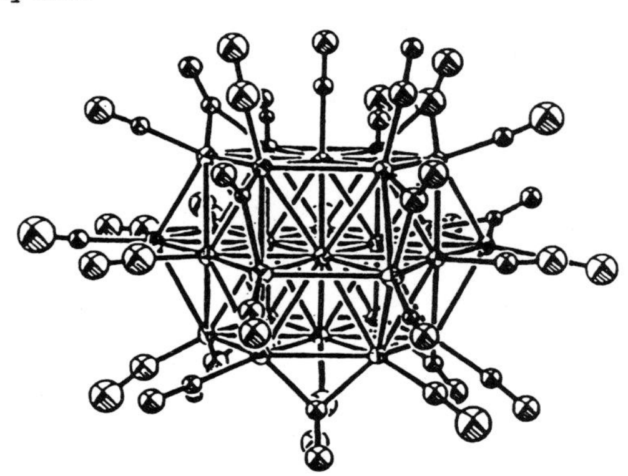

Figure 228.
Perspective view of $[Pt_{26}(CO)_{23}(\mu_2\text{-}CO)_9]^{2-}$ (from Ref. 394).

Also this complex undergoes a rich series of closely-spaced electron transfers, the redox potentials of which are reported in Table 45 |394|.

9.4. Octatridecanuclear Binary Complexes

9.4.1. $[Pt_{38}(CO)_{44}]^{2-}$. Figure 229 shows the structure of the metal core of the dianion $[Pt_{38}(CO)_{44}]^{2-}$ |394|.

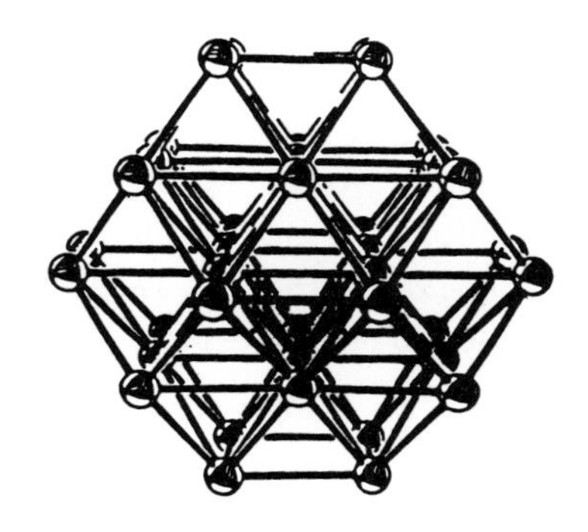

Figure 229. *Perspective view of the Pt_{38} core of $[Pt_{38}(CO)_{44}]^{2-}$ (from Ref. 394).*

It can be seen as the previous Pt_{24} assembly plunged in between two platinum-centred hexagonal faces.

As illustrated in Figure 230, like the Pt_{26} cluster, the present dianion undergoes a rich, but not easily resolvable, sequence of electron transfer steps |393|.

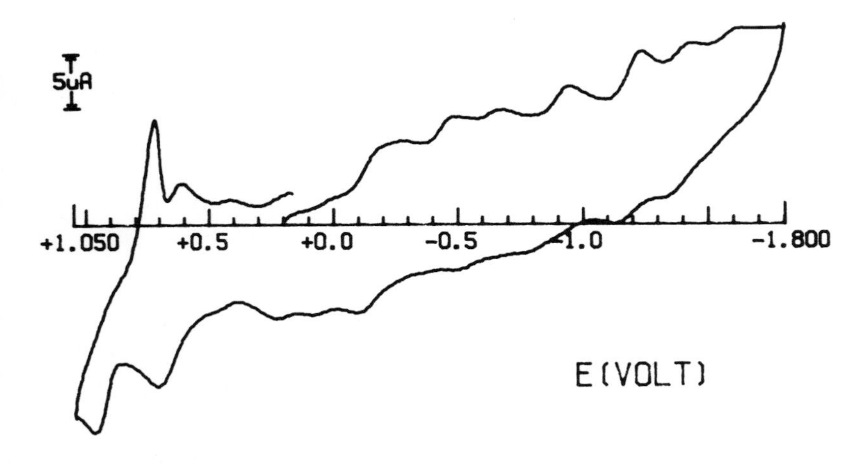

Figure 230. *Cyclic voltammogram recorded at a platinum electrode on a THF solution of $[Pt_{38}(CO)_{44}]^{2-}$. Scan rate 0.2 Vs^{-1}.*

The redox potentials for the high-nuclearity platinum clusters here described are compiled in Table 45.

TABLE 45. *Redox potentials (in Volt, vs. S.C.E.) for the redox changes exhibited by high nuclearity platinum-carbonyl cluster compounds*

complex	E°' 0/-	E°' -/2-	E°' 2-/3-	E°' 3-/4-	E°' 4-/5-	E°' 5-/6-	E°' 6-/7-	E°' 7-/8-	solvent	Ref
$[Pt_{19}]$			+0.05	-0.21	-1.08	-1.20	-1.98	-2.15	MeCN	393
$[Pt_{24}]$	+0.85	+0.60	-0.25	-0.55	-1.00	-1.25			CH_2Cl_2	394
		+0.28	-0.27	-0.52	-1.02	-1.17	-1.92		MeCN	394
$[Pt_{26}]$	+0.85	+0.55	-0.40	-0.40	-1.00	-1.00			CH_2Cl_2	394
			-0.22	-0.22	-0.88	-0.88	-1.32	-1.32	MeCN	394
$[Pt_{38}]$	+0.95	+0.70	-0.35	-0.45	-0.75	-0.90	-1.35		CH_2Cl_2	394

10. RHENIUM CLUSTERS

10.1. Tetranuclear Mixed-Ligand Complexes

10.1.1. $Re_4(CO)_{12}(Cl)_4$. The tetranuclear complex $Re_4(CO)_{12}(Cl)_4$ has assigned, on the basis of spectroscopic evidence |396|, a pseudocubane structure similar to that known for $Re_4(CO)_{12}(SMe)_4$ |397,398|, Figure 231, as well as for $Re_4(CO)_{12}(OH)_4$ |399|.

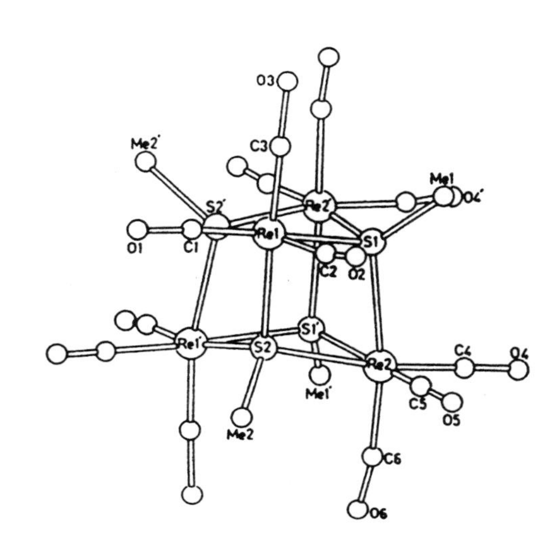

Figure 231. *Molecular structure of $Re_4(CO)_{12}(SMe)_4$. Mean Re-S bond distance, 2.51 Å. Mean Re...Re nonbonding length, 3.90 Å (from Ref. 397).*

The geometry can be thought to arise from two interpenetrating tetrahedra (Re_4 and S_4, respectively), having a common centre.

In dichloromethane solution, where decomposition of $Re_4(CO)_{12}(Cl)_4$ does not occur, two irreversible oxidation processes take place at Ep = +1.85 V and + 2.3 V, respectively |396|. This datum points out the notable stability towards redox reactions as well as the lack of flexibility of the actual complex.

10.2. Heptanuclear Carbido Complexes

10.2.1. $[Re_7(CO)_{21}C]^{3-}$. The heptarhenium trianion $[Re_7(CO)_{21}C]^{3-}$ has a geometry in which one triangular face of an octahedral Re_6C core is capped by a triply-bridging $Re(CO)_3$ fragment, Figure 232 |400|. Within the octahedron, the Re-Re bond distances range from 2.95 Å to 3.09 Å, with a mean value of 3.01 Å; the $Re-C_{(carbide)}$

distance has a mean value of 2.13 Å. In the capping unit, the Re-Re distance is 2.93 Å. The overall mean Re-Re bond length is 2.99 Å.

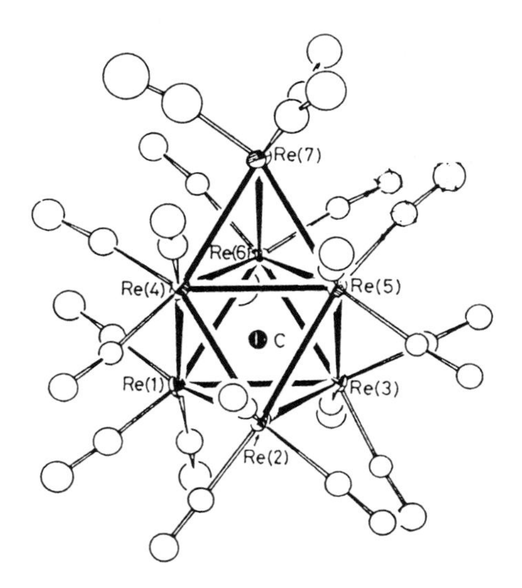

Figure 232.
View of the trianion $[Re_7(CO)_{21}C]^{3-}$ (from Ref. 400).

It has been briefly reported that the 98-electron system $[Re_7(CO)_{21}C]^{3-}$ undergoes, in dichloromethane solution, two quasireversible one-electron removals according to |401|:

$$[Re_7(CO)_{21}C]^{3-} \underset{+e}{\overset{-e}{\rightleftharpoons}} [Re_7(CO)_{21}C]^{2-} \underset{+e}{\overset{-e}{\rightleftharpoons}} [Re_7(CO)_{21}C]^{-}$$

first step: $E°' = +0.04$ V ($\Delta Ep_{0.1\ Vs^{-1}} = 60$ mV); second step: $E°' = +0.38$ V ($\Delta Ep_{0.1\ Vs^{-1}} = 70$ mV)

10.3. Octanuclear Carbido Complexes

10.3.1. $[Re_8(CO)_{24}C]^{2-}$. In closed correlation with $[Re_7(CO)_{21}C]^{3-}$, the dianion $[Re_8(CO)_{24}C]^{2-}$ possesses a structure in which two opposite triangular faces of the octahedral Re_6C core are capped by two $Re(CO)_3$ fragments, Figure 233 |402|. Within the octahedron, the Re-Re bond distances have a mean value of 2.99 Å, with a mean $Re-C_{(carbide)}$ distance of 2.12 Å. In the capping units, the mean

Re-Re distance is 2.97 Å. The overall mean Re-Re bond length, 2.98 Å, is only slightly smaller than the corresponding one in the monocapped $[Re_7(CO)_{21}C]^{3-}$.

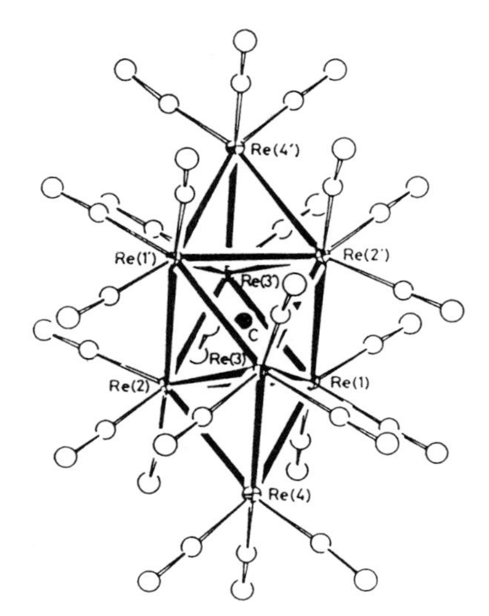

Figure 233. *Perspective view of the dianion* $[Re_8(CO)_{24}C]^{2-}$ *(from Ref. 402).*

It has been briefly reported that in dichloromethane solution, the 110-electron dianion $[Re_8(CO)_{24}C]^{2-}$ undergoes two subsequent one-electron oxidation steps, displaying more or less marked features of chemical reversibility ($E°'_{2-/-}$ = +0.80 V, $\Delta Ep_{(0.1\ Vs^{-1})}$ = 80 mV; $E°'_{-/0}$ = +1.05 V, $\Delta Ep_{(0.1\ Vs^{-1})}$ = 120 mV) |401,403|.

11. REFERENCES

1. K.Wade,
Adv.Inorg.Chem.Radiochem., *18* (1976) 1.
2. P.Chini, G.Longoni, and V.G.Albano,
Adv.Organomet.Chem., *14* (1976) 325.
3. E.L.Muetterties, T.N.Rhodin, E.Band, C.F.Brucker, and W.R.Pretzer,
Chem.Rev., *79* (1979) 91.
4. G.Ciani and A.Sironi,
J.Organomet.Chem., 197 (1980) 233.
5. P.Chini,
J.Organomet.Chem., *200* (1980) 37.
6. B.F.G.Johnson and J.Lewis,
Adv.Inorg.Chem.Radiochem., *24* (1981) 225.
7. M.C.Manning and W.C.Trogler,
Coord.Chem.Rev., *38* (1981) 89.
8. J.S.Bradley,
Adv.Organomet.Chem., *22* (1983) 1.
9. K.P.Hall and D.M.P.Mingos,
Progr.Inorg.Chem., *32* (1984) 237.
10. E.Sappa, A.Tiripicchio, and P.Braunstein,
Coord.Chem.Rev., *65* (1985) 219.
11. I.D.Salter,
Adv.Organomet.Chem., *29* (1989) 249.
12. R.D.Adams,
Chem.Rev., *89* (1989) 1703.
13. S.Harris,
Polyhedron, *8* (1989) 2843.
14. H.Vahrenkamp,
Adv.Organomet.Chem., *21* (1983) 169.
15. R.D.Adams and I.T.Horvath,
Progr.Inorg.Chem., *33* (1985) 127.
16. D.F.Shriver and M.J.Sailor,
Acc.Chem.Res., *21* (1988) 374.
17. P.Braunstein and J.Rose,
Stereochemistry of Organometallic and Inorganic Compounds, I.Bernal, Ed.; Elsevier, Amsterdam, 1989; Vol.3, p.3.
18. D.M.P.Mingos,
Acc.Chem.Res., *17* (1984) 311.

19. B.K.Teo, G.Longoni, and F.R.K.Chung,
Inorg.Chem., *23* (1984) 1257.
20. M.McPartlin,
Polyhedron, *3* (1984) 1279.
21. S.M.Owen,
Polyhedron, 7 (1988) 253.
22. D.Osella and P.R.Raithby,
Stereochemistry of Organometallic and Inorganic Compounds,
I.Bernal, Ed.; Elsevier, Amsterdam, 1989; Vol.3, p.304.
23. P.Lemoine,
Coord.Chem.Rev., *47* (1982) 55.
24. W.E.Geiger,
Progr.Inorg.Chem., *33* (1985) 275.
25. W.E.Geiger and N.G.Connelly,
Adv.Organomet.Chem., *24* (1987) 87.
26. P.Lemoine,
Coord.Chem.Rev., *83* (1988) 169.
27. S.R.Drake,
Polyhedron, *9* (1990) 455.
28. P.Zanello,
Coord.Chem.Rev., *83* (1988) 199.
29. P.Zanello,
Coord.Chem.Rev., *87* (1988) 1.
30. P.Zanello
Struct.Bonding (Berlin), *79* (1992) 101.
31. C.H.Wei and L.F.Dahl,
J.Am.Chem.Soc., *88* (1966) 1821.
32. C.H.Wei and L.F.Dahl,
J.Am.Chem.Soc., *91* (1969) 1351.
33. F.A.Cotton and J.M.Troup,
J.Am.Chem.Soc., *96* (1976) 4155.
34. R.E.Dessy, R.B.King, and M.Waldrop,
J.Am.Chem.Soc., *88* (1966) 5112.
35. A.M.Bond, P.A.Dawson, B.M.Peake, B.H.Robinson, and J.Simpson,
Inorg.Chem., *16* (1977) 2199.
36. D.Miholova, J.Klima, and A.A.Vlcek,
Inorg.Chim.Acta, *27* (1978) L67.
37. D.Miholova, J.Fiedler, and A.A.Vlcek,
J.Electroanal.Chem., *143* (1983) 195.

38. P.A.Dawson, B.M.Peake, B.H.Robinson, and J.Simpson, *Inorg.Chem.*, *19* (1980) 465.
39. D.R.Tyler, R.A.Levenson, and H.B.Gray, *J.Am.Chem.Soc.*, *100* (1978) 7888.
40. B.E.R.Schilling and R.Hoffmann, *J.Am.Chem.Soc.*, *101* (1979) 3456.
41. F.Y.-K.Lo, G.Longoni, P.Chini, L.D.Lower, and L.F.Dahl, *J.Am.Chem.Soc.*, *102* (1980) 7691.
42. C.Amatore, J.-N.Verpeaux, and P.J.Krusic, *Organometallics*, *7* (1988) 2426.
43. P.J.Krusic, J.San Filippo,Jr., B.Hutchinson, R.L.Hance, and L.M.Daniels, *J.Am.Chem.Soc.*, *103* (1981) 2129.
44. L.F.Dahl and J.F.Blount, *Inorg.Chem.*, *4* (1965) 1373.
45. P.J.Krusic, *J.Am.Chem.Soc.*, *103* (1981) 2131.
46. D.J.Dahm and R.A.Jacobson, *J.Am.Chem.Soc.*, *90* (1968) 5106.
47. R.E.Benfield, P.D.Gavens, B.F.G.Johnson, M.J.Mays, S.Aime, L.Milone, and D.Osella, *J.Chem.Soc., Dalton Trans.*, (1981) 1535.
48. G.Raper and W.S.McDonald, *J.Chem.Soc.*(A), (1971) 3430.
49. J.F.Blount, L.F.Dahl, C.Hoogzand, and W.Hübel, *J.Am.Chem.Soc.*, *88* (1966) 292.
50. D.Osella, R.Gobetto, P.Montangero, P.Zanello, and A.Cinquantini, *Organometallics*, *5* (1986) 1247.
51. K.Knoll, G.Huttner, L.Zsolnai, and O.Orama, *J.Organomet.Chem.*, *327* (1987) 379.
52. B.Eber, D.Buchholz, G.Huttner, Th.Fässler, W.Imhof, M.Fritz, J.C.Jochims, J.C.Daran, and Y.Jeannin, *J.Organomet.Chem.*, *401* (1991) 49.
53. A.Ceriotti, L.Resconi, F.Demartin, G.Longoni, M.Manassero, and M.Sansoni, *J.Organomet.Chem.*, *249* (1983) C35.
54. F.T.Al-Ani, D.L.Hughes, and C.J.Pickett, *J.Organomet.Chem.*, *307* (1986) C31.

55. G.Longoni and P.Zanello, unpublished results.
56. U.Honrath and H.Vahrenkamp, *Z.Naturforsh.*, *B39* (1984) 545.
57. C.H.Wei and L.F.Dahl, *Inorg.Chem.*, *4* (1965) 493.
58. P.Hübener and E.Weiss, *Cryst.Struct.Commun.*, *11* (1982) 331.
59. T.Madach and H.Vahrenkamp, *Chem.Ber.*, *114* (1981) 505.
60. T.M.Bockman and J.K.Kochi, *J.Am.Chem.Soc.*, *109* (1987) 7725.
61. A.Darchen, C.Mahé, and H.Patin, *Nouv.J.Chim.*, *6* (1982) 539.
62. L.F.Dahl and P.W.Sutton, *Inorg.Chem.*, *2* (1963) 1067.
63. A.Benoit, J.-Y.Le Marouille, C.Mahé, and H.Patin, *J.Organomet.Chem.*, *233* (1982) C51.
64. A.B.Rives, Y.Xiao-Zeng, and R.F.Fenske, *Inorg.Chem.*, *21* (1982) 2286.
65. L.Markò, T.Madach, and H.Vahrenkamp, *J.Organomet.Chem.*, *190* (1980) C67.
66. L.Markò, B.Markò-Monostory, T.Madach, and H.Vahrenkamp, *Angew.Chem.Int.Ed.Engl.*, *19* (1980) 226.
67. R.J.Haines, J.A.De Beer, and P.Greatrex, *J.Organomet.Chem.*, *85* (1975) 89.
68. A.Winter, L.Zsolnai, and G.Huttner, *J.Organomet.Chem.*, *234* (1982) 337.
69. E.Lindner, G.A.Weiss, W.Hiller, and R.Fawzi, *J.Organomet.Chem.*, *255* (1983) 245.
70. D.Buchholz, G.Huttner, L.Zsolnai, and W.Imhof, *J.Organomet.Chem.*, *377* (1989) 25.
71. B.Eber, G.Huttner, D.Günauer, W.Imhof, and L.Zsolnai, *J.Organomet.Chem.*, *414* (1991) 361.
72. B.Eber, G.Huttner, L.Zsolnai, and W.Imhof, *J.Organomet.Chem.*, *402* (1991) 221.
73. S.L.Cook, J.Evans, L.R.Gray, and M.Webster, *J.Organomet.Chem.*, *236* (1982) 367.

74. H.H.Ohst and J.K.Kochi, *Inorg.Chem.*, *25* (1986) 2066.
75. H.H.Ohst and J.K.Kochi, *J.Am.Chem.Soc.*, *108* (1986) 2897.
76. W.Clegg, G.M.Sheldrick, D.Stalke, S.Bhaduri, and H.K.Khwaja, *Acta Cryst.*, *C40* (1984) 2045.
77. R.P.Dodge and V.Schomaker, *J.Organomet.Chem.*, *3* (1965) 274.
78. B.Heim, J.C.Daran, Y.Jeannin, B.Eber, G.Huttner, and W.Imhof, *J.Organomet.Chem.*, *441* (1992) 81.
79. R.E.Dessy and R.L.Pohl, *J.Am.Chem.Soc.*, *90* (1968) 1995.
80. D.Osella, G.Arman, M.Botta, R.Gobetto, F.Laschi, and P.Zanello, *Organometallics*, *8* (1989) 620.
81. D.Osella, G.Arman, R.Gobetto, F.Laschi, P.Zanello, S.Ayrton, V.Goodfellow, C.E.Housecroft, and S.M.Owen, *Organometallics*, *8* (1989) 2689.
82. S.Aime, M.Botta, R.Gobetto, and D.Osella, *Inorg.Chim.Acta*, *115* (1986) 129.
83. K.H.Pannell, A.J.Mayr, and D.VanDerveer, *J.Am.Chem.Soc.*, *105* (1983) 6186.
84. J.R.Fish, T.Malinski, A.J.Mayr, and K.H.Pannell, *Inorg.Chim.Acta*, *150* (1988) 249.
85. R.J.Doedens and L.F.Dahl, *J.Am.Chem.Soc.*, *88* (1966) 4847.
86. G.van Buskirk, G.B.Knobler, and H.D.Kaesz, *Organometallics*, *4* (1985) 149.
87. R.F.Boehme and P.Coppens, *Acta Cryst.*, *B37* (1981) 1914.
88. J.H.Davis, M.A.Beno, J.M.Williams, J.Zimmil, M.Tachikawa, and E.L. Muetterties, *Proc.Natl.Acad.Sci. U.S.A.*, *78* (1981) 668.
89. P.L.Bogdan, C.Woodcock, and D.F.Shriver, *Organometallics*, *6* (1987) 1377.
90. M.Tachikawa and E.L.Muetterties, *J.Am.Chem.Soc.*, *102* (1980) 4541.

91. S.D.Wijeyesekera, R.Hoffmann, and C.N.Wilker, *Organometallics, 3* (1984) 962.
92. S.Harris and J.S.Bradley, *Organometallics, 3* (1984) 1086.
93. J.A.Hriljac, S.Harris, and D.F.Shriver, *Inorg.Chem., 27* (1988) 816.
94. S.Harris, M.L.Blohm, and W.L.Gladfelter, *Inorg.Chem., 28* (1989) 2290.
95. J.S.Bradley, G.B.Ansell, M.E.Leonowicz, and E.W.Hill, *J.Am.Chem.Soc., 103* (1981) 4968.
96. J.S.Bradley, S.Harris, J.M.Newsam, E.W.Hill, S.Leta, and M.A.Modrick, *Organometallics, 6* (1987) 2060, and references therein.
97. J.Wang, A.M.Crespi, M.Sabat, S.Harris, C.Woodcock, and D.F.Shriver, *Inorg.Chem., 28* (1989) 697.
98. D.E.Fjare and W.L.Gladfelter, *Inorg.Chem., 20* (1981) 3533.
99. P.Zanello and T.R.Spalding, work in progress.
100. A.Gourdon and Y.Jeannin, *J.Organomet.Chem., 440* (1992) 353.
101. J.A.Ferguson and T.J.Meyer, *J.Chem.Soc., Chem.Commun.,* (1971) 623.
102. J.A.Ferguson and T.J.Meyer, *J.Am.Chem.Soc., 94* (1972) 3409.
103. M.A.Neuman, Trinh-Toan, and L.F.Dahl, *J.Am.Chem.Soc., 94* (1972) 3383.
104. Trinh-Toan, W.P.Fehlhammer, and L.F.Dahl, *J.Am.Chem.Soc., 94* (1972) 3389.
105. C.R.Bock and M.S.Wrighton, *Inorg.Chem., 16* (1977) 1309.
106. M.E.Barr, B.R.Adams, R.R.Weller, and L.F.Dahl, *J.Am.Chem.Soc., 113* (1991) 3052.
107. K.S.Bose, E.Sinn, and B.A.Averill, *Organometallics, 3* (1984) 1126.
108. L.L.Nelson, F.Y.-K.Lo, A.D.Rae, and L.F.Dahl, *J.Organomet.Chem., 225* (1982) 309.

109. F.T.Al-Ani and C.J.Pickett, *J.Chem.Soc., Dalton Trans.,* (1988) 2329.
110. J.-F.Halet, R.Hoffmann, and J.-Y.Saillard, *Inorg.Chem., 24* (1985) 1695.
111. H.Vahrenkamp and D.Wolters, *Organometallics, 1* (1982) 87.
112. J.T.Jaeger, S.Aime, and H.Vahrenkamp, *Organometallics, 5* (1986) 245.
113. J.T.Jaeger, J.S.Field, D.Collins, G.P.Speck, B.M.Peake, J.Hähnle, and H. Vahrenkamp, *Organometallics,* 7 (1988) 1753.
114. H.Vahrenkamp and D.Wolters, *J.Organomet.Chem., 224* (1982) C17.
115. H.Vahrenkamp, E.J.Wucherer, and D.Wolters, *Chem.Ber., 116* (1983) 1219.
116. J.-F.Halet and J.-Y.Saillard, *Nouv.J.Chim., 11* (1987) 315.
117. J.T.Jaeger and H.Vahrenkamp, *Organometallics,* 7 (1988) 1746.
118. R.R.Gagnè, C.A.Koval, T.J.Smith, and M.C.Cimolino, *J.Am.Chem.Soc., 101* (1979) 4571.
119. T.Feng, P.Lau, W.Imhof, and G.Huttner, *J.Organomet.Chem., 414* (1991) 89.
120. K.Knoll, T.Fässler, and G.Huttner, *J.Organomet.Chem., 332* (1987) 309.
121. A.Gourdon and Y.Jeannin, *J.Organomet.Chem., 290* (1985) 199.
122. S.R.Drake, B.F.G.Johnson, and J.Lewis, *J.Chem.Soc., Chem.Commun.,* (1988) 1033.
123. E.H.Braye, L.F.Dahl, W.Hübel, and L.Wampler, *J.Am.Chem.Soc., 84* (1962) 4633.
124. C.G.Cooke and M.J.Mays, *J.Organomet.Chem., 88* (1975) 231.
125. J.W.Kolis, F.Basolo, and D.F.Shriver, *J.Am.Chem.Soc., 104* (1982) 5626.
126. M.Tachikawa, J.Stein, E.L.Muetterties, R.G.Teller, M.A.Beno, E.Gebert, and J.M.Williams, *J.Am.Chem.Soc., 102* (1980) 6648.

127. R.Hourihane, T.R.Spalding, G.Ferguson, T.Deeney, and P.Zanello,
J.Chem.Soc., Dalton Trans., (1993) 43.
128. P.L.Bogdan, M.Sabat, S.A.Sunshine, C.Woodcock, and D.F.Shriver,
Inorg. Chem., 27 (1988) 1904.
129. A.Gourdon and Y.Jeannin,
J.Organomet.Chem., 388 (1990) 195.
130. M.R.Churchill, J.Wormald, J.Knight, and M.J.Mays,
J.Am.Chem.Soc., 93 (1971) 3073.
131. M.R.Churchill and J.Wormald,
J.Chem.Soc., Dalton Trans., (1974) 2410.
132. J.Rimmelin, P.Lemoine, M.Gross, R.Mathieu, and D.de Montauzon,
J.Organomet.Chem., 309 (1986) 355.
133. M.Tachikawa, A.C.Sievert, E.L.Muetterties, M.R.Thompson, C.S.Day, and V.W.Day,
J.Am.Chem.Soc., 102 (1980) 1725.
134. G.L.Lilley, E.Sinn, and B.A.Averill,
Inorg.Chem., 25 (1986) 1073.
135. R.Mason and A.I.M.Rae,
J.Chem.Soc.(A), (1968) 778.
136. M.R.Churchill, F.J.Hollander, and J.P.Hutchinson,
Inorg.Chem., 16 (1977) 2655.
137. P.Zanello, S.Aime, and D.Osella,
Organometallics, 3 (1984) 1374.
138. J.C.Cyr, J.A.De Gray, D.K.Gosser, E.S.Lee, and P.H.Rieger,
Organometallics, 4 (1985) 950.
139. A.J.Downard, B.H.Robinson, J.Simpson, and A.M.Bond,
J.Organomet.Chem., 320 (1987) 363.
140. J.E.Cyr and P.H.Rieger,
Organometallics, 10 (1991) 2153.
141. B.Delley, M.C.Manning, D.E.Ellis, J.Berkowitz, and W.C.Trogler,
Inorg.Chem., 21 (1982) 2247, and references therein.
142. C.G.Nagel, J.C.Bricker, D.G.Alway, and S.G.Shore,
J.Organomet.Chem., 219 (1981) C9.
143. A.A.Bhattacharyya, C.G.Nagel, and S.G.Shore,
Organometallics, 2 (1983) 1187.

144. B.F.G.Johnson, J.Lewis, P.R.Raithby, and G.Süss, *J.Chem.Soc., Dalton Trans.,* (1979) 1356.
145. S.K.Malik and A.Poé, *Inorg.Chem., 17* (1978) 1484, and references therein.
146. M.I.Bruce, J.G.Matisons, and B.K.Nicholson, *J.Organomet.Chem., 247* (1983) 321, and references therein.
147. E.J.Forbes, N.Goodhand, D.L.Jones, and T.A.Hamor, *J.Organomet.Chem., 182* (1979) 143.
148. M.I.Bruce, J.G.Matisons, B.W.Skelton, and A.H.White, *J.Chem.Soc., Dalton Trans.,* (1983) 2375.
149. M.R.Churchill, T.P.Duggan, J.B.Keister, and J.W.Ziller, *Acta Cryst., C43* (1987) 203.
150. Z.A.Rahman, L.R.Beanan, L.M.Bavaro, S.P.Modi, J.B.Keister, and M.R.Churchill. *J.Organomet.Chem., 263* (1984) 75.
151. M.R.Churchill, C.H.Lake, W.G.Feighery, and J.B.Keister, *Organometallics, 10* (1991) 2384.
152. W.G.Feighery, R.D.Allendoerfer, and J.B.Keister, *Organometallics, 9* (1990) 2424.
153. A.J.P.Domingos, B.F.G.Johnson, and J.Lewis, *J.Organomet.Chem., 36* (1972) C43.
154. S.Aime, R.Gobetto, L.Milone, D.Osella, L.Violano, A.J.Arce, and Y.De Sanctis, *Organometallics, 10* (1991) 2854.
155. D.Osella, O.Gambino, C.Nervi, M.Ravera, M.V.Russo, and G.Infante, *Gazz.Chim.Ital.,* in the press.
156. G.Gervasio, D.Osella, and M.Valle, *Inorg.Chem., 15* (1976) 1221.
157. M.Evans, M.Hursthouse, E.W.Randall, E.Rosenberg, L.Milone, and M.Valle, *J.Chem.Soc., Chem.Commun.,* (1972) 545.
158. M.Catti, G.Gervasio, and S.A.Mason, *J.Chem.Soc., Dalton Trans.,* (1977) 60.
159. E.Rosemberg, S.Aime, L.Milone, E.Sappa, A.Tiripicchio, and A.M.Manotti Lanfredi, *J.Chem.Soc., Dalton Trans.,* (1981) 2023.
160. D.Osella, R.Gobetto, L.Milone, P.Zanello, and S.Mangani, *Organometallics, 9* (1990) 2167.

161. S.Aime, D.Osella, A.J.Deeming, A.J.Arce, M.B.Hursthouse, and H.M.Dawes,
J.Chem.Soc., Dalton Trans., (1986) 1459.
162. J.S.Field, R.J.Haines, and D.N.Smit,
J.Organomet.Chem., 224 (1982) C49.
163. J.S.Field, R.J.Haines, and D.N.Smit,
J.Chem.Soc., Dalton Trans., (1988) 1315.
164. S.Aime, G.Nicola, D.Osella, A.M.Manotti Lanfredi, and A.Tiripicchio,
Inorg.Chim.Acta, 85 (1984) 161.
165. B.F.G.Johnson, J.Lewis, and K.T.Schorpp,
J.Organomet.Chem., 91 (1975) C13.
166. B.F.G.Johnson, J.Lewis, B.E.Reichert, K.T.Schorpp, and G.M.Sheldrick,
J.Chem.Soc., Dalton Trans., (1977) 1417.
167. P.F.Jackson, B.F.G.Johnson, J.Lewis, P.R.Raithby, G.J.Will, M.McPartlin, and W.J.H.Nelson,
J.Chem.Soc., Chem.Commun., (1980) 1190.
168. D.Osella and P.Zanello,
unpublished results.
169. B.F.G.Johnson, J.Lewis, W.J.H.Nelson, J.N.Nicholls, J.Puga, P.R.Raithby, M.J.Rosales, M.Schröder, and M.D.Vargas,
J.Chem.Soc., Dalton Trans., (1983) 2447.
170. J.Wang, M.Sabat, L.J.Lyons, and D.F.Shriver,
Inorg.Chem.,30 (1991) 382.
171. D.H.Farrar, P.F.Jackson, B.F.G.Johnson, J.Lewis, J.N.Nicholls, and M.McPartlin,
J.Chem.Soc., Chem.Commun., (1981) 415.
172. B.F.G.Johnson, J.Lewis, J.N.Nicholls, J.Puga, P.R.Raithby, M.J.Rosales, M.McPartlin, and W.Clegg,
J.Chem.Soc., Dalton Trans., (1983) 277.
173. A.G.Cowie, B.F.G.Johnson, J.Lewis, and P.R.Raithby,
J.Organomet.Chem., 306 (1986) C63.
174. B.F.G.Johnson, J.Lewis, J.N.Nicholls, I.A.Oxton, P.R.Raithby, and M.J.Rosales,
J.Chem.Soc., Chem.Commun., (1982) 289.
175. M.I.Bruce, M.L.Williams, J.M.Patrick, and A.H.White,
J.Chem.Soc., Dalton Trans., (1985) 1229.

176. M.I.Bruce, M.J.Liddell, and E.R.T.Tiekink,
J.Organomet.Chem., *391* (1990) 81.
177. C.R.Eady, B.F.G.Johnson, J.Lewis, M.C.Malatesta, P.Machin, and M.McPartlin,
J.Chem.Soc., Chem.Commun., (1976) 945.
178. P.F.Jackson, B.F.G.Johnson, J.Lewis, M.McPartlin, and W.J.H.Nelson,
J.Chem.Soc., Chem.Commun., (1979) 735.
179. C.R.Eady, P.F.Jackson, B.F.G.Johnson, J.Lewis, M.C.Malatesta, M.McPartlin, and W.J.H.Nelson,
J.Chem.Soc., Dalton Trans., (1980) 383.
180. C.-M.T.Hayward and J.R.Shapley,
Inorg.Chem., *21* (1982) 3816.
181. B.Tulyathan and W.E.Geiger,
J.Am.Chem.Soc., *107* (1985) 5960.
182. S.D.Wijeyesekera and R.Hoffmann,
Organometallics, *3* (1984) 949.
183. B.F.G.Johnson, J.Lewis, S.W.Sankey, K.Wong, M.McPartlin, and W.J.H.Nelson,
J.Organomet.Chem., *191* (1980) C3.
184. J.S.Bradley, G.B.Ansell, and E.W.Hill,
J.Organomet.Chem., *184* (1980) C33.
185. G.B.Ansell and J.S.Bradley,
Acta Cryst., *B36* (1980) 726.
186. S.R.Drake, B.F.G.Johnson, and J.Lewis,
J.Chem.Soc., Dalton Trans., (1989) 243.
187. S.C.Brown, J.Evans, and M.Webster,
J.Chem.Soc., Dalton Trans., (1981) 2263.
188. S.R.Drake, B.F.G.Johnson, J.Lewis, G.Conole, and M.McPartlin,
J.Chem.Soc., Dalton Trans., (1990) 995.
189. A.Sirigu, M.Bianchi, and E.Benedetti,
Chem.Commun., (1969) 596.
190. B.M.Peake, B.H.Robinson, J.Simpson, and D.J.Watson,
J.Chem.Soc., Chem.Commun., (1974) 945.
191. E.R.Corey and L.F.Dahl,
Inorg.Chem., *1* (1962) 521.
192. M.R.Churchill and B.G. DeBoer,
Inorg.Chem., *16* (1977) 878.

193. M.R.Churchill, F.J.Hollander, and J.P.Hutchinson, *Inorg.Chem., 16* (1977) 2697.
194. V.F.Allen, R.Mason, and P.B.Hitchcock, *J.Organomet.Chem., 140* (1977) 297.
195. A.G.Orpen, A.V.Rivera, E.G.Bryan, D.Pippard, G.M.Sheldrick, and K.D.Rouse, *J.Chem.Soc., Chem.Commun.,* (1978) 723.
196. R.W.Broach and J.M.Williams, *Inorg.Chem., 18* (1979) 314.
197. D.Osella, E.Stein, G.Nervi, P.Zanello, F.Laschi, A.Cinquantini, E. Rosenberg, and J.Fiedler, *Organometallics, 10* (1991) 1929.
198. J.R.Shapley, J.B.Keister, M.R.Churchill, and B.G.DeBoer, *J.Am.Chem.Soc., 97* (1975) 4145.
199. M.R.Churchill and B.G.DeBoer, *Inorg.Chem., 16* (1977) 2397.
200. A.J.Deeming, S.Donovan-Mtunzi, S.E.Kabir, and P.J.Manning, *J.Chem.Soc., Dalton Trans.,* (1985) 1037, and references therein.
201. R.F.Alex, F.W.B.Einstein, R.H.Jones, and R.K.Pomeroy, *Inorg.Chem., 26* (1987) 3175, and references therein.
202. R.E.Benfield, B.F.Johnson, P.R.Raithby, and G.M.Sheldrick, *Acta Cryst., B34* (1978) 666.
203. B.F.G.Johnson, J.Lewis, B.E.Reichert, and K.T.Shorpp, *J.Chem.Soc., Dalton Trans.,* (1976) 1403.
204. D.Osella, M.Ravera, A.K.Smith, A.J.Mathews, and P.Zanello, *J.Organomet.Chem., 423* (1992) 255.
205. A.W.Coleman, D.F.Jones, P.H.Dixneuf, C.Brisson, J.-J.Bonnet, and G.Lavigne, *Inorg.Chem., 23* (1984) 952.
206. J.A.Glucas, D.F.Foster, M.M.Harding, and A.K.Smith, *J.Chem.Soc., Chem.Commun.,* (1984) 949.
207. C.G.Pierpont, *Inorg.Chem., 16* (1977) 636.
208. M.P.Brown, P.A.Dolby, M.M.Harding, A.J.Mathews, A.K.Smith, D.Osella, M.Arbrun, R.Gobetto, P.R.Raithby, and P.Zanello, *J.Chem.Soc., Dalton Trans.,* (1993) 827.

209. J.A.Glucas, P.A.Dolby, M.M.Harding, and A.K.Smith, *J.Chem.Soc., Chem.Commun.*, (1987) 1829.
210. B.F.G.Johnson, J.Lewis, P.R.Raithby, and C.Zuccaro, *Acta Cryst., B37* (1981) 1728.
211. B.F.G.Johnson, J.Lewis, P.R.Raithby, and C.Zuccaro, *Acta Cryst., B34* (1978) 3765.
212. B.F.G.Johnson, J.Lewis, P.R.Raithby, G.M.Sheldrick, and G.Süss, *J.Organomet.Chem., 162* (1978) 179.
213. B.F.G.Johnson, J.Lewis, P.R.Raithby, and C.Zuccaro, *J.Chem.Soc., Dalton Trans.*, (1980) 717.
214. B.F.G.Johnson, J.Lewis, P.R.Raithby, G.M.Sheldrick, K.Wong, and M.McPartlin, *J.Chem.Soc., Dalton Trans.*, (1978) 673.
215. D.Braga, B.F.G.Johnson, J.Lewis, J.M.Mace, M.McPartlin, J.Puga, W.J.H. Nelson, P.R.Raithby, and K.H.Whitmire, *J.Chem.Soc., Chem.Commun.*, (1982) 1081.
216. M.A.Collins, B.F.G.Johnson, J.Lewis, J.M.Mace, J.Morris, M.McPartlin, W.J.H. Nelson, J.Puga, and P.R.Raithby, *J.Chem.Soc., Chem.Commun.*, (1983) 689.
217. B.F.G.Johnson, J.Lewis, W.J.H.Nelson, J.Puga, P.R.Raithby, M.Schröder, and K.H.Whitmire, *J.Chem.Soc., Chem.Commun.*, (1982) 610.
218. C.R.Eady, J.J.Guy, B.F.G.Johnson, J.Lewis, M.C.Malatesta, and G.M.Sheldrick, *J.Chem.Soc., Chem.Commun.*, (1976) 807.
219. J.J.Guy and G.M.Sheldrick, *Acta Cryst., B34* (1978) 1722.
220. C.R.Eady, B.F.G.Johnson, J.Lewis, B.E.Reichert, and G.M.Sheldrick, *J.Chem.Soc., Chem.Commun.*, (1976) 271.
221. B.E.Reichert and G.M.Sheldrick, *Acta Cryst., B33* (1977) 173.
222. P.F.Jackson, B.F.G.Johnson, J.Lewis, J.N.Nicholls, M.McPartlin, and W.J.H. Nelson, *J.Chem.Soc., Chem.Commun.*, (1980) 564.
223. R.Mason, K.M.Thomas, and D.M.P.Mingos, *J.Am.Chem.Soc., 95* (1973) 3802.

224. C.R.Eady, B.F.G.Johnson, and J.Lewis,
J.Chem.Soc., Chem.Commun., (1976) 302.
225. M.McPartlin, C.R.Eady, B.F.G.Johnson, and J.Lewis,
J.Chem.Soc., Chem.Commun., (1976) 883.
226. S.B.Colbran, C.M.Hay, B.F.G.Johnson, F.J.Lahoz, J.Lewis, and P.R.Raithby,
J.Chem.Soc., Chem.Commun., (1986) 1766.
227. S.B.Colbran, F.J.Lahoz, P.R.Raithby, J.Lewis, B.F.G.Johnson, and C.J.Cardin,
J.Chem.Soc., Dalton Trans., (1988) 173.
228. M.H.Barley, C.E.Anson, B.F.G.Johnson, and J.Lewis,
J.Organomet.Chem., 339 (1988) 151.
229. R.J.Goudsmit, J.G.Jeffrey, B.F.G.Johnson, J.Lewis, R.C.S.McQueen, A.J.Sanders, and J.-C.Liu,
J.Chem.Soc., Chem.Commun., (1986) 24.
230. R.J.Goudsmit, B.F.G.Johnson, J.Lewis, P.R.Raithby, and K.H.Whitmire,
J.Chem.Soc., Chem.Commun., (1982) 640.
231. A.G.Orpen and G.M.Sheldrick,
Acta Cryst., B34 (1978) 1989.
232. C.Couture, D.H.Farrar, M.P.Gòmez-Sal, B.F.G.Johnson, R.A.Kamarudin, J.Lewis, and P.R.Raithby,
Acta Cryst., C42 (1986) 163.
233. D.G.Evans and D.M.P.Mingos,
Organometallics, 2 (1983) 435.
234. J.G.Jeffrey, B.F.G.Johnson, J.Lewis, P.R.Raithby, and D.A.Welch,
J.Chem.Soc., Chem.Commun., (1986) 318.
235. C.R.Eady, B.F.G.Johnson, J.Lewis, R.Mason, P.B.Hitchcock, and K.M.Thomas,
J.Chem.Soc., Chem.Commun., (1977) 385.
236. P.F.Jackson, B.F.G.Johnson, J.Lewis, M.McPartlin, and W.J.H.Nelson,
J.Chem.Soc., Chem.Commun., (1980) 224.
237. P.F.Jackson, B.F.G.Johnson, J.Lewis, W.J.H.Nelson, and M.McPartlin,
J.Chem.Soc., Dalton Trans., (1982) 2099.

238. D.H.Farrar, P.G.Jackson, B.F.G.Johnson, J.Lewis, W.J.H.Nelson, M.D.Vargas, and M.McPartlin, *J.Chem.Soc., Chem.Commun.*, (1981) 1009.
239. S.R.Drake, B.F.G.Johnson, J.Lewis, and R.C.S.McQueen, *J.Chem.Soc., Dalton Trans.*, (1987) 1051.
240. M.H.Barley, S.R.Drake, B.F.G.Johnson, and J.Lewis, *J.Chem.Soc., Chem.Commun.*, (1987) 1657.
241. S.R.Drake, M.H.Barley, B.F.G.Johnson, and J.Lewis, *Organometallics*, 7 (1988) 806.
242. M.H.Barley, C.E.Anson, B.F.G.Johnson, and J.Lewis, *J.Organomet.Chem.*, *339* (1988) 151.
243. D.W.Bullett, *Chem.Phys.Lett.*, *135* (1987) 373.
244. D.Braga, J.Lewis, B.F.G.Johnson, M.McPartlin, W.J.H.Nelson, and M.D.Vargas, *J.Chem.Soc., Chem.Commun.*, (1983) 241.
245. C.H.Wei and L.F.Dahl, *Inorg.Chem.*, *6* (1967) 1229.
246. C.E.Strouse and L.F.Dahl, *J.Am.Chem.Soc.*, *93* (1971) 6032.
247. H.Patin, G.Mignani, C.Mahé, J.-Y.Le Marouille, A.Benoit, D.Grandjeau, and G.Levesque, *J.Organomet.Chem.*, *208* (1981) C39.
248. C.Mahé, H.Patin, A.Benoit, and J.-Y.Le Marouille, *J.Organomet.Chem.*, *216* (1981) C15.
249. A.Benoit, A.Darchen, J.-Y.Le Marouille, C.Mahé, and H.Patin, *Organometallics*, *2* (1983) 555.
250. B.R.Penfold and B.H.Robinson, *Acc.Chem.Res.*, *6* (1973) 73.
251. D.Seyferth, *Adv.Organomet.Chem.*, *14* (1976) 98.
252. G.Schmid, *Angew.Chem.Int.Ed.Engl.*, *17* (1978) 392.
253. P.W.Sutton and L.F.Dahl, *J.Am.Chem.Soc.*, *89* (1967) 261.
254. P.Leung, P.Coppens, R.K.McMullan, and T.F.Koetzle, *Acta Cryst.*, *B37* (1981) 1347.
255. M.Ahlgrén, T.T.Pakkanen, and I.Tahvanainen, *J.Organomet.Chem.*, *323* (1987) 91.

256. G.H.Worth, B.H.Robinson, and J.Simpson,
Organometallics, *11* (1992) 501.
257. A.M.Bond, B.M.Peake, B.H.Robinson, J.Simpson, and D.J.Watson,
Inorg.Chem., *16* (1977) 410.
258. K.Hinkelmann, J.Heinze, H.-T.Schacht, J.S.Field, and
H.Vahrenkamp,
J.Am.Chem.Soc., *111* (1989) 5078.
259. A.J.Downard, B.H.Robinson, and J.Simpson,
Organometallics, *5* (1986) 1140.
260. J.C.Kotz, J.V.Petersen, and R.Reed,
J.Organomet.Chem., *120* (1976) 433.
261. P.N.Lindsay, B.M.Peake, B.H.Robinson, J.Simpson, V.Honrath,
H.Vahrenkamp, and A.M.Bond,
Organometallics, *3* (1984) 413.
262. (a) G.H.Worth, B.H.Robinson, and J.Simpson,
Organometallics, *11* (1992) 3863.
(b) D.Osella, O.Gambino, C.Nervi, M.Ravera, and
D.Bertolino,
Inorg.Chim.Acta, *206* (1993) 155.
263. B.M.Peake, B.H.Robinson, J.Simpson, and D.J.Watson,
Inorg.Chem., *16* (1977) 405.
264. M.D.Brice and B.R.Penfold,
Inorg.Chem., *11* (1972) 1381.
265. G.Allegra and S.Valle,
Acta Cryst., *B25* (1969) 107.
266. R.J.Dellaca, B.R.Penfold, B.H.Robinson, W.T.Robinson,
and J.L.Spencer,
Inorg.Chem., *9* (1970) 2204.
267. G.H.Worth, B.H.Robinson, and J.Simpson,
J.Organomet.Chem., *387* (1990) 337.
268. M.D.Brice, B.R.Penfold, W.T.Robinson, and S.R.Taylor,
Inorg.Chem., *9* (1970) 362.
269. T.W.Matheson and B.R.Penfold,
Acta Cryst., *B33* (1977) 1980.
270. A.J.Downard, B.H.Robinson, and J.Simpson,
Organometallics, *5* (1986) 1122.
271. (a) D.N.Duffy, M.M.Kassis, and A.D.Rae,
Acta Cryst., *C47* (1991) 2054.

(b) G.Balavoine, J.Collin, J.J.Bonnet, and G.Lavigne, *J.Organomet.Chem.*, *280* (1985) 429.
(c) A.J.Downard, B.H.Robinson, and J.Simpson, *J.Organomet.Chem.*, *447* (1993) 281.
(d) M.-J.Don, M.G.Richmond, W.H.Watson, M.Krawiec, and R.P.Kashyap, *J.Organomet.Chem.*, *418* (1991) 231.

272. P.A.Dawson, B.H.Robinson, and J.Simpson, *J.Chem.Soc., Dalton Trans.*, (1979) 1762.
273. T.W.Matheson, B.H.Robinson, and W.S.Tham, *J.Chem.Soc.*(A), (1971) 1457.
274. A.M.Bond, P.A.Dawson, B.M.Peake, P.H.Rieger, B.H.Robinson, and J.Simpson, *Inorg.Chem.*, *18* (1979) 1413.
275. G.J.Bezems, P.H.Rieger, and S.Visco, *J.Chem.Soc., Chem.Commun.*, (1981) 265.
276. A.J.Donward, B.H.Robinson, and J.Simpson, *Organometallics*, *5* (1986) 1132.
277. H.Beurich, F.Richter, and H.Vahrenkamp, *Acta Cryst.*, *B38* (1982) 3012.
278. H.Beurich, T.Madach, F.Richter, and H.Vahrenkamp, *Angew.Chem.Int.Ed.Engl.*, *18* (1979) 690.
279. U.Honrath, L.S.-Thang, and H.Vahrenkamp, *Chem.Ber.*, *118* (1985) 132.
280. W.L.Olson, A.M.Stacy, and L.F.Dahl, *J.Am.Chem.Soc.*, *108* (1986) 7646.
281. C.E.Barnes, D.L.Staley, A.L.Rheingold, and D.C.Johnson, *J.Am.Chem.Soc.*, *111* (1989) 4992.
282. W.L.Olson and L.F.Dahl, *J.Am.Chem.Soc.*, *108* (1986) 7657.
283. V.A.Uchtman and L.F.Dahl, *J.Am.Chem.Soc.*, *91* (1969) 3763.
284. P.D.Frisch and L.F.Dahl, *J.Am.Chem.Soc.*, *94* (1972) 5082.
285. R.L.Bedard, A.D.Rae, and L.F.Dahl, *J.Am.Chem.Soc.*, *108* (1986) 5924.
286. R.L.Bedard and L.F.Dahl, *J.Am.Chem.Soc.*, *108* (1986) 5933.

287. M.S.Ziebarth and L.F.Dahl,
J.Am.Chem.Soc., *112* (1990) 2411.
288. W.I.Bailey, Jr., F.A.Cotton, J.D.Jamerson, and B.W.S.Kolthammar,
Inorg.Chem., *21* (1982) 3131.
289. J.M.Mevs, T.Gennett, and W.E.Geiger,
Organometallics, *10* (1991) 1229.
290. P.Corradini,
J.Chem.Phys., *31* (1959) 1676.
291. C.H.Wei and L.F.Dahl,
J.Am.Chem.Soc., *88* (1966) 1821.
292. C.H.Wei and L.F.Dahl,
J.Am.Chem.Soc., *91* (1969) 1351.
293. C.H.Wei,
Inorg.Chem., *8* (1969) 2384.
294. F.H.Carré, F.A.Cotton, and B.A.Frenz,
Inorg.Chem., *15* (1976) 380.
295. J.Rimmelin, P.Lemoine, M.Gross, and D.de Mountauzon,
Nouv.J.Chim., *7* (1983) 453.
296. J.Rimmelin, P.Lemoine, M.Gross, A.A.Bahsoun, and J.A.Osborn,
Nouv.J.Chim., *9* (1985) 181.
297. A.Cinquantini, P.Zanello, R.Della Pergola, L.Garlaschelli, and S.Martinengo,
J.Organomet.Chem., *412* (1991) 215.
298. G.F.Holland, D.E.Ellis, and W.C.Trogler,
J.Am.Chem.Soc., *108* (1986) 1884.
299. G.F.Holland, D.E.Ellis, D.R.Tyler, H.B.Gray, and W.C.Trogler,
J.Am.Chem.Soc., *109* (1987) 4276.
300. E.Keller and H.Vahrenkamp,
Chem.Ber., *114* (1981) 1111.
301. P.H.Bird and A.R.Fraser,
J.Organomet.Chem., *73* (1974) 103, and references therein.
302. V.S.Kaganovich and M.I.Rubinskaya,
J.Organomet.Chem., *344* (1988) 383.
303. M.Elian, M.M.L.Chen, D.M.P.Mingos, and R.Hoffmann,
Inorg.Chem., *15* (1976) 1148.
304. A.A.Arduini, A.A.Bahsoun, J.A.Osborn, and C.Voelker,
Angew.Chem.Int.Ed.Engl., *19* (1980) 1024.

305. A.A.Bahsoun, J.A.Osborn, C.Voelker, J.J.Bonnet, and G.Lavigne,
Organometallics, 1 (1982) 1114.
306. D.J.Darensbourg, D.J.Zalewski, and T.Delord,
Organometallics, 3 (1984) 1210.
307. D.J.Darensbourg, D.J.Zalewski, A.L.Rheingold, and R.L.Durney,
Inorg.Chem., 25 (1986) 3281.
308. J.R.Kennedy, P.Selz, A.L.Rheingold, W.C.Trogler, and F.Basolo,
J.Am.Chem.Soc., 111 (1989) 3615.
309. C.H.Wei and L.F.Dahl,
Cryst.Struct.Comm., 4 (1975) 583, and references therein.
310. W.Heiber and T.Kruck,
Chem.Ber., 95 (1962) 2027.
311. R.C.Ryan and L.F.Dahl,
J.Am.Chem.Soc., 97 (1975) 6904.
312. E.Lindner, G.A.Weiss, W.Hiller, and R.Fawzi,
J.Organomet.Chem., 312 (1986) 365.
313. R.C.Ryan, C.U.Pittman,Jr., J.P.O'Connor, and L.F.Dahl,
J.Organomet.Chem., 193 (1980) 247.
314. A.M.Arif, A.H.Cowley, M.Pakulski, M.Hursthouse, and A.Karauloz,
Organometallics, 4 (1985) 2227.
315. M.G.Richmond and J.K.Kochi,
Inorg.Chem., 25 (1986) 656.
316. M.G.Richmond and J.K.Kochi,
Inorg.Chem., 26 (1987) 541.
317. M.G.Richmond and J.K.Kochi,
Inorg.Chem., 25 (1986) 1334.
318. M.G.Richmond and J.K.Kochi,
Inorg.Chim.Acta, 126 (1987) 83.
319. M.G.Richmond and J.K.Kochi,
Organometallics, 6 (1987) 254.
320. C.L.Shulman, M.G.Richmond, W.H.Watson, and A.Nagl,
J.Organomet.Chem., 368 (1989) 367.
321. M.G.Richmond, J.D.Korp, and J.K.Kochi,
J.Chem.Soc., Chem.Commun., (1985) 1102.
322. U.Krüerke and W.Hübel,
Chem.Ber., 94 (1961) 2829.

323. L.F.Dahl and D.L.Smith,
J.Am.Chem.Soc., *84* (1962) 2450.
324. G.Gervasio, R.Rossetti, and P.L.Stanghellini,
Organometallics, *4* (1985) 1612.
325. D.Osella, M.Ravera, C.Nervi, C.Housecroft, P.Raithby, P.Zanello, and F.Laschi,
Organometallics, *10* (1991) 3253.
326. R.Rumin, P.Courtot, J.E.Guerchais, F.Y.Pétillon, L.Manojlovic-Muir, and K.W.Muir,
J.Organomet.Chem., *301* (1986) C1.
327. R.Rumin, L.Manojlovic-Muir, K.W.Muir, and F.Y.Pétillon,
Organometallics, *17* (1988) 375.
328. F.Y.Pétillon, R.Rumin, and J.Talarmin,
J.Organomet.Chem., *346* (1988) 111.
329. V.Albano, P.Chini, and V.Scatturin,
Chem.Commun., (1968) 163.
330. V.Albano, P.Chini, and V.Scatturin,
J.Organomet.Chem., *15* (1968) 423.
331. S.Martinengo, D.Strumolo, P.Chini, V.G.Albano, and D.Braga,
J.Chem.Soc., Dalton Trans., (1985) 35.
332. V.G.Albano, P.Chini, G.Ciani, M.Sansoni, D.Strumolo, B.T.Heaton, and S.Martinengo,
J.Am.Chem.Soc., *98* (1976) 5027.
333. V.G.Albano, P.Chini, G.Ciani, M.Sansoni, and S.Martinengo,
J.Chem.Soc., Dalton Trans., (1980) 163.
334. T.Beringhelli, F.Morazzoni, and D.Strumolo,
J.Organomet.Chem., *236* (1982) 109.
335. V.G.Albano, P.Chini, G.Ciani, S.Martinengo, and M.Sansoni,
J.Chem.Soc., Dalton Trans., (1978) 463.
336. V.G.Albano, D.Braga, G.Ciani, and S.Martinengo,
J.Organomet.Chem., *213* (1981) 293.
337. V.G.Albano, D.Braga, P.Chini, G.Ciani, and S.Martinengo,
J.Chem.Soc., Dalton Trans., (1982) 645.
338. P.Zanello and L.Garlaschelli,
unpublished results.
339. V.G.Albano, D.Braga, A.Fumagalli, and S.Martinengo,
J.Chem.Soc., Dalton Trans., (1985) 237.
340. E.F.Paulus, E.O.Fisher, H.P.Fritz, H.Schuster-Woldan,
J.Organomet.Chem., *10* (1967) P3.

341. O.S.Mills and E.F.Paulus,
J.Organomet.Chem., 10 (1967) 331.
342. F.Faraone, S.Lo Schiavo, G.Bruno, P.Piraino, and G.Bombieri,
J.Chem.Soc., Dalton Trans., (1983) 1813.
343. Von H.Miessner,
Z.Anorg.Allg.Chem., 505 (1983) 187.
344. A.Fumagalli, T.F.Koetzle, F.Takusagawa, P.Chini, S.Martinengo, and B.T.Heaton,
J.Am.Chem.Soc., 102 (1980) 1740.
345. A.Bonny, T.J.Crane, N.A.P.Kane-Maguire,
Inorg.Chim.Acta, 65 (1982) L83.
346. S.Martinengo and P.Chini,
Gazz.Chim.Ital., 102 (1972) 344.
347. V.G.Albano, M.Sansoni, P.Chini, and S.Martinengo,
J.Chem.Soc., Dalton Trans., (1973) 651.
348. P.Chini and S.Martinengo,
Inorg.Chim.Acta, 3 (1969) 299.
349. V.G.Albano and P.L.Bellon,
J.Organomet.Chem., 19 (1969) 405.
350. E.R.Corey, L.F.Dahl, and W.Beck,
J.Am.Chem.Soc., 85 (1963) 1202.
351. V.G.Albano, D.Braga, P.Chini, D.Strumolo, and S.Martinengo,
J.Chem.Soc., Dalton Trans., (1983) 249.
352. P.Zanello and L.Garlaschelli,
work in progress.
353. V.G.Albano, D.Braga, D.Strumolo, C.Seregni, and S.Martinengo,
J.Chem.Soc., Dalton Trans., (1985) 309.
354. D.Strumolo, C.Seregni, S.Martinengo, V.Albano, and D.Braga,
J.Organomet.Chem., 252 (1983) C93.
355. J.Devillers, J.-J.Bonnet, D.de Montauzon, J.Galy, and R.Poilblanc,
Inorg.Chem., 19 (1980) 154.
356. P.Lemoine, M.Gross, D.de Montauzon, and R.Poilblanc,
Inorg.Chim.Acta, 71 (1983) 15.
357. M.R.Churchill and J.P.Hutchinson,
Inorg.Chem., 17 (1978) 3528.

358. P.Chini, G.Ciani, L.Garlaschelli, M.Manassero, S.Martinengo, A.Sironi, and F.Canziani, *J.Organomet.Chem.*, *152* (1978) C35.
359. G.Ciani, M.Manassero, and A.Sironi, *J.Organomet.Chem.*, *199* (1980) 271.
360. L.Garlaschelli, S.Martinengo, P.Chini, F.Canziani, and R.Bau, *J.Organomet.Chem.*, *213* (1981) 379.
361. R.Bau, M.Y.Chiang, C.Y.Wei, L.Garlaschelli, S.Martinengo, and T.F.Koetzle, *Inorg.Chem.*, *23* (1984) 4758.
362. V.Albano, P.L.Bellon, and V.Scatturin, *Chem.Commun.*, (1967) 730.
363. M.R.Churchill and J.P.Hutchinson, *Inorg.Chem.*, *19* (1980) 2765.
364. A.Strawczynski, R.Ros, R.Roulet, F.Grepioni, and D.Braga, *Helv.Chim.Acta*, *71* (1988) 1885.
365. G.Ciani, M.Manassero, V.Albano, F.Canziani, G.Giordano, S.Martinengo, and P.Chini, *J.Organomet.Chem.*, *150* (1978) C17.
366. M.H.Harding, B.S.Nicholls, and A.K.Smith, *Acta Cryst.*, *C40* (1984) 790.
367. D.J.Darensbourg and B.J.Baldwin-Zuschke, *Inorg.Chem.*, *20* (1981) 3846.
368. M.R.Churchill and J.P.Hutchinson, *Inorg.Chem.*, *18* (1979) 2451.
369. G.F.Stuntz and J.R.Shapley, *J.Organomet.Chem.*, *213* (1981) 389.
370. R.E.Benfield and B.F.G.Johnson, *J.Chem.Soc., Dalton Trans.*, (1980) 1743.
371. R.Della Pergola, L.Garlaschelli, S.Martinengo, F.Demartin, M.Manassero, and M.Sansoni, *Gazz.Chim.Ital.*, *117* (1987) 245.
372. M.P.Brown, D.Burns, M.M.Harding, S.Maginn, and A.K.Smith, *Inorg.Chim.Acta*, *162* (1989) 287.
373. A.Cinquantini and P.Zanello, unpublished results.
374. J.A.Clucas, M.M.Harding, B.S.Nicholls, and A.K.Smith, *J.Chem.Soc., Chem.Commun.*, (1984) 319.

375. F.Demartin, M.Manassero, M.Sansoni, L.Garlaschelli, S.Martinengo, and F.Canziani, *J.Chem.Soc., Chem.Commun.*, (1980) 903.
376. R.Della Pergola, L.Garlaschelli, S.Martinengo, F.Demartin, M.Manassero, N.Masciocchi, R.Bau, and D.Zhao, *J.Organomet.Chem.*, *396* (1990) 385.
377. R.Della Pergola, F.Demartin, L.Garlaschelli, M.Manassero, S.Martinengo, N.Masciocchi, and P.Zanello, *Inorg.Chem.*, *32* (1993) 3670.
378. R.Della Pergola, L.Garlaschelli, M.Manassero, N.Masciocchi, and P.Zanello, *Angew.Chem.Int.Ed.Engl.*, *32* (1993) 1347.
379. E.O.Fischer and C.Palm, *Chem.Ber.*, *91* (1958) 1725.
380. A.A.Hock and O.S.Mills, *Advances in the Chemistry of Coordination Compounds*, S.Kirschner, ed; Macmillan, New York, 1961, p.640.
381. J.J.Maj, A.D.Rae, and L.F.Dahl, *J.Am.Chem.Soc.*, *104* (1982) 3054.
382. L.R.Byers, V.A.Uchtman, and L.F.Dahl, *J.Am.Chem.Soc.*, *103* (1981) 1942.
383. T.E.North, J.B.Thoden, B.Spencer, A.Bjarnason, and L.F.Dahl, *Organometallics*, *11* (1992) 4326.
384. M.J.Bennett, F.A.Cotton, and B.H.C.Winquist, *J.Am.Chem.Soc.*, *89* (1967) 5366.
385. M.Bochmann, I.Hawkins, M.B.Hurtshouse, and R.L.Short, *J.Organomet.Chem.*, *332* (1987) 361.
386. M.Bochmann, I.Hawkins, L.J.Yellowlees, M.B.Hursthouse, and R.L.Short, *Polyhedron*, *8* (1989) 1351.
387. L.D.Lower and L.F.Dahl, *J.Am.Chem.Soc.*, *98* (1976) 5046.
388. K.C.C.Kharas and L.F.Dahl, *Adv.Chem.Phys.*, *70* (1988) 1.
389. D.F.Rieck, J.A.Gavney, R.L.Norman, R.K.Hayashi, and L.F.Dahl, *J.Am.Chem.Soc.*, *114* (1992) 10369.
390. A.Ceriotti, G.Longoni, M.Manassero, M.Perego, and M.Sansoni, *Inorg.Chem.*, *24* (1985) 117.

391. P.Zanello and L.Garlaschelli, unpublished results.
392. D.M.Washecheck, E.J.Wucherer, L.F.Dahl, A.Ceriotti, G.Longoni, M.Manassero, M.Sansoni, and P.Chini, *J.Am.Chem.Soc., 101* (1979) 6110.
393. P.Zanello, A.Ceriotti, and L.Garlaschelli, unpublished results.
394. J.D.Roth, G.J.Lewis, L.K.Safford, X.Jiang, L.F.Dahl, and M.J.Weaver, *J.Am.Chem.Soc., 114* (1992) 6159.
395. G.J.Lewis, J.D.Roth, R.A.Montag, L.K.Safford, X.Gao, S.-C.Chang, L.F.Dahl, and M.J.Weaver, *J.Am.Chem.Soc., 112* (1990) 2831.
396. B.J.Brisdon, D.A.Edwards, I.M.Towell, G.A.Moehring, and R.A.Walton, *J.Chem.Soc., Dalton Trans.,* (1988) 245.
397. E.W.Abel, W.Harrison, R.A.N.McLean, W.C.Marsh, and J.Trotter, *Chem.Commun.,* (1970) 1531.
398. W.Harrison, C.Marsh, and J.Trotter, *J.Chem.Soc., Dalton Trans.,* (1972) 1009.
399. E.Nuber, F.Oberdorfer, and M.L.Ziegler, *Acta Cryst., B37* (1981) 2062.
400. G.Ciani, G.D'Alfonso, M.Freni, P.Romiti, and A.Sironi, *J.Chem.Soc., Chem.Commun.,* (1982) 339.
401. C.-M.T.Hayward and J.R.Shapley, *Organometallics,* 7 (1988) 448.
402. G.Ciani, G.D'Alfonso, M.Freni, P.Romiti, and A.Sironi, *J.Chem.Soc., Chem.Commun.,* (1982) 705.
403. T.J.Henly, J.R.Shapley, A.L.Rheingold, and S.J.Gelb, *Organometallics,* 7 (1988) 441.

ACKNOWLEDGEMENTS

We are indebted to Mr. V.Bindi, Mr. V.Castelli, Mrs. G.Montomoli, Mrs. M.Pietrobono, and Mr. M.Porcù for their skilful assistance in preparing the manuscript.

We gratefully acknowledge receipt of the permission to reproduce figures from journals of:

- *AMERICAN CHEMICAL SOCIETY* (Inorganic Chemistry; Journal of the American Chemical Society; Organometallics);
- *ELSEVIER SEQUOIA S.A.* (Inorganica Chimica Acta; Journal of Organometallic Chemistry);
- *GAUTHIER-VILLARS* (Nouveau Journal de Chimie);
- *INTERNATIONAL UNION OF CRYSTALLOGRAPHY* (Acta Crystallographica);
- *PERGAMON PRESS PLC* (Polyhedron);
- *ROYAL SOCIETY OF CHEMISTRY* (Journal of Chemical Society; Chemical Communications; Journal of Chemical Society, Dalton Transactions);
- *VCH PUBLISHERS* (Angewandte Chemie; Chemische Berichte).

PREPARATION, PROPERTIES, AND STEREOCHEMICAL ASPECTS OF INCLUSION COMPOUNDS OF ORGANOMETALLIC COMPLEXES WITH CYCLODEXTRINS

A. Harada

Preparation, Properties, and Stereochemical Aspects of Inclusion Compounds of Organometallic Complexes with Cyclodextrins

Akira HARADA
Department of Macromolecular Science, Faculty of Science,
Osaka University, Japan

1. INTRODUCTION

Since ferrocene was discovered, a great number of organometallic complexes have been synthesized and shown to have unique structures and properties [1]. Molecular complexes in which two or more molecules are weakly bound by non-covalent bonds have received much attention because of their characteristic chemical and physical properties in relation to the functions of biologically important substances [2]. There are numerous reports on molecular complexes with both organic and inorganic compounds, but fewer on molecular complexes with organometallic complexes. However, recently, the number of papers on molecular complexes of organometallic complexes has been increasing [3-9]. Inclusion compounds of organotransition metal complexes with crown ethers [10] and cryptands [11] have been reported. However, in these cases, their guest complexes have been limited to complexes with protic ligands such as amines, water, or acetonitrile ligand in the first coordination sphere and only a part of the ligands has been found to interact with the hosts. Protonated polyammonium macrocycles act as efficient hosts toward transition metal complexes that carry cyanide anions in their first coordination sphere [12]. Reports on inclusion compounds of neutral transition metal complexes without protic ligands, such as metallocenes and π-arene complexes with synthetic macrocycles, have appeared only recently. These coordination modes may be possible only when the host molecules have a hydrophobic cavity to accommodate these lipophilic ligands. One of the most promising candidates for such a host is cyclodextrin.

2. CYCLODEXTRINS

Cyclodextrins (CDs) (cycloamyloses) are naturally occurring macrocyclic molecules consisting of a minimum of six D(+)-glucopyranose units attached by α-(1,4) linkages [13,14], Figure 1.

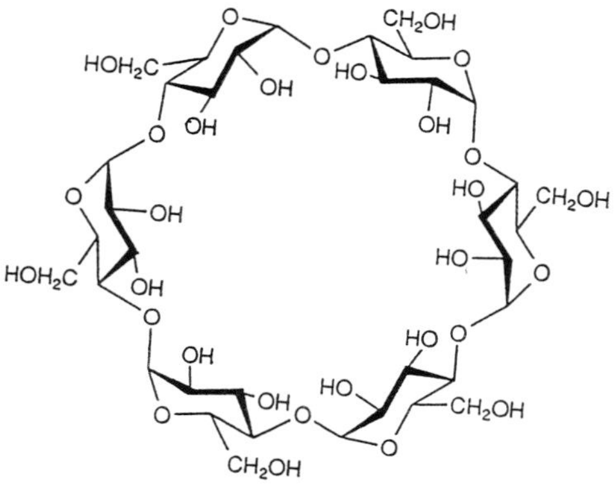

Fig. 1. The structure of α-cyclodextrin.

These compounds are called α-(6), β-(7), and γ-(8) cyclodextrin, respectively, Table 1. They are produced by the action of the amylase of *Bacillus macerans.* X-ray structural studies have shown that the secondary hydroxyl groups on the C-2 and C-3 of the glucose units are located on one side of the torus, whereas the primary hydroxyl groups are located on the opposite side of the torus [15]. Therefore, the outsides of cyclodextrins are hydrophilic and soluble in water, while the interior of the cavity consists of C-H groups and glucoside oxygens. So the interior of the cavities of cyclodextrins is relatively apolar compared to water. Cyclodextrins are V shaped, the secondary hydroxyl side being more open than the primary hydroxyl side.

Table 1. *Structural Characteristics of Cyclodextrins*

CD	Number of Glucose	Internal Diameter (Å)	Depth (Å)	Molecular Weight
α	6	4.5	6.7	972
β	7	7.0	7	1135
γ	8	8.5	7	1297

Cyclodextrins are well-known host molecules and have been studied as a model of enzymes for selective catalysis [8, 13]. Both in the solid state and in solution, cyclodextrins form inclusion complexes with various compounds, ranging from nonpolar hydrocarbons to small ions, and even rare gases. One requirement for the formation of inclusion complexes is that the guest molecule should be complementary to the cyclodextrin cavity. In inclusion complexes, the molar ratio of cyclodextrin to the guest molecule is usually stoichiometric one-to-one. Some of the inclusion complexes have been studied by single crystal X-ray diffraction and found that their guest molecules are included into the cyclodextrin cavities. Recently there have been some reports on the interaction between organometallic complexes and cyclodextrins.

3. FERROCENE AND ITS DERIVATIVES

3. 1. Ferrocene

Ferrocene is a typical organotransition metal complex and it is more stable than any other organotransition metal complexes. In 1975, Siegel and Breslow reported that ferrocene forms a 1:1 adduct with β-CD in both *N, N*-dimethylformamide and dimethyl sulfoxide

[16]. The acylation of β-CD by bound substrate has been studied as a model of serine acylase enzymes. The acylation of β-CD by the *p*-nitrophenyl ester of ferrocene cinnamic acid is accelerated over 50,000-fold compared with hydrolysis by buffer alone [17,18]. One enantiomer of the *p*-nitrophenyl ester of (*E*)-3-(carboxymethylene) 1,2-ferrocenecyclopentene acylates β-CD 5,900,000 times as fast in aqueous dimethyl sulfoxide as it hydrolyzes under the same conditions [19]. The actual rate achieved is comparable with or even exceeds the acylation of the enzyme chymotrypsin by *p*-nitrophenyl acetate. Breslow *et al.* proposed that β-CD forms complexes with substrates as intermediates [20,21]. Moreover, they reported that the *p*-nitrophenyl ester of (*E*)ruthenocene acrylic acid reacts in a complex to acylate β-CD, but with poorer binding and lower rate constants than with the corresponding ferrocene derivatives. Osa *et al.* reported the effect of β-CD on the electron transfer of ferrocenecarboxylic acid in solution [22]. However, inclusion compounds of ferrocene with cyclodextirns were not isolated.

Harada *et al.* prepared a range of crystalline inclusion compounds of ferrocene and its derivatives with cyclodextrins [23]. Ferrocenes are almost insoluble in water, so inclusion compounds could not be obtained by co-crystallization from aqueous solution, which is the usual method for obtaining inclusion complexes of water-soluble compounds. Therefore, Harada *et al.* tested the following methods for preparing CD-ferrocene inclusion compounds. **Method A**: An aqueous solution of cyclodextrin was added to an alcoholic solution of ferrocene with stirring and heating. **Method B**: An aqueous alcoholic solution (40 % ethanol) of CD was mixed with an aqueous alcoholic solution (60 % ethanol) of ferrocene, and then the added water and/or ethanol was partially removed by evaporation. **Method C**: Fine crystals of ferrocene were added to an aqueous solution of CD at 60 °C with stirring. The product obtained was washed thoroughly with water to remove the remaining cyclodextrin and with tetrahydrofuran to remove nonincluded molecules, and then recrystallized from water or aqueous alcohol. Of these three methods, Method C was found to be the best and gave β-CD-ferrocene inclusion compounds quantitatively. Method A gave a lower yield, while Method B afforded higher yields, but the Fe content of the complex was relatively low, Table 2.

α-CD-acetylferrocene and γ-CD-acetylferrocene complexes could not be obtained by method A, B, but C. The inclusion compounds of ferrocene with CD were characterized by elemental analyses, and IR, UV, and ^{1}H-NMR spectra. Stoichiometries were determined by elemental analyses, including atomic absorption analysis of the Fe contents, ^{1}H-NMR and absorption spectra. The results showed that all the complexes obtained were stoichiometric compounds and have definite CD:guest ratios depending on the combinations of host and guest [24]. Table 2 shows the results on the preparation of inclusion compounds of CDs with ferrocene.

Table 2. *Synthesis of Inclusion Compounds of Ferrocene with Cyclodextrins*

CD	Molar ratio of CD/Fc	Method	Yield %	Product: Fe % Found	Product: Fe % Calcd.	Product: CD:Fc
α	2:1	C	68	2.6	2.6	2:1
α	1:1	C	75	2.5	2.6	2:1
α	1:4	C	70	2.5	2.6	2:1
β	2:1	B	85	3.9	4.2	1:1
β	1:1	A	56	4.2	4.2	1:1
β	1:4	A	36	4.4	4.2	1:1
β	1:4	C	100	4.4	4.2	1:1
γ	1:1	C	59	3.6	3.8	1:1
γ	1:5	C	67	3.6	3.8	1:1

As illustrated in Scheme 1, β-CD formed 1:1 inclusion compounds with ferrocene, regardless of the molar ratio of the host to guest compound used. This result indicates that the interaction between cyclodextrin and ferrocene is a real inclusion phenomenon. γ-CD also formed 1:1 complexes, whereas α-CD formed 2:1 (α-CD :

guest) complexes in high yields even when α-CD was treated with excess ferrocene [25].

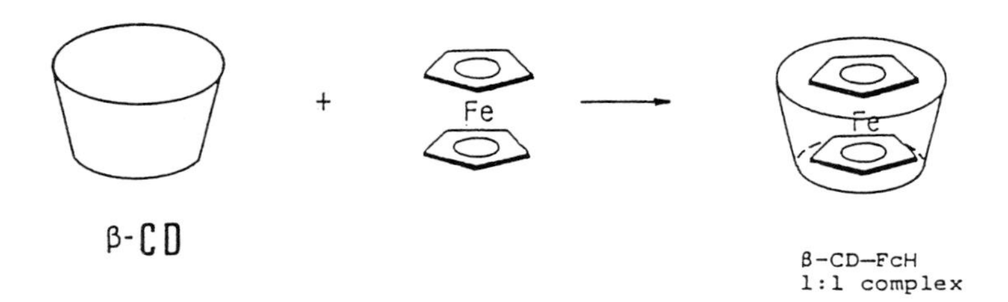

Scheme 1.

3. 2. Ferrocene Derivatives

Table 3 shows the results of the preparations of inclusion compounds of the substituted ferrocene derivatives shown in Scheme 2.

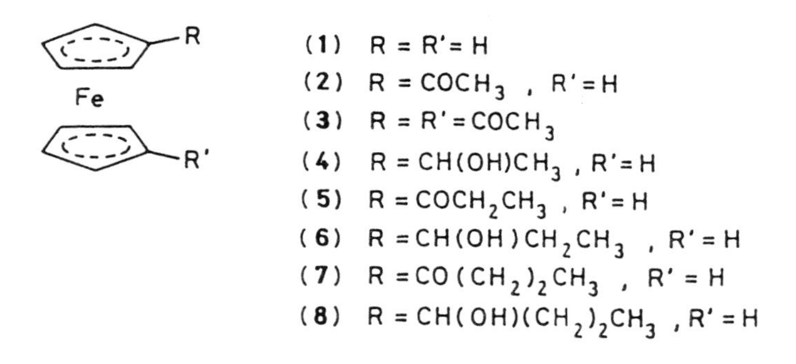

Scheme 2.

Table 3. *Synthesis of Inclusion Compounds of Ferrocene Derivatives with CDs*

CD	Guest	Molar ratio of CD/guest	Yield %	Product Fe % Found	Fe % Calcd.	CD:guest
α	2	1:2	73	2.5	2.6	2:1
β	2	1:2	100	4.1	4.1	1:1
γ	2	1:2	74	4.0	3.7	1:1
α	3	1:2	0	-	-	-
β	3	1:2	74	3.9	4.0	1:1
γ	3	1:2	90	3.7	3.6	1:1
α	4	1:2	80	2.6	2.6	2:1
β	4	1:2	100	4.5	4.1	1:1
γ	4	1:2	100	4.1	3.7	1:1
α	5	1:2	72	2.8	2.6	2:1
β	5	1:2	100	3.5	4.0	1:1
γ	5	1:2	98	3.5	3.6	1:1
α	6	1:2	46	3.5	2.6	2:1
β	6	1:2	100	3.6	4.0	1:1
γ	6	1:2	100	2.8	3.6	1:1
α	7	1:2	18	2.7	2.5	2:1
β	7	1:2	100	3.6	4.0	1:1
γ	7	1:2	100	3.6	3.6	1:1
α	8	1:2	67	2.8	2.5	2:1
β	8	1:2	100	3.5	4.0	1:1
γ	8	1:2	100	2.9	3.6	1:1
α	9	1:2	0	-	-	-
β	9	1:2	100	3.7	4.0	1:1
γ	9	1:2	100	3.1	3.66	1:1

β-CD and γ-CD formed 1:1 complexes with both mono- and disubstituted derivatives in high yields. For example, β-CD and γ-CD complexes with α-hydroxyalkyl ferrocene and oxybis-ethylidene ferrocene were obtained almost quantitatively. α-CD did not form complexes with 1,1-disubstituted derivatives, but it formed complexes with monosubstituted derivatives according to 2:1 (CD: guest) stoichiometry. The selective inclusion of CD could be exploited in the separation of ferrocene derivatives. Mono- and di-α-hydroxyethyl ferrocene, for example, could be successfully separated each other by formation of complexes with α-CD: monosubstituted ferrocene could be extracted from the inclusion compounds of β-CD, while the disubstituted derivative, which is not included, could be obtained by extraction of the residue.

3. 3. Properties

Inclusion compounds of ferrocene with β-CD and γ-CD are so stable that they can be recrystallized from water to give yellow crystals. β-CD-ferrocene complexes are stable to heat, not liberating ferrocene on heating at 100 °C *in vacuo*. Figure 2 shows the results of thermogravimetric analysis of the β-CD-ferrocene complex, free ferrocene, and a mixture of β-CD and ferrocene. The components in the mixture behaved independently: ferrocene sublimed below 80 °C, β-CD melted and decomposed at about 300 °C, and the complex did not change at about 80 °C, was stable up to 200 °C and dissociated at about the melting point of the CD component. These results indicate that the ferrocene molecule is tightly fixed within the β-CD cavity.

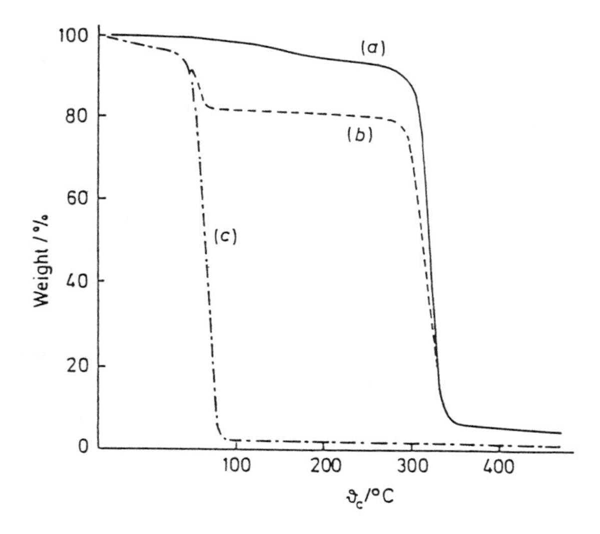

Fig. 2. Thermogravimetric analysis of: (a) β-CD-ferrocene complex; (b) a mixture of β-CD and ferrocene; (c) ferrocene (from Ref. 25).

Figure 3 shows the absorption and circular dichroism spectra of ferrocene in the presence of β-CD and γ-CD in ethylene glycol solution. β-CD caused marked decrease in the absorption spectrum of ferrocene at about 440 nm and showed a large positive Cotton effect at about 460 nm with a small negative Cotton effect at about 340 nm. γ-CD caused a smaller but definite decrease in the absorption spectrum of ferrocene and showed a negative Cotton effect at about 460 nm. According to calculations, an electric dipole moment on the axis of the cyclodextrin gives a positive Cotton effect, whereas one perpendicular to the axis gives a negative one. Thus, the above results indicate that the orientation of the ferrocene molecule in the cavity of β-CD is different from that in the γ-CD cavity.

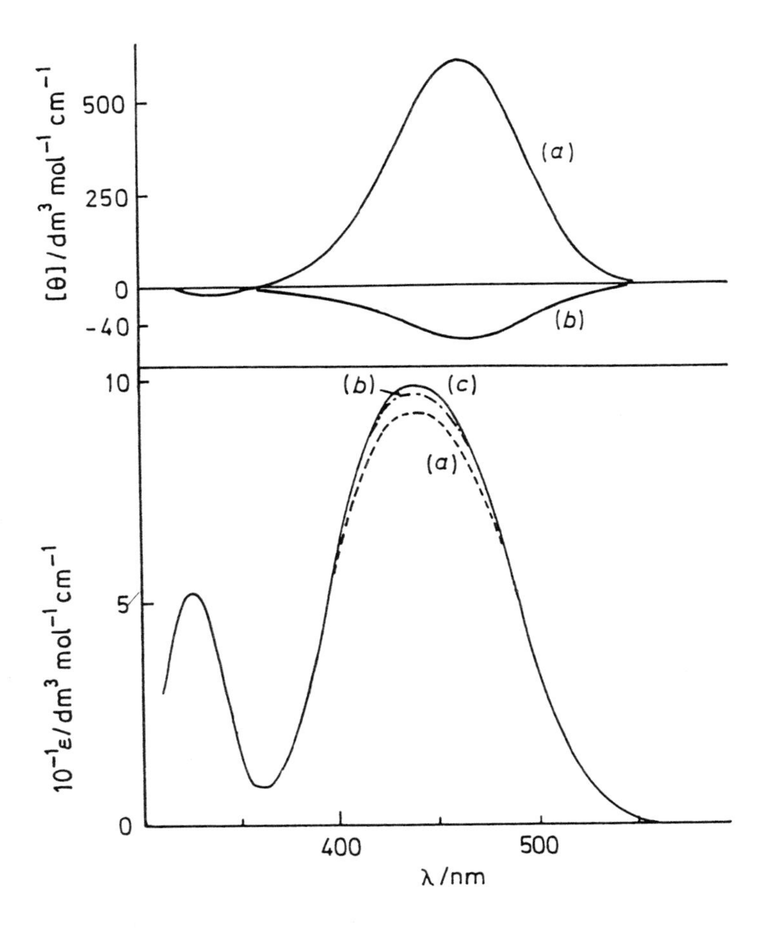

Fig. 3. Absorption and circular dichroism spectra of ethylene glycol solutions of ferrocene in the presence of β-CD (a) and γ-CD (b); [CD]=[ferrocene]=10^{-2} M. (c) In the absence of cyclodextrin (from Ref. 25).

3. 4. Structures

Molecular models of the CDs and ferrocene indicate that a ferrocene molecule could fit well into a β-CD cavity by axial inclusion, while a γ-CD cavity is large enough to accommodate a ferrocene molecule equatorially. The ferrocene molecule is too large to be completely included in a single α-CD cavity, which can accommodate only one cyclopentadienyl ring. A ferrocene molecule appears to fit well into the cavity formed by two molecules of α-CD. Proposed structures for the complexes are shown in Figure 4. The proposals for the dispositions of ferrocene within the cavities of β-CD and γ-CD have been substantiated by solid state ^2H NMR spectroscopic measurements [26] and by comparing the induced circular dichroism of 1,4,7,10,13-penta[13]ferrocenophane in the presence of β-CD with that of ferrocene in β-CD or γ-CD [27]. The interactions between some ferrocene derivatives with β-CD in solutions have been studied by cyclic voltammetry and the association constants have been determined [28].

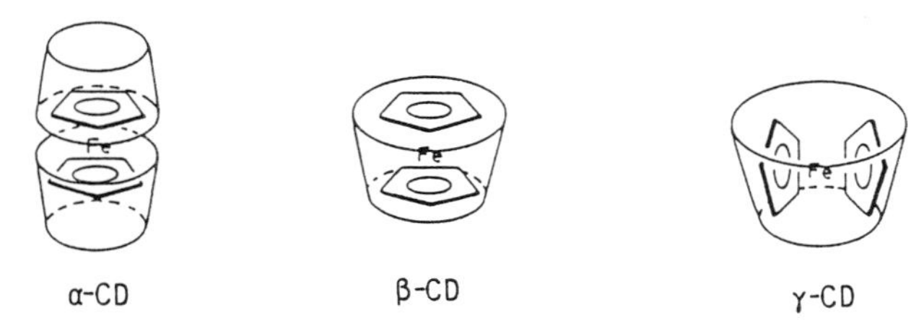

Fig. 4. Proposed structures of the inclusion compounds of cyclodextrin-ferrocene.

The crystal structure of the α-cyclodextrin-ferrocene (2:1) inclusion compound is shown in Figure 5. The ferrocene molecule with approximate D_{5d} symmetry is encapsulated by the dimer of the α-cyclodextrins in a tail-to-tail orientation and inclined by 42 ° with respect to the six-fold axes of the α-cyclodextrins of the dimer [29]. The overall structure of this α-CD-ferrocene (2:1) in the solid state is similar to that optimized in the gas phase by molecular mechanics calculations [30-32].

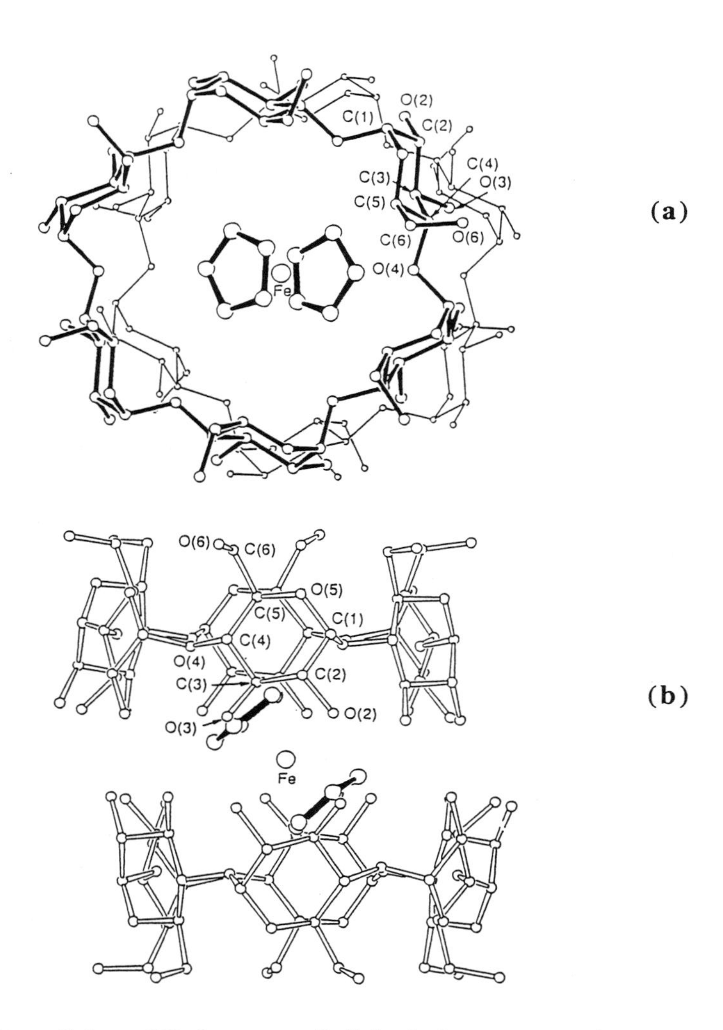

Fig. 5. View of the α-CD-ferrocene (2:1) inclusion compounds. (a) Perpendicular to the O(4) ether atoms planes of the macrocycles; (b) along a local two-fold axis which goes through the Fe atom (from Ref. 29).

3. 5. Binding Modes

The complexes of alkyldimethyl(ferrocenylmethyl)ammonium salts, where the alkyl group is methyl, heptyl or hexadecyl, with α-, β-, and γ-cyclodextrins were studied by electrochemical methods and high-field ^{1}H-NMR spectroscopy [33]. Methyl and heptyl complexes were found to form 1:1 complexes with α-CD, β-CD, and γ-CD, respectively. Hexadecyl complexes formed a 1:2 complex with α-CD and 1:1 complexes with β-CD and γ-CD. α-CD interacted with the aliphatic regions of the derivatives, while β-CD and γ-CD interacted with the ferrocene subunit. The dual mode of CD binding to these derivatives was used to build small supramolecular aggregates in which a ferrocene derivative directs several CD hosts to bind to different molecular regions. This was exemplified by the isolation of a quaternary complex between the hexadecyl complex and α-CD as well as β-CD, which exhibited the 1:2:1 stoichiometric ratio predicted in terms of the interactions of individual components of this ferrocene derivative with α-CD and β-CD, respectively.

Inclusion compounds of alkyldicarbonyl(η^5-cyclopentadienyl)iron with cyclodextrins were prepared [34]. In the solid state, carbon monoxide is inserted into the Fe-R bond. The insertion reactions greatly depend on the type of cyclodextrin used. The rate of the insertion reaction decreases sharply with the decrease in cavity size of the cyclodextrin (γ-CD > β-CD > α-CD).

The formations of adducts of ferrocenecarboxylic acid with β-CD and chemically modified derivatives have been studied by circular dichroism spectroscopy [35-39]. Dodecamethyl ferrocene formed complexes with γ-CD, although it did not form complexes with α-CD and β-CD. Pentamethylcyclopentadienyl group is too bulky to fit α-CD or β-CD cavities. The stoichiometry of the complex of dodecamethyl ferrrocene with γ-CD is two-to-one (2γ-CD:one guest).

4. CYCLOPENTADIENYL(ARENE) COMPLEXES

The mixed sandwich complex $[(\eta^5\text{-}C_5H_5)Fe(\eta^6\text{-}C_6H_6)]PF_6$ forms crystalline 2:1 host:guest inclusion complexes with α- and β-cyclodextrin, respectively. The crystal structure of the α-cyclodextrin adduct has been determined by single-crystal X-ray diffraction [40]. As illustrated in Figure 6, two adjacent α-CD molecules form a head-to-tail dimer by intermolecular hydrogen bonding across their secondary hydroxyl faces. The $[(\eta^5\text{-}C_5H_5)Fe(\eta^6\text{-}C_6H_6)]$ cation is encapsulated within the cavity of the dimer and is tilted at an angle of 39 ° (C_5H_5)/40 ° (C_6H_6) to the mean plane of the α-CD molecules. The cyclopentadienyl and the benzene rings are parallel. The PF_6 anion is located in the center of the primary hydroxyl side of the α-CD molecule that includes the benzene ring.

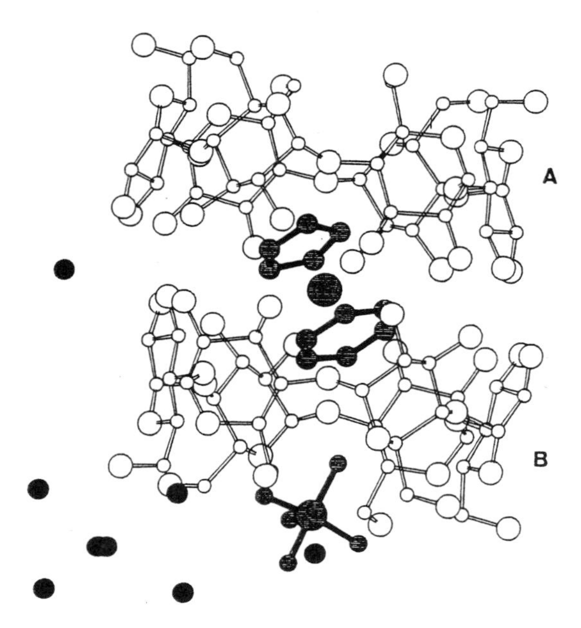

Fig. 6. Side-on view of the head-to-head type molecular structure of $[CpFe(C_6H_6)PF_6]2\alpha\text{-CD}\cdot 8H_2O$ (from Ref. 40).

The dimers are stacked along the crystallographic *c* axis to form the channel-type structure shown in Figure 7 [40].

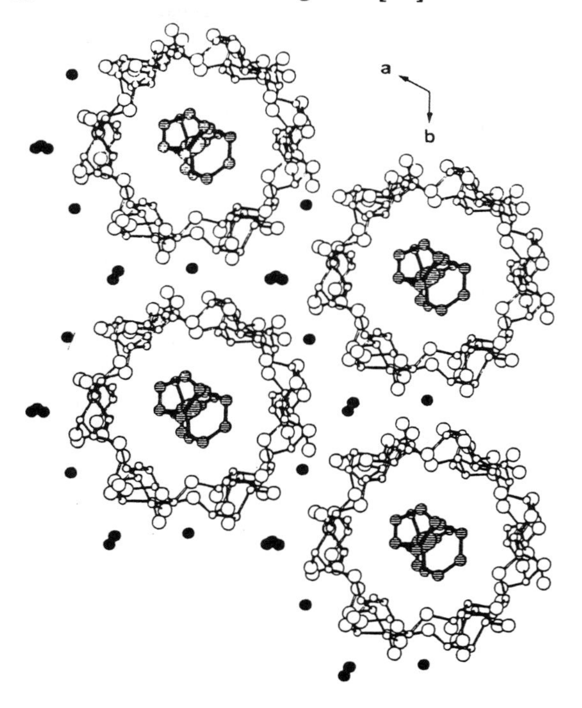

Fig. 7. Crystal structure viewed down the *c* axis showing the hexagonal array of the CD channels (from Ref. 40).

5. (η^6-ARENE)TRICARBONYLCHROMIUM

5. 1. Preparation

Inclusion compounds of η^6-arene transition-metal complexes with cyclodextrins (β-CD and γ-CD) have been prepared [41,42]. One-to-one inclusion compounds were obtained in high yields in a crystalline state by treatment of β- and γ-CDs with (η^6-arene)chromium tricarbonyl complexes. The formation of the inclusion compounds is selective.

Table 4 shows the results of the preparation of inclusion compounds of (π-arene)tricarbonyl chromium complexes with CDs.

Table 4. *Preparation of Inclusion Compounds of (Arene) tricarbonyl-chromium Complexes with Cyclodextrins*

CD	Guest Arene	Yield %	Product Cr % Found	Product Cr % Calcd.	Product CD:guest
α	Benzene	0	-	-	-
β	Benzene	91	3.7	3.9	1:1
γ	Benzene	21	3.5	3.4	1:1
α	Toluene	0	-	-	
β	Toluene	87	4.3	3.8	1:1
γ	Toluene	90	3.7	3.4	1:1
α	*o*-Xylene	0	-	-	-
β	*o*-Xylene	56	3.8	3.8	1:1
γ	*o*-Xylene	96	3.8	3.4	1:1
α	*m*-Xylene	0	-	-	-
β	*m*-Xylene	trace	3.7	3.8	1:1
γ	*m*-Xylene	84	4.2	3.4	1:1
α	*p*-Xylene	0	-	-	-
β	*p*-Xylene	trace	3.9	3.8	1:1
γ	*p*-Xylene	81	3.7	3.4	1:1
α	Mesitylene	0	-	-	-
β	Mesitylene	0	-	-	-
γ	Mesitylene	61	3.7	3.3	1:1
α	Hexamethylbenzene	0	-	-	-
β	Hexamethylbenzene	0	-	-	-
γ	Hexamethylbenzene	98	3.1	3.3	1:1
α	*o*-Methoxytoluene	0	-	-	-
β	*o*-Methoxytoluene	57	3.7	3.7	1:1
γ	*o*-Methoxytoluene	98	3.7	3.3	1:1

β-CD forms 1:1 inclusion compounds with tricarbonylchromium complexes of benzene-, toluene-, and *o*-xylene in high yields, but only in trace amounts with *m*- and *p*-xylene derivatives. However, β-CD does not form inclusion compounds with mesitylene-, or hexamethylbenzene chromium tricarbonyls. γ-CD forms 1:1 inclusion compounds with all the arene chromium complexes tested. γ-CD formed an inclusion compound with (benzene) tricarbonylchromium in 21 % yield. The inclusion compound of the γ-CD-tricarbonyl(hexamethylbenzene)chromium complex was obtained almost quantitatively. It should be noted that the (benzene) tricarbonylchromium complex was selectively included in β-CD, while the tricarbonylchromium complexes of mesitylene and hexamethylbenzene were specifically included in γ-CD. The aromatic ring of the complexes may be included in the cavity of the CDs.

Of the tricarbonylchromium complexes of *o*-, *m*-, and *p*-xylenes, only the tricarbonyl-(*o*-xylene)chromium complex was selectively included in β-CD. β-CD does not show such selectivity with uncomplexed *o*-, *m*-, and *p*-xylenes. α-CD does not form inclusion compounds with any arene chromium tricarbonyl complexes, although α-CD forms inclusion compounds with uncomplexed benzene, toluene, and xylenes.

All the inclusion compounds of cyclodextrins are stoichiometric 1:1 complexes even when cyclodextrin is treated with two-fold molar excess of (arene)tricarbonylchromium complexes.

5. 2. Properties

The inclusion compounds are thermally stable. The inclusion compound of β-CD with the (benzene)tricarbonylchromium complex, for example, did not liberate the guest when heated at 100 °C *in vacuo*, while under these conditions nonincluded (benzene) tricarbonylchromium complex sublimed. The inclusion compound of the (benzene) tricarbonylchromium complex with β-CD was stable even when exposed to sunlight for a prolonged period, while that with γ-CD was unstable. The IR spectra of the inclusion compound of γ-CD showed loss of a carbonyl ligand after irradiation with visible light for several days, while the spectrum of the β-CD complex did not change under the same conditions. These results indicate that β-

CD binds the complex more strongly than γ-CD and stabilizes the complex.

5. 3. Binding Modes

From the circular dichroism spectra of their inclusion compounds, the chromium complexes are supposed to be included in the cavities of cyclodextrins, according to the assemblies illustrated in Figure 8.

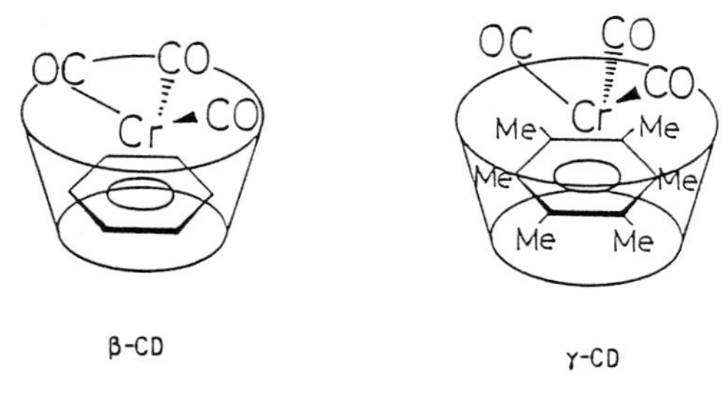

Fig. 8. Proposed structures of the 1:1 complex of (arene) tricarbonylchromium with β- and γ-cyclodextrins, respectively.

The IR spectra of the inclusion compounds show that the carbonyl stretching absorption bands are sharp. These spectra are different from those of the complexes in the solid state, which show broad bands, probably due to intermolecular interactions, and are similar to those of the complexes in solutions. These results indicate that the guest complexes are isolated from each other in the cavity as in solution.

Judging from the molecular dimensions of α-CD and (arene)tricarbonylchromium complexes, (arene)tricarbonylchromium complexes are too large to be included in the α-CD cavity. Uncomplexed benzene, toluene, and xylenes can be included in the α-CD cavity axially, but the chromium tricarbonyl group hinders inclusion of the whole guest molecule because the tricarbonyl chromium group is larger than the α-CD cavity. Actually α-CD does not form inclusion compounds with chromium hexacarbonyl. However, α-CD forms a 2:1(α-CD:guest) inclusion complex with (η^5-cyclopentadienyl)-tricarbonylmanganese in high yield. This is because the α-CD cavity can accommodate the cyclopentadienyl ring.

According to CPK molecular models, tricarbonyl chromium complexes of benzene and toluene appear to fit well into the β-CD cavity with axial orientation, as shown in Figure 8, whereas tricarbonylchromium complexes of mesitylene and hexamethylbenzene are too large to be included in the β-CD cavity[41,43] The γ-CD cavity is large enough to accommodate tricarbonylchromium complexes of xylene, mesitylene, and hexamethylbenzene, but is too large for the (benzene) tricarbonylchromium complex. The *o*-xylene complex, which is more compact than the *m*- or *p*-xylene complex, fits well into the β-CD cavity.

6. OLEFIN COMPLEXES

6. 1. Preparation

Cyclodextrins have been found to be efficient catalysts for the oxidation of olefins using palladium chloride. The ternary complexes consisting of CD, olefin, and metal are suggested to be intermediates in this oxidation reaction. Harada *et al.* isolated inclusion compounds of olefin metal complexes with cyclodextrins [44,45]. They chose cyclo-octa-1,5-diene(COD) and norbornadiene(NBD) as second sphere ligands because they form relatively stable complexes with transition metals.

Table 5 shows the results of the preparation of inclusion compounds of CD-olefin-metal complexes. β-CD formed a 2:1 (CD:guest) inclusion compounds with bis(μ-chloro)bis(cycloocta-1,5-diene)-dirhodium, $[(Rh(\mu\text{-}Cl)(COD)]_2$, and bis(μ-chloro)bis(norborna diene)-dirhodium, $[(Rh(\mu\text{-}Cl)(NBD)]_2$, in high yields. γ-CD formed 2:1 inclusion compounds with $[(Rh(\mu\text{-}Cl)(COD)]_2$ in 29 % yield, but did not form an inclusion compound with $[(Rh(\mu\text{-}Cl)(NBD)]_2$, which is smaller than $[(Rh(\mu\text{-}Cl)(COD)]_2$. It is noteworthy that $[(Rh(\mu\text{-}Cl)(NBD)]_2$ was specifically included only in the β-CD cavity. β-CD also formed inclusion compounds with dihalo(cycloocta-1,5-diene)-platinum, $Pt(COD)X_2$(X=Cl, Br, I), in high yields. γ-CD formed an inclusion compound with $Pt(COD)I_2$ quantitatively, while it did not form an inclusion compound with $Pt(COD)Cl_2$, and with $Pt(COD)Br_2$ it formed an inclusion compound in low yield, suggesting that γ-CD appears to distinguish the size of the guest compounds, especially their halogen part.

Table 5. *Preparation of Inclusion Compounds of Cyclic Olefin Metal Complexes*

Guest	CD	Yield %	Product Metal % Found	Product Metal % Calcd.	CD:guest
$[(Rh(\mu\text{-}Cl)(COD)]_2$	α	0	-	-	-
	β	90	7.9	7.4	2:1
	γ	29	6.5	6.6	2:1
$[(Rh(\mu\text{-}Cl)(NBD)]_2$	α	0	-	-	-
	β	100	7.9	7.9	2:1
	γ	0	-	-	-
$[(Rh(\mu\text{-}Br)(COD)]_2$	α	0	-	-	-
	β	70	7.5	7.2	2:1
	γ	58	7.0	6.5	2:1
$[(Rh(\mu\text{-}I)(COD)]_2$	α	0	-	-	-
	β	91	6.5	6.9	2:1
	γ	60	6.6	6.3	2:1
$Pt(COD)Cl_2$	α	0	-	-	-
	β	99	4.9[a]	4.6[a]	1:1
	γ	0	-	-	-
$Pt(COD)Br_2$	α	0	-	-	-
	β	100	9.9[a]	10.0[a]	1:1
	γ	14	11.1[a]	9.1[a]	1:1
$Pt(COD)I_2$	α	0	-	-	-
	β	90	17.2[a]	14.9[a]	1:1
	γ	100	13.9[a]	13.7[a]	1:1

[a] Halogen %.

6. 2. Structures

α-CD did not form inclusion compounds with any neutral transition metal complexes of cyclo-octa-1,5-diene. Stoddart *et al.* reported the preparation and structure of the ionic complex α-CD-[Rh(COD)(NH_3)$_2$][PF_6]6H_2O [46,47]. As shown in Figure 9, [Rh(COD)(NH_3)$_2$] is positioned almost exactly over the center of the torus in α-CD. One of the two ethylene groups is inserted, and the NH_3 groups are hydrogen-bonded with the secondary hydroxyl groups of α-CD. This hydrogen bond should be the driving force for the formation of the adduct.

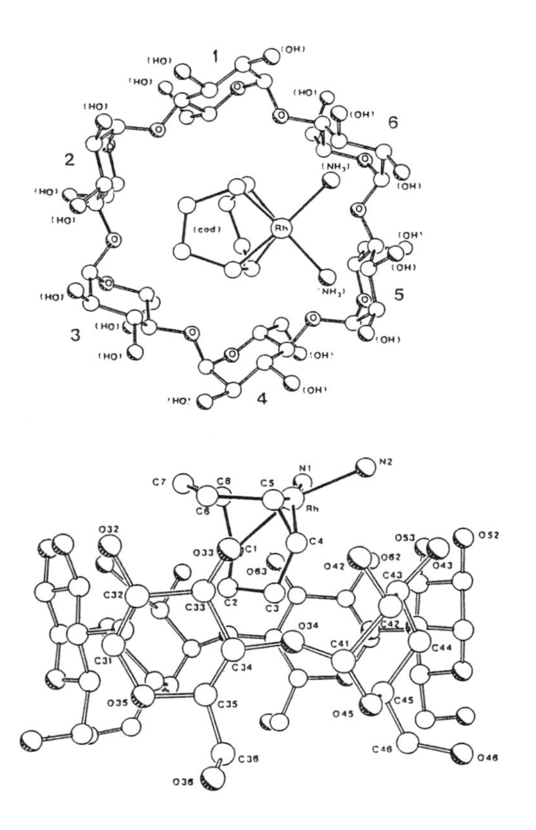

Fig. 9. View of the skeletal representation of [Rh(COD](NH3)2-α-CD]+ (from Ref. 46).

There was no sign of an interaction between α-CD and [(Rh(μ-Cl)(COD)]2. No solubilizing effects were observed. No change in the absorption spectrum and no induced Cotton effects were observed with the neutral complex [(Rh(μ-Cl)(COD)]2. Examination of Corey-Pauling-Koltun (C.P.K.) models showed that these olefin-metal complexes are a little too large to be fully included in the α-CD cavity. Although [(Rh(μ-Cl)(COD)]2 is almost insoluble in water, the inclusion compound of [(Rh(μ-Cl)(COD)]2 with β-CD is soluble.

Figure 10 shows the absorption spectrum of [(Rh(μ-Cl)(COD)]2 solubilized by β-CD in aqueous solution. [(Rh(μ-Cl)(COD)]2 was found to be strongly solubilized by β-CD. The spectrum of the inclusion compound was different from that of the nonincluded [(Rh(μ-Cl)(COD)]2: a new peak was seen at 400 nm, indicating that the electronic structure of [(Rh(μ-Cl)(COD)]2 was modified by its inclusion in the β-CD cavity. This may be due to the influence of outer-sphere coordination or some participation of the OH group of CD in the coordination.

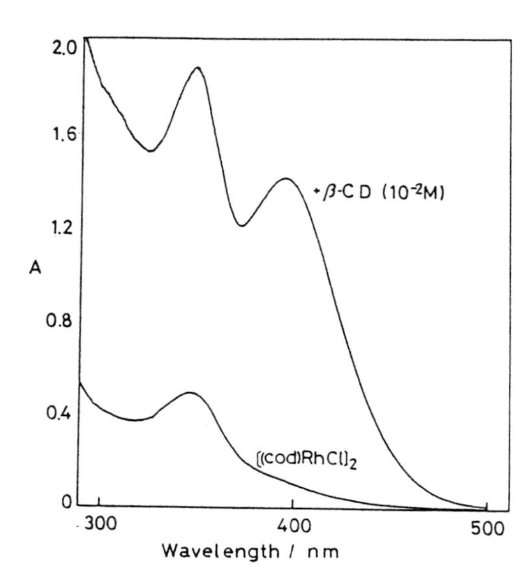

Fig. 10. Absorption spectra of [(COD]RhCl]2 in the presence of β-CD (from Ref. 45).

Figure 11 shows the circular dichroism spectra of $[(Rh(\mu\text{-}Cl)(COD)]_2$ in the presence of β-CD and γ-CD. A large induced Cotton effect was observed in the presence of β-CD, while a very weak induced Cotton effect was observed with γ-CD. These results indicate that the guest compound is well-inserted in the β-CD ring.

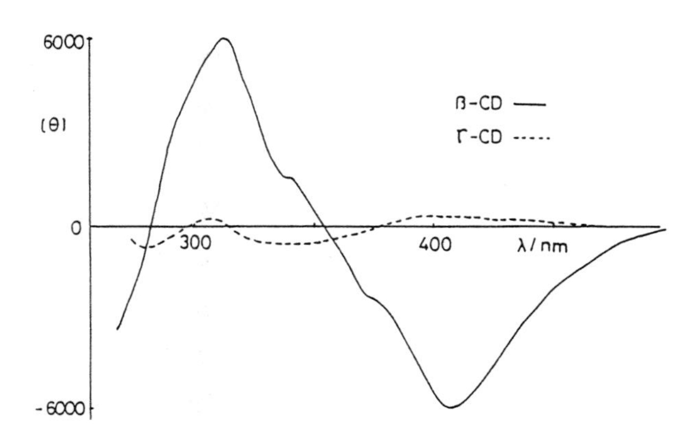

Fig. 11. Circular dichroism spectra of β-CD-$[(COD)RhCl]_2$ (from Ref. 45).

Figure 12 shows the ^1H NMR spectra of β-CD-$[(Rh(\mu\text{-}Cl)(COD)]_2$ [48]. The spectrum in D_2O displayed two sets of resonances for apparently two different CD species. One of them was superimposable on that of CD, and was that of free β–CD, while the other, which could represent that of complexed CD, showed large shifts of the signals for H1, H2, and H5 of β-CD to lower frequencies and a shift of H3 to a higher frequency.

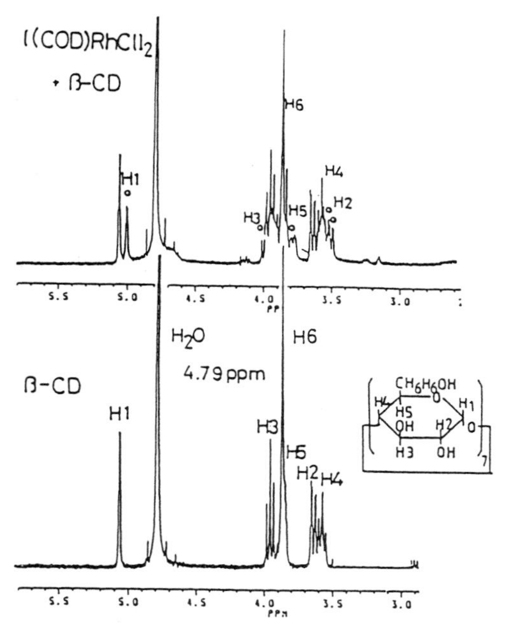

Fig. 12. ^1H NMR spectra of β-CD-$[(COD)RhCl]_2$ (from Ref. 48).

6. 3. Properties

The inclusion compounds are thermally stable, and thermogravimetric measurements indicated that the complexes are stabilized by inclusion in cyclodextrin cavities. Proposed structures of the inclusion compounds are shown in Figure 13.

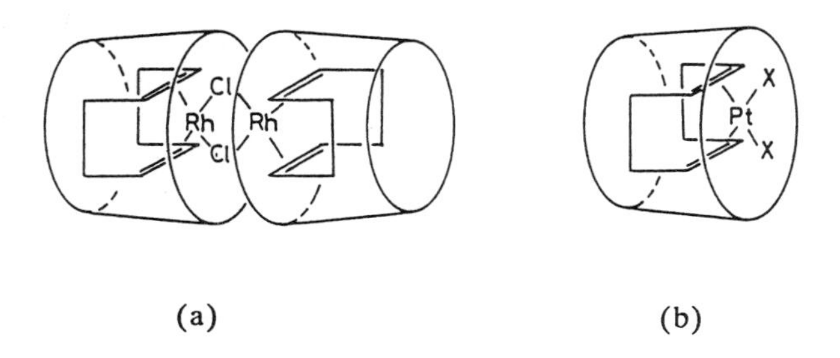

Fig. 13. Proposed structures of the complexes of β-CD with $[(COD)RhCl]_2$ (a), and $(COD)PtX_2$ (b).

Carboplatin, a new analog of *cis*-platin, is specific in forming a 1:1 adduct with β-CD in aqueous solution [49-51]. The changes in standard enthalpy (-25.3 KJ mol^{-1}) and entropy (-42 $JK^{-1}mol^{-1}$) for the "reaction" between carboplatin and β-CD indicate that the adduct formation is an entropy-driven process. The X-ray crystal structure of the α-CD complex of carboplatin has been reported, Figure 14.

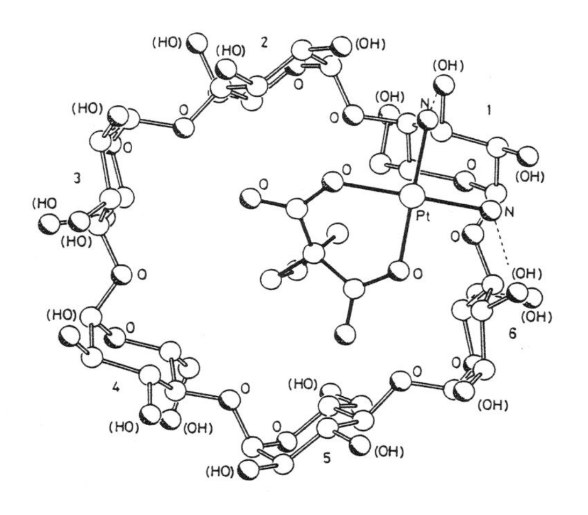

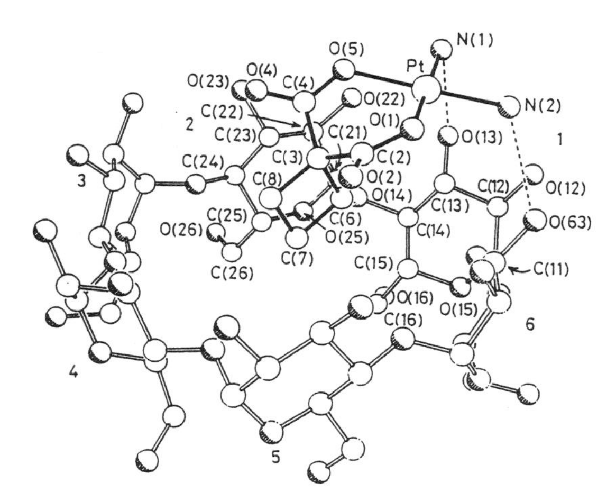

Fig. 14. View of the structure of [Pt(NH3)2(Cyclobutane-1,1-dicarboxylatodiammineplatinum)]-α-CD (from Ref. 51).

7. π-ALLYLPALLADIUM COMPLEXES

π-Allylpalladium complexes which are frequently used for a variety of organic syntheses form 1:1 inclusion compounds with β-CD and γ-CD. Table 6 shows the results of the preparation of inclusion compounds of π-allylpalladium complexes with cyclodextrins. β-CD formed inclusion compounds in high yields with all the π-allylpalladium complexes tested [52,53]. The inclusion compounds of $[(\pi\text{-allyl})PdCl]_2$ and $[(\pi\text{-allyl})PdBr]_2$ with γ-CD were obtained quantitatively. β-CD formed inclusion compounds with di-μ-chlorobis(2-methyl-η^3-allyl)dipalladium in high yield, but did not form inclusion compounds with di-μ-chlorobis(1-methyl-η^3-allyl)dipalladium. β-CD also formed 1:1 inclusion compounds with chloro- and bromo-bridged complexes in moderate yields but not with the iodo-bridged complex. β-CD is apparently able to distinguish not only π-allyl ligands but also the size of the bridging halogens. α-CD did not form inclusion compounds with any π-palladium complexes tested. This is reasonable because the examination of CPK models shows that these allylpalladium complexes are too large to fit in the α-CD cavity, Figure 15. π-Allylpalladium complexes are stabilized when they are included in the cavity of β-CD. Thermogravimetric analysis showed that π-allylpalladium decomposes at about 150 °C when it is not included, but at 200 °C when it is included in β-CD.

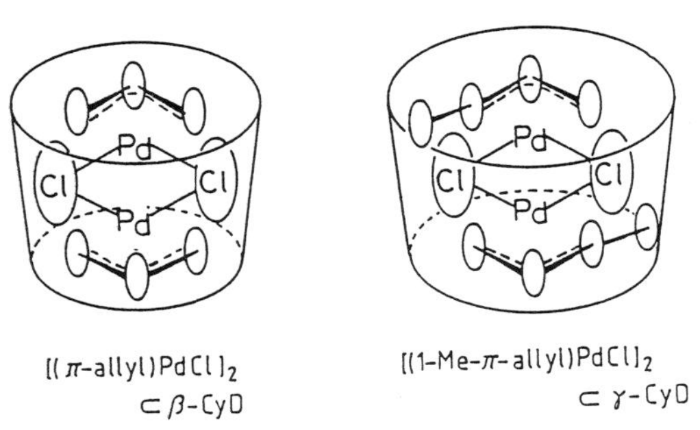

Fig. 15. Proposed structures of the inclusion compounds of $[(\pi\text{-allyl})PdCl]_2$.

Table 6. *Preparation of Inclusion Compounds of η^3-Allylpalladium Complexes with Cyclodextrin*

Guest	CD	CD:guest	Product		
			Yield[a]	Pd %	
			%	Found	Calcd.
$[(\eta^3\text{-allyl})PdCl]_2$	α	1:2	0	-	-
	β	1:1	21	-	-
	β	1:2	48	12.4	13.2
	γ	1:1	68	-	-
	γ	1:2	100	11.6	11.9
$[1\text{-}3\text{-}\eta^3(2\text{-methallyl})PdCl]_2$	α	1:2	0	-	-
	β	1:1	61	-	-
	β	1:2	95	14.8	13.0
	γ	1:1	58	-	-
	γ	1:2	63	9.6	11.7
$[1\text{-}3\text{-}\eta^3(2\text{-butenyl})PdCl]_2$	α	1:2	0	-	-
	β	1:2	0	-	-
	γ	1:2	96	7.6	11.7
$[(\eta^3\text{-allyl})PdBr]_2$	α	1:2	0	-	-
	β	1:2	23	12.7	12.5
	γ	1:2	100	13.2	11.3
$[(\eta^3\text{-allyl})PdI]_2$	α	1:2	0	-	-
	β	1:2	0	-	-
	γ	1:2	77	12.5	10.8

[a] Calculated from cyclodextrin

8. CARBORANE

Cyclodextrins form inclusion complexes with main-group compounds. β-CD and γ-CD form one-to-one inclusion complexes in crystalline states in high yields with *o*-carborane (1,2-dicarbododecaborane) on their treatment with *o*-carborane [54]. α-CD formed a two-to-one (CD:guest) complex on sonication and a one-to-one complex on crystallization from water-propan-2-ol. Results on the preparation of inclusion complexes of *o*-carborane with cyclodextrins are shown in Table 7. β-CD formed inclusion complexes quantitatively. The inclusion complexes are sufficiently stable to be crystallized from water-propan-2-ol. They are also thermally stable, and do not liberate *o*-carborane when heated at 200 °C *in vacuo.* Computer simulation of the molecular structure indicated that *o*-carborane fits well into the β-CD cavity, as shown in Figure 16. However, the *o*-carborane molecule is too large to be completely included in a single α-CD cavity, but fits well into the cavity formed by two α-CD molecules.

Table 7. *Preparation of Inclusion Compounds of o-Carborane*

CD	Molar ratio CD:guest	Method A Yield(%)	Method A CD:guest	Method B Yield(%)	Method B CD:guest
α	1:1	74	2:1	88	1:1
α	2:1	58	2:1	88	1:1
β	1:1	35	1:1	100	1:1
β	2:1			100	1:1
γ	1:1	54	1:1	83	1:1
γ	2:1			92	1:1

Method A: by sonication; Method B: by crystallization from water-propan-1-ol.

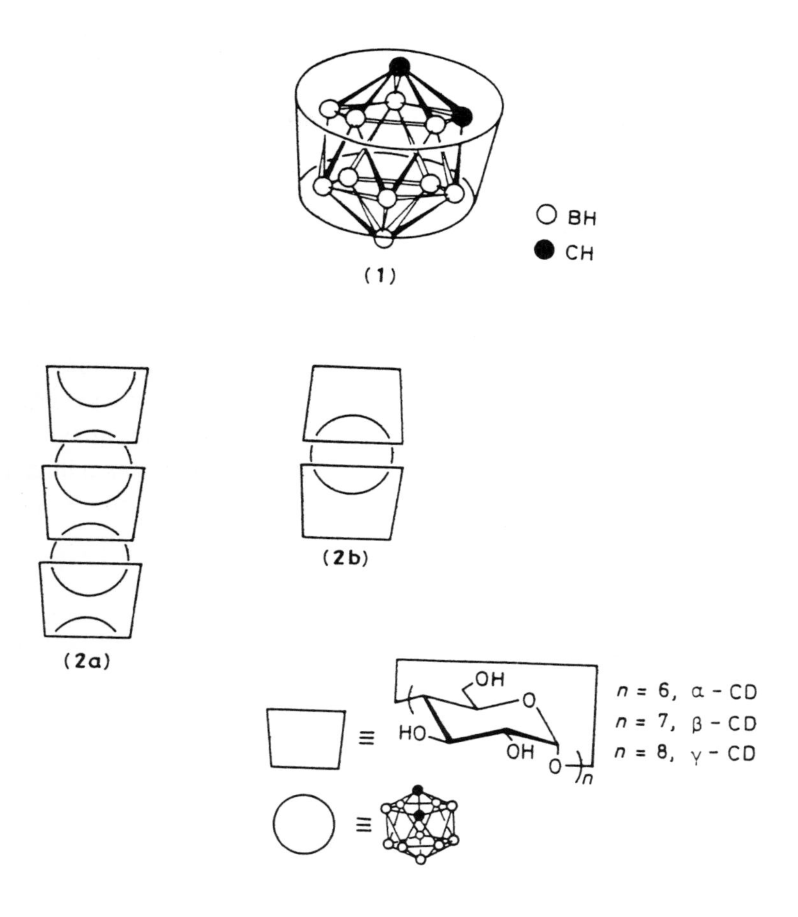

Fig. 16. Proposed structures of cyclodextrin-carborane complexes (from Ref. 54). β-CD-carborane complex (**1**), α-CD-carborane complex crystallized from water-propanol (**2a**) and α-CD-carborane complex obtained by sonication (**2 b**).

9. METAL CARBONYL COMPLEXES

9.1. Mononuclear Metal Carbonyl Complexes

Cyclodextrins form inclusion compounds with some mononuclear metal carbonyl complexes [55]. Table 8 shows the results of the preparation of inclusion compounds with some mononuclear metal carbonyl complexes. Cyclodextrins did not form inclusion compounds with chromium hexacarbonyl, molybdenum hexacarbonyl, and tungsten hexacarbonyl complexes. However, cyclodextrins form inclusion compounds with iron pentacarbonyl complex and cobalt nitrosyl tricarbonyl complex. α-CD forms inclusion compounds with iron pentacarbonyl and cobalt nitrosyl tricarbonyl complexes with 2:1 (2CD:guest) stoichiometry, while β- and γ-CD gave 1:1 complex with these metal carbonyl complexes, Figure 17.

Table 8. *Preparation of Inclusion Compounds of Mononuclear Metalcarbonyl Complexes*

Guest	CD	Yield(%)	CD:guest
$Co(NO)(CO)_3$	α	25	2:1
	β	28	1:1
	γ	32	1:1
$Fe(CO)_5$	α	14	2:1
	β	27	1:1
	γ	40	1:1
$Cr(CO)_6$	α	0	-
	β	0	-
	γ	0	-

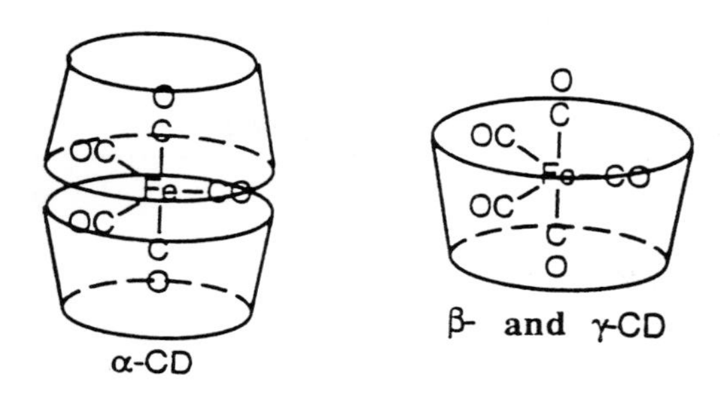

Fig. 17. Proposed structures of inclusion compounds between mononuclear metal carbonyl complexes and cyclodextrins.

9. 2. Binuclear Metal Carbonyl Complexes

γ-CD formed inclusion compounds with binuclear metal carbonyl complexes in high yields, although α-CD did not form complexes with binuclear complexes [55]. Table 9 shows the results of the preparation of inclusion compounds of CDs with binuclear metal carbonyl complexes. Trimethylphosphine or trimethylphosphite derivatives of cobalt carbonyl complexes, which are more stable than cobalt carbonyl complexes, form complexes only with γ-CD. They have cylindrical shape and are large enough to fit γ-CD cavity, Figure 18.

Table 9. *Preparation of Inlclusion Compounds of Binuclear Metal carbonyl Complexes with Cyclodextrins*

Guest	CD	Yield(%)	CD:guest
$Co_2(CO)_8$	α	0	-
	β	43	1:1
	γ	74	1:1
$Mn_2(CO)_{10}$	α	0	-
	β	0	-
	γ	85	1:1
$[Co(CO)_3PMe_3]_2$	α	0	-
	β	0	-
	γ	48	1:1
$[Co(CO)_3PEt_3]_2$	α	0	-
	β	0	-
	γ	68	1:1
$[Co(CO)_3P(OMe)_3]_2$	α	0	-
	β	0	-
	γ	89	1:1
$[Co(CO)_3P(n\text{-}Bu)_3]_2$	α	0	-
	β	0	-
	γ	0	-

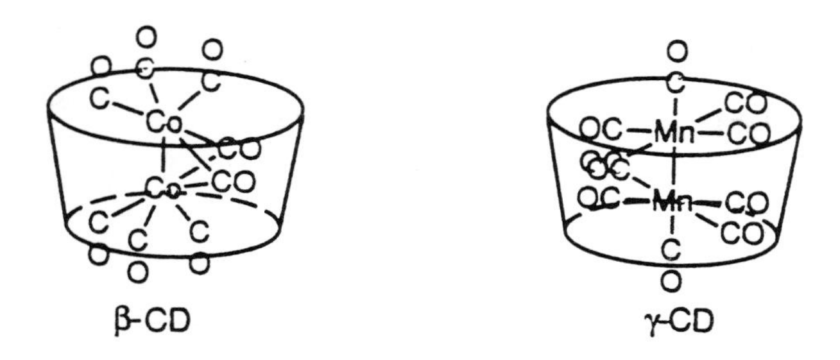

Fig. 18. Proposed structures of inclusion compounds of binuclear metal carbonyl complexes with cyclodextrins.

9. 3. Cobalt Cluster Complexes

One-to-one inclusion complexes were obtained in high yields by the treatments of di-, tri-, and tetranuclear cobalt complexes with γ-CD, but α-CD and β-CD did not form inclusion complexes with the cluster complexes [56]. Table 10 shows the results of the preparation of the inclusion compounds between the cobalt complexes shown in Scheme 3 and γ-CD. γ-CD formed inclusion compounds in high yields with the cobalt cluster complexes (**1**), (**2**), and the dinuclear complex (**4**), but it did not form an inclusion compound with complex (**3**). All the inclusion compounds obtained are stoichiometrically one-to-one inclusion compounds.

Cobalt cluster complexes are stabilized when they are included in the cavity of γ-CD. Thermogravimetric analysis shows that when not included, the cluster complex (**1**) decomposed at 120 °C, but when included in the cavity of γ-CD, it decomposed gradually above 200 °C.

Scheme 3.

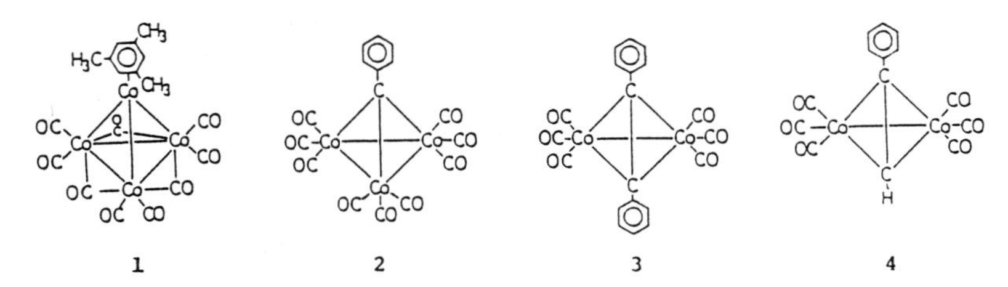

Table 10. *Preparation of Inclusion Compounds of Cobalt Complexes*

Complex	Yield(%)		
	α-CD	β-CD	γ-CD
(1)	0	0	74
(2)	0	0	67
(3)	0	0	0
(4)	0	0	73

Cobalt clusters included in the cyclodextrin cavity could be recovered by heating aqueous solutions of the inclusion complexes and then extracting the solution with organic solvents.

Judging from the molecular dimensions of cyclodextrins and the cobalt cluster complexes, these metal fragments can fit in the cavity of γ-CD, but they are too large to fit in the cavities of α-CD and β-CD. Complex (**3**) has a phenyl group at each end which may hinder inclusion of the whole guest molecule. The proposed structure of the inclusion compound of cluster complex (**1**) is shown in Figure 19.

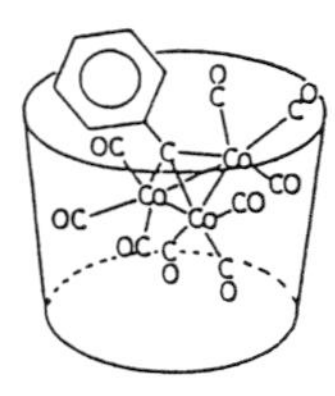

Fig. 19. Proposed structure of the inclusion compound formed by the cobalt cluster complex (**1**) with γ-CD.

9. 4. Cyclopentadienyl Metal Carbonyl Complexes

Cyclodextrins form inclusion compounds with ferrocenes as described in Section 3. They also form inclusion compounds with some mono- and bi-nuclear metal carbonyl complexes as described above. Cyclodextrins have been found to form inclusion compounds with some cyclopentadienyl metal carbonyl complexes with high selectivity [55]. Table 11 shows the results of the preparation of inclusion compounds of the cyclopentadienyl metal carbonyl complexes shown in Sheme 4 with cyclodextrins. Cyclopentadienyl iron carbonyl alkyl complexes, such as methyl, ethyl, propyl complexes form inclusion compounds in high yields. Acyl complexes and sulfinate complexes also gave inclusion compounds in high yields. The inclusion complexes of these cyclopentadienyl metal carbonyl complexes with α-CD are 2:1 (2CD:guest), while those with β-CD and γ-CD are 1:1. The binuclear complex (**1**) did not form inclusion compounds with α-CD, but it forms complexes with β-CD and γ-CD. β-CD gave a 2:1(CD:guest) complex and γ-CD gave a 1:1 complex with the binuclear complex (**1**).

Scheme 4.

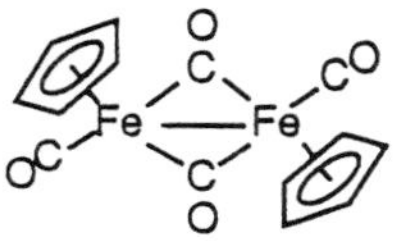

(1)

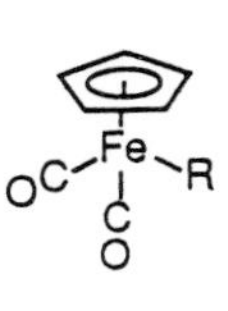

(2) R=CH_3
(3) =CH_2CH_3
(4) =$CH_2CH_2CH_3$
(5) =$COCH_3$
(6) =$COCH_2CH_3$
(7) =$COCH_2CH_2CH_3$
(8) =SO_2CH_3
(9) =$SO_2CH_2CH_3$
(10) =$SO_2CH_2CH_2CH_3$

Table 11. *Preparation of Inclusion Compounds of Cyclopentadienyl Metal Carbonyl Complexes with Cyclodextrins*

Guest	CD	Yield(%)	CD:guest
1	α	0	-
1	β	78	2:1
1	γ	53	1:1
2	α	99	2:1
2	β	80	1:1
2	γ	100	1:1
3	α	53	2:1
3	β	76	1:1
3	γ	92	1:1
4	α	66	2:1
4	β	52	1:1
4	γ	74	1:1
5	α	67	2:1
5	β	80	1:1
5	γ	99	1:1
6	α	72	2:1
6	β	79	1:1
6	γ	84	1:1
7	α	53	2:1
7	β	89	1:1
7	γ	77	1:1
8	α	52	2:1
8	β	60	1:1
8	γ	91	1:1
9	α	61	2:1
9	β	78	1:1
9	γ	79	1:1
10	α	70	2:1
10	β	83	1:1
10	γ	56	1:1

9. 5. Pentamethylcyclopentadienyl Metal Carbonyl Complexes

Table 12 shows the results of the preparation of inclusion compounds of the pentamethylcyclopentadienyl metal carbonyl complexes shown in Scheme 5. The binuclear complex (**1**) forms inclusion compounds only with γ-CD, as dodecamethylferrocene does. Pentamethyl-cyclopentadienyl iron carbonyl alkyl complexes, such as ethyl, propyl, bromo complexes did not form complexes with α-CD, but they form inclusion compounds with β-CD and γ-CD. All the γ-CD inclusion compounds show two-to-one (CD:guest) stoichiometry. β-CD inclusion compounds with methyl and bromo complexes are 1:1 and those of ethyl and propyl complexes are 2:1 (CD:guest).

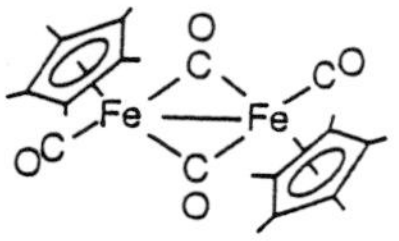

(1)

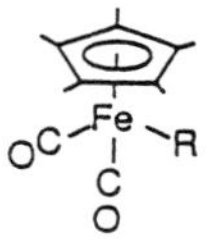

(2) R=CH_3
(3) =CH_2CH_3
(4) =Br

Scheme 5.

Table 12. *Preparation of Inclusion Compounds of Pentamethyl-cyclopentadienyl Metal Carbonyl Complexes with Cyclodextrins*

Guest	CD	Yield(%)	CD:guest
1	α	0	-
1	β	0	-
1	γ	72	1:1
2	α	0	-
2	β	88	1:1
2	γ	66	1:1
3	α	0	-
3	β	6	2:1
3	γ	23	1:1
4	α	0	-
4	β	78	1:1
4	γ	64	1:1

10. PHOSPHINE COMPLEXES

A large number of transition metal complexes bearing hydrophobic trialkylphosphine ligands are used for catalysis. Single crystals of the adduct of β-CD and *trans*-[Pt(PMe3)Cl2(NH3)] have been obtained and analyzed by X-ray crystallography [57]. As illustrated in Figure 20, the crystal structure of the one-to-one adduct indicates that the PMe3 ligand of the complex is inserted into and bound to the narrow primary hydroxy-group-bearing face of the β-CD torus.

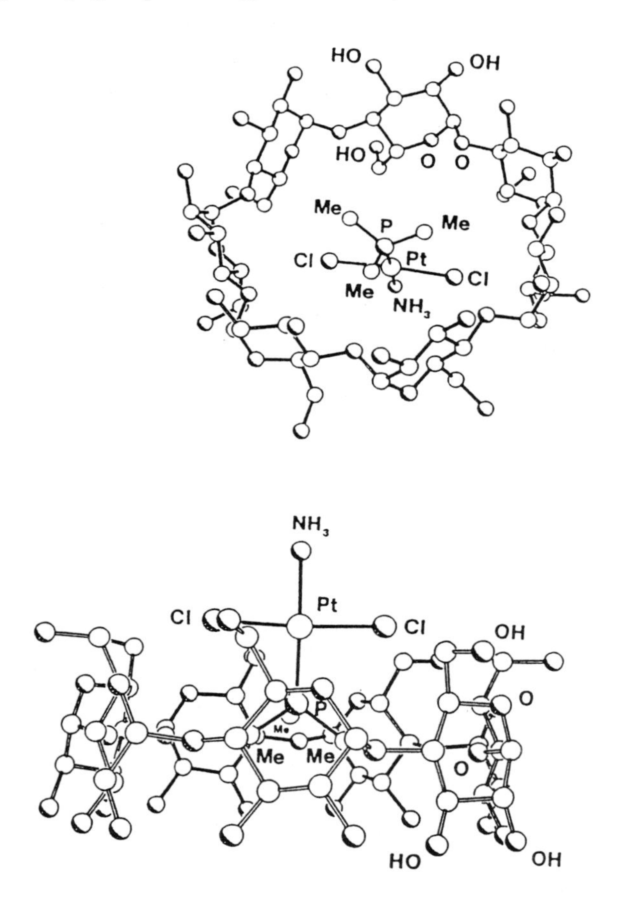

Fig. 20. X-Ray crystal structure of the inclusion complex of [Pt(PMe3)Cl2(NH3)] with β-CD (from Ref. 57).

11. OPTICAL RESOLUTION

During studies on the preparation of a series of inclusion compounds of organotransition metal complexes with cyclodextrins, α-CD was found to be very effective for optical resolution of 1-ferrocenylethanol by fractional precipitation. 1-Ferrocenylethanol and its analogues were completely resolved into their enantiomers by liquid chromatography on a polyamide column with an aqueous solution of α-cyclodextrin as a mobile phase [58,59]. Elution diagrams of 1-ferrocenylethanol and 1-ferrocenylpropanol are shown in Figure 21. This system should be useful for the resolution because it is simple and so can be used on a preparative scale.

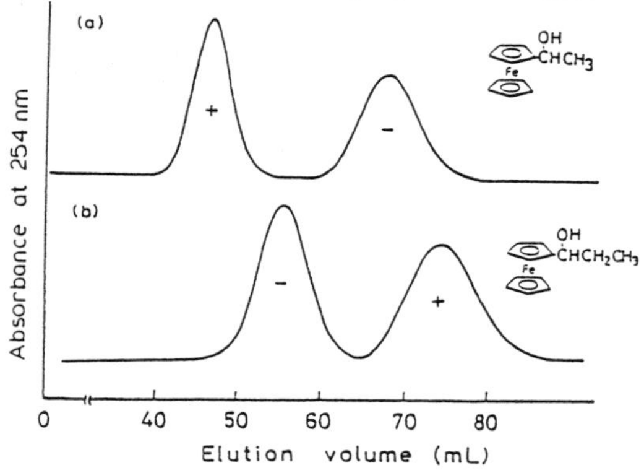

Fig. 21. Chromatographic separation of the enantiomers of (a) 1-ferrocenylethanol and (b) 1-ferrocenylpropanol using an aqueous solution of α-CD as eluent.

Recently, liquid chromatographic separation of enantiomers including racemic metallocenes using a α-cyclodextrin-bonded silica gel has been reported [60]. Enantiomeric pairs of ruthenocene, osmocene, and ferrocene derivatives were resolved by HPLC [61].

Membrane mediated separation of the enantiomers of (1-ferrocenylethyl)thiophenol using α-CD and β-CD as active transporting agents has also been reported.

Asymmetric reduction of acetylferrocene in the presence of cyclodextrins or the inclusion complexes gave the asymmetric ferrocenylethanol [62,63].

12. INCLUSION COMPLEXES IN SOLUTIONS

Cyclodextrins have the ability to solubilize organometallic complexes in water. For example, olefin-metal complexes are insoluble in water, but on addition of β-CD in water the olefin-metal complexes are included in CD and become soluble in water. During studies on the preparation of inclusion compounds of organometallic complexes with cyclodextrins, Harada *et al.* found that the inclusion complexes are formed not only in water but in alcohol solutions [64]. It has been thought that cyclodextrins form inclusion compounds only in aqueous solution. Therefore hydrophobic interactions are thought to play an essential role in the formation of the inclusion complexes. Breslow *et al.* reported that cyclodextrins form inclusion complexes in dimethylsulfoxide and dimethylformamide. However, Harada *et al.* found that cyclodextrins form inclusion complexes much more efficiently in ethylene glycol or monoglym solutions than in dimethylsulfoxide or dimethylformamide.

13. INVERSE PHASE TRANSFER CATALYSIS

The Wacker process, in which acetaldehyde is produced by oxidation of ethylene using palladium catalysts, is important industrially. However, olefins with long hydrocarbon chains cannot be oxidized by this process due to their insolubility in water. Cyclodextrins have the ability to solubilize insoluble organic compounds in water. Harada *et al.* applied CD to this system and found that the oxidation of long chain olefins proceeds smoothly affording methyl ketones in high yields.

Terminal olefins are oxidized to ketones in high yields under mild conditions using palladium chloride and cyclodextrins as catalysts in two-phase systems[65-67]. As shown in Figure 22, the yield of ketones depends on the substrates and high yields are obtained with substrates having C8-C10 structures. α-CD is the most effective for this purpose followed by β-CD and γ-CD.

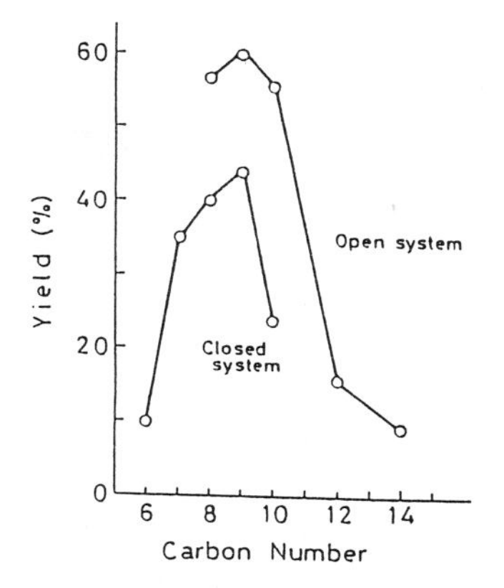

Fig. 22. Effects of the number of carbon atoms in the oxidation of olefins to ketones on the yields; 70 °C, 8h.

As schematized in Figure 23, a ternary complex, that is, a simultaneous assembly of CD, olefin, and metal fragment, is proposed to be an intermediate of this reaction [68-72].

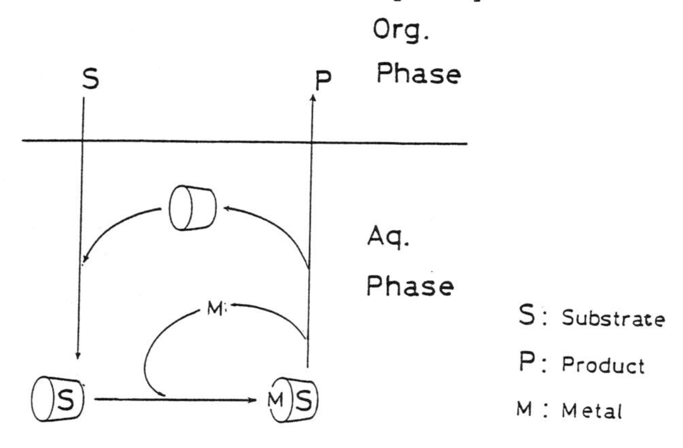

Fig. 23. Proposed mechanism for the inverse phase-transfer catalysis by cyclodextrin.

14. CONCLUSION

As described here, cyclodextrins form inclusion compounds with some organometallic complexes efficiently and selectively. Thus cyclodextrins should be useful receptors for organometallic complexes and should be applicable to a variety of processes.

References

1. F. G. Stone, G.Wilkinson, and E. W. Abel, *Comprehensive Organometallic Chemistry,* Pergamon Press, Oxford, 1982.
2. F. Vogtle, *Host Guest Complex Chemistry I,* Springer, Berlin, 1981.
3. H. M. Colquhoun, J. F. Stoddart, and D. J. Williams, *Angew. Chem., Int. Ed. Engl., 25* (1986) 487.
4. J. F. Stoddart and R. Zarzycki, *Recl. Trav. Chim. Pays-Bas, 107* (1988) 515.
5. A. Harada, *Kobunshi Kako, 37* (1988) 82.
6. A. Harada, *Chemical Industry, 39* (1988) 925.
7. A. Harada, *J. Synth. Org. Chem. Jpn., 48* (1990) 517.
8. A. Harada, *Hyomen, 28* (1990) 857.
9. I. Sokolov, *Metalloorg. Chim., 1* (1988) 25.
10. H. M. Colquhoun, J. F. Stoddart, and D. J. Williams, *J. Chem. Soc., Chem. Commun.,* (1981) 847.
11. D. R. Alston, A. M. Z. Slain, J. F. Stoddart, and D. J. Williams, *Angew. Chem., Int. Ed. Engl., 23* (1984) 821.
12. B. Dietrich, M. W. Hosseini, J. M. Lehn, and R. B. Sessions, *J. Am. Chem. Soc., 103* (1983) 1282.
13. M. L. Bender and M. Komiyama, *Cyclodextrin Chemistry,* Springer-Verlag, New York, 1977.
14. J. Szejtli, *Cyclodextrins and Their Inclusion Complexes,* Akademiai Kiado, Budapest, 1982.
15. W. Saenger, *Angew. Chem., Int. Ed. Engl., 19* (1980) 344.
16. B. Siegel and R. Breslow, *J. Am. Chem. Soc., 97* (1975) 6869.
17. M. F. Czarniecki and R. Breslow, *J. Am. Chem. Soc., 100* (1978) 7771.
18. R. Breslow, M. F. Czarniecki, J. Emert, and H. Hamaguchi, *J. Am. Chem. Soc., 102,*(1980) 762.

19. G. Trainor and R. Breslow,
J. Am. Chem. Soc., 103 (1981) 154.
20. R. Breslow, G. Trainor, and A. Ueno,
J. Am. Chem. Soc., 105 (1983) 2739.
21. W. J. Noble, S. Srivastava, R. Breslow, and G. Trainor,
J. Am. Chem. Soc., 105 (1983) 2745.
22. T. Matsue, K. Kobayashi, and T. Osa,
Nippon kagaku Kaishi, (1983) 303.
23. A. Harada and S. Takahashi,
J. Chem. Soc., Chem. Commun., (1984) 645.
24. A. Harada and S. Takahashi,
J. Incl. Phenom., 2 (1984) 791.
25. A. Harada, Y. Hu, S. Yamamoto, and S. Takahashi,
J. Chem. Soc., Dalton Trans., (1988) 729.
26. N. J. Clayden, C. M. Dobson, S. J. Heyes, and P. J. Wiseman,
J. Incl. Phenom., 5 (1987) 65.
27. N. Kobayshi and M. Opallo,
J. Chem. Soc., Chem. Commun., (1990) 477.
28. T. Matsue, D. H. Evance, T. Osa, and N. Kobayashi,
J. Am. Chem. Soc., 107 (1985) 3411.
29. Y. Odagaki, K. Hirotsu, T. Higuchi, A. Harada, and S. Takahashi,
J. Chem. Soc., Perkin Trans., (1990) 1230.
30. F. M. Menger and S. J. Sherrod,
J. Am. Chem. Soc., 110 (1988) 8606.
31. M. J. Sherrod,
Carbohydrate Research, 192 (1989) 17.
32. H-J. Tjiem, M. Brandl, and R. Breslow,
J. Am. Chem. Soc., 110 (1988) 8612.
33. R. Isnin, C. Salam, and A. Kaifer,
J. *Org. Chem., 56* (1991) 35.
34. M. Shimada, A. Harada, and S. Takahashi,
J. Chem. Soc., Chem. Commun., (1991) 263.
35. A. Ueno, F.Moriwaki, T. Osa, F. Hamada, and K. Murai,
Tetrahedron Lett., 26 (1985) 899.
36. N. Kobayashi and T. Osa,
Chem Lett., (1986) 421.
37. T. Matsue, T. Kato, U. Akiba, and T. Osa,
Chem. Lett., (1986) 843.
38. A. Ueno, F. Hamada, T. Tomokazu, and T. Osa,
Macromol. Chem., Rapid Commun., 6 (1985) 231.
39. A. Ueno, F. Moriwaki, T. Osa, F. Hamada, and K. Murai,
Chem. Pharm. Bull., 34 (1986) 438.

40. B. Klingert and G. Rihs,
Organometallics, 9 (1990) 1135.
41. A. Harada, K. Saeki, and S. takahashi,
Chem Lett., (1985) 1157.
42. A. Harada, K. Saeki, and S. Takahashi,
Organometallics, 8 (1989) 730.
43. C. A. Chang, H. Abdel-Aziz, N. Melchor, Q. Wu, K. H. Pannell, and D. W. Armstrong,
J. Chromatogr., 347 (1985) 51.
44. A. Harada and S. Takahashi,
J. Chem. Soc., Chem. Commun., (1986) 1229.
45. A. Harada, S. Yamamoto, and S. Takahashi,
Organometallics, 8 (1989) 2560.
46. D. R. Alston, A. M. Z. Slawin, J. F. Stoddart, and D. J. Williams,
Angew. Chem., Int. Ed. Engl., 24 (1985) 786.
47. D. R. Alston, P. R. Ashton, T. H. Lillley, J. F. Stoddart, and R. Zarzycki,
Carbohydr. Res., 192 (1989) 259.
48. A. Harada and S. Takahashi,
J. *Macromol. Sci. Chem., A26* (1989) 373.
49. D. R. Alston, J. F. Stoddart, and R. Zarzycki,
Tetrahedron Lett., 29 (1988) 2103.
50. D. R. Alston, T. H. Lilley, and J. F. Stoddart,
J. Chem. Soc., Chem. Commun., (1985) 1600.
51. D. R. Ashton, A. M. Z. Slawin, and J. F. Stoddart,
J. Chem. Soc., Chem. Commun., (1985) 1602.
52. A. Harada, M. Takeuchi, and S. Takahashi,
Chem. Lett., (1986) 1893.
53. A. Harada, M. Takeuchi, and S. Takahashi,
Bull. Chem. Soc., Jpn., 61 (1988) 4367.
54. A. Harada and S. Takahashi,
J. *Chem. Soc., Chem. Commun.,* (1988) 1352.
55. M. Shimada,
Doctoral Thesis, Osaka University, 1992.
56. A. Harada, M. Shimada, and S. Takahashi,
Chem. Lett., (1989) 275.
57. D. R. Alston, A. M. Z. Slawin, J. F. Stoddart, D. J. Williams, and R. Zarzycki,
Angew. Chem., Int. Ed., Engl., 27 (1988) 1184.
58. A. Harada, K. Saeki, and S. Takahashi,
J. Incl. Phenom., 5 (1987) 601.
59. A. Harada, K. Saeki, and S. Takahashi,
Carbohydr. Res., 192 (1988) 1.
60. W. L. Hinze, T. E. Riel, D. W. Armstrong, W. DeMond, A. Alak, and T. Ward,
Anal. Chem., 57 (1985) 237.

61. D. W. Armstrong, W. Demond, and P. Bronislov,
Anal. Chem., 57 (1985) 481.
62. R. Fornasier, F. Reniero, P. Scrimin, and U. Tonellato,
J. Org. Chem., 50 (1985) 3209.
63. Y. Kawajiri and N. Motohashi,
J. Chem. Soc., Chem. Commun., (1989) 1336.
64. A. Harada and S. Takahashi,
Chem Lett., (1984) 2089.
65. A. Harada, Y. Hu, and S. Takahashi,
Chem. Lett., (1986) 2083.
66. H. A. Zahalka, K. Januszkiewicz, and H. Alper,
J. Mol. Cat., 35 (1986) 249.
67. H. Alper,
Pure & Appl. Chem., 60 (1988) 35.
68. Y. Hu, M. Uno, A. Harada, and S. Takahashi,
Chem. Lett., (1990) 797.
69. Y. Hu, M. Uno, A. Harada, and S. Takahashi,
Bull. Chem. Soc., Jpn., in press.
70. Y. Hu, A. Harada, and S. Takahashi,
J. Mol. Cat., 60 (1990) L13.
71. Y. Hu, A. Harada, and S. Takahashi,
Synt. Commun., 18 (1988) 1607.
72. T. Joh, A. Harada, and S. Takahashi,
Memo. Inst. Sci. Ind. Res., 46 (1989) 37.

METAL-NITROXYL INTERACTIONS AS PROBES OF STEREOCHEMISTRY

S.S. Eaton and G.R. Eaton

METAL-NITROXYL INTERACTIONS AS PROBES OF STEREOCHEMISTRY

S.S. EATON AND G.R. EATON

1. INTRODUCTION

Electron paramagnetic resonance spectroscopy (EPR) of systems containing a nitroxyl radical interacting with a paramagnetic transition metal ion has been used to probe molecular stereochemistry in a variety of systems. The interaction between two unpaired electrons is analogous to the interaction between two nuclear spins - the spin-spin coupling can give rise to additional lines in the spectrum. With sufficient data for systems of known geometry, the spin-spin coupling constants can be interpreted in terms of molecular properties, including stereochemistry. Stereochemistry is used here in the broad sense that includes molecular conformations, spatial relationships between subunits, and isomers of transition metal complexes.

Electron spin-spin interaction between two paramagnetic centers can be divided into two components: (i) a through-bond contribution, *the exchange interaction*; (ii) a through-space contribution, *the dipolar interaction* [1]. The exchange interaction in EPR is analogous to the more-familiar isotropic nuclear spin-spin coupling constant in NMR. The exchange coupling constant is designated as J. The splitting of the EPR transitions due to exchange is observed in both rapidly tumbling and rigid lattice spectra. The magnitude of the exchange interaction is determined by orbital overlap and thus is sensitive to changes in molecular conformation. Due to the larger magnetic moment for an unpaired electron spin than for a nuclear spin, electron-electron coupling can be observed through larger numbers of bonds than is typical for nuclear-nuclear couplings. Thus, changes in J reflect conformational changes of multi-atom linkages. In this Chapter values of J are given in units of cm^{-1}. The conversion between units of gauss, that were used in some of the papers cited, and cm^{-1} is $J\ (cm^{-1}) = 4.668 \times 10^{-5}\ cm^{-1}\ g_{avg}\ J\ (G)$, where g_{avg} = the average of the metal and nitroxyl g values.

The dipolar interaction is an anisotropic through-space interaction. In frozen solution or single crystal spectra, the dipolar interaction causes splitting of the EPR transitions. For paramagnetic centers with anisotropic g- and A-values in rigid lattices, the anisotropy of the dipolar interaction can be used to obtain detailed geometrical information [2]. Since the trace of the dipolar interaction matrix is zero, splitting of continuous wave (CW) EPR spectra due to dipolar interaction is averaged to zero by rapid tumbling.

The impact of electron spin-spin coupling on a CW EPR spectrum depends upon the relaxation rate (the reciprocal of the relaxation time) for the faster relaxing paramagnetic center. If the relaxation rate T_1^{-1} (in hz) is slower than the spin-spin coupling (expressed in hz), and the coupling is greater than the linewidth of the EPR signal, resolved splittings are observed in the EPR spectra. In this regime the magnitude of the spin-spin coupling constant can be determined by analysis of the resolved splittings. As T_1^{-1} increases and becomes comparable in magnitude to the splitting, the lines broaden. For relaxation rates that are faster than the spin-spin coupling, the splittings are collapsed and information concerning the spin-spin interaction is obtained from measurements of relaxation rates rather than by analysis of resolved splittings. The impact of the electron spin relaxation rate in these spin-coupled systems is analogous to the effect of a dynamic process in an NMR experiment. The focus of this review is on cases in which the electron spin relaxation rates are sufficiently slow that resolved splittings in the CW spectra are observed and can be analyzed to evaluate the spin-spin coupling constant. It is anticipated that changes in electron spin relaxation rates due to electron-electron spin-spin coupling will play an increasingly important role in future studies of stereochemistry in metal-nitroxyl systems.

The emphasis of this Chapter is on the stereochemical information that can be obtained from the analysis of electron spin-spin interaction in discrete molecules. The analysis of EPR spectra to determine the exchange and dipolar interactions are discussed in the original papers. Related reviews include a literature survey of metal nitroxyl interactions [3] and discussions of long-range electron-electron exchange interaction [1], and the methodology for interpretation of splittings in EPR spectra that result from spin-spin interaction [2]. This Chapter does not include a discussion of the stereochemical aspects of the much stronger spin-spin interaction in polymeric metal-nitroxyl systems that is being utilized to design materials with particular magnetic properties [4]. The properties of metal complexes of imidazoline nitroxides have been summarized recently [5]. The stereochemistry of nitroxides is reviewed in Ref. 6.

2. SPIN-SPIN INTERACTION IN RAPIDLY TUMBLING SYSTEMS

When a molecule tumbles rapidly, anisotropic contributions to the spin-spin interaction are averaged and the isotropic terms are retained. The isotropic exchange interaction between two unpaired electrons produces EPR spectra that are analogous to AB spectra in NMR [7]. The EPR spectra for most organic radicals and many paramagnetic transition metals exhibit nuclear hyperfine splitting due to interaction of the unpaired electron with nuclear spins. Electron-electron interaction in systems with nuclear hyperfine splitting produces an AB pattern for each combination of nuclear spin states. As in NMR, the spectra can be analyzed by considering sub-spectra. For example, the fluid solution EPR spectrum of a complex with electron spin-spin coupling between a vanadyl ion and a nitroxyl spin label consists of 8x3 AB patterns due to interaction of the unpaired electrons with the spins of the vanadium ($I = 7/2$) and nitroxyl nitrogen ($I = 1$) nuclei. The ratio of J to the separation between the energy levels for the two interacting spins is different for each of the subspectra and spectra typically are second-order. Due to the large number of overlapping lines in the spectra, computer simulation often is needed to analyze the spectra. It is convenient to refer to transitions as "inner" or "outer" lines of the AB pattern. The splitting pattern for a rapidly tumbling molecule is independent of the sign of J, so only the magnitude of J is obtained by analysis of the fluid solution spectrum.

2.1 **Coordination Equilibria**

Since the magnitude of J is dependent on spin delocalization and orbital overlap, the interaction between an unpaired electron on a transition metal and an organic radical that is part of a ligand, is strongly dependent upon the mode of coordination of the ligand. This sensitivity of J to the mode of coordination of a spin-labeled ligand makes spin-spin interaction a powerful tool for detection of isomers and for studies of coordination equilibria.

2.1.1 Isomers. The EPR spectra of copper(II) complexes **I** (n=2,3) showed the presence of two isomers with populations that were solvent and temperature-dependent [8-10].

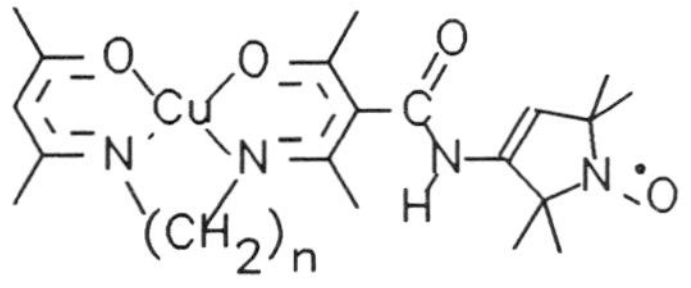

n = 2 Ia

n = 3 Ib

The values of J in toluene solution, at room temperature, were: 0.0488 cm^{-1} (55%) and 0.30 cm^{-1} (45%) for **Ia**; 0.0305 cm^{-1} (10%) and 0.26 cm^{-1} (90%) for **Ib**. As the value of J increases, the positions of the inner lines of the AB pattern become less sensitive to changes in the value of J, so to determine accurately the value of J it is important to observe the outer lines. At 9 GHz the intensities of the outer lines of the AB patterns with J = 0.30 cm^{-1} and 0.26 cm^{-1} were so small that the lines were difficult to detect. At 35 GHz the separation between the copper and nitroxyl electron spin energy levels is greater than at 9 GHz, which increases the intensities of the outer lines and facilitates their detection [10]. The populations of the isomers were obtained by computer simulation of the spectra at 9 and 35 GHz. The value of J for **Ia**, at room temperature, varied from 0.0450

cm^{-1} in THF to 0.0497 cm^{-1} in DMSO. In toluene solution J decreased from 0.0502 cm^{-1} at -20°C to 0.0451 cm^{-1} at 119°C [8]. For **Ib** the room temperature values of J varied from 0.0306 cm^{-1} in toluene to 0.0397 cm^{-1} in pyridine [9]. Although the solvent and temperature dependence of the values of J are significant, the range of values for the isomer with the smaller value of J is small compared with the large difference in J between the two isomers. The variations as a function of solvent and temperature are attributed to changes in chelate ring conformations and axial ligation of the metal. The assignment of the isomers with the larger values of J to the copper(II) complexes **IIa,b** was based on three observations.

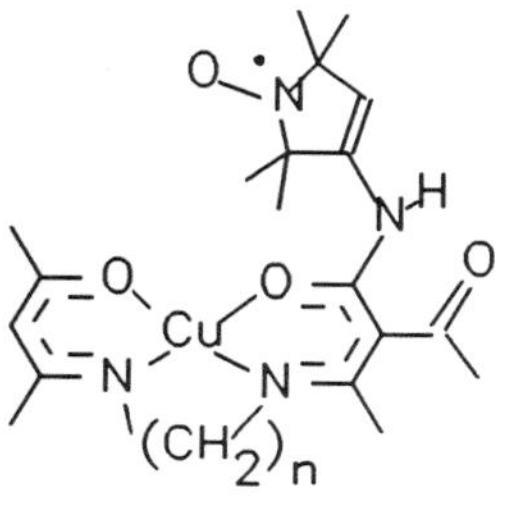

n = 2 IIa

n = 3 IIb

(1) The Cu(II) g- and A-values for the isomers with the larger values of J were close to those for the isomers with the smaller values of J, which indicated that the copper coordination sphere remained approximately planar with N_2O_2 donors.

(2) In **IIa,b**, the bonding pathway between the metal and nitroxyl is two bonds shorter than in **Ia,b**, but there are still three bonds between the copper and the nitroxyl ring. Values of J obtained for similar compounds suggest that J of the order of 0.20 to 0.30 cm^{-1} is plausible for **IIa,b**.

(3) The populations of the isomers with the larger values of J were increased in hydrogen bonding solvents.

The isomerization represented by **Ia,b** <—> **IIa,b** had been proposed previously, but had not been detected by other spectroscopic techniques [11,12].

In the EPR spectra of vanadyl bis(hexafluoroacetylacetonate) complexes of spin-labeled pyridines, VO(hfac)$_2$L, a single isomer was observed [13]. When the same spin-labeled pyridines were bound to vanadyl bis(trifluoroacetyl-acetonate), VO(tfac)$_2$L, two isomers were observed with approximately equal populations and with values of J that differed by about 20% [13,14]. For the same nitroxyl substituent, the values of J were 3 to 5 times larger when the substituent was attached at the pyridine para position than when it was at the meta position. If the pyridine is bound cis to the vanadyl oxygen with the pyridine plane parallel to the V=O bond, the geometry is optimum for back π bonding from the metal d_{xy} orbital that contains the unpaired electron to the pyridine π^* LUMO [13]. Unpaired electron spin density in the π^* LUMO would contribute to much larger values of J for the *para*-spin-labeled pyridine than for the *meta*-spin-labeled pyridine, which is consistent with the experimental data. If the pyridine binds trans to the vanadyl oxygen, no interaction between the d_{xy} orbital and the pyridine π orbitals would be expected, which would result in much smaller values of J than for the cis isomer. The 20% difference in the values of J for the two isomers of VO(tfac)$_2$L is too small to be consistent with cis and trans isomers with respect to the V=O bond. In addition, if the isomerism were cis/trans, at least some population of the second isomer would be expected for VO(hfac)$_2$L. The conclusion that the spin-labeled pyridine is coordinated cis to the V=O bond for all of the complexes is consistent with previous IR studies of pyridine coordination to VO(tfac)$_2$ [15,16]. It is more reasonable to attribute the two values of J to isomers **III** and **IV**, both of which have L cis to the V=O bond.

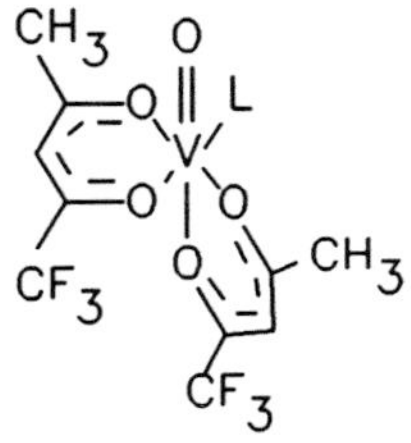

III 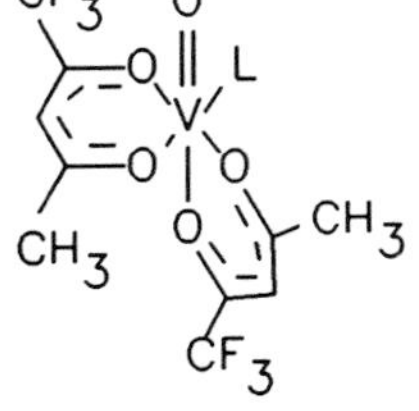   IV

Coordination of L cis to the V=O bond requires displacement of one end of a β-diketonate ligand. Since the $CF_3C=O$ oxygen is a poorer donor than the $CH_3C=O$ oxygen, it is more readily displaced from the equatorial plane. The pyridine ligand can bind trans to a $CF_3C=O$ (**III**) or trans to a $CH_3C=O$ (**IV**). This type of isomerism requires an unsymmetrical β-diketonate, so it is consistent with the observation of a single isomer for all of the $VO(hfac)_2$ adducts.

When spin-labeled pyridines were coordinated to $Cu(hfac)_2$, a single isomer was observed when the nitroxyl substituent was attached at the para position, but two isomers were observed for several ligands in which the substituent was attached at the meta position [13]. This isomerism is assigned to the presence of conformations **V** and **VI**. Steric interaction between the bulky R group and the hexafluoroacetylacetonate ring in **V** is expected to cause a greater displacement of the pyridine below the equatorial plane in **V** than in **VI**, resulting in different values of J.

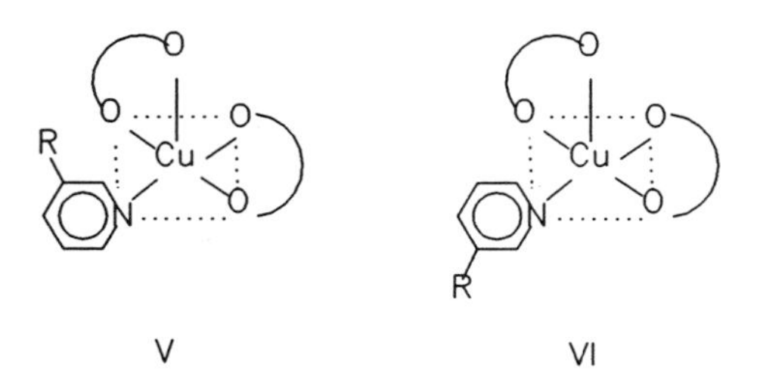

2.1.2 *Equilibria*. For pyridine-based ligands with a nitroxyl substituent, coordination to a metal can occur through the pyridine nitrogen or nitroxyl oxygen [17,18]. When the pyridine nitrogen is coordinated, J typically is small enough that the value can be determined from splittings in the EPR spectra. However, when the nitroxyl oxygen is directly coordinated to the metal, J is so large that EPR spectra are not observed for the spin-coupled system in fluid solution. (When J is large and antiferromagnetic, the spin system is diamagnetic. When J is large and ferromagnetic, the dipolar interaction is so large that incomplete averaging of the dipolar splitting results in spectra that are too broad to observe under normal operating conditions.) In addition, for some ligands, including **VII** and **VIII**, the coordination of the nitroxyl oxygen to one metal and coordination of the pyridine nitrogen to a second metal produces a ligand-bridged metal-metal dimer, **IX**. Quantitation of the EPR signal for the spin-coupled system in conjunction with visible spectroscopy has been used to determine equilibrium constants for these modes of coordination [17,18].

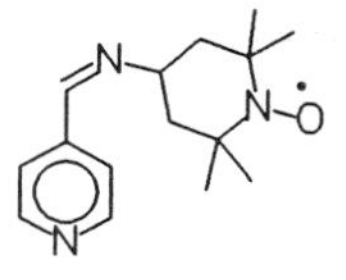

VII 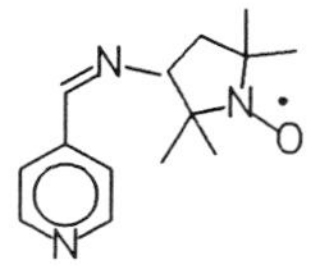 VIII

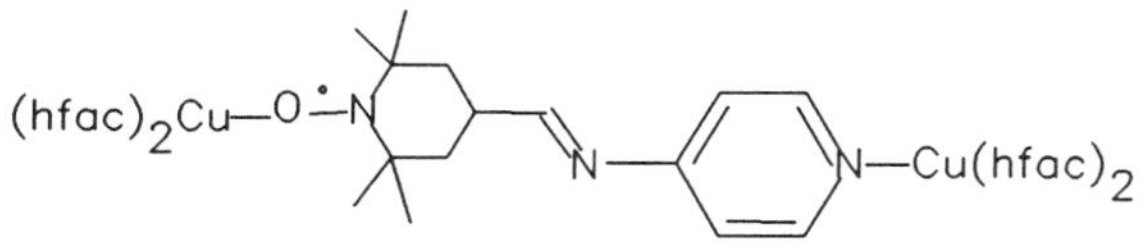

(hfac)$_2$Cu—O—N ... N—Cu(hfac)$_2$

IX

The EPR spectrum of dinitroxyl **X** is a multi-line signal due to nitroxyl-nitroxyl interaction with J approximately equal to the nitroxyl nitrogen hyperfine splitting.

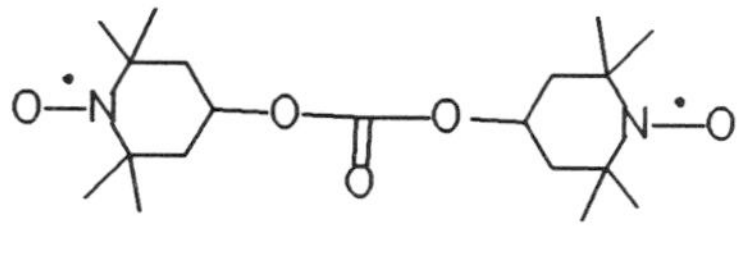

X 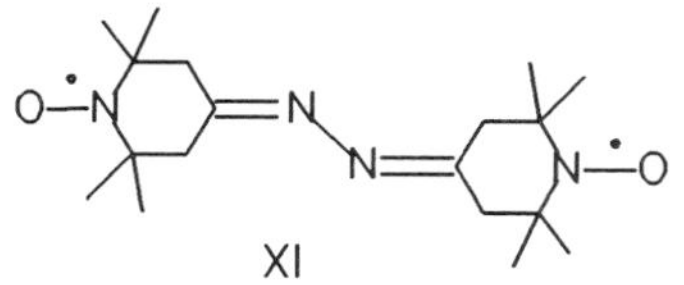 XI

For dinitroxyl **XI** the spectrum is a 5-line pattern due to J >> the nitroxyl nitrogen hyperfine splitting. Coordination of one of the nitroxyl oxygens of **X** or **XI** to $Cu(hfac)_2$, $Cu(tfac)_2$, $VO(hfac)_2$, or $VO(tfac)_2$ results in sufficiently strong electron spin-spin coupling between the metal and the coordinated nitroxyl that no EPR spectrum is observed for these two spins in fluid solution. The EPR spectrum of the complex reduces to the simple 3-line spectrum of a mono-nitroxyl [19]. The distinctive change in the EPR spectrum permitted measurement of the metal-nitroxyl binding constants.

$Ni(hfac)_2$ is square planar and diamagnetic. Coordination of a bidentate nitrogenous base generates a 6-coordinate paramagnetic complex. The fluid solution EPR spectrum of **XII** is a characteristic 3 line nitroxyl signal [20].

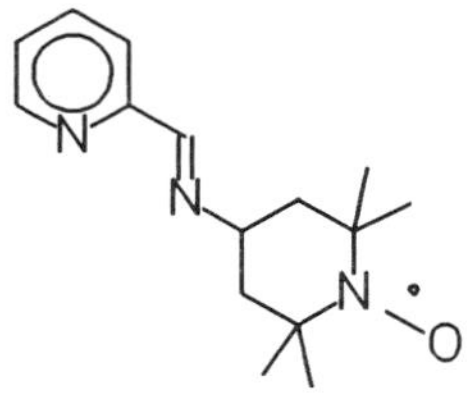

XII

When **XII** binds to $Ni(hfac)_2$ there is a strong nickel-nitroxyl interaction that broadens the fluid-solution nitroxyl EPR signal so extensively that the EPR signal for the complex is not observed at the spectrometer settings appropriate for **XII**. Quantitation of the EPR signal as a function of the ratio of $Ni(hfac)_2$ and **XII** concentrations gave an equilibrium binding constant of $(2 \pm 1) \times 10^4$ M^{-1} at 21°C, in toluene solution [20]. Similarly, the nickel(II) is diamagnetic in the square-planar spin-labeled complexes **XIII**, **XIV**, and **XV** and the nitroxyl-nitroxyl interaction is weak, so the EPR spectra are 3-line nitroxyl signals.

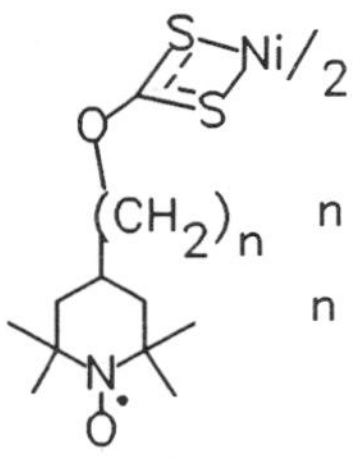

n = 0 XIII

n = 1 XIV

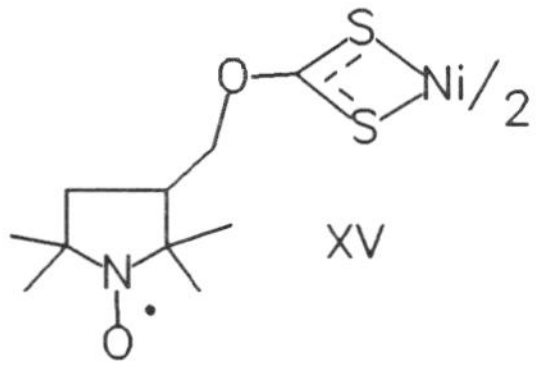

XV

Coordination of 2,2'-bipyridine (bipy) forms **XIII**·bipy, **XIV**·bipy, and **XV**·bipy in which the nickel is paramagnetic and the nickel-nitroxyl interaction is sufficiently strong that the nitroxyl EPR signal is too broad to observe at the spectrometer settings required for **XIII**, **XIV**, and **XV**. Quantitation of the EPR signal for **XIII**, **XIV**, or **XV** as a function of added bipy gave equilibrium binding constants between 3 x 10^3 and 2 x 10^4 M^{-1} for the three complexes, in $CHCl_3$ solution, at 21°C [20]. The fact that the loss of EPR signal was not the result of chemical destruction of the nitroxyl was demonstrated by observation of the characteristic frozen solution EPR spectra due to nickel-nitroxyl interaction [21] and by checks on the reversibility of the equilibria.

2.2 **Metal-ligand bonding**

Crystal structures of first-row transition metal complexes of bis(pyrazolyl)-amines, **XVI**, have shown that the complexes adopt a range of conformations from no bonding interaction between the metal and the central amine nitrogen to metal-amine nitrogen distances of 2.12-2.19 Å (22-25).

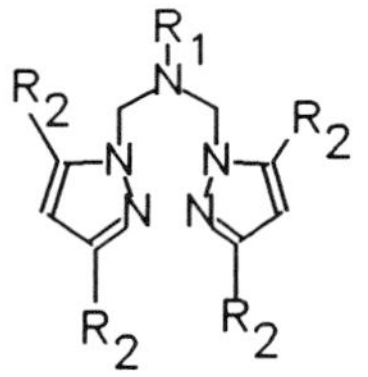

XVI

At X-band the value of J for Cu(**XVI**)Y_2 (Y = Cl^-, Br^-; R_1 = nitroxyl; R_2 = H) was large relative to the separation between the copper and nitroxyl energy levels and the copper and nitroxyl inner lines of the AB pattern were superimposed [26]. At Q-band the larger energy level separation permitted observation of separate copper and nitroxyl inner lines with a separation that indicated J = 0.20 cm^{-1}. The X-ray crystal structure of Cu(**XVI**)Cl_2 (R_1 = nitroxyl, R_2 = H) gave a Cu-amine nitrogen distance of 2.195 Å, which is substantially longer than the Cu-pyrazole nitrogen distance of 1.97 Å. When R_2 was CH_3 instead of H and the rest of the molecule was unchanged, the values of J were 0.0050 to 0.0800 cm^{-1} and were strongly dependent on solvent and temperature. The much smaller value of J for R_2 = CH_3 indicates that replacement of H by CH_3 resulted in a weakening of the bond between the copper and the amine nitrogen. The resulting conformation was quite flexible and was subject to variation as a function of solvent and temperature. As the temperature decreased the value of J increased, which suggested that as the sample cooled, the copper-amine nitrogen bond in the complex with R_2 = CH_3 became similar to that for R_2 = H. Analysis of the intensity of the half-field transition (as discussed below) in the frozen solution spectrum of Cu(**XVI**)Cl_2 (R_1 = nitroxyl, R_2 = CH_3) gave a copper-nitroxyl distance of 6.5 Å [26]. In the crystal structure of Cu(**XVI**)Cl_2 (R_1 = nitroxyl, R_2 = H) the through-space distances between the copper and the nitroxyl nitrogen and oxygen were 5.694 and 6.913 Å, respectively, for an average of 6.3 Å. The similarity between this value and the interspin distance obtained from the frozen solution EPR spectrum of Cu(**XVI**)Cl_2 (R_1 = nitroxyl, R_2 = CH_3) suggests that the two structures are similar.

2.3 **Ligand Conformations**

Electron spin-spin coupling was observed in the EPR spectra of the copper(II) complexes of spin-labeled porphyrins **XVII** - **XXII** which permitted examination of dynamic processes that interconvert conformations of the linkages between the porphyrin and nitroxyl rings [27,28].

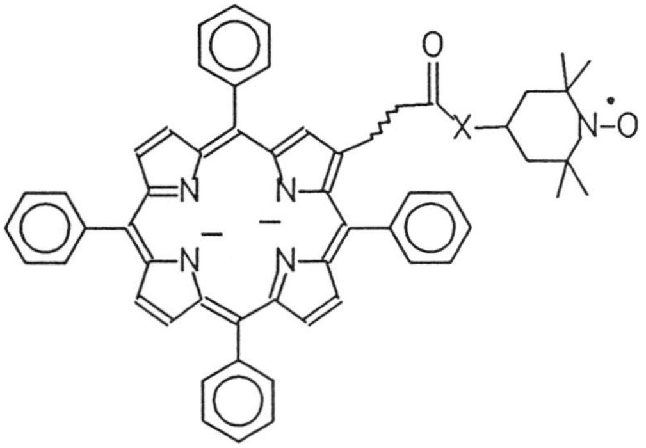

∿∿	X	
trans	NH	XVII
satur.	NH	XVIII
cis	NH	XIX
trans	O	XX
satur.	O	XXI
cis	O	XXII

In complex **XVII**, which has a trans olefin and an amide group between the porphyrin and nitroxyl rings, a single conformation was observed in $CHCl_3$, toluene, pyridine or combinations of these solvents, at temperatures between -60°C and room temperature. The value of J increased on cooling, but the temperature dependence was similar to what has been observed in other relatively rigid systems. Wojcik and Witanowski [29] studied halogenated acrylamides and concluded that for the trans olefins there is a single conformer that has a Z conformation of the C-C single bond adjacent to the olefin. The barrier to rotation about the amide C-N bond in acrylamides has been observed to be 12-16 kcal/mole, depending on the substituents on the olefin [30], so the rotation about the amide C-N bond in **XVII** is expected to be slow on the EPR timescale, at room temperature. By analogy with polypeptide amide linkages, it is expected that the piperidine ring is oriented cis to the carbonyl. These analogies suggest that **XXIII** is the sole conformation of **XVII** and that the structure is relatively rigid. Structure **XXIII** gives an "extended W plan" conformation for the bonds between the porphyrin and nitroxyl rings.

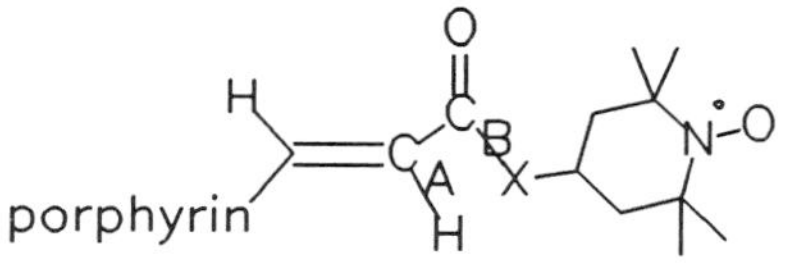

XXIII

This bond geometry in σ-bonded systems frequently has been observed to give rise to large long-range nuclear-nuclear couplings in NMR spectra and electron-nuclear couplings in EPR spectra [31,32]. The expectation of larger values of J for this geometry is consistent with the observation that the trans olefin gives larger values of J than the cis olefin, which cannot adopt an extended W-plan conformation (Table 1).

In contrast to the behavior of Cu**XVII**, which has a trans olefin and an amide in the linkage between the porphyrin and nitroxyl rings, two conformations were observed in the EPR spectra of Cu**XX**, which has a trans olefin and an ester in the linkage between the porphyrin and nitroxyl rings. One isomer exhibited a value of J that was similar to that for Cu**XVII** and the second isomer had a smaller value of J. On the basis of the coalescence temperature of the signals for the two species, the barrier to interconversion was estimated to be 6.3 kcal/mole [28]. The barrier to rotation is higher in chloroform or bromoform solution than in toluene due to specific interaction between the nitroxyl and the halocarbon [27,28]. The two isomers were assigned to E and Z conformations with respect to the C_a-C_b bond adjacent to the olefin (**XXIII**).

TABLE 1

Values of the spin-spin coupling constant, J, for spin-labeled copper porphyrins **XVII - XXII** in 2:1 toluene:pyridine[a]

		C-C linkage		
X	temperature	trans	saturated	cis
		XVII	**XVIII**	**XIX**
NH (amide)	22	24.5	13.4	3.4
	-60	30.6	3	6
			21	
		XX	**XXI**	**XXII**
O (ester)	22	19	12	4
	-60	11	6	4
		36	25	

[a]*Data taken from reference 28. Values of J in units of 10^{-4} cm^{-1}.*

At temperatures below -45°C, two conformations were observed for complexes Cu**XVIII** and Cu**XXI**, which have a saturated CH_2-CH_2 linkage and an ester or amide between the porphyrin and nitroxyl rings. The barrier to interconversion of these conformations (*ca.* 5 kcal/mole) is sufficiently low that it is difficult to assign the dynamic process to slowing of rotation about a particular bond. However, the larger of the J values observed for Cu**XVIII** and Cu**XXI** (0.0021 and 0.0025 cm^{-1}, respectively) are sufficiently similar to the values observed for the conformation sketched in **XXIII** to suggest that these conformations also correspond to an extended W plan. The order of increasing J values at room temperature in the complexes with saturated linkages, cis < saturated < trans, is consistent with rapid motions between extended W-plan conformations similar to that of the trans linkage, in which J is relatively large, and conformations more like that for the cis linkage, in which J is small.

At room temperature two species were observed in the EPR spectra of **XXIV** and **XXV** [3].

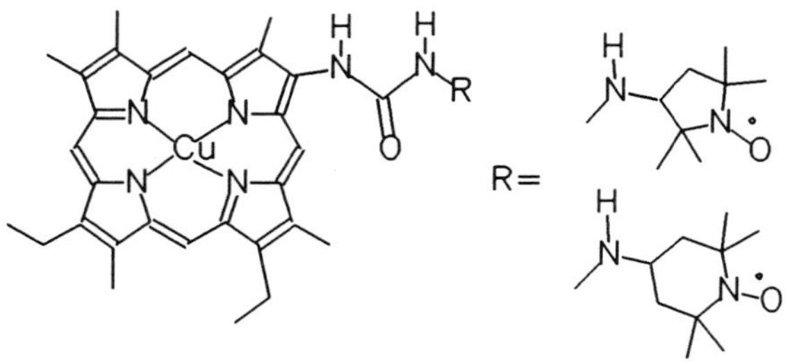

For **XXIV** the values of J for the two isomers differed by a factor of about 10; for **XXV** the values of J differed by about a factor of 4. The conformation with the larger value of J was present in higher concentration in pyridine solution than in CH_2Cl_2. Since isomers were observed for both the 5- and 6-membered nitroxyl rings, it is unlikely that they are due to differences in nitroxyl ring conformations. Isomers were not observed for the analogs of **XXIV** and **XXV** in which the urea linkage was replaced by an amide linkage, so the isomerism is attributed to different conformations of the urea linkage [33].

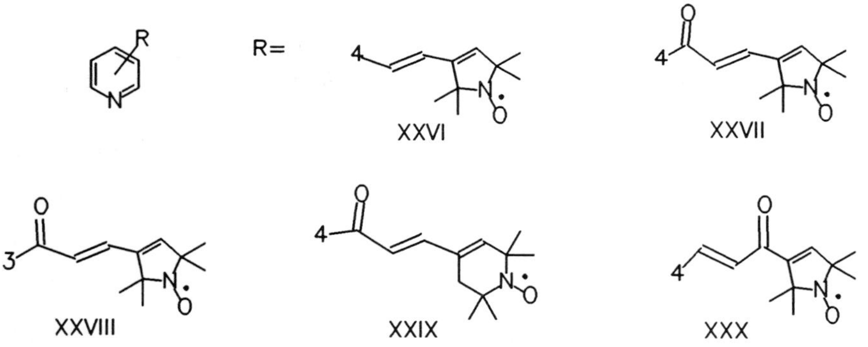

R=

XXVI XXVII

XXVIII XXIX XXX

The ^{1}H NMR spectrum of **XXVI** that had been reduced to an hydroxylamine with diphenylhydrazine showed an AB pattern for the olefinic protons with J = 16 Hz, which is consistent with a trans olefin [34]. No peaks characteristic of a cis olefin were observed. The Q-band spectrum of Cu(hfac)$_2$**XXVI** showed four components with different values of J [34]. The two major components had J = 0.18 and 0.075 cm^{-1} and were present in a ratio of about 5:1. The populations of conformations with J = 0.045 and 0.030 cm^{-1} were less than about 2%. Isomers of **XXVI** could be due to cis-trans isomerism about the olefinic bond or rotamers with respect to the single bond between the olefin and the nitroxyl ring. Combination of the two effects could give rise to four species, as observed in the EPR spectra. On the basis of the ^{1}H NMR data, the two major components in the EPR spectra are assigned to isomers with a trans conformation of the olefin. The two minor components are assigned to isomers with a cis conformation of the olefin. The second site for isomerism is the central bond of the diene. The barrier to rotation around the central bond in 1,3-butadiene is about 4.0 kcal/mole [35]. The separations between the peaks in the spin-coupled EPR spectrum of Cu(hfac)$_2$**XXVI** are sufficiently large that the absence of dynamic averaging at room temperature only requires an activation energy greater than about 5 kcal/mole. The bulky substituents in **XXVI** may cause the barrier to rotation about the central bond of this diene to be greater than in 1,3-butadiene. However, much higher barriers would be required to prevent averaging on the NMR time scale. Thus it is plausible that these conformers would be averaged in the NMR spectra, but not in the EPR spectra. In 1,3-butadiene the population of the trans isomer is about 100 times the population of the minor isomer [36], which is thought to be a skew or bent conformation rather than cis [37]. Although the trans isomer is likely to be favored for **XXVI**, the ratio of trans to skew is likely to be smaller for **XXVI** than for 1,3-butadiene due to steric interaction in the trans isomer between the 2,2-dimethyl protons of the nitroxyl ring and the olefinic proton. It seems plausible that the major isomer of **XXVI**, which also gives the largest value of J when coordinated to $Cu(hfac)_2$, is trans olefin with a trans conformation of the central bond of the diolefin. The value of J for this isomer is expected to be larger than for the other isomers. The fully trans linkage is expected to optimize both the σ and π contributions to the exchange interaction since the σ contribution is greatest through a W-plan geometry [31] and the π contribution is greatest for the planar conjugated conformation. It is proposed that the second-most abundant isomer is a trans olefin with a skew geometry with respect to the central bond of the diolefin. These assignments would give a trans to skew ratio of 5 for **XXVI** bound to $Cu(hfac)_2$. The value of J for the skew conformation is less than the value for the fully trans conformation by about a factor of 2.5, due to the less favorable geometry for both σ and π interaction.

The EPR spectra for $Cu(hfac)_2L$ (L = **XXVII**, **XXVIII**, **XXIX**, **XXX**) were temperature dependent [34]. At room temperature the nitroxyl signal for $Cu(hfac)_2$**XXVIII** was a broadened three-line spectrum. Below 0°C the spectrum split into two doublets of nitroxyl triplets with values of J that differed by about a factor of 8 (0.0009 and 0.00070 cm^{-1}). The conformation with the smaller value of J had the greater abundance (about 3:1) and dominated the frozen-solution spectrum. Since the averaged signal at room temperature had a smaller splitting than that observed for either of the slow-exchange signals, the signs of J must be opposite for the two conformations. The room temperature nitroxyl EPR spectrum for $Cu(hfac)_2$**XXIX** also was a broadened triplet and split into two doublets of nitroxyl triplets below -35°C. The behavior of $Cu(hfac)_2$**XXIX** was similar to that observed for $Cu(hfac)_2$**XXVIII**; the two values of J differed by about a factor of 10 (0.0005 and 0.0050 cm^{-1}), the conformation with the smaller value of J had the greater population (about 2.5:1), and the signs of J for the two conformations were opposite. The room temperature spectra of the nitroxyl signals for $Cu(hfac)_2L$ (L = **XXVII, XXX**) were similar to each other, with a broad doublet of triplets due to J = 0.0058 and 0.0080 cm^{-1}, respectively [68]. Below -60°C the nitroxyl signal for $Cu(hfac)_2$**XXVII** split into two doublets with J = 0.0080 and 0.0029 cm^{-1}. For L = **XXX** two doublets of triplets with J = 0.0110 and 0.0037 cm^{-1} were observed at temperatures below 10°C. For these two ligands the conformation with the larger value of J had the greater population and the signs of J are the same for the two conformations.

The barriers to isomerization in $Cu(hfac)_2L$ (L = **XXVII, XXVIII, XXIX, XXX**) were estimated to be about 5 kcal/mole by assuming two-site exchange and calculating $\Delta G^{\neq}$ from the coalescence temperature and the average separation between the interchanging nitroxyl signals [34]. The values were slightly higher for L = **XXVII, XXX** than for L = **XXVIII, XXIX**. It is proposed that this is the barrier to rotation about one of the single bonds in the pyridine-nitroxyl linkage. These ligands have a carbonyl group in the pyridine-nitroxyl linkage in addition to the olefin in **XXVI**. In **XXVII, XXVIII, XXIX**, the carbonyl group is between the olefin and the pyridine ring. Conformers in this system could be due to slowing of rotation of the bond between the two double bonds or the bond between the olefin and the carbonyl. In **XXX** the carbonyl group is adjacent to the nitroxyl ring instead of adjacent to the pyridine ring. In this case there is not a diolefin, so isomerism must be due to slow rotation about the bond(s) to the carbonyl group. At -50°C two conformations also were observed in the $VO(tfac)_2$ complexes of **XXVIII** and **XXX** [68]. Due to the larger number of nuclear hyperfine lines for vanadyl ion (I = 7/2) than for Cu(II) (I = 3/2), it is more difficult to detect isomers with values of J of the order of 0.01 cm^{-1} for the vanadyl complexes than for the copper complexes.

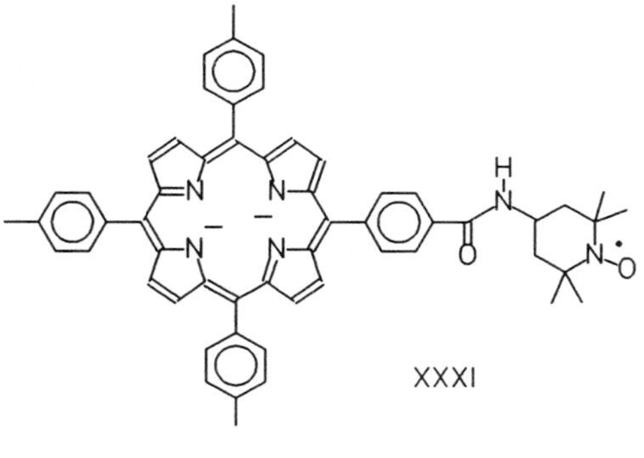

XXXI

The nitroxyl electron spin relaxation rates for high spin Fe(III), high spin Mn(III) and low spin Fe(III) complexes of spin-labeled porphyrins **XVII**, **XIX**, and **XXXI** were measured in toluene solution at room temperature [38]. The impact of the metal on the nitroxyl relaxation rates decreased as the distance between the metal and the nitroxyl increased and decreased in the order: high-spin Fe(III) > high spin Mn(III) > low-spin Fe(III). These trends are consistent with predominantly dipolar contributions to the interaction. The greater interaction for the high-spin Fe(III) complex of **XVII**, which contains a trans olefinic linkage, than for the complex of **XIX**, which contains a cis olefinic linkage, suggests that exchange contributions to the electron spin-spin interaction may also be significant.

2.4 **Weak Orbital Overlaps**

An important question in areas as diverse as electron transport, fluoresence resonance energy transfer and NMR coupling constants is the significance of weak orbital overlaps that are sometimes called "through-space" interactions. These are not interactions through the primary bonding pathways. Rather, these are interactions due to the spatial proximity of orbitals on atoms that are not bonded to (*i.e.* adjacent to, in an HMO sense) each other. When such orbital overlaps permit spin-spin interaction between the paramagnetic centers, the result is stronger exchange coupling than would have been expected through the primary bonding pathway. Since these weak orbital overlaps are strongly dependent on molecular conformation, changes in the exchange interaction due to weak orbital overlap can be used to monitor conformational changes.

The values of J obtained from the EPR spectra of Cu**XXXII** in a variety of solvents were 0.0003 to 0.0005 cm^{-1} [27]. The trans olefinic linkage in Cu**XXXII** positions the nitroxyl beside the porphyrin plane. When the trans olefinic linkage is replaced by a cis olefin (Cu**XXXIII**) the nitroxyl is positioned above the porphyrin plane.

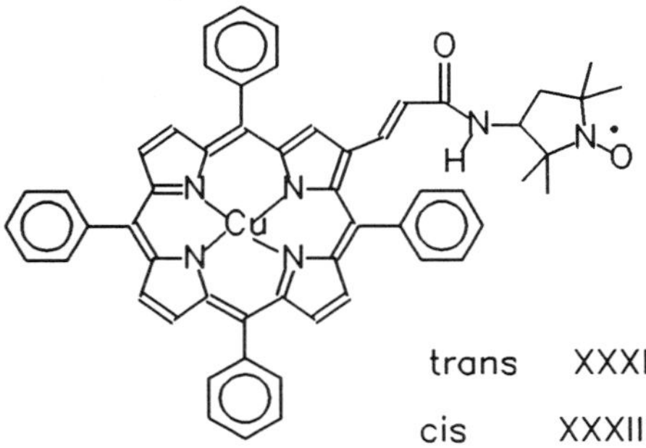

trans XXXII
cis XXXIII

In THF or pyridine solutions, the solvent coordinates to the copper and the value of J for Cu**XXXIII** is 0.0005 to 0.00010 cm^{-1}, but in the non-coordinating solvent $CHCl_3$ the value of J is 0.050 cm^{-1}. It is proposed that in the absence of a coordinating solvent, there is a weak orbital overlap between the amide oxygen in the cis isomer and the copper orbitals or the porphyrin π orbitals, which provides a pathway for stronger copper-nitroxyl interaction than occurs through the covalent bonds of the porphyrin-sidechain-nitroxyl system. Coordination of solvent to the copper causes the substituent to adopt a conformation that does not give an efficient weak orbital overlap pathway for copper-nitroxyl interaction.

Similarly, a series of copper(II) tetraphenylporphyrins has been studied in which a spin label is attached to one phenyl ring [39]. Cu**XXXIV** and Cu**XXXV** are two members of this series.

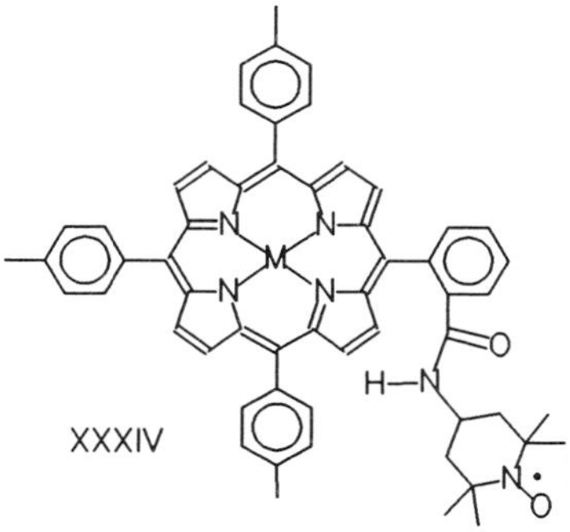

XXXIV 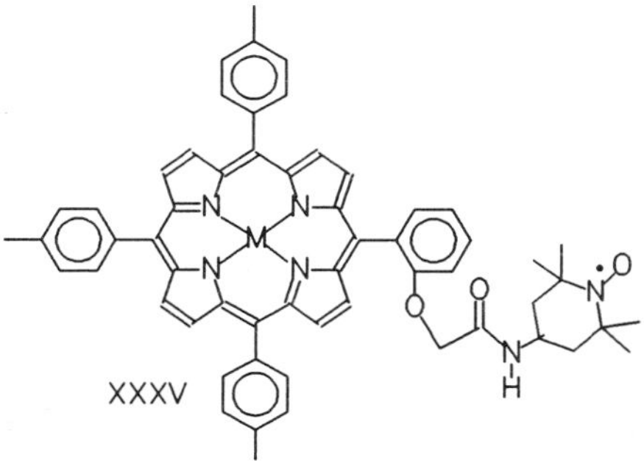 XXXV

When the substituent was at the meta or para position, the largest value of J was 0.0003 cm^{-1} and the values of J were weakly dependent on solvent. When the substituent was attached to the ortho position of the phenyl ring, the values of J in non-coordinating solvents were 0.030 and 0.10 cm^{-1}, depending upon the nature of the porphyrin-nitroxyl linkage and the nitroxyl ring size. These results indicate that there was a pathway for spin-spin interaction that was only accessible to the ortho isomers, presumably involving direct interaction of orbitals from the substituent with the metalloporphyrin orbitals. For amide (as in Cu**XXXIV**) or ester linkages, J fell to values between 0 and 0.010 cm^{-1} in pyridine solution. However, for an ether linkage between the ortho position of the phenyl ring and the nitroxyl, the value of J was about 0.10 cm^{-1}, independent of solvent coordination. The observation that the solvent dependence of J was dependent on the nature of the substituent indicated that the changes in J were not due solely to overall changes in the electron spin distribution in the porphyrin orbitals as a result of pyridine coordination. Rather, the changes in J are attributed to changes in the conformations of the amide and ester linkages as a result of axial ligand coordination. The conformation in the presence of axial ligand does not provide a weak orbital overlap pathway for metal-nitroxyl interaction. Conformational changes of the ether linkage as a result of axial ligand appear to be smaller than for the amides and esters.

For Ag(II)**XXXIV** the values of J in solutions of non-coordinating solvents were about 0.10 cm^{-1}. J decreased to about 0.010 cm^{-1} in pyridine solution [40].

This change in J is attributed to a conformational change of the amide substituent, analogous to what was observed in the copper(II) complex. Additional information about the conformational change was obtained by analysis of frozen solution spectra as discussed below in Section 3.1.3.

2.5 **Molecular Flexibility**

In systems with flexible linkages between the metal and the nitroxyl, there is a wider distribution in values of J than for systems with rigid linkages, and the values of J tend to be more strongly solvent and temperature dependent. The changes in J by 10 - 20% discussed above for **Ia,b** as a function of temperature and solvent are typical of the variations in J observed for complexes with relatively rigid geometries. The temperature dependence of J for copper(II) and vanadyl complexes of spin-labeled pyridines that have a CH_2 group in the linkage between the pyridine and the nitroxyl is larger than when the linkage consists only of an amide, ester, or Schiff base. For ligands with a CH_2 group, changes in J by up to a factor of 2 or 3 between 20° and -60° have been observed [14]. This strong temperature dependence of J reflects changes in the populations of rotamers and the strong effect of conformation on the value of J.

In the spin-labeled copper complexes **XXXVI** and **XXXVII** the nitroxyl is attached *via* an amide linkage to the α-carbon of an ethylene or propylene bridge [41].

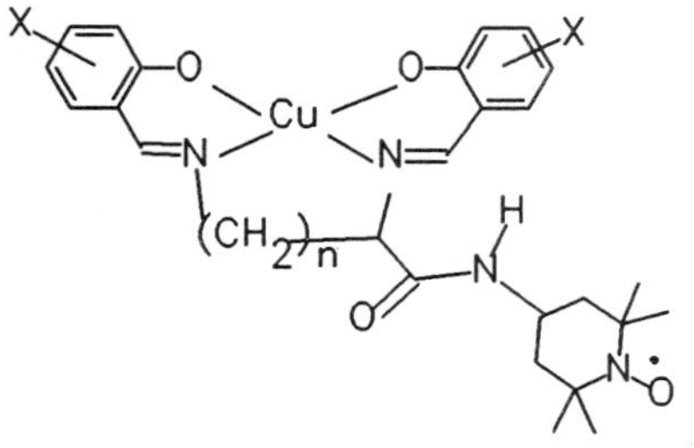

n = 2 XXXVI

n = 3 XXXVII

Since the copper-nitroxyl interaction in these complexes is expected to have a large σ contribution, the value of J is expected to be strongly dependent on the conformation of the methylene or propylene bridge. For **XXXVI** (X = H, 4-OMe), the values of J as a function of solvent increased in the order CH_2Cl_2 < $CHBr_3$ < $CHCl_3$, which parallels the order of increasing hydrogen-bonding ability of the solvent. Addition of the strongly hydrogen-bonding solvent CF_3CH_2OH to each of the chlorinated solvents caused further increases in J. X-ray crystal structures have shown that $CHCl_3$ and p-nitrophenol form hydrogen-bonded adducts with Cu(salen) [42,43]. The hydrogen bonding is to one of the coordinated oxygens of the ligand. Hydrogen bond formation decreases the "buckling" of the ethylene bridge and moves the bridge away from a *gauche* conformation [42,43]. The effects of solvent on the values of J for **XXXVI** suggest that hydrogen bonding may also be important in fluid solution. Only one component was observed in fluid solution; so, it is likely that the rate of exchange between free and hydrogen-bonded forms is fast on the EPR timescale. If the solution structures are similar to the structures in the crystals, the increase in J as a result of hydrogen bonding is due to a decrease in the buckling of the ethylene bridge. The values of J for **XXXVI** were between 0.0004 and 0.0020 cm^{-1} and increased as the temperature was decreased.

For **XXXVII** (X = H, 4-OMe, 3-OEt, 5-Cl) the values of J were between 0.0014 and 0.0075 cm^{-1} [41]. The value of J decreased as the hydrogen bonding ability of the solvent increased and decreased as the temperature was decreased. Both of these trends are the opposite of those observed for **XXXVI**. One possibility is that the values of J have opposite signs for the ethylene- and propylene-bridged complexes. Alternatively, the conformational changes may have opposite effects for the two series of complexes. The larger values of J for **XXXVII** than for **XXXVI** indicate significant differences in the conformations for the Cu-N-CH-amide linkages in the two series of complexes. Coordination of pyridine or THF to either **XXXVI** or **XXXVII** caused a large decrease in J, which again emphasizes the sensitivity of J to conformational changes.

The positions of the outer lines of an AB pattern are more sensitive to changes in the value of J than are the positions of the inner lines. The inability to detect the outer lines at either X-band or Q-band for a spin-labeled copper porphyrin with J ~ 0.050 to 0.060 cm^{-1} indicated a range of J values that was attributed to the presence, in solution, of a distribution of conformations that were interconverting slowly on the EPR timescale [27].

3. INTERACTION IN A RIGID LATTICE

In frozen solution or single crystal spectra of metal-nitroxyl complexes, both the exchange and dipolar interactions contribute to splittings in the EPR spectra. The two contributions can be separated by analysis of the orientation dependence of the splitting [1,3,44]. The magnitude of the dipolar interaction is dependent on the interspin distance, r. The exchange contribution, J, is dependent on molecular conformation as discussed for the fluid solution data. In fluid solution only the magnitude of J is measured. In single crystal studies, since the signs of the components of the dipolar interaction tensor are known, the sign of J impacts the orientation dependence of the spin-spin interaction. In frozen solution spectra, if the magnitudes of J and the dipolar interaction are comparable, the sign of J can be determined. In the studies discussed below, the exchange Hamiltonian was $-J\ S_1 \cdot S_2$. In this sign convention a negative sign of J indicates antiferromagnetic coupling, with the singlet energy level lower than the triplet. The separation between the singlet and triplet levels is J.

3.1 **Metal-nitroxyl interaction in frozen solution**

In frozen solution all possible orientations of the molecule with respect to the direction of the magnetic field are present. The EPR spectra are the superposition of the splittings arising from all of the orientations. If there are anisotropic g and/or A values for the interacting spins and if the frozen solution spectra are well resolved, it is possible to define the orientation of the interspin vector relative to the axes of g and/or A values and thereby obtain conformational information [1,3,44].

The intensity of the half-field transition is determined by the anisotropic contributions to the spin-spin interaction. For systems in which J is small enough that the anisotropic exchange contribution is negligible, the intensity of the half-field transitions, compared with the intensity of the allowed transitions, can be used to determine the interspin distance [1,45,46]. The lineshape of the half-field transition can be analyzed to determine the relative orientations of the interspin vector and hyperfine axes.

3.1.1 *Geometry of Transition Metal Complexes.* In the spin-labeled copper complexes **XXXVIII** [47], **XXXIX** [48], and **XL**$(py)_2$ [49], the copper-nitroxyl spin-spin exchange interaction is much larger than the differences between the energy levels of the copper and the nitroxyl unpaired electrons.

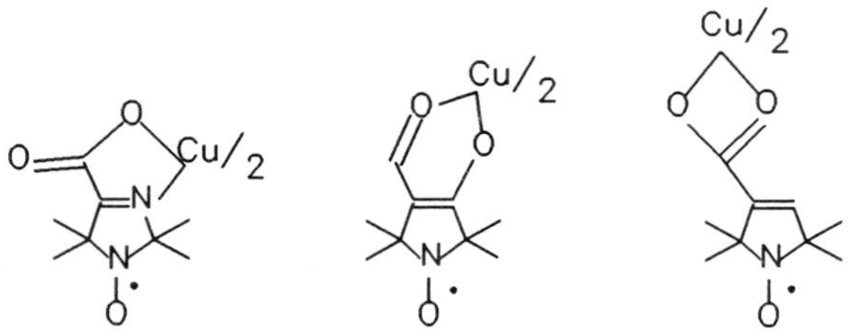

XXXVIII XXXIX XL

The g- and A-values for the complexes are the averages of those observed for the copper and nitroxyl in the absence of interaction. The turning points in the frozen solution spectra of these complexes were analyzed to show that, in each case, the nitroxyls were in an approximately planar array around the copper with copper-nitroxyl distances of 3.9 Å, 6.0 Å, and 7.3 Å for **XXXVIII** [47], **XXXIX** [48], and **XL**$(py)_2$ [49], respectively.

The EPR spectra of the allowed and half-field transitions for **XLI** and **XLII** are superpositions of spectra for several conformations with different values of r and J [46]. The dominant species had r = *ca.* 5 and 7 Å. The relative intensity of the half-field transition corresponded to a copper-nitroxyl distance of 5.7 Å. Since the relative intensity of the half-field transition depends on r^{-6}, the average is weighted in favor of molecules with smaller values of r.

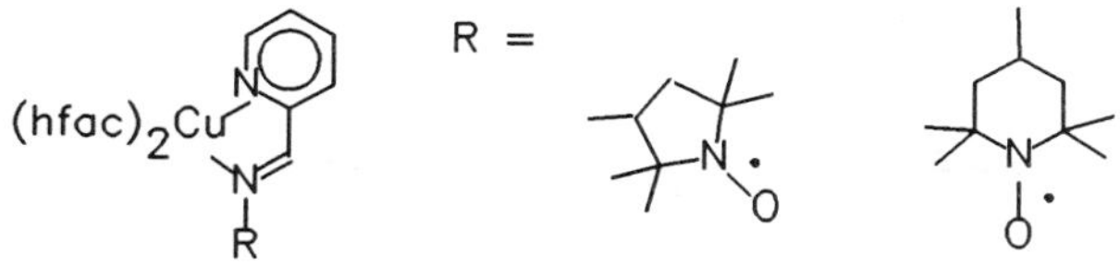

R =

XLI XLII

The crystal structure of a closely related compound reported copper to nitroxyl-oxygen distances of 5.99 to 6.0 Å [50].

Analysis of the relative intensity of the half-field transition for **XLIII** gave a copper-nitroxyl distance of 9.8 Å. Simulation of the lineshape of the half-field transition indicated that the angle between the z-axis of the Cu(II) hyperfine tensor and the interspin vector was about 45° [46]. The nitroxyl hyperfine splitting was not resolved in the allowed transitions so it was not possible to determine the orientation of the nitroxyl hyperfine axes relative to the copper axes.

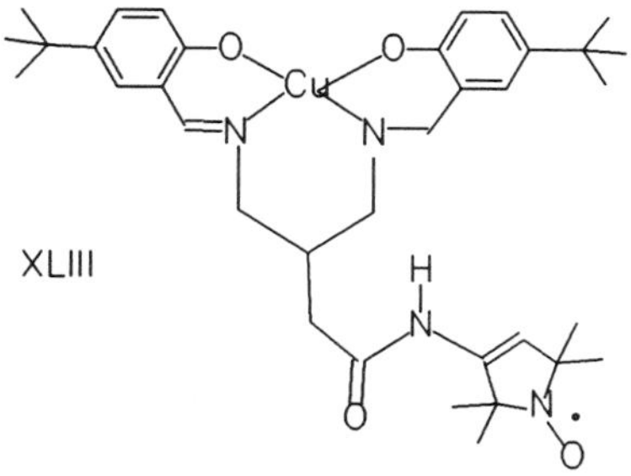

XLIII

The EPR spectrum of the Cu(II) complex of spin-labeled EDTA derivative **XLIV**, in a frozen solution of 1:1 water:propylene glycol, [51] had g values similar to those reported for Cu(H_2EDTA)(H_2O) [52]. A single crystal X-ray structure of Cu(H_2EDTA)(H_2O) showed that the coordinated water was in an equatorial position, one protonated carboxylate group was weakly coordinated in an axial position, and the second protonated carboxylate group was attached to the nitrogen that was bound weakly in the second axial position [53].

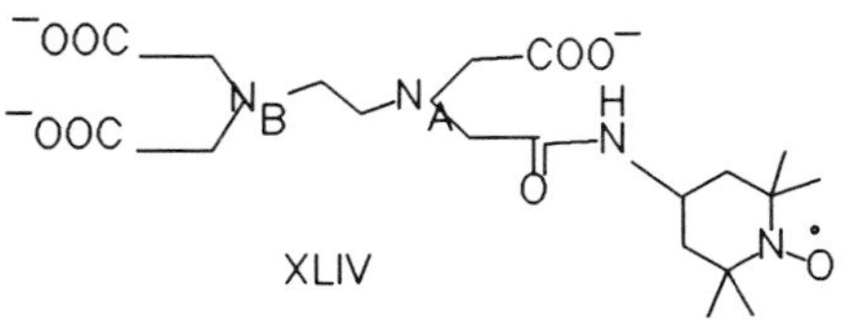

XLIV

By analogy, it was proposed that for CuH**XLIV**(H_2O), N_A is in an axial position, the water molecule is in an equatorial position and the protonated carboxylate is in the second axial position. The exchange contribution to the interaction between the copper and nitroxyl in CuH**XLIV**(H_2O) would be *via* N_A. Since the unpaired electron spin density is primarily in the xy plane, coordination of N_A in an axial position is expected to result in a very weak exchange interaction. Stronger interaction is expected for N_A bound in an equatorial position. The nitroxyl signal in the frozen solution spectrum of CuH**XLIV**(H_2O) was simulated with a dipolar interaction at r = 11 Å and J = 0. The negligible exchange interaction is consistent with the proposed structure in which N_A is axially coordinated. The relative intensity of the half-field transition was strongly concentration dependent, which indicated that intermolecular contributions were larger than intramolecular contributions and that the interspin distance was relatively long, which is consistent with the analysis of the lineshape of the nitroxyl signal. In contrast to the behavior of the Cu(II) complex, analysis of the frozen solution EPR spectrum of the vanadyl complex of **XLIV** gave J = -0.014 cm^{-1} [51]. It is proposed that N_A is coordinated in an equatorial position and that there is overlap of the vanadyl d_{xy} orbital containing the unpaired electron and orbitals on N_A, which provides a pathway for electron-electron interaction.

The nitroxyl region of the fluid solution EPR spectra of **XLV** and **XLVI** were three-line signals with linewidths that indicated J < 0.0001 cm^{-1} and < 0.0002 cm^{-1}, respectively [54]. In frozen solution the nitroxyl signal was split into a well-resolved multi-line pattern due to interaction between the Mn(II) and nitroxyl spins [55].

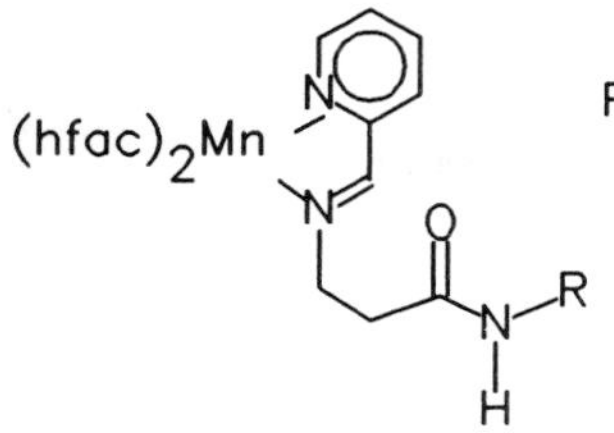

(hfac)$_2$Mn R = XLV 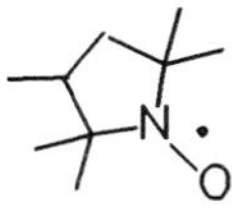 XLVI

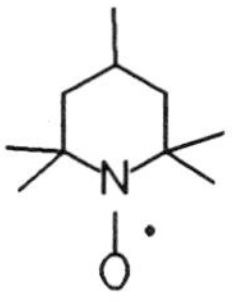

Since the values of J in fluid solution were small and there was no evidence of temperature-dependence of the value of J, simulations of the frozen solution spectra were performed assuming only a dipolar interaction between the two centers. The simulations indicated Mn(II)-nitroxyl distances of 9.7 Å and 9.0 Å for **XLV** and **XLVI**, respectively. The simulations were not improved by adding a small Mn(II)-nitroxyl exchange interaction. The shorter distance for **XLVI** than for **XLV** is consistent with the smaller dimensions of the 5-membered nitroxyl ring than of the 6-membered nitroxyl ring.

3.1.2 *Ligand Conformations*. Four-coordinate Co(II) tetra(*p*-trifluoromethylphenyl)porphyrin binds one mole of pyridine to form a five-coordinate complex [56]. Resolved electron spin-spin splitting was observed in the frozen solution EPR spectra for a series of spin-labeled pyridines. A single conformation was observed for several ligands including **VII**. Two conformations with resolvable differences in J were observed for the Co(II) porphyrin complexes with **XLVII** - **XLIX** [56].

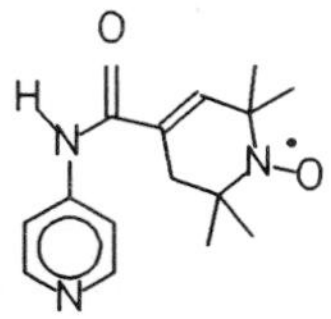

XLVII 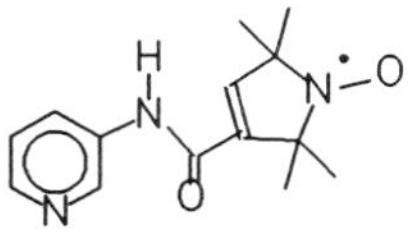 XLVIII 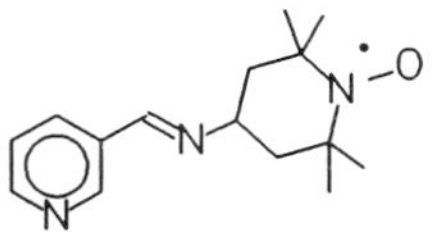 XLIX

3.1.3 *Weak orbital overlaps*. In the silver(II) complex of **XXXIV** the value of J in fluid solution was about 0.10 cm^{-1} in non-coordinating solvents and about 0.0010 cm^{-1} in the coordinating solvent pyridine [40,57]. Analysis of frozen solution spectra indicated that the decrease in J upon pyridine coordination was accompanied by about a 2 Å increase in the silver-nitroxyl distance [57]. The interspin distance was about 9 Å in the presence of coordinating solvent and 7 Å in the absence of coordinating solvent. This result confirms the proposal based on the changes in J in fluid solution, that coordination of an axial ligand caused a change in the conformation of the substituent at the ortho position of the phenyl ring.

In frozen solution, spectra of low-spin Fe**XXXIV**(1-Me-imidazole)$_2^+$, Fe**XXXIV**(imidazole)$_2^+$, and related complexes in which a nitroxyl is attached to one ortho-position of a tetraphenylporphyrin, two conformations were observed [58]. For Fe**XXXIV**(1-Me-imidazole)$_2^+$ and Fe**XXXIV**(imidazole)$_2^+$ the values of J for the strongly interacting conformations were ± 0.22 and ± 0.28 cm^{-1}, respectively, which are the largest values observed to date for metal-nitroxyl interaction in an ortho-substituted metalloporphyrin. The populations of the conformations with the larger values of J were greater in toluene solution than in $CHCl_3$ and were greater for the imidazole complex than for the 1-Me-imidazole complex. The presence of strongly and weakly interacting conformations for the low-spin Fe(III) complexes is similar to the behavior of the Cu(II) and Ag(II) complexes of the same spin-labeled porphyrin [39,40,57]. It is proposed that interaction between the carbonyl oxygen of the amide linkage and electron spin density in the porphyrin pyrrole ring provides a weak overlap pathway for electron spin-spin interaction. The magnitude of this interaction is strongly dependent on conformation.

At 6 to 12 K in non-coordinating solvents, multiple conformations were observed in the EPR spectra of Fe**XXXIV**Y (Y = F^-, Cl^-, Br^-) and in the spectra of high spin Fe(III) coordinated to other porphyrins in which the spin label was attached to an ortho position of one phenyl ring [59]. Conversion to the six-coordinate bis-DMSO complex resulted in large reductions in the magnitude of the spin-spin interaction, similar to the effect of axial ligand coordination on Ag(II)**XXXIV**.

At 6 K, several conformations were clearly resolved in the spectra of Fe**XXXV**Y (Y = F^-, Cl^-, Br^-) [60]. The electron-electron interaction in one of the conformations resulted in a splitting of both the nitroxyl signal at g = 2 and the Fe(III) signal at g = 6 due to r = 8 Å and J = $\pm 0.04 \pm 0.005$ cm^{-1}. Spin-spin interaction was stronger in a second conformation. The conformations responsible for these different strengths of spin-spin interaction may be similar to those observed for the Cu(II) and low spin Fe(III) complexes of ortho spin-labeled tetraphenylporphyrins. As the temperature was increased, the resolved spin-spin splitting in the spectra of Fe**XXXV**Y collapsed due to increasing relaxation rates of the Fe(III) unpaired electrons. These spectra illustrate the point, made in the introduction to this Chapter, that resolved spin-spin splitting can only be observed if the metal electron spin relaxation rate is long relative to the spin-spin splitting. Due to the increase of the Fe(III) electron spin relaxation rate with temperature, resolved spin-spin splittings were not observed for spin-labeled Fe(III) porphyrins in fluid solution [75].

3.2 **Metal-nitroxyl interactions in a single crystal**

When a metal-nitroxyl complex is doped into a single crystal of a diamagnetic host, rotation of the crystal about three orthogonal axes permits analysis of the full orientation dependence of the spin-spin interaction. Although this procedure is tedious, it provides a detailed picture of the relative orientations of the interspin vector and the axes of the g- and A-matrices for the two paramagnetic centers.

Copper(II) and silver(II) complexes of spin-labeled porphyrin **XVII** and copper(II), silver(II), and vanadyl complexes of spin-labeled porphyrin **XIX** were doped into $ZnTPP(THF)_2$ by slow evaporation of a THF solution containing a 99:1 ratio of ZnTPP to spin-labeled metalloporphyrin [61-63]. The axis definitions for the analysis of the data and definition of the relative orientations of the interspin vector and the axes for the g- and A-values are shown in Figure 1.

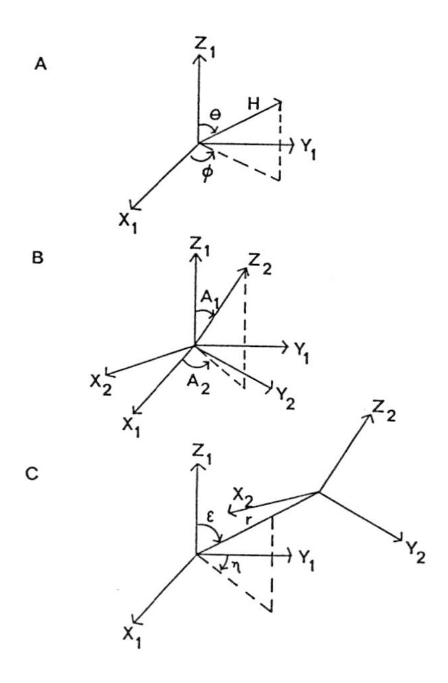

Figure 1. *Definitions of angles that relate the orientation of the magnetic field, the nitroxyl hyperfine tensor, and the interspin vector to the axes of the metal hyperfine tensor (electron 1). A) The angles* Θ *and* ϕ *define the orientation of the magnetic field. B) The angles* A_1 *and* A_2 *define the orientation of the z axis of electron 2 (nitroxyl). C) The angles* ϵ *and* η *define the orientation of the interspin vector, r. Throughout the text, r is used to denote the magnitude of the interspin vector.*

An X-ray crystal structure of the host lattice confirmed the coordination of two moles of THF and an analysis of the "free space" in the lattice showed that to accommodate a nitroxyl-substituted metalloporphyrin, THF must be lost from a neighboring unsubstituted $ZnTPP(THF)_2$ moiety [64]. Due to the presence of eight equivalent pyrrole hydrogens (sites for attachment of the substituent in the spin-labeled porphyrins) and the possibility that the substituent was located above or below the porphyrin plane, there were 16 possible locations of the substituent. Sites related by inversion symmetry are indistinguishable by single crystal EPR (x,y,z is equivalent to -x,-y,-z). For example, ϵ = 120°, η = -50° is indistinguishable from ϵ = 60°, η = 130°. For each of the spin-labeled complexes four locations of the nitroxyl side-chain were identified. Values of ϵ were arbitrarily constrained to fall between 0° and 180° and values of η were constrained to fall between -90° and +90°. Sites with different values of η correspond to location of the spin-labeled substituent at different pyrrole carbons. Since the g and A-values for CuTPP, VOTPP, AgTPP are axial, they do not provide a unique definition of the x and y axes, so the values of η and A_2 are relative to faces of the $ZnTPP(THF)_2$ crystal. Values of ϵ less than 90° correspond to locations of the substituent above the porphyrin plane and values of ϵ greater than 90° correspond to the locations of the substituent below the porphyrin plane. The experimental results are summarized in Table 2. Packing of the spin-labeled porphyrin into the host lattice resulted in the presence of distinct conformations with different values of J and r for the different sites.

TABLE 2

Orientation and Interaction Parameters for Spin-labeled Metal Complexes Studied in a Single Crystal of $ZnTPP(THF)_2$[a]

Complex	Site	A_1	A_2	ϵ	η	r[b]	J[b]	pop
CuXVII	1	173	85	77	-75	13.7	-21	26
	2	175	85	82	-27	15.0	-12	35
	3	173	85	68	10	13.0	1	4
	4	55	-35	85	45	15.5	-30	35
AgXVII	1	172	84	120	-50	11.5	-33	16
	2	70	28	102	-10	13.8	-26	34
	3	172	84	115	-10	11.8	-35	34
	4	172	84	107	39	13.0	-18	16
CuXIX	1	127	117	53	-55	10.8	7	22
	2	127	117	63	-16	12.5	0	44
	3	127	117	45	42	10.2	10	17
	4	127	117	155	80	9.2	-4	17
AgXIX	1	55	50	46	-70	10.3	-6	20
	2	41	50	63	-25	12.0	5	20
	3	41	50	120	25	11.5	8	40
	4	41	50	34	60	9.5	-12	20

Complex	Site	A_1	A_2	ϵ	η	r^b	J^b	pop
VOXIX	1	158	92	82	-80	13.6	0	29
	2	158	92	54	-50	10.0	4	14
	3	158	92	66	-9	11.0	-3	7
	4	158	92	49	28	9.8	-2	7
XXXII	1	30	60	117	-50	11.5	-42	14
	2	15	60	78	-10	13.5	-40	14
	3	90	25	105	30	13.0	9	14
	4	25	120	62	35	11.5	-24	27
	5	20	70	100	84	13.5	-20	31
XXXIII	1	63	-1	145	35	9.3	7	8
	2	68	-2	126	76	10.2	-2	46
	3	23	-18	144	77	9.4	35	23
	4	32	-27	146	88	9.3	22	23
L	1	10	45	83	-45	14	-68	14
	2	10	45	83	-40	14	-81	14
	3	152	-20	128	-40	10.5	37	14
	4	60	130	115	20	11	0	30
	5	150	-20	130	50	10	22	14
	6	160	45	55	50	10.5	-48	14

[a] Data taken from references 61-63, 65, 66; values of r are in A and values of J are in units of 10^{-4} cm^{-1}. $1x10^{-4}$ cm^{-1} is approximately 1 G in these systems. Pop is the % population based on the total concentration of paramagnetic metal. The total population of spin-labeled metal complex for some of the compounds is less than 100% due to due loss of nitroxyl during the slow crystal-growth process.

In attempts to simplify the distance-dependence of the exchange interaction it has been stated that J decays exponentially with the distance between the two paramagnetic centers [67,68]. Although this general trend may have some utility in describing through-space orbital overlap, it is not useful in describing the conformational effects on interaction through multi-atom bonding pathways. Inspection of the data in Table 2 clearly shows that there was no correlation between the values of J and r. The important variable is orbital interactions affected by the conformation of the intervening bonds rather than the through-space distance between the two centers. The values of J and orientations of the nitroxyl axes were different from those observed in frozen solution, which indicates the strong influence of the host lattice on the molecular geometry [61].

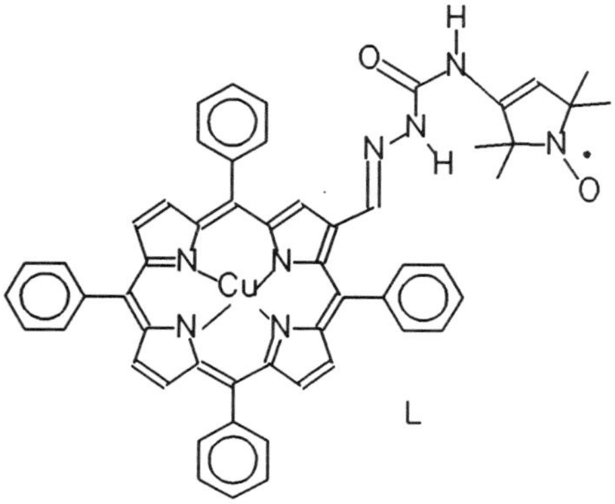

For the complexes of **XVII** and **XIX**, larger values of r correlate with increasing ϵ values, but the values for VO**XIX** fall on a different curve than the values for the Cu(II) or Ag(II) complexes of **XVII** or **XIX** (Figure 2) [61-63]. In these comparisons, the parameter of interest was the orientation of the substituent with respect to the porphyrin plane, independent of whether it was above or below the plane, so values of ϵ greater than 90° were set equal to 180-ϵ. For each of the complexes the values of A_1 and A_2 were approximately constant which indicated that the orientation of the nitroxyl ring relative to the metal hyperfine axes was similar for the multiple sites. The correlation between the values of r and ϵ showed that as the substituent was forced closer to the porphyrin plane, the porphyrin-nitroxyl linkage exhibited a more extended conformation. In the crystal structure of vanadyl octaethylporphyrin, the vanadium atom is 0.54 Å above the mean porphyrin plane [69]. If the conformations of the spin-labeled side chains are about the same for the copper, silver, and vanadyl complexes and the vanadyl ion is above the porphyrin plane, location of the substituent on the same side of the plane as the vanadium would give values of r that are shorter than for the Cu(II) or Ag(II) complexes. Locations on the opposite side of the plane would result in larger values of r for the vanadyl complex than for the Cu(II) or Ag(II) complexes. The consistently smaller values of r for the vanadyl complex than for the Ag(II) or Cu(II) complexes suggests that the substituent preferentially locates on the same side of the porphyrin plane as the vanadyl oxygen. The spatial requirements of the vanadyl oxygen may create a larger pocket on that side of the porphyrin plane that facilitates accommodation of the bulky nitroxyl side-chain.

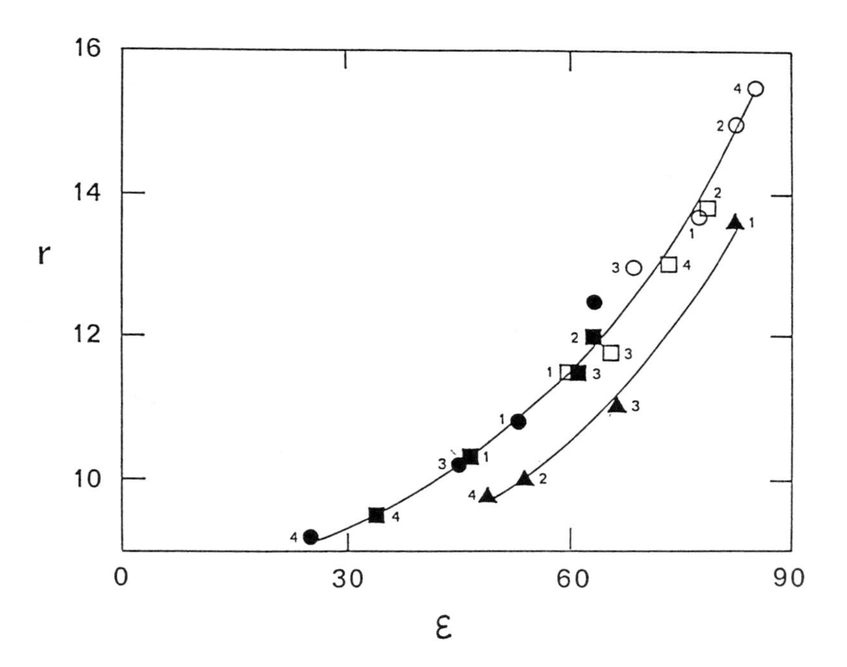

Figure 2. *Plots of the values of r and ε observed for spin-labeled complexes CuXVII (○), CuXIX (●), AgXVII (□), AgXIX (■), and VOXIX (▲) doped into ZnTPP(THF)$_2$. Each data point is labeled with the number of the site to which it refers. (Figure reproduced by permission of the American Chemical Society).*

For the major sites observed for Cu**XVII** the values of J ranged from -12 to -30 x 10^{-4} cm^{-1}. A minor component was observed with a J of 1 x 10^{-4}, but its small population (4%) suggests that it represents an unfavorable conformation of the molecule. For Ag**XVII** the values of J ranged from -18 to -35 x 10^{-4} cm^{-1}. The olefinic linkage in **XVII** is trans. For this linkage the value of J was consistently negative. For the complexes of **XIX**, which has a cis olefinic linkage, the values of J were smaller than for the trans linkage and the sign of J was either positive or negative. The larger values of J for the trans isomer than for the cis isomer is consistent with results obtained on these complexes in fluid solution [27,28,40]. In proton NMR it has been found that the three-bond coupling constants for olefinic protons are consistently greater for the trans isomer than for the cis isomer, with the sign of J positive for the trans isomer and either positive or negative for the cis isomer [70,71]. In comparing the signs of J from NMR and from these EPR studies, it should be noted that the exchange term in NMR is routinely given as J $S_1 \cdot S_2$, but the EPR studies used -J $S_1 \cdot S_2$ so the sign conventions are opposite. For a variety of other nuclei it has also been observed that 3- and 4-bond spin-spin couplings across olefinic linkages are greater for the trans isomer than for the cis isomer. Since the EPR couplings are through 10 bonds, there are undoubtedly many factors in addition to the conformation of the olefin that influence the magnitude of J. However, it appears that the effect of the olefinic linkage on the magnitude of the EPR coupling constants parallels the effect of the same linkage of NMR coupling constants. It is anticipated that further studies of EPR electron-electron coupling constants will become as important a stereochemical tool in EPR as nuclear coupling constants are in NMR.

The data obtained for spin-labeled copper complexes **XXXII** and **XXXIII**, doped into a single crystal of $ZnTPP(THF)_2$, [65] is given in Table 2. **XXXII** and **XXXIII** have a saturated 5-membered ring nitroxyl (pyrrolidine) in place of the 6-membered ring (piperidine) in Cu**XVII** and Cu**XIX**, respectively. The correlation between the values of r and ϵ for **XXXII** and **XXXIII** is similar to that observed for the copper and silver complexes of **XVII** and **XIX** [63,65]. The values of J for **XXXII**, which has the trans olefinic linkage, ranged from -42 to $+9 \times 10^{-4}$ cm^{-1} while the values for the cis isomer ranged from -2 to $+35 \times 10^{-4}$ cm^{-1}. The ranges of values for both isomers are greater than was observed for the closely related complexes with 6-membered nitroxyl rings, although the values are of the same order of magnitude. There is also a much greater variety of values of A_1 and A_2, the angles that define the orientation of the nitroxyl hyperfine tensor with respect to the copper hyperfine axes, for the 5-membered nitroxyl ring than for the 6-membered ring. Since the acrylamide linkage that attaches the nitroxyl ring to the porphyrin was the same for **XVII**, **XXXII** and for **XIX**, **XXXIII**, the differences may be due to the nitroxyl ring rather than the intervening linkage. The range of conformations accessible to the pyrrolidine ring may have a greater impact on the magnitude of the exchange interaction than do the range of conformations accessible to the piperidine ring. X-ray crystal structures of five-membered nitroxyl rings have shown that the pyrrolidine ring adopts a half-chair conformation with considerable variation in the bond angles [72,73]. These bond-angle variations may have a large impact on J. The largest values of J observed for the pyrrolidine nitroxyl were greater than for the piperidine nitroxyl, but for other conformations the inequality is reversed.

The values of J for **L**, in fluid solution at room temperature, were strongly solvent dependent ranging from 0.0013 cm^{-1} in chloroform to 0.0023 cm^{-1} in toluene and 0.0035 cm^{-1} in pyridine [33]. The diversity of J values indicated substantial conformational flexibility. The lines in the EPR spectra remained sharp at temperatures as low as -60°C, which indicated that any dynamic processes occurring in solution, that interconverted conformations with different values of J, had activation energies less than about 8 kcal/mole. **L** was doped into $ZnTPP(THF)_2$ to permit a more detailed examination of the conformations and the parameters obtained for it are included in Table 2. Although the urea moiety is likely to be all trans, the imine could be either cis or trans [66]. On a plot of ϵ vs. r, the values of **L** fall close to the line obtained for the Cu(II) and

Ag(II) complexes of **XVII** and **XIX** [66]. Some offset is expected because in **L** there is an additional atom in the metal-nitroxyl linkage compared to the linkages in **XVII** and **XIX**. The similarity in the correlation between ϵ and r values indicates that for all the substituents, the values of both ϵ and r increase as the sidechain is forced closer to the porphyrin plane by the steric constraints of the host lattice. The average value of r for the trans isomers of **MXVII** were longer than for the cis isomers of **MXIX**, although there is some overlap of the two sets (Table 2). The additional atom in the linkage between the porphyrin and the nitroxyl for **L** than for **MXVII** and **MXIX** is expected to cause some increase in r. For **L**, the values of r and the positions on the plot of r *vs.* ϵ indicate a trans conformation of the imine linkage at sites 1 and 2 and a cis conformation at sites 3-6. The values of J for sites 1 and 2 of **L** were large and negative, which is consistent with the conclusion that the conformations at these two sites are trans isomers. The values of J for sites 3-5 are smaller than for sites 1 and 2, which is consistent with the assignment as cis isomers. The large negative value of J for site 6 suggests a trans conformation in spite of the relatively short r. Inspection of molecular models indicated that the values of r for sites 1 and 2 (14 Å) were consistent only with an all-trans conformation of the imine-urea linkage, but that the values of r for sites 3-5 were consistent with either an all-trans conformation or a conformation containing a single cis linkage and the other linkages trans. It was proposed that both cis and trans isomers of the imine are present in the crystal. The magnitude of the values of J observed in solution are more similar to the values observed in the crystal for the cis imine. Since the barrier to interconversion of the cis and trans isomers is likely to be greater than 8 kcal/mole, it is likely that only one isomer is present in fluid solution [33]. The solvent dependence may reflect small changes in torsional angles. Interactions in the lattice may stabilize the trans isomer.

3.3 Comparison of conformations present in fluid solution and rigid matrices

One question that frequently arises is whether molecular conformations remain constant as the phase is changed from solution to solid state. Since the electron-electron coupling constants are sensitive to molecular geometry, comparison of values obtained in different phases provides insight on this question.

The values of J have been determined for the Cu(II) and Ag(II) complexes of **XVII** and the Cu(II), Ag(II), and vanadyl complexes of **XIX** in fluid solution, frozen solution, and doped into a single crystal of $ZnTPP(THF)_2$. The values are summarized in Table 3.

TABLE 3

Values of J[a] for spin-labeled metalloporphyrins in fluid solution, frozen solution, and a doped single crystal

Complex	J fluid soln.[b]	J frozen soln.[c]	J single crystl.[d]
Cu**XVII**	18 to 25[e]	-40[f]	-30 to -12[g]
Ag**XVII**	33 to 44[h]	-55[i]	-35 to -18[j]
Cu**XIX**	0 to 4[e]	<5[k]	-4 to +10[g]
Ag**XIX**	6 to 10[h]	~10[i]	-6 to +12[j]
VO**XIX**	<1[h]	<5[k]	-3 to +4[l]
XXXII	3 to 5[e]	m	-30 to +1[n]
XXXIII	10 to 550[e]	m	-2 to +35[n]

[a]In units of 10^{-4} cm^{-1}. If a sign is not given with the value, the sign is not known; [b]at room temperature in a variety of solvents; [c]toluene-THF glass at ca. -180°; [d]doped into $ZnTPP(THF)_2$; [e]reference 28; [f]reference 44; [g]reference 61; [h]reference 40; [i]reference 57; [j]reference 63; [k]spin-spin splitting is less than the line width of the nitroxyl lines; [l]reference 62; [m]spectra are poorly resolved; [n]reference 65.

For most of the complexes a range of values of J was observed in fluid solution and in the single crystals. Multiple contributions may also be present in the frozen solution, but overlapping spectra in frozen solution are more difficult to analyze because components with sharper lines dominate the spectra [44]. In all three phases, the magnitude of the exchange interaction increases in the order $V^{IV}O$ < Cu(II) < Ag(II), which agrees well with other measures of spin delocalization [18]. As noted in Section 3.1.2, the ZnTPP$(THF)_2$ host lattice appears to enforce a range of geometries for the porphyrin to nitroxyl linkage, which resulted in a variety of values of J and r for the same spin-labeled complex. As a result of the interaction with the host, there are substantial differences between the conformations of the molecules in solution and in the doped single crystal, which are reflected in differences in the values of J. The agreement between the fluid solution and frozen solution data is improved if a single solvent is examined and the temperature dependence of the value of J is considered [57]. The values of J in frozen solution for Cu**XVII** and Ag**XVII** agree closely with the values of J in fluid solution extrapolated to the freezing point of the solvent. The agreement between the values of J obtained in fluid solution and in the ZnTPP crystal for **XXXII** and **XXXIII** that contain the pyrrolidine nitroxyl ring is not as good as for the complexes with piperidine rings. Apparently, the single-crystal host preferentially stabilizes conformations of the pyrrolidine ring that are different from the preferred conformations in solution. In fluid solution the value of J for cis isomer **XXXIII** was much smaller in coordinating solvents than in non-coordinating solvents [28]. Since the single crystal was grown from THF solution, the zinc atom in ZnTPP$(THF)_2$ is coordinated to THF. The values of J observed in the crystal are more similar to the values observed in coordinating solvents, which suggests that the conformations of **XXXIII** observed in the crystal are more similar to the ones observed in coordinating solvents than in non-coordinating solvents.

The influence of the temperature at which the sample is immobilized on the values of J was examined by adsorbing Ag**XVII** or Ag**XIX** on imbiber beads at room temperature [57]. The values of J on imbiber beads were in good agreement with the values obtained in the CH_2Cl_2 solution from which the compound had been adsorbed. When the immobilized sample of Ag**XVII** was cooled, the value of J increased and at -180°C the value of J on imbiber beads was similar to that observed in frozen solution. For these relatively rigid molecules the change in temperature from *ca.* 20° to -180° had a greater impact on the value of J than the impact of the immobilization process for these complexes.

Complexes of spin-labeled pyridines with $Cu(hfac)_2$ and $VO(tfac)_2$ were immobilized on imbiber beads [74]. A comparison of the values of J in fluid solution and on imbiber beads is given in Table 4 for the Cu(II) complexes and in Table 5 for the vanadyl complexes. For most of the complexes, the values of J, including the temperature dependence of J, were similar in fluid solution and on imbiber beads. The similarity of J values indicates that the conformations of the linkages between the metal and the nitroxyl were largely unperturbed by immobilization.

TABLE 4

Values of J^a for $Cu(hfac)_2L$ in fluid solution and on imbiber beads

	fluid solution[b]		imbiber beads	
L			20°	-180°
VII	43 (20°)	60 (-60°)	-40	-62
XLIX	36,90 (20°)[c]	36, 121 (-60°)[c]	90[d]	110[d]
XLVIII	77 (20°)	170 (-25°)	120	85, 260[c]
XLVII	155 (20°)	175 (-25°)	150	170
LI	475 (20°)	310, 485 (-70°)[c]	450	400
VIII	595 (20°)	455 (-70°)	500[e]	500[e]

[a]*In units of* 10^{-4} cm^{-1} *for data taken from reference 74. If the sign of J is not given, the sign is not known;* [b]*the temperature in °C at which the data were obtained is given in parentheses;* [c]*two components were observed in the spectra;* [d]*a second component was observed with weaker spin-spin interaction, but overlap with the spectrum of spin-labeled ligand that was not coordinated prevented analysis of the parameters for the second component;* [e]*the outer lines of the AB patterns were not observed, so there is substantially greater uncertainty in this values than for values which were obtained from spectra in which the full AB pattern was observed.*

TABLE 5

Values of J[a] for $VO(tfac)_2$ in fluid solution and on imbiber beads

	fluid solution[b]		imbiber beads	
L			20°	-180°
VII	106, 125 (20°)[c]	120, 139 (-60°)[c]	100	115
XLIX	32 (20°)	37 (-40°)	-33	-40
VIII	1670 (20°)		2000[d]	2000[d]
LI	400 (20°)		380	380
XLVII	347, 412 (20°)[c]		400[d]	400[d]
XLVIII	83 (20°)	79 (-40°)	90	90

[a] *In units of 10^{-4} cm^{-1} for data taken from reference 74. The values of r for most of these complexes were too large, compared with the values of J, to permit determination of the sign of J;* [b]*the temperatures in °C at which data were obtained are given in parentheses;* [c]*two components were observed in the spectra;* [d]*the outer lines of the AB patterns were not observed so there is substantially greater uncertainty in this value than for values which were obtained from spectra in which the full AB pattern was observed.*

LI

Analysis of the frozen solution EPR spectra for a series of spin-labeled copper complexes gave values of J that were similar to those obtained in fluid solution, which indicated that the molecular conformations were largely unchanged by freezing the solution [44]. The interspin distances were between 8 Å and 13.5 Å.

4. SUMMARY AND PROGNOSIS

This contribution largely emphasizes results from our laboratory, since the metal-nitroxyl complexes we have studied have provided the largest set of resolved spin-spin interactions from which stereochemical information can be extracted. The approach revealed in this work is general. So long as relaxation times and other contributions to line widths allow resolution of the spin-spin interaction effects, one could study radical-radical, metal-radical, or metal-metal interactions. Eventually, a large enough body of information on electron-electron spin-spin interactions will accumulate to permit the facile application of this coupling information in the way that proton-proton couplings are used in applications of proton NMR to stereochemical problems. The use of relaxation times in systems in which the spin-spin interaction does not result in resolved splittings is at a much more primitive level of development.

The results described here have significance beyond the direct application of EPR spin-spin interactions to stereochemical problems. There is an expansive field of long-range interactions in science, of which long-range coupling in EPR is only a small part. The fundamental information obtained from electron-electron spin-spin interaction will inform fields as seemingly disparate as scanning tunneling microscopy and photosynthesis. A few references are provided here as an entry into possible applications in scanning tunneling microscopy [76,77], photosynthesis [78,79], electron transfer in proteins [80,81], interactions in non-conjugated π-systems [82], and electron transport *via* saturated hydrocarbon bridges [83], and other organic moieties [84,85].

5. ACKNOWLEDGMENT

The financial support of this work by the National Institutes of Health (Grant GM21156) is gratefully acknowledged. The names of our co-workers who have contributed to these studies are given on the papers cited.

6. REFERENCES

1. G.R. Eaton and S.S. Eaton, *Acc. Chem. Res. 21* (1988) 107-113.
2. G.R. Eaton and S.S. Eaton, *Biol. Magn. Reson. 8* (1989) 339-397.
3. S.S. Eaton and G.R. Eaton, *Coord. Chem. Rev. 83* (1988) 29-72.

4. A. Caneschi, D. Gatteschi, R. Sessoli, and P. Rey, *Acc. Chem. Res. 22* (1989) 392-398.
5. *Imidazoline Nitroxides*, L.B. Volodarsky, ed., vol I and II, CRC Press, Inc., Boca Raton, FL, 1988.
6. E. Breuer, H.G. Aurich, and A. Nielsen, *Nitrones, nitronates, and nitroxides*, John Wiley, Chichester, 1989.
7. S.S. Eaton, D.L. DuBois, and G.R. Eaton, *J. Magn. Reson. 32* (1978) 251-263.
8. D.L. DuBois, G.R. Eaton, and S.S. Eaton, *J. Am. Chem. Soc. 101* (1979) 2624-2627.
9. D.L. DuBois, G. R. Eaton, and S.S. Eaton, *Inorg. Chem. 18,* (1979) 75-79.
10. S.S. Eaton, K.M. More, D.L. DuBois, P.M. Boymel, and G.R. Eaton, *J. Magn. Reson. 41* (1980) 150-157.
11. P.N. Howells, J.W. Kenney, J.H. Nelson, and R.H. Henry, *Inorg. Chem. 15* (1976) 124-129.
12. R.P. Eckberg, J.H. Nelson, J.W. Kenney, and P. N. Howells, *Inorg. Chem. 16* (1977) 3128-3132.
13. B.M. Sawant, A.L.W. Shroyer, G.R. Eaton, and S.S. Eaton, *Inorg. Chem. 21* (1982) 1093-1101.
14. J.K. More, K.M. More, G.R. Eaton, and S.S. Eaton, *Inorg. Chem. 21* (1982) 2455-2460.
15. M.R. Caira, J.M. Haigh, L.R. Nasimbeni, *J. Inorg. Nucl. Chem. 34* (1972) 3171-3176.
16. N.S. Al-Naimi, A.R. Karaghouli, S.M. Aliwi, M.G. Jalhoom, *J.Inorg. Nucl. Chem. 36* (1974) 283-288.
17. P.M. Boymel, G.R. Eaton, and S.S. Eaton, *Inorg. Chem. 19* (1980) 727-735.
18. P.M. Boymel, G.A. Braden, G.R. Eaton, and S.S. Eaton, *Inorg. Chem. 19* (1980) 735-739.
19. B.M. Sawant, G.R. Eaton, and S.S. Eaton, *J. Magn. Reson. 45* (1981) 162-169.
20. S. Hafid, G.R. Eaton, and S.S. Eaton, *J. Magn. Reson. 51* (1983) 470-476.
21. P.H. Smith, G.R. Eaton, and S.S. Eaton, *J. Am. Chem. Soc. 106* (1984) 1986-1991.
22. W.L. Driessen, *Recl. Trav. Chim. Pays-Bas 101* (1982) 441-443.
23. J.W.F.M. Shoonhoven, W.L. Driessen, J. Reedijk, and G.C. Verschoor, *J. Chem. Soc. Dalton Trans* (1984) 1053-1058.
24. J.B.J. Veldhuis, W.L. Driessen, and J. Reedijk, *J. Chem. Soc. Dalton Trans.* (1986) 537-541.
25. H.L. Blonk, W.L. Driessen, and J. Reedijk, *J. Chem. Soc. Dalton Trans.* (1985) 1699-1705.

26. J.H. Riebenspies, O.P. Anderson, S.S. Eaton, K.M. More, and G. R. Eaton, *Inorg. Chem. 26* (1987) 132-137.
27. K.M. More, S.S. Eaton, and G.R. Eaton, *Inorg. Chem. 20* (1981) 2641-2647.
28. K.M. More, G.R. Eaton, and S.S. Eaton, *Can. J. Chem. 60* (1982) 1392-1401.
29. J. Wojcik, M. Witanowski and G.A. Webb, *J. Mol. Struct. 49* (1978) 249-257.
30. R.F. Hobson and L. W. Reeves, *J. Magn. Reson. 10* (1973) 243-252.
31. M. Barfield and B. Chakrabarti, *Chem. Rev. 69* (1969) 757-778.
32. F.W. King, *Chem. Rev. 76* (1976) 157-186.
33. K.M. More, G.R. Eaton, and S.S. Eaton, *Inorg. Chem. 22* (1983) 934-939.
34. K.M. More, G.R. Eaton, S.S. Eaton, O.H. Hankovszky, and K. Hideg, *Inorg. Chem. 28* (1989) 1734-1743.
35. M.E. Squillacote, R.S. Sheridan, O.L. Chapman, and F.A.L. Anet, *J. Am. Chem. Soc. 101* (1979) 3657-3658.
36. P.W. Mui and E. Grunwald, *J. Am. Chem. Soc. 104* (1982) 6562-6566.
37. J. Breulet, T.J. Lee, and H.F. Schaefer, *J. Am. Chem. Soc. 106* (1984) 6250-6253.
38. K.M. More, G.R. Eaton, and S.S. Eaton, *Inorg. Chem. 24* (1985), 3820-3823.
39. K.M. More, B.M. Sawant, G.R. Eaton, and S.S. Eaton, *Inorg. Chem. 20* (1981) 3354-3362.
40. K.M. More, S.S. Eaton, and G.R. Eaton, *J. Am. Chem. Soc. 103* (1981) 1087-1090.
41. K.M. More, G.R. Eaton, and S.S. Eaton, *Inorg. Chem. 23* (1984) 1165-1170.
42. E.N. Baker, D. Hall, T.N. Waters, *J. Chem. Soc. A* (1970) 400-405.
43. E.N. Baker, D. Hall, T.N. Waters, *J. Chem. Soc. A* (1970) 406-409.
44. S.S. Eaton, K.M. More, B.M. Sawant, P.M. Boymel, and G.R. Eaton, *J. Magn. Reson. 52* (1983) 435-449.
45. S.S. Eaton and G.R. Eaton, *J. Am. Chem. Soc. 104* (1982) 5002-5003.
46. S.S. Eaton, K.M. More, B. M. Sawant, and G.R. Eaton, *J. Am. Chem. Soc. 105* (1983) 6560-6567.

47. R.Z. Sagdeev, Yu.N. Molin, R.A. Sadikov, L.B. Volodarsky, and G. A. Kutikova, *J. Magn. Reson. 9* (1973) 13-26.
48. J.-C. Espie, J. Laugier, R. Ramasseul, A. Rassat, and P. Rey, *Nouv. J. Chem. 4* (1980) 205-209.
49. D.P. Dalal, S.S. Eaton, G.R. Eaton, *J. Magn. Reson. 42* (1981) 277-286.
50. M.K. Guseinova and S.D. Mamedov, *Zh. Strukt. Khim. 19* (1978) 515-521 (p. 445-450 in translation).
51. K.M. More, G.R. Eaton, and S.S. Eaton, *Inorg. Chem. 25* (1986) 2638-2646.
52. B.J. Hathaway, M.J. Bew, D.E. Billing, R.J. Dudley, and P. Nicholls, *J. Chem. Soc. A* (1969) 2312-2318.
53. F.S. Stephens, *J. Chem. Soc. A* (1969) 1723-1734.
54. J.K. More, K.M. More, G.R. Eaton, and S.S. Eaton, *J. Am. Chem. Soc. 106* (1984) 5395-5402.
55. K.M. More, G.R. Eaton, and S.S. Eaton, *J. Magn. Reson. 63* (1985) 151-167.
56. S.S. Eaton, P.M. Boymel, B.M. Sawant, J.K. More, and G.R. Eaton, *J. Magn. Reson. 56* (1984) 183-199.
57. K.M. More, G.R. Eaton, and S.S. Eaton, *Inorg. Chem. 23* (1984) 4084-4087.
58. L. Fielding, K.M. More, G.R. Eaton, and S.S. Eaton, *J. Am. Chem. Soc. 108* (1986) 618-625.
59. L. Fielding, K.M. More, G.R. Eaton, and S.S. Eaton, *Inorg. Chem. 29* (1989) 3127-3131.
60. L. Fielding, K.M. More, G.R. Eaton, and S.S. Eaton, *J. Am. Chem. Soc. 108* (1986) 8194-8196.
61. R. Damoder, K.M. More, G.R. Eaton, and S.S. Eaton, *J. Am. Chem. Soc. 105* (1983) 2147-2154.
62. R. Damoder, K.M. More, G.R. Eaton, and S.S. Eaton, *Inorg. Chem. 22* (1983) 2836-2841.
63. R. Damdoder, K.M. More, G.R. Eaton, and S.S. Eaton, *Inorg. Chem. 22* (1983) 3738-3744.
64. C.K. Schauer, O.P. Anderson, S.S. Eaton, and G.R. Eaton, *Inorg. Chem. 24* (1985) 4082-4086.
65. R. Damoder, K.M. More, G.R. Eaton, and S.S. Eaton, *Inorg. Chem. 23* (1984) 1320-1326.
66. R. Damoder, K.M. More, G.R. Eaton, and S.S. Eaton, *Inorg. Chem. 23* (1984) 1326-1330.
67. R.E. Coffman and G.R. Buettner, *J. Phys. Chem. 83* (1979) 2392-2400.
68. J.M. McNally and R.W. Kreilick, *Chem. Phys. Lett. 79* (1981) 534-540.

69. F.S. Molinaro and J.A. Ibers,
Inorg. Chem. 15 (1976) 2278-2283.
70. J.W. Emsley, J. Feeney, L.H. Sutcliffe,
"*High Resolution NMR Spectroscopy*", vol 2,
Pergamon Press, Oxford, 1966, pp. 727ff.
71. S. Sternhell,
Quart. Rev. Chem. Soc. 23 (1969) 236-270.
72. (a) B. Chion and J. Lajzerowicz,
Cryst. Struct. Commun. 7 (1978), 395-398.
(b) B. Chion, J. Lajzerowicz, D. Bordeaux, A. Collet, and J. Jacques,
J. Phys. Chem. 82 (1978) 2682-2688.
73. M.K. Guseinova, S.D. Mamedov, J.R. Amiraslanov, and T. M. Kutovaga,
Zh. Strukt. Khim. 19 (1978) 97-101.
74. K.M. More, G.R. Eaton, and S.S. Eaton,
J. Magn. Reson. 59 (1984), 497-505.
75. L. Fielding, K.M. More, G.R. Eaton, and S.S. Eaton,
Inorg. Chem. 26 (1987) 856-861.
76. G. Binnig and H. Rohrer,
IBM J. Res. Develop. 30 (1986) 355-369.
77. M. Thompson and V. Elings,
American Laboratory, April (1991) 36-42.
78. D.E. Budil, P. Gast, C.-H. Chang, M. Schiffer, and J.R. Norris,
Ann. Rev. Phys. Chem. 38 (1987) 561-583.
79. Y. Hu and S. Mukamel,
Chem. Phys. Lett. 160 (1989) 410-416.
80. B.E. Bowler, A.L. Raphael, and H.B. Gray,
Prog. Inorg. Chem. 38 (1990) 259-322.
81. P. Siddarth and R.A. Marcus,
J. Phys. Chem. 94 (1990) 8430-8434.
82. R. Gleiter and W. Shafer,
Acc. Chem. Res. 23 (1990) 369-375.
83. J.W. Verhoeven,
Pure & Appl. Chem. 62 (1990) 1585-1596.
84. G.L. Closs and J.R. Miller,
Science 240 (1988) 440-447.
85. A.P.L. Rendell, G.B. Bacskay, and N.S. Hush,
J. Amer. Chem. Soc. 110 (1988) 8343-8354.

RECENT DEVELOPMENTS ON STEREOCONTROL VIA (η^6-ARENE)CHROMIUM COMPLEXES

M. Uemura

RECENT DEVELOPMENTS ON STEREOCONTROL *VIA* (η^6-ARENE)CHROMIUM COMPLEXES

MOTOKAZU UEMURA

Faculty of Science, Osaka City University,
Sugimoto 3-3-138, Sumiyoshi-ku, Osaka 558, Japan

1. INTRODUCTION

Arene rings are known to form stable complexes with many transition metals; in particular, arene-chromium tricarbonyl species are attracting increasing attention because of their practical applications.[1] (η^6-Arene)chromium complexes are readily prepared by several convenient methods.[2] Various solvents have been used ranging from the arene itself to inert as well as polar solvents[3] in thermal reaction conditions with $Cr(CO)_6$. Recent studies[4] have indicated that the most suitable solvent medium for a wide range of high yield syntheses is an approximately 10/1 mixture of dibutyl ether and THF. This modification largely overcomes previous problems such as loss of $Cr(CO)_6$ *via* sublimation.[5] Another useful method for preparation of these complexes is by ligand transfer reaction from other ligands as naphthalene or acetonitrile on chromium. In addition to their ease of preparation, (η^6-arene)chromium complexes are air-stable, crystalline solids, easily characterized by spectroscopic methods, and easily purified by chromatographic techniques or recrystallization. The arene chromium complexes have some characteristic properties because of the strong electron-withdrawing ability and steric effect of the $Cr(CO)_3$ group. The chemical consequences of the chromium complexation are are summarized in Fig. 1.[6]

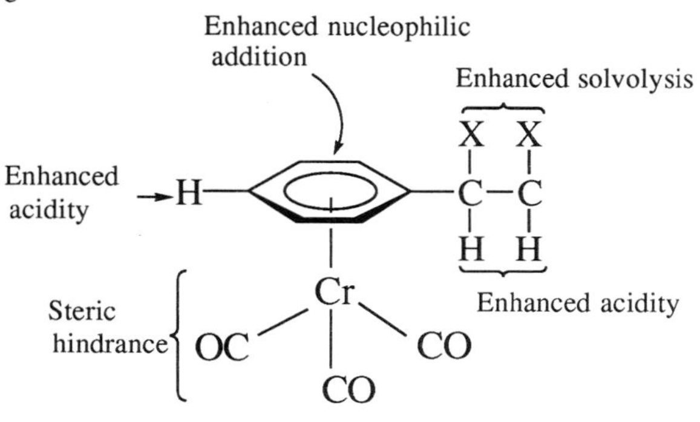

Fig. 1. Effect of $Cr(CO)_3$ Complexation of Arene

They involve: (1) activation of the aromatic ring to nucleophilic addition; (2) enhancement of the acidity of aromatic hydrogens; (3) stabilization of carbanions at the benzylic position; (4) stabilization of carbocations at the benzylic position; (5) steric hindrance of the arene face co-ordinated to the metal fragment. All of these effects have been exploited in organic syntheses. This article focuses on our recent development of stereocontrol utilizing the characteristic properties of (arene)chromium complexes.

2. STEREOCONTROL IN CYCLIC ARENE CHROMIUM COMPLEXES

It is well known that two possible diastereomeric chromium complexes, the *exo*- and *endo*-isomers, are obtained as a mixture by chromium complexation of the cyclic arene compounds (Scheme 1); the ratio of diastereomers depends on the steric and electronic effects of the substituents R as well as the reaction conditions. Stereoselective synthesis of each diastereomer of this type of complexes is important for organic syntheses such as natural product synthesis.

Scheme 1

$Cr(CO)_6$

$Cr(CO)_3$ R *exo*-

$Cr(CO)_3$ R *endo*-

The arene-$Cr(CO)_3$ unit can impart the unique property to stabilize both carbanions and carbocations at the benzylic position (Fig. 1). The formation of benzylic anions and cations are rendered more facile with respect to uncomplexed arenes, and these two contrasting properties have found widespread applications in organic syntheses.[1] The chromium tricarbonyl moiety can also function as a useful stereochemical template. Therefore, nucleophilic or electrophilic attack at the reactive center of an alicyclic ring condensed to an aromatic moiety, *e.g.*, indane or tetralin derivatives, always occurs stereoselectively in an *exo* fashion.

We have explored these interesting properties for the stereoselective synthesis of both *exo*- and *endo*-isomers from a common compound.[7] For example, as illustrated in Scheme 2, (1-*exo*-methyl-1-*endo*-tetralol)$Cr(CO)_3$ (**2**) obtained from (1-tetralone)$Cr(CO)_3$ (**1**) and MeLi, afforded (1-*endo*-methyltetralin)$Cr(CO)_3$ (**3**) by ionic hydrogenolysis[8] with an excess of triethylsilane and trifluoroacetic acid. On the other hand, the *endo* acetate complex **4**, prepared by hydride reduction of **1** and following acetylation, was converted to (1-*exo*-methyltetralin)$Cr(CO)_3$ (**5**) by treatment with trimethylaluminum. The key step in these reactions is the stereoselective *exo* attack of nucleophiles to $Cr(CO)_3$-stabilized carbocations. The results are summarized in Table 1.

Scheme 2

Table 1. Stereoselective Carbon-Carbon Bond Formation *via* Carbocations

Nucleophile (Lewis Acid)

4: R^1=Ac, R^2=H
7: R^1=R^2=H
2: R^1=H, R^2=Me

5: R^2=H, Nu=Me
6: R^2=H, Nu=Et
8: R^2=Nu=Me
9: R^2=H, Nu=$CH_2CH=CH_2$
10: R^2=H, Nu=CH=C=CH_2

entry	complex	nucleophile	Lewis acid	product	yield (%)
1	**4**	Me_3Al	none	**5**	99
2	**4**	Et_3Al	none	**6**	60
3	**7**	Me_3Al	$TiCl_4$	**5**	60
4	**2**	Me_3Al	$TiCl_4$	**8**	92
5	**4**	Et_2Zn	$TiCl_4$	**6**	62
6	**7**	Et_2Zn	$TiCl_4$	**6**	60
7	**4**	$CH_2=CHCH_2SiMe_3$	$TiCl_4$	**9**	95
8	**4**	$CH≡CCH_2SiMe_3$	$TiCl_4$	**10**	76

Although the free benzylic hydroxyl of (η^6-arene)chromium complexes could not be substituted by trialkylaluminum alone, the addition of $TiCl_4$ gave *exo*-alkyl chromium complexes smoothly (see Table 1). In the reaction with Et_2Zn, both the hydroxyl and the corresponding acetate groups afforded *exo*-ethylated chromium complexes only in the presence of $TiCl_4$. Therefore, different alkyl groups can be introduced stepwise at the different benzylic positions depending on the reaction conditions and the structure of substrates. Thus, the benzylic acetoxyl group of complex **11** was initially replaced by reaction with Et_3Al alone to give the ethylated complex **12**, in which the tertiary benzylic alcohol was subsequently substituted with Me_3Al in the presence of $TiCl_4$ (Scheme 3).

Scheme 3

Et, OAc, MeO, HO, Me, $Cr(CO)_3$ **11** —Et_3Al, 98%→ Et, Et, MeO, HO, Me, $Cr(CO)_3$ **12** —Me_3Al, 68%→ Et, Et, MeO, Me, Me, $Cr(CO)_3$ **13**

Like other carbon nucleophiles, allyl- and propargyl trimethylsilanes,[9] electron-rich aromatics and dicarbonyl compounds[10] and enolsilyl ethers[11] can be used for the trapping[12] of $Cr(CO)_3$-stabilized carbocations generated by reaction with Lewis acids.

An important feature of these carbon-carbon bond forming reactions is its wide flexibility for the preparation of a variety of (η^6-arene)chromium complexes. In particular, it is a very efficient method for the stereoselective construction of benzylic quarternary carbon without competitive elimination reactions. The stereochemistry of the quarternary carbon at the benzylic position depends only on the reaction order of the different nucleophiles. As shownin Scheme 4, complex **2** gave (1-*exo*-allyl-1-*endo*-methyltetralin)$Cr(CO)_3$ **(14)** by reaction with allyl trimethylsilane and $TiCl_4$. On the other hand, the diastereoisomeric (1-*exo*-methyl-1-*endo*-allyltetralin)$Cr(CO)_3$ **(16)** was synthesized from complex **1** by reaction with allylmagnesium chloride and subsequently with Me_3Al and $TiCl_4$. Since (1-tetralone)$Cr(CO)_3$ is available in its optically active form,[13] an asymmetric synthesis of complexes **14** and **16**, or decomplexed arenes, could be performed.

Scheme 4

$Cr(CO)_3$, O **1** —MeLi→ $Cr(CO)_3$, HO, Me **2** —$SiMe_3$ (allyl), $TiCl_4$, 99%→ $Cr(CO)_3$, Me **14**

1 —MgCl (allyl)→ $Cr(CO)_3$, HO **15** —Me_3Al, $TiCl_4$, 70%→ $Cr(CO)_3$, Me **16**

3. DIASTEREOSELECTIVE SYNTHESIS OF (*ORTHO*-SUBSTITUTED BENZYLALCOHOL)CHROMIUM COMPLEXES

As shown in Scheme 5, it is a further interesting problem with few precedents that each diastereomeric chromium complex of *o*- or *m*-substituted arenes with a chiral center on the side chain is stereoselectively synthesized. Since the benzylic acetoxyl or the hydroxyl group coordinated with $Cr(CO)_3$ can be easily substituted by a variety of carbon nucleophiles, it is a useful method to synthesize stereoselectively each diastereomer of *o*- or *m*-substituted benzyl alcohol chromium complexes in acyclic series for the preparation of highly functionalized arenes.

Scheme 5

The high stereoselectivity has been previously pointed out in the reaction of nucleophilic reagents with *o*-alkoxyphenyl ketone or aldehyde chromium complexes.[14] For example, the reaction of (*o*-methoxybenzaldehyde)$Cr(CO)_3$ **(17)** with alkyllithium gave exclusively (*S**,*S**)[15]-(*o*-methoxyphenylalkylcarbinol)$Cr(CO)_3$ **(19)** (Scheme 6, Table 2).[16] On the other hand, (*o*-methoxyphenyl alkyl ketone)$Cr(CO)_3$ **(18)** afforded predominantly the diastereomeric (*S**,*R**)[15]-complex **20** by $LiAlH_4$ reduction.

Scheme 6

a: R=Me, b: $R=CHMe_2$

Table 2. Stereoselective Addition of Nucleophiles to Chromium Complexed Benzylic Ketone

complex	nucleophile	**19 : 20**	yield (%)
17	MeLi	98 : 2	66
17	Me_2CHLi	99 : 1	44
18a	$LiAlH_4$	2 : 98	74
18b	$LiAlH_4$	1 : 99	84

Various transition state models have been proposed in order to rationalize these high selectivities. An easy to handle approach with many reactants may now be proposed.[17] Thus, as illustrated in

Scheme 6, the high diastereoselectivity of these reactions should result from the stereoselective *exo* addition of nucleophiles to the carbonyl group of these complexes, the carbonyl oxygen being placed away from the *ortho* methoxy substituent (*anti* conformation) because of a dipole effect.

The nomenclature utilized here is that described in the literature.[17,18] If the bonds from the chromium atom to the arene ring are regarded as single bonds, all the carbons of the *ortho*- or *meta*-disubstituted ring will be chiral centers. The enantiomer A of the complex will thus be called 1*S* (or 2*R*), and the enantiomer B will be 1*R* as shown in Fig. 2. The symbol * indicates a racemate.

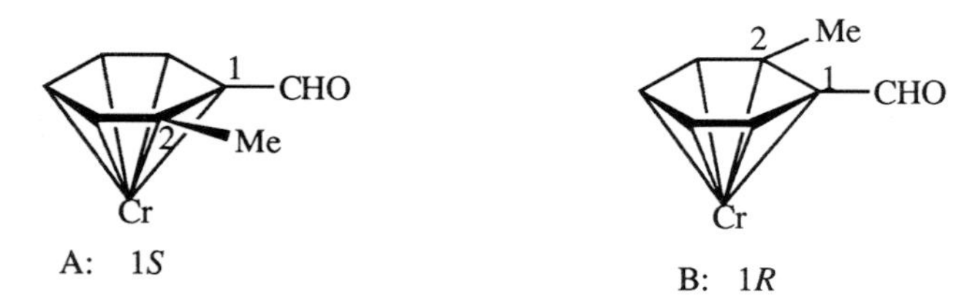

Fig. 2. Nomenclature of Enantiomers

However, the high stereoselectivity of the reactions shown in Scheme 6 is limited only to the case of chromium complexes having substituents such as OCH_3, CH_3, CF_3 etc. at the *ortho* position to the carbonyl group. Therefore, it becomes necessary to develop another useful method for the general preparation of these chromium complexes.

We next turned our attention to chromium complexation of *o*- or *m*-substituted secondary benzylalcohols. $Cr(CO)_3$ complexation[16] of 1-(*o*-methoxyphenyl)-ethylalcohol (**21**) (R=Me) with $Cr(CO)_6$ in butyl ether, heptane and THF (10/1/1) at 130-140°C for 24h afforded an easily separable mixture of (*S**,*S**)-isomer **19** (R=Me) and (*S**,*R**)-isomer **20** (R=Me) in a ratio of 70~80:30~20. The complexation by ligand transfer from (naphthalene)$Cr(CO)_3$ resulted in much higher selectivity (91:9), as described later. The α-ethyl compound gave a higher ratio of the corresponding (*S**,*S**)-complex, even under thermal conditions, while the α-allyl derivative afforded a 1:1 diastereomeric mixture (Scheme 7).

Scheme 7

R, OH, OMe — **21** → [$Cr(CO)_6$; Bu_2O/THF; 120~130°C] → **19** (*S**,*S**) ($Cr(CO)_3$; R, OH, H, OMe) + **20** (*S**,*R**) ($Cr(CO)_3$; R, H, OH, OMe)

R	**19 : 20**	yield (%)
Me	78 : 22	65
Et	92 : 8	80
$CH_2CH=CH_2$	50 : 50	56

This diastereoselective complexation to the arene nucleus would result from attachment of LnCr to the same face[19] using the benzylic hydroxyl group for an anchoring interaction between chromium and the oxygen atom (conformation C), in which the steric hindrance between R and the *o*-substituent is minimized (Scheme 8). The alternative conformation D, which leads to the (*S**,*R**)complex, is affected by severe steric interactions. In the case of the allyl derivatives **21** (R=allyl), the double bond as well as the hydroxyl group can also coordinate to chromium to give a 1:1 mixture of diastereomeric chromium complexes.

Scheme 8

OH R X Y C CrLn O H $Cr(CO)_3$ S*,S* D S*,R*

This direct diastereoselective complexation can be extended to the stereoselective synthesis[16,20] of each diastereomeric chromium complex of *meta* substituted secondary benzylalcohol derivatives in which a sterically bulky, and easily removable, trimethylsilyl group is temporarily introduced at either *ortho* position to the hydroxyalkyl group for getting higher selectivity in the complexation reaction. For example, as illustrated in Scheme 9 and Table 3, the reaction of 1-(2-trimethylsilyl-3-methoxyphenyl)-ethylalcohol (**22**) (R=Me) with 1 eq of $Cr(CO)_6$ under thermal conditions, followed by de-trimethylsilylation, gave predominantly the (*S**,*S**)-complex **24** (R=Me), as expected. On the other hand, the 6-trimethylsilyl compound **23** (R=Me) afforded the diastereomeric (*S**,*R**)-complex **25** (R=Me) as the major product under the same reaction sequences. However, the α-isopropyl-2-trimethylsilyl compound **22** (R=^{i}Pr) gave only a very low yield of the expected (*S**,*S**)-complex under thermal conditions, accompanied by other complexes (vide infra).

Thermal complexation of *o*-trimethylsilyl secondary benzylalcohol derivatives with $Cr(CO)_6$, sometimes gave different arene-chromium complexes in various ratios depending on reaction conditions. For example, reaction of **26** with an excess of $Cr(CO)_6$, during longer reaction times, gave unexpectedly the cyclic siloxane chromium complexes **28** and **29** as major products (Scheme 10, Table 4). Similarly, **22** (R=i-Pr) afforded the corresponding cyclic siloxane complexes as a mixture of *endo*- and *exo*-isopropyl derivatives with an excess of $Cr(CO)_6$ under severe reaction conditions. Also, treatment of complex **27** with $Cr(CO)_6$ for 2 days at 140°C gave a mixture of complexes **28** and

29. The formation of cyclic siloxane chromium complexes under these severe thermal conditions might result from an overall reaction path, in which the initially formed chromium complex **27** is converted to the *endo*-methyl complex **28**, by an intramolecular attack of the benzylic alkoxide anion to the silicone atom, and then the *endo*-methyl complex **28** is in turn converted to the less sterically hindered *exo*-methyl complex **29** under equilibrium conditions.[21]

Scheme 9

OH R SiMe$_3$ OMe **22** —1) A or B, 2)→ R OH H MeO $Cr(CO)_3$ **24** (S^*,S^*)

R H OH MeO $Cr(CO)_3$ **25** (S^*,R^*) ←1) A or B, 2)— OH R SiMe$_3$ OMe **23**

a R = Me, b R = $CHMe_2$

1) (A); 1 eq of $Cr(CO)_6$, (B); (naphthalene)$Cr(CO)_3$, 2) n-$Bu_4N^+F^-$

Table 3. Complexation of compounds **22** and **23** under thermal or kinetic conditions

compound	reagent (A or B)	**24 : 25**	yield (%)
22a	A	98 : 2	69
22b	A	100 : 0	10
23a	A	5 : 95	66
23b	A	4 : 96	77
22a	B	100 : 0	97
22b	B	100 : 0	88
23a	B	2 : 98	82
23b	B	0 : 100	85

Since this thermal complexation with $Cr(CO)_6$ reaches the equilibrium between the diastereomers on prolonged reaction times, a kinetic complexation under milder conditions is required to obtain higher diastereoselectivity and better yield. The ligand transfer from (naphthalene)$Cr(CO)_3$ to another arene reported by Kündig *et al*[22] proceeds in ether at lower temperature and seems to be promising for our purposes. The naphthalene chromium bond is labile, and the naphthalene ligand undergoes facile slippage leaving free a coordination site (η^6 --η^4) for an incoming ligand. This exchange reaction is accelerated by Lewis bases and donor solvents (*e.g.* THF). In fact, reaction of **22** with (naphthalene)$Cr(CO)_3$ in ether containing 1 eq of THF, at 70°C for 4h in a sealed tube, gave exclusively the (S^*,S^*)-chromium complex **24** after de-trimethylsilylation,in a better yield, and with a higher selectivity than the thermal complexation does (see Table 3). The more sterically bulky α-isopropyl compound **22** (R=^{i}Pr) gave only complex **24** (R=^{i}Pr), in high yields, without any diastereomeric contamination. Similarly, this arene exchange reaction of the regioisomeric trimethylsilyl compounds **23** gave the other diastereomeric (S^*,R^*)-complex **25** in good yield under the same reaction sequences (Scheme 9, Table 3).

Scheme 10

26 27 28 29

Table 4. Complexation of compound **26** under thermal condition

entry	eq. of $Cr(CO)_6$	time (h)	**27 : 28 : 29**	yield (%)
1	1.3	24	78 : 20 : 2	85
2	1.7	40	25 : 67 : 8	90
3	2.0	50	10 : 65 : 25	85
4	3.0	60	5 : 55 : 40	80

Such a diastereoselective $Cr(CO)_3$ complexation method contrasts with the diastereoselective *ortho* lithiation of chromium complexes with a chiral center at the benzylic position, which is here illustrated in Scheme 11, and which was developed by Heppert's[23] and Davies's[24] groups. Tricarbonyl(*N,N*-dimethyl-α-phenethylamine)chromium (**30**) is treated with t-BuLi to give a stereodefined, *ortho* lithiated complex with high selectivity. Deprotonation should occur at H(s) *via* a co-ordination of lithium with the amine according to the illustrated conformation.

Scheme 11

BuLi

X=NMe$_2$, OCH$_2$OMe

30 31

4. STEREOSPECIFIC CARBON-CARBON BOND FORMATION *VIA* $Cr(CO)_3$-STABILIZED CARBOCATIONS IN ACYCLIC SERIES

In the chromium complexes of cyclic compounds, carbon-carbon bond forming reactions *via* $Cr(CO)_3$-stabilized carbocations proceed stereoselectively to give *exo*-substituted complexes as discussed in Section 2. In the acyclic systems, the stereochemical course of S_N-1 type substitution is a much more interesting subject, since chromium complexes with chiral center at the benzylic position could be stereoselectively synthesized as discussed in Section 3. As shown in Scheme 12, reaction of the (*S**,*R**)-tricarbonyl(α-methyl-*o*-methoxybenzylacetate)chromium **32** with Et_3Al gave the single substitution complex **34** in 89% yield.[16,20] On the other hand, the diastereomeric acetate complex **33** afforded the other diastereomeric substitution product **35** without formation of complex **34**, under the same reaction conditions. Similarly, reaction of the (*S**,*S**)-(α-ethyl-*o*-methoxybenzylacetate)$Cr(CO)_3$ (**36**) with Me_3Al gave the single coupling product **34**, as well as the stereoisomeric complex **37** afforded **35** as the sole product.

Scheme 12

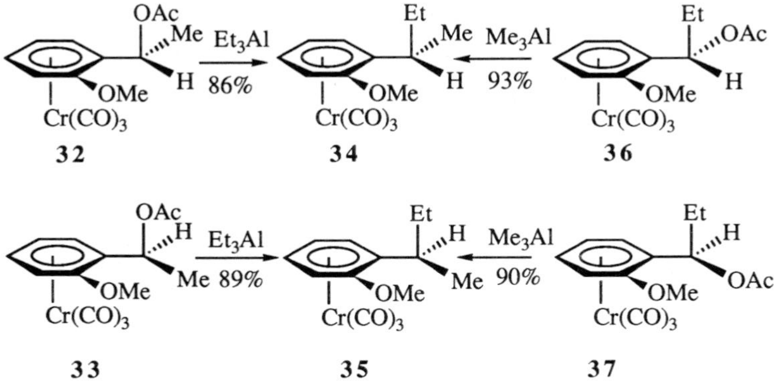

32 →(Et_3Al, 86%) 34 ←(Me_3Al, 93%) 36

33 →(Et_3Al, 89%) 35 ←(Me_3Al, 90%) 37

In a similar manner, the *meta*-substituted complexes **38** and **40** afforded stereospecifically **39** (Nu=Et,or $CH_2CH=CH_2$) and **41** (Nu=Et, or $CH_2CH=CH_2$), respectively, by reaction with Et_3Al alone or with allyl trimethylsilane in the presence of $BF_3{\cdot}OEt_2$ (Scheme 13).

Scheme 13

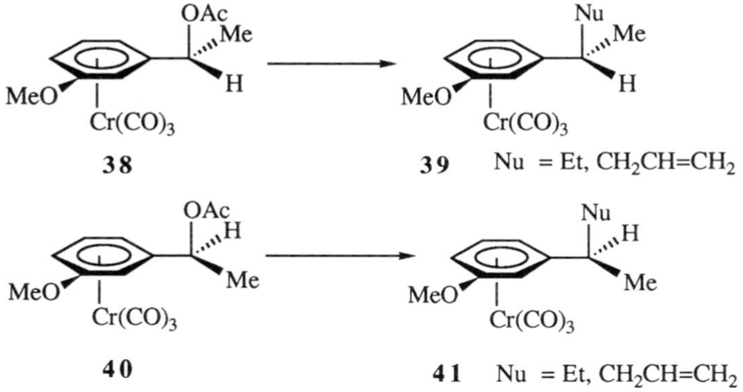

38 → 39 Nu = Et, $CH_2CH=CH_2$

40 → 41 Nu = Et, $CH_2CH=CH_2$

These results indicate that the substitution reaction *via* $Cr(CO)_3$-stabilized carbocations takes place **stereospecifically even in the acyclic systems**. There is, however, a question whether this stereospecific reaction occurs at the benzylic position with stereochemical retention or inversion. Jaouen *et al*[12c] reported that replacement of α-carbocations with some hetero-atom nucleophiles proceeds with retention of the configuration at the benzylic position. This result was also ascertained by using a carbon nucleophile with an optically active chromium compound, as follows (Scheme 14).

Complexation of (*S*)-*o*-methoxy-α-phenethylalcohol[25], $[\alpha]_D$= –70°, gave predominantly the (*S*,*S*)-complex **43**, $[\alpha]_D$= –188°, described in Section 3. After SiO_2 separation, the major (*S*,*S*)-complex **43** was converted into (*R*)-3-(*o*-methoxyphenyl)-n-butyric acid[26] (**46**), $[\alpha]_D$= –16°, by reaction with allyl trimethylsilane, followed by oxidation of the double bond. The minor (*R*,*S*)-complex **44**, $[\alpha]_D$=369°, also afforded the same (*R*)-carboxylic acid **46**. Therefore, carbon-carbon bond forming reactions *via* $Cr(CO)_3$-stabilized carbocations, proceed with **stereochemical retention** at the benzylic position.

Scheme 14

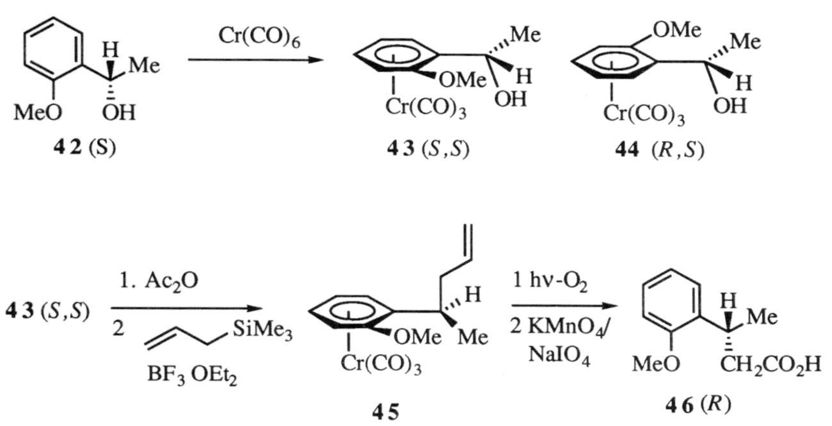

42 (S) → $Cr(CO)_6$ → 43 (*S*,*S*) + 44 (*R*,*S*)

43 (*S*,*S*) → 1. Ac_2O; 2. allyl-$SiMe_3$, $BF_3 \cdot OEt_2$ → 45 → 1 hν-O_2; 2 $KMnO_4$/$NaIO_4$ → 46 (*R*)

The mechanism of stereochemical retention in this S_N-1 reaction is illustrated in Scheme 15. The acetoxyl (or hydroxyl) group of the diastereomeric chromium complexes **47** and **48** is eliminated from the *exo*-side owing to an anchimeric assistance of the $Cr(CO)_3$ group[27] to generate the carbocation

Scheme 15

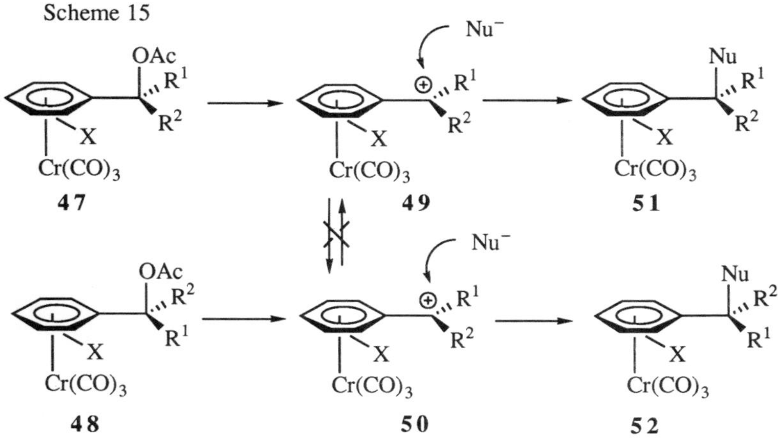

47 → 49 (+ Nu^-) → 51

48 → 50 (+ Nu^-) → 52

species **49** and **50**, respectively. Nucleophiles attack these carbocations from the *exo*-side to give stereospecifically the coupling products **51** and **52**, without equilibration of the carbocations *via* a free rotation about the C(1)-C(Ar) bond. Inhibition of this interconversion between the carbocations has been analyzed by NMR techniques.[28]

The diastereoselective $Cr(CO)_3$ complexation reaction and the subsequent stereospecific carbon-carbon bond formation, as mentioned previously, could be applied to the stereoselective synthesis of spiro-sesquiterpenoids acorenone (**53**) and acorenone B (**54**) by a combination of regioselective, intramolecular, nucleophilic additions[1,6] to $Cr(CO)_3$-anisole derivatives. Semmelhack and Yamashita reported[29] an elegant synthesis of these spiro terpenoids utilizing (arene)chromium complexes. However, no stereoselective complexation step has been included in their route (Scheme 16).

53 acorenone

54 acorenone B

Scheme 16

55 → $Cr(CO)_6$ → **56** + **57**

56: 1. LDA; 2. CF_3CO_2H; 3. NH_4OH; 4. H^+ → **58** → **54**

57 → **59** → **53**

Stereoselective synthesis of the chromium complexes **57** and **56**, which are key intermediates to acorenone and acorenone B, could be easily realized by the previously mentioned method. 6-Trimethylsilyl-3-methoxybenzaldehyde was reacted with isopropyllithium to afford **60**. The ligand transfer reaction of **60** with (naphthalene)chromium, followed by de-trimethylsilylation with (n-$Bu)_4NF$, gave exclusively the (S*,R*)-chromium complex **61**, as discussed above. Regioselective introduction of a methyl group at C-4 position was achieved by direct lithiation,[30] followed by quenching with MeI, in 95% yield, without formation of regioisomerically methylated complexes. This regioselective lithiation at C-4 position of (3-methoxybenzylalcohol)chromium complexes[30] and related chromium complexes contrasts with the results of lithiation (at C-2 position) of the chromium-free parent arenes.[31] After acetylation of the methylated complex, compound **62** was reacted with ethyl 2-trimethylsilyl-3-butenoate[33] in the presence of $BF_3 \cdot OEt_2$; subsequently, a double bond of the coupling product was hydrogenated with PtO_2 to afford the single diastereomeric chromium complex **63**, in 66% overall yield. The conversion of the ethyl ester complex **63** to the nitrile complex **56** was achieved by treatment with dimethylaluminum amide[33] in refluxing xylene in 45% yield, along with a 20% yield of the amide complex **64** (Scheme 17). The amide complex **64** has been also transformed into the nitrile complex **56** by reaction with Me_2AlNH_2 under the same conditions. The compound **56** had already been converted to acorenone B.[31] Similarly, the regioisomeric 2-trimethylsilyl compound **65** was stereoselectively converted to acorenone by the same reaction sequences (Scheme 18).

Scheme 17

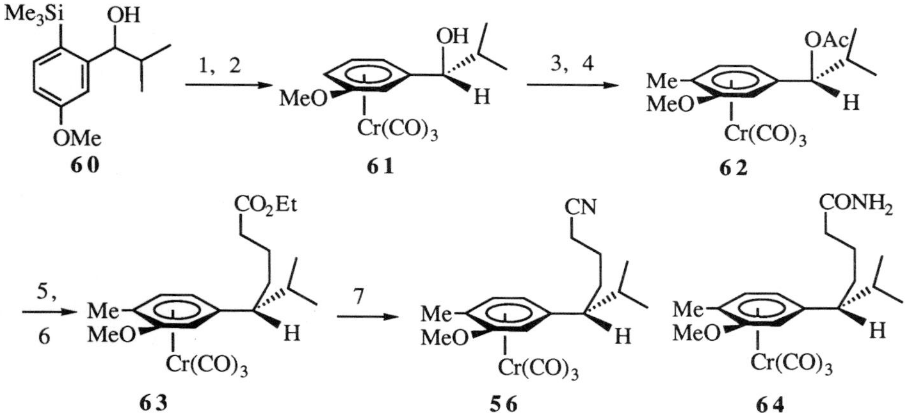

Reagents;
1) (naphthalene)$Cr(CO)_3$, 2) $n\text{-}Bu_4N^+F^-$, 3) n-BuLi/TMEDA, then MeI/HMPA, 4) Ac_2O/pyr
5) $CH_2{=}CHCH(SiMe_3)CO_2Et/BF_3 \cdot OEt_2$, 6) H_2/PtO_2, 7) Me_2AlNH_2

Scheme 18

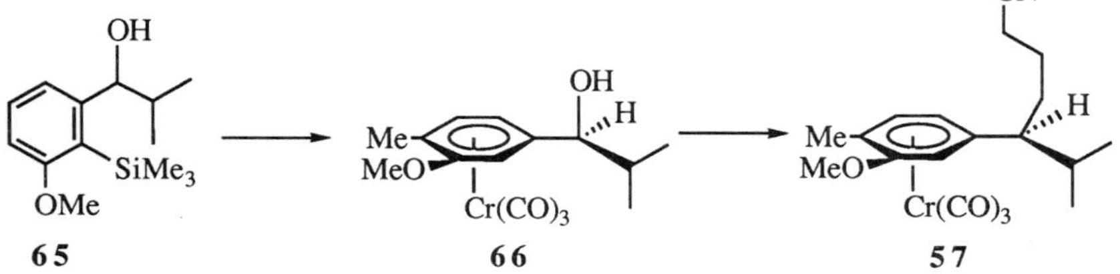

5. DIASTEREOSELECTIVE $Cr(CO)_3$ COMPLEXATION ASSISTED BY A REMOTE HYDROXYL GROUP

Diastereoselective $Cr(CO)_3$ complexation of *ortho*-substituted secondary benzylalcohol derivatives is described in Section 3. The chromium complexes of *o*-alkoxyphenyl alkyl ketone bearing a chiral center at C-n position in an alkyl side chain can be effectively employed to control the stereochemical relationship between C-1 and C-n positions (Scheme 19), since the benzylic carbonyl of these chromium complexes can be converted into each diastereomeric alkyl substitution product by stereoselective addition of nucleophiles, followed by stereospecific carbon-carbon bond forming reaction *via* $Cr(CO)_3$-stabilized carbocations, as discussed in Sections 3 and 4. In this Section, we wish to discuss the diastereoselective chromium complexation of *ortho*-alkoxyphenyl alkane derivatives having two functional groups, an acetal at C-1 (benzylic) and a hydroxyl at C-2 or C-3 positions in order to develop a remote stereocontrol.

Scheme 19

Chromium complexation of 3-hydroxy-1-(*o*-methoxyphenyl)pentan-1-one was unsuccessful with either $Cr(CO)_6$ or (naphthalene)chromium. Usually, the direct chromium complexation of *o*-alkoxyphenyl ketone derivative is difficult. For instance, 8-methoxy-1-tetralone and 5,8-dimethoxy-1-tetralone did not afford the corresponding (arene)chromium complexes, while 5- or 7-methoxy-1-tetralone gave the corresponding arene chromium complexes in good yields.[30,34] Failure of direct chromium complexation of these types of ketones may be attributed to an intramolecular coordination of chromium with two proximal oxygen atoms of the benzylic ketone and *o*-methoxyl groups. Such a difficulty was easily overcome by protection of the carbonyl group, as follows. The complexation of the ethylene acetal compound **67** with $Cr(CO)_6$ under thermal conditions gave a mixture of the easily separable diastereomers (*S**,*R**)-**68** and (*S**,*S**)-**69**, in a ratio of 55:45. The ligand transfer reaction of **67** with (naphthalene)chromium under milder conditions resulted in a much better selectivity (ratio 89:11) as shown in Scheme 20.[35]

Scheme 20

67 **68** (*S**,*R**) **69** (*S**,*S**)

$Cr(CO)_3L$	reaction conditions	**68 : 69**	yield (%)
$Cr(CO)_6$	butyl ether/heptane/THF (10 : 1 : 1); 24h	55 : 45	75
(naphthalene)$Cr(CO)_3$	ethyl ether/1 eq of THF 70°C, 4h	89 :11	85

In contrast with the ethyleneacetal compounds, the corresponding propyleneacetal and 2,2-dimethylpropyleneacetal derivatives **70** and **71** afforded no diastereoselectivity. Also, 1-(*o*-methoxyphenyl)-3-hydroxybutane (**72**), having no acetal function at the benzylic position, gave the corresponding diastereomeric complexes in a ratio of 61:39 by ligand transfer reaction. However, the *gem*-dimethyl compound **73** afforded exclusively the (S^*,R^*)-complex (92:8 ratio). The steric effect at the benzylic position is important for the high diastereoselectivity.

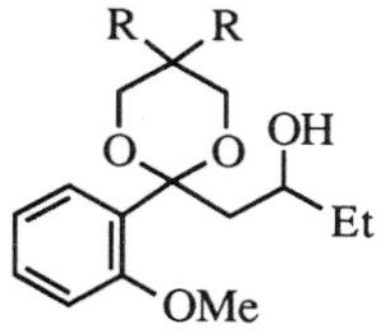

ratio

70; R = H (65 : 35)

71; R = Me (57 : 43)

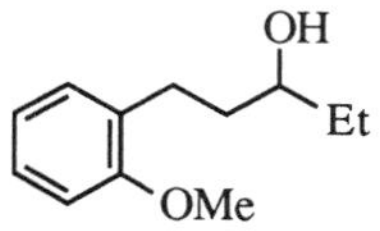

ratio

72; (61 : 39)

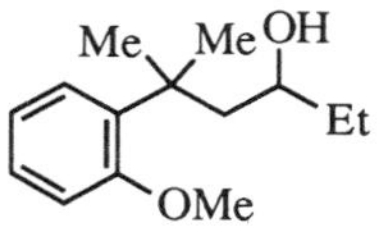

ratio

73; (92 : 8)

The results[35,36] of the ligand transfer reaction with (naphthalene)chromium are summarized in Table 5.

Table 5. Complexation with (naphthalene)$Cr(CO)_3$

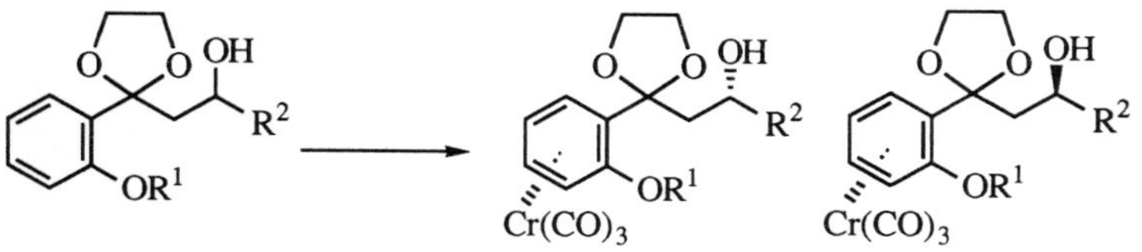

74 → **75** + **76**

entry	R^1	R^2	**75 : 76**	yield (%)
1	Me	Et	89 : 11	85
2	Me	^{i}Pr	96 : 4	89
3	Me	(E)-CH=CHMe	92 : 8	73
4	Me	(Z)-CH=CHMe	92 : 8	82
5	Me	(E)-C(Me)=CHMe	98 : 2	85
6	Et	Et	94 : 6	95
7	^{i}Pr	Et	94 : 6	95
8	^{i}Pr	(E)-CH=CHMe	94 : 6	80

The selectivity of this reaction increases with increasing steric bulkiness of both the alkyl groups (R^1 and R^2). The Thorpe-Ingold effect[37] may be operative for the high diastereoselective chromium complexation. The hydroxyl group in the side chain plays also an important role for high selectivity; as a matter of fact, the corresponding acetoxyl and dimethylamino compounds gave lower diastereoselectivity. It is noteworthy that the selectivity of $Cr(CO)_3$ complexation in these types of arene compounds is independent of the presence/absence of a double bond at C-4,5 position; that is, the chromium atom is not coordinated by the double bond in the complexation process (Table 5, entries

3,4,5 and 8). These results are in contrast with that of *o*-alkoxyphenyl compounds bearing a hydroxyl at C-1 and a double bond at C-3,4 position on the side chain (Scheme 7).

The high diastereoselectivity in the ligand exchange reaction of compounds **74** to afford the (*S**,*R**)-complexes **75** could be explained by $Cr(CO)_3$ complexation *via* a transition state E, in which the ethyleneacetal (or *gem* dimethyl) group is removed from the *o*-alkoxy group by rotation around the C(1)-C(Ar) bond due to steric effects, and the R^2 group is positioned to avoid severe steric interaction with the hydrogen atoms of ethyleneacetal (or *gem* dimethyl) group. The alternative intermediate F, which leads to the diastereoisomeric chromium complex **76**, requires a significant steric interaction between R^2 and the hydrogen atoms (Scheme 21).

Scheme 21

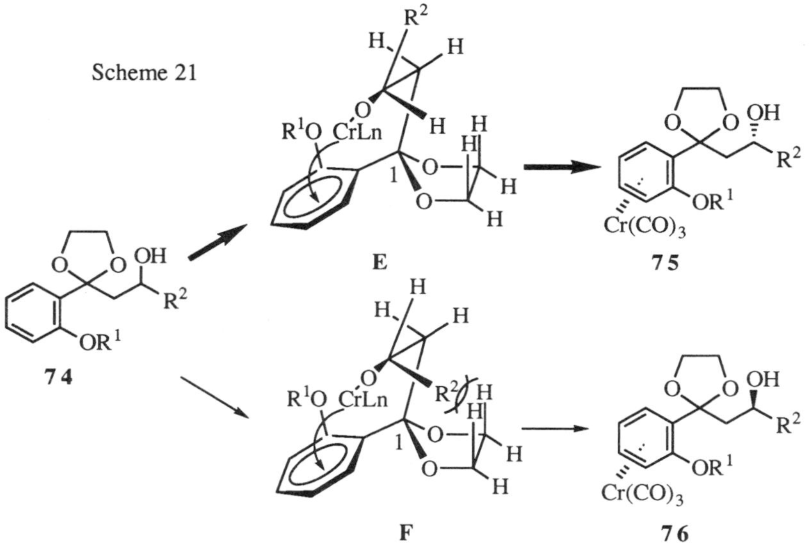

We next examined the diastereoselectivity of the ligand exchange reaction with homologous arene compounds possessing a hydroxyl at C-2 or C-4 position. In the reaction with the C-2 hydroxylated arene compounds[38] **77**, the (*S**,*S**)-chromium complexes **78** were predominantly obtained (Table 6). The stereochemistry of the major chromium complexation product was determined by X-ray crystallography. The transition state G leading to complex **78** is similar to those of *ortho*-substituted secondary benzylic alcohol derivatives (see Scheme 8). Both the alkoxyl and alkyl group (R^2) interactions are minimized by the spacial arrangement of each group.

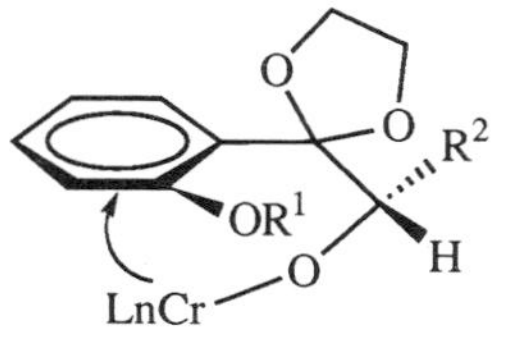

Table 6. Complexation with (naphthalene)$Cr(CO)_3$

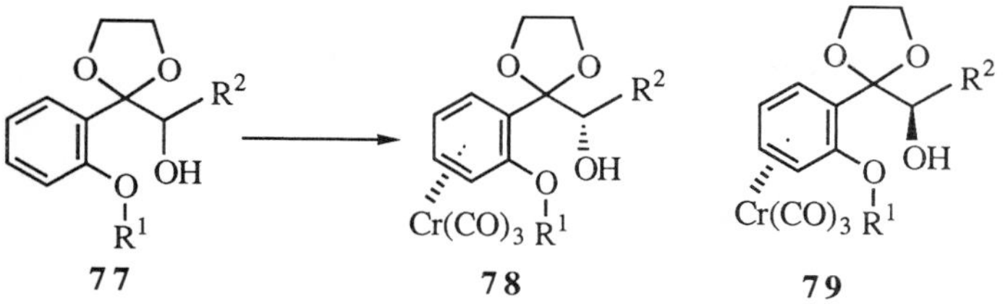

entry	R^1	R^2	**78 : 79**	yield (%)
1	Me	Me	85 : 15	80
2	Me	n-Bu	85 : 15	83
3	i-Pr	Me	92 : 8	88

Arene compounds with a hydroxyl at the C-4 position showed no diastereoselectivity on ligand transfer reaction. In these cases, a longer distance between the hydroxyl and the acetal moiety lowered the steric limitation in the conformation of the alkyl group at the chiral center (C-4) on the side chain.

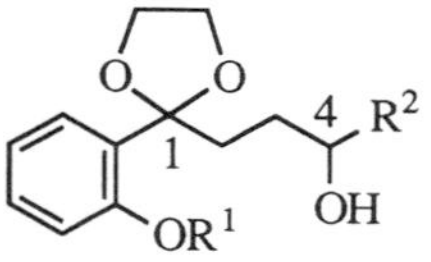

R^1 = Me, ^{i}Pr; R^2 = Et, ^{i}Pr

ratio of diastereomer, 1 : 1
yield, 60 ~ 75%

6. DIASTEREOSELECTIVE ALDOL REACTION OF (*ORTHO*-SUBSTITUTED ACETOPHENONE)CHROMIUM COMPLEXES

In Section 5, we discussed the diastereoselective $Cr(CO)_3$ complexation of *o*-alkoxyarenes with ethylene acetal at C-1 and a hydroxyl at C-3 position by ligand transfer reaction (Table 5). However, such a chromium complexation method is affected by the following disadvantages: (1) each arene compound to be complexed with (naphthalene)chromium should be prepared independently; (2) the preparation in optically active forms is not easy. Since compounds **75** are the equivalents of the metal-complexed aldol, diastereoselective aldol reaction of (*o*-alkoxyacetophenone)$Cr(CO)_3$ would provide a versatile method to overcome the actual disadvantages. Excellent methods have been explored for high stereoselectivity utilizing α-substituted enolates[39], while the aldol reaction of chiral enolates lacking an α-substituent is still a challenging problem for the achievement of high stereoselection.[40] Recently, transition metal-based aldol reactions have become of interest towards synthetic applications and mechanistic studies.[41] We have studied the aldol reaction of (*o*-substituted acetophenone)chromium complexes.[42]

Treatment of tricarbonyl(*o*-methoxyacetophenone)chromium (**80**) (L=CO) with LDA in ether, followed by addition of propionaldehyde, gave no satisfactory results with respect to both yield (less than 10%) and diastereoselectivity. The low yield of the aldol condensation would be attributed to the remarkable stabilization of the $Cr(CO)_3$-complexed enolate anion because of the strong π-acceptor ability of the CO ligands. Therefore, it would be desirable to find solutions to the general unreactivity of these enolates anions. One conceptually straightforward solution to this problem would be to increase the reactivity of either the electrophiles or the chromium complexed enolate anions. It should be possible to enhance the reactivity of enolate anions by replacing one of the three CO ligands with a more electron-releasing phosphine ligand on the chromium. Dicarbonyl(triphenylphosphine)(*o*-methoxyacetophenone)chromium has been prepared by photo-induced reaction of the corresponding tricarbonyl complex with triphenylphosphine.[43] Really, as shown in Table 7, aldol condensation of lithium enolate of the phosphine complex **80** (L=PPh_3, $P(OPh)_3$) with propionaldehyde was found to give the expected quantitative yield, but, unfortunately, the diastereoselectivity was still very low. Other metal enolates transferred from lithium enolates also resulted in low selectivity.

Table 7. Aldol reaction *via* lithium enolate

L	**81 : 82**	yield (%)
CO	55 : 45	10
PPh_3	60 : 40	99
$P(OPh)_3$	60 : 40	99

After several attempts, the aldol reaction of boron enolates of chromium complexes was found to produce satisfactory results (Table 8). The boron enolate generated from tricarbonyl(*o*-methoxyacetophenone)chromium (**83**), dibutylboryl trifluoromethanesulfonate and diisopropylethylamine in ether at –78 °C gave the (*S**,*S**)-chromium complex **84** (R^1=Et) and the (*S**,*R**)-complex **85** (R^1=Et) in a 90:10 ratio after treatment with propionaldehyde.

Table 8. Aldol reaction *via* boron enolate

entry	R^2BOTf	R^1CHO	**84 : 85**	yield (%)
1	$(n\text{-}Bu)_2BOTf$	EtCHO	90 : 10	80
2	$(c\text{-}pentyl)_2BOTf$	EtCHO	91 : 9	45
3	9-BBNOTf	EtCHO	92 : 8	35
4	$(n\text{-}Bu)_2BOTf$	C_6H_5CHO	92 : 8	75
5	$(n\text{-}Bu)_2BOTf$	Me_2CHCHO	95 : 5	81

The relative stereochemistry of the major product **84** was determined by comparison of the corresponding acetate complex with the stereo-defined, authentic acetate complex, derived from the minor product of $Cr(CO)_3$ complexation reaction, as shown in Scheme 20. These two reactions (aldol and complexation reactions) give different stereoisomers of chromium complexes as major products, and supplement each other in organic synthesis. The use of other dialkylboryl reagents resulted in lower yield or lower selectivity (Table 8).

The nature of both the aromatic substituents and the ligands of chromium have a significant influence on the stereoselectivity of the aldol reaction. (*o*-Methylacetophenone)$Cr(CO)_2L$ gave low diastereoselection in the aldol reaction (Table 9, entries 3,7). Therefore, the alkoxyl substituent at the *ortho* position is required for high selectivity. Interestingly, the complexes with monophosphine dicarbonyl(*o*-methoxyacetophenone)chromium exhibited slightly lower selectivities compared with that of the corresponding tricarbonyl complex (entries 4,5,8), while (2,4-dimethoxyacetophenone)$Cr(CO)_2L$ (L=PPh_3, $P(OPh)_3$) complexes were completely unselective (entries 6,10).

Table 9. Aldol reaction of complex **86** *via* boron enolate

$Cr(CO)_2L$ complex **86** (O, Me, X, Y) —1. R_2BOTf/DPEA, 2. EtCHO→ **87** (O, OH, Et, X, Y) + **88** (O, OH, Et, X, Y)

entry	L	X	Y	**87 : 88**	yield (%)
1	CO	OPr^i	H	85 : 15	75
2	CO	OMe	OMe	90 : 10	85
3	CO	Me	H	60 : 40*	45
4	Ph_3P	OMe	H	88 : 12	50
5	$(MeO)_3P$	OMe	H	86 : 14	60
6	Ph_3P	OMe	OMe	50 : 50	30
7	Ph_3P	Me	H	50 : 50	40
8	$(PhO)_3P$	OMe	H	88 : 12	85
9	$(PhO)_3P$	OPr^i	H	85 : 15	60
10	$(PhO)_3P$	OMe	OMe	50 : 50	80

* The ratio may be reversed.

The following transition states would be assumed for the aldol reaction of chromium complexes (Scheme 22). (*o*-Methoxyacetophenone)$Cr(CO)_3$ (**83**) exists in an *anti*-form[14] with respect to the *o*-alkoxy oxygen and the carbonyl oxygen, as discussed previously. Coordination of the boryl with the ketone oxygen, in the *anti*-form of complex **83**, followed by deprotonation with amine would generate *anti*-boron enolate. The *anti* enolate should attack a *re*-face of the aldehyde from an *exo* side of the chromium ligand through a cyclic twist boat transition state H, giving the (S^*,S^*)-condensation complexes **84**.

Scheme 22

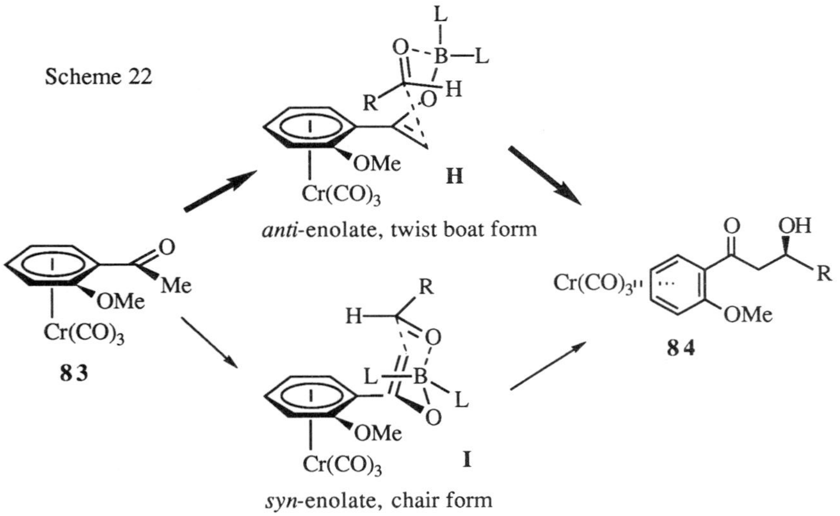

Recent *ab initio* calculations[44] and asymmetric induction[45] in simple boron enolate transition state suggest that a twist boat is easily accessible provided that no α-substituent, *i.e.*, methyl ketone derivatives, are present. However, it still remains the possibility that *syn* boron enolate is generated to form the cyclic chair transition state I. At this point, it cannot be determined whether only one conformation of the enolate is readily accessible or both the *anti*- and the *syn* enolates rapidly interconvert under the conditions in which the enolate is generated and the aldol condensation is carried out. No diastereoselection from the boron enolate aldol reactions with $[2,4(MeO)_2C_6H_3COCH_3]Cr(CO)_2L$ (L=PPh_3, $P(OPh)_3$) may be attributed to the conformation of the $Cr(CO)_2L$ group and the steric effects.

Aldol reaction of (*o*-alkoxyacetophenone)chromium and $Cr(CO)_3$ complexation of *o*-alkoxyarene compounds having ethylene acetal and hydroxyl can be useful reactions in view of further synthetic applications, since the major products in both reactions are diastereomers at C-3 position.

7. DIASTEREOSELECTIVE 1,4-ADDITION OF ORGANOCOPPER REAGENTS TO (*ORTHO*-SUBSTITUTED PHENYL (*E*)-ENONE)CHROMIUM COMPLEXES

The conjugate addition of organocopper reagents to α,β-unsaturated enones and enoates is now a fundamental tool in carbon-carbon bond forming processes. Interest in these reactions has recently extended in the direction of stereoselectivity and the related subject of reaction mechanism. Accordingly, asymmetric versions have attracted widespread attention during the last few years, and valuable chiral auxiliary compounds have been developed for asymmetric 1,4-addition reactions.[46] We discuss the diastereoselective 1,4-conjugate additions to (enone)chromium complexes in this chapter.[47]

The requisite (*E*)-enone complexes **89** were easily obtained from the aldol condensation products **84** of (*o*-substituted acetophenone)$Cr(CO)_3$ by dehydration with *p*-TsOH in refluxing benzene. Reaction of the racemic ((*E*)-enone)chromium complex **89** (R^1=Me, X=OMe) with $Bu^n{}_2CuLi$ in ether at –78°C for 1h gave a diastereomeric mixture of the 1,4-addition products of the Ar(1*S**,2*R**),3'(*R**) complex **90** (R^1=Me, R^2=Bu^n, X=OMe) and the Ar(1*S**,2*R**),3'(*S**) complex **91** (R^1=Me, R^2=Bu^n, X=OMe) in a ratio of 76:24. The other reaction results are shown in Table 10. Addition of Me_3SiCl to the reaction mixture slightly enhanced the formation of complex **90**, but the yield of the addition products was low and accompanied by recovery of the starting material. On the other hand, $Bu^nCu{\cdot}BF_3$ reagent[48] afforded predominantly the corresponding diastereoisomeric 1,4-addition product **91** (X=OMe, R^1=Me, R^2=Bu^n) in a ratio of 92:8 (entry 2). The relative stereochemistry between the newly created stereogenic center and the $Cr(CO)_3$-complexed arene ring of the addition product **91** (R^1=Me, R^2=Bu^n, X=OMe) was determined as the Ar(1*S**,2*R**),3'(*S**)-configuration by comparison with the optical rotation value of the authentic sample.[49] While the diastereoselectivity was low in the reaction of $Bu^n{}_2CuLi$ with *o*-methoxy complex (entry 1), the use of the corresponding sterically more bulky *o*-isopropoxy substituted complex resulted in a marked increase in the selectivity under the same reaction conditions (entry 7). In general, Gilman reagents produced predominantly **90** in the 1,4-addition reaction to the *ortho* alkoxy substituted enone complexes, while $RCu{\cdot}BF_3$ reagents gave mainly complexes **91**. With the *o*-isopropoxy substituted complex, the diastereoselectivity in the conjugated addition with the Gilman reagents was superior to that of the *o*-methoxy substituted complex due to steric effects. Instead of the *o*-alkoxy substituents, the (*o*-methyl substituted enone)chromium complex gave predominantly the Ar(1*S**,2*R**),3'(*S**)-addition product **91** by reactions with both $Bu^n{}_2CuLi$ and $BuCu{\cdot}BF_3$ reagents (entries 9,10).

Table 10. Diastereoselective 1,4-addition of organocopper reagents

entry	complex **89**	[R^2-Cu] reagent	**90 : 91**	yield (%)
1	X=OMe, R^1=Me	$Bu^n{}_2CuLi$	76 : 24	98
2	X=OMe, R^1=Me	$Bu^nCu{\cdot}BF_3$	8 : 92	96
3	X=OMe, R^1=Me	$Bu^n{}_2CuCNLi_2$	80 : 20	98
4	X=OMe, R^1=Me	$Bu^nCuPBu_3{\cdot}BF_3$	50 : 50	97
5	X=OMe, R^1=Me	$Bu^n{}_2CuLi/Me_3SiCl$	82 : 18	30
6	X=OMe, R^1=Bu^n	$MeCu{\cdot}BF_3$	10 : 90	88
7	X=OPr^i, R^1=Me	$Bu^n{}_2CuLi$	92 : 8	93
8	X=OPr^i, R^1=Me	$Bu^nCu{\cdot}BF_3$	48 : 52	94
9	X=Me, R^1=Me	$Bu^n{}_2CuLi$	20 : 80	98
10	X=Me, R^1=Me	$Bu^nCu{\cdot}BF_3$	3 : 97	85

Next, the diastereoselective conjugate addition reactions of the organocopper reagents with optically active arene chromium complexes were examined for the remote stereocontrol in the side chain (Scheme 23). Reaction of the Ar(1*R*,2*S*)-*o*-methoxy chromium complex (–)-**92** with ((*S*)-2-methyl-3-*t*-butoxypropyl)copper-borontrifluoride[50] (**93**) gave exclusively the addition product (–)-**94** with (*R*) configuration at the C-3 position in a ratio of >99:1. On the other hand, the corresponding Ar(1*S*,2*R*)-chromium complex (+)-**92** was reacted with the same (*S*) copper reagent **93** to a diastereomeric mixture of **95** in a 66:34 ratio. These results show that the former reaction is a matched case, while the latter combination is a mismatched pair. Similarly, the Ar(1*R*,2*S*)-*o*-isopropoxy chromium complex (–)-**96** was reacted with bis [(*S*)-2-methyl-3-butoxypropyl]cuprate (**97**) to provide the 3(*S*)-addition product **98** with 97% selectivity, but the corresponding (+)-complex **96** afforded mismatched selectivity with the same (*S*)-diorganocuprate **97**. The obtained complexes **94** and **98** can be applied for 1,3,5-stereocontrol in optically active forms as it will be mentioned in the following Section.

Scheme 23

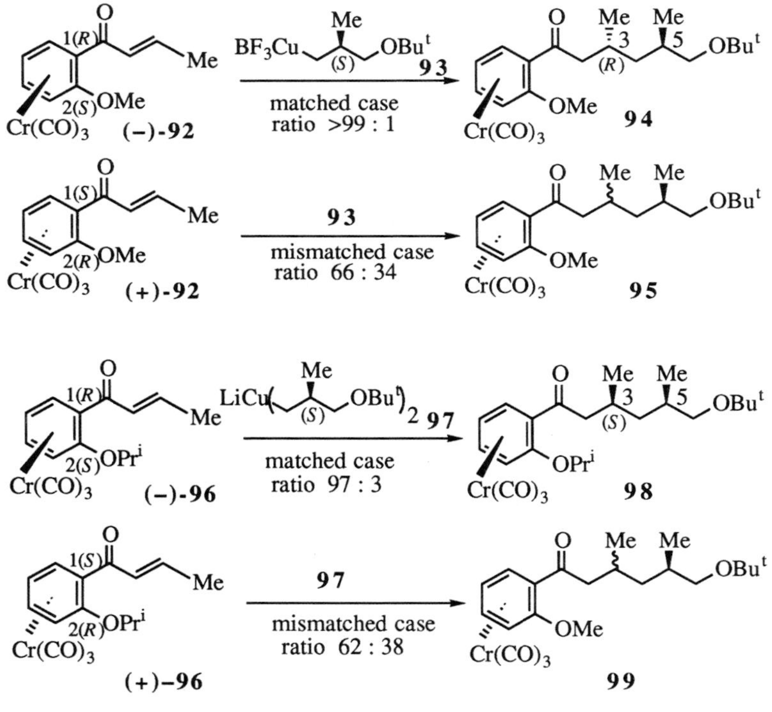

The observed π-face discrimination of the conjugate addition of the organocopper reagents would contribute to the conformation of the (*o*-substituted phenyl (*E*)-enone)chromium complexes. The X-ray crystallographic structures of [*o*-methoxyphenyl (*E*)-propenylketone]$Cr(CO)_3$ (**89**, X=OMe, R^1=Me) and [*o*-methylphenyl (*E*)-propenylketone]$Cr(CO)_3$ (**89**, X=Me, R^1=Me) are illustrated in Figures 3 and 4, respectively. The conformation of the benzylic carbonyl group of the *o*-methoxy substituted complex exists in an *anti*-form (OMe/C=O) and s-*cisoid* (C=O/C_α=C_β) structure, while the corresponding *o*-methyl substituted complex is featured as a *syn* disposed (Me/C=C), s-*cisoid* conformation.[51] The *syn* form in the *o*-methyl substituted complex would contribute to steric hindrance between the *o*-methyl and the propenyl groups.[51] Since the enone moiety in the *o*-methoxy complex is nearly coplanar with the arene, the Gilman reagents attack the *si* face at the C_β position in the s-*cisoid* and *anti*-disposed conformation from *exo* side, giving the addition product **90**. With the *o*-Me complex, the C_β carbon and methyl group of the enone moiety are proximal to the $Cr(CO)_3$ group and one face of the double bond is partially blocked by the *o*-methyl substituent. Therefore, the Gilman reagents attack the *re* face at the β-position in the s-*cisoid* and *syn*-disposed form, giving the complex **91**. On the other hand, the reversal of diastereoselectivity in the reaction of $RCu{\cdot}BF_3$ with *o*-OMe complex would contribute to a change of the conformation. The alternative s-*transoid* and *anti*-disposed form could be postulated as due to the steric effect of the coordination of the benzylic ketone with the Lewis acid, in which the *re* face at the C_β-position is attacked, giving the complex **91**. The diastereoselectivity in the reaction of the *o*-Me complex with $RCu{\cdot}BF_3$ is curious. At this point, it cannot be determined whether the conformation (*syn*- and s-*cisoid*) in the *o*-Me complex is unchanged or changed to the *anti* and s-*transoid* form, where both transition states produced the same complex **91**. Further investigations should be undertaken in order to find an explanation for a relationship between selectivities and conformations.

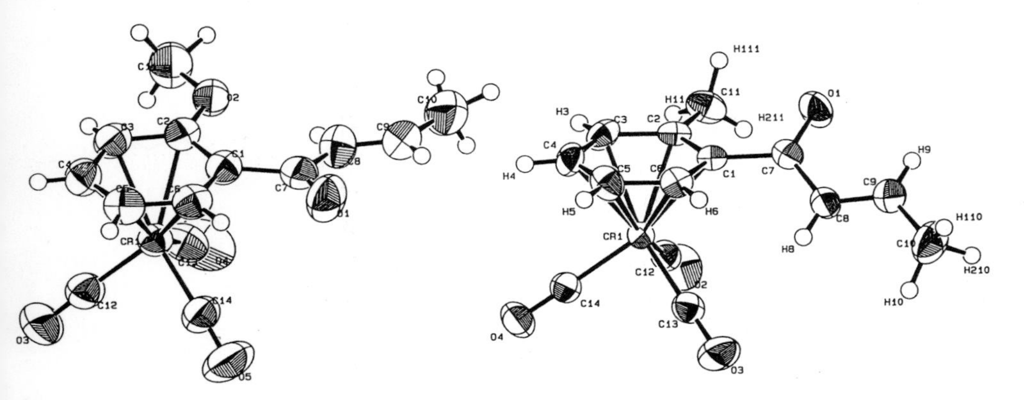

Fig. 3. Molecular structure of **89** (X=OMe, R=Me). Fig. 4. Molecular structure of **89** (X=Me, R=Me).

8. REMOTE STEREOCONTROL *VIA* (ARENE)CHROMIUM COMPLEXES

One of the most exciting challenges in synthetic methodology is the control of stereochemistry in conformationally flexible acyclic systems. In recent years, many efforts have been spent in the exploration of stereoselective reactions in acyclic precursors, and excellent methods have been developed for the diastereo- or enantioselective reactions between adjacent carbon atoms (1,2 relationships).[52] However, general approaches to the stereoselective construction of remote chiral relationships in acyclic systems have been less investigated and they still constitute a challenging problems in organic synthesis.[53]

We have described the synthesis of both diastereomeric chromium complexes of *o*-alkoxy arenes with benzylic ketone (or acetal) and C-3 hydroxyl group in the side chain by the $Cr(CO)_3$ complexation method, or aldol reaction. Also, the carbonyl group of the $Cr(CO)_3$-complexed benzylic ketones could be converted into both diastereomeric alkyl substituted chiral centers by utilization of nucleophilic addition to the carbonyl, and subsequent substitution reactions *via* carbocations. 1,3-Stereocontrol *via* (arene)chromium complexes is as follows (Scheme 24).

Protection of the hydroxyl group of complex **75** (R^1=Me,R^2=Et) and subsequent hydrolysis of the ethylene acetal gave complex **100** in 55% overall yield. Reduction of complex **100** with $LiAlH_4$ followed by acetylation afforded the (S^*,R^*,R^*)-chromium complex **101**. Treatment with Me_3Al gave the single diastereomeric 1,3-*syn* complex **102**. On the other hand, reaction of **100** with MeLi and subsequent hydrogenolysis afforded the other diastereomeric 1,3-*anti* complex **104**.

Scheme 24

OAc OBn
Et
OMe
$Cr(CO)_3$
101
Me_3Al
Me OBn
Et
OMe
$Cr(CO)_3$
102
O OBn
Et
OMe
$Cr(CO)_3$
100
1. $LiAlH_4$
2. Ac_2O
MeLi
HO Me OBn
Et
OMe
$Cr(CO)_3$
103
Et_3SiH
$BF_3 \cdot OEt_2$
Me OBn
Et
OMe
$Cr(CO)_3$
104

A saturated aliphatic moiety possessing methyl branches at 1,3,5- or 1,5-positions is a common feature in acyclic natural products such as α-tocopherol[54] or insect pheromons.[55] In order to explore the synthetic possibility of the 1,5-chiral systems *via* (arene)chromium complexes, we turned our attention to an efficient diastereoselective chirality transfer from C-3 to C-5 positions in allylic alcohol chromium complexes. For effective chirality transfer, Pd(0) catalyzed coupling reactions and the Claisen or related sigmatropic rearrangements occupy a prominent position among the available techniques for acyclic carbon-carbon bond formation of functionalized allylic systems. The wide-spread application of the Pd(0) coupling reaction[56] or the Claisen rearrangement, and its many variants,[57] to the synthesis of acyclic targets can be attributed to the reliability of the reported experimental procedures, the versatile disposition of the products with regard to further transformation, and the predictable stereochemical outcome of the electrocyclic event.

According to Scheme 25, the (*E*)-allylic acetate chromium complex **105** was treated with dimethyl sodiomalonate in the presence of 10 mol% of Pd(0) catalyst in THF, at room temperature, to give the single diastereomeric complex **106** in 95% yield without the formation of regio- and stereoisomeric coupling products. It is particularly noteworthy for synthetic applications that the malonate anion attacked regioselectively at the distal carbon (C-5 position) of the $Cr(CO)_3$-arene complex through the intermediate π-allyl palladium complex. This is probably due to steric and/or electronic effects of the arene and ethyleneacetal chromium complexes, and the stereochemical course of the coupling proceeds with net retention.[56,58] Similarly, the (*Z*)-allyl acetate chromium complex **107** under the same conditions produced the single coupling product **108** of complementary stereochemistry at C-5 position (5*S**), with an *E* double bond.

Scheme 25

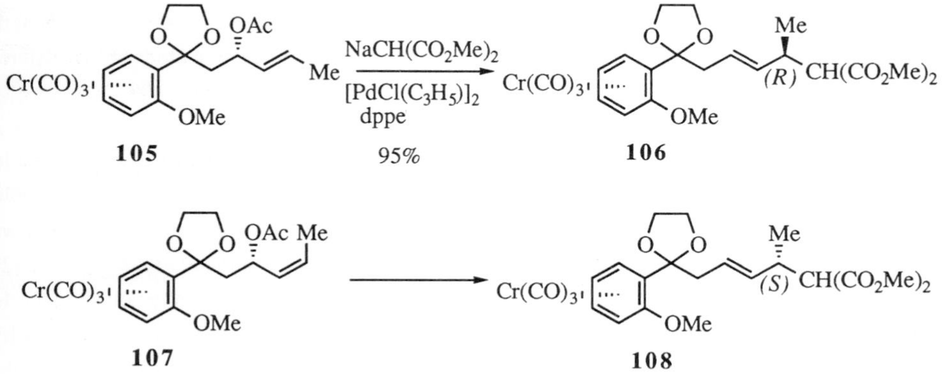

Next, the enolate Claisen rearrangement was examined for chirality transfer reactions. As shown in Scheme 26, treatment of the (*E*)-O-methyl glycolate chromium complex **109** with LDA in THF at –78°C, followed by addition of trimethylsilyl chloride and subsequent warming to room temperature,

Scheme 26

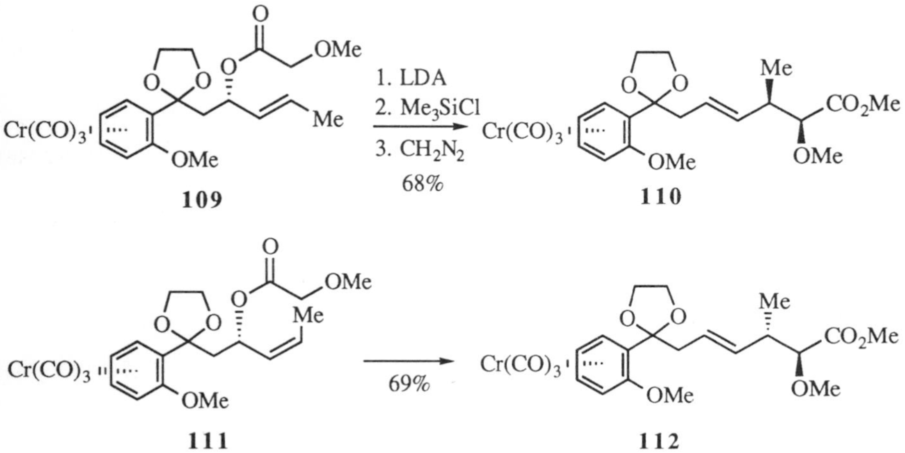

gave a rearranged silyl ester which was converted into the single 5,6-*syn* diastereomeric chromium complex **110** (Ar*S**,5*R**,6*S**) by the treatment with CH_2N_2, in 68% overall yield. The formation of the vicinal *syn* diastereomer from the (*E*)-complex **109** was expected from the assumption that the intermediate enolate exists as the cyclic chelate geometry, and the successive sigmatropic rearrangement proceeds *via* a chairlike six-center transition state.[59] On the other hand, starting from the (*Z*)-olefin complex **111**, the 5,6-*anti* complex **112** (Ar*S**,5*S**,6*S**) was obtained as a single rearranged product. Extremely high diastereoselectivity in the enolate Claisen rearrangement would be attributed to the steric bulkiness of the the $Cr(CO)_3$-complexed arene and the acetal ring.

Since the stereoselective chirality transfer to the C-5 position on the side chain of (arene)chromium complexes was achieved, the next requirement for 1,5-stereocontrol is the stereoselective conversion of the acetal to an alkyl substituted methine group. Both the diastereomers with alkyl substituent at the benzylic position could be stereoselectively synthesized by a change of reaction order of two different nucleophiles (*e.g.*, hydride and carbanions).

Conversion of the acetal group in complex **106**, or **108**, to the ketone group was carried out by an oxidative deacetalization[60] with trityl tetrafluoroborate to give the β,γ-unsaturated ketone complex **113a** without isomerization of the double bond to α,β-position (Scheme 27). Reduction of complex **113a**, with $NaBH_4$, followed by acetylation, gave the single acetoxyl complex **114a**, which was converted to the 1,5-*anti* dimethyl complex **115a** (Ar*S**,1*R**,5*R**) by treatment with Me_3Al, in 72% overall yield. On the other hand, reaction with MeLi at –78°C, followed by hydrogenolysis, afforded stereoselectively the 1,5-*syn* dimethyl complex **117a** (Ar*S**,1*S**,5*R**). Similarly, complex **113b** with 5(*S**) configuration was stereoselectively converted into the two diastereomeric dimethyl complexes **115b** and **117b**, respectively.

Scheme 27

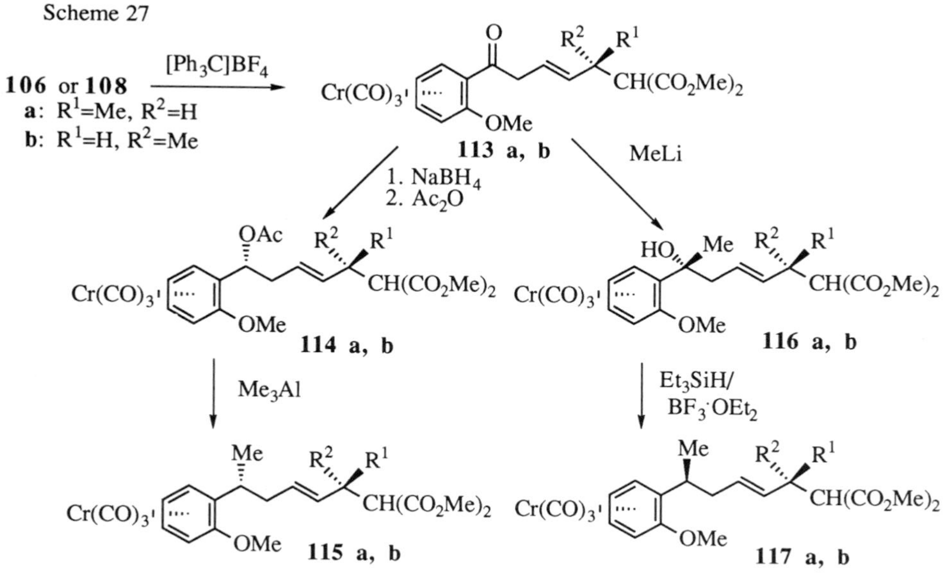

106 or 108 —[Ph3C]BF4→ 113 a, b
a: R1=Me, R2=H
b: R1=H, R2=Me

113 a, b —1. NaBH4, 2. Ac2O→ 114 a, b —Me3Al→ 115 a, b
113 a, b —MeLi→ 116 a, b —Et3SiH/BF3·OEt2→ 117 a, b

Cr(CO)3 · OMe · OAc · HO · Me · R2 · R1 · CH(CO2Me)2

Likewise, the enolate Claisen rearrangement products **110** and **112** can be employed in the stereoselective synthesis of acyclic systems containing 1,5,6-chiral centers using the above mentioned methods. Also, 1,4-conjugate addition product **94** can be converted to 1,3,5-stereodefined compounds in optically active forms by the same reaction sequences.

These reaction schemes incorporate significant flexibility with regard to the stereochemistry of the remote centers since a simple variation of the allylic olefin geometry would result in an effective stereocontrol of the remote chiral centers bearing various alkyl substituents. The results of such reaction sequences are very useful for the construction of the 1,5-pattern of alkyl substituents, a relationship reminiscent of naturally occurring acyclic or macrocyclic systems. Thus, the 1,5-*syn* dimethyl chromium complexes **115b** or **117a** were converted to the methyl 2(*R**),6(*R**),10-trimethylundecanoate **121** as shown in Scheme 28.

Scheme 28

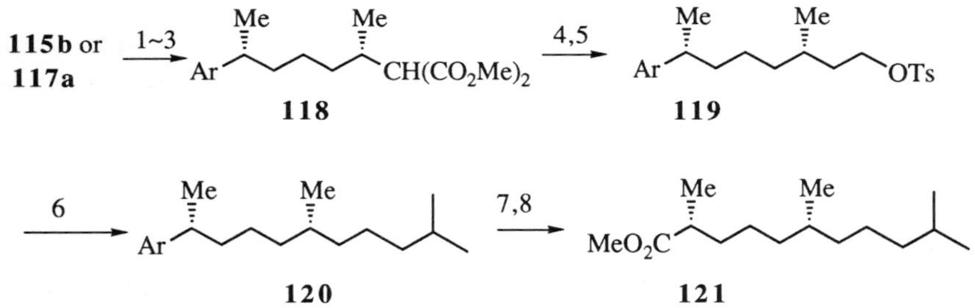

Reagents:
1. hv-O_2, 2. H_2/Pd-C, 3. KOAc/DMSO, 4. $LiAlH_4$, 5. TsCl/pyr, 6. $(i\text{-}Bu)_2CuLi$
7. $RuCl_3/NaIO_4$, 8. CH_2N_2

It is well known that $Cr(CO)_3$-complexed anisole derivatives can be used as precursors of substituted cyclohexenone derivatives by regioselective nucleophilic addition to the aromatic ring, as developed by Semmelhack.[1,6] Complex **122** was converted to cyclohexenone compound **123** by treatment with 2-lithio-2-methyl-1,3-dithiane, subsequent protonation, demetallation and final acidic hydrolysis, although the overall yield was not optimized (10% yield) (Scheme 29). Compound **123**, with four chiral centers, is useful for highly functionalized chiral acyclic systems.

Scheme 29

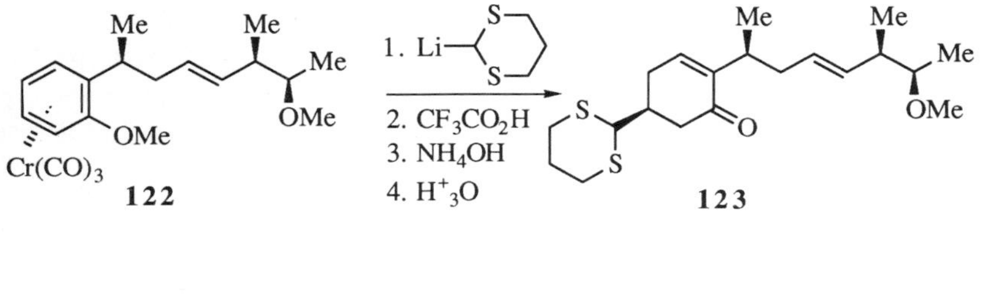

9. DIASTEREOSELECTIVE ADDITION OF CROTYL METAL TO $Cr(CO)_3$-COMPLEXED AROMATIC KETONES

Considerable attention has been focused on stereoselective carbon-carbon bond forming processes in acyclic and other conformationally flexible molecules, particularly by aldol addition[39b] and related reactions of crotyl metal reagents[61] with carbonyl compounds. Crotyl metal addition reactions are comparable or even superior to the aldol reactions, since the adducts, being homoallylic alcohols, can be transformed into aldol derivatives or other functional groups. Generally, a high degree of diastereoselectivity has been achieved in the reactions with **aldehydes** to form *anti*- and *syn*-adducts, respectively. The stereochemical outcome depends on the geometry of the double bond of crotyl metal, the nature of the metal and reaction conditions. However, much less selectivity[62] is observed in the reaction of **ketones** with crotyl metals (Scheme 30), because of the smaller difference in the steric size between both the groups attached to the ketone carbonyl. Recently, Seebach's[63] and Reetz's[64] groups have reported that crotyltitanium reagents react with ketones to yield *anti*-adducts with high selectivity. Proceeding toward the synthesis of dihydroxyserrulatic acid (**137**), we have examined diastereoselective additions of crotylmetals to aromatic ketones. In order to obtain high selectivity, the aromatic part was modified temporarily by sterically bulkier groups, *e.g.*, transition metal arene complexes.[65]

Scheme 30

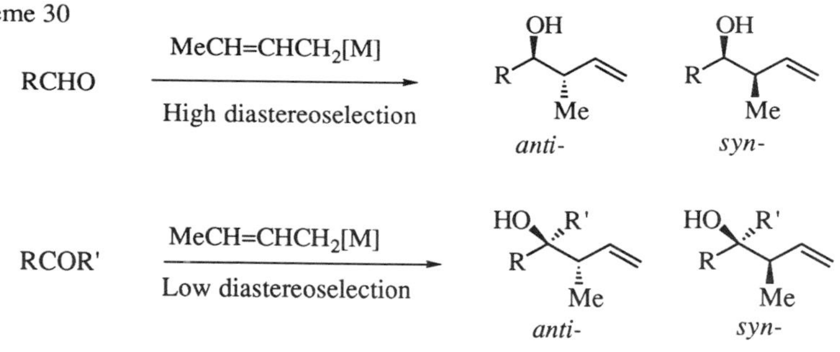

1-Tetralone gave no satisfactory diastereoselectivity with the Seebach's or Reetz's crotyl titanium reagents;[63,64] also, reaction of (1-tetralone)chromium complex with crotylmagnesium chloride gave no selectivity. However, reaction with the crotyl Grignard or lithium reagent, in the presence of 1 eq of trialkylaluminums, afforded the *anti*-adduct in sufficiently high selectivity, without formation of the regioisomer by α-attack of the crotyl metal group (Table 11).[65] The addition of Et_3Al showed a particularly high *anti*-selectivity. Since 1-tetralone itself without $Cr(CO)_3$ complexation gave the *anti*- and *syn*-adducts in a 2: 1 ratio under the same conditions, the $Cr(CO)_3$ complexation apparently increases the *anti*-selectivity in this reaction. Chromium complexes of acetophenone, propiophenone and isobutyrophenone also gave *anti* adducts predominantly with the aluminum "ate" complex. In contrast with the $Cr(CO)_3$-complexed aromatic ketones, the chromium complex of benzaldehyde exhibited no selectivity with this aluminum "ate" complex.[65]

Table 11. Diastereoselective addition of crotylmetal to complex **124**

$Cr(CO)_3$ — **124** (OMe) $\xrightarrow[\text{Additive}]{MeCH=CHCH_2[M]}$ **125** + **126**

entry	M	additive	**125 : 126**	yield (%)
1	MgCl	none	50 : 50	75
2	MgCl	Me_3Al	80 : 20	75
3	MgCl	Et_3Al	93 : 7	73
4	MgCl	$(i\text{-}Bu)_3Al$	85 : 15	65
5	MgCl	5 eq of Et_3Al	66 : 34	80
6	MgCl	Et_3B	28 : 72	30
7	MgCl	$(i\text{-}PrO)_4Ti$	64 : 36	30
8	Li	Et_3Al	94 : 6	60

This reaction presumably proceeds *via* a six-membered chair transition state, in which the smaller group on the ketone carbonyl function occupies a pseudo-axial position and the double bond of the "ate" complex exists as *E*-form. Although the cyclic penta-coordinate transition state of aluminum "ate" complex may be curious, a number of other reactions has been proposed to proceed *via* the "ate" complex mechanism.[66] Interestingly, the combination of Et_3B and crotyl Grignard reagent produced predominantly *syn*-adducts. An alternative transition state should be assumed in the case of the reaction of crotyl Grignard, Et_3B and chromium complexed aromatic ketones.

The hydrogenolysis of the benzylic hydroxyl group of the *anti*-adduct **125** with CH_3CO_2H/Et_3SiH gave smoothly the chromium complex **127** with inversion at the benzylic position (Scheme 31), in which the stereochemistry between the ring and the adjacent carbon (C-4 and C-11 in terpenoid numbering) is consistent with that of dihydroxyserrulatic acid (vide infra).

Scheme 31

125 $\xrightarrow[CF_3CO_2H]{Et_3SiH}$ **127**

10. [2,3]-WITTIG SIGMATROPIC REARRANGEMENT OF $Cr(CO)_3$-COMPLEXED BENZYL CROTYL ETHERS

The [2,3]-Wittig sigmatropic rearrangement has become an efficient method for acyclic stereocontrol.[67] It has already been reported[68] that the Wittig rearrangement of benzyl (*Z*)-crotyl ether provides an extremely high *syn* stereoselection, whereas the corresponding (*E*)-substrate gave poor stereoselectivity (Scheme 32). The mechanism of stereoselection in the [2,3]-Wittig rearrangement has been rationalized in terms of pseudo 1,3-diaxial interaction and gauche interaction in the enveloped five-membered transition state.[67,69] Since the extent of stereoselectivity is influenced by the steric bulkiness of the substituents, the modification of an aromatic ring to a sterically bulkier group, *e.g.*, by a temporary chromium complexation, is of interest in synthetic applications and mechanistic studies of the [2,3]-Wittig rearrangement.[70]

Scheme 32

LDA

	anti-		*syn-*
(*E*)-substrate	50	:	50
(*Z*)-substrate	5	:	95

As compiled in Table 12, treatment of (benzyl (*E*)-crotyl ether)$Cr(CO)_3$ (**128a**) with lithium diisopropylamide in THF at –78°C for 7h afforded a diastereomeric mixture of the *syn*-**129** and the *anti*-**130** derivatives, in a ratio of 95:5, after demetallation by exposure to sunlight. The high *syn* selectivity of (*E*)-substrate is in contrast with the results from the reaction of the chromium-free parent compound,[66] and can be explained as follows. The coordination of the sterically bulky $Cr(CO)_3$ would greatly enhance the gauche interaction between Ar(Cr) and methyl groups in the transition state K. Therefore, the other transition state J, having a pseudo axial orientation of Ar(Cr), would be preferred for the rearrangement to *syn*-**129**. Similarly, the bulkier substrate **128b** with triphenylphosphine ligand rearranges smoothly to give *syn*-**129** in good yield, but with lower selectivity. On the other hand, the corresponding (*Z*)-crotyl chromium complex **128c** gives a 1:1 diastereomeric mixture.

Table 12. [2,3]-Wittig sigmatropic rearrangement of (benzyl crotylether)$Cr(CO)_2L$

1 LDA, 2 hν-O_2

128 → **129** + **130**

complex	geometry (purity)	**129 : 130**	yield (%)
128a, L=CO	E (96%)	95 : 5	69
128b, L=PPh_3	E (96%)	88 : 12	95
128c, L=CO	Z (88%)	48 : 52	40

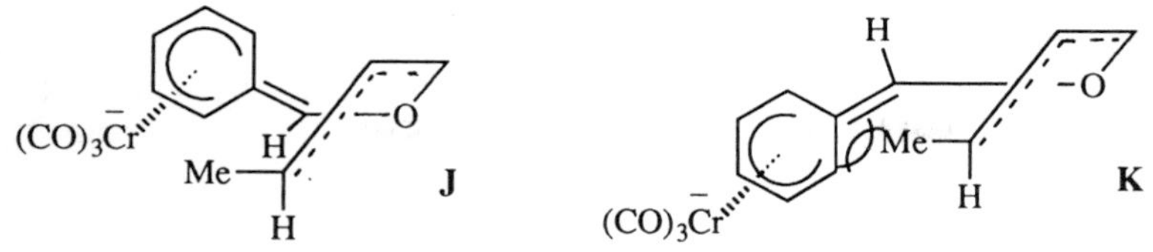

(CO)$_3\bar{C}r$ H Me H **J**

H Me H (CO)$_3\bar{C}r$ **K**

The [2,3]-Wittig rearrangement of (di-substituted arene)chromium complexes is also of interest in view of the stereochemistry of the rearranged chromium complexes, which can exist in four diastereomeric forms. The reaction of (*o*-methoxybenzyl (*E*)-crotyl ether)Cr(CO)$_3$ **(131)** (E>99.3%) with n-BuLi in the presence of TMEDA at –78°C gives the two diastereomeric chromium complexes *syn*-product **132** (Ar*S**,1*S**,2*R**) and *anti*-product **133** (Ar*S**,1*S**,2*S**) in a ratio of 97:3 (Scheme 33), without formation of the other two diastereomers. The formation of **132** and **133** can be explained in terms of an *exo* deprotonation[71] from the sterically most favourable conformation **134**, in which the methoxyl group is oriented *anti* to the side-chain ether moiety, followed by an *exo* attack of the rearranging double bond on the benzylic carbanion of the conformation **135** (Scheme 34). Furthermore, the chromium complex **131** can be employed as a "template" for highly asymmetric induction in the optically active form in the Wittig rearrangement.

Scheme 33

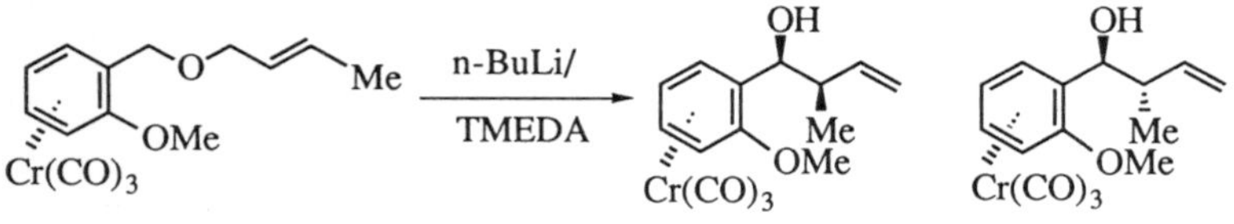

O Me OMe Cr(CO)$_3$ **131** — n-BuLi/ TMEDA → OH Me OMe Cr(CO)$_3$ **132** ; OH Me OMe Cr(CO)$_3$ **133**

ratio (93 : 7)

Scheme 34

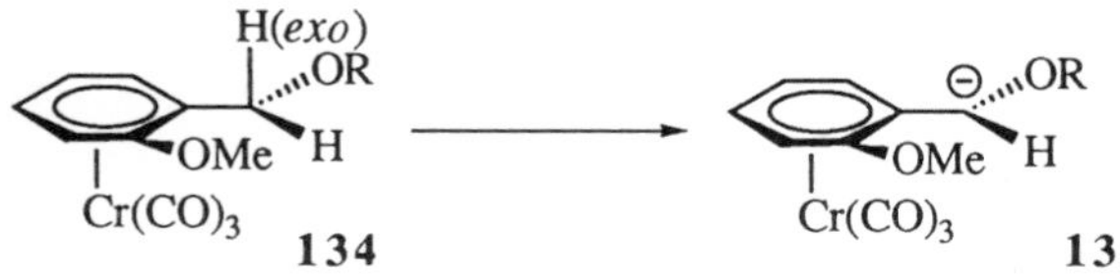

H(*exo*) OR OMe H Cr(CO)$_3$ **134** → ⊖ OR OMe H Cr(CO)$_3$ **135**

A high *syn* stereoselectivity in the Wittig rearrangement is also evident in chromium complexes with an alkyl substituent at the benzylic position. For example, as illustrated in Scheme 35, treatment of *endo*-(1-(*E*)-crotyloxy-5-methoxy-tetralin)Cr(CO)$_3$ **(136)** with n-BuLi gives predominantly the major rearranged chromium complex **126**, the product ratio being in marked contrast with that obtained from the reaction of (5-methoxy-1-tetralone)Cr(CO)$_3$ and crotylaluminum "ate" complex, as described above (see Table 11). The rearranged *syn* chromium complex **126** has suitable stereochemistry for the synthesis of pseudopterosins and related compounds, **138, 139**.

Scheme 35

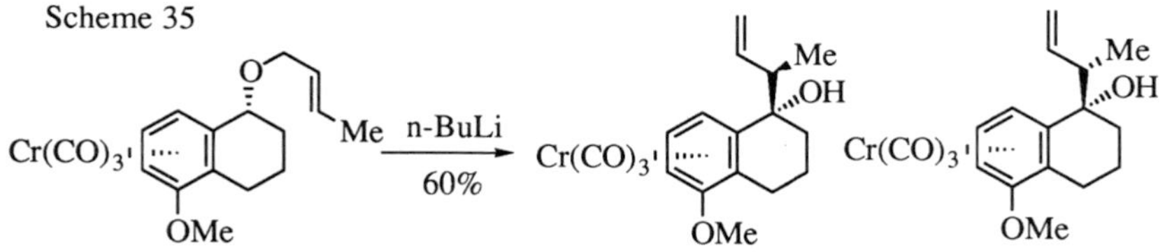

11. TOTAL SYNTHESIS OF (±)-DIHYDROXYSERRULATIC ACID *VIA* (ARENE)CHROMIUM COMPLEXES

Serrulatane class diterpenoids, such as dihydroxyserrulatic acid[72] **(137)**, seco-pseudopterosins[73] A-D **(138)**, pseudopterosins[74] A-D **(139)** and related compounds,[75] have been isolated from the leaves of *Eremophila serrulate*, a viscid shrub and marine sea whip *Pseudopterogorgia elisabethae*. Some of these natural products possess potent anti-inflammatory and analgesic properties.[76] For a general stereo- and regioselective synthetic method of these terpenoids from tetralin derivatives, the following three tactical problems need to be solved: (1) stereocontrol at C-4 and C-11 positions; (2) *trans* arrangement of two benzylic substituents (at C-1 and C-4); (3) introduction of C-1 unit at 6-position (*meta* position to the electron-donating C-8 OH group). Since these three questions have already been solved by utilizing (arene)chromium complexes, a highly selective synthesis of (±)-dihydroxyserrulatic acid is described in this Section.[77]

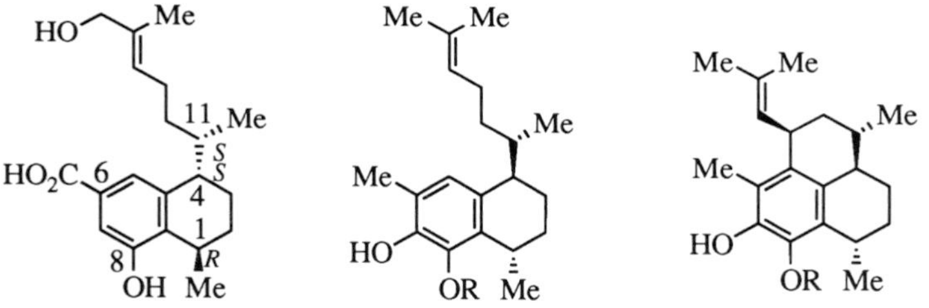

137 Dihydroxyserrulatic acid

138 Seco-pseudopterosin A; R=arabinose

139 Psedopterosin A; R=β-xylose

Mono ethyleneacetal of dihydro-1,4-naphthoquinone **(140)**[78] was converted to the corresponding (arene)chromium complex **141** using thermal conditions and $Cr(CO)_6$. Complex **141** was treated with crotylaluminum "ate" complex to afford the stereochemically desirable *anti* adduct **142** (in C-4/C-11 relationship) and a C-11 stereo-isomeric compound (ratio 85~90:15~10), as above described. Stereoselective conversion of the acetal group to the *exo* methyl derivative **143** was achieved in 61% overall yield, in four steps, as shown in Scheme 36. An ionic hydrogenolysis of the benzylic hydroxyl of complex **143** produced the C-4 *endo*-substituted complex **144** in 56% yield, along with a dehydration product at C-3/C-4 position (18% yield). The stereochemical outcome at the benzylic positions resulted in an *exo* attack of the reagents to $Cr(CO)_3$-stabilized carbocations. Therefore, the relative configuration at the C-1, C-4 and C-11 positions of dihydroxyserrulatic acid (**137**) is now satisfied in the chromium complex **144**. Next requirement for the total synthesis of **137** is the introduction of a carboxyl group at C-6 position. Although the introduction of a proper substituent at *meta* position to an electron-donating group is difficult by electrophilic reactions, this problem could be solved by utilizing (arene)chromium complexes in either of the following two methods. Direct lithiation[79] of (phenyl triisopropylsilyl ether)chromium complex or nucleophilic addition[1,6] of (anisole)chromium complex are well known to occur at the *meta* position of donor groups with high regioselectivity. The conversion of methoxyl in complex **144** to triisopropylsilyl ether failed; so, we next tried nucleophilic addition to complex **144**. However, reaction of complex **144** with 2-lithio-1,3-dithiane, and subsequent oxidative demetallation with I_2, gave an undesirable mixture of *ortho*- and *meta*-substituted compounds (ratio 3:1) in less than 10% yield. This low yield, and the unexpected formation of *ortho* isomer as a major product, may be attributed to the conformation of the three carbonyl ligands with respect to the arene ring. In (anisole)$Cr(CO)_3$, the carbonyl ligands prefer to assume a *syn*-eclipsed conformation with respect to OMe; therefore, nucleophiles attack at the *meta* position, which is also eclipsed with respect to another CO ligand, owing to the balance of charge and frontier orbital control.[80,81] However, the carbonyl ligands in complex **144** would avoid a severe steric interaction with the *endo*-oriented butenyl group at C-4 and could not adopt the *syn*-eclipsed conformation **147**. In this context, the *exo*-orientation of the large butenyl group seems to be essential for *meta*-introduction of nucleophiles, and we have confirmed this proposal with model chromium complexes, as follows.

Scheme 36

140 → (1) → **141** → (2) → **142** → (3~6) → **143** → (7) → **144** → (8,9) → **145**

Reagent:
1) $Cr(CO)_6$ (80%), 2) $MeCH{=}CHCH_2MgCl/Et_3Al$ (81%), 3) 1N-HCl (90%), 4) $NaBH_4$ (90%), 5) Ac_2O (95%), 6) Me_3Al (75%), 7) Et_3SiH/BF_3OEt_2 (56%), 8) hν-O_2 (95%), 9) $Cr(CO)_6$ (60%)

146

147

As illustrated in Scheme 37, reaction of (1-*exo*-isopropyl-5-methoxytetralin)$Cr(CO)_3$ (**148**) with 2-lithio-1,3-dithiane, followed by demetallation, gave, as expected, the *meta* substituted arene with high selectivity, while the corresponding *endo*-isopropyl complex **149** afforded, predominantly, an *ortho*-substituted compound in low yield. The olefinic complex **150** produced mainly *meta*-substituted compounds. The reactions of (arene)chromium complexes with lithio-dithiane are governed by kinetic control,[81] while other nucleophiles (carbanions stabilized by nitrile or cyanohydrine) gave different regioselectivities in the nucleophilic addition depending on the selected conditions.[82]

Scheme 37

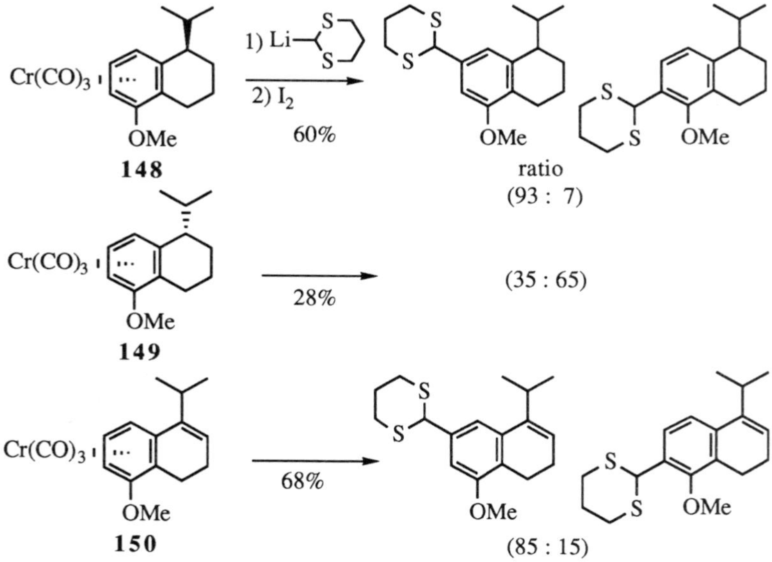

1) Li–(1,3-dithian-2-yl) 2) I_2 60%

148 → ratio (93 : 7)

149 → 28% (35 : 65)

150 → 68% (85 : 15)

Therefore, we next attempted to invert the face of the chromium complexation from **144** to **145** in order to achieve *meta* nucleophilic addition. Complex **145** could adopt the *syn*-eclipsed structure **147** because it is free from the steric interaction between the butenyl group and CO ligands. On oxidative demetallation, and subsequent re-complexation with $Cr(CO)_6$, **144** gave the inverted 4-*exo*-substituted complex **145** in 60% yield, but still accompanied by the diastereoisomer **144** in 20% yield (Scheme

36). In this route, the $Cr(CO)_6$ reagent has to be used repeatedly. Therefore, we turned our efforts to an alternative, simpler, method for the synthesis of the key intermediate **145**.

According to Scheme 38, the *endo* acetate complex **151**, obtained from complex **141**, was reacted with (*E*)-crotyl trimethylsilane[83] in the presence of boron trifluoride etherate to afford the stereochemically desirable C-4-*exo*-substituted tetralone complex **152** in 72% yield, accompanied by C-11 stereoisomer (24%). The corresponding (Z)-crotyl silane produced a 1:1 diastereomeric mixture. Stereoselective introduction of *endo*-methyl at C-1 position was straightforward. Treatment of complex **152** with MeLi, followed by hydrogenolysis of the resulting carbinol, produced the 1-*endo*-4-*exo*-substituted tetralin complex **145** in 45% overall yield. Nucleophilic addition of dithianyl carbanion to complex **145**, followed by oxidative demetallation, gave the 6-dithianylated tetralin **153**, as expected, in 50% yield, without detectable amount of regioisomers. The acetate compound **154**, derived from **153** by de-methylation and acetylation, was converted to the coupling product **155** by reaction with methyl β-bromomethacrylate in the presence of palladium catalyst after a hydroboration.[84] Reduction of the ester group in **155** followed by acetylation and subsequent hydrolysis of 1,3-dithianyl group produced the aldehyde complex **156**. Compound **156** was oxidized[85] to the phenolic methylester **157**, which was successfully converted to (±)-dihydroxyserrulatic acid (**137**) by basic hydrolysis.

Scheme 38

141 →(1, 2) **151** →(3) **152** →(4,5) 145 →(6~8) **153** R=Me, **154** R=Ac →(9) **155** →(10~12) **156** →(13) **157**

Reagent:
1) $LiAlH_4$ (95%), 2) Ac_2O (98%), 3) (E)-$MeCH{=}CHCH_2SiMe_3/BF_3{\cdot}OEt_2$ (72%), 4) MeLi (60%), 5) Et_3SiH/CF_3CO_2H (75%), 6) 2-lithio-1,3-dithian then I_2 (50%), 7) EtSH/NaH/DMF (95%), 8) Ac_2O (96%), 9) 9-BBN, then (E)-methyl β-bromomethacrylate/$PdCl_2$(dppf)/K_2CO_3/H_2O (77%), 10) Dibal (90%), 11) Ac_2O (98%), 12) $HgO/BF_3{\cdot}OEt_2/H_2O$ (69%), 13) $NaCN/MnO_2$/MeOH/AcOH (85%)

12. OPTICALLY ACTIVE (ARENE)CHROMIUM COMPLEXES AND THEIR UTILITY IN CATALYTIC ASYMMETRIC SYNTHESIS

(η^6-Arene)chromium complexes can exist in two enantiomeric forms when the phenyl ring is *ortho*- or *meta*-disubstituted. This planar chirality can be transferred into new central chirality with a high enantiomeric excess by means of appropriate organic reactions. The synthesis of optically active forms is an important subject for further development of (arene)chromium complexes. An excellent review by A. Solladié-Cavallo has appeared.[17] The racemic acids **158** and amines **159** have been resolved through crystallization of diastereomeric ammonium salts obtained with optically pure chiral amines[86] or acids.[87] More versatile complexes, as the aldehyde complexes **160**, have been easily resolved by column chromatography of the diastereomeric derivatives obtained with (*S*)-(-)-5-(α-phenylethyl)-semioxamazide[88] or (*S*)-(+)-valilol.[89] Reductive resolution of substituted benzaldehyde chromium complexes is achieved by bakers yeast or HLADH to produce (*R*)-aldehyde.[90]

$Cr(CO)_3$ (arene with X and R)

158 X=COOH
159 X=NH_2
160 X=CHO

We also investigated the kinetic resolution of (*ortho*-substituted benzylalcohol)chromium complexes by asymmetric esterification with a lipase.[91] Reaction of racemic (*o*-methylbenzyl alcohol) $Cr(CO)_3$ with isopropenyl acetate in the presence of the lipase from *Pseudomonas* sp. (Amano P) gave the corresponding (*S*)-acetate, 98%ee, 48% yield, remaining the (*R*)-alcohol, 100%ee, in 47% yield. *o*-Methoxy and *o*-TMS-substituted substrates were also *trans*acylated enantioselectively with Amano AK and Toyobo Type A (both from *Pseudomonas* sp.) as shown in Table 13.

Table 13. Kinetic resolution by lipase

161 → (lipase, isopropenyl acetate) → **162** (*S*, OAc) + **163** (*R*, OH)

X	lipase	acetate **162** % ee (yield %)	alcohol **163** % ee (yield %)
Me	Amano P	98 (48)	100 (47)
OMe	Amano K	97 (47)	95 (46)
$SiMe_3$	lipoprotein Type A	84 (45)	85 (48)

Similarly, optically active (η^6-arene)chromium complexes can be prepared by several other methods, *e.g.*, diastereoselective *ortho* lithiation[92] of the (chiral benzaldehyde acetal)$Cr(CO)_3$, or diastereoselective chromium complexation[93] of the chiral benzaldehyde aminoacetal. However, the employment of a stoichiometric amount of the chiral reagents is required for preparation of the chiral (arene)chromium complexes. We have next examined the *catalytic* asymmetric synthesis of optically active (arene)chromium complex *via* asymmetric cross-coupling reaction of a *meso* (arene)chromium complex in the presence of a chiral palladium catalyst.

An oxidative addition of the C-Cl bond of the arene compounds to palladium(0) is usually difficult. However, this limiting step is favoured by the coordination of an electron-withdrawing $Cr(CO)_3$ group to the arene ring.[94] The selective mono-substitution of one of the enantiotopic chlorine atoms of (*o*-dichlorobenzene)chromium complex (**164**) would result in the formation of optically active (η^6-arene)-chromium complexes.[95] Reaction of **164** with tributyl(vinyl)stannane catalyzed by palladium complexes coordinated with chiral ligands gave good yields of the mono-coupling product **165**, though in a racemic form. In the presence of chiral monophosphine ligand, the reaction of the vinylstannane gave the di-coupling product **166** as a major product. Since no enantioselectivity was observed in the coupling reaction with vinylstannane reagent, the coupling reaction with other vinylmetals was investigated for the achievement of high enantioselectivity. Althogh vinylmagnesium bromide gave low yield of the coupling products, the corresponding zinc reagent afforded 42% ee of the (–)-mono coupling product **165** in 44% yield catalyzed by 10 mol% of di-μ-chlorobis(π-allyl)palladium(II) and 12 mol% of (*S*)-(*R*)-PPFA. The asymmetric reaction with vinylboric acid catalyzed by palladium in the presence of PPFA in aqueous thallium hydroxide gave 38% ee of the mono coupling product. Comparable enantioselectivities were obtained in the reactions with substituted vinylboric acids. The ratio of mono- to di-coupling products was improved with α-methylvinylboric acid in the presence of PPFA, probably due to stereoelectronic effects of the electron-donating methyl group. Reaction with the corresponding boric ester gave similar enantioselectivity. The results are summarized in Table 14.

Table 14. Catalytic asymmetric cross-coupling reaction of *meso* (dichlorobenzene)$Cr(CO)_3$

$Cr(CO)_3$ (Cl, Cl) **164** —Vinylic Metal (RM), $[PdCl(\pi\text{-}C_3H_5)]_2$, Chiral Ligand (L*)→ $Cr(CO)_3$ ((1*S*)Cl, (2*R*)R) **165** (+ $Cr(CO)_3$ (R, R) **166**)

entry	vinylic metal	ligand (L*)	°C, hrs	ratio of **165** : **166** (yield %)	% ee of **165**
1[a]	CH_2=$CHSnBu_3$	(*R*)-BINAP	40, 18	75 : 25 (80)	0
2[a]	CH_2=$CHSnBu_3$	(*S*)-(*R*)-PPFA	40, 18	87 : 13 (46)	0
3[a]	CH_2=CHMgBr	(*S*)-(*R*)-PPFA	50, 48	75 : 25 (8)	–
4[a]	CH_2=CHZnCl	(*S*)-(*R*)-PPFA	40, 18	67 : 23 (56)	42
5[a]	CH_2=CHZnCl	(*R*)-MeO-MOP	0, 18	24 : 76 (93)	0
6[b]	CH_2=$CHB(OH)_2$	(*S*)-(*R*)-PPFA	23, 48	73 : 27 (59)	38
7[b]	CH_2=$CMeB(OH)_2$	(*S*)-(*R*)-PPFA	27, 48	95 : 5 (64)	44
8[b]	CH_2=$CMeB(OH)_2$	(*R*)-BINAP	35, 48	47 : 53 (93)	25
9[b]	CH_2=$CMe[O(CH_2)_3O]$	(*S*)-(*R*)-PPFA	35, 48	76 : 24 (63)	39
10[b]	(*E*)-BuCH=$CH_2B(OH)_2$	(*S*)-(*R*)-PPFA	27, 18	77 : 23 (53)	44

[a] THF was used as solvent. [b] Reaction with vinylic boric acids was carried out in the presence of 3 eq of 0.4M TlOH in aqueous THF solution.

Thus, we have succeeded, for the first time, in the catalytic asymmetric synthesis of optically active molecules whose chirality is based on planar chirality due to 1,2-unsymmetrically substituted (η^6-arene)chromium structure. The stereochemical outcome in the present catalytic asymmetric cross-coupling reaction should be determined at the oxidative addition step where one of the enantiotopic carbon-chlorine bonds reacts selectively with a chiral palladium(0) species forming palladium-carbon bond. The loss of enantioselectivity observed with the vinylstannane may imply that the metal fragment of the vinylation reagents also plays an important role on the oxidative addition.

The development of catalytic enantioselective reactions is now recognized as one of the most important and challenging problems in organic synthesis. Nucleophilic addition of organometallic reagents to carbonyl substrates constitutes one of the most fundamental operations. Modification of the organometallic compounds by chiral, nonracemic auxiliaries offers a general way to create optically active alcohols, and the catalytic version in particular provides maximum synthetic efficiency. The use of organozinc compounds allows catalytic enantioselective alkylation of aldehydes leading to a diverse array of secondary alcohols of high optical purity. Several groups have reported that the chiral amino alcohols not only accelerate the alkylation of aldehydes with dialkylzinc compounds but also govern the stereochemical outcome in an absolute sense.[96] Chiral β-amino alcohols have been most effective for the preparation of secondary alcohols with high optical purity, presumably owing to the formation of cyclic five-membered transition states by coordination of the zinc metal with two heteroatoms. We have examined the effectiveness of optically active (1,2-disubstituted arene)chromium complexes having δ- or γ-amino alcohols as the chiral ligands in catalytic asymmetric reactions.

The chiral ligands **167~172** used in this work were stereoselectively prepared from commercially available (*R*)-α-phenethylamine.[97] (η^6-Arene)chromium complexes with an electron-donating phosphine or phosphite ligand **171** were synthesized from the corresponding tricarbonyl ligand by irradiation in benzene solution with a high pressure mercury lamp. The chromium complexes **173** and **174** were obtained from resolved (o-*N*,*N*-dimethylamino benzaldehyde)$Cr(CO)_3$.

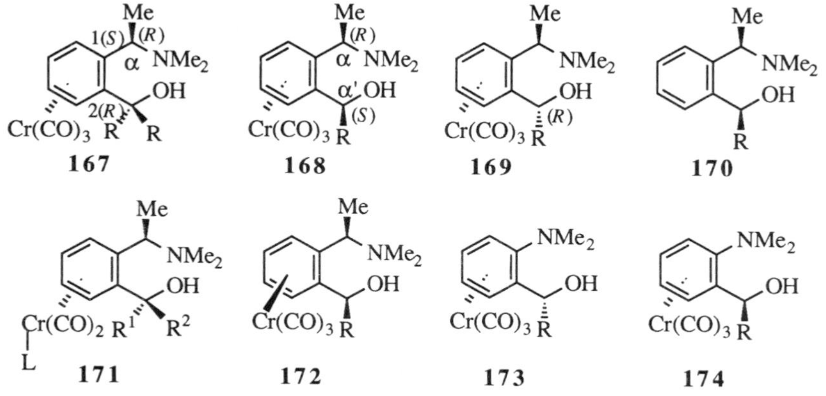

167 **168** **169** **170**

171 **172** **173** **174**

Table 15. Catalytic asymmetric ethylation of benzaldehyde

PhCHO → (Et_2Zn, 5mol% of (arene)chromium complex, toluene, 0°C) → PhCH(OH)Et

entry	complex	yield (%)	% ee	absolute config.
1	**167** (R=H)	70	15	*S*
2	**167** (R=Et)	89	95	*S*
3	**167** (R=C_6H_5)	83	93	*S*
4	**168** (R=Me)	83	63	*S*
5	**168** (R=Et)	87	93	*S*
6	**168** (R=*i*-Pr)	78	81	*S*
7	**168** (R=C_6H_5)	82	83	*S*
8	**168** (R=*o*-$CF_3C_6H_4$)	97	89	*S*
9	**169** (R=Et)	87	50	*S*
10	**170** (R=Et)	69	24	*S*
11	**171** (R^1=R^2=Et, L=PPh_3)	99	97	*S*
12	**171** (R^1=R^2=C_6H_5, L=PPh_3)	99	94	*S*
13	**171** (R^1=R^2=Et, L=$P(OPh)_3$)	96	97	*S*
14	**171** (R^1=H, R^2=Et, L=PPh_3)	97	96	*S*
15	**172** (R=Et)	82	29	*R*
16	**173** (R=Me)	58	11	*S*
17	**174** (R=C_6H_5)	56	12	*S*

Reaction of the diethylzinc with benzaldehyde was carried out in the presence of the chiral (1,2-disubstituted arene)chromium complexes in order to examine the structural effect of the complexes upon enantioselectivity (Table 15).[98] The complex **167** having ethyl or phenyl group without the chiral center at the α'-position gave the α(*S*)-phenylpropyl alcohol in high enantioselectivity.[98] These chiral ligands afforded high enantiomeric excess in the asymmetric ethylation of benzaldehyde with diethylzinc, regardless of the δ-amino alcohols. However, the unsubstituted complex **167** (R=H) at the α'-position exhibited very low selectivity. The effect of chirality at the benzylic alcohol moiety in the ligands was next examined in the enantioselective ethylation of benzaldehyde. Complexes **168** with the α'(*S*)-configuration gave a high degree of enantioselectivity in the catalytic ethylation of benzaldehyde giving the α(*S*)-phenylpropyl alcohol. Particularly, the (*S*)-ethyl substituent complex was very effective. The complexes with methyl or sterically bulky substituents, *e.g.* aromatics and isopropyl, resulted in lower enantioselectivities. The corresponding chromium free chiral amino alcohol **170** (R=Et) resulted in less efficient 24% ee giving the (*S*)-product (entry 10). Thus, $Cr(CO)_3$ complexation of the ligands and a proper steric size of the substituents at the alcohol moiety have a large effect on the enantioselectivity in the asymmetric ethylation of benzaldehyde.

On the other hand, stereoisomeric α'(*R*)-chromium complexes **169** are less effective in producing the same (*S*)-alcohol (entry 9). Since both the stereoisomeric chromium complexes **168** and **169** gave the same (*S*)-alcohol, the direction of the asymmetric induction is independent of the chirality of alcohol moiety in the chromium complexes. From these results, the chromium complexes with the (*S*)-configuration or disubstituents at the benzylic hydroxyl position are significant for the achievement of high enantioselectivity. In contrast with the Ar(1*S*,2*R*) complexes **168**, the face-inverted Ar(1*R*,2*S*)-chromium complex **172** gave the (*R*)-alcohol in 25% ee (entry 15). Thus, the face chirality of the (arene)chromium complexes determines the absolute configuration of the product.

The effect of electron-donating ligands on the chromium was examined in the catalytic asymmetric ethylation. The enantioselectivities and yields of the product with the chromium complex **171** possessing triphenylphosphine or triphenylphosphite increased (entries 11-14).

The proposed transition state **175** for the enantioselective addition of diethylzinc to benzaldehyde with chromium complexes **168** is postulated as follows. The zinc alkoxide is formed as a seven-membered ring in the *exo*-configuration to avoid severe steric interaction with the $Cr(CO)_3$ and α'-benzylic substituents. The ethylation can be interpreted in terms of a six-membered cyclic transition state with an equatorial orientation of the phenyl ring of the benzaldehyde, attacking the *si*-face of the aldehyde.

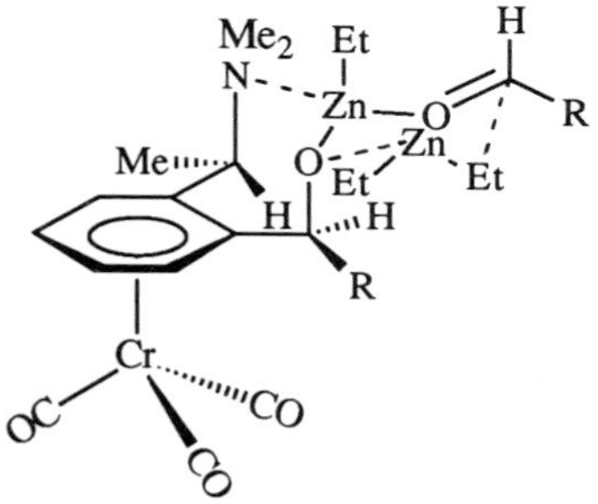

175

Unfortunately, both the stereoisomeric homologous γ-amino alcohol (arene)chromium complexes as chiral ligands, which would form six-membered transition state, resulted in poor enantioselectivities (entries 16,17).

In conclusion, the direction of $Cr(CO)_3$ complexation, the chirality of benzylic alcohol moiety, and the coordinating ring size of the chiral (1,2-disubstituted arene)chromium complexes are important factors in the catalytic asymmetric ethylation of benzaldehyde.

Enantioselective conjugate additions of organometallic reagents to prochiral enones afford synthetically useful optically active β-substituted ketones. Rapid progress has recently been made in the field by modifying the organocopper reagents with nontransferable chiral ligands. In some cases, the enantioselective conjugate additions of organocopper reagents have been achieved by addition of chiral ligands. However, most studies have employed stoichiometric conditions. A more attractive approach would affect enantioselective conjugate additions by use of catalytic amounts of chiral ligands. It was recently reported[99] that the chiral β-amino alcohols and related compounds catalyze the conjugate addition of dialkylzincs to enones in the presence of transition metal. We have examined the reactivities and enantioselectivities in the asymmetric conjugate addition of diethylzinc with the above cited (1,2-disubstituted arene)chromium complexes as chiral ligands.[100]

Conjugate addition reactions were carried out in acetonitrile at –30°C with 2 equiv. of diethylzinc in the presence of Ni(II) catalyst, generated *in situ* by mixing $Ni(acac)_2$ and the chiral (arene)$Cr(CO)_3$ (Table 16). The diethyl complex **167** (R=Et), the most efficient ligand for the asymmetric ethylation of benzaldehyde, gave a racemic 1,3-diphenylpentan-1-one under catalytic conditions with 5 mol % of $Ni(acac)_2$ and 50 mol % of the chromium complex (entry 1). Use of stoichiometric amounts of the Ni(II) and complex **167** (R=Et) produced the (*R*)-conjugate addition product in 37% ee (entry 2). The asymmetric induction is highly dependent on the amount of the chiral catalyst. The corresponding diphenyl-substituted complex **167** ($R=C_6H_5$) resulted in higher selectivity. Among the (S)-configuration complexes, the phenyl-substituted complex **168** ($R=C_6H_5$) is an efficient chiral ligand for conjugate addition in both catalytic (62% ee) and stoichiometric (78% ee) reactions.

Table 16. Enantioselective conjugate addition to chalcone catalyzed by Ni(II)/(arene)$Cr(CO)_3$ system

Et_2Zn/$Ni(acac)_2$, (arene)$Cr(CO)_3$, CH_3CN

entry	(arene)$Cr(CO)_3$	mol % $Ni(acac)_2$	ratio of Ni(II):(arene)$Cr(CO)_3$	yield (%)	% ee (config)
1	**167** (R=Et)	5	1 : 10	92	0
2	**167** (R=Et)	100	1 : 1	70	37 (*R*)
3	**167** ($R=C_6H_5$)	10	1 : 10	83	44 (*R*)
4	**167** ($R=C_6H_5$)	100	1 : 1	71	62 (*R*)
5	**168** (R=Et)	5	1 : 10	91	43 (*R*)
6	**168** (R=Et)	100	1 : 1	72	53 (*R*)
7	**168** ($R=C_6H_5$)	5	1 : 10	90	62 (*R*)
8	**168** ($R=C_6H_5$)	100	1 : 1	70	78 (*R*)

Since it is possible to modify chiral ligands, it seems that there is considerable opportunity for further improvement in this class of chiral (arene)chromium complexes.

13. ACKNOWLEDGMENTS This work was mainly supported by a Grant-in-Aid from the ministry of Japanese Education. A number of graduate students, whose names appear in the papers cited, have contributed to our work in this area. In particular, the contributions made by T. Minami, K. Isobe, T. Kobayashi, H. Nishimura, and R. Miyake are gratefully acknowledged. I would also like to thank prof. Y. Hayashi for helpful and insightful discussions.

REFERENCES AND NOTES

1 For reviews:
(a) M. F. Semmelhack,
Ann. N. Y. Acad. Sci., **295** (1977) 36.
(b) G. Jaouen,
Transition Metal Organometallics in Organic Synthesis. Vol. 2. *Arene Complexes in Organic Synthesis*; H. Alper ed, Academic Press, New York, **1978**, p. 65.
(c) J. P. Collman, L. S. Hegedus, J. R. Norton and R. G. Finke,
Princeples and Applications of Organotransition Metal Chemistry, University Science Books, Mill Valley, Calf., **1987**, p. 921.
(d) S. G. Davies,
Organotransition Metal Chemistry, Applications to Organic Synthesis, Pergamon Press, Oxford, **1982**.
(e) A. J. Pearson,
Metallo-Organic Chemistry, John Wiley, Chichester, **1985**, p. 348.
(f) R. Davis and L. A. P. Kane-Maquire,
Comprehensive Organometallic Chemistry, Chromium Compounds with η^2-η^8 *Carbon Ligands*; G. Wilkinson ed, Pergamon Press, Oxford, **1982**, Vol. 3, p. 953.
(g) E. P. Kündig,
Pure Appl. Chem., **57** (1985) 1855.

2 (a) N. Nicholls and M. C. Whiting,
J. Chem. Soc., **(1959)** 551.
(b) E. O. Fischer and K. Ofele,
Z. Naturforsch. Teil B, **13** (1958) 458.
(c) E. O. Fischer, K. Ofele, H. Esseler, W. Frohlich, J. P. Mortensen and W. Semmlinger,
Chem. Ber., **91** (1958) 2763.
(d) G. Natta, R. Ercoli, F. Calderazzo and E. Santambrogio,
Chim. Ind. (Milan), **40** (1958) 1003.

3 (a) W. E. Siverthorn,
Adv. Organomet. Chem., **13** (1975) 47.
(b) M. D. Rausch,
J. Org. Chem., **39** (1974) 1787.
(c) W. P. Anderson, N. Hsu, C. W. Stanger Jr. and B. Munson,
J. Organomet. Chem., **69** (1974) 249.
(d) G. A. Moser and M. R. Rausch,
Synth. React. Inorg. Metal. Org. Chem., **4** (1974) 38.

4 (a) S. Top and G. Jaouen,
J. Organomet. Chem., **182** (1979) 381.
(b) C. A. L. Mahaffy and P. L. Pauson,
Inorg. Synth. **19** (1979) 154.

5 W. Strohmeier,
Chem. Ber., **94** (1961) 2490.

6 This figure is adapted from excellent literature
(a) M. F. Semmelhack, G. R. Clark, G. L. Garcia, J. J. Harrison, Y. Thebtaranonth, W. Wulff, and A. Yamashita,
Tetrahedron, **37** (1981) 3957.
(b) see ref. 1(c).

7 M. Uemura, K. Isobe, and Y. Hayashi,
Tetrahedron Lett., **26** (1985) 767.

8 D. N. Kursanov, Z. N. Parnes, and N. M. Lion,
Synthesis, **(1974)** 633.

9 M. Uemura, T. Kobayashi, and Y. Hayashi,
Synthesis, **(1986)** 386.

10 M. Uemura, T. Minami, and Y. Hayashi,
J. Organomet. Chem., **299** (1986) 119.

11 M. T. Reetz and M. Sauerwald,
Tetrahedron Lett., **24** (1983) 2387.

12 Also, O-, N-, and S-nucleophiles other than carbon nucleophiles could be easily trapped with $Cr(CO)_3$-stabilized carbocations:

(a) S. Top, A. Meyer, and G. Jaouen,
Tetrahedron Lett., **(1979)** 3537.
(b) S. Top and G. Jaouen,
J. Chem. Soc., Chem. Commun., **(1979)** 224.
(c) Idem.,
J. Org. Chem., **46** (1981) 78.
(d) S. Top, G. Jaouen, and McGlinchey,
J. Chem. Soc., Chem. Commun., **(1980)** 1110.

13 (a) G. Jaouen and R. Debard,
Tetrahedron Lett., **(1971)** 1015.
(b) A. Meyer and G. Jaouen,
J. Chem. Soc., Chem. Commun., **(1974)** 787.
(c) G. Jaouen and A. Meyer,
J. Am. Chem. Soc., **97** (1975) 4667.

14 (a) A. Meyer and R. Dabard,
J. Organomet. Chem., **36** (1972) C38.
(b) J. Besancon, J. Tirouflet, A. Card, and Y. Dusausoy,
Ibid., **59** (1873) 267.
(c) A. Solladié-Cavallo and J. Suffert,
Tetrahedron Lett., **25** (1984) 1897.
(d) Idem.,
Synthesis, **(1985)** 659.
(e) A. Solladié-Cavallo, D. Farkhani, A. C. Dreyfus, and F. Sanch,
Bull. Chem. Soc. Chim. Fr., **(1986)** 906.
(g) A. Solladié-Cavallo, G. Lapitajs, P. Buchert, A. Klein, S. Colonna, and A. Manfredi,
J. Organomet. Chem., **330** (1987) 357.

15 The symbol (*S**,*S**) represents an enantiomeric mixture of (*S*,*S*) and (*R*,*R*)-chromium complexes. The picture is shown in one enantiomeric form for clarity. The first symbol indicates the configuration of the aromatic part co-ordinated with $Cr(CO)_3$, and the second one shows that of the chiral center at the benzylic position.

16 M. Uemura, T. Kobayashi, K. Isobe, T. Minami, and Y. Hayashi,
J. Org. Chem., **51** (1986) 2859.

17 A. Solladié-Cavallo,
Advances in Metal-Organic Chemistry, Vol. 1, "*Chiral Arene Chromium Complexes*"; L. S. Liebeskind ed, JAI press, Greenwich, **1989**, p. 99.

18 (a) K. Schlögl,
Topics in Stereochemistry, Vol. 1; E. L. Eliel and N. L. Allingers, Eds, Wiley-Interscience, New York, **1967**.
(b) G. Paiaro and A. Panunzi,
J. Am. Chem. Soc., **86** (1964) 5148.

19 Chromium complexation of 1-hydroxyindane with (trispyridine)$Cr(CO)_3$ in the presence of boron trifluoride etherate gave endo-hydroxy complex, but the yield was very low:
D. E. Gracey, W. R. Jackson, W. B. Jennings, and T. R. B. Mitchell,
J. Chem. Soc., (B), **(1969)**,1204.

20 M. Uemura, T. Kobayashi, T. Minami, and Y. Hayashi,
Tetrahedron Lett., **27** (1986) 2479.

21 D. E. Gracey, W. R. Jackson, C. H. McMullen, and N. Thompson,
J. Chem. Soc., (B), **(1969)** 1167.

22 (a) E. P. Kündig, C. Perret, S. Spichinger, and G. Bernadienelli,
J. Orgnomet. Chem., **286** (1985) 183.
(b) E. P. Kündig, V. Desobry, C. Grivet, B. Rudolph, and S. Spichinger,
Organometallics, **6** (1987) 4505.
(c) T. G. Traylor, M. J. Goldberg,
Ibid., **6** (1987) 2551.

23 J. A. Heppert, J. Aube, M. E. Thomas-Miller, M. L. Milligan, and F. Takusagawa,
Organometallics, **9** (1990) 727.

24 J. Blagg, S. G. Davies, C. L. Goodfellow, and K. H. Sutton,
J. Chem. Soc., Perkin Trans 1, **(1987)** 1805.

25 Optical pure (*S*)-alcohol **42** was prepared by fractional recrystallization of brucine salt of an adduct from (*DL*)-compound and phthalic anhydride. Optical purity was determined by ^{1}H-NMR of Mosher ester.

26 A. I. Meyers, R. K. Smith, and C. E. Wintten,
J. Org. Chem., **44** (1979) 2250.

27 W. S. Trahanovsky and R. J. Card,
J. Am. Chem. Soc., **94** (1972) 2897.

28 M. Acampora, A. Ceccon, M. Dalfarra, G. Giacometti, and G. Rigatti,
J. Chem. Soc. Perkin Trans 2, **(1977)** 483

29 M. F. Semmelhack and A. Yamashita,
J. Am. Chem. Soc., **102** (1980) 5924.

30 (a) M. Uemura, N. Nishikawa, and Y. Hayashi,
Tetrahedron Lett., **21** (1980) 5924.
(b) M. Uemura, N. Nishikawa, K. Take, M. Ohnishi, K. Hirotsu, T. Higuchi, and Y. Hayashi,
J. Org. Chem., **48** (1983) 2349.
(c) M. Uemura, K. Take, K. Isobe, T. Minami, and Y. Hayashi,
Tetrahedron, **41** (1985) 5771.

31 (a) M. Uemura, S. Tokuyama, and T. Sakan,
Chem. Lett., **(1975)** 1195.
(b) H. O. House, R. C. Strincland, and E. J. Zaiko,
J. Org. Chem., **41** (1976) 2401.
(c) B. M. Trost, G. Rivers, and J. M. Gold,
Ibid., **45** (1980) 1835.
(d) M. R. Winkle and R. C. Ronald,
Tetrahedron, **39** (1983) 2031.

32 P. Albaugh-Robertson and J. A. Katzenellenbogen,
J. Org. Chem., **48** (1983) 5288.

33 J. L. Wood, N. A. Khatri, and S. M. Weinreb,
Tetrahedron Lett., **(1979)** 4907.

34 (a) M. Uemura, T. Minami, and Y. Hayashi,
J. Chem. Soc., Chem. Commun., **(1984)** 1193.
(b) M. Uemura, K. Isobe, and Y. Hayashi,
Chem. Lett., **(1985)** 91.

35 M. Uemura, T. Minami, and Y. Hayashi,
J. Am. Chem. Soc., **109** (1987) 5277.

36 (a) M. Uemura, T. Minami, K. Hirotsu, and Y. Hayashi,
J. Org. Chem., **54** (1989) 469.
(b) M. Uemura, T. Minami, and Y. Hayashi,
Tetrahedron Lett., **29** (1988) 6271.

37 (a) D. F. DeTar and N. P. Luthra,
J. Am. Chem. Soc., **102** (1980) 4505.
(b) C. Exon and P. Magnus,
Ibid., **105** (1983) 2477.

38 These compounds **77** were prepared by the following literature method:
T. Cubigny, M. Larcheveque, and H. Normant,
Synthesis, **(1978)** 857.

39 For reviews:
(a) D. A. Evans, J. V. Nelson, and T. R. Taber,
Top. Stereochem., **13** (1982) 1.
(b) C. H. Heathcock,
Asymmetric Synthesis, J. D. Morrison ed, Academic Press, New York, **1984**, Vol. 3, p. 111.
(c) S. Masamune, W. Choy, J. C. Petersen, and L. R. Sita,
Angew. Chem. Int. Ed. Engl., **24** (1983) 1.

40 (a) M. Braun,
Angew. Chem. Int. Ed. Engl., **26** (1987) 24.
(b) R. Devant, U. Mahler, and M. Braun,
Chem. Ber., **121** (1988) 397.

41 (a) L. S. Liebeskind, M. E. Welker, and V. Goedken,

J. Am. Chem. Soc., **106** (1984) 441.
(b) L. S. Liebeskind, M. E. Velker, and R. W. Fengel,
Ibid., **108** (1986) 6328.
(c) S. G. Davies, I. M. Dorder-Hedgecook, and P. Warner,
J. Chem. Soc., Chem. Commun., **(1984)** 956.
(d) S. G. Davies, I. M. Dorder-Hedgecook, P. Warner, R. H. Jones, and K. Prout,
J. Organomet. Chem., **285** (1985) 213.
(e) I. Ojima and H. B. Kwon,
J. Am. Chem. Soc., **110** (1988) 5617.
(f) W. D. Wulff and S. R. Gilbertson,
Ibid., **107** (1985) 503.
(g) E. R. Burkhardt, J. J. Doney, R. G. Bergman, and C. H. Heathcock,
Ibid., **109** (1987) 2022.
(h) G. A. Slough, R. G. Bergman, and C. H. Heathcock,
Ibid., **111** (1989) 938.

42 M. Uemura, T. Minami, and Y. Hayashi,
Tetrahedron Lett., **30** (1989) 6383; *Ibid.*, **31** (1990) 2218.

43 G. Jaouen and R. Dabard,
J. Organomet. Chem., **72** (1974) 377.

44 Y. Li, M. N. Paddon-Row, and K. N. Houk,
J. Am. Chem. Soc., **110** (1988) 3684.

45 I. Paterson and J. M. Goodman,
Tetrahedron Lett., **30** (1989) 937.

46 (a) G. H. Posner,
Org. React. 19 (1972) 1.
(b) B. H. Lipshutz, R. S. Wihelm, J. A. Kozlowsky,
Tetrahedron, **40** (1984) 5005.
(c) Y. Yamamoto,
Angew. Chem. Int. Ed. Engl. **25** (1986) 947.
(d) B. H. Lipshutz,
Synthesis, **(1987)** 325.
(e) W. Oppolzer,
Tetrahedron, **43**, (1987) 1969.
(f) W. Oppolzer, A. J. Kingma, G. Poli,
Ibid., **45** (1989) 479.
(g) E. J. Corey, F. J. Hannon, N. W. Boaz,
Ibid., **45** (1989) 545.

47 M. Uemura, H. Oda, T. Minami, M. Shiro, Y. Hayashi,
Organometallics, **11** (1992) 3705;
M. Uemura, H. Oda, T. Minami, Y. Hayashi,
Tetrahedron Lett. **32** (1991) 4565.

48 Y. Yamamoto, K. Maruyama,
J. Am. Chem. Soc. **100** (1978) 3240;
Yamamoto's reagent ($RCu{\cdot}BF_3$) is recognized as a composition of $RCu^{I}Li{+}BF_3$,
B. H. Lipshutz, E. L. Ellsworth, S. H. Dimock,
J. Am. Chem. Soc. **112** (1990) 5869.

49 The absolute configuration was determined by rotation value of *o*-methoxyphenyl 2-methylhexyl ketone;
W. Oppolzer, H. Loher,
Helv. Chim. Acta, **64** (1981) 2808.

50 Optically active ((*S*)-2-methyl-3-*tert*-butoxypropyl)copper-boron trifluoride (**93**) was prepared via (*S*)-(+)-3-butoxy-2-methyl-1-bromopropane from methyl (*S*)-(+)-3-hydroxyisobutyrate;
(a) H. Nagaoka, Y. Kishi,
Tetrahedron, **37** (1981) 3873.
(b) N. C. Cohen, W. F. Eichel, R. J. Lopresti, C. Newkom, G. Saucy,
J. Org. Chem. **41** (1976) 3505.

51 (a) A. Solladié-Cavallo,
Chiral-Arene-Chromium-Carbonyl Complexes in Asymmetric Synthesis, in *Advances in Metal-Organic Chemistry*; L. S. Liebeskind ed, JAI Press, Greenwich, **1989** Vol. 1, p. 99.

(b) J. Besançon, J. Tirouflet,
Bull. Soc. Chim. Fr. **(1969)** 861.
(c) J. Besançon, G. Tainturier, J. Tirouflet,
Ibid. **(1971)** 1804.
52 *Asymmetric Synthesis*, J. Morrison ed, Academic Press, New York, Vol. 3, **1984**.
53 A recent representative 1,5-stereocontrol:
(a) C. H. Heathcock, B. L. Finkelstein, E. J. Jarvi, P. A. Radel, and C. R. Hadley,
J. Org. Chem., **53** (1988) 1922.
(b) H. Takaya, T. Ohta, N. Sayo, H. Kumobayashi, S. Akutagawa, S. Inoue, I. Kasahara, and R. Noyori,
J. Am. Chem.Soc., **109** (1987) 1596.
(c) T. Harada, T. Hayashiya, I. Wada, N. Iwa-ake, and A. Oku,
Ibid., **109** (1987) 527.
(d) K. Tomooka, T. Okinaga, K. Suzuki, and G. Tsuchihashi,
Tetrahedron Lett., **28** (1987) 6335.
(e) T. Ibuka, T. Nakano, S. Nishi, and Y. Yamamoto,
J. Am. Chem. Soc., **108** (1986) 7420.
(f) C. H. Heathcock and P. A. Radel,
J. Org. Chem., **51** (1986) 4322.
54 (a) K. K. Chan, N. C. Cohen, J. P. DeNoble, A. C. Specian, Jr., and G. Saucey,
J. Org. Chem., **41** (1976) 3497.
(b) P. Schudel, H. Mayer, and O. Isler:
Vitamines, J. Morrison ed, Academic Press, New York, 1972, Vol. V, p.168.
55 (a) D. M. Jewett, F. Matsumura and H. C. Coppel,
Science, **192** (1976) 51.
(b) T. Suzuki,
Agric. Biol. Chem., **45** (1981) 1357.
(c) S. Bystrom, H. E. Hogerg, and T. Norin,
Tetrahedron, **37** (1981) 2249.
56 For reviews:
(a) B. M. Trost,
Acc. Chem. Res., **13** (1980) 385.
(b) J. Tsuji, *Organic Synthesis with Palladium Compounds*, Springer-Verlag, New York, **1980**.
(c) J. Tsuji,
Pure Appl. Chem., **54** (1982) 197.
57 (a) S. J. Rhodes and N. R. Raulines,
Org. React., **22** (1975) 1.
(b) R. K. Hill,
Asymmetric Synthesis, J. D. Morrison ed, Academic Press, New York, Vol. 3, Ch. 8, **1984**.
(c) F. E. Ziegler,
Acc. Chem. Res., **10** (1977) 227.
58 T. Hayashi, A. Yamamoto, and T. Hagihara,
J. Org. Chem., **51** (1986) 723 and references cited therein.
59 T. J. Gould, M. Balestra, M. D. Wittman, J. A. Gary, L. T. Rossano, and J. Kallmerten,
J. Org. Chem., **52** (1987) 861.
60 D. H. R. Barton, P. D. Magnus, G. Smith, and D. Zurr,
J. Chem. Soc., Chem. Commun., **1971**, 861.
61 (a) R. W. Hoffman,
Angew. Chem. Int. Ed. Engl., **21** (1982) 555.
(b) Y. Yamamoto,
Acc. Chem. Res., **20** (1987) 243.
(c) Y. Yamamoto and K. Maruyama,
Heterocycles, **18** (1982) 357.
62 R. E. Sjoholm,
Acta. Chem. Scand., **44** (1990) 82.
63 D. Seebach and L. Widler,
Helv. Chim. Acta., **65** (1982) 1972.
64 M. T. Reetz, R. Steinbach, J. Westermann, R. Peter, and B. Wenderoth,

Chem. Ber., **118** (1985) 1441.
65 M. Uemura, T. Minami, K. Isobe, T. Kobayashi, and Y. Hayashi,
Tetrahedron Lett., **27** (1986) 967.
66 (a) Y. Yamamoto, H. Yatagai, and K. Maruyama,
J. Am. Chem. Soc., **103** (1981) 1969.
(b) M. T. Reetz and B. Wenderoth,
Tetrahedron Lett., **23** (1982) 5259.
67 T. Nakai and K. Mikami,
Chem. Rev., **86** (1986) 885.
68 K. Mikami, Y. Kimura, N. Kishi, and T. Nakai,
J. Org. Chem., **48** (1983) 279.
69 R. W. Hoffmann,
Angew. Chem. Int. Ed. Engl., **18** (1979) 563.
70 (a) M. Uemura, H. Nishimura, and Y. Hayashi,
J. Organomet. Chem., **376** (1989) C3.
(b) J. Brocard, M. Mahmoudi, L. Pelinski, and L. Maciejewski,
Tetrahedron Lett., **30** (1989) 2549.
71 J. Blagg, S. G. Davies, N. J. Holman, C. A. Laughton, and B. E. Mobbs,
J. Chem. Soc., Perkin Trans. 2, **(1986)** 1581.
72 (a) K. D. Croft, E. L. Ghisalberti, P. R. Jefferies, C. L. Raston, A. H. White, and S. R. Hall,
Tetrahedron, **33** (1977) 1475.
(b) K. D. Croft, E. L. Ghisalberti, P. R. Jefferies, and A. D. Stuart,
Aust. J. Chem., **32** (1979) 2079.
73 S. A. Look and W. Fenical,
Tetrahedron, **43** (1987) 3363.
74 S. A. Look, W. Fenical, G. K. Matsumoto, and J. Clardy,
J. Org. Chem., **51** (1986) 5140.
75 (a) S. R. Hall, C. L. Raston, B. W. Skelton, and A. H. White,
J. Chem. Soc., Perkin Trans. 2, **(1981)** 1467.
(b) C. A. Harris, M. T. Burch, and W. Fenical,
Tetrahedron Lett., **29** (1988) 4361.
76 S. A. Look, W. Fenical, R. S. Jacobs, and J. Clardy,
Proc. Natl. Acad. Sci., U.S.A. **83** (1986) 6238.
77 (a) M. Uemura, H. Nishimura, and Y. Hayashi,
Tetrahedron Lett., **31** (1990) 2319.
(b) M. Uemura, H. Nishimura, T. Minami and Y. Hayashi,
J. Am. Chem. Soc. **113** (1991) 5402.
78 D. J. Crouse, M. M. Wheeler, M. Goemann, P. S. Tobin, S. K. Basu, and D. M. S. Wheeler,
J. Org. Chem., **46** (1981) 1814.
79 (a) N. F. Masters and D. A. Widdowson,
J. Chem. Soc., Chem. Commun., **(1983)** 955.
(b) P. J. Beswick, S. J. Leash, N. F. Masters, and D. A. Widdowson,
Ibid., (1984) 46.
(c) J. M. Clough, I. S. Mann, and D. A. Widdowson,
Tetrahedron Lett., **28** (1987) 2645.
(d) M. Fukui, T. Ikeda, and T. Oishi,
Ibid., 23 (1982) 1605.
(e) Idem.,
Chem. Pharm. Bull., **31** (1983) 466.
80 (a) A. Solladié-Cavallo,
Polyhedron, **11** (1985) 901.
(b) T. A. Albright and B. Carpenter,
Inorg. Chem., **19** (1980) 3092.
81 (a) E. P. Kündig, V. Desobry, D. P. Simmons, and E. Wenger,
J. Am. Chem. Soc., **111** (1989) 1804.
(b) B. Ohlson and C. Ullenius,
J. Organomet. Chem., **350** (1988) 35.
(c) W. R. Jackson, I. D. Rae, M. G. Wong, M. F. Semmelhack, and J. N. Garcia,
J. Chem. Soc., Chem. Commun., **(1982)** 1359.

(d) M. F. Semmelhack, J. L. Garcia, D. Cortes, R. Farina, R. Hong, and B. K. Carpenter, *Organometallics*, **2** (1983) 467.
82 The regioselectivity in the nucleophilic addition was dependent on the nature of nucleophiles: M. Uemura, T. Minami, Y. Shinoda, H. Nishimura, M. Shiro and Y. Hayashi, *J. Organomet. Chem.* **406** (1991) 371.
83 T. Hayashi, K. Kabeta, I. Hamachi, and M. Kumada, *Tetrahedron Lett.*, **24** (1983) 2865.
84 N. Miyaura, T. Ishiyama, H. Sasaki, M. Ishikawa, M. Satoh, and A. Suzuki, *J. Am. Chem. Soc.*, **111** (1989) 314.
85 E. J. Corey, N. W. Gilman, and B. E. Gahem, *J. Am. Chem. Soc.*, **90** (1968) 5616.
86 (a) A. Mandelbraum, Z. Neuwirth, and N. Cais, *Inorg. Chem.*, **2** (1963) 902.
(b) R. Dabard, A. Mayer, and G. Jaouen, C. R. *Acad. Sci., Paris, Ser. C*, **268** (1969) 201.
(c) H. Fak, K. Schlögl, and W. Steyrer, *Monatsch. Chem.*, **97** (1966) 1029.
87 S. Rosca and C. D. Nenitzescu, *Rev. Roum. Chim.*, **15** (1970) 259.
88 (a) A. Solladié-Cavallo, G. Solladié, and E. Tsamo, *Inorg. Synth.*, **23** (1985) 85.
(b) Idem., *J. Org. Chem.*, **44** (1979) 4189.
89 (a) S. G. Davies and C. L. Goodfellow, *J. Chem. Soc., Perkin Trans, 1*, **(1989)** 192.
(b) Idem. *Synlett*, **1** (1989) 59.
90 (a) S. Top, G. Jaouen, J. Gillois, C. Baldoli, and S. Maiorana, *J. Chem. Soc., Chem. Commun.*, **(1988)** 1284.
(b) Y. Yamazaki and K. Hosono, *Tetrahedron Lett.*, **30** (1989) 5313.
91 K. Nakamura, K. Ishihara, A. Ohno, M. Uemura, H. Nishimura, and Y. Hayashi, *Tetrahedron Lett.*, **31** (1990) 3603.
92 (a) Y. Kondo, J. R. Green, and J. Ho, *J. Org. Chem.* **56** (1991) 7199.
(b) J. Aube, J. A. Heppert, M. L. Milligan, M. J. Smith, and P. Zenk, *Ibid.* **57** (1992) 3563.
93 A. Alexakis, P. Mangeney, I. Marek, F. Rose-Munch, E. Rose, A. Semra, and F. Robert, *J. Am. Chem. Soc.* **114** (92) 8288.
94 (a) W. J. Scot, *J. Chem. Soc., Chem. Commun.* **(1987)** 1755.
(b) J-F. Carpentier, Y. Castanet, J. Brocard, A. Mortreux, and F. Petit, *Tetrahedron Lett.* **32** (1991) 4705; *Ibid.* **33** (1992) 2001.
(c) J. M. Clough, I. S. Mann, and D. A. Widdowson, *Ibid.* **23** (1987) 2645.
(d) R. Mutin, C. Lucas, J. Thivolle-Cazat, V. Dufaud, F. Dany, and J. M. Basset, *J. Chem. Soc., Chem. Commun.* **(1988)** 896.
(e) F. Dany, R. Mutin, C. Lucas, V. Dufaud, J. Thivolle-Cazat, and J. M. Basset, *J. Mol. Catal.* **51** (1989) L15.
(f) V. Dufaud, J. Thivolle-CaZat, J. M. Basset, R. Mathieu, J. Jaud, and J. Waissermann, *Organometallics*, **10** (1991) 4005.
95 M. Uemura, H. Nishimura, and T. Hayashi, *Tetrahedron Lett.* **34** (1993) 107.
96 For some representative examples:
(a) R. Noyori and M. Kitamura, *Angew. Chem. Int. Ed. Engl.* **30** (1991) 49.
(b) N. Ogini and T. Omi, *Tetrahedron Lett.* **25** (1984) 2823.
N. Oguni, Y. Matsuda, and T. Kaneko,

J. Am. Chem. Soc. **110** (1988) 7877.
(c) M. Kitamura, S. Okada, S. Suga, and R. Noyori,
Ibid. **111** (1989) 4028.
(d) K. Soai, A. Ookawa, K. Ogawa, and T. Kaba,
Ibid. **109** (1987) 7111.
(e) E. J. Corey, P. W. YuenF. J. Hannon, and D. A. Wierda,
J. Org. Chem. **55** (1990) 784.
(f) K. Takano, H. Ushio, and H. Suzuki,
J. Chem. Soc., Chem. Commun. **(1989)** 1700.
(g) M. Watanabe, S. Araki, Y. Butsugan, and M. Uemura,
J. Org. Chem. **56** (1991) 2218.

97 M. Uemura, R. Miyake, M. Shiro, and Y. Hayashi,
Tetrahedron Lett. **32** (1991) 4569.

98 (a) M. Uemura, R. Miyake, and Y. Hayashi,
J. Chem. Soc., Chem. Commun. **(1991)** 1696.
(b) M. Uemura, R. Miyake, K. Nakayama, M. Shiro, and Y. Hayashi,
J. Org. Chem. **58** (1993) 1238.

99 (a) K. Soai, T. Hayasaka and S. Ugajin,
J. Chem. Soc., Chem. Commun. **(1989)** 516.
(b) K. Soai, M. Okuda, and M. Okamoto,
Tetrahedron Lett. **32** (1991) 95.
(c) C. Bolm and M. Ewald,
Ibid. **31** (1990) 5011.
(d) J. F. G. A. Jansen and B. Feringa,
J. Org. Chem. **55** (1990) 4168.
(e) K-H. Ahn, R. B. Klassen, and S. Lippard,
Organometallics, **9** (1990) 3178.

100 M. Uemura, R. Miyake, K. Nakayama, and Y. Hayashi,
Tetrahedron Asymmetry, **3** (1992) 713.

THE STEREOCHEMISTRY OF PALLADIUM CATALYZED CYCLIZATION REACTIONS

A. Heumann

ABBREVIATIONS

Ac	acetyl
AcOH	acetic Acid
Ar	aryl
BINAP	1,1'[binaphthyl]-2,2'[bis (diphenylphosphine)]
Bu	butyl
Bz	benzyl
bzq	benzoquinone
COD	1,5-cyclooctadiene
c/t	*cis-trans* ratio
d	day
DBA	dibenzylideneacetone
de	diastereomeric excess
DMAD	dimethyl acetylenedicarboxylate
DME	1,2-dimethoxyethane
DMF	N,N-dimethylformamide
DMSO	dimethylsulfoxide
dppe	DIPHOS, 1,2-bis(diphenylphosphino)ethane
dppm	bis(diphenylphosphino)methane
E	CO_2Me
E'	CO_2Et
ee	enantiomeric excess
eq	equivalent
Et	ethyl
EtOH	ethanol
h	hour
Hex	hexyl
HMPA	hexamethylphosphoramide
Me	methyl
MeOH	methanol
Nu	nucleophile
Ph	phenyl
Pr	propyl
r.t.	room temperature
TCPC	tetracarbomethoxypalladacyclopentadiene
THF	tetrahydrofuran
Ts	tosyl, *p*-toluenesulphonyl

THE STEREOCHEMISTRY OF PALLADIUM CATALYZED CYCLIZATION REACTIONS

Andreas Heumann

1. Introduction

The development of methods for the construction of cyclic systems is an important challenge in organic synthesis. These systems are widespread in nature (alkaloids, terpenes, hormones, pheromones,etc.) and largely represented amongst fine chemicals or pharmaceuticals. Therefore, it is not astonishing that new, especially selective synthetic methods in organic and inorganic chemistry are immediately applied to the preparation of cyclic and polycyclic compounds.

Since the discovery of the Wacker process in the late fifties [1], the organic chemistry of palladium has seen a really explosive expansion. Excellent books by Maitlis [2], Hartley [3], Henry [4], Tsuji [5], Segnitz [6], and Heck [7] are well suited to document this fact. The ability of transition metals to bind or to coordinate different functional groups at the same metal center, a prerequisite for metal-catalyzed reactions, is certainly the most useful feature for cycloaddition and cyclization reactions. Beyond the 'simple' fact of bringing two ends of a chain together a metal catalyst, and predominantly palladium, is often the reagent of choice when regio- and, more importantly, stereoselective arrangements are the main objectives of the operation. A further attractive point is the possibility of reactions with catalytic amounts of palladium reagents, palladium(0) and palladium(II), salts or complexes. In the latter case an additional oxidant is necessary when the palladium is reduced during the operation. This article is therefore centered on reactions in which palladium reagents may offer simple and elegant solutions for the stereoselective preparation of carbo- and heterocyclic compounds. Though arbitrary like any organisation of a complex theme, the division into 2 parts seemed useful **Part A: Addition to carbon-carbon double bonds (π-route)** and **Part B: Addition to π-allyl intermediates and Cascade reactions**. This organisation is based on the fact that most of the reactions are addition or insertion reactions to π- or π-allyl-palladium complexes with the intermediate formation of carbon palladium or heteroatom palladium bonds.

2. Addition to carbon-carbon double bonds (π-route)

The observation that the combination of water, ethylene and palladium chloride leads to the formal addition of one oxygen atom to the olefin and the formation of acetaldehyde is one century old [8]. The development of a catalytic reaction (addition of $CuCl_2/O_2$ as a reoxidation system) to an industrial process by Wacker Chemie [1] stimulated more academic investigations [4, 9, 10]. The mechanism was studied [11], and has since been discussed [12]. The important step is the nucleophilic addition of water to the coordinated ethylene:

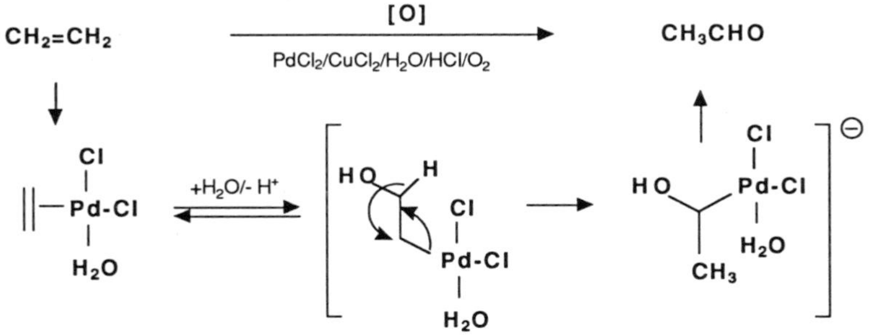

The subsequent hydride shift is a second important characteristic of this transformation; it can be observed in numerous palladium catalyzed reactions [13]. The last palladium hydride elimination step (ß-elimination) liberates the organic substrate and the palladium catalyst. This kind of activation of the olefinic double bond towards nucleophilic attack [14], probably well characterized by the German expression Umpolung [15], is a general phenomenon and can be observed with a great number of carbon and heteroatom centered nucleophiles. By isolating and characterizing such palladium-σ complexes [16], it has been shown with more complex olefins that the hydroxypalladation of alkenes is a *trans* process.

More closely connected to the original ethylene oxidation [1] is Stille's reaction [17a], another very elegant demonstration of the *trans* stereochemistry. It makes use of the fact that palladium-carbon-σ bonds react with carbon monoxide [17b] and, more importantly, are transformed with complete retention of configuration at carbon, prior to any other reaction (hydride shift, for example). Intramolecular alcoholysis after rotation around the C–C axis leads to the ß-lactone **1**.

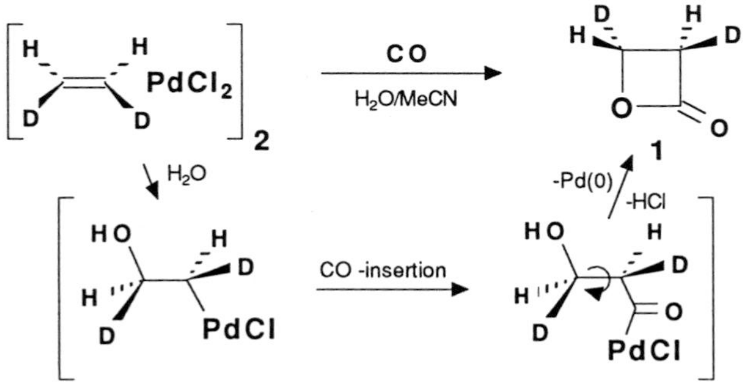

This completely stereospecific carbonylation/cyclization reaction demonstrates several important features which are often encountered in palladium catalyzed organic transformations: (i) the reactivity of the intermediate is readily modified and subsequent reactions are possible (substitution, insertion of small molecules, transposition.etc.); (ii) the addition of the metal and the nucleophile is completely stereo- and (if regioisomers are possible) regioselective; (iii) the termination process may switch from ß-elimination to other basic reactions like reductive elimination, the latter frequently with well defined stereochemistry. Thus, not surprisingly, intramolecular versions of these C-C and C-heteroatom forming reactions have become an extremely valuable tool for building up all kind of cyclic and polycyclic systems.

2.1 Formation of carbon-oxygen bonds

The replacement of water by alcohols or carboxylic acids in the palladium catalyzed oxidation of olefins [18] gives a direct access to saturated aliphatic ethers, or esters, from simple olefins. The first intramolecular version of these reactions, the synthesis of benzofuran derivatives from 2-allylphenol, has been described by a Japanese group [19]; an enantioselective version [20] will be discussed later.

OH R $PdCl_2$ O R

Unsaturated aliphatic alcohols (alk-4-en-1-ol compounds) are less reactive; nevertheless, tetrahydrofuran derivatives are formed in moderate yield [21].

R^1 R^2 OH $Pd(OAc)_2$ 0.1eq $Cu(OAc)_2xH_2O$ 1eq MeOH, H_2O, O_2, r.t. H Ph O

2 R^1=H, R^2=Ph **3** 40%

H Ph O H Pd **4-π** H Ph O Pd **4-σ**

Both terminal groups, phenyl at C_1 and a methyl substituent at the double bond are necessary to ensure the complete diastereoselectivity observed during the reaction of **2** (R_1=H,R_2=Ph). The phenyl group directs the OH-attack towards the alkene, *trans* to palladium, in a way that steric interactions between Ph and the vinylic H are minimized, the result being the *trans* configuration in **4-σ** and **3**. If the phenyl group is replaced by Me in **2** (R_1=H,R_2=Me) or if two substituents are present at C_1 (Me,Et, Ph) this stereochemical discrimination is partly cancelled and a partial to nearly complete loss of selectivity is observed (*cis/trans* ratio of 25/75 and 60/40, respectively). A double substitution at the terminal double bond directs the cyclization to the six-membered ring 5.6-dihydropyran **6**.

Ph OH
5
Ph O
6 45%

A Markovnikov type selectivity during the alkoxy palladation step may be the explanation for this different regioselectivity.

In a very recent communication, Semmelhack and coworkers [22] studied the alkoxy-palladation of alk-5-en-1-ol compounds. The sole oxacyclic system formed is a six-membered ring. Wacker type catalysts, e.g. $PdCl_2$(0.2 eq.), $CuCl_2$ (3 eq.), or $Pd(OAc)_2$ (1.1 mol-eq.) in MeOH at 24°C, without the copper chloride, are efficient systems for the cyclization processes, however leading to regio-isomers, and chlorinated compounds in the case of $PdCl_2$.

R 2 1 5 H OH
7
R=H
$Pd(OAc)_2$ 1.1eq.
solvent
H O H
8 95% (c/t=18:1)
39% (c/t=19:1)
+
O
9 0% (solvent: DMSO)
47% (solvent: HOAc)
R=Me
1,2-elimination
'hydride shift'
Me
H O H
10 96% (c/t=31:1)
Pd
H O H
11
R' O
12

A study of solvent effects revealed DMSO to be the solvent of choice in order to eliminate any rearrangement. The excellent yields and the high *cis*-diastereoselectivity observed in **8** or **10** were promising features for an application to the formation of the tetronomycin synthon **14**.

13 — $Pd(OAc)_2$ 1.1eq., DMSO-24h-24°C → **14** 72%

Exomethylene compounds, such as **12** have not been observed, even with starting material including unsubstituted terminal double bonds.

Substituents at the allylic position of unsaturated alcohols may also change the reactivity pattern. Two Japanese groups [23, 24] reported on a reaction which is formally an α,α'- addition to the double bond:

15 R^1=alkyl,COOMe,OBz,Ts; R^2=H,alkyl — $PdCl_2$ 0.1 - 0.5eq., reoxidant-solvent →

16 R=H, 43-87%, solvent: wet DMF (ref. 23)

17 R=Me, 17-95%, solvent: MeOH (ref. 24)

The stereochemistry at C_1 and C_2 is not affected during the transformation of **15**. This has been used for the preparation of enantiomerically pure (protected) desoxyribose (*erythro*-**15** → *trans*-**16**; R=H, R^1=OBz, R^2=CH_2OCOPh). The mechanism of this useful reaction has not yet been clarified. A substituent at C_2 (carbon or heteroatom) as well as a free OH group seem to be necessary. The regiochemistry observed is the opposite to the one of 'normal' palladium catalyzed oxidations of terminal alkenes [25] where the Pd attacks the terminal unsaturated carbon. After a rearrangement (hydrogen shift) a second substitution takes place at C_6. An explanation could be that the substituent at C_3 induces a 'terminal' oxidation to an aldehyde (and to the lactol), a reaction which is known in palladium(II) chemistry [26]. Another mechanism is depicted in Scheme I.

Scheme I

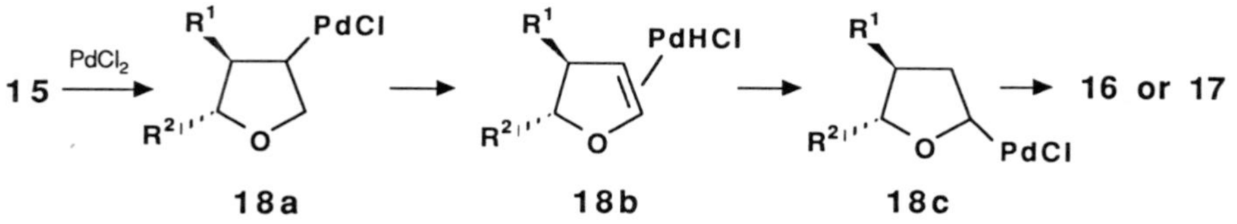

The authors of reference [24] disagree with such a possibility and discuss a mechanism where the first step is supposed to be the addition of Pd(II) and methanol, followed by rearrangement and cyclization. However it seems more reasonable to assume that the internal OH-group, much better located for the alkoxypalladation step, should react first.

The same kind of intermediate **18**, which is unfavorable according to Balwin's rules [27] (endo-trig cyclization), should be involved in the following reaction [28]:

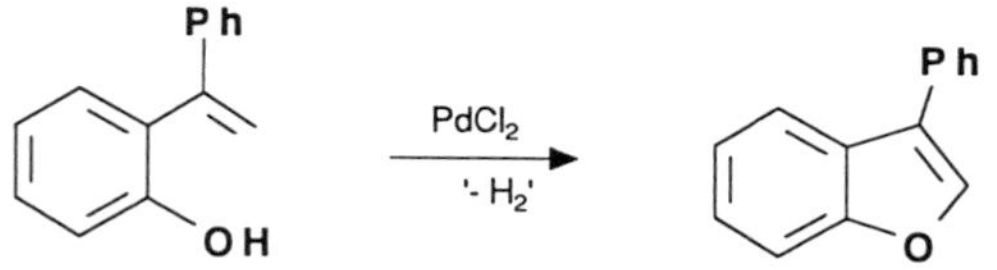

The influence of the substituents at C_2 remains unexplained, and indeed another example of this kind of regioselectivity has been reported [29] under Heck type reaction conditions [7, 30].

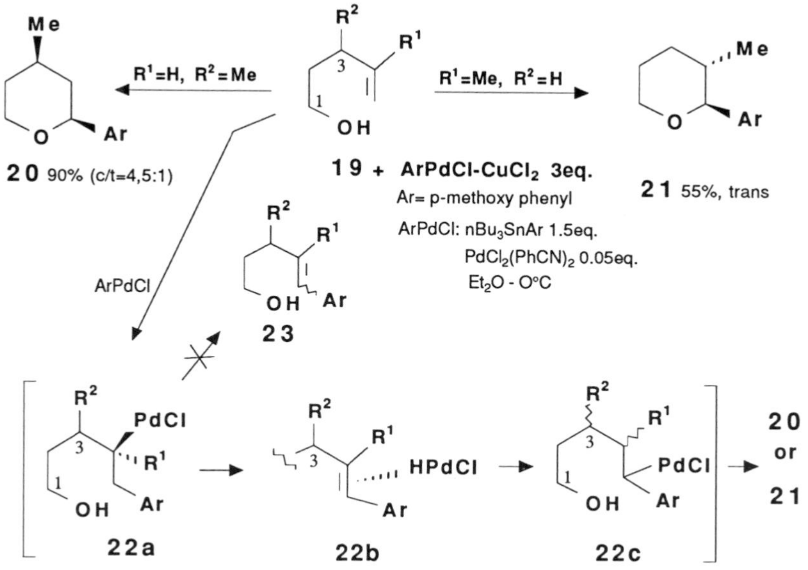

Now, contrary to compound **15**, which leads to non cyclized oxidation products in the absence of a substitutent at carbon-2, substitution at C_2 (or not) does not alter the reaction pathway in reactions of **19.** The reason for the α,α'-addition to the terminal unsaturated carbon, instead of the usual 1.2-addition (Heck reaction), has to be found in the particular system applied for the formation of the aryl palladium from nBu$_3$SnAr and $PdCl_2(PhCN)_2$. Indeed, the first step, a normal Heck type 1.2-addition of the ArPd moiety to the double bond, leads to the first organopalladium intermediate. It has been suggested that the polarity of the solvent may be responsible for the unusual course of the reaction. In the unpolar solvent (Et_2O), the 'normal' reductive elimination of the 1.2-addition complex **22a** to **23** is not operative (intermediate too stable to undergo heterolytic cleavage); and, thus, the transposition to **22c** becomes a favorable process. The new termination step consists in a S_N2 replacement of Pd by the oxygen nucleophile, followed by regioselective cyclization. It should be mentioned that only an O-methylated aryl group exclusively leads to cyclized products. Another transposition, a formal 1.3-addition, not leading to cyclization is observed with unsubstituted phenyl-PdCl as intermediate. The factors controlling the stereochemistry of the tetrahydropyrans, mainly *cis* configuration in 1.3 and *trans* in 1.2-disubstituted compounds **20** and **21**, respectiveley, are not yet elucidated.

The carbonyl insertion is a well studied property of palladium alkyl complexes [2]. In combination with the nucleophilic addition to the coordinated double bond, an intramolecular alkoxycarbonylation has been developed by Semmelhack's group [31, 32, 34]. Initially introduced as a cyclization step in the synthesis of naphthoquinone antibiotics [31], this reaction has been extensively studied as summarized in Schemes II and III.

Scheme II

C$_3$or C$_4$-chain

R= 1 or 2 substituents
substituents at a,b, or c

and/or

+ CO/MeOH

conditions:

$PdCl_2$ 0.05 eq.;$CuCl_2$ 2eq.
CO/MeOH, r.t.

CO_2Me and/or CO_2Me

These examples give evidence of the remarkable degree of diastereoselectivity which can be attained in many cases. A new termination step, the CO insertion into the intermediate C-Pd σ-bond, seems to be beneficial for the reaction profile also. Two different groups are thus added to the double bond in a 1.2-addition. The following parameters have been studied: (i) the influence of the chain length: 3 or 4 carbons between OH and the double bond; (ii) the effect of different substituents R on the stereoselectivity, *i.e.* the attack from one distinct diastereoface; (iii) the ring size of the oxacyclic product. In the case of a C_3-chain tetrahydrofurans and pyrans are formed [32].

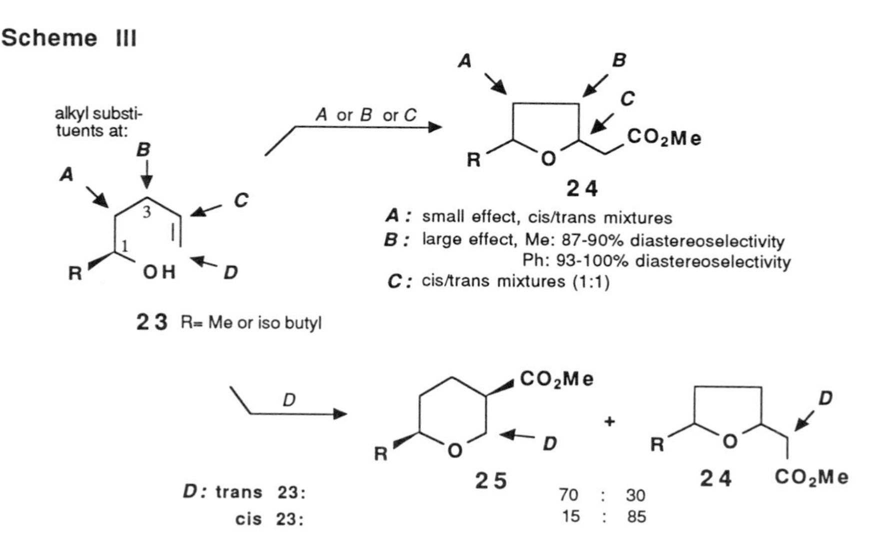

The importance of substituents (alkyl or aryl) is manifest for ring-size preferences and for the stereochemistry of the products: substituents at positions *A*,*B*, or *C* give only five membered rings **24**; at *D* position they give both five- and six-membered rings **24** and **25**. Regarding the stereochemistry at *A* and *C* , the effects are small and mixtures of *cis* and *trans* disubstituted cyclic ethers are obtained. This is also the case for **24** when position *D* is substituted; in contrast, the pyran **25** is formed with selectivity in this reaction. In addition, the ratio of 5 to 6 membered rings depends on the configuration of the (internal) double bond: *trans* substituted olefins give more pyran **25**; however, *cis*-olefin **23**-*D*, now, predominantly leads to mixtures of tetrahydrofurans. A remarkably large effect has been found when position *B* is substituted by a methyl (<90% diastereoselectivity) or phenyl (93-100% diastereoselectivity).

This remarkable directing effect of an alkyl or aryl group at the allylic position (C_3) is rationalized on the basis of steric interactions in the intermediate palladium-π complex **26a** or **26b**. The conformation with less non-bonded interactions **26a** is favoured so much that only the reaction to **24a** takes place, and the formation of **24b** (*cis*-1.4 relationship) is completely suppressed. The prediction that the latter *cis*-type cyclic compound may be formed from the alkenol with opposite stereochemistry at C_3 has been demonstrated with **23b**, which reacts to **24b** with nearly the same diastereoselectivity. The preference of six- or five-membered ring formation from *trans* or *cis* olefins, respectively, (case *D* in Scheme III) is readily understandable with the same model, as well as the lack of *cis*/*trans* selectivities in case *A* or *B*, where sterically discriminating interactions are obviously lacking.

Ph 3 1 Me OH **23a** — $PdCl_2/CuCl_2$, CO/MeOH → Me OH R² R¹ **26a** $PdCl_2$ — R^1= Ph, R^2=H → Ph Me O **24a** CO_2Me

Ph 3 1 Me OH **23b** → Me OH R² R¹ $PdCl_2$ **26b** — R^1= H, R^2=Ph → Ph Me O **24b** CO_2Me

Comparable results, *i.e.* highly regio- and stereoselective cyclizations have been obtained by another research group, starting from alkenols containing more than two substituents, or even when only one substituent (at C_3) is present in the olefin [33]. Usually five-membered rings are formed with the exception of Z-substituted olefins (**27**) or olefins with two terminal methyl groups at the double bond. One example may illustrate both the influence on stereo- and regiochemistry:

Me 3 1 OH **27** — $PdCl_2/CuCl_2$, CO/MeOH → Me 3 4 1 O CO_2Me Me **28** + Me CO_2Me 4 5 O Me **29** — 1 : 4 (75%)

$PdCl_2$ Me O Me ← OH Me $PdCl_2$ → Me $PdCl_2$ O Me

Semmelhack's concept of a conformational control in the transition state, minimizing non-bonded interactions in the different intermediated palladium complexes, finds here an elegant confirmation: the combination of Z configuration at the double bond and substituent at C3 suppresses completely the (unfavorable) formation of the pyran ring.

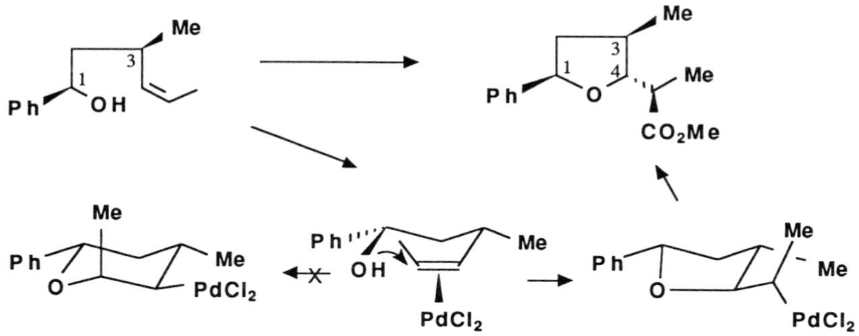

Even compounds with a trisubstituted double bond, usually quite unreactive to palladium, have reacted; however, the original $PdCl_2$ had to be activated by addition of triethylamine in order to start with an (more reactive) olefin-palladium-amine monomer.

In the case of a four carbon chain (see also Scheme II) tetrahydropyran derivatives are the exclusive products [31], sometimes with very high *cis/trans* selectivities (>97% *cis*)[32]:

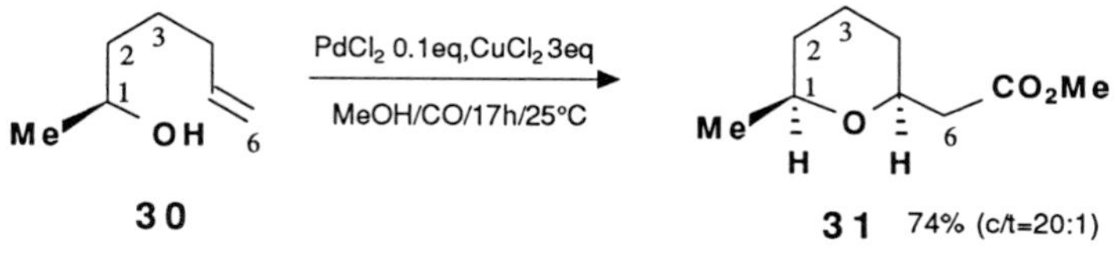

$PdCl_2$ 0.1eq, $CuCl_2$ 3eq

MeOH/CO/17h/25°C

30

31 74% (c/t=20:1)

The influence of different substitution patterns has been studied in this system, too. The more important reactions concern those where C3-C4 is part of a ring system (aromatic, quinoid or cyclohexane). It turned out that, only when the latter carbons are part of a saturated chain, high stereoselectivies are possible for both the substituents of the heterocyclic ring and the ring junction. Consistent with this observation, the key intermediate **33** has been prepared in the synthesis of the natural antibiotic frenolicin.

32 cis
cis/trans (82:18)

87%
3h

33 1 isomer
2 isomers (81:19)

In a preliminary study [34], the possibility was tested to accomplish two successive cyclizations, with two alcoholic groups present in the starting olefin.

$Pd(OAc)_2$ 1eq
THF

CO

34

35

36 (lactone ring:c/t=4.8:1)

This direct alkoxycarbonylation/carbonylation proceeds with moderate *cis*-stereoselectivity (4.8:1), irrespective of the stereochemistry of the starting diol. The usual carbonylation reagents ($PdCl_2$ cat., $CuCl_2$) were not active in this case, and stoichiometric amounts of $Pd(OAc)_2$ are necessary. This kind of 'cascade reactions' will be discussed in a separate chapter.

Sometimes the oxacyclic target molecules require configurations, which are not those observed during alkoxycarbonylation reactions. This is the case in the tetrahydropyran subunit of the polyether nigericin. Bartlett [35] tried an interesting approach directed to reverse the *cis*-selectivity, when tetrahydrofurans are formed from hexenols.

* + CO
MeOH

37 R=H,H
* Semmelhack's cond. R=CHR'
(ref. 32,34)

38 39% (7.9 : 1) 39
<94% (2.1 : 1)

The originally observed high *cis*-selectivity in diequatorial 2.6-disubstituted tetrahydropyran **38** (R=H,H) could be reduced from ca 8:1 to about 2:1 when the unsaturated alcohol **37** (R= H,H) was reacted in a ketal form **37** (R=CHR'); a complete reversion to predominant *trans* selectivity, however, does not seem possible.

The principle that substitution of the nucleophile, *i.e.* the OH-group, has an influence on the selectivity of the cylization-carbonylation-reaction has already been put in evidence on γ-allenic alcohols [36]. Silylether **40** can be cyclized to substituted acrylates either by Hg(II) or Pd(II) salts. In the case of mercury, the reaction can be directed to high (15:1) *cis*/*trans* selectivities provided the trialkyl-silicon group is properly chosen, the palladium exclusively serves to catalyze the final carbonylation step. With palladium as cyclization catalyst, the reaction takes place easily; however, the siloxyether is hydrolysed before the nucleophilic attack at the olefinic bond occurs, and in all cases the non-selectivity of the free hydroxyallene is observed. A brillant demonstration of improved stereoselectivity in these systems has been reported recently by the same group [37]. The considerable *cis*/*trans* ratio higher than 49:1 can be achieved in the formation of compound **42b**, a precursor of a nactin antibiotic subunit, by both a 'two step protocol' (mercuration/palladation) and silyl ethers being the requirements for the high stereoselectivity.

1. $Hg(OCOCF_3)_2$ 2. Pd(II)-CO; CO

40a R = Me, R' = $SiMe_2{}^tBu$

40b R = alk-CH(OR')-CH_2, R' = $SiMe_2{}^tBu$

41 M=Hg, M=Pd

42a (44-90%; c/t= 15:1)

42b (>70%; c/t= >49:1)

Pd(II)-CO: $PdCl_2$ 0.1eq./$CuCl_2$ 3eq./CO/MeOH

Enantioselective oxidation of olefins are of particular interest [38] and the chiral palladium catalyzed cyclization of alkenyl phenols is an interesting approach to this fundamental problem [20]. In this reaction, already mentioned before [19], the ordinary palladium salt is replaced by the chiral π-allyl palladium complex, easily prepared from $PdCl_2$ and (1S)(-) ß-pinene. Palladium maintains the oxidation state II and, in combination with $Cu(OAc)_2/O_2$ in methanol, it still constitutes a catalytic oxidation system. Apparently no dissociation of the chiral allylic ligand takes place during the oxidative cylization, and Hosokawa, Murahashi and coworkers [20] have proposed the following reaction scheme.

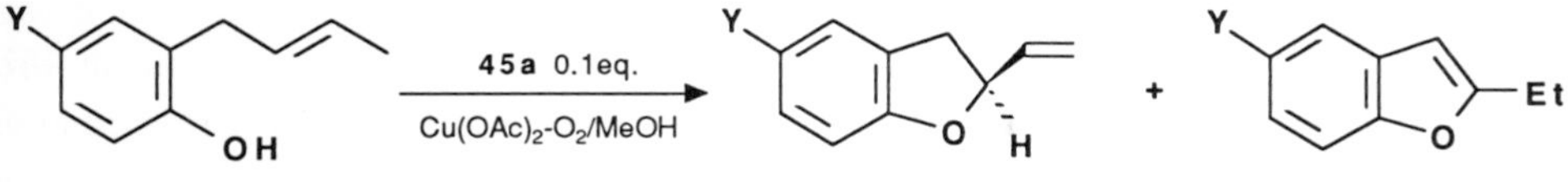

45a 0.1eq.
$Cu(OAc)_2$-O_2/MeOH

43 Y=OMe, alkyl, H, Cl, COMe

(S)-(+) 44 83 : 17 (81%)
Y=H, ee <18%
Y=OMe, ee <26%

The major product formed from the 2-butenylphenol **43** (high yield and fairly high regioselectivity), is the chiral 2.3-dihydrobenzofuran **44**. In an extensive study both the substrates (substituents at the aromatic ring) and the chiral allylic part in the palladium catalyst have been investigated. With electron-donating substituents (Y=OMe), the optical induction can be increased from originally 18% to 26%. In the chiral palladium complex **45**, substituents at C_4 in the pinene part modify the amount and the orientation of the induction. When no substitutent is present in **45a** (X=H), the (S)(+) isomer of **44** will be formed predominantly (ee_{max} 26%); with **45a** (X=OAc, syn to the palladium) as chiral catalyst, the formation of the opposite enantiomer (R)(-) **44** is favoured.

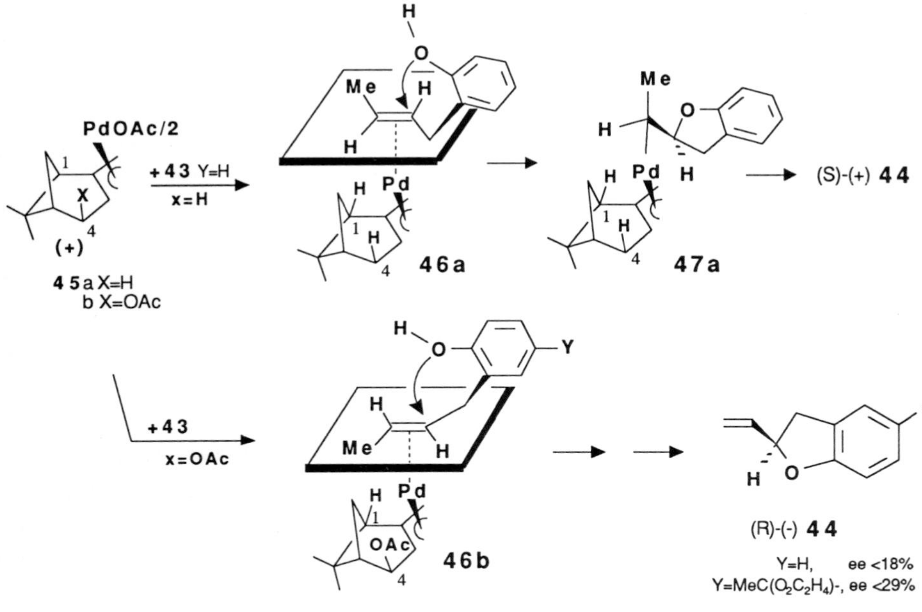

PdOAc/2
+43 Y=H
x=H
(+)
45a X=H
b X=OAc
46a
47a
(S)-(+) 44
+43
x=OAc
46b
(R)-(-) 44
Y=H, ee <18%
Y=MeC($O_2C_2H_4$)-, ee <29%

Steric interactions in the different intermediates are the controlling factors for enantioselectivity. According to the *trans* attack of the oxygen nucleophile at the coordinated olefin, and keeping in mind that the pinanyl ligand remains coordinated throughout the reaction, the diastereomers **46a** and **46b** have been proposed as the enantio-controlling species. Minimum steric interactions in **46a** are achieved when the methyl group in the olefinic part is situated in the 'pocket' existing over C_1-H in the pinane. The predominance of the opposite intermediate **46b** is a consequence of the blocking of the front side corner of the acetyl pinanyl ligand by an acetate bridge between the syn acetyl group and the palladium. Consequently, the location of the bulky phenoxy part is achieved as far as possible from the C_4-OAc group. The influence of aromatic substituents changes with the polarity of Y and is rather electronic in nature.

This unique chiral catalyst system only seems to be efficient in reactions on alkenylphenols having an E-configuration; the corresponding Z-isomer reacts with a chiral discrimination close to zero (ee 0.7%). The use of the catalyst **45a** in other chiral oxidation reactions has not been reported or has failed until now, as in the case of non conjugated dienes [39].

2.2 Formation of carbon-nitrogen bonds

Nitrogen centered nucleophiles, such as amines, amides, carbamates etc. easily add to coordinated olefins. This is of great importance since many natural bases (alkaloids, aminoacids..) or synthetic pharmacologically active substances (neurotoxic compounds, psychopharmaca, antibiotica..) contain the carbon-nitrogen bond. As in the addition of oxygen nucleophiles, the addition is a *trans* process. This promising stereocontrolling element is somewhat obscured by the fact that elevated basicity and nucleophilicity of amines usually complicate the reactions: stoichiometric reaction conditions with isolation of the π- or σ-complexes are one major drawback, another being the competing coordination to palladium instead of addition to the double bond. Nevertheless, in many cases low temperatures and/or substitution of the amine are remedies. The formation of nitrogen containing cyclic systems, *via* amine addition to palladium coordinated olefins, has been reviewed [40]. Two reaction paths have essentially been developed: the cyclization of o-allylanilines and olefinic tosylamides. One example may be representative [41]. Under catalytic conditions only five-membered rings are formed easily (**48** *cis* $\rightarrow$ **49**). The configuration of the substituents in the aliphatic or alicyclic part of the substrate usually is not affected during the reaction. Nevertheless, this configuration exerts a strong influence on the regiochemical outcome of the reaction. *Trans* 1.2-disubstituted alicyclic compounds do not react in some cases; or, as the consequence of ring strain in *trans* fused azabicyclic products, cyclize to six-membered rings (**48** *trans* → **50**).

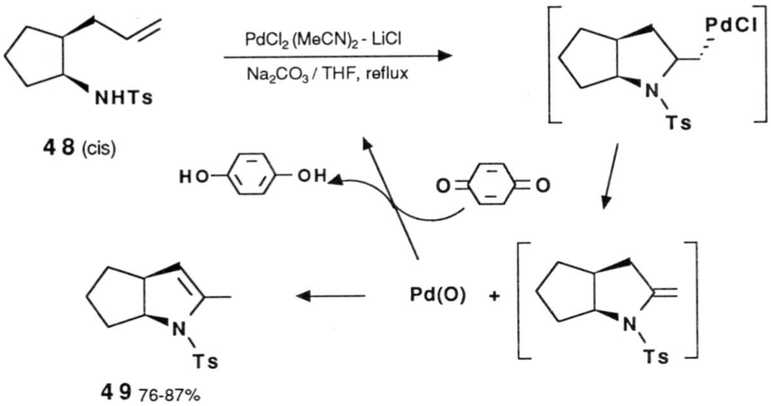

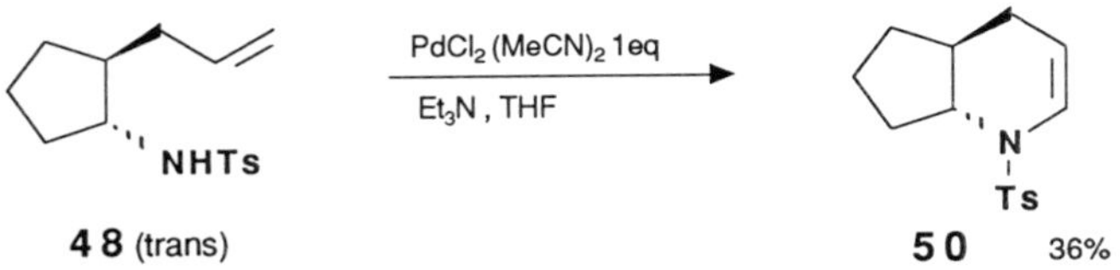

Though Pugin and Venanzi [42] could catalytically cyclize primary aminoalkenes to form five- and six-membered imines under the conditions of the Wacker Process in water, the necessity for using stoichiometric quantities of palladium in aminations is more or less a rule. Moreover, reactions are often complicated by amine replacement of the olefin ligand from the metal instead of adding to the olefin. Possibilities to circumvent these problems are to work: (i) at low temperatures; (ii) in strong acidic medium [43].

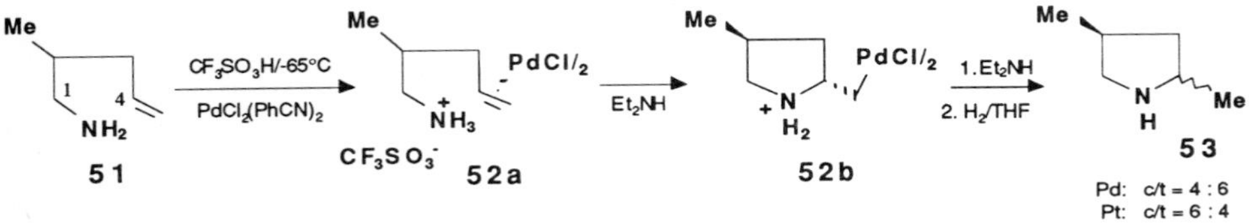

Under these conditions no olefin displacement, no isomerization and no ß-hydride abstraction are observed: two moles of *sec* amine are necessary to remove the protons from the nitrogen fragment. According to the position of the double bond, 2-

methyl substituted pyrrolidines and piperidines are formed. The formation of pyrrolidine itself does not seem possible since 4-aminobutene does not cyclize. With substituents in the carbon chain in **51** only poor selectivites are achieved to the cyclic amines **53** (c/t=1:1.5). Also, platinum salts ($PtCl_4^{2-}$) are active as catalysts in these stoichiometric aminometallation. However, an interesting difference is observable: whereas palladium prefers to form 5-membered rings and *trans* configuration, platinum gives rise to the opposite selectivities, piperidine ring and *cis*-configuration. The preference for the cyclization to five-membered rings may be the consequence of different charge distribution induced by Pt or Pd at the coordinated olefin part of the metal complexes. The difference in *cis/trans* ratios are attributed to small differences in the conformational energies of the metal complexes (**52a**) prior to cyclization. The termination step has to be a stoichiometric reduction of the primary palladium-σ-complex (**52b**).

The particular behaviour of unsaturated amines towards palladium catalyzed reactions is demonstrated, once again, by the absence of any cyclized product from 1-aza-4-cyclooctene **54** [44]. This classical model system for transannular reactivity is however easily transformed to the pyrrolizidine **55** with electrophiles such as HgX_2 or Br_2.

R^+ ; $PdCl_2(PhCN)_2$; HN ; R ; N ; **54** ; **55** ; R^+ = Br,HgX

It is clear from the following example that transannular reactions in medium rings not only are a favorable process, but can even proceed with remarkable selectivities, provided the adequate reactive intermediate is present within the molecule.

$PdCl_2(PhCN)_2$; CH_2Cl_2 - r.t. ; N–CO_2Et ; **56** ; **57** (71-83%) ; $-PdCl_2$; CO_2Et ; **58** ; $PdCl_2$; EtO ; (Pd?) ; Cl ; **59** ; ^+PdCl ; –Pd-Cl ; **60a**

In a veritable multistep rearrangement sequence N-carbethoxy-8-azabicyclo[5.1.0]oct-3-ene **56** is transformed into N-carbethoxy nortropidine **57** [45] (49 hours, 74-83% yield). Coordination of $PdCl_2$ to the double bond at the opposite side of the labile aziridine ring and selective ring opening by cleavage at C_1 (inversion of configuration) occurs instantaneously. The formally negatively charged nitrogen is supposed to be stabilized by the delocalized carbalkoxy group together with the coordination of a second palladium(II). Intramolecular *trans* addition of this nucleophile to the coordinated alkene creates the new C-N bond in the bicyclic tropane skeleton. The final, less well documented, 1.3-PdCl elimination has to proceed *via* hydride shift and recreates the double bond. The total process is a palladium catalyzed isomerization, the finally uncomplexed $PdCl_2$, stoichiometric or catalytic, being recovered unchanged. No intermediate of the four step mechanism can actually be observed by spectroscopic methods, but this mechanism is backed by: (i) comparison with the reactivity of saturated aziridines (π-complex); (ii) trapping of **59** with aqueous cyanide, formation of **61**; (iii) the interception the palladium-σ-complex with CO/MeOH, dppe or $NaBH_4$, in order to isolate **62**, **60b**, and **63**, respectively.

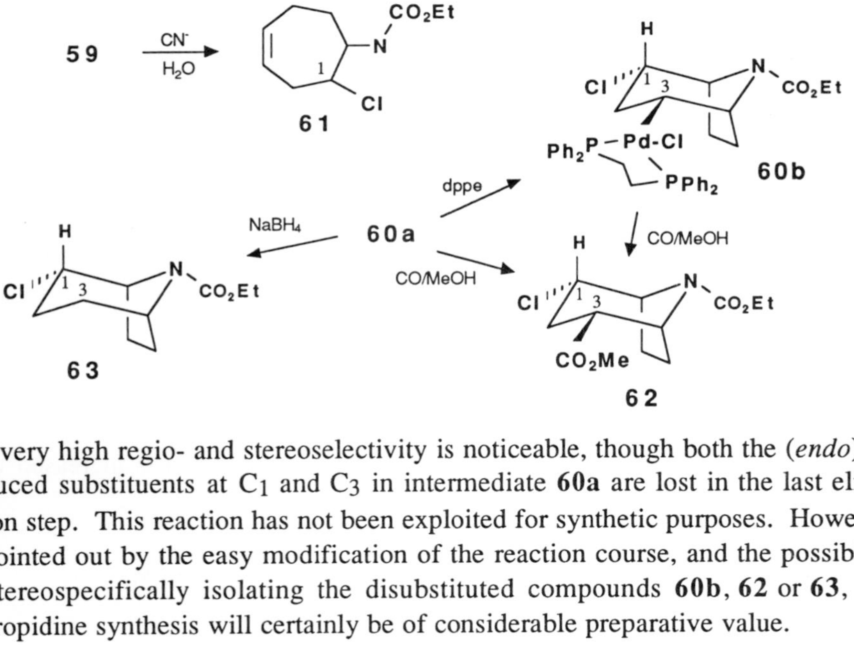

The very high regio- and stereoselectivity is noticeable, though both the (*endo*) introduced substituents at C_1 and C_3 in intermediate **60a** are lost in the last elimination step. This reaction has not been exploited for synthetic purposes. However, as pointed out by the easy modification of the reaction course, and the possibility of stereospecifically isolating the disubstituted compounds **60b**, **62** or **63**, this nortropidine synthesis will certainly be of considerable preparative value.

In some respects , the regio- and stereochemistry in palladation reactions of *sec.* amines or amides is comparable to the reaction of alcohols or acids. The aza Heck type cyclization of alkenyl amides to 5- and 6-membered rings [46] follows the same reaction pathway than the corresponding alcohol transformation (**19** → **20**) [29]. Again α,α'-addition to the olefin of nucleophile and aryl group takes place, when the modified Heck arylation conditions (ArHgCl or $ArSn(nBu)_3$ and $PdCl_2$ in diethyl ether) are used.

R
3
1
NHTs
64
+ ArPdCl-$CuCl_2$ 4eq.
Ar= p-methoxy phenyl
ArPdCl: nBu_3SnAr 1.5eq.
$PdCl_2(PhCN)_2$ 0.05eq.
Et_2O - 0°C
R
N
Ts
OMe
65a R=OAc, 77%, c/t=1:1)
65b R=Me, 58%, c/t=4.2:1)

Only modest to low *cis/trans* selectivities are possible; however, the good yield (77% in case of **65a** R=OAc) and the stability of the allylic acetoxy group under the reaction conditions should be noticed. Palladium catalyzed allylic substitution [47] or elimination reactions [48] apparently seem to be unfavored in the nonpolar (Et_2O) reaction medium. Homoallylic amides lead to the pyrrolizidine system.

The cyclization of allenic amines, amides and carbamates has been reported by Gallagher's group [49]. Depending upon the chain length, five-membered rings are formed. Whereas with a substituent at the allenic group the palladium catalyzed amino-alkoxycarbonylation leads to a 1:1 E/Z mixture, a carboethoxy group at C_1 directs the ring closure to predominantly *trans* stereochemistry.

EtO
O
1
NH
R
C
66
$PdCl_2$ 0.1eq, $CuCl_2$ 3eq
CO/MeOH, r.t.
EtO
O
N
R
$PdCl/_2$
67
EtO
O
N
R
CO_2Me
68 R=Bz: 55%
'predominantly trans'
R=Ts: 68% (add.of Na_2CO_3)

6-Membered rings are formed less easily than azacyclopentanes, and the addition of a base is necessary. Now the E/Z selectivity in the product is increased when the allene part is substituted.

NH
C
Ph
69
$PdCl_2$ 0.1eq, $CuCl_2$ 3eq
CO/MeOH, r.t.
N
Ph
CO_2Me
70 (52%, E/Z=6:1)

Spontaneous cyclization [50] takes place when the allyloxycarbonyl function is removed in **71** by palladium(0) catalysis.

71 → 72 85%

Pd$(PPh_3)_4$, Ph_3P, CH_2Cl_2
Bu-CH(Et)-COOH r.t.,48h

2.3 Formation of carbon-carbon bonds

2.3.1 Carbon nucleophiles or Heck type reactions

Stabilized carbon nucleophiles, such as dialkyl sodiomalonates, are of main importance in palladium(0) catalyzed allylic substitution reactions [47]. It has been shown [51] that they also may easily be added to palladium(II) coordinated olefins. The addition across the unsaturated linkage proceeds likewise with oxygen or nitrogen nucleophiles, in a *trans* fashion; however, an intramolecular stabilization of the palladium σ-complex is indispensable. This can be performed by allylic electron-donor ligands, an amino [51,52] or sulfido group.

$N(CH_3)_2$ + HNu + Li_2PdCl_4 → (THF/25°C, 81-95%) Nu, PdCl/2, $N(CH_3)_2$

HNu=MeOH, (ref 52)
Nu=$^-CHE_2$, (ref 51 and 53)

An intramolecular version of this carbopalladation provides an interesting entry to regiospecifically constructed 5- to 7-membered rings [53].

73-E → (Li_2PdCl_4, O°C; o,5h, base) → 74trans 93%
73-Z → 90% → 74cis
74trans, 74cis → $+H_2$ → 75

E=CO_2Me

When 5-membered rings are to be formed (in a rapid reaction: 0°C, 0.5h) the palladium complexes **74** are readily isolated. The ring junction is directly related to the geometry of the allylic starting compound **73E** or **73Z**. The organic product is formed by hydrogenation to amine **75**. Sulfide diesters react less easily (low yields), higher temperatures are needed (25°C,15h), however, with identical stereo- and regioselectivities. The kinetic difference to the allylamines can be attributed to the higher electron donating properties of the sulfide ligand, which reduces the charge density in the palladium, and leads to lower polarization of the olefin.

When higher rings are involved (6- ,7-membered ring) the palladacyclic system is too unstable to be isolated. A slight modification of the work-up procedure permits, after reduction of the metal complex, to isolate either the amine in the presence of 4Å molecular sieves, or an aldehyde; the latter *via* acetic acid hydrolysis, and most probably with the intermediate formation of an imminium cation. Work-up without any precaution leads to a mixture of amine **78** and aldehyde **79** in a 3:1 ratio.

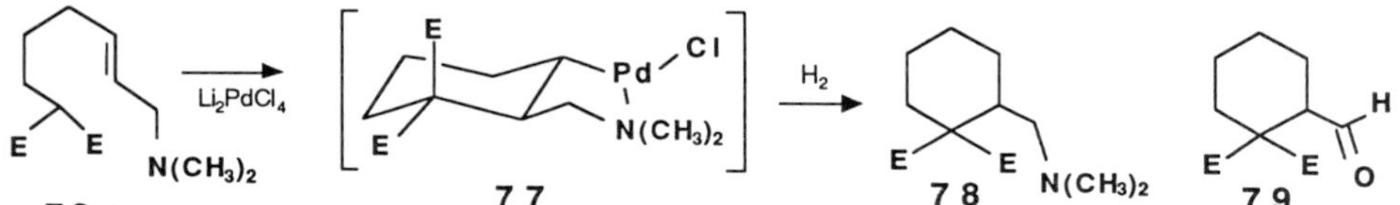

$E=CO_2Me$
presence of mol sieves 4Å 91%
1. 6h/25°C 2. HOAc 93%

Different C-nucleophiles can be used, for example, enolates. Depending on the position of the keto group, exo- or endocyclic products are formed. In the latter reaction it is of interest that the seven-membered ring **81** is more easily formed than the cyclohexanone derivative **82**.

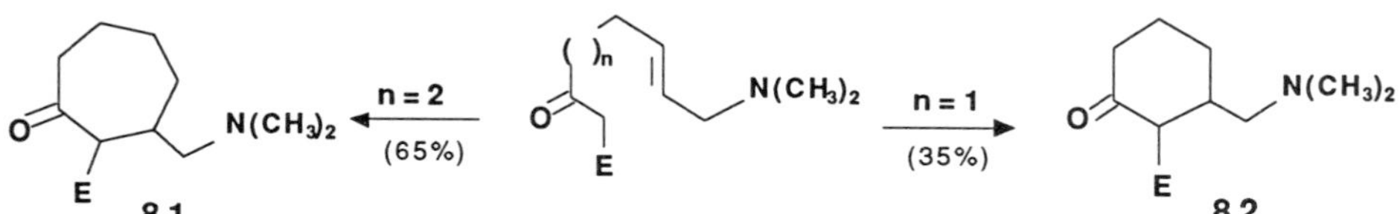

cond.: 1. Li_2PdCl_4 1eq., THF/CH_2Cl_2 1:1, KOtBu 1.1eq, 1.5h-r.t.
2. H_2

The stereochemistry of aminoketones **81** and **82** has not been determined. Sulfides are too unreactive to be cyclized in this case.

The addition of carbon nucleophiles to olefins may be performed catalytically under Heck type reaction conditions. A vinyl- or phenyl-palladium(II) species is preformed and added to the double bond. The ring forming step is now an S_N2 type substitution of an intermediate palladium σ-complex by the malonate type nucleophile [54]. No stabilizing amino or sulfido group is required, as in Holton's reactions [51, 53]; however, the reactions seem to be limited to the formation of cyclopentane rings.

n=3

83 *(n= 4 or1: no cyclization)*

+ PhI 75min/85°C-dppe

$Pd(dba)_2$ 0.04eq/NaH/DMSO

84

Pd° + 105min

85 57%

86 75%

A quite unique way to create a carbon nucleophile is the following palladium mediated diazo insertion reaction [55]: when an olefinic diazo ketone **87** is heated in nitromethane with $PdCl_2(PhCN)_2/LiBF_4$ in the presence of 2,2'-azobis(2-methyl-propionitrile), cyclopentenone **89** is formed. The mechanism is rationalized as an addition-elimination process *via* the (not isolated) intermediate palladacycle **88**.

$PdCl_2(PhCN)_2$ 0.005eq

MeCN, 3h, reflux

87 $E=CO_2Me$

88

89 55%

In the reaction of the allylic cyclohexene derivative **90** two isomeric derivatives **91a** and **91b** are isolated. The favorable elimination from the intermediate palladium complex **92** leads to the enone **91a**. Nevertheless, the proximity of the axial hydrogen at C_3 to palladium makes the olefin formation possible in the six-membered ring and the unsaturation is retained in **91b** to nearly 30%. According to the stereochemistry of the intermediate **92**, it seems reasonable to attribute the stereochemistry drawn in **91b** (*cis* ring junction and *exo* configuration of the carboxylate group).

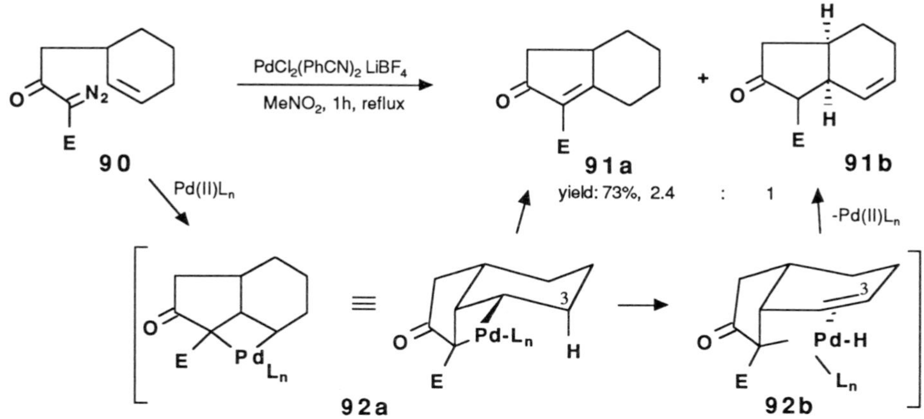

$PdCl_2(PhCN)_2$ $LiBF_4$, $MeNO_2$, 1h, reflux; **90** → **91a** + **91b**; yield: 73%, 2.4 : 1; **90** —Pd(II)L_n→ [**92a** → **92b**] —-Pd(II)L_n→ **91a**, **91b**

It remains to be clarified if in the first coordination-addition step(s) the carbene type carbon adds to the coordinated double bond, or if the carbene coordinates to palladium(II) and, in a consecutive step, adds to the double bond in a [2+2] cycloaddition reaction. We have already discussed the heteroatom (N,S) assisted addition of malonate type nucleophiles to the coordinated olefinic bond (*path A*);

Scheme IV

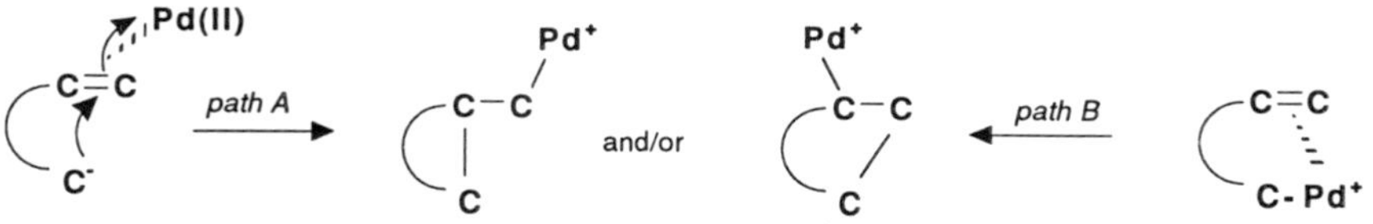

path A ... and/or ... *path B*

Now the conceptually related insertion of a double bond into the carbon-palladium σ-bond will be discussed (*path B*).

A possible method of generating a carbon-Pd σ-bond consists in the oxidative addition of C-X (X=halogen, ester, ether..) to a palladium(0) complex. In aliphatic or alicyclic systems the C-X group has to be activated, *e.g.* by a neighboring carboxylate or carboxamide group. If C-X is aromatic or vinylic in nature, the oxidative addition is more favorable and the palladium complex can be implied in vinylation or phenylation reactions (Heck reaction) [7, 30].

In oxidative addition reactions at 'activated' aliphatic or alicyclic substrates, palladium catalyzed cyclization reactions are known with homoallylic chloroformiates [56] and formamides [57]. Carbocyclic- and heterocyclic systems are thus available. α-Haloamides [58], α-haloketones [59], or α-haloesters [60] are cyclized in a similar way in the presence of catalytic amounts of $Pd(PPh_3)_4$ together with bis(1,8 dimethylamino)naphthalene (proton sponge) as a base. Five-, six- [58a] and seven-membered [58b] lactams are formed in low to fairly good yields and variable selectivity.

As an example [58a] of this kind of reactions, the easy (5 minutes, 50% yield) and highly stereoselective (*cis* configuration of **94**) cyclization of N-iodoacetyl-2-vinylpyrrolidine **93** may be representative. It is of synthetic interest since the three products **94, 95** and **96** are readily transformed into pyrrolizidine alkaloids *dl*-trachelantamidine **98c**, *dl*-isoretronecanol **98t**, *dl*-heliotridane (endo)-**99** and *dl*-pseudoheliotridane (*exo*)-**99**. The ratio of the bicyclic amides is solvent dependent: acetonitrile ensures the highest ratio of saturated products (**94:95:96** = 4:2:1). On the contrary, DMF or HMPA give rise to a high percentage of unsaturated amides (yield 24-28%), predominantly **96**. The question whether the selective formation of the *cis* substituted compound **94** is the direct consequence of a palladium σ-complex **97b** exhibiting identical configuration has not been answered; a nearly 1:1 ratio between saturated and unsaturated cyclization products cannot exclude a stereocontrol during the elimination step from *exo*/*endo*-**97b**, reductive elimination from *exo*-**97b** to **94**, or ß-elimination of Pd-H from *endo*-**97b** to form the alkenes **95** and **96**.

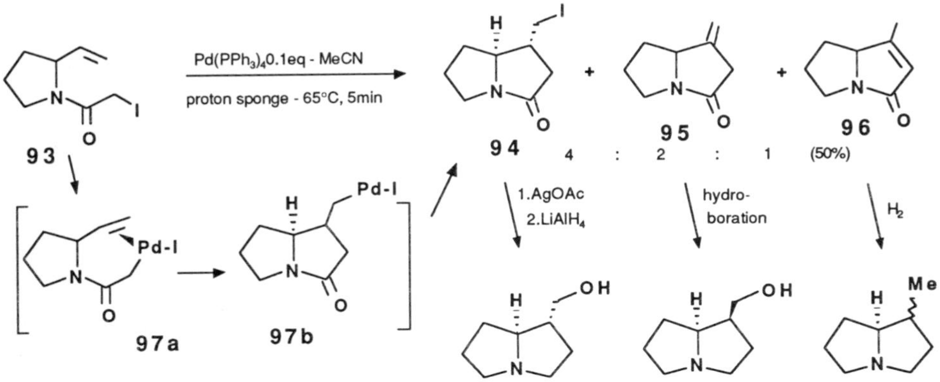

Pd(PPh$_3$)$_4$0.1eq - MeCN
proton sponge - 65°C, 5min
93 94 95 96
4 : 2 : 1 (50%)
Pd-I
97a 97b
1.AgOAc
2.LiAlH$_4$
hydro-
boration
H$_2$
OH OH Me
98c 98t 99

The mechanism proposed by Mori, Ban and coworkers [58, 59], in order to rationalize these palladium mediated isomerization reactions of unsaturated α-iodocarbonyls, involves the formation and insertion of a palladium σ-complex across an olefinic double bond (*e.g.*, **93** → **97a** → **97b**). In light of recent results, under free radical conditions, another mechanism has been suggested for these transformations [60].

(Bu$_3$Sn)$_2$ or Pd(dppe)$_2$

100 E=CO$_2$Me → **101**(c/t=1:1) + **102** (c/t=1:2) 93 : 7

The observation of virtual identical reaction mixtures independent of either ditin/hν or palladium(0) catalyst from the unsaturated iodo carbonyl compound **100** (and 2 other comparable olefins) seems to provide strong evidence that this "ene-halogeno-cyclization should be classed as a radical chain reaction that proceeds by an atom transfer mechanism" rather than an organometallic reaction. Indeed, the radical reaction (ditin) is well understood and the metal serves, in a dual manner, as a radical initiator and as a trap for iodine; the same rôle is now attributed to palladium(0). It remains to be clarified if these conclusions are valid for most of Mori-Ban reactions, including the cyclization of **93**. The formation of aminoenone **96,** the principal reaction product, at least in one case, is difficult to explain by an atom transfer process via the radical intermediate **103**. The two step elimination process from the palladium complex **97b** to **96**, however, is well precedented with organopalladium intermediates such as **104** [62] and **105** [63].

103 (?)→ **96** ← **97b** (Pd-I)   **104** (Pd-X, OAc)   **105** (Pd-X, OAc)

There may be a disagreement on oxidative addition and reductive elimination reactions of palladium iodide complexes [60, 61]; similar metal chloro complexes are much less doubtful, since they can be isolated, and the mechanism of intramolecular acetylene and olefin insertion reactions have been studied by Samsel and Norton [64].

The unsaturated α-carbonyl palladium complexes are readily formed by reactions of suitable chloroformiates with Pd(0)-phosphine. On heating, stable vinyl-complexes **107-L** can be isolated from the acetylene **106-L**. According to kinetic and ^{31}P-NMR measurements, a four-coordinate mechanism is most likely to occur *via* the acetylene coordination in **106**. The rate constant for the ligand displacement from **106-L** to **106** depends on the chain length. This emphasizes the importance of coordination geometries for the insertion.

Compared to the formation of palladium complex **108** (3 carbons between acetylene and oxygen), the rate constant k decreases on formation of compounds **106**, having a less mobile C_2-chain. In **106**, the acetylene group has to be placed in the unfavorable in-plane geometry with (at best) an angle of 45° with the coordination plane. Consequently, the global reaction from **106L** to **107-L** is rapid, though the coordination of **106-L** → **106** is slow. Indeed, in spite of the insertion reaction to **107**, and the formation of the five-membered ring, it is a favorable process. The opposite effect is observed with complex **108**, where the triple bond is easily coordinated in the favorable position, perpendicular to the coordination plane. Now, the reaction to a six-membered alkylidene complex has to go through an unfavorable transition state and will, evidently, be slower.

The corresponding olefin complexes **109** and **112** are likely to react in a very similar fashion. Apparently the σ-complex **110** is readily formed from the palladium phosphine butenyl chloroformiate complex **109**. The immediate ß-elimination of $Pd(PPh_3)_2$ and HCl leads to lactone **111**, exclusively. The longer chain complex **112** only reacted sluggishly to 'unidentifiable insertion products'. The 4-coordinate mechanism also seems to be operative (inhibition by added ligand) though reliable values for the rates could not be obtained.

PdL$_4$ or PdL$_3$ + chloroformiate/130°C

PPh$_3$-Pd-PPh$_3$ Cl **109** → toluene → [PPh$_3$-Pd-PPh$_3$ Cl **110**] → **111** + Pd(PPh$_3$)$_2$ + HCl

PPh$_3$-Pd-PPh$_3$ Cl **112** → 130°C, toluene → decomposition; no identifiable insertion products

Beside a clear demonstration of a four-coordinate mechanism prior to insertion reactions of C-Pd σ-bonds across the double and the triple bond, this study presented conclusive evidence for the relationship between reaction rates and the stability of intermediate σ–π-coordinated complexes. Therefore, the preference of five- over six-membered ring formation, so often encountered in palladium catalysis, can be related directly to the energy of the corresponding intermediates.

Beside the oxidative addition reaction, involving activated carbon-halogen bonds, the intermediate palladium σ-complex can also form in a metal exchange reaction. Alkylmercurials constitute easy accessible precursors, which readily react with palladium(II). The following palladium promoted cyclization of allylic 'alcohols' is a contribution of Larock and coworkers. Rather unselective in the case of cyclic allylic alcohols [65], the stoichiometric formation of butenolides from aliphatic α-chloromercurio acetate esters, such as **113**, is shown to proceed with extremely selective double bond isomerization [66].

HgCl **113** → Li$_2$PdCl$_4$, THF&DMF or HMPA → **114** (53-69%) + HX + Pd

PdX$_2$ → PdX → PdX → HPdX → PdX

Vinyl- or aryl palladium species are easily accessible by oxidative addition of palladium(0) to vinyl- or aryl halides (esters). The addition of these organometallic species across a double bond usually followed by metal hydride elimination, well known as the Heck Reaction [7, 30], is one of the most powerful carbon-carbon bond forming reaction in organic palladium chemistry. Intramolecular versions, aimed to the formation of carbo- and heterocyclic systems, have been extensively studied by several groups. In this report, only some recent results with relevant regio- and stereochemical information will be discussed.

The compatibility with a number of functional groups, makes the Heck reaction a powerful tool for heterocyclic synthesis. This is particularly true for azacyclic and -bicylic systems [67-71]. The following annelation reaction of *ortho*-iodobenzoic amide **115**, *via* aryl palladium intermediates, and leading to a 2:1 mixture of the olefinic regioisomers, is an illustrative example [67]. This contribution of Grigg's group shows several important attributes often met in these reactions. The usually observed unreactivity of palladium towards highly substituted double bonds is no more a rule in these reactions, and tertiarty carbon atoms are formed fairly easily. The addition of Ar-Pd across the double bond proceeds with *cis*-stereochemistry; however, the elimination step is often accompanied by rather unselective isomerization reactions. The usual ring size of cyclized compounds is the five- or six-membered ring, in some cases also the seven-membered ring [70].

Pd(OAc)$_2$/PPh$_3$/DMF

K$_2$CO$_3$/NEt$_4$Cl/30°C

115 **116a** 2 : 1 **116b**

72%

117a **117b**

A different termination reaction takes place in the presence of anionic species, and in the 'tandem-cyclization-anion capture ' process, the intermediate palladium species, such as **117b**, is intercepted by hydride species [69a] or anionic carbon groups [69b, c]. In the example shown below, the *cis-exo* addition to the norbornene double bond in **118** leads to a polycyclic amide **119** (R=H or -CCPh), the anion source being either HCOOH/piperidine (hydride) or phenylacetylene (carbon nucleophile) *via* the organopalladated intermediates **119** (R=Pd-I and subsequently R=Pd-H or Pd-CC-Ph).

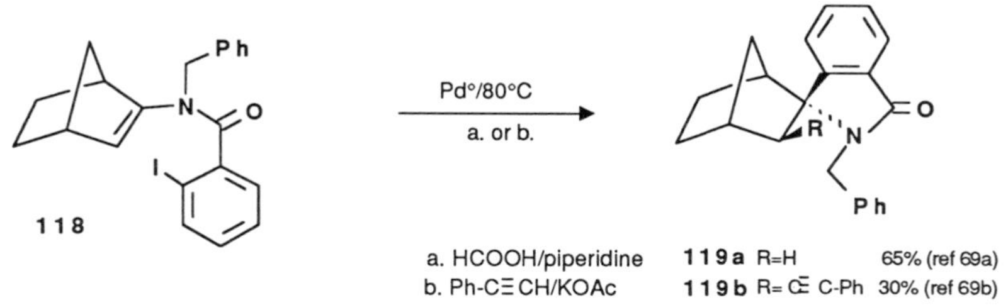

Pd°/80°C

a. or b.

a. HCOOH/piperidine | **119a** R=H | 65% (ref 69a)
b. Ph-C≡CH/KOAc | **119b** R= C≡ C-Ph | 30% (ref 69b)

Now, the formation of the quaternary carbon (Markovnikov type addition) leads to the formation of spiro compounds.

Scheme V

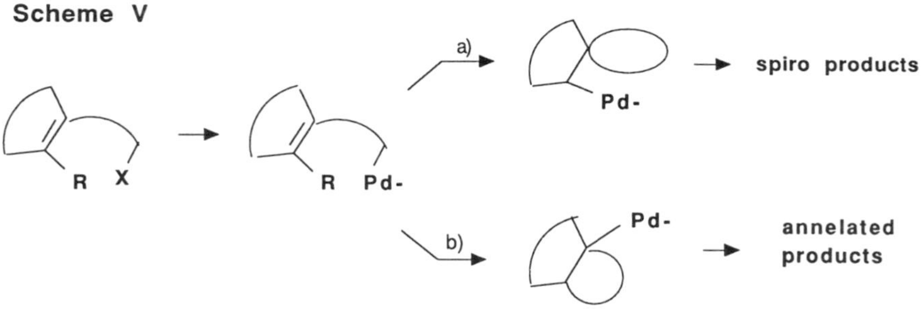

a) → spiro products

b) → annelated products

It is noteworthy that the latter reaction (a) is predominant over the annelation reaction (b), even in the case of tetrasubstituted olefins [70].

An interesting application of these spirocyclizations has been reported by Overman and coworkers [71] during their studies towards the synthesis of gelsemium alkaloids. The pentacyclic spirooxindole derivative **121** is formed as the major isomer in the intramolecular alkene arylation of **120**. Only minor influences on the stereoselectivity are exerted from the substituents R at the amido group (best result: R=Me, *endo vs. exo* attack, **121a:b** = 1.5-2.0 : 1).

120 E=CO$_2$Me → Pd(PPh$_3$)$_4$, MeCN-Et$_3$N, 82°C → 121a + 121b

R=Me (81%) 1.5-2 : 1

Similar C-C bond forming transformations are useful for the obtaining of oxygen containing heterocycles, mainly hydrofurans or hydropyrans. As for cyclic olefins spiro-, annelated-, or bridgehead bicyclic compounds may be obtained from unsaturated alkyl or aryl ethers [70, 72]. In a typical example, the highly selective *cis* connection in the oxacyclic ring as well as the complete absence of bridgehead olefins should be remarked, the latter despite the possibility of conjugation with the aromatic nucleus.

122 → 123 → (-PdH) 124a + 124b

cond.:PdCl$_2$(PPh$_3$)$_2$ 0.05eq NEt$_3$-MeCN, reflux

124a (74%) ratio 1 : 1 124b

Inside the cyclohexane ring, the commonly observed hydrogen shift takes place, apparent by the equivalent formation of the two olefins **124a** and **124b**, in a 1:1 ratio [72].

Considering the formation of pure carbocyclic systems by the intramolecular Heck reaction, the parallel and simultaneous investigations of three different research groups, the ones of Overman [70], Negishi [73], and Larock [74], emphasize its importance. A number of bicyclic and tricyclic systems have been synthesized by these groups. The regio- and stereochemical features are perfectly similar to the ones that have been found in the cyclization to aza- and oxa cyclic compounds. The balance of spiro- *vs.* fused cyclization seems delicate (Scheme V) as shown by the high yield transformation of olefins **124** and **126** [73c].

E'=CO_2Et

Pd $(PPh_3)_4$ 0.03eq, NEt_3-MeCN-THF, reflux,10h

124 → 125 86% (3-4 isomers)

idem, reflux,1h

126 → 127 70%, (>90%cis)
2 regio isomers (ratio 7:3)

With heteroatoms present in the side chain (nitrogen or oxygen), only the spirocyclization pathway is operative [70]. Nevertheless, the regio control seems total since there is no mention of mixtures of spiro- and annelated reaction products. The ease of selective formation of the bicyclo[3.3.1]nonane skeleton [75] is also noteworthy [70, 73b]. During the last Pd-H elimination step (e.g.) *exo* or *endo*-cyclic double bonds can be formed, depending on the regiochemistry of the palladation.

Pd°/L, NEt_3-MeCN-THF, reflux

128 → [129] → 130 70%,
olefin-isomeric ratio 9:1

At this stage also, the isomerization, *via* readdition-elimination of Pd-H species, takes place. Overman [70] found an elegant solution to this problem.

[Pd], 80°C- 2-4d

131 → 132a + 132b

[Pd] : $Pd(OAc)_2$ 0.03eq, DMF, KOAc, nBu_4NCl 1 : 5.1 (63%)
$Pd(OAc)_2$ 0.03eq, CH_3CN, Ag_2CO_3,PPh_3 >95 : 5 (72%)

By adding traps of hydrohalic acids, silver carbonate or silver nitrate, to the reaction mixture, the hydrogen migration is considerably suppressed. Larock's reaction of **131** shows the efficiency of this approach [74].

One of the major challenges in organic synthesis is the enantioselective, or, if this is possible, enantiospecific formation of carbon-carbon bonds. The highly stereoselective vinyl- or aryl palladation of olefins creates, in many cases, a chiral carbon from symmetrical sp^2 carbons. The necessity of stabilizing the intermediate organopalladium species by phosphine ligands enables access to diastereomeric intermediates by the use of a chiral phosphine, issued from the considerable 'chiral ligand pool' that is available today. The first approach to the study of chiral induction has been undertaken using the intramolecular Heck vinylation of the 1.4-cyclohexadiene alkenyl iodide **133** leading to an unsaturated chiral *cis*-decalin derivative **135** [76].

133 R=I → [Pd°] → R=I-PdL_n → **134** (I-PdL_n) → (±) **135** (68%)

achiral [Pd°]: $Pd(OAc)_2$ 0.03eq, DIPHOS, MeCN, Ag_2CO_3, 5h, 60°C

chiral [Pd°]: $Pd(OAc)_2$ 0.03eq, *L 0.09eq, cyclohexene 0.06eq, Ag_2CO_3 2eq, 1-methyl-2pyrrolidinone (solvent), 3h, 60°C add. of 133 (R=I), 37.5h, 60°C → (+) **135** (74%, ee: 46%)

*L = (R)-BINAP

Overman's silvercarbonate conditions, as well as a judicious choice of a 1.4-dienic structure in order to avoid any isomerization, are prerequisites for the formation of the single isomer (±) **135** in the achiral reaction. The enantioselectivity can be induced with a variety of chiral phosphines. Good optical yields, however, are only obtained with the C_2-symmetry, atropisomeric diphosphine (R)-BINAP. It is well known that for the Heck reaction optimum conditions have to be determined for nearly every reaction. The strong solvent effect on the degree of chiral induction (ee) in the 'decalin' **135** is therefore no surprise: DMSO (1% ee), THF (2% ee), CH_3CN (8% ee), 1,1,3,3-tetramethylurea (19% ee), HMPA (20% ee), N,N'-(dimethylpropylene)urea (23% ee) and, finally, 1-methyl-2-pyrrolidinone (33%). An additional increase to 46% ee, combined with the best chemical yield of 74%, could be achieved by carefully controlling the formation of the active palladium(0)-chiral ligand catalyst before adding vinyl iodide **133** (R=I) to the reaction mixture. This elegant asymmetric study on catalytic C-C bond formation is a further demonstration of the power of the Heck reaction in the tool box of modern, selective palladium catalyzed transformations. Moreover it is a nice demonstration how to upgrade a nearly inefficient reaction to one reaching high asymmetric induction.

Another interesting extension of intramolecular carbopalladation reactions to benzyl halides or esters, instead of vinyl- or aryl halides, has been reported by Negishi's group [77]. In many respects the features of the Heck reaction are found, such as high alkene regiospecificity, formation of five- to seven-membered rings, olefin isomerization, easy formation of spiro compounds and *cis* stereochemistry of the cyclized products. The possibility of a radical mechanism is excluded by comparison with stannation reactions (no Sn-promoted cyclization) and the ease of trapping palladium-carbon σ complexes with acrylate:

Cl

cat $Pd(PPh_3)_4$

NEt_3,MeCN
reflux, 2d

PdL_nCl

Me

E

Me

E

136

137

138 (55%)

The intramolecular trapping of intermediates **137** (homoallyl- instead of methyl group) converts the starting diene into a tricyclic system by double (cascade) cyclization.

2.3.2 Enyne and enallene reactions

This part describes mainly reactions in which the composition and number of atoms remains unchanged. The cyclization process is therefore a metal catalyzed isomerization reaction. These kind of reactions have proven extremely useful for organic syntyhesis, particularly in the case of Ene reactions [78], intramolecular Diels Alder cycloadditions [79] or metal catalyzed enyne/enallene cyclizations [80]. In a series of papers, B.M. Trost and coworkers have developed a palladium mediated version of the latter reaction [81-91], (Scheme VI). Starting from 1-6-enynes, this reaction is a general approach to highly substituted cyclopentane derivatives, all containing conjugated or 1.4-dienic double bonds. In one case, the cyclization of 1.7-enynes to cyclohexane rings has also been reported [89].

Scheme VI

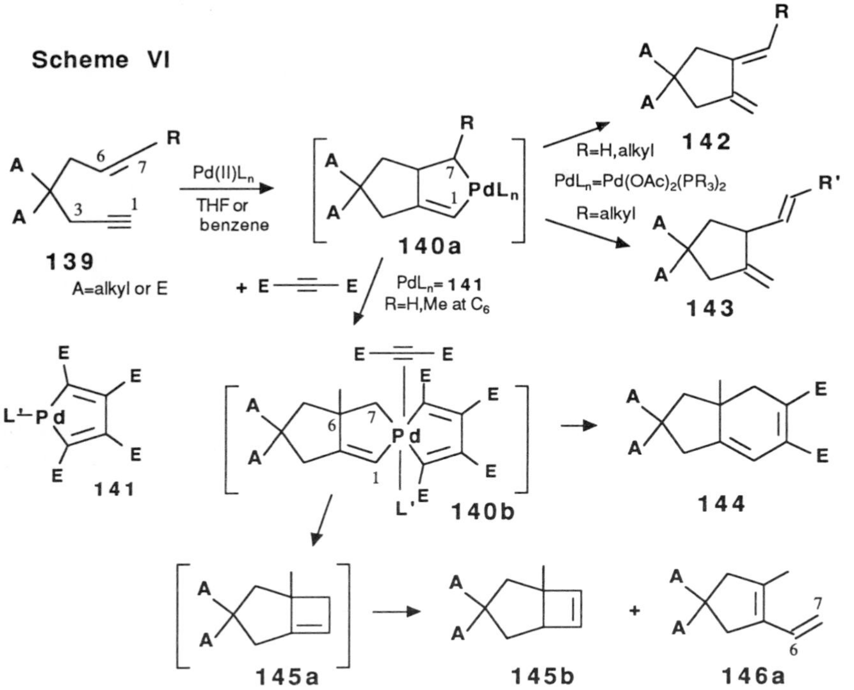

Depending upon the catalyst, ligands, substitution patterns and reaction conditions, four different types of dienes can be obtained under mild and catalytic reaction conditions. The usual products in (catalytic) $Pd(OAc)_2(PR_3)_2$ catalyzed reactions are 1.3-dienes **142** and 1.4-dienes **143**. A great effort had been devoted to the elucidation of the mechanism operative in these isomerizations [86, 87]. The formation of the cyclic dienes involves the intermediacy of metallacyclopentene **140a**, representing a Pd(IV) species, and subsequent hydrogen migrations. This hydride shifts can be suppressed by the less electron deficient chelated Pd complex TCPC (tetracarbomethoxy-palladacyclopentadiene) **141** and trapping of the new palladacyclopentene by dimethyl acetylenedicarboxylate (DMAD). The [2+2+2] cycloaddition products **144** are the sole non rearranged products in these isomerizations.

By the formation of **145b** and **146** the DMAD intermediate **140b** reveals novel metal catalyzed rearrangements of 1,6-enynes. Proceeding *via* 1,1 reductive elimination on palladium in **140b**, and forming the unstable cyclobutene intermediate **145a**, this new pathway becomes predominant when C_1 is substituted by the carboxylate group E. After labelling **139** specifically with deuterium or carbon-13 an unexpected carbon migration becomes evident:

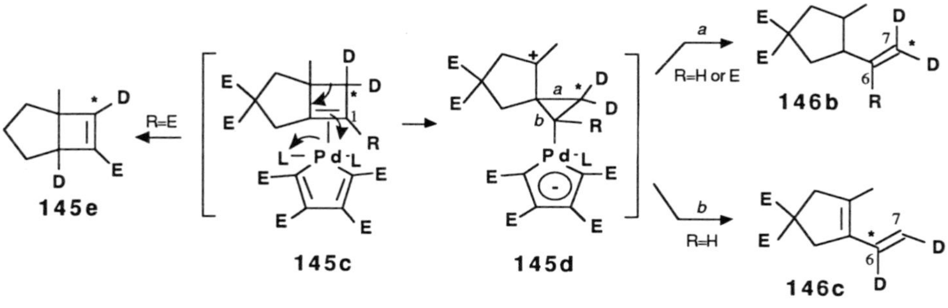

A rational explanation of the results with labeled 1,6-enynes, *i.e.* the formation of **145e**, together with the vinylcyclopentenes **146b** and **146c**, is the following: the coordinated palladium intermediate **145c** either reacts to the rearranged cyclobutene **145e** (R=E) or leads to a new anionic cyclopropylcarbinyl and palladacyclopentadienyl complex **145b**. Cyclopropane cleavage according to a or b has been established by the labels in **146b** and **146c**, respectively.

The regioselectivity of the formation of 1,3 *vs*. 1,4-dienes depends, to a great extent, on the different substituents at the enyne carbon chain. Carbon or oxygen atoms in allylic positions seem to favour the 1,3-diene route, whereas methyl or methylene groups in these positions (carbon-5 or carbon-8) give 1,4 isomerized dienes **143** (Scheme VI). A high preference for *trans* configuration is found in the vinylic part. The diastereoselectivities are good in the formation of 1.2-disubstituted cyclopentanes (*trans/cis* ratio=3:1 from E-substituted olefin), and excellent in the case of annelated bicyclic rings of type **148** [81, 85]:

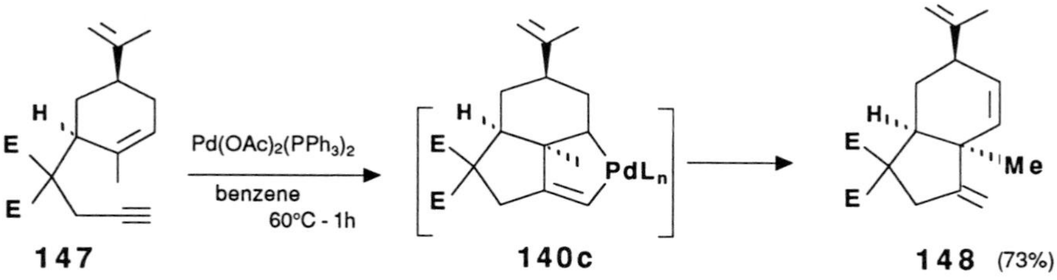

The intermediate formation of the [2+2] cycloaddition complex **140c** is the controlling factor for complete *cis*-stereoselectivity in the indane system. A *cis* or *trans* substituent in the cyclohexane ring, as well as trisubstitution of the double bond, has no important influence on the reaction; quaternary and vicinal quaternary centers are readily formed. Thus Trost's carbametallation compares favorably to the Heck type cyclisations, discussed before. The potentialities of the isomerization-cyclization procedure in natural product chemistry has immediately been exploited for the synthesis of the picrotoxane skeleton **150** [85]:

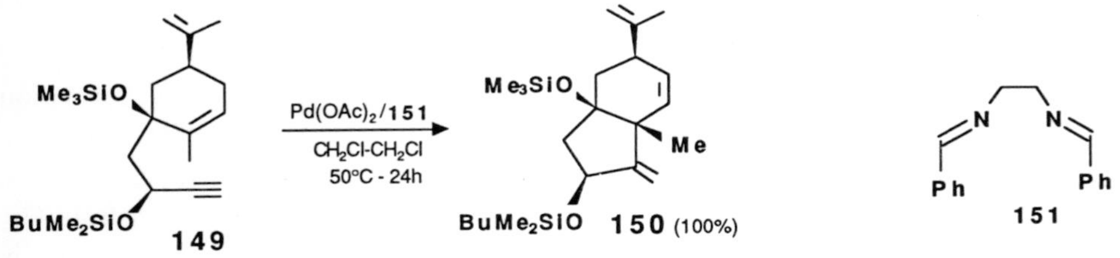

For this particular reaction a new, less reactive palladium(II) catalyst had to be found. The bis-imine **151**, stronger σ-donor than phosphines but still with some π-acceptor properties proved to be the ideal new ligand, together with palladium acetate, in order to raise the yield in the transformation of **149** from less than 10% (phosphine ligand) to nearly 100% (ligand: **151**). Very interesting perspectives for Trost's carbametallation lie in the application to heterocyclic syntheses.

Unlike in palladium(II) mediated oxyamination of alkenes [92], nitrogen is compatible with catalytic reaction conditions. With palladium acetate-phosphine in isomerization reactions, highly selective formation of five-membered azacyclic and azabicyclic systems are possible [84]. The unusually low 1,3 - 1,4 regioselectivity in the isomerization of **152** could be considerably improved by the chelating bis-imine ligand **151** (ratio **153/154** 1:2, L=PPh_3 and 1:9, L=N,N'-bis(phenylmethylene)-1,2-ethylenediamine **151**).

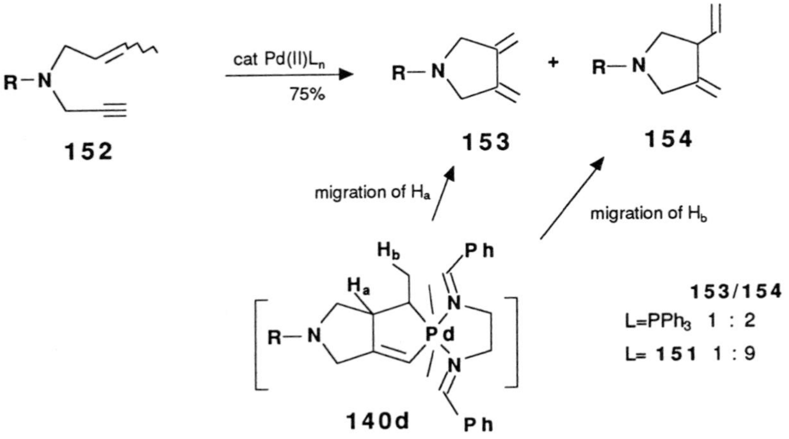

It has been suggested that the steric hindrance in the intermediate **140d** disfavours the migration of the most reactive hydrogen, H_a, which should lead to 1,3-dienes, in favour of the transposition of H_b, giving the 1,4-diene. Within the syntheses of different bicyclic alkaloids, the successful formation (50% yield) of the carbapenem nucleus is an outstanding result of these reactions [84].

The stereochemically induced ligand control during the final olefin formation step is perceptible even in the carbon chain R in **139** [83] (Scheme VI). On coordinating to the palladium(IV) in **140e**, the unsaturated side chain in **155** is able to fix the molecule in a conformation that prevents the hydrogen migration from C_8. Indeed, in this position no alignement of C_8-H and palladium for *cis*-migration is possible (four centered transition state) [93]. The consequence is the (nearly) exclusive migration of C_{13}-H and formation of **156b**. In the presence of (phosphine) ligands, the more mobile palladium complex **140f** is the intermediate, a situation that leads to the migration of C_{13}-H and C_8-H and, thus, to much lower preference of **156b** (ratio **156b/a** 2.9:1). This coordination control is further confirmed by the comparable non-selectivity in the case of saturated side chains. A different reaction control (conformationally induced minimized non-bonded interactions) is operative in the thermal reaction.

By obtaining **156a**, the products of the flash vacuum thermolysis (625°C) are nicely complementary to those of the metal catalyzed reactions [83].

The high degree of understanding of the isomerization-cyclization sequence of enynes has allowed the development of another, extremely simple, catalytic system. It is based on the idea that palladium(II) can be formed by oxidative addition of an acid (HX) to palladium(0). Actually the inverse reaction, the reductive elimination is the final step of palladium catalyzed oxidation reactions. However, the sensitivity of the Wacker Reaction, for example, to the concentration of strong acids (HCl, stabilization of HPdCl species) is well known. The stabilization of H-Pd-X species in the case of weaker acids, however, was unknown. It is a most important feature for enyne cyclizations that these HPdX complexes are stable enough, even with acids such as acetic acid (and a wide range of carboxylic acids) to promote very efficiently the cyclization of 1,6-enynes (and 1,7-enynes in some cases) in the presence of phosphines. In many respects, the reactions with the new catalyst are comparable to the former $Pd(OAc)_2$-phosphine reaction: similar high selectivities (1,3 or 1,4-dienes), soft conditions, good yields; on the contrary, substrates with a carboxylate group at the terminal acetylenic carbon do not cyclize. Though phosphine ligands are not obligatory, their presence seems to be beneficial for the reactions. The easy formation of 6-membered rings is noteworthy and last, but not least, the possibility to perform catalytic asymmetric induction in the presence of chiral carboxylic acids. Despite an unsatisfactory regioselectivity in favour of the (chiral) 1,4-diene, the asymmetric inductions are appreciable at least with (S) binaphthoic acid **159**. A new mechanism is operative with HPdX catalysis. It is a stepwise addition-cyclization-elimination sequence (and no concerted cycloaddition). In its asymmetric version, the enantioselective step is the insertion of the isopropylidene double bond into the vinylic carbon-palladium bond.

157 → $(dba)_3Pd_2xCHCl_3$, *RCOOH, PhH, r.t., PPh_3 → **158a** + **158b** (59-82%)

HPdX → **161a** → **161b** X=OCO*R

*RCOOH: (S)**159**, (S)**160**

	ratio **158a/b**	(ee)
S(-)-binaphthoic acid (61%)	3 : 1	(33%)
S(-)-Mosher's acid (60%)	3 : 1	(8%)
(+)-camphorcarboxylic acid (64%)	3.2 : 1	(2-5%)

The *cis*-stereospecificity of the insertion reaction is a favorable factor for efficient, asymmetric reaction control by the chiral acid. The four centered palladacyclic transition state becomes diastereomeric, and given the proximity of the chiral carboxylate ligand to the prochiral olefin, exhibits efficient chiral discrimination and a promising enantiomeric excess of 33%, at least in the case of the C_2 symmetrical binaphthoic acid **159**. The geometry of the chiral acid is decisive since monoacids, *e.g.* MTPA (2-methoxy-2-trifluoromethylphenylacetic acid) (S)**160**, Mosher's [94] widely used chiral analytical reagent, or other chiral 'cyclopentane' carboxylic acids don't give ee values greater than 10%.

The course of the 'HPdOAc'-catalyzed cyclization can be modified, reductively, by adding an excess of silanes, *e.g.* polymethylhydrosiloxane (PMHS). This particular hydride source intercepts the palladium acetate complex **161d** and transforms the original enyne double bond into a saturated carbon carbon bond. Under these conditions **149** reacts to the more saturated indane **162**, exhibiting the 'usual' hight selectivities known from the non reductive cyclizations.

Complementary to the enyne isomerisation the new variant allows the selective reduction of the double bond of the original enyne during the cyclization. This control of either olefin maintenance (**150**) or olefin hydrogenation (**162**) should proof extremely valuable, more precisely in polyenic compounds where chemoselective reduction is not an easy task.

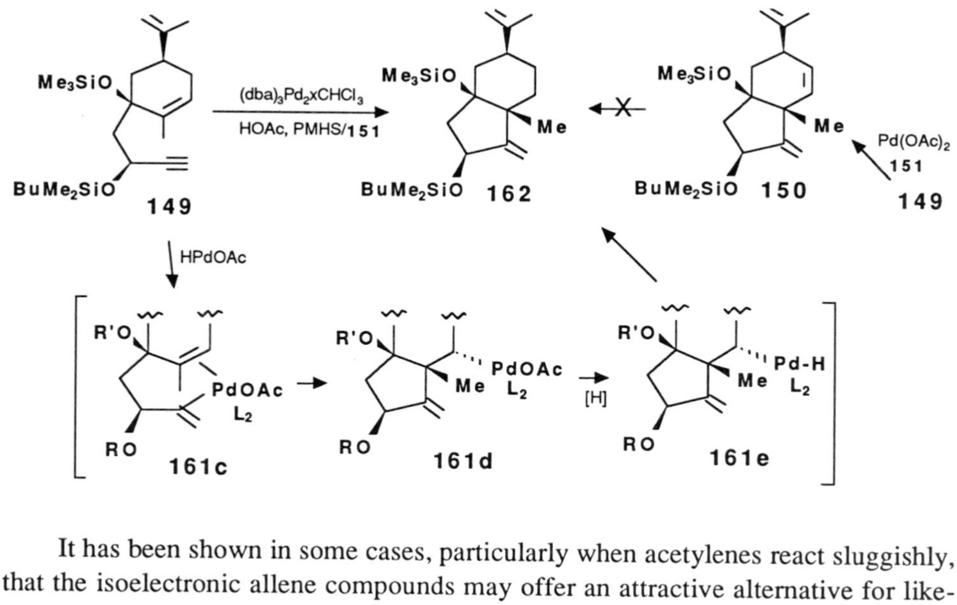

It has been shown in some cases, particularly when acetylenes react sluggishly, that the isoelectronic allene compounds may offer an attractive alternative for likewise selective isomerisation cyclization reactions [95]. The ten-membered ring allene **163** rearranges in the presence of a palladium(II) catalyst [96].

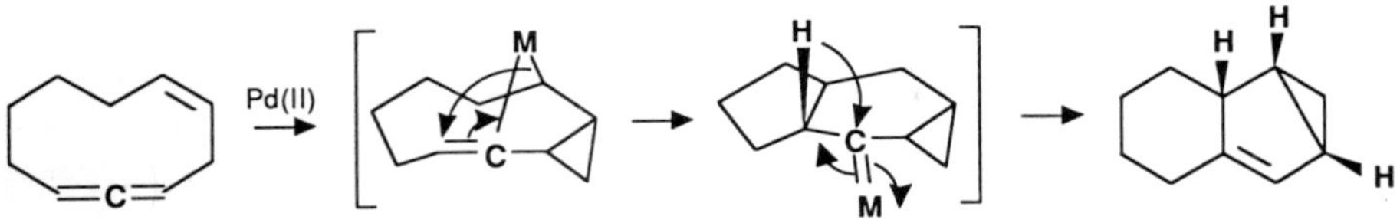

Pd(II)

163 **164**

The cyclic allene **163** is highly reactive and actually the formation of **164** is observed with a number of metals (Ag,Hg,Zn,Rh,Cu,Tl,Ru). In this work, which was published in 1983, several mechanisms are discussed. In the light of the recent mechanistic results of B. Trost [87], cycloaddition and carbene type palladium intermediates seems probable even if a strained bridgehead double bond is involved.

2.3.3 Organopalladation-olefin insertion reactions

In organopalladation-olefin insertion reactions the carbon-palladium bond is created by nucleophilic addition to the coordinated olefin. The insertion of a second double bond present in the molecule, into the C-Pd σ-bond, is the cyclization step. In this regards it compares to the isomerization cyclization reactions and the intramolecular Heck type reactions. Differently, during the last elimination step palladium is reduced; this step is frequently assisted by a second oxidant, thus recreating the original palladium(II) oxidant. Usually, the nucleophile is the solvent (carboxylic acids, alcohols or in rare cases water); but, non solvolytic reaction conditions are also possible (C-nucleophiles, solid carboxylic acids). These organopalladation reactions are the sole cyclizations where external, heteroatomic nucleophiles (one or two) are added to the hydrocarbon. This permits the controlled formation of up to 3 (or 4 if CO is incorporated) new chemical bonds in a single step.

Historically, these are the 'first' cyclization reactions with palladium. They may be performed under Wacker type conditions and they may make use of non conjugated dienes which form stable π- and π-σ-complexes with palladium(II). In several review articles [97, 98] and one monograph [16] the chemistry of intramolecular-coordination compounds of diolefins such as 1,5-cyclooctadiene (COD), dicyclopentadiene, norbornadiene, 1,5-hexadiene etc., is compilated. Interestingly, the 'unusual cyclization reaction' reported by an ICI research group in 1965 [99] applies to 1,5-hexadiene **165**, whose $PdCl_2$-di π complex **167a** was prepared in 1937 and published in the early fifties [100]; the chelating character of the diene in this complex has been confirmed by X-ray diffraction studies [101].

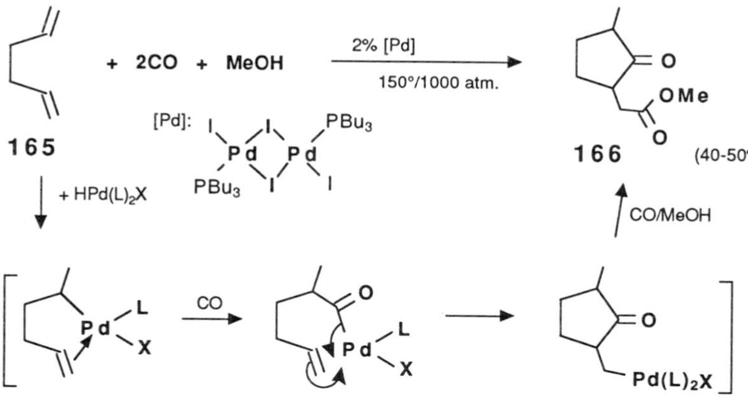

In this remarkable cyclization of **165**, three carbon-carbon and one carbon-oxygen bonds are formed in one step and under catalytic conditions. The stereochemistry of the product has not been studied and it seems probable that the unusual 'industrial' reaction conditions (150°C/1000 atmospheres) have prevented more detailed investigations. It is important to note that under lower carbon monoxide pressure (3-5 atm) the two double bonds react separately and cyclization does not take place with $PdCl_2$-$CuCl_2$ [102]. This reveals a subtle, pressure dependent competition between olefin and CO coordination at the diene, and the destruction of the chelated structure of the intermediate palladium complex(es).

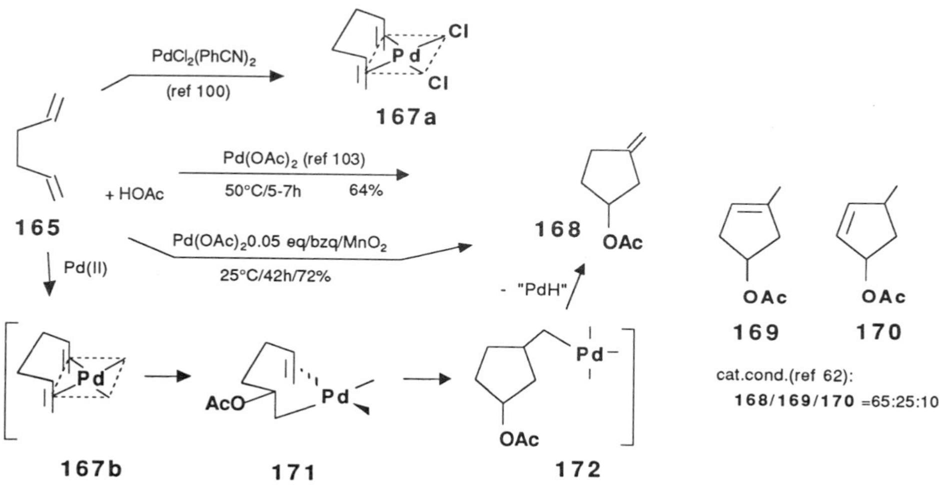

With (stoichiometric) palladium acetate, and carbon monoxide free conditions, chelation and cyclization is possible in acetic acid [103].

Only quite recently, conditions could be found which allow efficient, and variably selective cyclization under catalytic conditions [62]. The regeneration of the palladium acetate is easily performed by a combination of benzoquinone (catalytic, second oxidant) and manganese dioxide (stoichiometric, reoxidant of hydroquinone). This regeneration process is an electron transfer reaction *via* the quinone coordinated to palladium [104] and it allows an interesting approach to a multicomponent system, with molecular oxygen as an ultimate oxidant [105].

R

[Pd]
HOAc

AcO⁻

R=Ph

Ph OAc **174** (87, c/t = 1.29:1)

Ph OAc **175** (10)

Ph OAc **176** (3)

173 **167c**

[Pd]: $Pd(OAc)_2$ 0.05eq /bzq 0.2eq /MnO_2

R=CH_2COOH

COOH (?) AcO⁻ **167d**

HOOC OAc **177** (40%,t/c=7:3)

The catalytic reaction is less selective than the stoichiometric reaction [103] and 35% of isomerized olefins **169** and **170** are formed. Nevertheless, it is a preparative useful [106], general cyclization reaction. The regio- and diastereoselectivities with a substituent at C3 are as above. In the reaction of **173** (R=Ph), high regioselectivity (32:1) is combined with minor isomerization. That means that the substituent in **173** controls the nucleophilic attack in the π-complex **167c**: the nucleophile adds to the uncrowded vinyl group. The stereochemical control is low (nearly 1:1 mixture of *cis/trans* isomers **174**), which reflects the absence of any preferred (*exo/endo*) orientation of R in **167c**. With other substituents R, *e.g.* coordinating (CH_2)X groups, the *cis/trans* ratio in the final products **177** becomes more selective, however never exceeding a value of 7:3 (R=CH_2COOH) in favour of the *trans* product **177** [62].

Norbornadiene is one of the dienes which permitted to establish the stereochemistry of the three fundamental reactions that are involved in the palladation-cyclization reactions. Facilitated by the easy interpretation of the ^{1}H-NMR spectra of the bicyclic palladium compounds, the *trans* addition of nucleophile and palladium(II) to one double bond could be demonstrated unambiguously [107, 108].

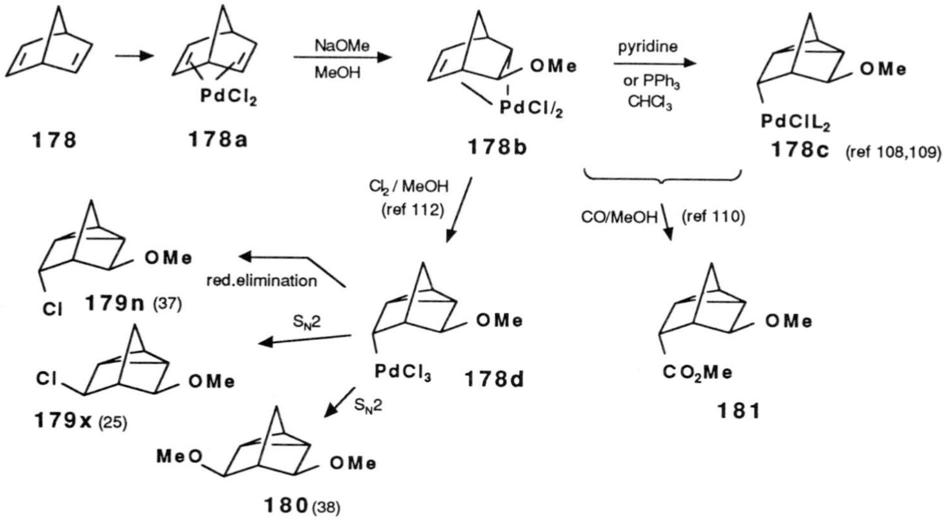

Bridge splitting and olefin displacement in **178b**, with pyridine or trialkyl phosphines, lead to cylization *via* (*cis*) insertion rection. The real character of this cyclization [108] has only been recognized some years later [109]. The carbonylation of the C-Pd bond in **178c**, proceeding with retention of configuration, constituted a chemical confirmation of the designed palladium complexes [110, 111]. With such Pd-σ complexes in hand, the oxidative cleavage of the carbon palladium bond could be studied readily [112].

The cleavage of **178b** with bromine or chlorine (halogenation), a reaction that leads to cyclized products **179** and **180**, is rationalized by two different mechanistic pathways. Oxidative addition of the halogen leads to the palladium(IV) intermediate **178d**. Reductive elimination (retention of configuration) is now in competition with S_N2 attack of the nucleophile (inversion of configuration). In apolar solvents (methylene chloride) reductive elimination is favoured (*e.g.* halogen Cl_2 *endo*/*exo*=83:17), whereas solvolytic conditions (MeOH) and addition of halide ions lead to more S_N2 cleavage. These results, obtained on isolated metal complexes, are also observable in a catalytic reaction under Wacker type conditions [113].

cat.$PdCl_2/CuCl_2$ NaOAc/HOAc

177 → 182n (56) + 182x (13) + 183 (13) y:85%

Now the copper chloride holds the rôle of both oxidant of Pd-C and the added halide, and seems responsible for the fairly unselective formation (comparable amounts of retention and inversion products) of three nortricyclenyl esters **182n**, **182x**, and **183**.

Carbon nucleophiles add to coordinated double bonds in a different way than heteroatoms, and it was the reaction on norbornadiene complexes that permitted the first demonstration of the *cis*-addition pathway [114, 115]:

$PdCl_2$ 178a → (Ph_2Hg, CH_2Cl_2) → [ClPd-Ph 184a] → Ph $PdCl/_2$ 184b (50%) → (pyridine) → Ph $PdClpy_2$ 184c

The phenylation [114] of the dichloro norbornadiene palladium **178a** is performed with diphenylmercury, which gave the *endo* substituted addition product in 50% yield. The cyclization to **184c** after addition of pyridine, is common with alkoxy or acetoxy complexes.

Beside norbornadiene, another chelating olefin, 1,5-cyclooctadiene, has been used extensively [97, 98] as a model for studies concerning palladium-olefin coordination and the reactivity towards nucleophiles. The $PdCl_2$-cyclooctadiene complex is easily formed, very stable and reactive enough to react *via* ligand exchange reactions. A cyclization was observed with oxo-palladium adducts in 1967 by irradiation of **187** in the presence of $Pd(OAc)_2$ [116]; however, the stereochemical relationship between the final product and the nucleophilic addition to the olefin was not realized immediately [117].

The acetoxy indene **186** (*exo* configuration of OAc), a formal *cis*-oxypalladation product, has been isolated from *cis,cis*-1,5 cyclooctadiene. This was in contradiction to the usual *trans*-organopalladation stereochemistry [118]. In reality, a *cis*/*trans* isomerization of the original diene **187**, or COD-$PdCl_2$, takes place before the cyclization, a fact that could be demonstrated by reaction of the *cis,trans* cyclooctadiene **185**. The light also behaves as a reoxidant by promoting the *cis*-insertion and the final carbon-palladium cleavage.

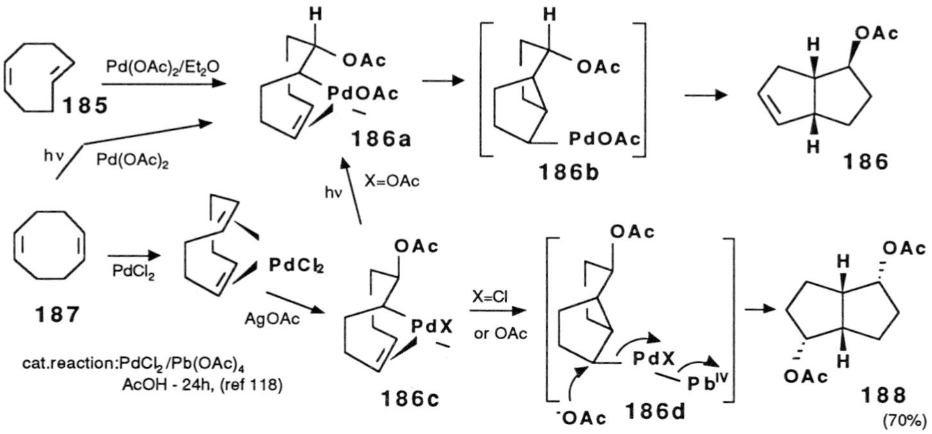

A complete catalytic cyclization of COD to *endo*-diacetate **188** has been reported from P.M. Henry's laboratory with the binary oxidation system $PdCl_2$-$Pb(OAc)_4$ [119]. This reaction is a nice demonstration of the synthetic value of palladium catalyzed oxidative cylization reactions. The incorporation of 2 oxygen nucleophiles is achieved under catalytic conditions and complete control of (*cis*) ring junction and (*endo*) configuration of both acetoxy groups. The uncommon reoxidant $Pb(OAc)_4$ is the determining reagent in the oxidative cleavage of the carbon-palladium bond in **186d** leading to the complete S_N2 type Pd eliminating step [119, 120].

Cyclooctadienes are useful starting materials for bridged bicyclic molecules [121]. When introducing *exo*methylene groups into the eight-membered ring, it is possible to gain entry into these cyclic systems by palladium catalysis [122].

The Wacker type catalyst ($PdCl_2$ 0.03 eq/$CuCl_2$ in buffered acetic acid) allows to cyclize 5-methylenecyclooctene **189** to the bridgehead substituted bicyclo[3.3.1]- and bicyclo[4.2.1]nonane derivatives **190 - 193** under extremely mild reaction conditions. Compared to an ionic cyclization of this type [123], the palladium mediated reaction is less regioselective since cyclization takes place at C_1 and C_2 with formation of both possible ring systems. On the contrary, the introduction of the different nucleophiles (chloride and acetate) is highly regio- and stereoselective (bridgehead position and *endo* configuration). Despite some similarities with the cyclization of N-carbethoxy-8-azabicyclo[5.1.0]oct-3-ene **56** [45], another mechanism seems operative in the case of the carbocyclization of **189** *via* the *exo*-bicylic palladium complexes **195** and the chloride substitution with inversion of configuration.

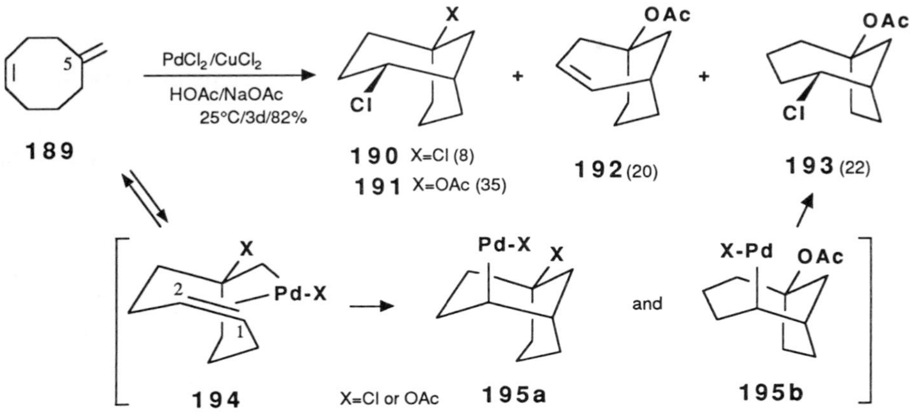

The observation of an olefinic substrate **192** (*trans* elimination) is an additional support of the latter intermediate.

Bis *exo*methylenecyclooctane **196** is likewise cyclized to bridged bicyclic compounds [124].

Now, the products carry substituents at both bridgehead positions. This formal double functionalization of tertiary carbon atoms is specific for metal catalysis since a similar reaction of **196** is not known under pure ionic conditions. The formation of **199** and **200** is rationalized in terms of the following transposition: as a result of the *cis* stereochemistry of olefin insertion into carbon-metal bonds, the new C-C bond can only be formed between C_9 and C_5. The intermediate palladium complex **198a** has a primary CH_2-Pd unit which is known to easily undergo

C-H or C-C transposition reactions [13]. Indeed, with Pd as a good leaving group, the reductive elimination step is accompanied by a Wagner Meerwein rearrangement to the bicyclo[4.3.1]decane compounds **199** and **200**. A subsequent base catalyzed fragmentation of chloroacetate **200**, which is the main product in most reactions, reveals the size of the newly formed nine membered ring.

The regioselective formation of **201** from **196** in two steps might be an interesting entry to cyclononane ring systems.

O B$^-$ O Cl t-BuOK/25°C hexane O ≡ O

200 **201**

The direct access to complicated systems with concomitant introduction of one or several functional groups is one of the major advantages of palladium catalyzed oxidative cyclization reactions. The tricyclic brendane skeleton is of theoretical and practical interest [125]; however, no straightforward synthesis had been found. Vinylnorbornene **202** (70% endo), readily accessible from butadiene and cyclopentadiene, is an ideal building block for the brendane series.

+ 2CO + $(CH_3)_2CHOH$ $PdCl_2(PPh_3)_2$(0.16%) PPh_3 / 80kg/cm^2

202 (yield: 1.41% turnover: 8-9) **203**

+ HCN $Pd(PPh_3)_4$(2 mol%) 130°C-8h

NC Pd L **204a** PdCN **204b** CN **205** yield:5.2%

Palladium catalyzed cyclizations to **205** are possible in the presence of HCN, as a carbon nucleophile [127], proceeding *via cis* migration. Treatment of **202** with carbon monoxide also leads to cylization; in this case, two CO units are incorporated during the cyclization [126]. Differently from these low yield reactions (**203** and **205** are formed in quantities of 1.41 and 5.2%, respectively), treatment of vinylnorbornene **202** with a Wacker type catalyst leads to substituted brendyl derivatives **207** with high yields and very high control of regio- and stereoselectivity of the different, newly formed, carbon-carbon or carbon-heteroatom bonds [63]. In order to obtain high isomeric purity in **207**, the best conditions are: (i) high chloride concentration (LiCl, $CuCl_2$); (ii) molecular oxygen as the ultimate oxydant [128].

$PdCl_2$-$CuCl_2$-LiCl-O_2, HOAc-NaOAc-80°C

202 → [206a (AcO, ClPd) → 206b (AcO, PdCl, Cl^-)] → 207 (AcO, Cl) (yield >75% selectivity >90%)

Another diene, *cis*-1,2-divinylcyclohexane **208**, (*cis*-DVCH), allows to observe of a very high degree of chemoselectivity combined with the total control of diastereoselectivity in the presence of palladium(II). Two different oxidation systems have been studied, both leading to acetoxy indanes.

208 → 209: $PdCl_2$-$CuCl_2$-70°C, HOAc-NaOAc-1h

208 → 210 (70%): $Pd(OAc)_2$ (1-5mol%)-25°C, HOAc-MnO_2 - benzoquinone; (1R*.6S*.7S*)

208 → [211 (H, OAc, PdX, H) → 105 (H, OAc, H, PdX)]; 'chloride free' → 210; $[Cl]^-$ → 209

$PdCl_2$-$CuCl_2$ give the rearranged indene **209** in a rapid reaction at 50°C (1hour) [63]. Cyclization with the three component catalyst $Pd(OAc)_2$-benzoquinone-MnO_2 proceeds much slower (42 hours at room temperature); however, it performs the *exo* acetoxy methylenindane **210** with complete stereoselectivity [62]. On studying this reaction in detail, it was possible to get more insight into the mechanism of the oxidative cyclization of 1,5-dienes. Second, the stereoselective formation of **210** under mild reaction conditions makes this transformation of *cis*-divinylcyclohexane **208** a model reaction for the investigation of enantioselective, catalytic oxidation of olefins with palladium.

A di-π complex of cis-1,2-divinylcyclohexane has been known since 1966 [129]; **212** is easily prepared from **208** and $PdCl_2(PhCN)_2$. No structural investigation has been reported. *Cis*-1,2-divinylcyclohexane has two different vinyl groups; one in the equatorial, the other in the axial position. In order to explain the high diastereocontrol during the formation of **209** or **210** it was of interest to determine the structure of **212**, and to study the nucleophilic attack to the π complex, one question being: which of the double bonds reacts preferentially?

By X-ray analysis it is possible to show that $PdCl_2$-*cis*-1,2-divinylcyclohexane **212** contains two crystallographically independent organometallic conformers [130].

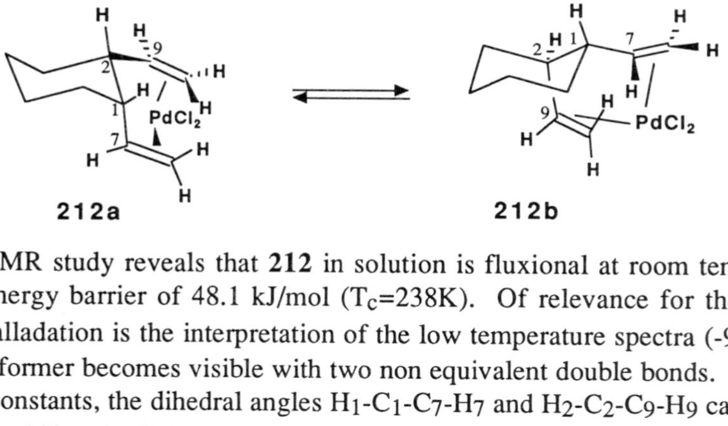

212a **212b**

The ^{1}H-NMR study reveals that **212** in solution is fluxional at room temperature with an energy barrier of 48.1 kJ/mol (T_c=238K). Of relevance for the mechanism of palladation is the interpretation of the low temperature spectra (-90°C). A single conformer becomes visible with two non equivalent double bonds. From the coupling constants, the dihedral angles H_1-C_1-C_7-H_7 and H_2-C_2-C_9-H_9 can be estimated to be 90° and 160°, respectively. This means that both the olefins are nearly parallel in the π complex **212**. The effect of coordination to palladium on the conformational change is small, a fact which suggests that the conformation of the non coordinated part (degenerate chair form) is much more important for the stability of **212**. The attack of OAc^- at either double bond would lead to the σ,π complex **211** and to the bicyclic acetates **209** and **210**.

When 'labelling' the cyclohexane ring, the diene becomes chiral and the unselective attack to the double bonds should result in the formation of two different acetoxyindanes **215** and **216**. The silylated *cis*-DVCH palladium chloride **214** is readily prepared from 9-(trimethylsilyl)-(Z,E)-1,5-cyclodecadiene **213** via Cope rearrangement. The free olefin is oxidatively cyclized under the reaction conditions to the unique reaction product **215**. This surprising result demonstrates that the nucleophilic attack, leading to oxidative cyclization, exclusively occurs at the equatorial vinyl group in *cis*-DVCH.

$(CH_3)_3Si$ **213** $\xrightarrow{PdCl_2(PhCN)_2}$ **214-PdCl$_2$** $\xrightarrow{dppe}$ **214**

214 $\xrightarrow[bzq/HOAc/r.t.]{Pd(OAc)_2/MnO_2}$ [**214**-Pd(OAc)$_2$] $\rightarrow$ **215** (not **216**)

The isolation of acetate complexes, the real intermediates in the catalytic cyclization reaction, failed when treating *cis*-DVCH-$PdCl_2$ **212** with silver acetate [131].

208 $\xrightarrow{DOAc_d,\ Pd(OAc)_2\ bzq}$ or **212** $\xrightarrow{AgOAc}$ **211** (X=OAc) $\rightarrow$ **209** + **210** + **217**

A di-π acetate complex could not be found; however, the σ,π complex **211** (X=OAc) is observable in the ^{1}H-NMR spectra. It is a short lived species which undergoes, spontaneously, subsequent insertion and elimination reactions. Contrary to the catalytic reaction(s), the high concentration of different palladium species, even in the presence of other oxidants (benzoquinone), renders the oxidative cyclization much less selective. The presence of the three isomers of unsaturated indane acetates **209**, **210** and **217** testifies this fact.

This observation highlights the fundamental problem of comparing catalytic and stoichiometric reactions, and the dilemma, when observing new, usually slow side reactions, which are suppressed under catalytic conditions provided the total amount and the number of active species is controlled by the (low) concentration of the catalyst.

The development of enantioselective reactions is one of the major objectives of actual organic chemistry. The cyclization of *cis*-DVCH meets the criteria of a reaction serving as a model reaction for enantioselective oxidation: high yield, high diastereoselectivity, soft and modulable reaction conditions, basic knowledge of the mechanism. According to our mechanistic study, we had established that the nucleophile added *trans* to the coordinated, equatorial double bond in **212**. *Cis*-DVCH-$PdCl_2$ is fluxional between the two enantiomeric conformers **212a** and **212b**, and the addition to the **212a** conformer would lead to (1R,6S,7S) esters; the reaction of the opposite conformer **212b** has to lead to the corresponding enantiomeric substrate with (1S,6R,7R) configuration.

The formation of chiral products from symmetrical starting material requires diastereomeric transition states. In DVCH cyclizations, this may be achieved with: (i) chiral palladium complexes; (ii) by using chiral nucleophiles. In both cases the asymmetric discrimination takes place during the nucleophilic attack to the (equatorial) double bond in the intermediate DVCH-$Pd(OAc)_2$.

The most attractive, completely catalytic, chiral palladium(II) catalyzed reaction, in the presence of (+)-(3,2,10-η-pinene)palladium(II) acetate **45a** [20] for example, failed to give chiral induction with **208**. On the other hand, the application of chiral carboxylic acids as nucleophiles could be easily realized, and we actually have obtained very promising results with a great variety of different chiral acids, in most cases derivatives of mandelic or lactic acid [132]. The optical inductions which could be observed are low to fairly good; they depend on the structure of the chiral acid and the reaction conditions. We present 2 examples, that show the complementary behaviour of the O-aryl substituted lactic acids **218** and **219**. The 'dichloro' acid **218** leads to the formation of (1R.6S.7S) isomer **220a** with a diastereomeric excess of 17%. A slight modification in the chiral acid, the replacement of the *ortho* Cl atom by a methyl group (**219**), changes the direction of induction, and the opposite isomer **221b** is isolated (de 17%). This means that the system is very sensitive to structural changes in the chiral acid, even if the modification is several atom distant from the chiral carbon.

208 + **218** R=Cl / **219** R=CH_3

+ **218** [Pd] → (1R,6S,7S) **220a**
(X= (R)-$OCOCH(CH_3)OC_6H_3Cl_2$
65%, de 17%

+ **219** [Pd] → **221b**
(X= (R)-$OCOCH(CH_3)OC_6H_3(Cl)CH_3$
31%, de 17%

[Pd]=$Pd(OAc)_2$(0.05eq)/MnO_2/p-benzoquinone
acetone - 20°C - 3-5 days

An interesting effect of powdered molecular sieves has been observed in these reactions. Irrespective of the nature of the arylated lactic acid (**218** or **219**), only the isomer with S configuration **220a** at the newly created chiral center is formed in the presence of the zeolites. At the same time the stereoselectivity considerably improves (up to 76% de) [133]. In contrast with other, titanium- [134-136] and aluminium-catalyzed [137] chiral reactions, that has been improved by the water scavenging properties of the sieves, only non dried sieves have an effect on the diastereoselectivity of the cyclization of **208**.

The cyclization of 1,6-dienes such as **222** under Wacker type reaction conditions are known in some cases [137]. Five- and six membered rings have been obtained according to the substitution pattern of the double bonds.

It can be concluded from the products that the palladation always takes place at the more substituted olefin part. This is surprising since the unsubstituted double bond should react first. Equilibria between different intermediate palladium complexes are supposed to occur. Likewise, in the cyclization of *cis*-DVCH it is conceivable that the non coordinated part of the molecule(s) controls the rate of the insertion (cyclization) step. Unfavourable interactions of the G groups are, most probably, in favour of **224a** or **226a**, and, subsequently, to cyclized chlorides **225** and **227**, respectively. The high $CuCl_2$ concentration (10 fold excess) increases the rate of oxidative chlorination of the σ complex **224b** or **226b**, and the surprising production of the unsaturated chloride **227**.

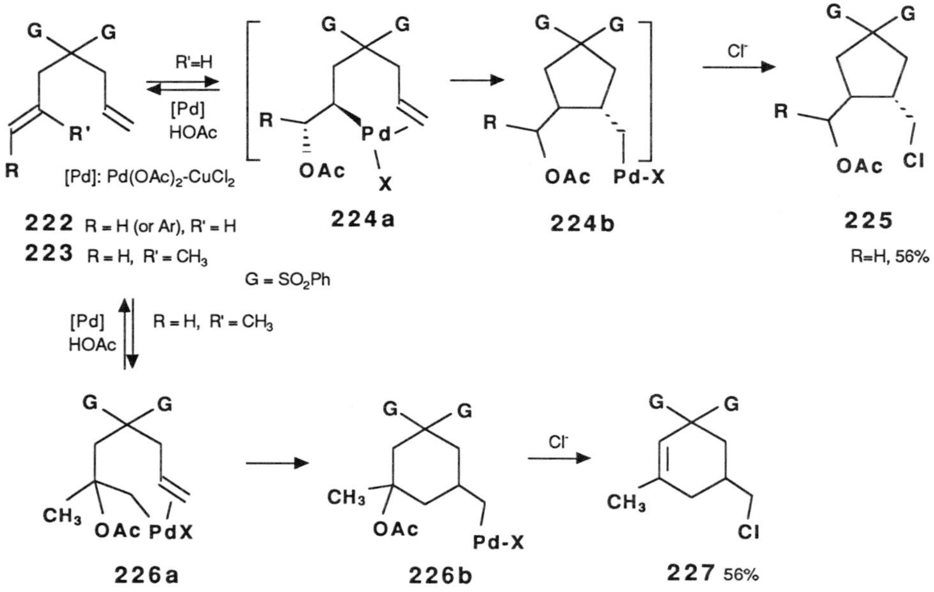

It should be emphasized that one of these dienes, more precisely **222** (R=H) permits palladation with carbon nucleophiles under Heck type conditions.

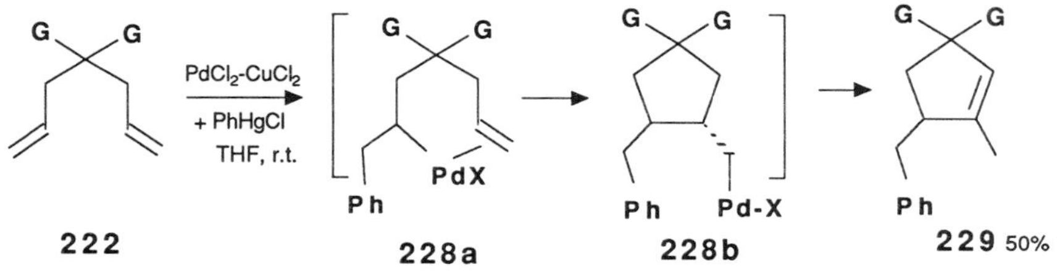

The final double bond isomerization is 'attended'; however, the double bond position is unusual and a directing effect is assumed to be due to the geminal sulphone grouping.

Note added after submission of this article

Several publications have appeared that deal with palladium catalyzed cyclization reactions, concerning the intramolecular [139] and chiral Heck Reactions [140 - 142] as well as the possibility to suppress isomerization with thallium(I) salts [143]. The stereo- and regiocontrol in the oxypalladation followed by CO insertion has been studied with unsaturated steroids [144] and new variations of the enyne reaction have been found with metathesis- [145] and alkylative cyclization [146].

Acknowledgement

I would like to thank my wife Carmen for her very patient support throughout the work on this article. I would also like to emphasize my gratitude to all persons involved in our own work, and whose names appear in the references. Especially, I got profit from a very pleasant and fruitful, still ongoing collaboration with several research groups in Stockholm and Uppsala (Sweden) and, for cyclization chemistry, particularly with Dr. Christina Moberg, Dr. Thomas Antonsson, Dr. Lori Sutin and Dr. Louise Tottie.

3 References

1. (a) J. Smidt,
   Chem.Ind. (London), (1962) 54.
   (b) J. Smidt, W. Hafner, R. Jira, R. Sieber, J. Sedlmeier and A. Sabel,
   Angew.Chem.Int.Ed.Engl., *1* (1962) 200.
2. P.M. Maitlis,
   The Organic Chemistry of Palladium, Vols 1, 2;
   Academic Press, New York, 1971.
3. F.R. Hartley,
   The Chemistry of Platinum and Palladium, Applied Science Publishers,
   London, 1973.
4. P.M. Henry,
   Palladium-Catalyzed Oxidation of Hydrocarbons, D.Reidel Publishing
   Co.: Dordrecht, 1980.
5. J. Tsuji,
   Organic Synthesis with Palladium Compounds, Springer Verlag,
   Berlin, 1980.
6. A. Segnitz,
   *Methoden zur Herstellung und Umwandlung von σ-Organo-palladium
   Verbindungen* in *Houben-Weyl Methoden der Organischen Chemie*,
   Bd.13, Teil 9b, Thieme Verlag, Stuttgart,1983.
7. R.F. Heck,
   Palladium Reagents in Organic Synthesis, Academic Press, New
   York,1985.
8. F.C. Philips,
   Am.Chem.J., *16* (1894), 255.
9. F.R. Hartley,
   Chem.Rev., *69* (1969) 799.
10. (a) P.M. Henry,
    Trans.N.Y.Acad.Sci., *33* (1971) 41.
    (b) R.F. Heck,
    Topics Curr.Chem., *16* (1971) 221.
11. J.E. Bäckvall, B. Åkermark and S.O. Ljunggren,
    J.Am.Chem.Soc., *101* (1979) 2411.
12. N. Gregor, K. Zaw and P.M. Henry,
    Organometallics, *3* (1984) 1251.

13. A. Heumann,
in *Catal.Met.Complexes (Metal Promoted Selectivity in Organic Synthesis)*, M. Graziani, A.J. Hubert and A.F. Noels (Eds), Kluwer Academic Publishers,Dordrecht, 1991, p. 133.

14. J.E. Bäckvall,
in *Reaction of Coordinated Ligands*, P.S. Praterman (Ed.), Plenum Press: New York, 1986.

15. D. Seebach,
Angew.Chem., 91 (1979) 259.

16. I. Omae,
Organometallic Intramolecular Coordination Compounds, Elsevier, Amsterdam, 1986.

17. (a). J.K. Stille and R. Divakaruni,
J.Am.Chem.Soc.,100 (1978) 1303.
and *J.Organomet.Chem., 169* (1979) 239.
(b) cf also J.K. Stille and D.E. James,
J.Am.Chem.Soc.,97 (1975) 674-676
and *J. Organomet.Chem., 108* (1976) 401.
(c) C. Narayana and M. Periasamy,
Synthesis, (1985) 253.

18. (a) L.S. Hegedus,
Tetrahedron, 40 (1984) 2415.
(b) J. Tamaru and Z. Yoshida ,
J.Organomet.Chem., 334 (1987) 213.

19. (a) T. Hosokawa, K. Maeda, K. Koga and I. Moritani,
Tetrahedron Lett., (1973) 739.
(b) T. Hosokawa, H. Ohkata and I. Moritani,
Bull.Chem.Soc.Jpn., 48 (1975) 1533.
(c) T. Hosokawa, S. Yamashita,.S.-I. Murahashi and A. Sonoda,
Bull. Chem.Soc.Jpn., 49 (1976) 3662.
(d) T. Hosokawa, S. Miyagi, S.-I. Murahashi and A. Sonoda,
J.Org.Chem., 43 (1978) 2752.

20. (a) T. Hosokawa, S. Miyagi, S.-I. Murahashi and A. Sonoda,
J.Chem.Soc.Chem. Commun., (1978) 687.
(b) T. Hosokawa, U. Tetsuyuki and S.-I. Murahashi,
J.Chem.Soc.Chem.Commun., (1979) 475.
(c) T. Hosokawa, U. Tetsuyuki, S.Inui and S.-I. Murahashi,
J.Am.Chem.Soc., 103 (1981) 2318.
(d) T. Hosokawa, C. Okuda and S.-I. Murahashi,
J.Org.Chem., 50 (1985) 1282.

(e) T. Hosokawa, Y. Imada and S.-I. Murahashi, *Bull.Chem.Soc.Jpn.*, *58* (1985) 3282.
(f) T. Hosokawa, Y. Imada and S.-I. Murahashi, *Bull.Chem Soc.Jpn.*, *59* (1986) 2191.
(g) T. Hosokawa, T. Kono, T. Shinohara and S.-I. Murahashi, *J.Organomet.Chem.*, *370* (1989) C13.
(h) T. Hosokawa, and S.-I. Murahashi, *Accounts Chem.Res.*, *23* (1990) 49.

21. T. Hosokawa, M. Hirata, S.-I. Murahashi and A. Sonoda, *Tetrahedron Lett.*, (1976) 1821.
22. M.F. Semmelhack, C.R. Kim, W. Dobler and M. Meier, *Tetrahedron Lett.*, *30* (1989) 4925.
23. J. Nokami, H.Ogawa, S. Miyamoto, T. Mandai, S. Wakabayashi and J. Tsuji, *Tetrahedron Lett.*, *29* (1988) 5181.
24. S. Igarashi, Y. Haruta, M. Ozawa, Y. Nishide, H. Kinoshita and K. Inamota, *Chem.Lett.*, (1989) 737.
25. J. Tsuji, *Synthesis*, (1984) 369.
26. B. Feringa, *J.Chem.Soc.Chem.Commun.*, (1986) 909.
27. (a) J.E. Baldwin, *J.Chem.Soc.Chem.Commun.*, (1976) 734
(b) J.E. Baldwin, J. Cutting, W. Dupont, L. Cruse, L. Silberman and R.C.Thomas, *J.Chem.Soc. Chem.Commun.*, (1976) 736.
28. G. Casiraghi, G. Casnati, G. Puglia, G. Sartori and G. Terenghi, *Synthesis*, (1977) 122.
29. Y. Tamaru, M. Hojo, H.Higashimura and Z. Yoshida, *Angew.Chem.Int.Ed. Engl.*, *25* (1986) 735.
30. (a) R.F.Heck, Topics Curr.Chem. 16 (1971) 221.
(b) R.F.Heck, *Org.React.(N.Y.)*, *27* (1982) 345.
(c) H.U. Reissig, *Nachr.Chem.Tech.Lab.*, 34 (1986) 1066.
(d) G.D. Daves, Jr. and A. Hallberg, *Chem.Rev.*, *89* (1989) 1433.

31. (a) M.F. Semmelhack, L. Keller, T. Sato and E. Spiess, *J.Org.Chem.*, *47* (1982) 4382.
(b) M.F. Semmelhack, J.J. Bozell, T.Sato, M. Wulff, E. Spiess and A. Zask, *J.Am.Chem.Soc.*, *104* (1982) 5850.
(c) M.F. Semmelhack and A. Zask, *J.Am.Chem.Soc.*, *105* (1983) 2034.
32. (a) M.F. Semmelhack and C. Bodurow, *J.Am.Chem.Soc.*, *106* (1984) 1496.
(b) M.F. Semmelhack and Nan Zhang, *J.Org.Chem.*, *54* (1989) 4483.
33. M. McCormick, R. Monahan III, J. Soria, D. Goldsmith and D. Liotta, *J.Org.Chem.*, *54* (1989) 4485.
34. M.F. Semmelhack, C. Bodurow and M. Baum, *Tetrahedron Lett.*, *25* (1984) 3171.
35. C.P. Holmes and P.A.Bartlett, *J.Org.Chem.*, *54* (1989) 98.
36. R.D. Walkup and G. Park, *Tetrahedron Lett.*, *28* (1987) 1023.
37. R.D. Walkup and G. Park, *J.Am.Chem.Soc.*, *112* (1990) 1597.
38. (a) J.D. Mosher, (Ed.) *Asymmetric Synthesis* Vol 1 - , Analytical Methods, Academic Press, New York,1983.
(b) B. Bosnich, (Ed.) *Asymmetric Catalysis*, NATO ASI Series E: Applied Science 103, Martinus Nyhoff, Dordrecht, 1986.
39. A. Heumann, unpublished results, cf footnote in ref 132.
40. (a) L.S. Hegedus, *J.Mol.Catal.*, *19* (1983) 201.
(b) L.S. Hegedus, *Angew.Chem.Int.Ed.Engl.*, *27* (1988) 1113.
(c) L.S. Hegedus, P.A. Weider, T.A.Mulhern, H. Asada and S. D'Andrea, *Gazz.Chim.Ital.*, *116* (1986) 213.
41. (a) L.S. Hegedus and J.M. McKearin, *J.Am.Chem.Soc.*, *104* (1982) 2444.
(b) L.S. Hegedus, M.S. Holden and J.M. McKearin,. *Organic Syntheses*, *62* (1984) 48.

42. B. Pugin and L.M. Venanzi,
J.Am.Chem.Soc., *105* (1983) 6877.
43. B. Pugin and L.M. Venanzi,
J.Organomet.Chem., *214* (1981) 125.
44. S.R. Wilson and R.A. Sawicki,
J.Org.Chem., *44* (1979) 287.
45. G.R. Wiger and M.F. Rettig,
J.Am.Chem.Soc., *98* (1976) 4168.
46. Y. Tamaru, M. Hojo, S. Kawamura and Z. Yoshida,
J.Org.Chem., *51* (1986) 4089.
47. (a) B.M. Trost,
Accounts Chem.Res., *13* (1980) 385.
(b) J. Tsuji,
Tetrahedron, *42* (1986) 4361.
48. J. Tsuji, T. Yamakawa, M. Kaito and T. Mandai,
Tetrahedron Lett., (1978) 2075.
49. (a) D. Lathbury, P. Vernon and T. Gallagher,
Tetrahedron Lett., *27* (1986) 6009.
(b) T. Gallagher, I.W. Davies, S.W. Jones, D. Lathbury, M.F. Mahon, K.C. Molloy, R.W. Shaw and P. Vernon,
J.Chem.Soc.Perkin Trans.1, (1992) 433.
50. S.F. Martin and C.L. Campbell,
Tetrahedron Lett., *28* (1987) 503.
51. (a) R.A. Holton and R.A. Kjonaas,
J.Am.Chem.Soc. 99 (1977) 4177.
(b) R.A. Holton and R.A. Kjonaas,
J.Organomet.Chem., *142* (1977) C15.
52. A.C. Cope, J.M. Kliegman and E.C. Friedrich,
J.Am.Chem.Soc., *89* (1967) 287.
53. R.A. Holton and J.R. Zoeller,
J.Am.Chem.Soc., *107* (1985) 2124.
54. G. Fournet, G. Balme and J. Goré,
Tetrahedron Lett., *30* (1989) 69.
55. D.F. Taber, J.C. Amedio and R.G. Sherril,
J.Org.Chem., *51* (1986) 3382.
56. F. Henin and J.P. Pete,
Tetrahedron Lett., *24* (1983) 4687.
57. F. Henin J. Muzart and J.P. Pete,
Tetrahedron Lett., *27* (1986) 6339.

58. (a) M. Mori, I. Oda and Y. Ban,
Tetrahedron Lett., *23* (1982) 5315.
(b) M. Mori, N. Kanda, I. Oda and Y. Ban,
Tetrahedron, *41* (1985) 5465.
(c) M. Mori, N. Kanda and Y. Ban,
J.Chem.Soc.Chem.Commun., (1986) 1375.
(d) M. Mori, N. Kanda, Y. Ban and K. Aoe,
J.Chem.Soc.Chem. Commun., (1988) 12.
59. (a) M. Mori, Y. Kubo and Y. Ban,
Tetrahedron Lett., *26* (1985) 1519.
(b) M. Mori, Y. Kubo and Y. Ban,
Tetrahedron, *44* (1988) 4321.
60. (a) D.P. Curran and C.-T. Chang,
Tetrahedron Lett., *31* (1990) 933.
(b) cf also D.P. Curran, M.-H. Chen, E. Spletzer, C.M. Seong and C.-T. Chan,
J.Am. Chem.Soc., *111* (1989) 8872.
61. *For a discussion of the activation of Csp^3X bonds:* M. Chanon,
Bull.Soc.Chim. France II, (1982) 197.
62. (a) T. Antonsson, A. Heumann and C. Moberg,
J.Chem.Soc.Chem.Commun., (1986) 518.
(b) T. Antonsson, C. Moberg, L. Tottie and A. Heumann,
J.Org.Chem., *54* (1989) 4914.
63. A. Heumann, M. Reglier and B.Waegell,
Angew.Chem.Int.Ed.Engl., *18* (1979) 866.
64. E.G. Samsel and J.R. Norton,
J.Am.Chem.Soc., 106 (1984) 5505.
65. R.C. Larock and D.E. Stinn,
Tetrahedron Lett., *30* (1989) 2767.
66. R.C. Larock, D.E. Stinn and M.-Y. Kuo,
Tetrahedron Lett., *31* (1990) 17.
67. (a) R. Grigg, V. Sridharan, P. Stevenson and T. Worakun,
J.Chem.Soc.Chem. Commun., (1986) 1697.
(b) R. Grigg, V. Sridharan, P. Stevenson and S. Sukirthalingam,
Tetrahedron, *45* (1989) 3557.
(c) R. Grigg, V. Santhakumar, V. Sridharan, P. Stevenson, A. Teasdale, M. Thornton-Pett and T. Worakun,
Tetrahedron, *47* (1991) 9703.
68. R. Grigg, P. Stevenson and T. Worakun,
Tetrahedron, *44* (1984) 2033.

69. (a) B. Burns, R. Grigg, P. Ratananukul, V. Sridharan, P. Stevenson and T. Worakun,
Tetrahedron Lett., *29* (1988) 4329
(b) R. Grigg and V. Sridharan,.
Tetrahedron Lett., 30 (1989) 1139
(c) B. Burns, R. Grigg, P. Ratananukul, V. Sridharan, P. Stevenson, S. Sukirthalingam and T. Worakun,
Tetrahedron Lett., *29* (1988) 5565
(d) B. Burns, R. Grigg, V. Sridharan, P. Stevenson, S. Sukirthalingam and T. Worakun,
Tetrahedron Lett., *30* (1989) 1135.

70. M.M. Abelman, T. Oh and L.E. Overman,
J.Org.Chem., *52* (1987) 4130.

71. W.G. Earley, T. Oh and L.E. Overman,
Tetrahedron Lett., *31* (1988) 3785.

72. E. Negishi, T. Nguyen, B. O'Connor, J.M. Evans and A. Silveira,Jr.,
Heterocycles, *28* (1989) 55.

73. (a) B. O'Connor, Y.Zhang and E. Negishi,
Tetrahedron Lett., *29* (1988) 3903
(b) Y. Zhang, B. O'Connor and E. Negishi,
J.Org.Chem., *53* (1988) 5588
(c) E. Negishi, Y. Zhang and B. O'Connor,
Tetrahedron Lett., *29* (1988) 2915.

74. R.C. Larock, H.Song, B.E.Baker and W.H. Gong,
Tetrahedron Lett., *29* (1988) 2919-.

75. J.A. Peters,
Synthesis, (1979) 321.

76. Y. Sato, M. Sodeoka and M. Shibasaki,
J.Org.Chem., *54* (1989) 4738.

77. G. Wu, F. Lamaty and E. Negishi,
J.Org.Chem., *54* (1989) 2507.

78. cf. J.M. Takacs and L.G. Anderson,
J.Am.Chem.Soc., *109* (1987) 2200.

79. D.F. Taber,
Intramolecular Diels-Alder and Alder Ene Reactions, Springer Verlag, Berlin 1984.

80. Nickel-chromium:
a) enynes: B.M. Trost and J.M. Tour,
J.Am.Chem.Soc., *109* (1987) 5268

(b) enallenes: B.M. Trost and J.M. Tour, *J.Am.Chem.Soc.*, *110* (1988) 5231.

81. B.M. Trost and M. Lautens, *J.Am.Chem.Soc.*, *107* (1985) 1781.

82. B.M. Trost and J.Y.L. Chung, *J.Am.Chem.Soc.*, *107* (1985) 4586.

83. B.M. Trost and M. Lautens, *Tetrahedron Lett.*, *26* (1985) 4887.

84. B.M. Trost and S.-F. Chen, *J.Am.Chem.Soc.*, *108* (1986) 6053.

85. B.M. Trost and D.J. Jebaratnam, *Tetrahedron Lett.*, *28* (1987) 1611.

86. B.M. Trost and G.J. Tanoury, *J.Am.Chem.Soc.*, *109* (1987) 4753.

87. B.M. Trost and G.J. Tanoury, *J.Am.Chem.Soc.*, *110* (1988) 1636.

88. B.M. Trost, P.A. Hipskind, J.Y.L. Chung and C. Chan, *Angew.Chem.Int Ed. Engl.*, *28* (1989) 1502.

89. B.M. Trost, D.C. Lee and F. Rise, *Tetrahedron Lett.*, *30* (1989) 651.

90. B.M. Trost and F. Rise, *J.Am.Chem.Soc.*, *109* (1987) 3161.

91. B.M. Trost and R. Braslau, *Tetrahedron Lett.*, *29* (1988) 1231.

92. *cf.* J.E. Bäckvall and E.E. Björkmann, *J.Org.Chem.*, *45* (1980) 2893.

93. G. Henrici-Olivé and S. Olivé, *Fortschr.Chem.Forsch.*, *67* (1976) 107.

94. (a) J.A. Dale and H.S. Mosher, *J.Am.Chem.Soc.*, *95* (1973) 512.
(b) G.R. Sullivan, J.A. Dale and H.S. Mosher, *J.Org.Chem.*, *38* (1973) 2143.

95. B.M. Trost and K.M. Matsuda, *J.Am.Chem.Soc.*, *110* (1988) 5233.

96. R.W. Thiess, J.L. Boop, M. Schiedler, D.C. Zimmermann and T.H. LaPage, *J.Org.Chem.*, *48* (1983) 2021.

97. I. Omae, *Angew.Chem.Int.Ed.Engl.*, *21* (1982) 889.

98. I. Omae,
Coord.Chem.Rev., *51* (1983)1.

99. (a) S. Brewis and P.R. Hughes,
Chem.Comm., (1965) 489.
(b) S. Brewis and P.R. Hughes,
Chem.Comm., (1967) 71.

100. K.A. Jensen,
Acta Chem.Scand., 7 (1953) 866.

101. (a) I.A. Zakharova, G.A. Kukina, T.S. Kuli-Zade, I.I. Moiseev, G.Y. Pek and M.A. Porai-Koshits,
Russ.J.Inorg.Chem., *11* (1966) 1364
(b) I.A. Zakharova, L.A. Leiters and V.T. Aleksanyan,
J.Organomet.Chem., *72* (1974) 283.

102. J.K. Stille and R. Divakaruni,
J.Org.Chem., *44* (1979) 3474.

103. (a) T. Matsuda, T. Mitsuyasu and Y. Nakamura,
Kogyo Kagaku Zasshi, *72* (1969) 1751
(b) N. Adachi, K. Kikukawa, M. Takagi and T. Matsuda,
Bull.Chem.Soc.Jpn., *48* (1975) 521.

104. J.E. Bäckvall and A. Gogoll,
Tetrahedron Lett., *29* (1988) 2243.

105. (a) J.E. Bäckvall, A.K. Awasthi and Z.D. Renko,
J.Am.Chem.Soc., *109* (1987) 4750.
(b) J.E. Bäckvall, R.B. Hopkins, H. Grennberg, M.M. Mader and A.K. Awasthi,
J.Am.Chem.Soc., *112* (1990) 5160.

106. (a) T. Antonsson, C. Malmberg and C. Moberg,
Tetrahedron Lett., *29* (1988) 5973.
(b) C. Moberg, K. Nordström and P. Helquist,
Synthesis, (1992) 685.

107. J.K. Stille and R.A. Morgan,
J.Am.Chem.Soc., *88* (1966) 5135.

108. M. Green and R.I. Hancock,
J.Chem.Soc.(A), (1967) 2054.

109. D.R. Coulson,
J.Am.Chem.Soc., *91* (1969) 200.

110. (a) J.K. Stille and L.F. Hines,
J.Am.Chem.Soc., *92* (1970) 1798
(b) L.F. Hines and J.K. Stille,
J.Am.Chem.Soc., *94* (1972) 485.

111. catalytic carbonylation of norbornadiene:
C.B. Anderson and R. Markovic,
J.Serb.Chem.Soc., *50* (1985) 125.

112. P.K. Wong and J.K. Stille,
J.Organomet.Chem., *70* (1974) 121.

113. A. Heumann and B. Waegell,
Nouv.J.Chim., *1* (1977) 277.

114. (a) A. Segnitz. P.M. Bailey and P.M. Maitlis,
J.Chem.Soc.Chem.Commun., (1973) 698
(b) A. Segnitz, E. Kelly, S.H. Taylor, and P.M. Maitlis,
J.Organomet.Chem., *124* (1977) 113.

115. E. Vedejs and P.D. Weeks,
J.Chem.Soc.Chem.Commun., (1974) 223.

116. C.B. Anderson and B.J. Burreson,
Chem.Ind., (1967) 620.

117. M. Akbarzadeh and C.B. Anderson,
J.Organomet.Chem., *197* (1980) C5.

118. C.B. Anderson and B.J. Burreson,
J.Organomet.Chem., 7 (1966) 181.

119. P.M. Henry, M. Davies, G. Ferguson, S. Phillips and R. Restivo,
J.Chem. Soc.Chem.Commun., (1974) 112.

120. (a) S.K. Chung and A.I. Scott,
Tetrahedron Lett., (1975) 49
(b) K.Y. Chernyuk, V.I. Mel'nikova and K.K. Pivnitskii,
Zh.Org.Khim., *18* (1982) 577.

121. G. Haufe, *Z.Chem.*, *19* (1979) 170.

122. A. Heumann, M. Reglier and B.Waegell,
Angew.Chem.Int.Ed.Engl., *18* (1979) 867.

123. Formation of bicyclo[3.3.1]nonanes under Friedel Craft's conditions:
A. Heumann and W. Kraus,
Tetrahedron, *34* (1978) 405.

124. A. Heumann, M. Reglier and B.Waegell,
Tetrahedron Lett., *24* (1983) 1971.

125. A. Nickon, H.R. Kwasnik, C.T. Mathew, T.D. Swartz, R.O. Williams and J.B. DiGiorgio,
J.Org.Chem., *43* (1978) 3904.

126. Y. Kobori and T. Takezono,
Eur.Pat.Appl., EP 170,227 (1986); *CA 105*: 191437s (1986).

127. E.S. Brown and E.A. Rick,
Abstracts Am.Chem.Soc.Pet.Div.Chem.Reprints, *14* (1969) B29.

128. A. Heumann, S. Kaldy and A. Tenaglia, *J.Chem.Soc.Chem.Commun.*, (1993) 420.

129. (a) J.C. Trebellas, J.R. Olechowski, and H.B. Jonassen, *J.Organomet.Chem.*, *6* (1966) 412
(b) P. Heimbach and M. Molin, *J.Organomet.Chem.*, *49* (1973) 477.

130. C. Moberg, L. Sutin, I. Csöregh and A. Heumann, *Organometallics*, *9* (1990) 974.

131. C. Moberg, L. Sutin and A. Heumann, *Acta Chem.Scand.*, *45* (1991) 77.

132. A. Heumann and C. Moberg, *J.Chem.Soc.Chem.Commun.*, (1988) 1516.

133. (a) A. Heumann, L. Tottie and C. Moberg, *J.Chem.Soc.Chem.Commun.*, (1991) 218.
(b) L. Tottie, P. Baeckström, C. Moberg, J. Tegenfeldt and A. Heumann, *J.Org.Chem.*, *57* (1992) 6579.

134. Sharples epoxidation:
(a) R.M. Hanson and K.B. Sharpless, *J.Org.Chem.*, *51* (1986) 1922
(b) Y. Gao, R.M. Hanson, J.M. Klunder, S.Y. Ko, H. Masamune and K.B. Sharpless, *J.Am.Chem.Soc.*, *109* (1987) 5765.

135. Diels-Alder cycloaddition:
K. Narasaka, H. Tanaka and F. Kanai, *Bull.Chem.Soc.Jpn.*, *64* (1991) 387.

136. Glyoxylate-ene reaction:
(a) K. Mikami, M. Terada and T. Nakai, *J.Chem.Soc. Chem.Commun.*, (1990) 1623.
(b) M. Terada, K. Mikami and T. Nakai, *J.Am.Chem.Soc.*, *112* (1990) 3949.

137. Ene reaction of aldehydes with alkenes:
K. Maruoka, Y. Hoshino, T. Shirasaki and H. Yamamoto, *Tetrahedron Lett.*, *29* (1988) 3967.

138. B.M. Trost and K. Burgess, *J.Chem.Soc.Chem.Commun.*, (1985) 1084.

139. R. Grigg, V. Sridharan, P. Stevenson, S. Sukirthalingam and T. Worakun, *Tetrahedron*, *46* (1990) 4003.

140. (a) Y. Sato, M. Sodeoka and M. Shibasaki, *Chem.Lett.*, (1990) 1953.

(b) K. Kagechika and M. Shibasaki,
J.Org.Chem., 56 (1991) 4093.
(c) Y. Sato, S. Watanabe and M. Shibasaki,
Tetrahedron Lett., *33* (1992) 2589.
(d) Y. Sato, T. Honda and M. Shibasaki,
Tetrahedron Lett., *33* (1992) 2593.
141. (a) R. Grigg, M.J.R. Dorrity, J.F. Malone, W.D.J.A. Norbert and V. Sridharan,
Tetrahedron Lett., *31* (1990) 3075.
142. A. Ashimori and L.E. Overman,
J.Org.Chem., 57 (1992) 4571.
143. R. Grigg, V. Loganathan, V. Santhakumar, V. Sridharan, and A. Teasdale,
Tetrahedron Lett., *32* (1991) 687.
144. P. Kocovsky and M. Pour,
J.Org.Chem., *55* (1990) 5580.
145. B.M. Trost and M.K. Trost,
J.Am.Chem.Soc., *113* (1991) 1850.
146. (a) B.M. Trost, W. Pfrengle, H. Urabe and J. Dumas,
J.Am.Chem.Soc., *114* (1992) 1923.
(b) B.M. Trost and J. Dumas,
J.Am.Chem.Soc., *114* (1992) 1924.

SUBJECT INDEX

(grouped by chapter)

SUBJECT INDEX
(grouped by chapter)

Chapter 1 (pp. 3-160)

Chapter 2 (pp. 163-408)

Chapter 3 (pp. 411-455)

Chapter 4 (pp. 459-504)

Chapter 5 (pp. 507-555)

Chapter 6 (pp. 559-623)